Lecture Notes in Computer Science   10954

Commenced Publication in 1973
Founding and Former Series Editors:
Gerhard Goos, Juris Hartmanis, and Jan van Leeuwen

More information about this series at http://www.springer.com/series/7409

De-Shuang Huang · Vitoantonio Bevilacqua
Prashan Premaratne · Phalguni Gupta (Eds.)

Intelligent Computing Theories and Application

14th International Conference, ICIC 2018
Wuhan, China, August 15–18, 2018
Proceedings, Part I

 Springer

Editors
De-Shuang Huang
Tongji University
Shanghai
China

Vitoantonio Bevilacqua
Polytechnic of Bari
Bari
Italy

Prashan Premaratne
University of Wollongong
North Wollongong, NSW
Australia

Phalguni Gupta
Indian Institute of Technology Kanpur
Kanpur
India

ISSN 0302-9743   ISSN 1611-3349   (electronic)
Lecture Notes in Computer Science
ISBN 978-3-319-95929-0   ISBN 978-3-319-95930-6   (eBook)
https://doi.org/10.1007/978-3-319-95930-6

Library of Congress Control Number: 2018947576

LNCS Sublibrary: SL3 – Information Systems and Applications, incl. Internet/Web, and HCI

This Springer imprint is published by the registered company Springer International Publishing AG
part of Springer Nature
The registered company address is: Gewerbestrasse 11, 6330 Cham, Switzerland

Preface

The International Conference on Intelligent Computing (ICIC) was started to provide an annual forum dedicated to the emerging and challenging topics in artificial intelligence, machine learning, pattern recognition, bioinformatics, and computational biology. It aims to bring together researchers and practitioners from both academia and industry to share ideas, problems, and solutions related to the multifaceted aspects of intelligent computing.

ICIC 2018, held in Wuhan, China, August 15–18, 2018, constituted the 14th International Conference on Intelligent Computing. It built upon the success of ICIC 2017, ICIC 2016, ICIC 2015, ICIC 2014, ICIC 2013, ICIC 2012, ICIC 2011, ICIC 2010, ICIC 2009, ICIC 2008, ICIC 2007, ICIC 2006, and ICIC 2005 that were held in Liverpool, UK, Lanzhou, Fuzhou, Taiyuan, Nanning, Huangshan, Zhengzhou, Changsha, China, Ulsan, Korea, Shanghai, Qingdao, Kunming, and Hefei, China, respectively.

This year, the conference concentrated mainly on the theories and methodologies as well as the emerging applications of intelligent computing. Its aim was to unify the picture of contemporary intelligent computing techniques as an integral concept that highlights the trends in advanced computational intelligence and bridges theoretical research with applications. Therefore, the theme for this conference was "Advanced Intelligent Computing Technology and Applications." Papers focused on this theme were solicited, addressing theories, methodologies, and applications in science and technology.

ICIC 2018 received 632 submissions from 19 countries and regions. All papers went through a rigorous peer-review procedure and each paper received at least three reviews. Based on the review reports, the Program Committee finally selected 275 high-quality papers for presentation at ICIC 2018, included in three volumes of proceedings published by Springer: two volumes of *Lecture Notes in Computer Science* (LNCS) and one volume of *Lecture Notes in Artificial Intelligence* (LNAI).

This volume of *Lecture Notes in Computer Science* (LNCS) includes 90 papers.

The organizers of ICIC 2018, including Tongji University, Wuhan University of Science and Technology, and Wuhan Institute of Technology, made an enormous effort to ensure the success of the conference. We hereby would like to thank the members of the Program Committee and the referees for their collective effort in reviewing and soliciting the papers. We would like to thank Alfred Hofmann, executive editor at Springer, for his frank and helpful advice and guidance throughout and for his continuous support in publishing the proceedings. In particular, we would like to thank all the authors for contributing their papers. Without the high-quality submissions from the authors, the success of the conference would not have been possible. Finally, we are especially grateful to the IEEE Computational Intelligence Society, the International Neural Network Society, and the National Science Foundation of China for their sponsorship.

May 2018

De-Shuang Huang
Vitoantonio Bevilacqua
Prashan Premaratne
Phalguni Gupta

ICIC 2018 Organization

General Co-chairs

De-Shuang Huang, China
Huai-Yu Wu, China
Yanduo Zhang, China

Program Committee Co-chairs

Kang-Hyun Jo, Korea
Xiao-Long Zhang, China
Haihui Wang, China
Abir Hussain, UK

Organizing Committee Co-chairs

Hai-Dong Fu, China
Yuntao Wu, China
Bo Li, China

Award Committee Co-chairs

Juan Carlos Figueroa, Colombia
M. Michael Gromiha, India

Tutorial Chair

Vitoantonio Bevilacqua, Italy

Publication Co-chairs

Kyungsook Han, Korea
Phalguni Gupta, India

Workshop Co-chairs

Valeriya Gribova, Russia
Laurent Heutte, France
Xin Xu, China

Special Session Chair

Ling Wang, China

International Liaison Chair

Prashan Premaratne, Australia

Publicity Co-chairs

Hong Zhang, China
Michal Choras, Poland
Chun-Hou Zheng, China
Jair Cervantes Canales, Mexico

Sponsors and Exhibits Chair

Wenzheng Bao, Tongji University, China

Program Committee

Abir Hussain
Akhil Garg
Angelo Ciaramella
Ben Niu
Bin Liu
Bing Wang
Bingqiang Liu
Binhua Tang
Bo Li
Chunhou Zheng
Chunmei Liu
Chunyan Qiu
Dah-Jing Jwo
Daowen Qiu
Dong Wang
Dunwei Gong
Evi Syukur
Fanhuai Shi
Fei Han
Fei Luo
Fengfeng Zhou
Francesco Pappalardo
Gai-Ge Wang

Gaoxiang Ouyang
Haiying Ma
Han Zhang
Hao Lin
Hongbin Huang
Honghuang Lin
Hongjie Wu
Hongmin Cai
Hua Tang
Huiru Zheng
Jair Cervantes
Jian Huang
Jianbo Fan
Jiang Xie
Jiangning Song
Jianhua Xu
Jiansheng Wu
Jianyang Zeng
Jiawei Luo
Jing-Yan Wang
Jinwen Ma
Jin-Xing Liu
Ji-Xiang Du

José Alfredo Costa
Juan Carlos
    Figueroa-García
Junfeng Xia
Junhui Gao
Junqing Li
Ka-Chun Wong
Khalid Aamir
Kyungsook Han
Laurent Heutte
Le Zhang
Liang Gao
Lida Zhu
Ling Wang
Lining Xing
Lj Gong
Marzio Pennisi
Michael Gromiha Maria
    Siluvay
Michal Choras
Ming Li
Mohd Helmy Abd Wahab
Pei-Chann Chang

Ping Guo
Prashan Premaratne
Pu-Feng Du
Qi Zhao
Qingfeng Chen
Qinghua Jiang
Quan Zou
Rui Wang
Sabri Arik
Saiful Islam
Seeja K. R.
Shan Gao
Shanfeng Zhu
Shih-Hsin Chen
Shiliang Sun
Shitong Wang
Shuai Li Hong
Stefano Squartini
Sungshin Kim
Surya Prakash
Takashi Kuremoto
Tao Zeng
Tarık Veli Mumcu

Tianyong Hao
Valeriya Gribova
Vasily Aristarkhov
Vitoantonio Bevilacqua
Waqas Haider Khan
   Bangyal
Wei Chen
Wei Jiang
Wei Peng
Wei Wei
Wei-Chiang Hong
Weijia Jia
Weiwei Kong
Wen Zhang
Wenbin Liu
Wen-Sheng Chen
Wenyin Gong
Xiandong Meng
Xiaoheng Deng
Xiaoke Ma
Xiaolei Zhu
Xiaoping Liu
Xinguo Lu

Xingwen Liu
Xinyi Le
Xiwei Liu
Xuesong Wang
Xuesong Yan
Xu-Qing Tang
Yan Wu
Yan Zhang
Yi Xiong
Yong Wang
Yonggang Lu
Yongquan Zhou
Yoshinori Kuno
Young B. Park
Yuan-Nong Ye
Zhan-Li Sun
Zhao Liang
Zhendong Liu
Zhenran Jiang
Zhenyu Xuan
Zhihua Zhang

Additional Reviewers

Huijuan Zhu
Yizhong Zhou
Lixiang Hong
Yuan Wang
Mao Xiaodan
Ke Zeng
Xiongtao Zhang
Ning Lai
Shan Gao
Jia Liu
Ye Tang
Weiwei Cai
Yan Zhang
Zhang Yuanpeng
Han Zhu
Wei Jiang
Hong Peng
Wenyan Wang
Xiaodan Deng

Hongguan Liu
Hai-tao Li
Jialing Li
Kai Qian
Huichao Zhong
Huiyan Jiang
Lei Wang
Yuanyuan Wang
Biao Zhang
Ta Zhou
Wei Liao
Bin Qin
Jiazhou Chen
Mengze Du
Sheng Ding
Dongliang Qin
Syed Sadaf Ali
Zheng Chenc
Shang Xiang

Xia Lin
Yang Wu
Xiaoming Liu
Jing Lv
Lin Weizhong
Jun Li
Li Peng
Hongfei Bao
Zhaoqiang Chen
Ru Yang
Jiayao Wu
Dadong Dai
Guangdi Liu
Jiajia Miao
Xiuhong Yang
Xiwen Cai
Fan Li
Aysel Ersoy Yilmaz
Agata Giełczyk

Akila Ranjith
Xiao Yang
Cheng Liang
Alessio Ferone
José Alfredo Costa
Ambuj Srivastava
Mohamed Abdel-Basset
Angelo Ciaramella
Anthony Chefles
Antonino Staiano
Antonio Brunetti
Antonio Maratea
Antony Lam
Alfredo Pulvirenti
Areesha Anjum
Athar Ali Moinuddin
Mohd Ayyub Khan
Alfonso Zarco
Azis Ciayadi
Brendan Halloran
Bin Qian
Wenbin Song
Benjamin J. Lang
Bo Liu
Bin Liu
Bin Xin
Guanya Cai
Casey P. Shannon
Chao Dai
Chaowang Lan
Chaoyang Zhang
Zhang Chuanchao
Jair Cervantes
Bo Chen
Yueshan Cheng
Chen He
Zhen Chen
Chen Zhang
Li Cao
Claudio Loconsole
Cláudio R. M. Silva
Chunmei Liu
Yan Jiang
Claus Scholz
Yi Chen
Dhiya AL-Jumeily

Ling-Yun Dai
Dongbo Bu
Deming Lei
Deepak Ranjan Nayak
Dong Han
Xiaojun Ding
Domenico Buongiorno
Haizhou Wu
Pingjian Ding
Dongqing Wei
Yonghao Du
Yi Yao
Ekram Khan
Miao Jiajia
Ziqing Liu
Sergio Santos
Tomasz Andrysiak
Fengyi Song
Xiaomeng Fang
Farzana Bibi
Fatih Adıgüzel
Fang-Xiang Wu
Dongyi Fan
Chunmei Feng
Fengfeng Zhou
Pengmian Feng
Feng Wang
Feng Ye
Farid Garcia-Lamont
Frank Shi
Chien-Yuan Lai
Francesco Fontanella
Lei Shi
Francesca Nardone
Francesco Camastra
Francesco Pappalardo
Dongjie Fu
Fuhai Li
Hisato Fukuda
Fuyi Li
Gai-Ge Wang
Bo Gao
Fei Gao
Hongyun Gao
Jianzhao Gao
Gaoyuan Liang

Geethan Mendiz
Guanghui Li
Giacomo Donato
  Cascarano
Giorgio Valle
Giovanni Dimauro
Giulia Russo
Linting Guan
Ping Gong
Yanhui Gu
Gunjan Singh
Guohua Wu
Guohui Zhang
Guo-sheng Hao
Surendra M. Gupta
Sandesh Gupta
Gang Wang
Hafizul Fahri Hanafi
Haiming Tang
Fei Han
Hao Ge
Kai Zhao
Hangbin Wu
Hui Ding
Kan He
Bifang He
Xin He
Huajuan Huang
Jian Huang
Hao Lin
Ling Han
Qiu Xiao
Yefeng Li
Hongjie Wu
Hongjun Bai
Hongtao Lei
Haitao Zhang
Huakang Li
Jixia Huang
Pu Huang
Sheng-Jun Huang
Hailin Hu
Xuan Huo
Wan Hussain Wan Ishak
Haiying Wang
Il-Hwan Kim

Kamlesh Tiwari
M. Ikram Ullah Lali
Ilaria Bortone
H. M. Imran
Ingemar Bengtsson
Izharuddin Izharuddin
Jackson Gomes
Wu Zhang
Jiansheng Wu
Yu Hu
Jaya sudha
Jianbo Fan
Jiancheng Zhong
Enda Jiang
Jianfeng Pei
Jiao Zhang
Jie An
Jieyi Zhao
Jie Zhang
Jin Lu
Jing Li
Jingyu Hou
Joe Song
Jose Sergio Ruiz
Jiang Shu
Juntao Liu
Jiawen Lu
Jinzhi Lei
Kanoksak Wattanachote
Juanjuan Kang
Kunikazu Kobayashi
Takashi Komuro
Xiangzhen Kong
Kulandaisamy A.
Kunkun Peng
Vivek Kanhangad
Kang Xu
Kai Zheng
Kun Zhan
Wei Lan
Laura Yadira Domínguez
  Jalili
Xiangtao Chen
Leandro Pasa
Erchao Li
Guozheng Li

Liangfang Zhao
Jing Liang
Bo Li
Feng Li
Jianqiang Li
Lijun Quan
Junqing Li
Min Li
Liming Xie
Ping Li
Qingyang Li
Lisbeth Rodríguez
Shaohua Li
Shiyong Liu
Yang Li
Yixin Li
Zhe Li
Zepeng Li
Lulu Zuo
Fei Luo
Panpan Lu
Liangxu Liu
Weizhong Lu
Xiong Li
Junming Zhang
Shingo Mabu
Yasushi Mae
Malik Jahan Khan
Mansi Desai
Guoyong Mao
Marcial Guerra
  de Medeiros
Ma Wubin
Xiaomin Ma
Medha Pandey
Meng Ding
Muhammad Fahad
Haiying Ma
Mingzhang Yang
Wenwen Min
Mi-Xiao Hou
Mengjun Ming
Makoto Motoki
Naixia Mu
Marzio Pennisi
Yong Wang

Muhammad Asghar
  Nadeem
Nadir Subaşi
Nagarajan Raju
Davide Nardone
Nathan R. Cannon
Nicole Yunger Halpern
Ning Bao
Akio Nakamura
Zhichao Shi
Ruxin Zhao
Mohd Norzali Hj Mohd
Nor Surayahani Suriani
Wataru Ohyama
Kazunori Onoguchi
Aijia Ouyang
Paul Ross McWhirter
Jie Pan
Binbin Pan
Pengfei Cui
Pu-Feng Du
Iyyakutti Iyappan
  Ganapathi
Piyush Joshi
Prashan Premaratne
Peng Gang Sun
Puneet Gupta
Qinghua Jiang
Wangren Qiu
Qiuwei Li
Shi Qianqian
Zhi Xian Liu
Raghad AL-Shabandar
Rafał Kozik
Raffaele Montella
Woong-Hee Shin
Renjie Tan
Rodrigo A. Gutiérrez
Rozaida Ghazali
Prabakaran
Jue Ruan
Rui Wang
Ruoyao Ding
Ryuzo Okada
Kalpana Shankhwar
Liang Zhao

Sajjad Ahmed
Sakthivel Ramasamy
Shao-Lun Lee
Wei-Chiang Hong
Hongyan Sang
Jinhui Liu
Stephen Brierley
Haozhen Situ
Sonja Sonja
Jin-Xing Liu
Haoxiang Zhang
Sebastian Laskawiec
Shailendra Kumar
Junliang Shang
Guo Wei-Feng
Yu-Bo Sheng
Hongbo Shi
Nobutaka Shimada
Syeda Shira Moin
Xingjia Lu
Shoaib Malik
Feng Shu
Siqi Qiu
Boyu Zhou
Stefan Weigert
Sameena Naaz
Sobia Pervaiz
Somnath Dey
Sotanto Sotanto
Chao Wu
Yang Lei
Surya Prakash
Wei Su
Qi Li
Hotaka Takizawa
FuZhou Tang
Xiwei Tang
LiNa Chen
Yao Tuozhong
Qing Tian
Tianyi Zhou
Junbin Fang
Wei Xie
Shikui Tu
Umarani Jayaraman
Vahid Karimipour

Vasily Aristarkhov
Vitoantonio Bevilacqua
Valeriya Gribova
Guangchen Wang
Hong Wang
Haiyan Wang
Jingjing Wang
Ran Wang
Waqas Haider Bangyal
Pi-Jing Wei
Fangping Wan
Jue Wang
Minghua Wan
Qiaoyan Wen
Takashi Kuremoto
Chuge Wu
Jibing Wu
Jinglong Wu
Wei Wu
Xiuli Wu
Yahui Wu
Wenyin Gong
Zhanjun Wang
Xiaobing Tang
Xiangfu Zou
Xuefeng Cui
Lin Xia
Taihong Xiao
Xing Chen
Lining Xing
Jian Xiong
Yi Xiong
Xiaoke Ma
Guoliang Xu
Bingxiang Xu
Jianhua Xu
Xin Xu
Xuan Xiao
Takayoshi Yamashita
Atsushi Yamashita
Yang Yang
Zhengyu Yang
Ronggen Yang
Yaolai Wang
Yaping Yang
Yue Chen

Yongchun Zuo
Bei Ye
Yifei Qi
Yifei Sun
Yinglei Song
Ying Ling
Ying Shen
Yingying Qu
Lvjiang Yin
Yiping Liu
Wenjie Yi
Jianwei Yang
Yu-Jun Zheng
Yonggang Lu
Yan Li
Yuannong Ye
Yong Chen
Yongquan Zhou
Yong Zhang
Yuan Lin
Yuansheng Liu
Bin Yu
Fang Yu
Kumar Yugandhar
Liang Yu
Yumin Nie
Xu Yu
Yuyan Han
Yikuan Yu
Ying Wu
Ying Xu
Zhiyong Wang
Shaofei Zang
Chengxin Zhang
Zehui Cao
Tao Zeng
Shuaifang Zhang
Liye Zhang
Zhang Qinhu
Sai Zhang
Sen Zhang
Shan Zhang
Shao Ling Zhang
Wen Zhang
Wei Zhao
Bao Zhao

Zheng Tian
Zheng Sijia
Zhenyu Xuan
Fangqing Zhao
Zhipeng Cai
Xing Zhou
Xiong-Hui Zhou

Lida Zhu
Ping Zhu
Qi Zhu
Zhong-Yuan Zhang
Ziding Zhang
Junfei Zhao
Juan Zou

Quan Zou
Qian Zhu
Zunyan Xiong
Zeya Wang
Yatong Zhou
Shuyi Zhang
Zhongyi Zhou

Contents – Part I

An Efficient Elman Neural Networks Based on Improved Conjugate Gradient Method with Generalized Armijo Search

Mingyue Zhu[1], Tao Gao[2], Bingjie Zhang[1], Qingying Sun[1],
and Jian Wang[1(✉)]

[1] College of Science, China University of Petroleum, Qingdao 266580, China
mingyue_zhu16@163.com, bingjie_zhang_1993@163.com,
sunqingying01@163.com, wangjiann1@upc.edu.cn
[2] College of Information and Control Engineering,
China University of Petroleum, Qingdao 266580, China
gaotao_1989@126.com

Abstract. Elman neural network is a typical class of recurrent network model. Gradient descent method is the popular strategy to train Elman neural networks. However, the gradient descent method is inefficient owing to its linear convergence property. Based on the Generalized Armijo search technique, we propose a novel conjugate gradient method which speeds up the convergence rate in training Elman networks in this paper. A conjugate gradient coefficient is proposed in the algorithm, which constructs conjugate gradient direction with sufficient descent property. Numerical results demonstrate that this method is more stable and efficient than the existing training methods. In addition, simulation shows that, the error function has a monotonically decreasing property and the gradient norm of the corresponding function tends to zero.

Keywords: Elman · Conjugate gradient · Generalized Armijo search
Monotonicity

1 Introduction

Elman neural network is a typical recurrent neural network with one single hidden layer [1]. In the hidden layer, each neuron is connected to other neurons and itself [2]. This feedback mechanism contributes to recognizing and generating the spatial and temporal patterns [3]. In general, gradient descent learning algorithm has been widely used in training neural networks [4, 5]. To reduce computational complexity and accelerate the convergence speed, some efficient algorithms have been proposed, such as approximate gradient method of Elman networks [6]. Unfortunately, the indispensable reason of poor performance is that these algorithms essentially have first-order convergence properties. In order to overcome this obstacle, conjugate gradient (CG) and Newton methods are often used as the common replacement of the gradient descent method [7–9]. Compared with the gradient descent algorithm, Newton method is a more efficient algorithm. However, Newton method requires the second derivative information of the function, which involves the Hessian matrix and the matrix inversion operation. In this sense,

© Springer International Publishing AG, part of Springer Nature 2018
D.-S. Huang et al. (Eds.): ICIC 2018, LNCS 10954, pp. 1–7, 2018.
https://doi.org/10.1007/978-3-319-95930-6_1

the calculation of Newton method is more time-consuming. Compared with Newton method, conjugate gradient method has certain advantages. It does not require the second derivative information and still converges faster than the gradient method [10].

The classical CG methods include: Hestenes-Stiefel (HS) method [11], Fletcher-Reeves(FR) method [12], Polak-Ribiere-Polyak (PRP) method [13] etc. We note that the conjugate coefficient plays an important role in designing CG algorithm. And it is also a challenging work to determine the optimal coefficient. Recently, Rivaie et al. [14] have proposed an interesting CG coefficient. Motivated by this work, one main contribution of this paper is to look for a more efficient conjugate coefficient during training Elman neural networks.

In recent years, CG method becomes a popular choice in training neural networks [8, 9]. In [8], the PRP conjugate gradient algorithm was employed to train the feedforward neural network. In fact, a fixed learning rate would often lead to low training performance in training networks. As an improvement, the generalized Armijo search method [15] for feedforward neural network has been proposed in [9]. The main idea is to use the generalized Armijo search method to find out the best learning rate. Inspired by [9, 15, 16], an efficient learning algorithm based on generalized Armijo search method has been proposed with a novel conjugate coefficient. Then, we construct a CG Elman neural network model by employing the optimal coefficient and the generalized Armijo search method. For given problems, numerical simulation demonstrates that the proposed algorithm is more efficient than its counterparts. In addition, the decreasing monotonically property and the deterministic convergence of this presented algorithm have also been verified from an application manner.

The rest of the paper is organized as follows: in Sect. 2, we introduce the structure of the Elman network and its learning algorithm: the CG method with generalized Armijo search for Elman neural networks (ECGGA). The supporting numerical experiment is shown in Sect. 3. Finally, we give the conclusion in Sect. 4.

2 Structure of Elman Networks and Algorithms

Without loss of generality, we consider one typical Elman network which has m input, n hidden and one output neurons. $\{x^q, O^q\}_{q=1}^{Q} \subset R^m \times R$ is the training sample set, where x^q and O^q are the input and the corresponding target of the q-th sample, respectively. For convenience, we assume that $U_I = \left(u_{ij}^I\right)_{n \times m}$ to denote the connecting weight matrix of input layer and hidden layer, $U_L = \left(u_{ij}^L\right)_{n \times n}$ is the recurrent weight matrix. Let $v = (v_1, v_2, \cdots, v_n)^T \in R^n$ be the weight vector connecting the hidden layer and the output layer.

Two continuously differentiable functions are employed to be the activation functions, g and $f : R \rightarrow R$, of hidden and output layers, respectively. We introduce the following vector-valued function

$$G(\boldsymbol{b}) = (g(b_1), g(b_2), \cdots, g(b_n))^T, \forall \boldsymbol{b} \in R^n. \tag{1}$$

The hidden layer output, $\boldsymbol{y}^q$, corresponding to the total input is as follows:

$$\boldsymbol{y}^q = G(U_I x^q + U_L \boldsymbol{y}^{q-1}), \boldsymbol{y}^0 \equiv \boldsymbol{0}. \tag{2}$$

The corresponding network output is then expressed as

$$d^q = f(\boldsymbol{v} \cdot \boldsymbol{y}^q), q = 1, 2, \cdots, Q. \tag{3}$$

where $\cdot$ indicates the inner product of the two vectors.

For simplicity, all of the weights among the input, hidden and output layers are incorporated as a total vector $\boldsymbol{w} = (\boldsymbol{v}^T, \boldsymbol{u}^T)^T \in R^{n(n+m+1)}$ where $\boldsymbol{u} = vec(U_I, U_L)$. The error function of the network is defined as follows:

$$E(\boldsymbol{w}) = \frac{1}{2} \sum_{q=1}^{Q} (O^q - d^q)^2. \tag{4}$$

In order to facilitate the expression, we use some abbreviations to represent different algorithms for Elman networks. The gradient descent training algorithm for Elman neural networks (EGD). The CG training method for Elman neural networks (ECG), and the novel CG method with generalized Armijo search for Elman neural networks (ECGGA).

For any given initial weight vector $\boldsymbol{w}^0$, we introduce the ECGGA algorithm as follows: the weight iteration sequence is generated by the following formula

$$\boldsymbol{w}^{k+1} = \boldsymbol{w}^k + \eta^k \boldsymbol{p}^k, k \in N. \tag{5}$$

where $\eta^k \in (0, 1)$ is the learning rate of the k-th iteration. The learning rate is computed by carrying out the generalized Armijo search method. The conjugate gradient search direction, $\boldsymbol{p}^k$, is denoted as

$$\boldsymbol{p}^k = \begin{cases} -E_w^k, & k = 0, \\ -E_w^k + \beta^k \boldsymbol{p}^{k-1}, & k \geq 1, \end{cases} \tag{6}$$

where

$$E_w^k = \begin{pmatrix} E_v(\boldsymbol{w}^k) \\ E_u(\boldsymbol{w}^k) \end{pmatrix} = \begin{pmatrix} -\sum_{q=1}^{Q} (O^q - d^q) f'(\boldsymbol{v} \cdot \boldsymbol{y}^q) \boldsymbol{y}^q \\ -\sum_{q=1}^{Q} (O^q - d^q) f'(\boldsymbol{v} \cdot \boldsymbol{y}^q) \left(\frac{\partial \boldsymbol{y}^q}{\partial \boldsymbol{u}}\right)^T \boldsymbol{v} \end{pmatrix} \tag{7}$$

and β^k is the conjugate gradient coefficient.

$$\frac{\partial \boldsymbol{y}^q}{\partial \boldsymbol{u}} = G'\left(U\left(\begin{matrix} \boldsymbol{x}^q \\ \boldsymbol{y}^{q-1} \end{matrix}\right)\right)\left(\left(\begin{matrix} \boldsymbol{x}^q \\ \boldsymbol{y}^{q-1} \end{matrix}\right)^T \otimes \boldsymbol{I}_n + U_L \frac{\partial \boldsymbol{y}^{q-1}}{\partial \boldsymbol{u}}\right), \frac{\partial \boldsymbol{y}^0}{\partial \boldsymbol{u}} = \boldsymbol{0} \tag{8}$$

where the notation $\otimes$ presents the Kronecker product.

Inspired by this work [14], we propose another new CG coefficient, β^k, which is denoted as β^k_{MBQJ} as follows:

$$\beta^k_{MBQJ} = \frac{\mu \left(E^k_w\right)^T \left(\boldsymbol{p}^{k-1}\right)}{\left(\boldsymbol{p}^{k-1}\right)^T \left(\boldsymbol{p}^{k-1} - E^k_w\right)} \tag{9}$$

where $\mu \in (0,1)$ is a given constant.

For the generalized Armijo search technique, the learning rate η^k in (5) satisfies that

$$E\left(\boldsymbol{w}^k + \eta^k \boldsymbol{p}^k\right) \leq E\left(\boldsymbol{w}^k\right) + \mu_1 \eta^k \left(E^k_w\right)^T \boldsymbol{p}^k \tag{10}$$

and

$$\eta^k \geq \gamma_1 \text{ or } \eta^k \geq \gamma_2 \eta^k_* > 0, \tag{11}$$

where $\mu_1, \mu_2 \in (0,1)(\mu_1 \leq \mu_2)$, and η^k_* requires that

$$E\left(\boldsymbol{w}^k + \eta^k_* \boldsymbol{p}^k\right) > E\left(\boldsymbol{w}^k\right) + \mu_2 \eta^k_* \left(E^k_w\right)^T \boldsymbol{p}^k, \tag{12}$$

γ_1 and γ_2 are positive constant values.

3   Simulation

In this section, we simulate a binary classification problem to compare the performance of EGD, ECG and ECGGA. We can get Banknote authentication dataset in the UC Irvine machine learning repository [17]. Banknote data set has a total of 1,372 samples, each of which has four input variables and one output variable. We randomly selected 1,029 samples as training data set and the remaining 343 samples as the test data set. Thus, we set the Elman network with four input units, five hidden units and one output unit. In this experiment, we construct three identical network structures and set the same training parameters to compare the performance of the different algorithms EGD, ECG and ECGGA. The initial weights are randomly chosen in the interval $[-0.5, 0.5]$. The logistic function $f(b) = \frac{1}{1+e^{-b}}, (b \in R)$ as the activation functions of the hidden and output layers, respectively. Training process is terminated when the error E is less than 1.0×10^{-6} or the iterations are up to 4,500.

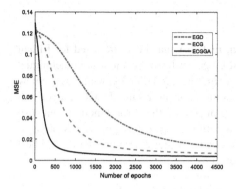

Fig. 1. Square error of Banknote problem

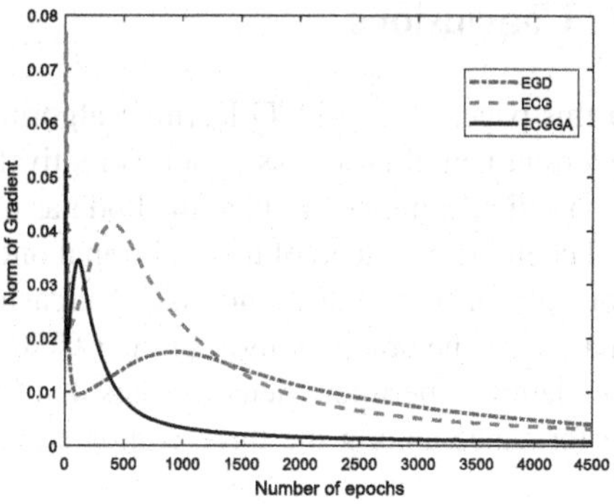

Fig. 2. Norm of gradient in Banknote problem

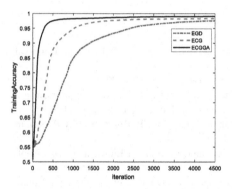

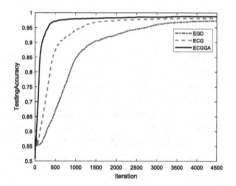

Fig. 3. Training accuracy of Banknote problem

Fig. 4. Testing accuracy of Banknote problem

From Fig. 1, we can see the monotone descent property of the error function. It is obvious that the error curve for the ECGGA algorithm converges faster than the other two counterparts. As shown in Fig. 2, the gradient of the error function tends to zero with the increase of iterations. This also demonstrates the convergent behavior of the proposed algorithm. Figure 3 shows the training accuracy curves on different algorithms. Obviously, the convergence rate of ECGGA algorithm is faster than that of the other two. Furthermore, its training accuracy is significantly higher than that of EGD and ECG. Figure 4 demonstrates the curves of testing accuracy on Banknote data set. We can clearly see that ECGGA algorithm is better than the other two methods.

4 Conclusions

In this paper, a novel CG learning algorithm, ECGGA, has been proposed for Elman recurrent neural networks which is motivated by generalized Armijo search technique. A specific conjugate gradient method has been employed in ECGGA, which uses a new CG coefficient instead of the traditional ones. Simulation shows that ECGGA algorithm not only ensures the monotonic decreasing property of the error function, but also guarantees the gradient norm of the ECGGA algorithm convergence to zero. Moreover, simulation experiment demonstrates that ECGGA has faster convergence rate and best performance among the three compared algorithms.

Acknowledgments. This work was supported in part by the National Natural Science Foundation of China (No. 61305075), the Natural Science Foundation of Shandong Province (Nos. ZR2015AL014, ZR201709220208) and the Fundamental Research Funds for the Central Universities (Nos. 15CX08011A, 18CX02036A).

References

1. Ooi, S.Y., Tan, S.C., Cheah, W.P.: Experimental Study of Elman Network in Temporal Classification. Springer, Singapore (2017)
2. Elman, J.L.: Finding structure in time. Cogn. Sci. **14**(2), 179–211 (1990)
3. Ahalt, S.C., Liu, X.M., Wang, D.L.: On temporal generalization of simple recurrent networks. Neural Netw. Official J. Int. Neural Network Soc. **9**(7), 1099–1118 (1996)
4. Williams, R.J., Zisper, D.: A Learning Algorithm for Continually Running Fully Recurrent Neural Networks. MIT Press **1**(2), 270–280 (1989)
5. Wu, W., Xu, D.P., Li, Z.X.: Convergence of gradient method for Elman networks. Appl. Math. Mech. **29**(9), 1231–1238 (2008)
6. Xu, D.P., Li, Z.X., Wu, W.: Convergence of approximated gradient method for Elman network. Neural Network World **18**(3), 171–180 (2008)
7. Güntürkün, R.: Using elman recurrent neural networks with conjugate gradient algorithm in determining the anesthetic the amount of anesthetic medicine to be applied. J. Med. Syst. **34**(4), 479–484 (2010)
8. Wang, J., Wu, W., Zurada, J.M.: Deterministic convergence of conjugate gradient method for feedforward neural networks. Neurocomputing **74**(14), 2368–2376 (2011)
9. Wang, J., Zhang, B.J., Sun, Z.Q., Hao, W.X., Sun, Q.Y.: A novel conjugate gradient method with generalized Armijo search for efficient training of feedforward neural networks. Neurocomputing **275**, 308–316 (2018)
10. Nocedal, J., Wright, S.J.: Numerical Optimization. Springer, New York (2006)
11. Hestenes, M.R., Steifel, E.: Method of Conjugate Gradients for Solving Linear Systems. National Bureau of Standards, Washington (1952)
12. Fletcher, R., Reeves, C.M.: Function minimization by conjugate gradients. Comput. J. **7**(2), 149–154 (1964)
13. Polak, E., Ribiere, G.: Note sur la convergence de methodes de directions conjures. Revue Francaise Information Recherche Operationnelle **16**(16), 35–43 (1969)

14. Rivaie, M., Fauzi, M., Mamat, M., Mohd, I.: A new class of nonlinear conjugate gradient coefficients with global convergence properties. AIP Conf. Proc. **1482**, 486–491 (2012)
15. Sun, Q.Y., Liu, X.H.: Global convergence results of a new three terms conjugate gradient method with generalized Armijo step size rule. Math. Numer. Sin. **26**(1), 25–36 (2004)
16. Dong, X., Yang, X., Huang, Y.: Global convergence of a new conjugate gradient method with Armijo search. J. Henan Normal Univ. **6**, 25–29 (2015)
17. UCI Machine Learning Repository. http://archive.ics.uci.edu/ml

Natural Calling Gesture Recognition in Crowded Environments

Aye Su Phyo$^{(\boxtimes)}$, Hisato Fukuda$^{(\boxtimes)}$, Antony Lam$^{(\boxtimes)}$,
Yoshinori Kobayashi$^{(\boxtimes)}$, and Yoshinori Kuno$^{(\boxtimes)}$

Graduate School of Science and Engineering, Saitama University, Saitama, Japan
{ayesuphyo, fukuda, antonylam,
kuno}@cv.ics.saitama-u.ac.jp,
yosinori@hci.ics.saitama-u.ac.jp

Abstract. Most existing gesture recognition algorithms use fixed postures in simple environments. In natural calling, the user may perform gestures in various positions and the environment may be occupied by many people with many hand motions. This paper presents an algorithm for natural calling gesture recognition by detecting gaze, the position of the hand-wrist, and fingertips. A challenge to solve is how to make the natural calling gesture recognition work in crowded environments with randomly moving objects. The approach taken here is to get the key-points of individual people using a real-time detector by using a camera and detect gaze and hand-wrist positions. Then, zooming into the hand-wrist part and getting the key-points of fingertips, we calculate the positions of the fingertips to recognize calling gestures. We tested the proposed approach in video under different conditions: from one person to over four people that sit and walk around. We obtained an average recognition accuracy of 87.12%, thus showing the effectiveness of our approach.

Keywords: Natural calling gesture · Gaze · Hand-wrist · Fingertip
Crowded environment · Openpose

1 Introduction

Many countries in the world are faced with the aging of its population, which needs help and care. One solution, service robots, may even replace home care staff (e.g. a caregiver) to take care of the elderly. For service robots to perform tasks like humans, service robots should have abilities such as recognizing faces, gestures, and signs [1].

Gestures, one of the most natural forms of communication, can be performed using any part of the body, from the arms, the head, as in a nod, or even the face. This work mainly focuses on natural calling hand gestures. Hand gesture recognition is a challenging problem in computer vision and is a topic of active research using sensor-based and vision-based methods [2]. The study of vision-based methods, which uses only a camera without the use of any extra devices, has made great progress in different areas. However, there are still challenges such as illumination changes and the background-foreground problem, where objects in the scene might even contain skin-like colors. Another issue is the presence of crowds moving around with many hand motions.

© Springer International Publishing AG, part of Springer Nature 2018
D.-S. Huang et al. (Eds.): ICIC 2018, LNCS 10954, pp. 8–14, 2018.
https://doi.org/10.1007/978-3-319-95930-6_2

In crowded environments, conventional methods might erroneously recognize hand movements as calling gestures. Based on these challenges, we propose an algorithm that works in crowded environments with randomly moving objects. In addition, the caller's position may be varied. Conventional methods would also typically check all the people in the scene. However, in our approach, only the people who gaze towards the camera with defined wrist positions are checked from the scene. Among all the body and hand motions, we perform zooming on the hand because it is difficult to detect details of the hands. And as for wrist positions, there may be many similar actions to calling gestures. Our approach uses the key-points of the fingertips to differentiate between whether something is a calling gesture or not.

The remainder of this paper is organized as follows: related work is reviewed in Sect. 2. In Sect. 3, our approach for gesture recognition is explained. Experiments and results are given in Sect. 4. Section 5 is our conclusion and future work.

2 Related Work

Applied to elderly care scenarios, a Kinect based approach to recognize calling gestures is proposed in [3] which used a skeleton-based recognition system to detect when the user is standing up, and an octree when the skeleton is not properly tracked. Canal et al. [4] applied gesture recognition methods for human-multirobot interaction using the Euclidean distance between the hand and the neck joints, the angle in the elbow joint, the distance between the hip and the hand joints, and the position of the hand joint. Grazia et al. [5] recognized ten gestures from visual signals used by the army to control a mobile robot performed by Neural Network (NN) classifiers using quaternions and angles. Gu et al. [6] used the Euler angle to recognize left arm gestures to control a Pioneer robot with four gestures (come, go, rise up, and sit down) using joint angles (left elbow yaw and roll left shoulder yaw and pitch). Some methods are based on detecting and counting which of the 5 fingers are extended. Zhou et al. [7] proposed an algorithm that extracts fingers from salient hand edges and high-level geometry features from hand parts. The state of fingers is determined by the distance of external contours of the fingers to the wrist position.

Most past work focuses on one person in a simple scene, however, in our approach, we address the issue of working in a cluttered and noisy environment, which includes moving people, hand motions, and more. Moreover, most of the works assume a fixed position for the person to perform gesture recognition and have defined start-points and end-points of the gestures. In our work, the person who performs the calling gesture can do so from various positions.

3 Process of Natural Calling Gesture Recognition

Our approach can be divided into some steps. We perform body key-point acquisition, detection of gaze and hand-wrist position, zooming in the hand-wrist part, hand fingertip key-point acquisition, and finally calling gesture recognition (Fig. 1).

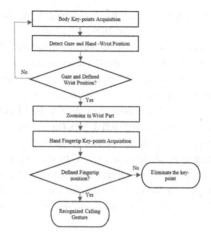

Fig. 1. Overall process of the proposed natural calling gesture recognition system

3.1 Body Key-Points Feature Acquisition for Gesture Recognition

We perform gesture recognition by first extracting body and hand key-points from the OpenPose detector [8]. The key-points can be extracted from RGB images or videos captured by camera. There are times, the hand part can be missed in crowded scenes.

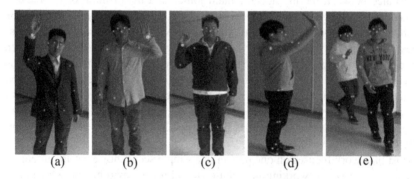

| (a) | (b) | (c) | (d) | (e) |

Fig. 2. Examples of calling gestures and similar hand motions with key-points

3.2 Detecting Gaze and Finding Hand-Wrist Position

To call someone, we must face the person and perform some hand calling gesture. Based on this, our work detects gaze and finds hand-wrist positions to avoid tracking of all people. In crowded environments, there are many people moving around with hand movements. By detecting gaze and hand-wrist positions first, we can reduce tracking other hand motions. Hand-wrist positions for calling are shown in Fig. 2. We do not consider Fig. 2d as calling gesture as the person does not face the camera. In Fig. 2e, the person in the back faces the camera and the hand-wrist is also in a raised position, and so, zoom in for closer inspection.

3.3   Zoom in Wrist Part

There are times when we may miss the hand part in a crowded scene because the hand area is too small. It is possible to get the fingertip positions if the hand is near the camera. For that reason, after detecting gaze and the defined hand-wrist position, our system zooms in on the hand-wrist part as in Fig. 3.

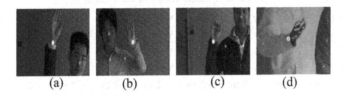

(a)   (b)   (c)   (d)

Fig. 3. Hand fingertip key-points features acquisition

3.4   Hand Fingertip Key-Points Features Acquisition

In the defined hand-wrist position, there are many hand motions. This could lead to the action being confused with a calling gesture. We observe that the pose of the fingertips of calling gestures are typically open. For that reason, we use the detailed key-points of the fingertips. After zooming in the hand-wrist part, the system calculates positions of the fingertips. Figures 3(a–c) show the fingertip positions of the calling gesture and Fig. 3d is not a calling gesture. The pose of the fingertips of calling gestures are typically open like in Figs. 4a and b. Although Figs. 4c and d are similar with the open-type poses, we do not consider these to be calling gestures.

(a)   (b)   (c)   (d)

Fig. 4. Example of positions of fingertips.

To classify calling gestures, our approach uses the x-coordinates of each fingertip and the angles from the base of the palm to the thumb and to the little finger. Firstly, we consider whether the order of the x-coordinate values of each fingertip (from the thumb to the little finger) is ascending or descending or not. If the order is either ascending or descending, the angle of the vector form the base of the palm to the thumb is between 30 and 70°, and the angle of the vector from the base of the palm to the little finger is between 0 to 45°, we consider that a calling gesture.

4 Experiments and Results

In our experiment, we captured hundreds of videos in several specific situations and evaluated our algorithm on them. Table 1 summarizes the characteristics of each situation, S and W denote sitting and walking respectively.

Table 1. Characteristics of each video

Video ID	V1– V10	V10– V20	V21– V30	V31– V40	V41– V50	V51– V60	V61– V70	V71– V80	V81– V100
Person	1	2	3	4	1	2	3	4	>4
Situation	S	S	S	S	W	W	W	W	S, W
Distance	~2 m	~2 m	~2 m	~2 m	~2 m	~2 m	~2 m	~2 m	~3 m
FPS	30	30	30	30	30	30	30	30	30

Fig. 5. Detected calling gestures from our algorithm in three different scenes.

Figure 5 shows sample results of the accuracy of our algorithm in three different scenes. The quantitative performance of our algorithm is shown in Table 2. To get the hand part clearly, we recorded the video at high resolution (1980×1080). We find that a body is usually correctly detected if it is within 2 m. With a gesture recognition accuracy of 74% in the case of sitting and walking with more than four people, recognition performance is notably worse. This can be explained by the varied motion of the people. If people walk fast, we cannot correctly detect the body key-points. If we cannot get the correct key-points, our algorithm reduces in performance. In addition, we may sometimes miss the hand fingertip key-points, and, if gaze detection fails, we miss the calling gesture.

Table 2. Performance of gesture recognition

Situation	Total no. of gestures	Accuracy rate	No. correctly identified gestures
Sitting one person	50	96%	48
Sitting two people	50	96%	48
Sitting three people	50	86%	43
Sitting four people	50	82%	41
Walking one person	50	96%	48
Walking two people	50	90%	45
Walking three people	50	82%	41
Walking four people	50	78%	39
Sitting and Walking more than four people	35	74%	26

5   Conclusion and Future Work

This paper introduces natural calling gesture recognition in crowded environments, for human-robot interaction. To detect users, this study used the key-points from the OpenPose real-time detector. Using these key-points, gaze detection and finding hand-wrist positions were performed. If our algorithm detects a gaze towards the camera and a defined hand-wrist position, it zooms into the hand-wrist part. After that, it finds the key-points of the hand's fingertips. From these key-points, our algorithm recognizes whether the gesture is for calling or not. We expect that many robotic applications will make effective use of the proposed algorithm for human robot interaction in crowded environments.

For future work, we will extend our work to real-time and to more complicated situations with service robots. Furthermore, we can use sound information (when possible) as well as vision to allow for better human-robot interaction. For instance, sound localization can be used with calling motions and would be especially helpful in cases where the person could be occluded in the crowded environment.

Acknowledgement. This work was supported by JSPS KAKENHI Grant Number JP26240038.

References

1. Lemaignan, S., Warnier, M., Sisbot, E.A., Clodic, A., Alami, R.: Artificial cognition for social human–robot interaction: an implementation. Artif. Intell. **247**, 45–69 (2017)
2. Escalera, S., Athitsos, V., Guyon, I.: Challenges in multimodal gesture recognition. J. Mach. Learn. Res. **17**(1), 2549–2602 (2016)
3. Zhao, X., Naguib, A.M., Lee, S.: Kinect based calling gesture recognition for taking order service of elderly care robot. In: The 23rd IEEE International Symposium on Robot and Human Interactive Communication 2014 RO-MAN, pp. 525–530, August 2014

4. Canal, G., Angulo, C., Escalera, S.: Gesture based human multirobot interaction. In: 2015 International Joint Conference on Neural Networks (IJCNN), pp. 1–8, July 2015
5. Cicirelli, G., Attolico, C., Guaragnella, C., D'Orazio, T.: A kinect-based gesture recognition approach for a natural human robot interface. Int. J. Adv. Rob. Syst. **12** (2015)
6. Gu, Y., Do, H., Ou, Y., Sheng, W.: Human gesture recognition through a kinect sensor. In: 2012 IEEE International Conference on Robotics and Biomimetics (ROBIO), pp. 1379–1384, December 2012
7. Zhou, Y., Jiang, G., Lin, Y.: A novel finger and hand pose estimation technique for real-time hand gesture recognition. Pattern Recogn. **49**, 102–114 (2016)
8. Hidalgo, G., Cao, Z., Simon, T., Wei, S.-E., Joo, H., Sheik, Y.: Openpose. https://github.com/CMU-Perceptual-Computing-Lab/openpose

An Adaptive Segmentation Algorithm for Degraded Chinese Rubbing Image Binarization Based on Background Estimation

Han Huang[1(✉)], Zhi-Kai Huang[2], Yong-Li Ma[2], and Ling-Ying Hou[2]

[1] Department of Mechanical and Biomedical Engineering,
City University of Hong Kong, Kowloon Tong, Hong Kong
huang.hanl@husky.neu.edu
[2] College of Mechanical and Electrical Engineering,
Nanchang Institute of Technology, Nanchang 330099, Jiangxi, China
huangzhik2001@163.com

Abstract. Image Segmentation plays an important role in image processing and analysis. In order to preserve strokes of a Chinese character while enhancing character details for degraded historical document image, we propose an adaptive segmentation algorithm for degraded historical document image binarization based on background estimation for non-uniform illumination images. The novelty of the proposed method is that find an optimal background estimation based on Blind/Referenceless Image Spatial QUality Evaluator. The proposed method has four steps: (i) preprocess using median filtering; (ii) extraction of the red color components; (iii) a morphological operation in order to find an optimal background estimation; and (iv) segmented binary image using Otsu's Thresholding. Experimental results demonstrate that it is capable of extracting more accurate segmentation of characters for degraded Chinese rubbing document image.

Keywords: Top-Hat transform
Blind/Referenceless Image Spatial QUality Evaluator · Background estimation
Mathematical morphology

1 Introduction

There is no other civilization just like China that preserve the memory of its own history and culture using rubbing which writing on stone (inscriptions), at the same time, rubbings are important components of ancient Chinese books, it is the main source for people to learn, study, and research history. The rubbing document digitization is a novel approach to promote and pass on Chinese traditional arts, as well as a new idea to protect stone relics. On the other way, many Chinese rubbing from the surviving ancient codices often lose their visual quality over time due to their storage and deterioration from moisture, infestation and many other factors. After those Chinese rubbings becoming digital image, the foreground and background are difficult to separate due to varying contrast, heavy noise and variable background intensity.

© Springer International Publishing AG, part of Springer Nature 2018
D.-S. Huang et al. (Eds.): ICIC 2018, LNCS 10954, pp. 15–24, 2018.
https://doi.org/10.1007/978-3-319-95930-6_3

In addition to the low contrast, background noise is often introduced due to texture of stone, deterioration of the paper material, etc. Therefore, extracting clean Chinese character from raw images of the ancient scripts is a challenging yet crucial step for any further automatic document image analysis tasks such as page layout analysis, character recognition, etc. In the acquisition process of the rubbing image, the non-uniform distribution of the brightness will affect the quality of images. To improve the quality of character segmentation, a great deal of methods and techniques have been studied. The most important preprocessing step of these methods is text binarization that converts the document images from a gray-scale or color image into a binary image, it is separate the foreground and background of ancient document image in way that the background information is represented by white pixels and the foreground by black ones.

One of the simplest and yet efficient image processing techniques which can be used to separate foreground and background layers of document images is thresholding [1]. Although many thresholding techniques that could be categorized as global [2–4] and local thresholding [5–7] algorithms, multi-thresholding methods [8–10] and adaptive thresholding techniques [11, 12]. Global thresholding is preferred when the images are having equal contrast spreading on background and foreground whose illumination is uniform and have a great difference between target and background. Local adaptive thresholding is used for restoring the foreground pixel from document image. Generally, the choice of a suitable algorithm for a degradation document image proved to be a very difficult procedure itself. Figure 2 shows an Example of a Chinese degraded rubbing image and its histogram. Due to the existing complex degradations, problems are still challenging. The effect of many experiments have indicated that the traditional method of uniform illumination of the weak target image processing is not acceptable, there are many disadvantages such as: the separation of target and background is incomplete or the loss of a valid target, low processing efficiency and so on.

In here, we present an new adaptive algorithm for Chinese rubbing image using the Blind/Referenceless Image Spatial QUality Evaluator (BRISQUE) which utilizes a Natural Scene Statistics (NSS) model framework of locally normalized luminance coefficients and quantifies 'naturalness' using the parameters of the model [13]. To correct low contrast of rubbing images, a morphological top-hat operator with disk-shaped structuring element and adaptive size of pixels was applied to the red component of the color Chinese rubbing image which utilizes the difference between structure elements. To reduce noise, a median filter is applied to the shade corrected image. The outline of this paper is as follows. We will briefly discuss about mathematical morphology strategy, followed Blind/Referenceless Image Spatial QUality Evaluator (BRISQE). We will later explain our algorithm followed by our experimental results and discuss them. Finally, the paper is concluded in the last section.

2   Related Work

Generally, degradation document image is occurring due to background noise or variation in contrast and illumination. A shading degradation illustrated document image is more common because camera documents are more susceptible to the lighting variation. There are some binarization document image methods have incorporated background estimation and normalization steps. In [16], the author proposed adaptive contrast technique which estimates document background surface using an iterative polynomial smoothing procedure, which estimates the shading variation and compensates the shading degradation based on the estimated shading variation. In [11] estimate the document background surface through an adaptive and iterative image averaging procedure. In [12], background estimation is applied along with image normalization based on background compensation. For badly illuminated documents, adaptive thresholding, which calculates a local threshold for each image pixel, is the traditional approach for the document binarization. In [17], the authors use a Support Vector Machine (SVM) to select an optimal global threshold for binarization of each image block, the entire image is further binarized by a locally adaptive thresholding method.

There are many algorithms that attempt to separate foreground (characters) from background in scanned text documents, but thresholding in one form or another remains one of the standard workhorses, such as Bernsen's adaptive thresholding that estimates a threshold for each pixel according to its local neighbors. In [18], the authors use local maximum and minimum to build a local contrast image. Then, a sliding window is applied across that image to determine local thresholds. In [19], a phase-based binarization model for ancient document images is proposed, develop a ground truth generation tool, called PhaseGT, to simplify and speed up the ground truth generation process for ancient document images. More recently, in [20], proposed an active contours evolving according to intrinsic geometric measures of the document image, the image contrast is defined by the local image maximum and minimum which is used to automatically generate the initialization map of our active contour model, finally, an average thresholding is also used to produce the final delineation and binarization. As we observed, most binarization methods are based on intuitive observation of the gray level distinctions between characters and backgrounds, regardless of the adaptive selection thresholding for the degraded document image. To overcome those difficulties, we propose an adaptive method that applies different techniques to separate characters from degraded document images.

3   Mathematical Morphology

The basic idea of image processing using mathematical morphology is that detect a target image by structural elements with a certain shape (a basic element that has a certain structural shape, such as a certain size of a rectangle, a circle, or a diamond, etc.). By examining the effectiveness of the structural elements in the image target area and the filling method, we can get relevant information about image morphology and structure, using them to achieve the purpose of image analysis and recognition.

3.1   The Influence of Structural Elements

The structure element is a key point for the morphological image processing, different structure elements determines the analysis and processing of various geometric information in the image, also determines the amount of the computation during the transformation of the data, so the analysis of the structural elements is an important content of image edge detection. In general, both the size and structure shape of the structural elements will affect the image edge detection. Small size structure elements have weak denoising ability, but they can detect precise edge details. Large size structure elements have stronger ability to denoise, but the edges detected are more rough. What's more, the structure elements of different shapes apply different induction ability to the edge of different images.

A grayscale image can be regarded as a set of two dimensional points, the expansion and corrosion operation can be expressed as follows:

$$(f \otimes S)(x, y) = \max\{g(x - k, y - l)|(k, l) \in S\} \tag{1}$$

$$(f \oplus S)(x, y) = \min\{g(x - k, y - l)|(k, l) \in S\} \tag{2}$$

The Top-hat algorithm can be divided into Top-hat algorithm and Bot-hat algorithm based on the different components of open operation and closed operation.

Apply Top-hat algorithm to the image and expressed as TH:

$$TH(x, y) = (f - f \circ S)(x, y) \tag{3}$$

Apply Bot-hat algorithm to the image and expressed as BH:

$$BH(x, y) = (f \bullet S - f)(x, y) \tag{4}$$

In the equation, $f(x, y)$ is the original grayscale image, $S(x, y)$ is the structural elements. Top-hat transforming extract foreground information through the difference between the original image and its open operation, and Bot-hat transforming suppress background information through the difference between the original image and its closed operation.

3.2 Blind/Referenceless Image Spatial QUality Evaluator

BRISQUE is a universal non-reference image quality evaluation algorithm based on spatial image statistical features. The algorithm is based on the following theoretical premise: the natural image has certain laws, and the visual characteristics of the human eye evolve with the law. In [13], Mittal et al. studies have found that the normalized luminance coefficient of natural images in the spatial domain has the characteristics of statistical properties and conforms to the unit Gauss distribution. This feature is affected by image distortion, and different distortions have different effects on the distribution. Based on the above research results, the author proposes a BRISQUE non-reference image quality evaluation algorithm based on the statistical features of the spatial domain.

For a given grayscale image with a size of M * N, the luminance normalization coefficient of each pixel is satisfied as follows:

$$\hat{I}(i,j) = \frac{I(i,j) - \mu(i,j)}{\sigma(i,j) + c} \tag{5}$$

$$\mu(i,j) = \sum_{k=-K}^{K} \sum_{l=-l}^{L} \omega_{k,l} I_{k,l}(i,j) \tag{6}$$

$$\sigma(i,j) = \sqrt{\sum_{k=-K}^{K} \sum_{l=-l}^{L} \omega_{k,l} (I_{k,l}(i,j) - \mu(i,j))^2} \tag{7}$$

Among those equations:

$i = 1, 2, \cdots, M$; $j = 1, 2, \cdots, N$; c is a constant, $c = 1$; $K = L = 3$; $\mu(i,j)$ and $\sigma(i,j)$ are mean value and standard deviation;

$\omega = (\omega_{k,j} | k = -K, -k+1, \ldots, K, l = -L, -L+1, \ldots, L)$ is sampling and normalization of a two-dimensional Gauss equation.

The BRISQUE algorithm uses the luminance normalization coefficient as the quality correlation feature to evaluate the image quality. The use of image features, compared to the other non-reference quality assessment, eliminates the need for a variety of complex transform. Therefore, the algorithm has the advantages of simpler calculation and time saving on the premise of similar accuracy. However, the decorrelation processing of the image brightness is equal to ignoring the effect of brightness on the quality of the test image.

4   Description of the Method

A robust algorithm was developed for segmenting Chinese Rubbing images from their backgrounds using color images. The proposed framework consists of four steps: (i) preprocess using median filtering (The median filtering allows a great deal of high spatial frequency detail to pass while remaining very effective at removing noise on images where less than half of the pixels in a smoothing neighborhood have been effected); (ii) extraction of the red color components; (iii) a morphological operation in order to find an optimal diameter of disk Thr^* if minimum BRISQUE is found; and (iv) segmented binary image using Otsu's Thresholding. The flowchart of the proposed segmentation method is represented in Fig. 1.

In this way, we get a binarization algorithm that can deal with document images suffering from uneven illumination and low contrast.

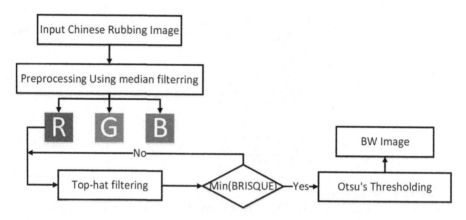

Fig. 1. Flowchart of the proposed segmentation algorithm (Color figure online)

5 Experimental Analysis and Results

To evaluate and test the proposed method, we employ dataset from Chinese Sone Rubbing collection in East Asian Library, University of California, Berkeley (http://ucblibrary4.berkeley.edu:8088/xtf3/search?rmode=stonerubbings&identifier=&title=&name=&text=&date=&startdate=15&subject=&height=&width=&material=&script=&enc_provenance=). We evaluate the performance of our system both quantitatively and qualitatively in terms of Jaccard similarity coefficient, FPR (false positive ratio) and FNR (false negative ratio), compared with classic Otsu's algorithms. Figure 2 shows a scan of a page from a Chinese historical book represent a brownish background that the rubbing image gets darker shows a low-contrast image with its histogram. The degraded Thai historical document images. Before we estimated the foreground, it was necessary to apply a median filter to remove the noise. The median filter considers each pixel in the image in turn and looks at its nearby neighbors to decide whether or not it is representative of its surroundings. The neighborhood was selected to be a 3×3 pixel square. This is a very small neighborhood compared to the size of the images, which are at the end 371×1260 pixels. After that we used function *imopen* in Matlab which performs morphological open on our estimated foreground.

We applied the binarization methods previously presented over our test collection composed of an old Chinese rubbing documents images having several types of degradations and structure complexity. We present in this section the results of applying the previous methods over the images of our collection. For the objective evaluation we have used three parameters for segmentation measurement that is Jaccard coefficient, false positive rate (FPR) and false negative rate (FNR) FPR shows the degree of under segmentation and FNR shows degree of over segmentation [15]. The Jaccard coefficient measures similarity between finite sample sets, and is defined as the size of the intersection divided by the size of the union of the sample sets, for binary images, it computes the intersection of binary images A and B divided by the union of A and B. The Jaccard coefficient can be calculated by using the following formula:

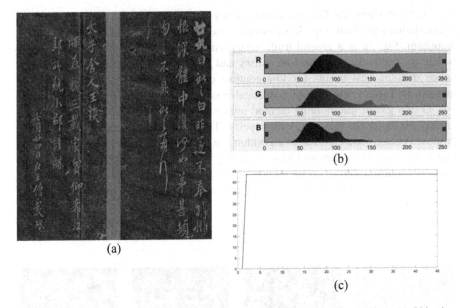

Fig. 2. (a) Example of a Chinese rubbing image [14]. (b) Histogram of document. (c) Objective function values.

$$J(A, B) = \frac{|A \cap B|}{|A \cup B|} \tag{8}$$

The FPR and FNR are defined as follow.

$$FPR = \frac{FP}{N} = \frac{FP}{FP + TN} \tag{9}$$

$$FNR = \frac{FN}{N} = \frac{FN}{FN + TN} \tag{10}$$

where *FP* is the number of false positives (white in the ground truth image and black in the binarize image), *FN* is the number of false negative, *TN* is the number of true negatives and $N = FP + TN$ is the total number of negatives. The performance of the segmentation algorithm is given in the Table 1.

Table 1. Quantitative measurement results of segmentation

	Jaccard	FPR	FNR
Our algorithm	0.6009	0.0348	0.3781
Ostu' algorithm	0.3948	0.7433	0.3188

Figure 3b show the Chinese character segmentation obtained with our adaptive segmentation algorithm, Fig. 3d shows the results by the obtained Ostu' segmentation algorithm, Fig. 3e is a ground truth binary image refers to the actual binary image which are manually eliminated all noises and degraded factors. In Table 1, we can notice that the best results for low contrast document image are obtained by our method. In Table 1, the Jaccard coefficient was significantly higher than Ostu' methods. Otsu's global thresholding method misses some text while classifies dark background pixels as the text pixels improperly. The experimental results show that the proposed background elimination algorithm can be achieved better precision and recall than Ostu' method for various Chinese rubbing image. It could performed well for low contrast Chinese rubbing image. However, since the threshold value is applied globally, it tends to overthreshold some of the weak handwriting resulting in broken handwriting.

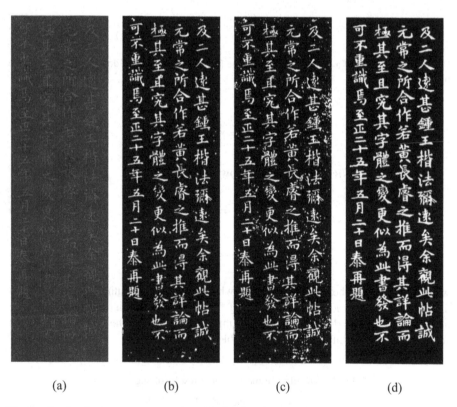

(a)   (b)   (c)   (d)

Fig. 3. Results for image segmentation. (a) Original image. (b) Segmentation image of our algorithm. (c) Segmentation image of Ostu's. (d) Ground truth binary image.

6 Conclusions

The Chinese rubbing images obtained through rubbing were featured by many fuzzy details, bad effect and so on, so it might lose more details in the traditional handling process. Pre-processing is an important stage in the image processing especially in the case of Chinese historic document image segmentation applications. An efficient image preprocessing algorithm will increase accuracy of the segmentation algorithms and reduces misclassification. This paper proposes an adaptive segmentation algorithm for degraded historical document image binarization. Subjective and objective evaluation methods are used to judge the efficiency of our algorithm. Experimental results demonstrate that the images background estimate by a morphological operation that are adaptive selection to find an optimal diameter of disk. However, since our background estimation algorithm does not take into consideration the relation of illumination in different scenes, it may introduce slight flickering for other applications in case that the scenes vary apparently. In the future, the method will be tested in OCR (Optical Character Recognition) application to test the readability of proposed method in degraded documents.

Acknowledgement. This work was supported by the National Natural Science Foundation of China (Grant No. 61472173), Natural Science Foundation of Jiangxi Province of China, No. 20161BAB202042, the grants from the Educational Commission of Jiangxi province of China, No. GJJ151134.

References

1. Khashman, A., Sekeroglu, B.: Document image binarisation using a supervised neural network. Int. J. Neural Syst. **18**(05), 405–418 (2008)
2. Huang, Z.K., Chau, K.W.: A new image thresholding method based on Gaussian mixture model. Appl. Math. Comput. **205**(2), 899–907 (2008)
3. Otsu, N.: A threshold selection method from gray-level histograms. IEEE Trans. Syst. Man Cybern. **9**(1), 62–66 (1979)
4. Solihin, Y., Leedham, C.G.: Integral ratio: a new class of global thresholding techniques for handwriting images. IEEE Trans. Pattern Anal. Mach. Intell. **21**(8), 761–768 (1999)
5. Sauvola, J., Pietikäinen, M.: Adaptive document image binarization. Pattern Recogn. **33**(2), 225–236 (2000)
6. Su, B., Lu, S., Tan, C.L.: Binarization of historical document images using the local maximum and minimum. In: Proceedings of the 9th IAPR International Workshop on Document Analysis Systems, pp. 159–166. ACM (2010)
7. Sezgin, M., Sankur, B.: Survey over image thresholding techniques and quantitative performance evaluation. J. Electron. Imaging **13**(1), 146–166 (2004)
8. O'Gorman, L.: Binarization and multithresholding of document images using connectivity. CVGIP Graph. Models Image Process. **56**(6), 494–506 (1994)
9. Huang, D.Y., Wang, C.H.: Optimal multi-level thresholding using a two-stage Otsu optimization approach. Pattern Recogn. Lett. **30**(3), 275–284 (2009)
10. Chen, S., Wang, M.: Seeking multi-thresholds directly from support vectors for image segmentation. Neurocomputing **67**, 335–344 (2005)

11. Gatos, B., Pratikakis, I., Perantonis, S.J.: Adaptive degraded document image binarization. Pattern Recogn. **39**(3), 317–327 (2006)
12. Moghaddam, R.F., Cheriet, M.: A multi-scale framework for adaptive binarization of degraded document images. Pattern Recogn. **43**(6), 2186–2198 (2010)
13. Mittal, A., Moorthy, A.K., Bovik, A.C.: No-reference image quality assessment in the spatial domain. IEEE Trans. Image Process. **21**(12), 4695–4708 (2012)
14. http://www.lib.berkeley.edu/EAL/stone/rubbings.html
15. Huang, Z.K., Li, Z.H., Huang, H., et al.: Comparison of different image denoising algorithms for Chinese calligraphy images. Neurocomputing **188**, 102–112 (2016)
16. Lu, S.J., Tan, C.L.: Binarization of badly illuminated document images through shading estimation and compensation. In: International Conference on Document Analysis and Recognition, pp. 312–316. IEEE (2007)
17. Xiong, W., Xu, J., Xiong, Z., et al.: Degraded historical document image binarization using local features and support vector machine (SVM). Optik **164**, 218–223 (2018)
18. Su, B., Lu, S., Tan, C.: Binarization of historical document images using the local maximum and minimum. In: Proceedings of the 9th IAPR International Workshop DAS, pp. 159–166 (2010)
19. Nafchi, H.Z., Moghaddam, R.F., Cheriet, M.: Phase-based binarization of ancient document images: model and applications. IEEE Trans. Image Process. Publ. IEEE Sig. Process. Soc. **23**(7), 2916–2930 (2014)
20. Hadjadj, Z., Meziane, A., Cheriet, M., et al.: An active contour based method for image binarization: application to degraded historical document images. In: Isko-Maghreb: Concepts and Tools for Knowledge Management, pp. 1–7. IEEE (2015)

Parameter Selection of Image Fog Removal Using Artificial Fish Swarm Algorithm

Fan Guo, Gonghao Lan, Xiaoming Xiao[✉], and Beiji Zou

School of Information Science and Engineering, Central South University,
Changsha 410083, Hunan, China
xmxiao@mail.csu.edu.cn

Abstract. Although image defogging is widely used in many working systems, existing defogging methods have some limitations due to the lack of enough information to solve the equation of fog formation model. To overcome the limitations, a novel defogging parameter selection algorithm based on artificial fish swarm algorithm (AFSA) is proposed in this paper. Two representative defogging algorithms are used to test the effectiveness of the method. The proposed method first selects the two main parameters and then optimizes them using the AFS algorithm. An assessment index of image defogging effect is used as the food concentration of the AFSA. Thus, these parameters may be adaptively and automatically adjusted for the defogging algorithms. A comparative study and qualitative evaluation demonstrate that better quality results are obtained by using the proposed method.

Keywords: AFSA · Defogging effect assessment · Parameter selection
Single image defogging

1 Introduction

Most automatic systems assume that the input images have clear visibility, therefore removing the effects of bad weather from these images is an inevitable task. In the past decades, extensive research efforts have been conducted to remove fog or haze from a single input image. Most of these methods [1–4] intend to recover scene radiance using the image degradation model that describe the formation of a foggy image. Tan [1] removed fog by maximizing the local contrast of the restored image. Nishino et al. [2] proposed a Bayesian probabilistic method that estimates the scene albedo and depth from a foggy image with energy minimization of a factorial Markov random field. He et al. [3] estimated the transmission map and the airlight of the degradation model using the dark channel prior. Tarel et al. [4] introduced an atmospheric veil to restore image visibility based on the fast median filter. However, these methods are controlled by a few parameters with fixed values that cannot be automatically adjusted for different foggy images.

In recent years, people are quite interested in automatic fog removal, which is useful in applications such as surveillance video [5], intelligent vehicles [6], and outdoor object recognition [7]. In this paper, we thus focus on the AFSA-based adaptive parameter adjustment for single image defogging.

© Springer International Publishing AG, part of Springer Nature 2018
D.-S. Huang et al. (Eds.): ICIC 2018, LNCS 10954, pp. 25–37, 2018.
https://doi.org/10.1007/978-3-319-95930-6_4

Artificial fish swarm algorithm (AFSA) was first presented by Li et al. [8]. This algorithm is a technique based on swarm behaviors that were inspired from social behaviors of fish swarm in nature. AFSA works based on population, random search, and behaviorism. The fish swims towards locations where food concentration is highest. As a typical application of behaviorism in artificial intelligence, AFSA can search for the global optimum [9]. It is widely agreed that AFSA can be used for digital image processing, such as image segmentation [10, 11], image enhancement and restoration [12], image matching [13], image quantization [14], etc. However, due to the lack of proper objective criterion of defogging effect as the fitness function, present defogging algorithms seldom use AFSA to effectively remove fog from a single image. Though the objective image quality evaluation methods have achieved some promising results, they are just applied to assess the quality of degraded image, such as image denosing results and deblurring results. The aim of defogging algorithm is to recover color and details of the scene from input foggy image. Unlike image quality assessment, the fog can not be addressed like a classic image noise or degradation which may be added and then removed. Meanwhile, there is no easy way to have a reference no-fog image, and the quality evaluation criteria of degraded image, such as the structural similarity (SSIM) [15], the peak signal-to-noise ratio (PSNR) [16], and the mean square error (MSE), are not suitable for assessing image defogging effects. This makes the problem of adaptive parameter adjustment of defogging algorithm not straightforward to solve.

The main contribution of this paper is we propose a way to select parameter values for single image defogging, which can help to overcome the limitations of existing defogging algorithms. Here, the defogging effect assessment index presented in our previous work [17] is taken as the food concentration of the proposed AFSA-based method. Thus, the parameter values can be adaptively and automatically adjusted for different input foggy images. A comparative study and qualitative evaluation demonstrate that the better quality results are obtained by using the proposed method.

2   Limitations of the Existing Defogging Methods

Most current defogging methods recover the scene radiance by solving the fog formation model. Since the model contains three unknown parameters and the solving process is an ill-posed inverse problem, it is thus inevitable to introduce many application-based parameters that used in various assumptions for image defogging. A large quantity of experimental results shows that the selection of the algorithm parameters has direct influence on the final defogging effect. However, there exists a major problem for the parameter setting in most defogging algorithms, i.e. the parameters always have fixed values in the defogging algorithms.

In our experiments, we find that the fixed parameter values caused that the fog removal algorithms just have good defogging effect for a certain kind of foggy image, and the algorithms may not work well for the images captured under other foggy conditions. For example, He's algorithm [3] has mainly three parameters to control: ω which alters the amount of haze kept at all depths, c the patch size for estimating transmission map, and t_0 restrict the transmission to a lower bound to make a small amount of fog preserve in very dense fog regions. All these parameters have fixed value suggested by the authors, such as the fog parameter ω, which is set to be 0.95 in the algorithm [3]. Our experimental results show that, if ω is adjusted downward, more fog will be kept, and vice versa. Using $\omega = 0.95$ keeps a slight amount of fog effect around at all depths. However, the experiments also show that ω sometimes needs to be decreased when an image contains substantial sky regions, otherwise the sky region may wind up having artifacts.

Besides, most defogging algorithms have introduced some parameters, which lead to user interaction and make the final defogging effect hard to control as well. For example, Tarel's algorithm [4] is controlled by five parameters in which p is the percentage of removed atmospheric veil, s_v the assumed maximum size of white objects in the image, b the white balance control for global or local process, s_i the maximum size of adapted smoothing to soften the noise amplified by the restoration, and g an extra factor during final gamma correction. It is obvious that the controllability can be greatly improved and the user-interaction can be also largely reduced if defogging algorithm has no more than two parameters. Therefore, distinguishing between main parameters which directly affect the results and other less important parameters which can be considered as fixed values, and then automatically adjusting their values are very important for the defogging algorithms.

3   The Proposed Parameter Value Selection Approach

3.1   Optimal Parameter Selection

The task of optimal parameter selection is to distinguish between two main parameters which directly affect the results and other less important parameters which can be considered as fixed values for various defogging methods. There are two ways to select the two main parameters: one is to analyze the related parameters from the perspective of physical mechanism, and the other is to tune one of algorithm parameters by fixing the rest to see whether the defogging results have significant change. Two representative fog removal methods [3, 4] are taken as examples to describe the parameter selection process in this paper. Since He's method [3] is recognized as one of the most effective ways to remove fog, and Tarel's method [4] is regarded as one of the fastest defogging algorithms at present.

For He's method [3], the dark channel prior with the haze imaging model is used to directly estimate the thickness of the haze and recover a high-quality haze-free image. In the method, a small amount of haze for distance objects is kept to make the final defogging results seem more natural and preserve the feeling of depth as well. There are two key parameters for the fog preservation purpose: ω $(0 < \omega < 1)$ and t_0 $(0 < t_0 < 1)$. Other parameters have less influence on the final defogging effect, and can thus be regarded as less important paramters. For Tarel's method [4], the method is based on median filter and consists in: atmospheric veil inference, image restoration and smoothing, tone mapping. The most important two steps that determine the final defogging effect are atmospheric veil inference and tone mapping. Since the value of p $(0 < p < 1)$ controls the amount of atmospheric veil that can be removed, this parameter is useful to compromise between highly restored visibility where colors may appear too dark, and less restored visibility where colors are clearer. The parameter g ($0 < g < 10$) is used to perform gamma correction to achieve more colorful result. Experimental results show that the larger the value of g, the clearer the defogging result is. Compared with the two parameters p and g, other parameters have less effect on the final defogging results.

3.2 Measurement of Defogging Effect

The CNC index, an effective defogging evaluation indicator proposed in our previous work [17] is used here to guide the parameter adjustment process. For the input foggy image $\mathbf{x}$ and its corresponding fog removal image $\mathbf{y}$, the CNC index is obtained after carrying out the following steps: (i) compute the rate e of visible edges after and before fog removal, (ii) calculate the image color naturalness index (CNI) and color colour-fulness index (CCI) to measure the color naturalness of the defogging image $\mathbf{y}$, and (iii) Combine the three components e, CNI and CCI to yield an overall defogging effect measure CNC. The mathematical formula of e can be expressed as:

$$e(\mathbf{x}, \mathbf{y}) = \frac{n(\mathbf{y})}{n(\mathbf{x})}. \tag{1}$$

where $n(\mathbf{x})$ and $n(\mathbf{y})$ denote the cardinal numbers of the set of visible edges in x and y, respectively. More details about the mathematical formula of CNI and CCI can be found in [18].

For the overall variation trend of the three indexes, the statistical results show that the peak of CNI curve stands for the most natural result, but it is not necessarily the best defogging effect. However, the best effect must have good naturalness (high CNI value). When the image is over enhanced, the color is distorted, and CNI goes down rapidly. For e and CCI, they achieve the best effect before reaching their peaks. When the image is overenhanced, the curves continue ascending. After reaching their peaks, these curves begin to go down. Therefore, if the uptrend of e and CCI (from their best effect points to their curve's peaks) can be largely counteracted by the downtrend of CNI, and the peak of CNC curve can be more close to the real best effect point. Meanwhile, the value variation of CNI is small, while that of e and CCI is relatively big. Thus, the effect of e and CCI on the CNC index needs to be weakened. The CNC index between image $\mathbf{x}$ and $\mathbf{y}$, *i.e.* the function CNC can be defined as

$$\text{CNC}(\mathbf{x}, \mathbf{y}) = e(\mathbf{x}, \mathbf{y})^{1/5} \cdot \text{CNI}(\mathbf{y}) + \text{CCI}(\mathbf{y})^{1/5} \cdot \text{CNI}(\mathbf{y}) \tag{2}$$

A good result is described by the large value of CNC. Therefore, the optimal value of the two main parameters of defogging algorithms can be obtained when the CNC index (2) achieves the largest value.

3.3 Parameter Value Selection Using Artificial Fish Swarm Algorithm

Artificial fish (AF) is a fictitious entity of true fish, which is used to carry on the analysis and explanation of problem. The AF realizes external perception by its vision. Suppose X is the current state of an AF, Visual is the visual distance, X_v is the visual position at some moment, and N is the population size. If the state at the visual position is better than the current state, it goes forward a step in this direction and arrives the X_{next} (X_i $(t +1)$) state; otherwise, continues an inspecting tour in the vision. Using the artificial fish swarm algorithm (AFSA), the two main parameters of various defogging algorithms can be automatically determined.

Specifically, the AF model includes two parts (variables and functions). The variables include: X (X_i (t)) is the current position of the AF, *step* is the moving step length, *visual* represents the visual distance, *try_number* is the try number, and δ is the crowd factor ($0 < \delta < 1$). The functions include the behaviors of the AF: AF_Prey, AF_Swarm, and AF_Follow. In each step of the optimization process, AF looks for locations with better fitness values in problem search space by performing the previous three behaviors based on algorithm procedure. The pseudo-code of the AFSA-based method that is used for image defogging is presented in Algorithm 1 (see Fig. 1).

Algorithm 1. The AFSA used for image defogging

Begin
  Initialize the population: N, visual, *step, maxgen*, δ, *try_number*.
  While running do
    For 1 to N do
      Calculate the CNC function value for the two parameters represented as an initial
fish swarm with the size of $2 \times N$
      Execute the basic fish behaviors
      Generate the new individuals
  End while
  The optimal solution is the selected parameter value for the defogging method.
  The enhanced image with the optimal parameter values is our final defogging result.
End

Fig. 1. Algorithm 1: The AFSA used for image defogging

The simple but effective AFSA-based parameter selection consists of the following steps:

(1) Fish swarm initialization

Each artificial fish in the swarm is described by a random array that generated within a given range. For He's defogging algorithm, suppose the size of fish swarm is N, ω $(0 < \omega < 1)$ and t_0 $(0 < t_0 < 1)$ are the two parameters that need to be optimized. Thus, the individual's status of each artificial fishes can be expressed as a vector $X = \{x_1, x_2\}$, in which the $x_i (i = 1, 2)$ represents ω and t_0. For Tarel's algorithm, p and g are the two key parameters, where $p \in (0, 1)$ and $g \in (1, 10)$. Similarly, the $x_i (i = 1, 2)$ here represents the parameters p and g. In initialization phase, the AFSA will produce an initial fish swarm with the size of $2 \times N$ for both algorithms, and each column represents the two parameters of an artificial fish for the defogging algorithms.

As the fitness value of the objective function, the food concentration of the current location about the artificial fish is actually the CNC index value for image defogging. Therefore, the food concentration can be expressed as $Y = CNC(X)$. The other parameters such as the visual field of the artificial fish, the maximum moving step, the maximum generation, the crowd factor and the maximum number of tries in every forage are expressed as *visual, step, maxgen, δ* and *try_number*. Note that the crowd factor $δ$ can be used to limit the fish swarm size of the artificial fish swarm so as to make more artificial fish individuals gather in the region with better state rather than the neighborhood with suboptimal state. For the proposed approach, the parameters of ASFA are set as following, $N = 10$, *visual* $= 1$, *step* $= 0.1$, *maxgen* $= 20$, $δ = 0.5$ and *try_number* $= 10$. In our experiment, we find that the parameter settings can make algorithm get better optimization performances.

(2) Prey behavior

This behavior is an individual behavior that each AF performs independently and performs a local search around itself. Every AF by performing this behavior attempts *try_number* times to move to a new position with better food concentration. Here, the better food concentration is corresponding to the higher CNC value. Suppose AF_i is in position X_i, then within the visual area S, it randomly chooses another position X_j. AF_i will swim towards X_j. This process can be written as:

$$X_j = (X_i + visual \times \text{rand}(0,1)) \times \frac{1}{2} \tag{3}$$

Since Y represents the food concentration, if $Y_j > Y_i$, position of AF_i is updated to the next step by Eq. (4).

$$X_i(t+1) = X_i(t) + \frac{X_j - X_i(t)}{\|X_j - X_i(t)\|} \times step \times \text{rand}(0, 1) \tag{4}$$

The above two steps are repeated *try_number* times. If AF_i could not move toward better positions, it moves with a random step in its visual using Eq. (5), and the food concentration of Y_{next} ($Y_i (t +1)$) state can also be obtained by means of CNC index.

$$X_i(t+1) = (X_i(t) + visual \times \text{rand}(0, 1)) \times \frac{1}{2} \tag{5}$$

(3) Swarm behavior

Fish usually assembles in groups to capture colonies and/or to avoid dangers. There are n_f neighbors within the visual area S. X_c is the center of those neighbors. AF_i will swim a random distance towards X_c. If $(n_f/n) < δ$ (where n is the number of artificial fishes) and $Y_c > Y_i$ which means that the companion center has more food (higher CNC value) and is not very crowded, it goes forward a step to the companion center (Eq. 6). Otherwise, AF_i will resume the behavior of Prey.

$$X_i(t+1) = X_i(t) + \frac{X_c - X_i(t)}{\|X_c - X_i(t)\|} \times step \times rand(0, 1) \qquad (6)$$

(4) Follow behavior

When some fishes find food through the moving process of swarm, their neighborhood partners tend to follow them to the best food concentration location. Let X_i be the AF current state, and it explores the companion X_j in the neighborhood ($\|X_j - X_i(t)\| <$ Visual), which has the greatest Y_j. If $Y_j > Y_i$ and $(n_f/n) < \delta$, which means that the companion X_j state has higher food concentration (higher CNC value) and the surrounding is not very crowded, it goes forward a step to the companion X_j. Otherwise, AF_i will resume the behavior of Prey.

$$X_i(t+1) = X_i(t) + \frac{X_j - X_i(t)}{\|X_j - X_i(t)\|} \times step \times rand(0, 1) \qquad (7)$$

The collective behaviors of Prey, Follow and Swarm of all fishes are simulated in each generation. The fishes will choose the behavior that has the best position (concentration). AFSA is independent on the initial condition. A termination criterion can be added for each specific problem. For the proposed method, when the maximum number of generations is reached, the AFSA computation is terminated. The optimal solution of X indicates the selected value for the two main parameters. The enhanced image with these parameter values is our final fog removal result. Figure 2 shows the optimal value (food concentration) variation for a test foggy imag. One can clearly see that the best food concentration value is 0.4762 for He's algorithm [3], and its

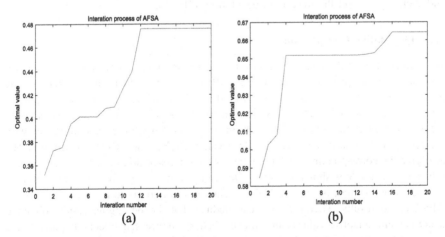

Fig. 2. The optimal value variation for a test foggy image. (a) He's result. (b) Tarel's result.

corresponding X value is (0.9977, 0.56) for the parameter value of (ω, t_0) in the method. For Tarel's algorithm [4], the best food concentration value is 0.6643, and its corresponding X value is (0.9765, 8.5) for the parameter value of (p, g). In our experiments, we find that the defogging results obtained by the two optimal parameter values seem quite promising in most cases. Thus, the effectives of the proposed approach can be verified.

4   Experimental Results

The publicly available dataset frida2 [19] is used to evaluate image defogging methods. This dataset contains synthetic no-fog images and associated foggy images for 66 diverse road scenes. The absolute difference (AD) on the images between defogged images and target images without fog is used as performance metrix, and good results are described by small value of AD. To verify the effectiveness and validity of the proposed parameter value selection method, three criteria have been considered: (i) generation number influence, (ii) qualitative comparison, (iii) quantitative evaluation, and (iv) time complexity. In the experiments, all the results are obtained by executing Matlab R2008a on a PC with 3.10 GHz Intel® CoreTM i5-2400 CPU.

4.1   Generation Number Evaluation

To evaluate the influence of the generation number *maxgen* used in the proposed method, some group experiments are performed by varying the generation number *maxgen* from 10 to 50. From the parameter values and the resulting AD metrics of a test image we can clearly see that the results are visually and statistically close (the value range of AD are [36.1573, 36.9429] for He's method, and [48.0486, 48.9980] for Tarel's method) when varying *maxgen* from 10 to 50. It demonstrates that the influence of the generation number is very limited in the proposed method. The experiments on other test images also confirm the observations. Thus, to make a tradeoff between speed and accuracy, we set the maximum generation to be 20.

4.2   Qualitative Comparison

To evaluate the AFSA-based method for selecting the most proper parameter values, the image dataset provide by Tarel et al. [19] is used to validate the accuracy of the parameter selection, since this dataset provides the original image without fog and the image with fog for 66 road scenes simultaneously, thus we can use the AD index to measure the defogging effects. For each image, we obtain the fog removal results with the default and the auto-adaptive parameter values for He's and Tarel's algorithms. We also give the corresponding no-fog images as the reference images for comparison. An illustrative example is shown in Fig. 3. One can clearly see that the far-away buildings or trees which cannot be seen in He's or Tarel's results become more obvious in our AFSA-based results. Therefore, we can deduce that the defogging results obtained using the auto-adaptive values can achieve a better enhancement effect compared to the results obtained using the default values for both defogging algorithms.

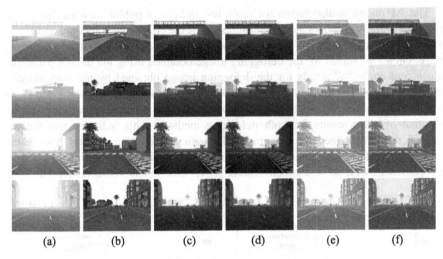

(a)   (b)   (c)   (d)   (e)   (f)

Fig. 3. Visual comparison of defogging results for public database frida2 [19]. (a) Foggy images. (b) Fog-free images. (c) He's results obtained using default parameter values ($\omega = 0.95$, $t_0 = 0.1$, $c = 3$). (d) He's results obtained using auto-adaptive parameter values. (e) Tarel's results obtained using default parameter values ($p = 0.95$, $b = 0.5$, $s_v = 9$, $s_i = 1$, $g = 1.3$). (f) Tarel's results obtained using auto-adaptive parameter values.

4.3   Quantitative Evaluation

To quantitatively assess the proposed parameter value selection method, we compute the AD index value for the images in Fig. 3, and the statistical results are shown in Tab. 1. One can notice that the AD value obtained by the auto-adaptive parameter is smaller than that of default parameters for both defogging algorithms, which means that the better defogging effect can be obtained by using the proposed method. This confirms our observations in Fig. 3.

Table 1. AD index between enhanced images and no-fog images for the test images in Fig. 3

Img	Method				
	Foggy image	He's method (default value)	He's method **(adaptive value)**	Tarel's method (default value)	Tarel's method **(adaptive value)**
Figure 3 (Row #1)	42.67	33.68	**31.65**	23.11	**16.42**
Figure 3 (Row #2)	55.73	34.46	**30.65**	38.67	**32.29**
Figure 3 (Row #3)	62.56	40.12	**32.66**	47.01	**27.92**
Figure 3 (Row #4)	66.20	31.05	**29.84**	47.27	**23.64**

The AD index is also tested for more test images in public database frida2 (66 images). Figure 4(a) shows the statistical results of the AD for He's method and Fig. 4 (b) shows the AD results for Tarel's method. In Fig. 4, symbol "♦" stands for foggy image, circle "♦" stands for the defogging image obtained using the default parameter values, and circle "◇" stands for the defogging image obtained by the auto-adaptive parameter values. The horizontal axes are the image number index and vertical axes are the AD index values. It is clear that the ADs of adaptive parameter results are smaller than that of other results for both defogging methods. This indicates that the fog removal results obtained by the proposed parameter selection method have better defogging effect for the public image database compared to the other results. This is also consistent with the assessment results of AD and human visual perception.

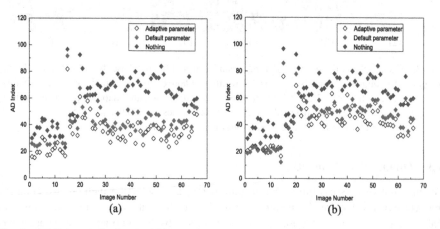

Fig. 4. AD index results for the 66 test images in public database frida2 [16]. (a) He's statistical results. (b) Tarel's statistical results.

4.4 Time Complexity

For image defogging algorithms, the computational efficiency is also an important issue. For the two representative defogging algorithms, the most time-consuming operation in He's defogging algorithm is using soft matting to obtain the refined transmission map [3]. Suppose the input image has a size of M × N, He's algorithm involves a (M × N) × (M × N) matting Laplacian matrix [20]. Therefore, the time complexity of He's algorithm TC_H is about $O(M^2 \times N^2)$. While the time complexity of Tarel's algorithm TC_T is about $O(MNs_v^2 \ln s_v)$, where s_v is the assumed maximum size of white objects in the image [4].

For the fitness function calculation, the time complexity of the CNC method TC_{CNC} is about $O\big((M/s)^2 \times (N/s)^2\big)$, where s is the subwindow size that used for patch segmentation. For AFS optimization algorithm, suppose the N artificial fishes compose a group in the target search space of two dimensions, m is the maximum generation and t_r is the maximum number of tries in every forage. Let the running time of each artificial fish in each iteration is T_f, the total running time of the proposed AFSA-based algorithm can be written as: $N \times m \times t_r \times T_f$. Thus, the time complexity of AFSA-based algorithm can be reflected by the number of artificial fishes and the running time of each fish in an iteration. Since the fish swarm populations is set to 10, the maximum number of iterations is 20 and the try number is 10 in our experiment, and the ideal defogging results are obtained by applying the determined optimal parameter values to defogging algorithms, we can thus deduce that the whole time complexity of the proposed method is about $10 \times 20 \times 10 \times (TC_H + TC_{CNC}) \times T_f + TC_H$ for He's algorithm, and about $10 \times 20 \times 10 \times (TC_T + TC_{CNC}) \times T_f + TC_T$ for Tarel's algorithm.

Generally, the average running time of the proposed parameter selection method for He's algorithm is about 50 min for an image with a size of 200×150, and about 350 s for that of Tarel's algorithm with the same size image in Matlab environment. This speed can be further improved by more efficient defogging methods or a GPU-based parallel algorithm. Besides, surveillance video defogging can also benefit much from the proposed method. That's because for surveillance video, once the optimal parameter values are determined by a image frame, the same values can be directly applied to a series of video frames to obtain the restored images whose fog density is similar in the corresponding original video frames. Therefore, the once-for-all strategy can save a lot of time for video defogging.

5 Conclusions

In this paper, a novel AFSA-based parameter value selection method was proposed. Different from the most defogging methods which generally fix the parameter values, the proposed approach can help defogging methods automatically select proper parameter values for different foggy images. In the proposed method, the two main parameters which directly affect the results are first distinguished from other less important parameters which can be considered as fixed values. Then, the two parameter values are adaptively determined by using the artificial fish swarm algorithm. The proposed parameter selection method has been applied to two representative defogging algorithms, which demonstrated the superior performance of the proposed scheme in both qualitatively and quantitatively. In the future, we will try to investigate the parameter value selection issue based on more advanced assessment index, since the CNC index may not be the best one to measure image defogging effect.

Acknowledgements. The authors would like to thank Erik Matlin and Kaelan Yee for providing He's source code, and Dr. Tarel and Dr. Hautiere for providing the Matlab code of their approach. This work was supported by the National Natural Science Foundation of China (61502537), Hunan Provincial Natural Science Foundation of China (2018JJ3681), and the National Undergraduate Programs for Innovation.

References

1. Tan R.T.: Visibility in bad weather from a single image. In: Proceedings of IEEE Conference on Computer Vision and Pattern Recognition, pp. 1–8. IEEE Computer Society, Anchorage (2008)
2. Nishino, K., Kratz, L., Lombardi, S.: Bayesian defogging. Int. J. Comput. Vision 98(3), 263–278 (2012)
3. He, K.M., Sun, J., Tang, X.O.: Single image haze removal using dark channel prior. IEEE Trans. Pattern Anal. Mach. Intell. 33(12), 2341–2353 (2011)
4. Tarel, J.P., Hautiere, N.: Fast visibility restoration from a single color or gray level image. In: Proceedings of IEEE International Conference on Computer Vision, pp. 2201–2208. IEEE Computer Society, Kyoto (2009)
5. Lagorio, A., Grosso, E., Tistarelli, M.: Automatic detection of adverse weather conditions in traffic scenes. In: Proceedings of IEEE Fifth International Conference on Advanced Video and Signal Based Surveillance, pp. 273–279. IEEE Computer Society, Santa Fe (2008)
6. Hautiere, N., Tarel, J.-P., Aubert, D.: Towards fog-free in-vehicle vision systems through contrast restoration. In: Proceedings of IEEE Conference on Computer Vision and Pattern Recognition, pp. 2374–2381. IEEE Computer Society, Minneapolis (2007)
7. Hautiere, N., Tarel, J.-P., Halmaoui, H., Bremond, R., Aubert, D.: Enhanced fog detection and free-space segmentation for car navigation. Mach. Vis. Appl. 25(3), 667–679 (2014)
8. Li, L.X., Shao, Z.J., Qian, J.X.: An optimizing method based on autonomous animate: fish swarm algorithm. In: Proceedings of system engineering theory and practice, pp. 32–38. IEEE Computer Society, Los Alamitos (2002)
9. Neshat, M., Sepidnam, G., Sargolzaei, M., Toosi, A.N.: Artificial fish swarm algorithm: a survey of the state-of-the-art, hybridization, combinatorial and indicative applications. Artif. Intell. Rev. 42(4), 965–997 (2014)
10. Ye, Z.W., Li, Q.Y., Zeng, M.D., Liu, W.: Image segmentation using thresholding and artificial fish-swarm algorithm. In: Proceedings of International Conference on Computer Science and Service System, pp. 1529–1532. IEEE Computer Society, Los Alamitos (2012)
11. Janaki, S.D., Geetha, K.: Automatic segmentation of lesion from breast DCE-MR image using artificial fish swarm optimization algorithm. Pol. J. Med. Phys. Eng. 23(2), 29–36 (2017)
12. Sui, D., He, F.: Image restoration algorithm based on artificial fish swarm micro decomposition of unknown priori pixel. Telkomnika 14(1), 187–194 (2016)
13. Zhu, J.L., Wang, Z.L., Liu, H.: Gray-scale image matching technology based on artificial fish swarm algorithm. Appl. Mech. Mater. 411–414, 1295–1298 (2013)
14. El-said, S.A.: Image quantization using improved artificial fish swarm algorithm. Soft. Comput. 19(9), 2667–2679 (2015)
15. Wang, Z., Bovik, A.C., Sheikh, H.R., Simoncelli, E.P.: Image quality assessment: from error visibility to structural similarity. IEEE Trans. Image Process. 13(4), 600–612 (2004)
16. Ji, Z.X., Chen, Q., Sun, Q.S., Xia, D.D.: A moment-based nonlocal-means algorithm for image denoising. Inf. Process. Lett. 109(23–24), 1238–1244 (2009)
17. Guo, F., Tang, J., Cai, Z.X.: Objective measurement for image defogging algorithms. J. Cent. South Univ. 21(1), 272–286 (2014)

18. Huang, K.Q., Wang, Q., Wu, Z.Y.: Natural color image enhancement and evaluation algorithm based on human visual system. Comput. Vis. Image Underst. **103**(1), 52–63 (2006)
19. Tarel, J.-P., Hautiere, N., Caraffa, L., Cord, A., Halmaoui, H., Gruyer, D.: Vision enhancement in homogeneous and heterogeneous fog. IEEE Intell. Transp. Syst. Mag. **4**(2), 6–20 (2012)
20. He, K.M.: Single image haze removal using dark channel prior. Ph.D. dissertation, The Chinese University of Hong Kong (2011)

On Bessel Structure Moment
for Images Retrieval

Zi-ping Ma$^{(\boxtimes)}$ and Jin-lin Ma

College of Information and Computational Science,
North Minzu University, Yinchuan 750021, China
maziping@tom.com

Abstract. This paper proposed a new Bessel Structure moments for image retrieval. The proposed method has rotation invariance and performs better than orthogonal Fourier-Mellin and Zernike moments in terms of represent global features. The experiments show that the feature descriptors extracting from the proposed algorithm perform better for image retrieval than conventional descriptors by comparing the retrieval accuracy with the same order.

Keywords: Image retrieval · Bessel Structure moments · Invariant descriptor

1 Introduction

Moments method have been widely applied in image processing and 3D shape recognition due to their ability to describe global features and their geometric invariance.

Among these moments, geometric moments are ones of the typical moments. However, because of lacking of orthogonality it is more difficult to achieve their reconstruction [1]. On the other hand, the typical orthogonal moments include orthogonal Fourier-Mellin moments [2], Zernike moments [3, 4], Legendre moments [5], Bessel moments [6] and so on. And orthogonal moments have some advantages such as their orthogonality, image representation and being reconstruction more easily. The important difference between them is the Integral kernel function. The integral kernel functions make moments orthogonal and robustness, which can overcome a certain degree of information redundancy. As one of the orthogonal moments, the discrete orthogonal moments have been introduced such as discrete Tchebiche [7], discrete Krawtchouk moments [8], discrete dual Hahn moments [9] and so on. The implement of the discrete orthogonal moments does not involve any numerical approximations [6].

In addition, some researchers have developed to constructing geometrical invariant features from different moment methods [10, 11]. Moreover, some researchers combined moment methods with other transformation space such as Fourier, Gabor, Radon, wavelet transformation and so on [12–15]. It is verified that these combined methods have better accuracy compared to Gabor, Radon and wavelet based methods and it require very low computational effort.

© Springer International Publishing AG, part of Springer Nature 2018
D.-S. Huang et al. (Eds.): ICIC 2018, LNCS 10954, pp. 38–45, 2018.
https://doi.org/10.1007/978-3-319-95930-6_5

Meanwhile due to their invariance Hu moments first are proposed in 1962 and are used widely in image analysis and pattern recognition. And some researchers have drawn greater attention on constructing complete set of geometric moment invariants [2, 16]. Different from the conventional methods, the rotation and translation invariants of Gaussian–Hermite moments are proposed by being defined with Hermite polynomials and a family of orthogonal moments. And the numerical experiments show that this method has better invariance to rotation and translation and comparably better noise robustness than the corresponding geometric invariants [16].

Bessel–Fourier moments are a new set of moments based on the Bessel function of the first kind. It has proved that they are more suitable than orthogonal Fourier-Mellin and Zernike moments [6].

Several years later, Xiao et al. presented an improved method of Bessel Fourier moments to obtain the rotation, scaling and translation invariant [17]. Meanwhile they extended their previous work and propose a new moment method in the Radon space, which can obtain rotation, scaling, translation and affine invariance [18].

In case of the imprecisions of Exponent-Fourier moments [19], Xiao et al. proposed the improvement of them which has the more precision and superiority of reconstruction [20]. Similarity, in order to solve the instability of proposed method in literature [20], a novel Improved Exponent–Fourier moments are redefined which can avoid the value of Exponent–Fourier moments infinity at the origin and has the superiority in image reconstruction [21].

This paper is organized as follows: In Sect. 2 we present an overview of Bessel function of first kind, Bessel-Fourier moments and propose Bessel Structure moments. Section 3 is conducted to numerical experiments. The conclusion is presented in Sect. 4.

2   Bessel Structure Moments

2.1   Bessel Fourier of First Kind

The definition of the Bessel function of first kind is as follows [22]:

$$J_v(x) = \sum_{k=0}^{\infty} \frac{(-1)^k}{k!\Gamma(v+k+1)}\left(\frac{x}{2}\right)^{v+2k} = \frac{(x/2)^v}{\Gamma(v+1)}{}_0F_1(v+1, -(x/2)^2) \qquad (1)$$

Where v is a real constant, $\Gamma(a)$ is the gamma function, ${}_0F_1$ is the generalized hypergeometric function. The Bessel function is the solution of the Bessel's equation.

$$y'' + \frac{1}{x}y' + (1 - \frac{v^2}{x^2})y = 0 \qquad (2)$$

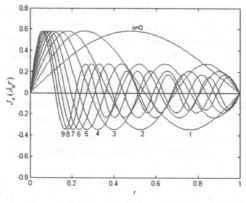

(a) The plot of Bessel polynomial function

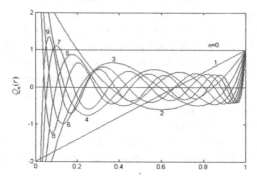

(b) The plot of Fourier-Mellin polynomial function

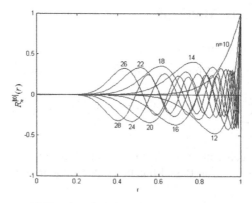

(c) The plot of Zernike polynomial function

Fig. 1. Curves of different polynomial functions in polar coordinates

And have the following recurrence relation:

$$J_{v-1}(x) + J_{v+1}(x) = \frac{2v}{x} J_v(x) \tag{3}$$

Figure 1 shows that plots of Bessel polynomial function, Fourier-Mellin polynomial function and Zernike polynomial function. From Fig. 1, it can be observed that Bessel polynomial function has more zeros than other polynomial functions, which indicates Bessel polynomial function can have corresponding better shape description ability theoretically [6].

2.2 Defining of Bessel Structure Moment

Structure Moment is proposed in reference [23]. The key concept of Structure Moment is to redefinition on transformed image by a linear transformation or nonlinear transformation using Moment method. This novel method not only can inherit merits from traditional moment methods but also can improve shape description ability of traditional moment methods.

Inspired by the idea of definition of Structure Moment, we defined Bessel Structure Moment as:

$$BSM_{nm} = \frac{1}{2\pi a_n} \int_0^{2\pi} \int_0^1 f^p(r, \theta) J_v(\lambda_n r) \exp(-jm\theta) r dr d\theta \tag{4}$$

Where $f(r, \theta)$ is the image in polar coordinate, n, m is the moment order, and $a_n = [J_{v+1}(\lambda_n)]^2/2$ is a normalization constant, $J_v(\lambda_n r)$ is the Bessel polynomial in r of degree n, λ_n is the n-th zero of $J_v(r)$ and p is the power of $f(r, \theta)$. Equation (4) can enlarge the information of $f(r, \theta)$.

2.3 Bessel Structure Moments' Rotation Invariants

Let $f'(r, \theta)$ be the rotated image of an origin image function $f(r, \theta)$ with the angle φ, and we have

$$f'(r, \theta) = f(r, \theta - \varphi) \tag{5}$$

Let $\theta' = \theta - \varphi$, we have $\theta = \theta' + \varphi$, Eq. (4) can be written as

$$\begin{aligned} BSM'_{nm} &= \frac{1}{2\pi a_n} \int_0^{2\pi} \int_0^1 f^p(r, \theta') J_v(\lambda_n r) \exp(-jm(\theta + \varphi)) r dr d\theta' \\ &= [\frac{1}{2\pi a_n} \int_0^{2\pi} \int_0^1 f^p(r, \theta') J_v(\lambda_n r) \exp(-jm\theta') r dr d\theta'] \exp(-jm\phi) \\ &= BSM_{nm} \exp(-jm\phi) \end{aligned} \tag{6}$$

Equation (6) shows that $|BSM_{nm}|$ equal to $|BSM'_{nm}|$, the Bessel Structure moments of rotation of the image by an angle of φ.

3 Experimental Results and Discussions

If we definite feature vectors of Bessel Structure moments as follows:

$$BSM = [bs_1, bs_2 \cdots, e_1, e_2] \tag{7}$$

Where the value of p is 2, and

$$e_1 = \frac{1}{(M \times N)^2} \sum_{m=1}^{M} \sum_{n=1}^{N} |BSM(m, n)|^2 \tag{8}$$

$$e_2 = \frac{1}{(M \times N)^2} \sum_{m=1}^{M} \sum_{n=1}^{N} |BSM(m, n)|^2 \log(|BSM(m, n)|^2) \tag{9}$$

We conducted retrieval experiments on three databases in order to verity retrieval capability of the proposed algorithm. We made D1, D2, D3 database, which are respectively chosen from illumination color (COL), illumination direction (ILL), and object viewpoint (VIEW) from the Amsterdam Library of Object Image (ALOI). There are respectively 12, 24, 72 color objective images. Each database consists of 50, 25, 16 images per category respectively.

We compared the proposed method (BSM) with other moments including Bessel Fourier moments (BFM in [6]), Zernike moments (ZM, in [13]) and orthogonality Fourier-Mellin moments (OFFM, in [14]). The Precision-Recall curves (PRC) on three databases are shown in Fig. 2. As shown in Fig. 2, the proposed method (BSM) achieved best results than other comparative methods in three databases. Especially the precision and recall figure is almost the ideal figure in subset D1.

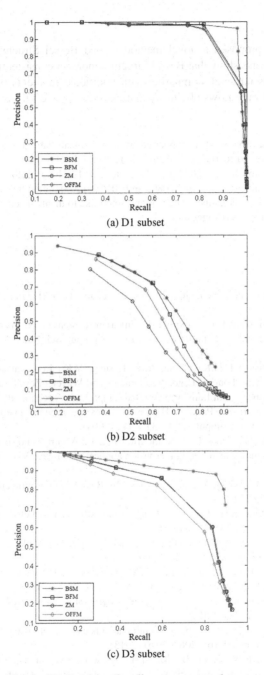

(a) D1 subset

(b) D2 subset

(c) D3 subset

Fig. 2. The Precision-Recall curves on test datasets.

4   Conclusion

In this paper we proposed a novel method named Bessel Structure moments. The numerical experiments show that Bessel Structure moment consistently performs better in terms of images retrieval, comparing with traditional moments of the same order. The good retrieval rate shows obviously that our descriptor is very are useful for color image retrieval.

Acknowledgements. This work is supported by the National Natural Science Foundation of China under grant nos. 61462002 and 61261043, Higher School Scientific Research Projects of Ningxia Province (No. NGY2016144), Education and Teaching Reform Project of North University of Nationalities (Nos. 2016JY0805 and 2016JY1205), Initial Scientific Research Fund of North University of Nationalities. The authors would like to thank the anonymous referees for their valuable comments and suggestions.

References

1. Hu, M.K.: Visual pattern recognition by moment invariants. IEEE Trans. Inf. Theory **8**, 179–182 (1962)
2. Stéphane, D., Mohamed, D., Ghorbel, F.: Invariant content-based image retrieval using a complete set of Fourier-Mellin descriptors. Int. J. Comput. Sci. Netw. Secur. **9**(7), 240–247 (2009)
3. Yadav, R.B., Nishchal, N.K., Gupta, A.K., Rastogi, V.K.: Retrieval and classification of objects using generic Fourier, Legendre moments and Wavelet Zernike moment descriptors and recognition using joint transform correlator. Opt. Laser Technol. **40**(3), 517–527 (2008)
4. Sim, D.-G., Kim, H.-K., Park, R.-H.: Invariant texture retrieval using modified Zernike moments. Image Vis. Comput. **22**(4), 331–342 (2004)
5. Papakostas, G.A., Karakasis, E.G., Koulouriotisb, D.E.: Accurate and speedy computation of image Legendre moments for computer vision applications. Image Vis. Comput. **28**(3), 414–423 (2010)
6. Xiao, B., Ma, J.-F., Wang, X.: Image analysis by Bessel-Fourier moments. Pattern Recogn. **43**(8), 2620–2629 (2010)
7. Mukundan, R., Ong, S.H., Lee, P.A.: Image analysis by Tchebichef moments. IEEE Trans. Image Process. **10**(9), 1357–1364 (2001)
8. Yap, P., Paramedran, R., Ong, S.H.: Image analysis by Krawtchouk moments. IEEE Trans. Image Process. **12**(11), 1367–1377 (2003)
9. Zhu, H.Q., Shu, H.Z., Liang, J., Luo, L.M., Coatrieux, J.L.: Image analysis by discrete orthogonal dual-Hahn moments. Pattern Recogn. Lett. **28**(13), 1688–1794 (2007)
10. Hoang, T.V., Tabbone, S.: Invariant pattern recognition using the RFM descriptor. Pattern Recogn. **45**(1), 271–284 (2012)
11. Dominguez, S.: Image analysis by moment invariants using a set of step-like basis functions. Pattern Recogn. Lett. **34**(16), 2065–2070 (2013)
12. Wang, X., Guo, F.-x., Xiao, B., Ma, J.-f.: Rotation invariant analysis and orientation estimation method for texture classification based on Radon transform and correlation analysis. J. Vis. Commun. Image Represent. **21**(1), 29–32 (2010)

13. Toharia, P., Robles, O.D., Rodríguez, Á., Pastor, L.: A study of Zernike invariants for content-based image retrieval. In: Mery, D., Rueda, L. (eds.) PSIVT 2007. LNCS, vol. 4872, pp. 944–957. Springer, Heidelberg (2007). https://doi.org/10.1007/978-3-540-77129-6_79
14. Sheng, Y., Shen, L.: Orthogonal Fourier-Mellin moments for invariant pattern recognition. Opt. Soc. Am. **11**(6), 1748–1757 (1994)
15. Novotni, M., Klein, R.: Shape retrieval using 3D Zernike descriptors. Comput. Aided Des. **36**(11), 1047–1062 (2004)
16. Yang, B., Kostková, J., Flusser, J., Suk, T.: Scale invariants from Gaussian-Hermite moments. Sig. Process. **132**, 77–84 (2017)
17. Xiao, B., Gang, L., Zhao, T., Xie, L.: Rotation, scaling and translation invariant texture recognition by Bessel-Fourier moments. Pattern Recogn. Image Anal. **26**(2), 302–308 (2016)
18. Xiao, B., Cui, J.-T., Qin, H.-X., Li, W.-S., Wang, G.-Y.: Moments and moment invariants in the Radon space. Pattern Recogn. **48**, 2772–2784 (2015)
19. Hai-tao, H., Zhang, Y.-d., Shao, C., Quan, J.: Orthogonal moments based on exponent functions: Exponent-Fouriermoments. Pattern Recogn. **47**(8), 2596–2606 (2014)
20. Xiao, B., Li, W.-s., et al.: Errata and comments on orthogonal moments based on exponent functions: Exponent–Fourier moments. Pattern Recogn. **48**, 1571–1573 (2015)
21. Hai-tao, H., Quan, J., Shao, C.: Errata and comments on "Errata and comments on Orthogonal moments based on exponent functions: Exponent-Fourier moments". Pattern Recogn. **52**, 471–476 (2016)
22. Ana, M.B., Beigi, I., Benitez, A.B., Chang, S.: MetaSEEk: a content-based meta-search engine for images. In: Proceedings of SPIE Storage and Retrieval for Image and Video Databases, vol. 3312, 118–128 (1997)
23. Li, Z., Zhang, Y., Hou, K., Li, H.: 3D polar-radius invariant moments and structure moment invariants. In: Wang, L., Chen, K., Ong, Y.S. (eds.) ICNC 2005. LNCS, vol. 3611, pp. 483–492. Springer, Heidelberg (2005). https://doi.org/10.1007/11539117_70

Optimizing Edge Weights for Distributed Inference with Gaussian Belief Propagation

Brendan Halloran$^{(\boxtimes)}$ (iD), Prashan Premaratne, and Peter James Vial

School of Electrical, Computer and Telecommunication Engineering,
University of Wollongong, Wollongong, NSW 2522, Australia
bh294@uowmail.edu.au

Abstract. Distributed processing is becoming more important in robotics as low-cost ad hoc networks provide a scalable and robust alternative to tradition centralized processing. Gaussian belief propagation (GaBP) is an effective message-passing algorithm for performing inference on distributed networks, however, its accuracy and convergence can be significantly decreased as networks have higher connectivity and loops. This paper presents two empirically derived methods for weighting the messages in GaBP to minimize error. The first method uses uniform weights based on the average node degree across the network, and the second uses weights determined by the degrees of the nodes at either end of an edge. Extensive simulations show that this results in greatly decreased error, with even greater effects as the network scales. Finally, we present a practical application of this algorithm in the form of a multi-robot localization problem, with our weighting system improving the accuracy of the solution.

Keywords: Gaussian belief propagation · Distributed algorithms
Markov random fields · Factor graphs · Localization

1 Introduction

Distributed processing is becoming more of an important paradigm as larger networks are being applied to various problems, particularly robotics and robotic vision. Traditional centralized solutions, where a single central processor takes data from each node, suffer from scalability issues due to communication bottlenecks and have a substantial risk of a single point of failure at the central node. Distributed algorithms alleviate these issues by moving the processing to the nodes themselves and using only local information with communication between direct neighbors, thereby taking better advantage of the communication structure and increasing scalability. These algorithms still need improvements, however, to achieve the same accuracy as centralized systems.

Belief propagation (BP) is a well-known message passing algorithm for performing efficient inference on graphical models [1, 2]. Whilst the original BP algorithm performed inference on discrete random variables, Gaussian belief propagation (GaBP) is an extension of BP to continuous random variables which are modelled as Gaussian densities [3, 4]. This allows many problems dealing with continuous variables, such as localization, to be interpreted as a distributed problem on a graph.

1.1   Related Work

Belief propagation was first proposed by Pearl for inference over discrete random variables, guaranteeing exact inference on acyclic graphs [1, 2]. It was later reformulated as *Generalized Belief Propagation* (GBP) for approximate inference on cyclic graphs [5]. It has also been shown that BP is equivalent to many similar *sum-product* algorithms, such as fast Fourier transforms, Kalman filtering, and average-consensus [6].

BP has been extended to work on continuous variables through *nonparametric belief propagation* (NBP) [7], which uses particle filtering, and *Gaussian belief propagation* [3, 4], which uses Gaussian densities. Although NBP can represent more complex non-Gaussian distributions, GaBP requires much less memory and processing power making it more suitable for low-cost hardware. GaBP has been applied to many problems such as parallel SVMs [8], MMSE estimators [9], Kalman filtering [10], linear least squares estimation [11], and multi-camera calibration [12]. NBP has also seen successful application to localization and tracking [13, 14], however GaBP has been shown to achieve similar results through iterative linearization of the problem [15]. Recently, GaBP has been applied to finite-element analysis, allowing greater parallelisation on GPUs [16] and high-performance computing applications [17]. BP also has wider applications to image processing, such as in image segmentation and recognition [18, 19].

Recent work has been done to understand the convergence conditions of GaBP, particularly analysing the information matrix [20, 21]. Other work has sought to improve the accuracy of the different BP algorithms, generally using different strategies for applying importance weights to the messages [7]. *Tree-reweighted BP* (TRW-BP) calculates the entire set of possible spanning-trees for a given cyclic graph then weights edges based on the probability that an edge appears in the set, calling these edge appearance probabilities (EAPs) [22]. It was later shown that uniform weights across all edges could achieve similar results by approximating the EAPs [23]. This has been successfully applied to NBP (TRW-NBP), however the empirical relationship derived was based only on a specific network size [24]. We are motivated to derive a more general weighting system for networks of varied sizes and connectivities.

1.2   Paper Contributions

To increase the accuracy of Gaussian belief propagation, we present the following improvements. Firstly, we have derived uniform message weights for minimizing root-mean-square error (RMSE), approximating the EAPs, by analyzing optimal weights for a range of small-world networks of differing sizes and connectivity. Next, we have proposed an alternative non-uniform message weight based on our prior derivation, which provides a further improvement to the algorithm accuracy. Finally, we have provided a practical example of this being used in a distributed inference problem through a demonstration of distributed localization.

The paper is organized as follows. Section 2 reviews the preliminaries of undirected graphical models. Section 3 describes GaBP and reweighted messages, and Sect. 4 explains our empirical approach to optimal weights. Section 5 provides an application of this algorithm to the localization problem, and finally, Sect. 6 concludes the paper.

2   Preliminaries

Consider a network of N nodes which are connected in an ad hoc manner. This network can be described by an undirected graph, or *Markov random field* (MRF), $\mathcal{G} = \{\mathcal{V}, \mathcal{E}\}$. Here, $\mathcal{V} = \{1, \ldots, N\}$ is the set of nodes and $\mathcal{E} \subseteq \mathcal{V} \times \mathcal{V}$ is the set of edges containing $(i,j) \in \mathcal{E}$ pairs where node i and j are connected. We denote the direct neighbours of node i as $\mathcal{N}_i = \{j \in \mathcal{V} : (i,j) \in \mathcal{E}\}$, which has degree $d_i = |\mathcal{N}_i|$. An example of a simple MRF is shown in Fig. 1.

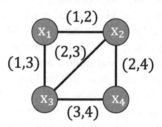

Fig. 1. An example Markov random field, with nodes and edges identified.

3   Gaussian Belief Propagation

Consider each node i having a true state given by some variable y_i. The random variable z_{ij} is the noisy observation by node i of either y_j itself or some constraint on y_i and y_j, where $j \in \mathcal{N}_i$. We wish to estimate y_i from all z_{ij} by marginalizing the density

$$p(y_i | z_{ij} : (i,j) \in \mathcal{E}) = \int_{(y_j, j \neq i)} p(y_1, \ldots, y_N | z_{ij} : (i,j) \in \mathcal{E}) dy_j.$$

The joint density can be factorised into node potentials ϕ_i and edge potentials ψ_{ij} for the MRF,

$$p(y_1, \ldots, y_N) \propto \prod_{i \in \mathcal{V}} \phi_i(y_i) \prod_{(i,j) \in \mathcal{E}} \psi_{ij}(y_i, y_j).$$

Belief propagation is an iterative algorithm which can estimate all y_i by passing messages $m_{i \to j}^t(y_j)$ from node i to j across edges $(i,j) \in \mathcal{E}$ at every iteration t.

$$m_{i \to j}^t(y_j) \propto \int_{y_i} \phi_i(y_i) \psi_{ij}(y_i, y_j) \prod_{(k \in \mathcal{N}_i \backslash j)} m_{k \to i}^{t-1}(y_i) dy_i \qquad (1)$$

In Eq. 1, $\mathcal{N}_i \backslash j$ means all neighbours of i except j. That is, a message from node i to node j combines the node potential with all incoming messages from the prior iteration from all neighbours except the node that will be receiving the message, node j. Figure 2 shows this process graphically for one iteration. We also compute a belief for each iteration t, which is the product of the node potential and all messages from that iteration. This value converges towards a global estimate for y_i, resulting in exact inference on trees and approximate inference otherwise.

$$b_i^t(y_i) \propto \phi_i(y_i) \prod_{k \in \mathcal{N}_i} m_{k \to i}^t(y_i) \tag{2}$$

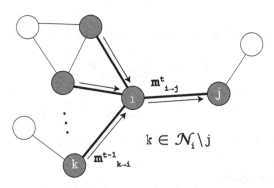

Fig. 2. Messages from node i to node j include all messages received at node i from neighbours excluding node j.

In GaBP, random variables are modelled as Gaussian densities. Therefore, two scalars each represent potentials, messages, and beliefs – mean and precision. The mean is the noisy measurement z_{ij} and precision P_{ij} is the inverse of its variance $\Sigma_{z_{ij}}$. The graph is then represented by an information matrix Ω, encoding the edges and precisions, and a measurement vector ξ, containing precision-weighted sums of measurements. For example, if each node has relative measurements of its neighbours, $z_{ij} = y_j - y_i + w_{ij}$, as well as an absolute measurement of itself, $z_{ii} = y_i + w_{ii}$, where w is some zero-mean additive Gaussian noise, then we would have

$$\Omega_{ij} = \begin{cases} (d_i + 1)P_{ij}, & i = j \\ -P_{ij}, & j \in \mathcal{N}_i \\ 0, & \textit{otherwise} \end{cases}$$

$$\xi_i = (d_i + 1)z_{ii}P_{ij} - \sum_{(j \in \mathcal{N}_i)} z_{ij}P_{ij}.$$

We can then reduce the messages of Eq. 1 to mean and precision values,

$$P^t_{i\setminus j} = \Omega_{ii} + \sum_{(k \in \mathcal{N}_i \setminus j)} P^{t-1}_{k \to i}$$

$$\mu^t_{i\setminus j} = \left(P^t_{i\setminus j}\right)^{-1} \left(\xi_i + \sum_{(k \in \mathcal{N}_i \setminus j)} \mu^{t-1}_{k \to i} P^{t-1}_{k \to i}\right)$$

$$P^t_{i \to j} = -\Omega^2_{ij} \left(P^t_{i\setminus j}\right)^{-1}$$

$$\mu^t_{i \to j} = -\Omega_{ij} \mu^t_{i\setminus j} \left(P^t_{i\setminus j}\right)^{-1}$$

From these messages, the beliefs can so be determined at each node using

$$P^t_i = \Omega_{ii} + \sum_{(k \in \mathcal{N}_i)} P^t_{k \to i}$$

$$\mu^t_i = \left(P^t_i\right)^{-1} \left(\xi_i + \sum_{(k \in \mathcal{N}_i)} \mu^t_{k \to i} P^t_{k \to i}\right)$$

with μ^t_i converging towards a global consensus value for y_i.

3.1 Message Reweighting

Wainwright et al. first showed how messages can be reweighted in *tree-reweighted belief propagation*, where weights were based on the probability that an edge appears on a spanning-tree subgraph, ρ_{ij} [22]. This method reduces RMSE in the solution but wasn't suitable for use in distributed networks as it required all possible spanning-trees to be determined to calculate the edge appearance probabilities. Savic et al. simplified this by approximating EAPs for lattice graphs and applied this to *nonparametric belief propagation* [23]. Here, we will apply these weights to the Gaussian-based equations.

Firstly, edge weights ρ_{ij} are applied to the messages for belief in Eq. 2,

$$b^t_i(y_i) \propto \phi_i(y_i) \prod_{k \in \mathcal{N}_i} \left(m^t_{k \to i}(y_i)\right)^{(\rho_{ij})}.$$

Then outgoing messages of Eq. 1 include this new reweighted belief and have the inverse weight applied to the edge potential, giving

$$m^t_{i \to j}(y_j) \propto \int_{y_i} \left(\psi_{ij}(y_i, y_j)\right)^{(1/\rho_{ij})} \frac{b^{t-1}_i(y_i)}{m^{t-1}_{j \to i}(y_i)} dy_i.$$

This is then converted to mean and precision for the Gaussian beliefs and messages.

$$P_i^t = \Omega_{ii} + \sum_{(k \in \mathcal{N}_i)} \left(\rho_{ij} P_{k \to i}^t \right)$$

$$\mu_i^t = \left(P_i^t \right)^{-1} \left(\xi_i + \sum_{(k \in \mathcal{N}_i)} \left(\mu_{k \to i}^t \rho_{ij} P_{k \to i}^t \right) \right)$$

$$P_{i \backslash j}^t = P_i^{t-1} - P_{j \to i}^{t-1}$$

$$\mu_{i \backslash j}^t = \left(P_{i \backslash j}^t \right)^{-1} \left(P_i^{t-1} \mu_i^{t-1} - \mu_{j \to i}^{t-1} P_{j \to i}^{t-1} \right)$$

$$P_{i \to j}^t = -\Omega_{ij}^2 \left(\rho_{ij}^2 P_{i \backslash j}^t \right)^{-1}$$

$$\mu_{i \to j}^t = -\Omega_{ij} \mu_{i \backslash j}^t \left(\rho_{ij} P_{i \backslash j}^t \right)^{-1}$$

In Savic's TRW-NBP, the weights are uniform for the entire graph. That is, $\rho_{ij} = \rho_{opt}$ for all edges $(i,j) \in \mathcal{E}$, with ρ_{opt} being determined by their empirical formula.

4   Optimizing Weights for Small World Networks

For TRW-NBP, Savic determined an empirical relationship between the optimal edge weights, with regard to RMSE, and the average node degree of the graph, n_d

$$\rho_{opt}(n_d) = \rho_0 e^{-k_\rho n_d} \tag{3}$$

where ρ_0 and k_ρ were constants. These constants were found empirically for a network of $N = 25$ nodes and were different for lattice and random graphs. In this section, optimal weights are analyzed for minimizing RMSE in GaBP and extended to more general small-world networks for a range of network sizes.

4.1   Small World Generation

A small-world network is an interpolation between lattice graphs, where nodes are connected in a regular pattern, and random graphs. Small-world networks are characterized by high clustering and having few hops between any two nodes. We chose to optimize our algorithm weights for small-world networks due to their prevalence in many naturally occurring systems. These networks can be generated with the Watts-Strogatz model, where a lattice graph has its edges rewired randomly with some probability [25]. This process is as follows:

1. For a graph of N nodes, it has connectivity or average node degree $n_d \in 2\mathbb{Z}$ with $N \gg n_d$ and a parameter representing the 'randomness' of the graph $0 \leq \beta \leq 1$.
2. The graph is initially connected as in a regular lattice where each node is connected to the $n_d/2$ neighbors either side. That is,

$$\mathcal{E} = \{(i,j) : 0 < |i - j| mod \left(N - \frac{n_d}{2} - 1\right) < \frac{n_d}{2}\}$$

3. For each node i, take each edge (i,j) where $j \in \mathcal{N}_i$ and $i < j$, and with probability β replace it with a random new edge (i,k) where $k \neq i$ and $k \notin \mathcal{N}_i$.

4.2   Uniform Optimal Weights

The main effect of the graph cycles on the GaBP solution is an increase of RMSE, with acyclic graphs performing exact inference. Using small-world networks, the effect of uniform message weights on RMSE was simulated on a variety of structures, with four network sizes N and β varying from 0 to 1 in steps of 0.1. For $N = 10$, small-worlds were generated of connectivities $1 \leq n_d/2 \leq 4$. For $N = 20$ and $N = 30$, connectivities were $1 \leq n_d/2 \leq 7$. For $N = 40$, connectivities were $2 \leq n_d/2 \leq 8$. For each size, β, and connectivity, 1000 small-world networks were generated. In each network a simple single variable problem was simulated with nodes assigned true values $y_i \in [1, 100]$ and measurements taken of themselves and neighbors affected by zero mean Gaussian noise of variance $\Sigma_{z_{ij}} = 0.5$. The maximum likelihood of y_i was estimated using GaBP with uniform weights ρ ranging from 0.01 to 1.0 in steps of 0.01, and the RMSE of these solutions was compared to the RMSE of the *unweighted* GaBP solution.

Figure 3a shows results for $N = 40$ and $\beta = 0.5$. As the connectivity increases the optimal weight decreases, whilst its effect on RMSE increases. For higher connectivities, most weights have some positive effect on the RMSE compared to the unweighted solution. It is only once the weights become much lower than the optimal weight that the error begins to increase rapidly, and the solution no longer converges. Figure 3b shows that the location of the optimal weight does not rely on the average node degree n_d alone, but also the graph size N. This is because the amount by which the connectivity of a cyclic graph exceeds that of a valid spanning-tree is dependent on N as well. Figure 4 shows the effect that different β values in the small-world generation have on the location of the optimal weight. In a network of $N = 40$ and $n_d = 12$, the optimal weights for each value of β cluster around $\rho = 0.25$ with no significant pattern. This result was similar for all graph sizes and connectivities that we tested.

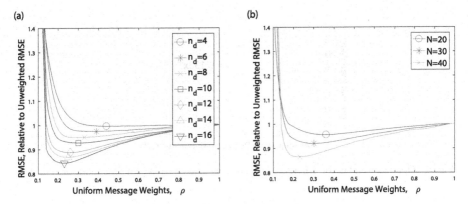

Fig. 3. Average RMSE relative to unweighted RMSE for different uniform weights and (a) connectivities with $N = 40$ and $\beta = 0.5$, and (b) network sizes for $\beta = 0.5$ and $n_d = 12$.

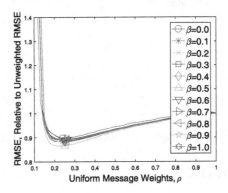

Fig. 4. Average RMSE relative to unweighted RMSE for different uniform message weights and β values in an $N = 40$ small world network of connectivity $n_d = 12$.

To explore these relationships further, the optimal weights for each graph size and connectivity were analyzed against n_d/N. This revealed a power relationship, as opposed to the exponential relationship of Eq. 3.

$$\rho_{opt}(n_d, N) = \rho_0 \cdot \left(\frac{n_d}{N}\right)^{-k_\rho} \tag{4}$$

Figure 5 shows this for $N = 40$, taking the average optimal weight across all β. This was repeated for all network sizes and revealed changing values of ρ_0 and k_ρ with N, shown in Fig. 6. Using these trends, the following empirical relationships were derived,

$$\rho_0(N) = a_\rho N^{-b_\rho}$$

$$k_\rho(N) = a_k N + b_k$$

From our data, we found $a_\rho = 2.9969$, $b_\rho = 0.845$, $a_k = 0.0026$, and $b_k = 0.444$.

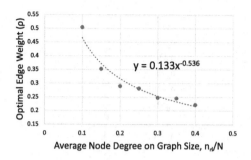

Fig. 5. Optimal weights for different connectivities against n_d/N, for a network of $N = 40$.

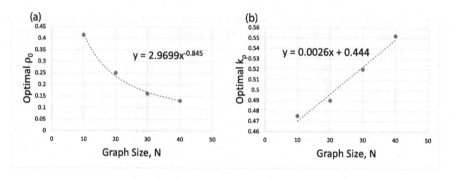

Fig. 6. For different graph sizes, optimal values of (a) ρ_0 and (b) k_ρ.

4.3   Non-uniform Weights

Next, the effect of non-uniform weights was explored. Considering that the heuristic of Eq. 4 aims to emulate the effect of EAP weights, individual node degrees can be used rather than the average node degree. Since the probability that a given edge is pruned in to make a spanning-tree is related to the degrees of the nodes at either end, we use the average of these degrees, d_{ij}, for edge specific weights given in Eq. 6.

$$d_{ij} = \frac{d_i + d_j}{2} \tag{5}$$

$$\rho_{opt}\left(d_{ij}, N\right) = \rho_0 \cdot \left(\frac{d_{ij}}{N}\right)^{-k_\rho} \tag{6}$$

4.4   Comparison of Weighting Methods

The effects of these different weighting methods were compared using the same graph sizes, connectivities and β as in the previous simulation, measuring the average RMSE relative to the unweighted solution, shown in Fig. 7. As a baseline, the ideal weights at the minima of the previous simulation, shown in Figs. 3 and 4, were included. As the uniform weighting heuristic of Eq. 4 aims to follow these minima as closely as possible, it is as expected that these two lines are close together. The non-uniform weighting heuristic, however, out-performed these weights. The heuristic of Eq. 3, from TRW-NBP, was also included for comparison. This performed slightly worse than our weights and resulted increased RMSE for low connectivities on the $N = 10$ graph.

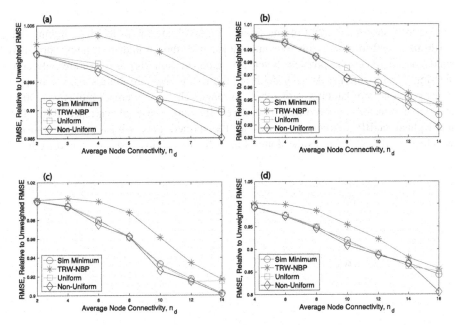

Fig. 7. Average RMSE, relative to RMSE of unweighted GaBP of 1000 trials, for different weighting systems and graph sizes of (a) N = 10, (b) N = 20, (c) N = 30, and (d) N = 40.

5   Application to Factor Graph Localization

As a practical example of this algorithm, our weights were tested on a linear multi-robot *Simultaneous Localization and Mapping* (SLAM) problem. SLAM uses measurements between robot poses and landmarks to localize the robots and map the landmarks. Graph SLAM represents this as *maximum a posteriori* optimization on a factor graph [26]. BP has been shown to efficiently solve graphical representations in localization [15], tracking [27], and SLAM [28]. The difference between an MRF and factor graph can be seen in Fig. 8a, with visual representation of graph SLAM shown in Fig. 8b. Nodes are the same as those on an MRF, whilst the constraints between nodes and priors are now shown as factors represented by squares.

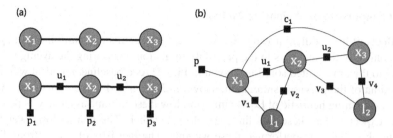

Fig. 8. Factor graphs, (a) the conversion from an MRF with constraints between nodes u and prior information p as factors, and (b) SLAM as a factor graph with x a pose, l a landmark, p the prior, u a constraint between poses, v a measurement of a landmark, and c a loop-closure.

Figure 9 shows an output of this simulation, involving robots moving randomly on a 100 m × 100 m plane with four landmarks, with the ground truth paths shown as solid lines. To keep the problem linear for demonstration purposes, we are only considering (x, y) positions. At each iteration, the robots measure the distance to its previous pose and any landmarks or robots in range. The measurement radius was 30 m and were affected by zero-mean additive Gaussian noise of variance 3 m. The unweighted GaBP solution is shown as the dashed line, whilst the non-uniform weighted GaBP is shown as the dotted line, which was closer to the ground truth.

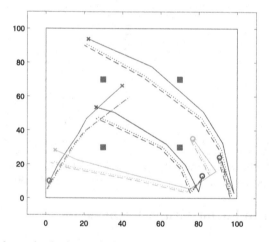

Fig. 9. Example result of using weighted GaBP to solve a 2D graph SLAM problem.

This simulation was repeated for different numbers of robots ranging from 1 to 6, for 1000 trials each, to measure the average reduction in RMSE compared to unweighted GaBP, shown in Fig. 10. As the number of robots increases, and therefore the graph size, the effect of the weighted messages on RMSE increases as well. The TRW weights, however, decrease in effectiveness as the graph size increases.

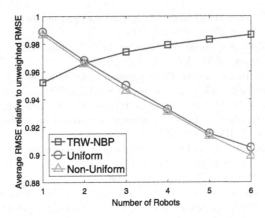

Fig. 10. Relative reduction in RMSE compared to unweighted GaBP for different numbers of robots, using TRW weights as well as our uniform and non-uniform weights.

6 Conclusions and Future Work

This paper has presented two weighting systems to reduce RMSE in Gaussian belief propagation on cyclic graphs. The first involves uniform weights based on graph size and average node degree, whilst the second uses non-uniform weights from the average degree at either end of an edge. These weighting systems were empirically derived by analyzing the RMSE minima for uniform weights on graphs of varied sizes and connectivities. We have shown that these weights effectively reduce the error in the GaBP solution, with more significant effects at higher connectivities. Finally, an example application was given in the form of multi-robot graph SLAM, with the non-uniform weights achieving a more accurate solution than unweighted GaBP. We will be continuing this work to better relate the optimal message weights to the intrinsic properties of the graph and will also investigate applying it to further robotics problems.

References

1. Pearl, J.: Probabilistic Reasoning in Intelligent Systems: Networks of Plausible Inference, 2nd edn. Morgan Kaufmann, San Francisco (1988)
2. Pearl, J.: Fusion, propagation, and structuring in belief networks. Artif. Intell. **29**(3), 241–288 (1986)
3. Weiss, Y., Freeman, W.T.: Correctness of belief propagation in Gaussian graphical models of arbitrary topology. In: Advances in Neural Information Processing Systems, pp. 673–679 (2000)
4. Bickson, D.: Gaussian belief propagation: theory and application. arXiv preprint arXiv:0811.2518 (2008)
5. Yedidia, J.S., Freeman, W.T., Weiss, Y.: Generalized belief propagation. In: Advances in Neural Information Processing Systems, pp. 689–695 (2001)

6. Kschischang, F.R., Frey, B.J., Loeliger, H.A.: Factor graphs and the sum-product algorithm. IEEE Trans. Inf. Theor. **47**(2), 498–519 (2001)
7. Sudderth, E.B., Ihler, A.T., Isard, M., Freeman, W.T., Willsky, A.S.: Nonparametric belief propagation. Commun. ACM **53**(10), 95–103 (2010)
8. Bickson, D., Yom-Tov, E., Dolev, D.: A gaussian belief propagation solver for large scale support vector machines. arXiv preprint arXiv:0810.1648 (2008)
9. Bickson, D., Dolev, D., Shental, O., Siegel, P.H., Wolf, J.K.: Gaussian belief propagation based multiuser detection. In: IEEE International Symposium on Information Theory, ISIT 2008, pp. 1878–1882. IEEE, July 2008
10. Bickson, D., Shental, O., Dolev, D.: Distributed Kalman filter via Gaussian belief propagation. In: 2008 46th Annual Allerton Conference on Communication, Control, and Computing, pp. 628–635. IEEE, September 2008
11. Shental, O., Bickson, D., Siegel, P.H., Wolf, J.K., Dolev, D.: Gaussian belief propagation for solving systems of linear equations: theory and application. arXiv preprint arXiv:0810.1119 (2008)
12. Halloran, B., Premaratne, P., Vial, P., Kadhim, I.: Distributed one dimensional calibration and localisation of a camera sensor network. In: Huang, D.-S., Jo, K.-H., Figueroa-García, J. C. (eds.) ICIC 2017. LNCS, vol. 10362, pp. 581–593. Springer, Cham (2017). https://doi.org/10.1007/978-3-319-63312-1_51
13. Savic, V., Zazo, S.: Cooperative localization in mobile networks using nonparametric variants of belief propagation. Ad Hoc Netw. **11**(1), 138–150 (2013)
14. Savic, V., Wymeersch, H.: Simultaneous localization and tracking via real-time nonparametric belief propagation. In: 2013 IEEE International Conference on Acoustics, Speech and Signal Processing (ICASSP), pp. 5180–5184. IEEE, May 2013
15. García-Fernández, Á.F., Svensson, L., Särkkä, S.: Cooperative localization using posterior linearization belief propagation. IEEE Trans. Veh. Technol. **67**(1), 832–836 (2018)
16. Hosseinidoust, Z., Giannacopoulos, D., Gross, W.J.: GPU optimization and implementation of Gaussian belief propagation algorithm. In: 2016 IEEE Conference on Electromagnetic Field Computation (CEFC). IEEE (2016)
17. El-Kurdi, Y., et al.: Acceleration of the finite-element gaussian belief propagation solver using minimum residual techniques. IEEE Trans. Magn. **52**(3), 1–4 (2016)
18. Yang, S., Premaratne, P., Vial, P.: Hand gesture recognition: an overview. In: 2013 5th IEEE International Conference on Broadband Network & Multimedia Technology (IC-BNMT). IEEE (2013)
19. Premaratne, P., Ajaz, S., Premaratne, M.: Hand gesture tracking and recognition system for control of consumer electronics. In: Huang, D.-S., Gan, Y., Gupta, P., Gromiha, M.M. (eds.) ICIC 2011. LNCS (LNAI), vol. 6839, pp. 588–593. Springer, Heidelberg (2012). https://doi.org/10.1007/978-3-642-25944-9_76
20. Du, J., Ma, S., Wu, Y.C., Kar, S., Moura, J.M.: Convergence analysis of distributed inference with vector-valued Gaussian belief propagation. arXiv preprint arXiv:1611.02010 (2016)
21. Du, J., Ma, S., Wu, Y.C., Kar, S., Moura, J.M.: Convergence analysis of the information matrix in Gaussian belief propagation. In: 2017 IEEE International Conference on Acoustics, Speech and Signal Processing (ICASSP), pp. 4074–4078. IEEE, March 2017
22. Wainwright, M.J., Jaakkola, T.S., Willsky, A.S.: Tree-reweighted belief propagation algorithms and approximate ML estimation by pseudo-moment matching. In: AISTATS, January 2003

23. Wymeersch, H., Penna, F., Savić, V.: Uniformly reweighted belief propagation: a factor graph approach. In: 2011 IEEE International Symposium on Information Theory Proceedings (ISIT), pp. 2000–2004. IEEE, July 2011

24. Savic, V., Wymeersch, H., Penna, F., Zazo, S.: Optimized edge appearance probability for cooperative localization based on tree-reweighted nonparametric belief propagation. In: 2011 IEEE International Conference on Acoustics, Speech and Signal Processing (ICASSP), pp. 3028–3031. IEEE, May 2011

25. Watts, D.J., Strogatz, S.H.: Collective dynamics of 'small-world' networks. Nature **393** (6684), 440 (1998)

26. Cadena, C., Carlone, L., Carrillo, H., Latif, Y., Scaramuzza, D., Neira, J., Reid, I., Leonard, J.J.: Past, present, and future of simultaneous localization and mapping: Toward the robust-perception age. IEEE Trans. Rob. **32**(6), 1309–1332 (2016)

27. Meyer, F., et al.: Message passing algorithms for scalable multitarget tracking. Proc. IEEE **106**(2), 221–259 (2018)

28. Leitinger, E., et al.: A scalable belief propagation algorithm for radio signal based SLAM. arXiv preprint arXiv:1801.04463 (2018)

Auto Position Control for Unmanned Underwater Vehicle Based on Double Sliding Mode Loops

Wei Jiang[1](✉), Jian Xu[2], Xiao-Feng Kang[2], and Liang-liang Wang[1]

[1] Department of Information Engineering, City College,
Wuhan University of Science and Technology, Wuhan, China
jiangweisky@qq.com
[2] College of Automation, Harbin Engineering University, Harbin, China

Abstract. To deal with the large speed jump and position overshoot when desired position changes suddenly for UUV in the horizontal plane, an auto position controller based on double sliding mode loops is proposed. In comparison with the conventional control approach, the virtual speed is adopted in the proposed controller designing. Double loops are contained in the designed controller, and sliding mode surface is both contained in the outer loop and inner loop. The outer loop generates the virtual speed of UUV to avoid the large speed jump, while the inner loop realizes the speed tracking and eliminate the error between virtual speed and real speed. To estimate the uncertainty of the UUV's motion model and the environment disturbance, an adaptive law is adopted. The stability of the proposed control method is proven based on Lyapunov stability theory, and the effectiveness of the proposed controller is demonstrated through simulation.

Keywords: UUV · Double sliding mode loops · Virtual speed
Auto position

1 Introduction

The purpose of dynamic positioning control is to maintain the desired position and attitude for vessels under environment disturbance [1, 2]. As one mode of dynamic positioning control, auto position control is to change the vessel's position automatically when the operator demands another desired position. Mostly auto position control is for surface vessels, while there are also some applications for unmanned underwater vehicle (UUV). In the UUV surface missions such as observation, recovery at a fixed point, UUV also needs to be auto position controlled. When the desired position changes, the auto position controller works by using the thrusters. The speed jump and the position overshoot occur when the desired position changes suddenly. Therefore, it is very essential to design a suitable controller.

The robustness of the controller for UUV should be taken into consideration, because the hydrodynamic parameters of UUV model may not be accurate and the environment disturbance can be a great factor to the control accuracy [3, 4]. Considering the parameters uncertainties, an adaptive PD controller for ROV dynamic

© Springer International Publishing AG, part of Springer Nature 2018
D.-S. Huang et al. (Eds.): ICIC 2018, LNCS 10954, pp. 60–71, 2018.
https://doi.org/10.1007/978-3-319-95930-6_7

positioning control is proposed in [5]. The simulation results and experiments show a good performance under the constant uncertainties. However, the overshoot of the position still exists, and the speed jump occurs. To solve position overshoot and speed jump problem for UUV, some other control algorithms have been proposed such as fuzzy control strategy [6], neural control algorithm [7], sliding mode method [8]. In [9], a tracking controller based on some fuzzy rules can reduce the speed jump when tracking error suddenly changes. These fuzzy rules should be designed based on lots of experimental results and experience of skilled engineers. But it is difficult to set those fuzzy rules for UUV control, since the experiments data or experience may be not enough. In [10], a three layer neural network algorithm is applied in tracking control for UUV. The neural network performs very well with a large initial position error. But the network needs to be training with a lot of experimental data, and it is not that easy to be realized in UUV real-time control system.

Therefore, it is very important to take the implementation of the control algorithm into consideration. As it is not difficult to be implemented, sliding mode control has been widely used in UUV control. Moreover, it is also not sensitive to the uncertainty of the motion model and the external disturbance. A drawback of sliding mode is the chattering caused by the switch function [11]. To solve the chattering problem in the practical application, Bing Sun proposed a chattering free sliding mode method by using an adaptive continuous term to replace the switch term [12]. However, when the desired position changes suddenly and position error is great, the speed jump and the position overshoot occur with conventional sliding mode controller. The real speed is contained in the conventional control law directly [13]. In [14] a virtual speed vector is adopted in the designed tracking controller based on back-stepping and dynamical sliding mode algorithm for UUV. The real speed in the control law is replaced by the generated virtual speed in tracking control for UUV. The simulation results are satisfactory, and the speed jump is avoided.

Concentrating on the position control problem of UUV, a controller based on double sliding mode loops is proposed in this paper that is inspired by the virtual speed idea in [14]. The outer loop generates smooth virtual speed and the inner loop guarantees accurate position control. The speed jump and position overshoot are overcome. Moreover, an adaptive law is adopted to estimate the uncertainty of the UUV's motion model and the environment disturbance. With the proposed controller, the anti-interference ability and dynamic response ability of the system are improved. This paper is organized as follows: In Sect. 2, the mathematical model of UUV is given. In Sect. 3, the controller based on double sliding mode loops is introduced. In Sect. 4, the stability of the system is analyzed. In Sect. 5, the effectiveness of the proposed controller is simulated. At last, the conclusion is given based on the previous contents.

2 Mathematical Model of UUV

The kinematic and dynamic models of UUV in the horizontal plane are presented briefly in this section. To describe the motion of UUV, two coordinate frames are defined: the earth fixed frame and the UUV frame. The two frames are shown in Fig. 1. To simplify the UUV motion model, the position (x, y) and heading ψ of UUV in the earth fixed frame are expressed in vector form by $\eta = [x \quad y \quad \psi]^T$, meanwhile the speed of UUV in the horizontal plane are also expressed in vector form by $\upsilon = [u \quad v \quad r]^T$.

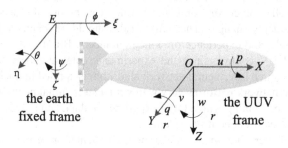

the earth
fixed frame

the UUV
frame

Fig. 1. Definition of the earth fixed frame and the UUV frame

The three motions are referred as the surge, sway and yaw of UUV. The transformation of speed vector between the UUV frame and the earth fixed frame is written in vector form as follows:

$$\dot{\eta} = J(\eta)\upsilon \tag{1}$$

Since the roll and pitch are neglected in the horizontal plane in this paper and only yaw is considered, the transformation matrix $J(\eta)$ can be rewritten as follows:

$$J(\eta) = J(\psi) = \begin{bmatrix} \cos\psi & -\sin\psi & 0 \\ \sin\psi & \cos\psi & 0 \\ 0 & 0 & 1 \end{bmatrix} \tag{2}$$

As some hydrodynamic terms can be neglected, the dynamic model of UUV in the earth-fixed frame can be simplified as follows:

$$\begin{cases} (m - X_{\dot{u}})\dot{u} = X_T - X_F - X_u u - X_{u|u|}u|u| \\ (m - Y_{\dot{v}})\dot{v} + (mx_G - Y_{\dot{r}})\dot{r} = Y_T - Y_F - Y_v v - Y_{v|v|}v|v| \\ (mx_G - N_{\dot{v}})\dot{v} + (I_z - N_{\dot{r}})\dot{r} = N_T - N_F - N_r r - N_{r|r|}r|r| \end{cases} \tag{3}$$

where state $u, v, r \in R^1$ denote the surge, sway, and yaw speed in the UUV frame. $m, X_{\dot{u}}, Y_{\dot{v}}, N_{\dot{r}} \in R^{3\times3}$ are the mass and added mass of UUV; $x_G, y_G \in R^1$ are the UUV gravity coordinate in the UUV frame; $I_z \in R^1$ is the moment of inertia about the earth-fixed z-axis. $X_T, Y_T, N_T \in R^{3\times1}$ are the control force and moment provided by the UUV thrusters, $X_F, Y_F, N_F \in R^{3\times1}$ are the environment disturbance.

To design the control law easily, the dynamic model can be rewritten in vector form as [15]:

$$M\dot{\upsilon} + D(\upsilon)\upsilon + \tau_f = \tau_T \tag{4}$$

where $M \in R^{3\times3}$ is the mass matrix, υ is the speed vector in UUV frame, $\tau_T \in R^{3\times1}$ is the control output vector, $\tau_f \in R^{3\times1}$ is the environment disturbance vector.

3 Controller Design

When mathematical model of UUV is established, the essence of auto position needs to be considered. The purpose of auto position for UUV is to maintain the desired position. When the desired position changes suddenly, UUV will be moved to the desired position by the thrusters with the controller in the horizontal plane. To avoid the speed jump and position overshoot, a good controller which are supposed to have a smooth and accurate control effect should be designed. The position of UUV could not be stably maintained if the speed varies dramatically, so the outer loop speed control is introduced, which makes the system rapidly restore stability when suffering from disturbance. And thus, the anti-interference ability and dynamic response ability of the system are improved. Inspired by the virtual speed idea in tracking control of UUV [14], the proposed auto position controller of UUV consists of double sliding mode loops. The outer loop generates the virtual speed based on the position error, and the inner loop calculates the control output based on the speed error. In order to realize the auto position control for UUV, the control output needs to be allocated to the main thrusters and horizontal auxiliary thrusters of UUV. The thruster allocation is very important in the controller design. The framework of the whole control system is shown in Fig. 2.

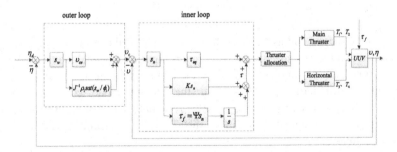

Fig. 2. The framework of auto position control system for "BSA-IV" UUV

3.1 Outer Loop

First of all, a sliding mode surface of the position error is defined:

$$s_w = \tilde{\eta} + K_1 \int_0^t \tilde{\eta} dt \tag{5}$$

where $\tilde{\eta} = \eta_d - \eta$, $\tilde{\eta}$ is the position control error, η_d is the desired position, and η is the real position. $K_1 \in R^{3 \times 3}$ is the constant matrix. Based on this sliding mode surface, the virtual speed vector is designed as follows:

$$v_c = J^{-1}(\eta)(\dot{\eta}_d + K_1 \tilde{\eta}) + J^{-1}(\eta)\rho_1 sat(s_w/\phi_1) \tag{6}$$

where v_c is the virtual speed vector, ρ_1 is a constant, sat is the saturation function, and the exact form is described as follows:

$$sat(s_w/\phi_1)_i = \begin{cases} \text{sgn}(s_{wi}/\phi_1), & if \left|\frac{s_{wi}}{\phi_1}\right| > 1 \\ s_{wi}/\phi_1, & else \left|\frac{s_{wi}}{\phi_1}\right| \le 1 \end{cases} \tag{7}$$

where ϕ_1 is the positive constant, and it is the thickness of the boundary layer. The purpose of using the boundary layer is to reduce the chattering problem caused by sliding mode switch function.

3.2 Inner Loop

To eliminate the error between the generated virtual speed and real speed of UUV, inner loop is designed with another sliding mode surface. This sliding mode surface related with speed error is designed as follows:

$$s_n = \tilde{v} + K_2 \int_0^t \tilde{v} dt \tag{8}$$

where s_n is the second sliding mode surface, $\tilde{v} = v_c - v$ is the error between generated virtual speed and the real speed, $K_2 \in R^{3 \times 3}$ is the constant diagonal matrix. Taking the dynamic model Eq. (4) into consideration, the control law consists of two parts: the equivalent control term $\tau_{eq} \in R^{3 \times 1}$ and the sliding mode compensation term $\tau_{wn} \in R^{3 \times 1}$. The purpose of adding $\tau_{wn} \in R^{3 \times 1}$ is to make sure that the system state will be converged to the designed sliding mode surface. The exact form of the control law is designed as follows:

$$\begin{cases} \tau = \tau_{eq} + \tau_{wn} \\ \tau_{eq} = M(\dot{v}_c + K_2 \tilde{v}) + D(v)v + \hat{\tau}_f \\ \tau_{wn} = K_3 s_n + \rho_2 sat(s_n/\phi_2) \end{cases} \tag{9}$$

where $K_3 \in R^{3 \times 3}$ is the diagonal matrix, ρ_2 is a constant, *sat* is the saturation function which is as same as the saturation in the outer loop. However, the environment disturbance vector in the control law is still unknown, and there are no sensors equipped in UUV to measure the disturbance. Therefore, an adaptive law is adopted to estimate the total environment disturbance and uncertainty of motion model of UUV. The adaptive law is designed as follows:

$$\dot{\hat{\tau}}_f = \Psi s_n \tag{10}$$

Where $\dot{\hat{\tau}}_f$ is the derivative of the estimated environment disturbance vector, and $\Psi \in R^{3 \times 3}$ is the constant diagonal matrix.

Table 1. The parameters of thrusters of "BSA-IV" UUV

Thruster type	Space/m	Min-Force/N	Max-Force/N
Main thruster	0.620	−1274	1274
Horizontal auxthruster	4.875	−768	768

3.3 Thruster Allocation of UUV

Thruster allocation is a very important step of auto position control for UUV, and the forces of the controller output in surge, sway and yaw need to be allocated to main thrusters and auxiliary thrusters of UUV to realize auto position. A thruster allocation method is introduced which is applied in "BSA-IV" UUV of Harbin Engineering University. The parameters of thrusters are shown in Table 1 and Fig. 3.

The controller output in surge, sway and yaw can be described as follows:

$$\tau = \begin{bmatrix} \tau_x & \tau_y & \tau_N \end{bmatrix}^T \tag{11}$$

The thruster allocation is designed as:

$$\begin{cases} T_1 = 0.5(\tau_x + \beta\tau_N/L_m), \ \gamma_1 T_{1\,min} \leq T_1 \leq \gamma_1 T_{1max} \\ T_2 = 0.5(\tau_x - \beta\tau_N/L_m), \ \gamma_1 T_{2\,min} \leq T_2 \leq \gamma_1 T_{2max} \\ T_3 = 0.5(\tau_y - (1-\beta)\tau_N/L_h), \ \gamma_2 T_{3\,min} \leq T_3 \leq \gamma_2 T_{3max} \\ T_4 = 0.5(\tau_y + (1-\beta)\tau_N/L_h), \ \gamma_2 T_{4min} \leq T_4 \leq \gamma_2 T_{4max} \end{cases} \tag{12}$$

where T_1, T_2, T_3, T_4 are left main thruster, right main thruster, bow horizontal auxiliary thruster, stern horizontal auxiliary thruster respectively, L_m, L_h are the space between main thrusters, space between horizontal auxiliary thrusters, τ_x, τ_y, τ_N represent the controller output in the three degrees of freedom in horizontal plane. β is the thruster allocate coefficient which is used to control the percentage of moment in heading provided by the main thrusters. T_{1max}, T_{2max} are the maximum force which are provided by two main thrusters and two horizontal auxiliary thrusters respectively. γ_1, γ_2 are the limit of the usage of the main thrusters, horizontal auxiliary thrusters respectively.

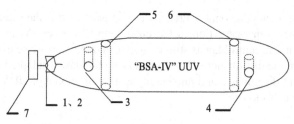

1、2: Main Thruster,   3、4: Horizontal aux thruster,
5、6: Vertical auxthruster,   7: rudder

Fig. 3. Configuration of thrusters of "BSA-IV" UUV

4 Stability Analysis

As the proposed control method is designed with double loops, the stability will be analyzed by using two Lyapunov functions. For the outer loop, let the first Lyapunov function candidate be:

$$V_1 = \frac{1}{2} s_w^T s_w \tag{13}$$

Then the time derivative of V_1 yields:

$$\dot{V}_1 = s_w^T \dot{s}_w = s_w^T (\dot{\tilde{\eta}} + K_1 \tilde{\eta}) = s_w^T (\dot{\eta}_d - \dot{\eta} + K_1 \tilde{\eta}) \tag{14}$$

Substitute Eq. (1) and Eq. (6) into Eq. (14), yields:

$$\begin{aligned} \dot{V}_1 &= s_w^T (\dot{\eta}_d - J(\eta) v_c + K_1 \tilde{\eta}) = s_w^T (\dot{\eta}_d - \dot{\eta}_d - K_1 \tilde{\eta} - \rho_1 sat(s_w/\phi_1) + K_1 \tilde{\eta}) \\ &= -s_w^T \rho_1 sat(s_w/\phi_1) \end{aligned} \tag{15}$$

According to Eq. (7), one obtains:

$$-s_{wi}^T \rho_1 sat(s_{wi}/\phi_1) = \begin{cases} -\rho_1 |s_{wi}| \le 0, & if \ |s_{wi}/\phi_1| > 1 \\ -s_{wi}^2/\phi_1, & else \ \left| \frac{s_{wi}}{\phi_1} \right| \le 1 \end{cases} \tag{16}$$

Hence, based on Eq. (16), it can be concluded:

$$\dot{V}_1 \le 0 \tag{17}$$

Therefore, the Lyapunov stability of the outer loop is proven. If s_w is bounded at $t = 0$, it remains bounded for $t > 0$. And then:

$$\lim_{t \to \infty} \int_0^t \dot{V}_1 dt = \lim_{t \to \infty} V_1 - V_1(0) < \infty \tag{18}$$

By Barbalat lemma:

$$\lim_{t \to \infty} V_1(t) = 0 \tag{19}$$

which implies:

$$\lim_{t \to \infty} s_w = 0 \tag{20}$$

It means that the system is stable and the position error asymptotically converges to zero. For inner loop, let the second Lyapunov function candidate be:

$$V_2 = \frac{1}{2}(s_n^T M s_n + \tilde{\tau}_f^T \Psi^{-1} \tilde{\tau}_f) \tag{21}$$

where $\tilde{\tau}_f = \hat{\tau}_f - \tau_f$ is the disturbance estimate error, and $\Psi \in R^{3 \times 3}$ is constant diagonal matrix used in the second sliding mode surface. Then the time derivative of V_2 yields:

$$\dot{V}_2 = s_n^T M \dot{s}_n + \tilde{\tau}_f^T \Psi^{-1} \dot{\tilde{\tau}}_f \tag{22}$$

Substitute Eq. (4) and Eq. (8) into Eq. (22), yields:

$$\begin{aligned}
\dot{V}_2 &= s_n^T M (\dot{v}_c - \dot{v} + K_2 \tilde{v}) + \tilde{\tau}_f^T \Psi^{-1} \dot{\tilde{\tau}}_f = s_n^T (M \dot{v}_c - M \dot{v} + M K_2 \tilde{v}) + \tilde{\tau}_f^T \Psi^{-1} \dot{\tilde{\tau}}_f \\
&= s_n^T [M \dot{v}_c - (\tau - D(v)v - \tau_f) + M K_2 \tilde{v}] + \tilde{\tau}_f^T \Psi^{-1} \dot{\tilde{\tau}}_f
\end{aligned} \tag{23}$$

Considering the designed control law Eq. (9), one obtains:

$$\dot{V}_2 = -s_n^T \tilde{\tau}_f - s_n^T K_3 s_n - s_n^T \rho_2 sat(s_n/\phi_2) + \tilde{\tau}_f^T \Psi^{-1} \dot{\tilde{\tau}}_f + \tilde{\tau}_f^T s_n \tag{24}$$

Then substitute the adaptive law Eq. (10) into Eq. (24), one obtains:

$$\dot{V}_2 = -s_n^T \tilde{\tau}_f - s_n^T K_3 s_n - s_n^T \rho_2 sat(s_n/\phi_2) + \tilde{\tau}_f^T s_n \tag{25}$$

Because $s_n^T \tilde{\tau}_f$ and $\tilde{\tau}_f^T s_n$ are both scalar, the results can be regarded as the same, then one gets:

$$\dot{V}_2 = -s_n^T K_3 s_n - s_n^T \rho_2 sat(s_n/\phi_2) \tag{26}$$

As the second term in Eq. (26) is as same as the Eq. (16), it can also be concluded:

$$\dot{V}_2 \leq 0 \tag{27}$$

Hence, the stability of the inner loop is proven. If s_n is bounded at $t = 0$, it remains bounded for $t > 0$. And then:

$$\lim_{t \to \infty} \int_0^t \dot{V}_2 dt = \lim_{t \to \infty} V_2 - V_2(0) < \infty \tag{28}$$

By Barbalat lemma:

$$\lim_{t \to \infty} V_2(t) = 0 \tag{29}$$

which implies:

$$\lim_{t \to \infty} s_n = 0 \tag{30}$$

It means that the system is stable and the speed error asymptotically converges to zero.

As the stability and asymptotic convergence of the outer loop and inner loop are both proven, the stability and asymptotic convergence of the whole control system for UUV is guaranteed.

5 Simulation

Simulation studies are performed with "BSA-IV" UUV. In order to illustrate the advantage of the proposed controller based on double sliding mode loop (DSMC), a conventional sliding mode controller (SMC) based on the reaching law is also adopted. These two controllers are simulated with an identical simulation case.

The hydrodynamic parameters of "BSA-IV" UUV are shown in Table 2.

Table 2. Hydrodynamic parameters of "BSA-IV" UUV

Parameter	Value	Quantity	Description
m	8000	kg	Mass
$X_{\dot{u}}$	−932.35	kg	Added mass
$Y_{\dot{v}}$	−5552.9	kg	Added mass
$N_{\dot{r}}$	−10412	kgm^2	Added moment
$X_{u\lvert u \rvert}$	182.1	Ns^2/m^2	Quadratic drag in u
$Y_{v\lvert v \rvert}$	762.6	Ns^2/m^2	Quadratic drag in v
$N_{r\lvert r \rvert}$	888.7	Ns^2	Quadratic drag in r
X_u	199.6	Ns/m	Linear drag in u
Y_v	111.2	Ns/m	Linear drag in v
N_r	101.1	Ns/m	Linear drag in r
x_G	0.22	m	Gravity position in x axis
y_G	0	m	Gravity position in y axis
I_Z	20012	kgm^2	Moment of inertia in z axis

The total simulation time is 200 s, and the initial states of UUV in horizontal plane are set as: $[X \quad Y \quad \psi] = [0\,m \quad 0\,m \quad 0°]$, $[u \quad v \quad r] = [0\,m/s \quad 0\,m/s \quad 0\,rad/s]$. The desired position and heading are $[X \quad Y \quad \psi] = [5\,m \quad 5\,m \quad 30°]$, $t \in [0, 50)$, $[0\,m \quad 0\,m \quad 0°]$, $t \in [50, 100)$, $[6\,m \quad 6\,m \quad 60°]$, $t \in [100, 200)$ $[u_e \quad v_e \quad r_e]$ $= [0\,m/s \quad 0\,m/s \quad 0\,rad/s]$. The parameters of the proposed controller are set as $K_1 = diag[0.3, 0.3, 0.3]$, $\rho_1 = 0.5$, $\phi_1 = \phi_2 = 0.1$ $K_2 = diag[1, 1, 1]$, $\Psi = diag[10, 10, 10]$, $\rho_2 = 0.7$, $K_3 = diag[600, 600, 600]$, $\gamma_1 = \gamma_2 = 0.8$, $\beta = 0$.

The environment disturbance vector is set as $\tau_f = [30\,N \quad 30\,N \quad 10\,N \cdot m]$. The simulation results are shown in Figs. 4, 5, 6 and 7. Simulation result of UUV auto position control is shown in Fig. 4. The UUV speed simulation result is shown in Fig. 5. The forces of main and horizontal thrusters are shown in Fig. 6. The auto position trajectory of UUV with SMC and DSMC are shown in Fig. 7.

In Fig. 4, the black line is the desired states, the red dotted line represents position control of sliding mode controller based on reaching law (SMC), and the blue solid line represents position control of the proposed controller based on double sliding mode loops (DSMC). It can be seen that the control effect of DSMC is better than the SMC. The adjust time in North and East of SMC is longer than DSMC. And there are also overshoots in North and East of SMC. When the desired values change in North, East and Heading, the control accuracy of DSMC is better than SMC. In Fig. 5, the dotted line represents speed change of SMC, and the solid line represents speed change of DSMC. The speed jump of SMC is bigger than DSMC.

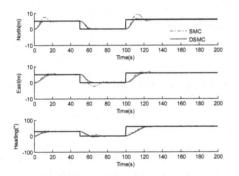

Fig. 4. Position of UUV with SMC/DSMC

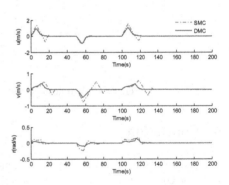

Fig. 5. Speed of UUV with SMC/DSMC

In Fig. 6, dotted line represents thrusters forces with SMC, while solid line represents thrusters force with DSMC. It can be seen that the thrusters are changing a litter more frequently with SMC than DSMC. The trajectory of UUV with SMC and DSMC in the first period $t \in [0, 50\,s)$ is shown in Fig. 7. Left represent the trajectory of UUV with SMC, right represent the trajectory of UUV with SMC. It can be seen that the trajectory is smoother with DSMC than that with SMC. And there are no overshoots in the horizontal plane with DSMC which is helpful in the auto position task of UUV.

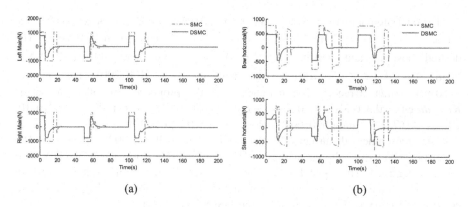

Fig. 6. Forces of main Thrusters with SMC: (a) Main thrusters, (b) Horizontal auxiliary thrusters

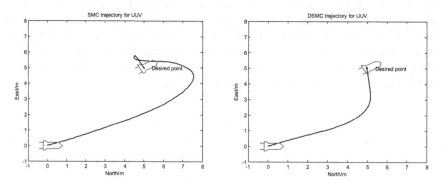

Fig. 7. UUV trajectory with SMC/DSMC in the first period

6 Conclusion

The present work addresses the auto position controller based on double sliding mode loop for UUV. The outer loop generates the virtual speed based on the position error, and the real speed will be replaced by the virtual speed in the controller designing. Meanwhile the error between the real speed and actual speed is eliminated in the inner loop. Since the sliding mode surface is contained in both loops, the state of UUV in the horizontal plane will be converged to the designed sliding mode surfaces. Based on the Lyapunov stability theory, the stability and asymptotic convergence of the double loops are both guaranteed. Therefore, the stability of the designed auto position controller for UUV is proven. Conclusion can be derived from the simulation results: the speed jump will be decreased with the proposed controller, and the overshoot in position in the horizontal plane will be eliminated. So the effectiveness of the proposed control method is demonstrated.

Acknowledgement. This work is supported by Science and Technology Research Project of Hubei Provincial Department of Education under Grant No. B2016433.

References

1. Shusheng, X., Xiaogong, L.: A closed loop hierarchical multi-sensor fusion algorithm for vessel dynamic positioning. Acta Electronica Sinica **42**(5), 512–516 (2014)
2. Fang, W., Ming, L., Feng, X.: Design and implementation of a triple-redundant dynamic positioning control system for deepwater drilling rigs. Appl. Ocean Res. **57**, 140–151 (2016)
3. Karthick, S., Kumar, K.S., Mohan, S.: Relative analysis of controller effectiveness for vertical plane control of an autonomous underwater vehicle. In: OCEANS 2016, No. 7485569 (2016)
4. Mohan, S., Singh, Y.: Task space position tracking control of an autonomous underwater vehicle with four tilting thrusters. In: OCEANS 2016, No. 7485382 (2016)
5. Hoang, N.Q., Kreuzer, E.: Adaptive PD-controller for positioning of a remotely operated vehicle close to an underwater structure: theory and experiments. Control Eng. Pract. **15**(4), 411–419 (2007)
6. Jeen-Shing, W., Lee, C.S.G.: Self-adaptive recurrent neuro-fuzzy control of an autonomous underwater vehicle. IEEE Trans. Robot. Autom. **19**(2), 283–295 (2003, 2006)
7. Liaw, H.C., Shirinzadeh, B.: Neural network motion tracking control of piezo-actuated flexure-based mechanisms for micro-/nanomanipulation. IEEE/ASME Trans. Mechatron. **14**(5), 517–527 (2009)
8. Healey, A.J., Lienard, D.: Multivariable sliding mode control for autonomous diving and steering of unmanned underwater vehicles. IEEE J. Oceanic Eng. **18**(3), 327–339 (1993)
9. Antonelli, G., Chiaverini, S., Fusco, G.: A fuzzy-logic-based approach for mobile robot path tracking. IEEE Trans. Fuzzy Syst. **15**(2), 211–221 (2007)
10. Van de Ven Pepijn, W.J., Flanagan, C., Toal, D.: Neural network control of underwater vehicles. Eng. Appl. Artif. Intell. **18**(5), 533–547 (2005)
11. Soylu, S., Buckham, B.J., Podhorodeski, R.P.: A chattering-free sliding-mode controller for underwater vehicles with fault-tolerant infinity-norm thrust allocation. Ocean Eng. **35**(16), 1647–1659 (2008)
12. Sun, B., Zhu, D.: A chattering-free sliding-mode control design and simulation of remotely operated vehicles. In: IEEE Control and Decision Conference, pp. 4173–4178 (2011). Chinese
13. Zhu, D., Sun, B.: The bio-inspired model based hybrid sliding-mode tracking control for unmanned underwater vehicles. Eng. Appl. Artif. Intell. **26**(10), 2260–2269 (2013)
14. Xu, J., Wang, M., Qiao, L.: Dynamical sliding mode control for the trajectory tracking of underactuated unmanned underwater vehicles. Ocean Eng. **105**, 54–63 (2015)
15. Fossen, T.I., Strand, J.P.: Passive nonlinear observer design for ships using Lyapunov methods: full-scale experiments with a supply vessel. Automatica **35**(1), 3–16 (1999)

Vision-Based Hardware-in-the-Loop-Simulation for Unmanned Aerial Vehicles

Khoa Dang Nguyen and Cheolkeun Ha(✉)

School of Mechanical and Automotive Engineering, University of Ulsan,
Ulsan 680-749, South Korea
{nguyendangkhoacntt, cheolkeun}@gmail.com

Abstract. This paper proposes a new configuration of a general hardware-in-the-loop-simulation (HILS) setup of Unmanned Aerial vehicle (UAV) especially for vision algorithm. In our setup, the Gazebo software is used to simulate six-degree-of-freedom (6 DOF) model and corresponding sensor readings such as the inertial measurement unit (IMU) and the camera for a quad-rotor UAV. Meanwhile, the flight control algorithm is performed on the Pixhawk hardware. The Raspberry hardware is installed the vision algorithms to estimate the position of the quad-rotor UAV for the landing task. The middle software named control application software (CAS) is developed to collect the communication between the Gazebo, Pixhawk and Raspberry by using the multithread architecture. Numerical implementation has been performed to prove effectiveness of the suggested HILS components approach.

Keywords: Vision · Raspberry · Vision HILS

1 Introduction

The Unmanned Aerial Vehicles (UAVs) have been well-known as an automatic robot helping people carry out the hard and dangerous works. Their applications attracted significant interests of researchers [1, 2]. Specially, the vision application was presented exciting, such as tracking a ground moving target [3], autonomous landing on a moving unmanned ground vehicle (UGV) [4], 3D mapping [5]. In these studies, the researchers used the realistic UAV for demonstrating their applications. The testing process may have high risk of property damage. To restrict the accidents and safety for human, the HILS is often performed for the UAVs system on the real time platform, which contains the hardware devices and the simulation softwares. The hardwares are used to install the flight control algorithms and the vision algorithm while the simulation software provides the 6DOF model and corresponding sensor readings of the UAV.

For some research associated with vision in HILS, the real camera is used to estimate the state of UAV based on the virtual image which is shown on the desktop monitor screen [6, 7], projector screen [8, 9]. Although the developed HILS systems showed some interesting results, their applicability is limited due to the cost and associated complicated components of the camera. Beside, the presented 3D information of scene is also the trouble of using the virtual image on the screen.

© Springer International Publishing AG, part of Springer Nature 2018
D.-S. Huang et al. (Eds.): ICIC 2018, LNCS 10954, pp. 72–83, 2018.
https://doi.org/10.1007/978-3-319-95930-6_8

In another approach, the Gazebo software [10] was utilized to simulate the 6DOF model of the UAV, virtual camera and virtual 3D environment workspace for the simulation, which uses the open dynamics engine to present a simulation in the real-time conditions. For example, Odelga et al. [11] combined the Odroid board with the Gazebo to present the operation of the UAV. Therein, the hardware board is installed for the flight control algorithm and the vision control algorithm. In this case, the hardware system may overload because the operating system (OS) must perform many processes at the same time. And the Odroid board lacks the basic sensor for a UAV such as IMU sensor, GPS sensor, and etc. Therefore, the real UAV needs to equip these sensors to provide state of UAV for the flight controllers. Consequently, the system becomes complex, and needs expensive solution. As another solution, the Pixhawk was used to setup the flight control algorithm, which is integrated with the full sensor for a quad-rotor UAV [12]. In this study, the vision algorithms were performed on the a desktop computer which is difficult to apply to the real UAV because the weight of desktop computer is too heavy to integrate on the UAV.

In the traditional HILS setup, the communication among the parts of HILS used either the direct connection or a middleware software. For direct connection [8, 13], any parts of HILS on the several platform can use the line transfer signal without the permission from the system. Subsequently, the HILS communication is conflicted. Therefore, the HILS performance is degraded. To overcome the limitation of direct connection, the middleware software named robot operating system (ROS) [12, 14] was used to ensure the performance connection in HILS. Nevertheless, this solution needs to integrate ROS to all parts of HILS. And the message signal must use ROS message format. Consequently, the HILS system becomes more complex, limiting its applicability.

The main contribution of this paper is to build a general HILS for testing the flight control algorithm and vision algorithm of the UAV on the real-time condition. Specifically, a quad-rotor UAV is chosen to present the operation HILS. First, the Gazebo software is used to construct the 6DOF model and the sensor models such as IMU and camera for the UAV. Second, the algorithms are robustly developed on the real-time platform. The flight control algorithm based on PX4 source code is performed on the Pixhawk board for the flight tracking position. And the vision algorithm is installed on the Raspberry board to estimate position of the quad-rotor UAV for autonomous landing. Third, the middleware software named CAS is developed to ensure the communication between the parts of HILS. The CAS is written in the QtCreator by C program. To improve the performance of HILS, the multithread architecture is applied in CAS in order to speed up signal transfer and manage state of the line transfer. Unlike ROS, the CAS does not need to integrate all parts of HILS so that the HILS does not require to specify the hardwares and softwares.

The remainder of this paper is organized as follows: Sect. 2 includes the problem description and system modeling of the quad-rotor UAV while the HILS proposed is introduced in Sect. 3. Numerical simulation is performed in Sect. 4 to illustrate the effectiveness of our HILS setup. Finally, some concluding remarks are given in Sect. 5.

2 Problem Description and System Modeling

The quad-rotor UAV is chosen to present the operation flights in the HILS development, which contains a body frame that is connected with four motors. The motors are configured to different rotating direction such as clockwise and counter clockwise as in Fig. 1. Let's define l_1 and l_2 are the distances along the axes to UAV's center of gravity.

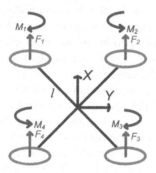

Fig. 1. Description of the quad-rotor UAV

The mathematical model of the quad-rotor UAV can be presented as: [15]

$$\begin{bmatrix} U_1 \\ U_2 \\ U_3 \\ U_4 \end{bmatrix} = \begin{bmatrix} K_f & K_f & K_f & K_f \\ -l_1K_f & -l_1K_f & l_1K_f & l_1K_f \\ l_2K_f & -l_2K_f & -l_2K_f & l_2K_f \\ -K_m & K_m & -K_m & K_m \end{bmatrix} \begin{bmatrix} \omega_1^2 \\ \omega_2^2 \\ \omega_3^2 \\ \omega_4^2 \end{bmatrix} \tag{1}$$

where U_1, U_2, U_3 and U_4 are the throttle, roll, pitch and yaw control input, respectively. ω_i is angular velocity of the ith motor; K_f, K_m are constant value of rotor thrust coefficient and rotor drag coefficient, respectively.

The motion of the quad-rotor UAV can be divided into two subsystems such as rotational subsystem roll (ϕ), pitch (θ), yaw (ψ) and translational subsystem (x, y, z). The dynamic model can be implemented as [16].

$$\ddot{\phi} = \frac{1}{I_{xx}}\left(U_2 - J_r\dot{\theta}\omega_r + \left(I_{yy} - I_{zz}\right)\dot{\Psi}\dot{\theta}\right) \tag{1}$$

$$\ddot{\theta} = \frac{1}{I_{yy}}\left(U_3 - J_r\dot{\phi}\omega_r + (I_{zz} - I_{xx})\dot{\phi}\dot{\Psi}\right) \tag{2}$$

$$\ddot{\Psi} = \frac{1}{I_{zz}}\left(U_4 + (I_{xx} - I_{yy})\dot{\theta}\dot{\phi}\right) \tag{3}$$

$$\ddot{x} = (cos(\phi)cos(\psi)sin(\theta) + sin(\phi)sin(\psi))\left(\frac{U_1}{m}\right) \qquad (4)$$

$$\ddot{y} = (cos(\phi)sin(\psi)sin(\theta) - sin(\phi)cos(\psi))\left(\frac{U_1}{m}\right) \qquad (5)$$

$$\ddot{z} = (cos(\phi)cos(\theta))\left(\frac{U_1}{m}\right) - g \qquad (6)$$

where $\omega_r = \omega_1 - \omega_2 + \omega_3 - \omega_4$; I_{xx}, I_{yy}, I_{zz} are the moments of inertia about the principle axes in the body frame; J_r denotes the inertia of the motor; m is the total mass of the quad-rotor UAV and g is the gravitational constant.

3 General HILS Development

The HILS is performed to test the effectiveness of the flight control algorithm for the UAV in the real-time platform. Based on the modeling of the quad-rotor UAV, the flight control is developed by using the PX4 open source, which is compiled and uploaded to the Pixhawk board. The Gazebo software is used to configure the dynamic model, the sensor model and the 3D visualization of the quad-rotor UAV. In the Gazebo, the camera model is also defined to generate the camera data for the vision algorithm which is installed on the Raspberry board. The CAS is used to collect the signals between the Gazebo, Pixhawk and Raspberry components. The general HILS is proposed as shown in Fig. 2.

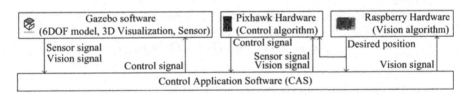

Fig. 2. Proposed HILS system

3.1 Software Development

In this section, the Gazebo software is developed to create the dynamic model, sensor model and 3D visualization for simulating the operation of the quad-rotor UAV. Based on Linux operating system, the Gazebo software is performed based on three components. First, the 3D CAD of the quad-rotor UAV is constructed in the Solidworks. Then the SW2URDF toolbox is used for the dynamic model of UAV in the Gazebo. The geometry of the quad-rotor UAV is presented in Fig. 3.

Second, the measurement state and control UAV are configured in the plugin through the libraries such as physics *(physics.hh)*, sensor *(sensor.hh)*, camera *(cameraSensor.hh)* in the Gazebo. The IMU sensor model is constructed with the

Fig. 3. 3D visualization of the quad-rotor UAV

information as the orientation, linear acceleration and angular velocity, which is used to estimate the state of the quad-rotor UAV. The camera sensor is built to provide the video data for the vision algorithms. Simultaneously, the angular velocity which comes from the controller in Pixhawk, is also used to set the force and moment for the quad-rotor UAV in the plugin. For example, the code in the list below describes for IMU, camera and control plugin.

```
//IMU
Gazebo::msgs::Quaternion* orientation = new Gazebo::msgs::Quaternion();
math::Vector3 angular_vel_I = link_->GetRelativeAngularVel();
 Eigen::Vector3d linear_acceleration_I(acc_I.x, acc_I.y, acc_I.z);
 Eigen::Vector3d angular_velocity_I(angular_vel_I.x, angular_vel_I.y,
                                 angular_vel_I.z);
//Camera
CameraPlugin::Load(_parent, _sdf);
for (int i=0; i<_fov ; ++i) {  int index = startingPix + i*_width;
        for (int j=0; j<_fov ; ++j) illum += _image[index+j];  }
msg.illuminance = illum/(_fov*_fov);
//Control
physics::Link_V parent_links = link_->GetParentJointsLinks();
 math::Pose pose_difference = link_->GetWorldCoGPose() -
                 parent_links.at(0)->GetWorldCoGPose();
```

Third, the above two components are integrated in a world file to make a full simulation in the Gazebo. A virtual environment simulation in the Gazebo software is shown in Fig. 4.

Fig. 4. Visualization of simulation in the Gazebo software

In HILS setup, the communication between the parts is an important work. Because if a number of packets in HILS is lost, the performance of the UAV will deteriorate. The middleware CAS is developed to ensure communicate between the components of HILS. The structure of CAS is shown in Fig. 5.

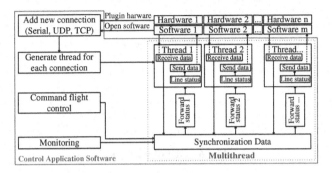

Fig. 5. Structure of the CAS

Herein, three modules are built in CAS. First, the connection module is used to open the interface with any parts of HILS such as the hardware (Pixhawk, Arduino, and etc.), and softwares (Gazebo, X-plane, Qgroundcontrol, and etc.). This interface is very diverse like the serial, ethernet (TCP/UDP), and USB connection. Second, each connection in HILS is created by several thread for the communication with other part based on the multithread architecture. The threads are synchronized with the signal and the state line transfer is checked in all threads by using '*line status*' flag to ensure that the system do not have the confliction. For distribution of the signal to each part in HILS, the '*forward status*' flag is used to allow/deny the signal to each thread. Third, the commands of flight control and monitoring are used to set some custom commands and to track the state of the UAV as well as the state of signal in the system. To easy with the user, GUI of the CAS is designed as shown in Fig. 6.

3.2   Hardware Development

The Pixhawk hardware board is used to perform the flight control algorithm for the quad-rotor UAV on the real time. The flight controller is developed based on the PX4 source, which is presented in Fig. 7.

To track a desired trajectory, the position control and attitude control based on PID control techniques are applied. The output of attitude controller is sent to the inverse function (IF) (obtain from Eq. (1)) to calculate the angular velocity for each motor. The feedback signal for each controller is provided by the IMU sensor model in the Gazebo [17, 18]. All controllers are written by C program which is compiled to make a firmware and then uploading to microchip of the Pixhawk hardware.

The Raspberry board is adopted to perform the vision algorithm in our HILS development. Being installed with the Linux (Ubuntu Mate 16.04 LTS), this board can run the C program which allows access and analysis data from a camera source.

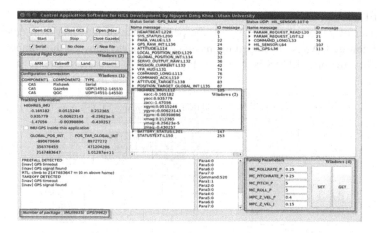

Fig. 6. GUI of the CAS

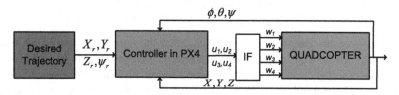

Fig. 7. Schema tracking flight controller of the quad-rotor UAV

Moreover, it provides the input/output ports such as the ethernet, USB and serial for easy communicate with other peripheral devices especially for the camera. A standalone system of the Raspberry board is configured as in Fig. 8.

Fig. 8. Standalone system of the Raspberry board

In our HILS setup, the camera signal is generated by the Gazebo simulation software, which is used for the input signal of the Raspberry board in order to test the vision algorithm.

4   Implementation of HILS and Results

In this section, the simulation has been carried out to demonstrate the effectiveness of our HILS setup which is constructed as in Fig. 9. The capability of HILS is appreciated in two flight scenarios such as the position control, and the autonomous landing using a vision algorithm. At initial time, the quad-rotor UAV is configured as in Fig. 10.

Fig. 9. Full HILS setup

Firstly, the communication in HILS is performed to check the activation of each part. Herein, the sensor signals and the actuator controls are transferred from the Gazebo and the Pixhawk, respectively, with a frequency as in Table 1.

The ground control station named QGroundControl (QGC) is added to HILS to evaluate the performance communication based on CAS as well as HILS setup. Two test cases such as the traditional communication HILS setup (denote as TS), and the HILS with CAS (denote as TCAS) are performed. As the result in Table 2, the CAS with the multithread architecture has better the performance communication than TS. This result proved that the CAS has high effectiveness in the HILS development.

Secondly, the quad-rotor UAV is tested with the desired trajectory (X_r, Y_r, Z_r, ψ_r) which is defined from a program in the Raspberry board with the desired values of $(X_r = 0, Y_r = 0, Z_r = 2.5, \psi_r = 0)$ for the takeoff, $(X_r = 8, Y_r = 5, Z_r = 6, \psi_r = 0)$ for moving to a position and $(X_r = 8, Y_r = 5, Z_r = 0, \psi_r = 0)$ for landing. At the first time, the quad-rotor UAV performs takeoff to 2.5 [m] and keeps waiting the desired signal from Raspberry. As result in Fig. 11, the Pixhawk receives the desired trajectory signal at time 0.4 [min], 1.5 [min] and 3.2 [min] to change position for the quad-rotor UAV. Herein, the UAV can move to the desired position with a small error. This test shows that the combination among the Gazebo, Pixhawk and Raspberry worked well.

Fig. 10. Workspace of the quad-rotor UAV in Gazebo software

Table 1. Setting frequency of the signal in HILS.

Name component	Frequency	Description
IMU sensor	500 Hz	HIL_SENSORS - Generation from Gazebo
Actuator control	400 Hz	HIL_CONTROLS - Generation from Pixhawk

Table 2. Frequency and ratio errors in simulations.

Name	IMU		Control actuator	
	Frequency source 500 Hz		Frequency source 400 Hz	
	Average Hz	Ratio error %	Average Hz	Ratio error %
TS	352.8847	29.4231	279.4541	30.1365
TCAS	431.8999	13.6200	342.6261	14.3435

Specially, the flight controllers in Pixhawk can ensure tracking position for the quad-rotor UAV. Other parts such as Gazebo and Raspberry can perform correctly the tasks in HILS setup.

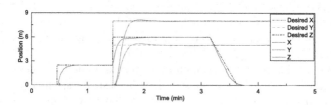

Fig. 11. Responses from the quad-rotor UAV

Thirdly, to evaluate the effectiveness of the Raspberry board, the vision algorithm is installed on this board to detect a marker pad [19]. In this test, a virtual camera in the Gazebo and a real camera are connected to the Raspberry board via the CAS and USB connection support, respectively. As the results in Fig. 12, in both simulation and real experimental cases, the vision algorithm installed on the Raspberry board can recognize the marker pad with high accuracy (red rectangle). And it can apply to the quad-rotor UAV to generate the position for the landing process.

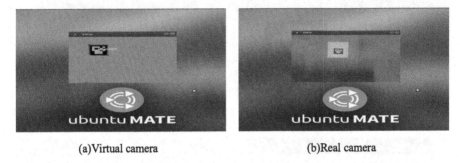

(a)Virtual camera   (b)Real camera

Fig. 12. Raspberry board recognizes the marker pad by using the vision algorithm (Color figure online)

Next, the vision algorithm is used to estimate the position for the autonomous landing of the quad-rotor UAV by using the marker pad [19]. The flight task is performed in two steps. Step 1, the quad-rotor UAV is takeoff to $(X_r = 0, Y_r = 0, Z_r = 2.5)$. Step 2, the vision algorithm is run to detect the landing marker and then generate desired position (X_r, Y_r, Z_r) for the controller in the Pixhawk. As the results in Figs. (13 and 14), the quad-rotor UAV can land to the marker pad with the small error. All results showed that the our HILS could provide well simulation for testing the flight control algorithm and vision algorithm of the quad-rotor UAV based on the Pixhawk, Gazebo, Raspberry and CAS.

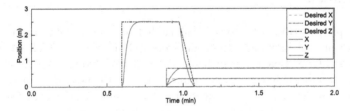

Fig. 13. Responses from the quad-rotor UAV in the landing task

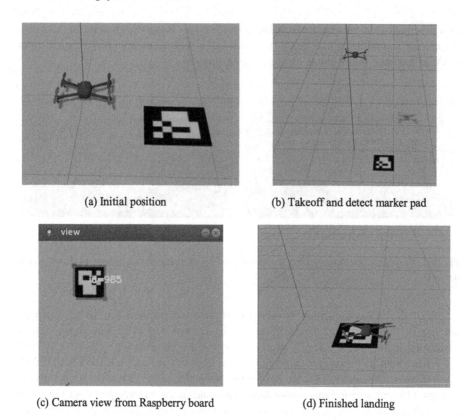

(a) Initial position   (b) Takeoff and detect marker pad

(c) Camera view from Raspberry board   (d) Finished landing

Fig. 14. Quad-rotor UAV landing to marker pad

5   Conclusion

This paper presents HILS setup to simulate the flight operation of the quad-rotor UAV by using vision. The proposed HILS could work well based on the Pixhawk, Gazebo, Raspberry and CAS. Herein, the Pixhawk and Raspberry provide the real time hardware to perform the flight control and vision algorithm. The Gazebo is used to present the dynamic model, sensor model and 3D visualization for the quad-rotor UAV. Meanwhile, the CAS is developed based on the multithread architecture to ensure the communication between the parts of HILS. As a result, our proposed techniques could establish the HILS better performed in a sense of high speed communication and camera application for the quad-rotor UAV.

References

1. Khosiawan, Y., Nielsen, I.: A system of UAV application in indoor environment. Prod. Manuf. Res. Open Access J. **4**(1), 2–22 (2016)
2. Nex, F., Remondino, F.: UAV for 3D mapping applications: a review. Appl. Geomatics **6**, 1–15 (2014)
3. Gomez-Balderas, J.E., Flores, G., Garcia, C.L.R., Lozano, R.: Tracking a ground moving target with a quadrotor using switching control. J. Intell. Rob. Syst. **70**, 65–78 (2013)
4. Hoang, T., Bayasgalan, E., Wang, Z., Tsechpenakis, G., Panagou, D.: Vision-based target tracking and autonomous landing of a quadrotor on a ground vehicle. In: American Control Conference, USA, pp. 5580–5585 (2017)
5. Fraundorfer, F., Heng, L., Honegger, D., Lee, G.H., Meier, L., Tanskanen, P., Pollefeys, M.: Vision-based autonomous mapping and exploration using a quadrotor UAV. In: IEEE/RSJ International Conference on Intelligent Robots and Systems, Portugal, pp. 4557–4564 (2012)
6. Duan, H., Zhang, Q.: Visual measurement in simulation environment for vision based UAV autonomous aerial refueling. IEEE Trans. Instrum. Meas. **64**, 2468–2480 (2015)
7. Mingu, K., Daewon, L., Jaemann, P., Chulwoo, P., Hyoun, J.K., Youdan, K.: Vision-based hardware-in-the loop simulation test of vision based net recovery for fixed wing unmanned aerial vehicle. In: Third Asia-Pacific International Symposium on Aerospace Technology (2011)
8. Trilaksono, B.R., Triadhitama, R., Adiprawita, W., Wibowo, A.: Hardware-in-the-loop simulation for visual target tracking of octorotor UAV. Aircr. Eng. Aerosp. Technol. **83**, 407–419 (2011)
9. Gans, N.R., Dixon, W.E., Lind, R., Kurdila, A.: A hardware in the loop simulation platform for vision-based control of unmanned air vehicles. Mechatronics **19**, 1043–1056 (2009)
10. Gazebo. http://Gazebosim.org/. Accessed 6 June 2017
11. Odelga, M., Stegagno, P., Bülthoff, H.H., Ahmad, A.: A setup for multi-UAV hardware-in-the-loop simulations. In: 2015 Workshop on Research, Education and Development of Unmanned Aerial Systems, Mexico, pp. 204–210 (2015)
12. Bu, Q., Wan, F., Xie, Z., Ren, Q., Zhang, J., Liu, S.: General simulation platform for vision based UAV testing. In: 2015 IEEE International Conference on Information and Automation, China, pp. 2512–2516 (2015)
13. Saunders, J., Beard, R.: UAS flight simulation with hardware-in-the-loop testing and vision generation. J. Intell. Rob. Syst. **57**, 407–415 (2010)
14. Lepej, P., Navarro, A.S., Sola, J.: A flexible hardware-in-the-loop architecture for UAVs. In: 2017 International Conference on Unmanned Aircraft Systems, USA, pp. 1751–1756 (2017)
15. Pounds, P., Mahony, R., Corke, P.: Modeling and control of a large quadrotor robot. Control Eng. Pract. **18**, 691–699 (2010)
16. Xiong, J.J., Zhang, G.B.: Global fast dynamic terminal sliding mode control for a quadrotor UAV. ISA Trans. **66**, 233–240 (2017)
17. Abellanosa, C.P., Lugpatan, R.P.J., Pascua, D.A.D.: Position estimation using inertial measurement unit (IMU) on a quadcopter in an enclosed environment. Int. J. Comput. Commun. Instrum. Eng. **3**, 332–336 (2016)
18. Zhang, T., Kang, M., Achtelik, M., Kühnlenz, K., Buss, M.: Autonomous hovering of a Vison/IMU guided quad-rotor. In: 2009 International Conference on Mechatronics and Automation, China, pp. 2870–2875 (2009)
19. Carreira, T.G.: Quadcopter automatic landing on a docking station. Master thesis, Instituto Superior Tecnico (IST) (2013)

Development of a New Hybrid Drone and Software-in-the-Loop Simulation Using PX4 Code

Khoa Dang Nguyen[1], Cheolkeun Ha[1(✉)], and Jong Tai Jang[2]

[1] School of Mechanical and Automotive Engineering, University of Ulsan,
Ulsan 680-749, South Korea
nguyendangkhoacntt@gmail.com, cheolkeun@gmail.com
[2] Future Aerospace Technology Team, Korea Aerospace Research Institute,
Daejeon, South Korea
jjt@kari.re.kr

Abstract. The fixed-wing vertical takeoff and landing unmanned aerial vehicles (UAVs) called the hybrid drone is a new type of aircraft that inherits the multi-rotor (MR) and the fixed-wing (FW) structure to use their strengths. Normally, the MR uses the forces and the moments generated from the four motors to drive roll and pitch angle while the FW uses the elevator and aileron control surfaces which have often complex structure and its difficult design. Therefore, this paper aims to develop a new design for the hybrid drone. By removing the complex components in FW, the angle controls of our UAV by using common basis on four motors of the MR structure are used. Furthermore, to improve the heading control performance, two extended side-motors are used. To verify the effectiveness of the hybrid drone configuration, the software-in-the-loop (SiTL) simulation is performed based on PX4 source, Gazebo simulation and ground control station named QGroundControl (QGC). Herein, PX4 is modified to design a flight control in which a desired trajectory from the QGC is set while the Gazebo is used to construct the dynamic model and 3D visualization for the hybrid drone. Numerical simulations have been performed to demonstrate the effectiveness of the our design approach.

Keywords: Hybrid drone · SiTL · PX4

1 Introduction

Nowadays, UAVs have been known as a hot topic in the military and civil applications. In the platform configuration, the UAV can be mainly classified into three categories, the fixed-wing types [1], the helicopter types [2] and the multi-rotor types [3]. The strength of FW types is high speed flight and fuel efficiency. However, it cannot perform to vertical takeoff and landing (VTOL) and hover in the flight itinerary. In the opposite points of view, the helicopter and multi-rotor types can provide the limitation on FW types, but the their speed is slower than the FW types.

To overcome the shortcomings of the traditional type platforms, a hybrid drone which belongs to the types between VTOL and FW are known as one of feasible

© Springer International Publishing AG, part of Springer Nature 2018
D.-S. Huang et al. (Eds.): ICIC 2018, LNCS 10954, pp. 84–93, 2018.
https://doi.org/10.1007/978-3-319-95930-6_9

solutions, which owns the advantages of both configurations [4–6]. Most of hybrid drones are designed based on characteristics of the multi-rotor platform and the fixed-wing platform. Normally, in hybrid drone the MR mode uses the motors only for the vertical takeoff and landing. Meanwhile, the FW mode uses several motors to make movement of the drone. Therein, the hybrid drone must change the activation motor among two modes. Furthermore, the FW mode uses the control surfaces as main function in the flight operation, which is not easy to assemble.

For any UAV development, the software-in-the-loop (SiTL)-simulation [7] has been known as a valuable simulation process to verify the designed drone model and the flight control algorithms installed in the drone. Some researchers build the flight control algorithms based on Matlab/Simulink [7] or PX4 source [8]. And the drone model is built based on X-Plane [9], FlightGear [10], JMAVSim [11], or Gazebo [12]. Among them, PX4 source and Gazebo simulation have more advantages due to following reasons. First, PX4 [13] is an open source that includes many libraries to drive unmanned aerial and ground vehicles. In addition, PX4 can be uploaded to an open hardware Pixhawk [14] which is a very popular hardware for UAV applications. Second, Gazebo simulation is also an open source, and it can provide the dynamic model of UAV, sensor model and 3D visualization. In particular, this software contains the open dynamics engine (ODE), which can present a system model robot with high accuracy in real-time conditions [15].

In this paper, a new hybrid drone is designed to improve the limitations of the traditional platforms. The UAV which have seven motors used four motors for the roll and pitch control, two motors at the rear for yaw control and one motor for push motion control. The motors are arranged in the position suitable on the frame. The flight controllers of the hybrid drone are verified by the SiTL simulation which is performed in Linux system. Herein, several controllers for UAV is developed based on the open source PX4. And the dynamics, 3D visualization and sensor model are simulated on the open software Gazebo.

The remainders of this paper are organized as follows: Sect. 2 includes the design and modeling of the hybrid drone while the control design scheme is introduced in Sect. 3. Numerical simulations are then performed in Sect. 4 to verify the effectiveness of the hybrid drone. Finally, some concluding remarks are given in Sect. 5.

2 Modeling and Flight Controller Design of the Hybrid Drone

2.1 Dynamic Model of Hybrid Drone

In this paper, a new hybrid drone is designed as shown in Fig. 1. Herein, the UAV is constructed with a body frame installing seven electric motors on top. Each motor attaches a propeller to produce forces and moments for motion flight of UAV. The rotation direction of the motor and the distance from center of mass to the motors are defined in Fig. 1.

Let's define F_i, M_i ($i = 1..7$) to be the force and moment of the ith motor, respectively. These can be defined as:

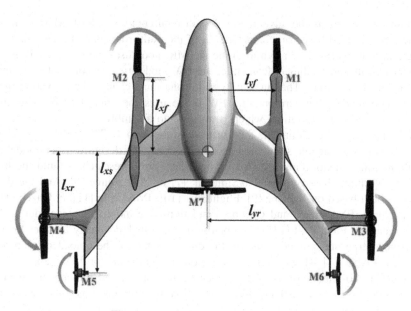

Fig. 1. Description of the hybrid drone

$$F_i = k_t \omega_i^2 \tag{1}$$

$$M_i = k_m \omega_i^2 \tag{2}$$

where ω_i is the angular velocity of the ith motor; k_t and k_m are constant values of rotor thrust coefficient and rotor torque coefficient, respectively.

The aerodynamic of UAV includes the lift (F_l), the drag (F_d) and the moments (M) that can be defined as:

$$F_l = C_l \frac{\rho}{2} s v^2 \tag{3}$$

$$F_d = C_d \frac{\rho}{2} s v^2 \tag{4}$$

$$M = C_m \frac{\rho}{2} s v^2 C_h \tag{5}$$

where ρ, s, v, C_h, C_l, C_d, C_m and are the fluid density, wing area, flight speed, chord, lift, drag, and moment coefficient, respectively.

Suppose that the control inputs of the hybrid drone are defined as follows: U_1, U_2, U_3, U_4 and U_5 are vertical thrust control, roll control, pitch control (using M_1, M_2, M_3 and M_4 motors) and yaw control (using M_5 and M_6 motors) and push control (using M_7 motor), respectively.

2.2 Design Control for Hybrid Drone

The hybrid drone is driven by two controllers: the attitude controller and the position controller. Two flights of multi-rotor mode and fixed-wing mode are combined to maintain operations of the hybrid drone. The control schematic for the hybrid drone is designed as shown in Fig. 2.

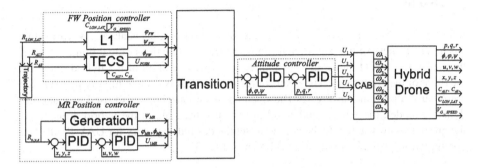

Fig. 2. Schema of control for the hybrid drone

The position controller of the hybrid drone is divided into two flight modes: the MR mode uses the PID controllers and the FW mode uses the *L1* [16] and the total energy control system (*TECS*) [17]. To make the full system controller in Fig. 2, the transition block is applied to decide which mode is selected for the hybrid drone. Then the output of the position controller is set to the desired input to the attitude controller. All flight modes use the same attitude controller structure to control the angular motion of the hybrid drone. This controller contains two control loops. Based on PID controller, the angle rate control and the angle control are applied to the inner loop and the outer loop control, respectively.

From this relationship between the pseudo-control inputs of U_i ($i = 1..5$) and the control motor inputs of M_i ($i = 1..7$), the control allocation block (CAB) can be obtained in Fig. 2. The output of the controller is used to compute the angle velocity of six motors through CAB consisting of the inverse function allocation (IFA) and the switch function allocation (SFA), which can be obtained from Eq. (6) and Eq. (7).

$$\begin{bmatrix} \omega_1^2 \\ \omega_2^2 \\ \omega_3^2 \\ \omega_4^2 \\ \omega_7^2 \end{bmatrix} = [\text{IFA}] \begin{bmatrix} U_1 \\ U_2 \\ U_3 \\ U_5 \end{bmatrix}, \ [\text{IFA}] = \begin{bmatrix} -1.3 & -4.2 & 7.1 & 0 \\ -1.3 & 4.2 & 7.1 & 0 \\ -1.3 & -4.2 & -7.1 & 0 \\ -1.3 & 4.2 & -7.1 & 0 \\ 0 & 0 & 0 & 50 \end{bmatrix} \quad (6)$$

Herein, the U_1 is determined by the transition block, which has two modes. For the FW mode, the U_1 is set to zero and for the MR mode, the U_1 is set by the position controller. The U_2 and U_3 are generated from the attitude controller.

Moreover, SFA for the U_4 can be defined as

$$\text{SFA} : \begin{cases} \omega_5^2 = 0 \, ; \ \omega_6^2 = -160.8 \times 10^2 \times U_4 & \text{if} \ \ U_4 < 0 \\ \omega_5^2 = 160.8 \times 10^2 \times U_4 \ ; \ \omega_6^2 = 0 & \text{if} \ \ U_4 > 0 \end{cases} \tag{7}$$

3   Software-in-the-Loop (SiTL) Simulation for the Hybrid Drone

In order to realize a real flight of UAV, the simulation process is always performed to ensure correctness of the control algorithms or the safe functions of real flight UAV. In this section, the SiTL simulation is applied to trying operation of the hybrid drone by the drive of the controllers. The open source named PX4 is used to modify the code for the control algorithms in Fig. 2. The open software Gazebo is also used to make the dynamic model, the sensor model and to visualize the state of the hybrid drone. And the ground control station QGC is used to set the desired trajectory for UAV flight. The configuration of SiTL simulation is shown in Fig. 3.

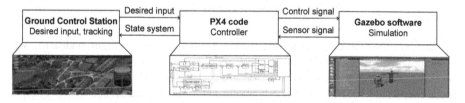

Fig. 3. Configuration of SiTL simulation

Firstly, the hybrid drone is drawn on the 3D CAD in Solidworks software to ensure the size and geometry parameters. The weight of the hybrid drone is configured as 0.5 kg. By using the SW2URDF, this CAD drawing is converted to the drone model on the Gazebo simulation as in Fig. 4.

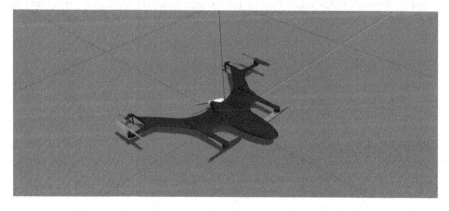

Fig. 4. Hybrid drone visualization in the Gazebo simulation

The model can work based on the plugin which is defined to set the velocity of the motors and to measure the states of UAV via the sensor models. All plugins in the Gazebo are written in C programs. For the feedback signal of the hybrid drone, the global positioning system (GPS) and inertial measurement unit (IMU) sensor model are used to provide the attitude and the position of the UAV.

Secondly, the control algorithms in Fig. 2 for the hybrid drone is written in the *fw_pos_control_ll, mc_pos_control, mc_fw_att_control, vtol_att_control* files in PX4 source for the fixed-wing position control, the multi-rotor position control, the attitude control and the transition, respectively.

Thirdly, the ground control station software named QGC software is used to set the desired input to the controller in PX4. This software allows setting the desired trajectory for the UAV and tracking the states of the drone system in the SiTL simulation developed.

Fourth, the connection between PX4 source code and Gazebo simulation are performed based on the UDP connection via the mavlink packet [18]. Therefore, the control signals and the sensor signals can be transferred flexibly in the SiTL simulation. All parts of the SiTL are developed in the Linux PC.

4 Numerical Simulation

In this section, the SiTL simulation has been carried out to evaluate the performance of the hybrid drone and the flight control algorithms in the SiTL simulation. Based on Linux operating system (OS) in PC, the PX4, Gazebo software and the ground control station QGC are integrated in the SiTL and applied as shown in Fig. 5.

Fig. 5. SiTL setup in the Linux OS

For the simulation, the desired flight trajectory of the hybrid drone is defined as in Fig. 6. Herein, the eight waypoints are selected by the click on the map of QGC, which are (T)-takeoff, MR mode to (2), (3)-transition to FW mode, FW mode to (4), (5)-transition to MR mode, MR mode to (6), (L)-landing. The desired airspeed for FW mode is defined as 11 (m/s). At the beginning of simulation, the MR mode is activated.

Fig. 6. Desired trajectory for the hybrid drone

As the results shown in Fig. 6, the hybrid drone tracks the desired flight trajectory by using the controllers in Fig. 2. At the first time, the hybrid drone in the MR mode performs takeoff flight with the altitude 25 (m) from (T) waypoint. Then it moves to (2) (3) (4) (5) (6) and landing at (L). The movement between (T)-(2), (5)-(6) and (6)-(L) are performed by MR mode. And (3)-(4) uses the FW mode. These modes are changed via the transition waypoint at (3) and (5). The results in Fig. 7 show that the hybrid drone is highly stable in all flights. The altitude and the yaw angle of the hybrid drone track accurately the desired input as in Figs. 7(a) and (b). Meanwhile, the airspeed in the FW mode is maintained around 11 (m/s) (Fig. 7(c)). The motor responses and the attitude responses show the proper motion between two modes of the UAV (Figs. 7(b) and (d)). All results in Fig. 7 prove convincingly that the newly designed hybrid drone in Fig. 1 can be successfully controlled by the flight control algorithms in Fig. 2 and the SiTL in Figs. 3 and 4 accurately performs the flight along the desired trajectory in the MR mode, FW mode and transition mode.

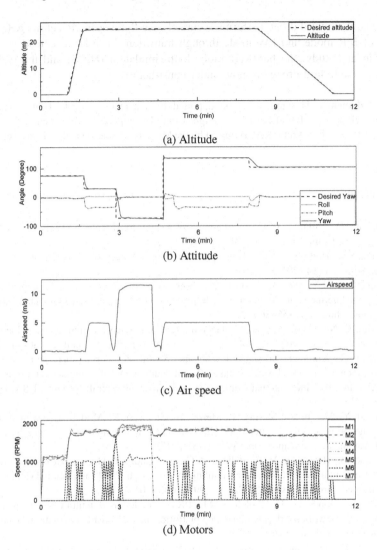

Fig. 7. Response of the hybrid drone

5 Conclusion

This paper presents a new design of the hybrid drone as a VTOL UAV. In this design configuration the four motors that are conventionally used in MR mode also play a role of control surfaces for pitch and roll control in FW mode. Especially a pair of side motors is used to control yaw motion which is quite different from many other VTOL UAVs. To verify operation of the hybrid drone, the SiTL is developed based on PX4 source, Gazebo and QGC. Herein, PX4 code is modified for the flight controllers of the hybrid drone and Gazebo provides the dynamic model of the hybrid drone. QGC is used to define the desired flight trajectory and to display the motion of the drone.

As results, the hybrid drone performs accurately the flight scenario set in QGC in two modes of MR mode and FW mode through transition. In order to extend the works shown in this study, the hardware-in-the-loop-simulation (HILS) and the real flight testbed of the hybrid drone are now under construction.

Acknowledgments. This research was supported from a major project, 'High Performance Multicopter/Propeller Hybrid Drone and Flight Control Computer Development' of the Korea Aerospace Research Institute (KARI) funded by the Ministry of Science and ICT, the Republic of Korea.

References

1. Arifianto, O., Farhood, M.: Optimal control of a small fixed-wing UAV about concatenated trajectories. Control Eng. Pract. **40**, 113–132 (2015)
2. Rivera, S.C., Rodriguez, M.T.: Helicopter modeling and study of the accelerated rotor. Adv. Eng. Softw., 1–14 (2017)
3. Zhao, Y., Cao, Y., Fan, Y.: Disturbance observer-based attitude control for a quadrotor. In: 2017 4th International Conference on Information, Cybernetics and Computational Social Systems, China, pp. 355–360 (2017)
4. Maxim, T., Nhu, V.N., Sangho, K., Jae-Woo, L.: Comprehensive preliminary sizing/resizing method for a fixed wing - VTOL electric UAV. Aerosp. Sci. Technol. **71**, 30–41 (2017)
5. Haowei, G., Ximin, L., Zexiang, L., Shaojie, S., Fu, Z.: Development and experimental verification of a hybrid vertical take-off and landing (VTOL) unmanned aerial vehicle (UAV). In: 2017 International Conference on Unmanned Aircraft Systems, USA, pp. 160–169 (2017)
6. Demitri, Y., Verling, S., Stastny, T., Melzer, A., Siegwart, R.: Model-based wind estimation for a hovering VTOL tailsitter UAV. In: 2017 IEEE International Conference on Robotics and Automation, Singapore, pp. 3945–3952 (2017)
7. Adriano, B., Helosman, V. F., Poliana, A.G., Alessandro, C.M.: Guidance software-in-the-loop simulation using X-Plane and simulink for UAVs. In: 2014 International Conference on Unmanned Aircraft Systems, USA, pp. 993–1002 (2014)
8. Meier, L., Honegger, D., Pollefeys, M.: PX4: a node-based multithreaded open source robotics framework for deeply embedded platforms. In: 2015 IEEE International Conference on Robotics and Automation, USA (2015)
9. Castrol, D.F.D., Santos, D.A.D.: A software-in-the-loop simulation scheme for position formation flight of multicopters. J. Aerosp. Technol. Manage. **8**(4), 431–440 (2016)
10. Zhang, J., Geng. Q., Fei, Q.: UAV flight control system modeling and simulation based on FlightGear. In: International Conference on Automatic Control and Artificial Intelligence, China, pp. 2231–2234 (2012)
11. Seung, H.C., Seok, W.H., Yong, H.M.: Hardware-in-the-loop simulation platform for image-based object tracking method using small UAV. In: 2016 IEEE/AIAA 35th Digital Avionics Systems Conference, USA (2016)
12. Odelga, M., Stegagno, P., Bulthoff, H.H., Ahmad, A.: A setup for multi-UAV hardware-in-the-loop simulations. In: 2015 Workshop on Research, Education and Development of Unmanned Aerial Systems, Mexico, pp. 204–210 (2015)

13. PX4 Pro Drone Autopilot. https://github.com/PX4/Firmware. Accessed 6 June 2017
14. PX4 autopilot. https://pixhawk.org/modules/pixhawk. Accessed 6 June 2017
15. Carlos, E.A., Nate, K., Ian, C., Hugo, B., Steven, P., John, H., Brian, G., Steffi, P., Jose, L.R., Justin, M., Eric, K., Gill, P.: Inside the virtual robotics challenge simulating real-time robotic disaster response. IEEE Trans. Autom. Sci. Eng. 12(2), 494–506 (2015)
16. Park, S., Deyst, J., How, J.: A new nonlinear guidance logic for trajectory tracking. In: Proceedings of the AIAA Conference on Guidance, Navigation, and Control, USA (2004)
17. Faleiro, L.P., Lambregts, A.A.: Analysis and tuning of a total energy control system control law using eigenstructure assignment. Aerosp. Sci. Technol. 3(3), 127–140 (1999)
18. Iulisloi, Z., Carlos, E.T.L., Janana, S., Edison, P.D.F.: Control platform for multiple unmanned aerial vehicles. IFAC-PapersOnLine 49, 36–41 (2016)

Research and Application of LIN Bus Converter

Li Huang[1(✉)] and Hui Huang[2]

[1] City College of Wuhan University of Science and Technology,
Wuhan College of Foreign Languages and Foreign Affairs,
Wuhan 430083, China
240113921@qq.com
[2] Central Southern China Electric Power Design Institute,
China Power Engineering Consulting Group Corporation, Wuhan 430070, China

Abstract. In order to improve the performance of vehicle communication system and reduce the cost of testing, a LIN bus converter has been designed. This paper discusses the design and implementation of interface converter based on LIN bus and USB bus.

Keywords: LIN bus · USB bus · BCM

1 Introduction

The LIN bus is widely used in the control of car door, steering wheel, seat, temperature control and engine cooling fan [1]. At present, the engineer mostly use special testing equipment to test the BCM (body control module) with LIN bus. But this kind of equipment have simple function, low efficiency and high cost. The USB interface has been used in most of computers. If the LIN bus and USB bus interconnection interface are designed, and the information in the BCM module is directly sent to the computer through the USB interface, then BCM can be tested with the help of computer's resources.

The LIN bus is based on a general UART/SCI interface which is easy to implement on both hardware and software [2]. Its low cost advantage makes some advanced mechanical electronic devices can be applied to the vehicle system. PIC18F2550 MCU is embedded with USB interface and the USART interface with LIN protocol. In this paper, a LIN and USB interface converter is designed with the MCU.

2 Overall Design of the USB-LIN Converter

The USB-LIN converter is consists of three major parts: the USB interface part of the microcontroller, the interface part of the LIN bus and the power isolation, as shown in Fig. 1. The 5 V power supply of the MCU comes from the computer's USB interface, and the power of the LIN transceiver comes from the 12 V battery of the car. The optocoupler PS9701 can realize the isolation of two power systems. It effectively protects the USB interface and BCM module of the computer.

© Springer International Publishing AG, part of Springer Nature 2018
D.-S. Huang et al. (Eds.): ICIC 2018, LNCS 10954, pp. 94–101, 2018.
https://doi.org/10.1007/978-3-319-95930-6_10

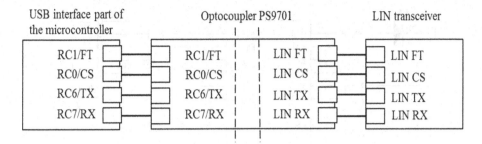

Fig. 1. Overall design of USB-LIN conversion interface

3 Hardware Design of USB-LIN Conversion Interface

3.1 Introduction of PIC18F2550

PIC18F2550 is an USB Microcontrollers with 24k bytes enhanced flash memory, embedded with 2048 bytes SRAM, and 1-Kbyte Dual Access RAM for USB. The USB protocol stack realizes 4 transmission modes, such as control, interrupt, isochronous and bulk transfer. In addition, the MCU is embedded with the enhanced USART module to support the LIN bus protocol. Therefore, the design of the peripheral circuit of LIN and USB conversion interface can be simplified by using this MCU.

3.2 Hardware Design of USB Interface

In Fig. 2, USBH is a USB interface, and the voltage of the USB port of the computer is 5 V + 5%, the current minimum is 100 mA, and the maximum is 500 mA. The power supply range of PIC18F2550 should be 4.2 V–5.5 V. In actual testing, the current consumed by this circuit is only 50 mA. Therefore, the power supply provided by the computer to the USB port can guarantee the normal operation of the circuit.

In this figure, these signals are connected to optical isolation circuit, such as RC1/FT, RC0/CS, RC6/TX and RC7/RX. The power supply is isolated from the USB port of computer and the LIN bus. RC1/FT is the LIN bus fault detection port, through which the ATA6625 (LIN interface chip) transfers the state of the LIN bus to PIC18F2550. RC0/CS is the chip selection signal of the LIN bus transceiver. RC6/TX is the sending signal of the LIN bus transceiver. RC7/RX is the receiving signal of the LIN bus transceiver. To improve the coupling waveform of the optical septum, the pull up resistance of 1 K is added to both the RC1/FT signal and the RC7/RX signal.

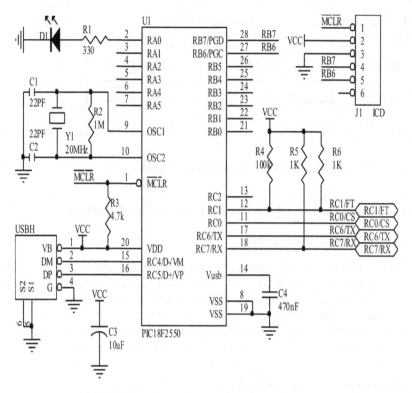

Fig. 2. Circuit of the USB interface with computer

3.3 Hardware Design of LIN Bus Interface

In Fig. 3, ATA6625 is a LIN transceiver [3], and this chip is designed in accordance with the LIN specification and the SAEJ2602 standard. The ESD stability of the chip is higher than 6 kV. ATA6625 includes a step-down regulator which can provide power for the MCU and other chips in the application.

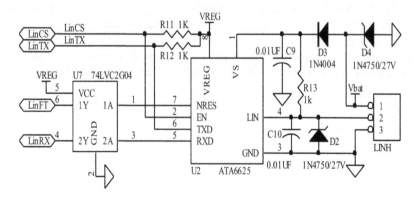

Fig. 3. Circuit of LIN bus interface

The LINH is the LIN bus interface in the figure, and Vbat is powered by 12 V battery. D3 and D4 prevent battery overload. The power of the LIN bus terminal is isolated from the USB port of computer. Data is exchanged between LIN bus transceiver and MCU through a photoelectric isolation chip of PS9701.

3.4   Hardware Design of Power System Isolation

In order to prevent the damage of the USB interface of computer, the photoelectric isolation is added between the 5 V power supply of the PC machine and the battery of the car. Considering that the baud rate upper limit of the LIN bus is 20 Kbps [4], ps9701 with higher speed optical isolation is used in the circuit. This photoelectric isolation rising delay is only 50 ns (Fig. 4).

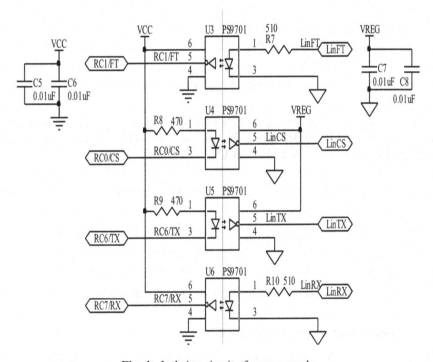

Fig. 4. Isolation circuit of power supply

4   Software Design of LIN Bus Communication Conversion

The overall program block diagram of USB-LIN interface is shown in Fig. 5. The whole program is divided into two parts, one is main program and the other is interrupt service program. The main task state machine of LIN bus is implemented in the timer interrupt service program, and is implemented from the task state machine on the LIN bus's rising edge interrupt service program. The slave state machine is implemented on the LIN bus in the rising edge interrupt service program.

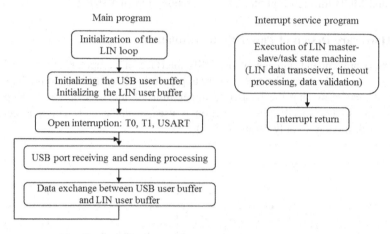

Fig. 5. Overall program flow chart of USB-LIN interface

The design of data buffer for USB-LIN interface is illustrated in Fig. 6. It includes data exchange on LIN node, data exchange on USB side, and data parsing and exchanging between LIN bus buffer and USB end buffer.

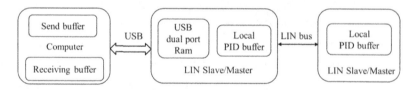

Fig. 6. Data exchange schematic of USB-LIN interface

4.1   Software Implementation of LIN Bus

The software part of the LIN bus includes the implementation of network management, realization of master and slave state machine, and the update of scheduling table and information binding table. The LIN protocol specifies the state transfer diagram of the master and slave task [5]. The core task of the main task is to produce the right message head. The mechanism of message header transmission is based on the time scheduling table. The scheduling principle of the LIN bus task is shown in Fig. 7. In this figure, Timer 1 is the slice timer setting as 1 ms for scheduling, timer 0 is timer for frame timeout, and its setting depends on the clock frequency of the node and the baud rate of the communication.

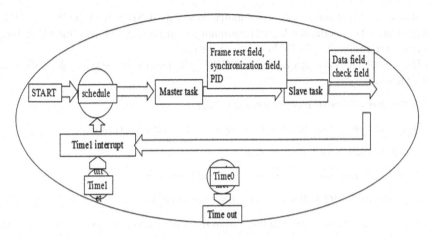

Fig. 7. Scheduling schematic diagram of LIN bus

A scheduler table gives the time slice for each frame, and the main task starts the transmission of the frame based on this table. The scheduling table consists of three items: the processing handle of the frame, the frame's protected identifier (PID) and the time slice for the frame.

An information binding table defines a node's response to a received frame with a particular protected identifier (PID). The master node can operate the data buffer of slave node, this means that the physics parameter from the node can be operated directly by master node.

The scheduler table and the information binding table are stored in the EEPROM of the PIC18F2550, and this two tables can be updated through USB of computer.

4.2   Software Implementation of USB Bus

In the CDC mode [6], the CDC subclass implements the protocol for virtual serial communication. Most operating systems have device drivers that support the CDC class, and can identify the CDC class devices, therefore it is only necessary to view the interface as a virtual serial port in computer. When the USB port of the PIC18F2550 works in the CDC mode, the highest communication rate is 1 Mbps. This MCU fully meets the requirements of LIN bus data exchange.

The USB interface of PIC18F2550 can be used in general mode or CDC mode to exchange data with computer. Although the general mode has a higher USB rate, the driver of computer is more complex. The USB-LIN interface just handles low-speed LIN communication, and the communication rate above 1 Mbps is not necessary. Therefore, the CDC mode is used in the USB interface of PIC18F2550.

In the CDC mode, the CDC subclass implements the protocol for virtual serial communication [7]. Most operating systems have device drivers that support the CDC class, and can identify the CDC class devices, therefore it is only necessary to view the interface as a virtual serial port in computer. When the USB port of the PIC18F2550r works in the CDC mode, the highest communication rate is 1 Mbps. This MCU fully meets the requirements of LIN bus data exchange.

The USB/CDC class firmware of PIC18F2550 mainly implements the following API function.

- void putrsUSBUSART(const rom char *data)

  Write the string from the program storage area to the USB port.

- void putsUSBUSART(char *data)

  Write the string from the data buffer to the USB port.

- void mUSBUSARTTxRom(rom byte *pData, byte len)

  A specific length of string is written from the program storage area to the USB port.

- void mUSBUSARTTxRam(byte *pData, byte len)

  A specific length of string is written from data buffer to the USB port.

- BOOL mUSBUSARTIsTxTrfReady(void)

  Check whether the CDC class has more data to send.

- byte getsUSBUSART(char *buffer, byte len)

  Copy the data from the USB/CDC buffer to the user buffer.

- byte mCDCGetRxLength(void)

Get the actual length of the data copied from the USB/CDC buffer to the user buffer. This function is called after the getsUSBUSART function is executed.

The function, mUSBUSARTIsTxTrfReady, is called to check whether the firmware is ready to send data to the USB port, then the following function are called, putrsUSBUSAR, putsUSBUSART, mUSBUSARTTxRam, and mUSBUSARTTxRom.

When calling getsUSBUSART, the length of the target buffer must be greater than the length of the USB/CDC buffer.

5 Conclusions

In this paper, the design methods of LIN bus and USB bus interface are discussed from the aspects of the overall system structure, hardware, and software design. The interface is used to connect the LIN bus interface with the auto body control module and the USB of computer. This design has been used in the testing of the BCM module and in the teaching and training of the LIN network protocol. Practice has proved that it is valuable.

Acknowledgements. This paper is supported by Hubei Provincial Department of Education Science and Technology Research Foundation of B2017589; College Teachers Guiding Students' innovation and entrepreneurship training program project of 2017.

References

1. Specks, J.W., Rajnäk, A.: LIN - protocol, development tools, and software interfaces for local interconnect networks in vehicles. In: VDI-Berichte, pp. 1227–1250 (2000)
2. Ruff, M.: Evolution of local interconnect network (LIN) solutions. IEEE Veh. Technol. Conf. 5(2), 3382–3389 (2003)
3. Xu, Y., Wang, J., Chen, W., Tao, J.: Application of LIN bus in vehicle network. In: IEEE International Conference on Vehicular Electronics & Safety, pp. 119–123 (2006)
4. Krastev, G.: Microcomputer protocol implementation at local interconnect network. Comput. Appl. 24(6), 114–116 (2004)
5. Shi, B.B.: Application research of automatic guided vehicle system based on LIN bus. Adv. Mater. Res. 267, 710–714 (2011)
6. Ye, Q.: Research and application of CAN and LIN bus in automobile network system. Int. Conf. Adv. Comput. Theor. Eng. 6(14), V6-150–V6-154 (2010)
7. Kenarsari Anhari, A.: Medium access control protocol design for in-vehicle power line communication. J. Clin. Res. Pediatr. Endocrinol. 5(1), 1022–1027 (2013)

Unmanned Ship Path Planning Based on RRT

Xinjia Chen[1], Yanxia Liu[1], Xiaobin Hong[2(✉)], Xinyong Wei[2],
and Yesheng Huang[2]

[1] School of Software Engineering, South China University of Technology,
Guangzhou, China
[2] School of Mechanical and Automotive Engineering,
South China University of Technology, Guangzhou, China
mexbhong@scut.edu.cn

Abstract. Path planning is a task of primary importance for unmanned ship, but current algorithms are complex and inefficient. In this paper, we propose a Rapidly-Exploring Random Tree algorithm (RRT) for path planning of unmanned ship, which can obtain an asymptotically optimal path planning in limited time. Moreover, an extension of RRT algorithm has been proposed to overcome the actual demand of multi-waypoint path planning for unmanned ship. The feasibility and effectiveness of the proposed algorithm was proved by simulation on MATLAB™ platform.

Keywords: Unmanned ship · Path planning · Rapidly-Exploring Random Tree

1 Introduction

With the rapid development of science and technology, the field of modern unmanned ship is constantly being exploited and utilized. Unmanned ship plays an important role in autonomous navigation, especially sea patrol and cargo delivery. The rise of artificial intelligence promotes the rapid development of driverless technology, which has led to many path planning algorithms to be proposed successively [1].

Traditional path planning algorithm is the main choice for unmanned ship path planning, such as artificial potential field method [2], Dijkstra search algorithm [3] and A* algorithm [4, 5]. These algorithms have the following shortcomings: Map modeling is complex; the heuristic ideas always complicated and real-time performance is poor. Whereas, the RRT algorithm applied in the mobile robot path planning is superior: The path points in RRT algorithm are simply generated by random sampling; map information requirement is simple; RRT algorithm can maintain the consistency with the dynamic constraint during the generation of path points, and a feasible path can be generated within limited time.

In this paper, we introduce the implementation of RRT-based path planning algorithm for unmanned ship. The definition of path planning problem in the area of unmanned ship is given in the first place. We then promote the multi-waypoint path planning method based on improved RRT and analyze the performance of RRT algorithm. Finally, by simulation experiments, the effectiveness of RRT path planning algorithm for unmanned ship is verified.

© Springer International Publishing AG, part of Springer Nature 2018
D.-S. Huang et al. (Eds.): ICIC 2018, LNCS 10954, pp. 102–110, 2018.
https://doi.org/10.1007/978-3-319-95930-6_11

2 Definition of Unmanned Ship Path Planning Problem

The main goal of this paper is to propose a method for unmanned ships to generate a valid path within limited time. We obtain sea surface information by S57 electronic chart [6]. Moreover, an improved path planning method based on RRT algorithm is proposed to solve the problem of unmanned ship multi-waypoint path planning problem [7].

The environment model for unmanned ships is constructed from S57 electronic chart, which conforms to the S57 international standard, encapsulated by the ISO8211 standard. S57 electronic chart contains the geographical location and object information of the sea area. We parse the S57 electronic chart to acquire sea surface information using the GDAL library.

For illustrative purposes, represent the map space as M, and obstacles and non-navigation areas as $M_{obstacle}$, and navigable areas as M_{free}, then $M_{obstacle} = M/M_{free}$. The definition of single-waypoint path planning for unmanned ships is to find a continuous sequence from the specified start point q_{start} to the end point q_{goal} in M_{free}. We assume that the unmanned ship must pass n spots, which represented as targets[n]. The problem of multi-waypoint path planning for unmanned ship is defined as finding a continuous sequence in M_{free} which passes all the targets.

3 Single-Waypoint Unmanned Ship Path Planning Method Based on RRT

Rapidly-exploring Random Tree (RRT) is an incremental sampling algorithm proposed by Lavalle [8], with good performance in practical applications [9] with only a few parameters.

Moreover, RRT is an efficient planning algorithm that can be applied in multidimensional space. By setting a start point as root node, RRT will generate a random extended tree stochastically, and the algorithm terminates when a leaf enters the target area or reaches the target point.

When using RRT algorithm to plan a path for unmanned ship, points on the path tree represent the positions where the unmanned ship arrives. The algorithm terminates when the unmanned ship can reach the final target point. Assume that M represents the map converted from S57 electronic chart. Random extended tree T that keeps the information of extended nodes and the edges between the points; the start point q_{start} and the end point q_{goal}; randomly created point q_{rand} in the map area; the nearest node q_{near} in the tree T to the q_{rand}; new extended node q_{new}.

The collision detection parameter t, which can be chosen according to the size of the actual obstacle, denotes that taking t points evenly between q_{near} and q_{new}, and judging whether the edge[q_{near}, q_{new}] is in the free space. Besides, in the process of tree extension, probability p defines the probability of a new extended point toward the target point. The number of point k defines the maximum number of iterations of the algorithm. Tree extension distance Δt of the extended leaf node defines the distance between the new point and the nearest node in the tree, which can be adjusted on the basis of the dynamic restraint of unmanned ship [10]. Parameters mentioned above are those who have the impact on RRT performance.

RRT Algorithm process is as follows:

RRT(q_{start}, q_{goal}, map)

---

1. T <- InitTree();
2. T <- InsertNode(q_{start});
3. for j=0 to k
4.   if rand()<p
5. q_{rand}= q_{goal};
6.   else q_{rand}=(randi(mapWidth),randi(mapLength));
7. q_{near} =Nearest(T,q_{rand});
8. q_{new} =NewPoint(q_{near} q_{rand}, Δt);
9.   if ObstacleFree(q_{near}, q_{new}) //collision check
10.   if GoalInNewEdge(q_{new},q_{rand}, q_{goal}) or q_{new} == q_{goal}
11.   return T;
12. return T

Algorithm steps:

Steps 1, 2: Initialize unique start node q_{start} for random extended tree T.

Step 3: Define the maximum number of iterations.

Steps 4, 5: Create a random point q_{rand} with probability p towards the q_{goal} direction.

Step 6: Create a random point q_{rand} with probability $(1 - p)$ towards any direction.

Step 7: Search node q_{near} for q_{rand} from T.

Step 8: A newly extended node q_{new} is generated by q_{near}, q_{rand} and distance Δt.

Step 9: Check whether there is any obstacle in q_{near} to q_{new}.

Step 10: The target is on the newly created edge or reach the target point

Step 11: Successfully find the path and returns tree T.

Step 12: After k times of iterations, q_{goal} is still not reached, and error information is returned.

New Node Extension: The point q_{rand} is randomly created in the finite space M, and the nearest node q_{near} to the q_{rand} is searched in the random extended tree T. Besides, q_{new} will be created by q_{near} and q_{rand}. The q_{new} (x, y) generation formula is as follows: (Fig. 1)

$$q_{new}(x, y) = q_{near} + \Delta t * \frac{q_{rand} - q_{near}}{\|q_{rand} - q_{near}\|} \tag{1}$$

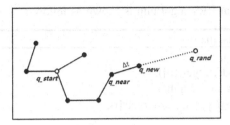

Fig. 1. New node extension

Collision Detection: To determine q_{new} is a legal extended point if there are any obstacles between q_{near} and q_{new}. Selecting some test points evenly between q_{near} and q_{new}, and then, q_{new} point is valid if all test points belong to M_{free} space (Fig. 2).

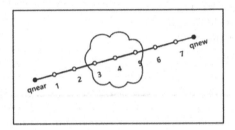

Fig. 2. Collision detection

4 Multi-waypoint Unmanned Ship Path Planning Method Based on RRT

The original RRT algorithm can quickly generate a sailing route in the case of a single waypoint. However, in the actual situation of unmanned ship, perhaps need to reach multiple preset waypoints. Current algorithm only solves the single-waypoint problem. In order to achieve multi-waypoint path planning, the target point is dynamically adjusted to waypoint, and the node information of the tree is dynamically refreshed during the growth process of RRT trees, so that the planned path passes through all the waypoints.

By setting the waypoints in a pair and consider them as root node of tree and the extension target position, makes the final path to pass all the waypoints. In this process, the extension process of the tree needs to utilize the existed node information, and the growth of the current tree is not related to tree nodes in the previous round, hence, the tree nodes need to be marked in each round to indicate whether it is an extensible node. Furthermore, to ensure that the final path includes all the waypoints, path smoothing algorithm is used for each segment of path.

The multi-waypoint RRT algorithm is shown as follows:

MultiRRT(targets, map)
1. T <- InitTree();
2. T <- InsertNode(targets[0]);
3. for i=1 to i<targets.length()
4. q_{start} =targets[i-1]; q_{goal} =targets[i];
5.   for j=0 to k
6.   if rand()<p
7. q_{rand}= q_{goal};
8.   else q_{rand}=(randi(mapWidth),randi(mapLength));
9. q_{near}=Nearest(T,q_{rand});
10. q_{new} =NewPoint(q_{near},q_{rand}, $\Delta\tau$);
11.   if ObstacleFree(q_{near}, q_{new})
12.   if GoalInNewEdge(q_{new},q_{rand}, q_{goal})
13.   or q_{new} == q_{goal}
14.   smooth(T,q_{start},q_{goal});
15.   refresh(T);
16.   break;
17.   break;
18. return T;

Step 3: Iterate by the number of waypoints.

Step 4: Dynamically change the start and end points in each iteration.

Step 14: After a path is generated in each cycle, invoking the path smoothing algorithm.

Step 15: Update the nodes information and mark the nodes generated in the current cycle.

Path Smoothing: There are a large number of redundant points in the path from q_{start} to q_{goal} generated by the RRT algorithm. We used a greedy approach to connect the farthest point with the start point and remove redundant points. Set the points sequence to be smoothed as path [1, n], and the smoothed path is path_smooth[1,n].

The path smooth algorithm is shown as follows:

Smooth(path)
1.path_smooth=path[1] //initialization
2.currentIndex = 1; //the start index to be smoothed
3.currentSmoothIndex = n; // the end index to be smoothed
4.while currentIndex < n :
5.   while currentIndex < currentSmoothIndex :
6.   if ObstacleFree(path[currentSmoothIndex], path[currentIndex])
7.   path_smooth = [path_smooth, path[currentSmoothIndex]];
8.   currentIndex = currentSmoothIndex;
9.   break;
10.   else   currentSmoothIndex = currentSmoothIndex - 1;
11.currentSmoothIndex = n;

The result of the path smoothing algorithm is as follows: (Fig. 3)

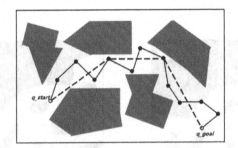

Fig. 3. Path smoothing

5   Simulation Experiments and Performance Analysis

5.1   Path Planning in Different Sea Environments Experiments

All simulation experiments in this paper running in MATLAB. Experiment 1 simulates the RRT path planning result of unmanned ship in the case of a simple sea environment. Experiment 2 simulates the path planning result under the situation of the sailing area is complex and has multiple corners. Experiment 3 simulates the result in the sea environment with multiple obstacles. RRT algorithm can still avoid obstacles accurately and quickly generate an effective path in all situations.

From Figs. 4, 5 and 6, it can be concluded that the RRT algorithm can solve a variety of sea environment situations, and can plan an asymptotically optimal sailing route within a reasonable time for different sea surface, which shows the excellent performance of the RRT algorithm for unmanned ship path planning.

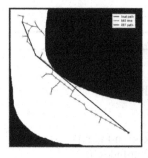  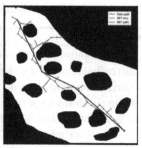

Fig. 4. Simple situation   **Fig. 5.** Complex situation   **Fig. 6.** Many obstacles situation

5.2   Multi-waypoint Path Planning Experiment

Experiment 4 simulates a common sea environment. The RRT algorithm calculates the result if the path only requires one waypoint. Experiment 5 is based on experiment four, presetting multiple points through which unmanned ship must pass, and using multi-waypoint RRT algorithm for path planning (Figs. 7 and 8).

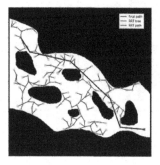   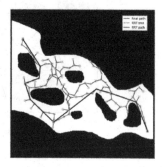

Fig. 7. One target position
situation

Fig. 8. Multi-waypoint
situation

The simulation results show that even if the limitation of multiple waypoints is added, a path can be obtained through a multi-waypoint RRT algorithm in a roughly similar time with experiment 4, and can be accurately smoothed. In short, multi-waypoint RRT algorithm is feasible and effective.

5.3   Path Smoothing Experiment

The path smoothing algorithm obtains shorter path by removing redundant points. The comparison of result is as Fig. 9, and the black line indicates the smoothed path. Compared with the original path, the length of the path can be effectively shortened.

Fig. 9. Path smoothing

5.4   A* Algorithm Comparison Experiment

The A*(A-Star) algorithm in path planning is an effective algorithm for finding the optimal path. It combines the advantages of the Best-First Search and Dijkstra algorithms to improve the efficiency.

We compare the RRT algorithm with the path planning classical A* algorithm in the same operating environment and the same input matrix as M [500, 500].

Fig. 10. A* algorithm

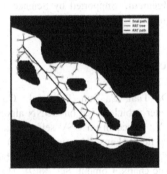

Fig. 11. RRT algorithm

The results of Experiment 6 and Experiment 7 are compared as follows:

It can be seen from the experiments, the results show in Figs. 10, 11 and Table 1, that the A* algorithm needs to detect a large number of points to find an optimal path,

Table 1. Algorithm performance analysis

Algorithm	Detection points	Time (s)	Path length
RRT	279	1.5	564
A*	8905	36	516

and the length of the final path of the RRT algorithm is slightly longer. Although the results of RRT algorithm is random, the number of created points is far less than that of the A* algorithm. Conversely, the rules for generating and detecting RRT algorithms are much simpler than other algorithms, even more, the efficiency can be increased by at least two orders of magnitude for real-time path planning of unmanned ship.

To sum up, RRT algorithm has many superiorities compared with other path planning algorithm, for instance, such as the artificial potential field method and ant colony optimization algorithms, need to set up complex heuristic rules, while RRT algorithm only extends through a simple random algorithm and can meet the dynamic constraints during the extension process.

6 Conclusion

This paper proposes the idea of applying the RRT algorithm to the field of unmanned ship path planning. And simulation experiments show that RRT can solve the problem of sea path planning in different situations. Based on the RRT algorithm, this paper proposes a multi-waypoint RRT algorithm to solve the problem of multi-waypoint path planning for unmanned ship. Finally, the path smoothing algorithm is used to shorten the path obtained by the RRT algorithm.

Acknowledgement. Supported by Science and Technology Project of Guangdong Province, China (Granted No. 2017B010118002).

References

1. Zhu, D., Tian, C., Sun, B., et al.: Complete coverage path planning of autonomous underwater vehicle based on GBNN algorithm. J. Intell. Rob. Syst., 1–13 (2018)
2. Zhang Y.: Research on USV self-avoidance navigation control system based on artificial potential field. Hainan University (2017)
3. Zhuang, J., Wan, L., Liao, Y., et al.: Global path planning of unmanned watercraft based on electronic charts. Comput. Sci. **38**(9), 211–214 (2011)
4. Chen, S., Liu, C., Huang, Z., et al.: AUV global path planning based on sparse A* algorithm. Torpedo Technol. **20**(4), 271–275 (2012)
5. Zammit, C., Kampen, E.J.V.: Comparison between A* and RRT algorithms for UAV path planning. In: Aiaa Guidance, Navigation, and Control Conference (2018)
6. IHO: IHO Transfer Standard for Digital Hydrographic Data, 3.1 edn., Publication S-57. International Hydrographic Bureau, Monaco (2000)
7. Liu, Y., Bucknall, R.: Efficient multi-task allocation and path planning for unmanned surface vehicle in support of ocean operations. Neurocomputing **275**, 1550–1566 (2018)
8. Lavalle, S.M.: Rapidly-exploring random trees: a new tool for path planning. Algorithmic Comput. Rob. New Dir., 293–308 (1998)
9. LaValle, S.M.: Planning Algorithms. Cambridge University Press, Cambridge (2006)
10. Du, Z., Wen, Y., Xiao, C., et al.: Motion planning for unmanned surface vehicle based on trajectory unit. Ocean Eng. **151**(151), 46–56 (2018)

A Natural Language Interaction Based Automatic Operating System for Industrial Robot

Yunhan Lin[1,2,3], Huasong Min[3(✉)], Haotian Zhou[3], and Mingyu Chen[3]

[1] College of Computer Science and Technology,
Wuhan University of Science and Technology, Wuhan, China
[2] Hubei Province Key Laboratory of Intelligent Information Processing
and Real-Time Industrial System, Wuhan, China
[3] Institute of Robotics and Intelligent Systems,
Wuhan University of Science and Technology, Wuhan, China
mhuasong@wust.edu.cn

Abstract. In this paper, an automatic operating system is designed for industrial robot using natural language. Some cutting-edge technologies, such as 3D visual perception, human-robot-environment interaction and auto-programming, are integrated into the design of automatic operating system for industrial robot. In particularly, a new "rule-scene" matching and interaction algorithm is proposed, which can realize the interaction among human, robot and environment, and can guide user to give the correct rule when the user's rule is wrong or invalid. Automatic operation is achieved through auto-programming and implementation after rules are defined correctly. In the experiment section, our designed system is applied to a fruits sorting platform as a demonstration. Experiment result proves the validity and practicability of our system.

Keywords: Natural language interaction · Automatic operation
Human-robot-environment interaction · Auto-programming · Industrial robot

1 Introduction

Industrial robots could previously only complete automatic operations using fixed programs, which strengthens specificity while reduces expansibility of robots. As more and more diverse requirements for automatic operation are put forward in industrial production, it becomes more and more difficult for previous industrial robots to meet requirements, so teaching-playback robots and robot programming languages have been developed [1]. Robot programming languages can be divided into action-level, object-level and task-level programming languages according to the levels of operation description [2]. Among them, a task-level programming language refers to a language showing a higher level than the former two languages and is the ideal programming language. This language allows users to directly give commands to the goals to be achieved for a task, with no need for regulating the details of each action of robots. As long as the initial environment model and the final working state are given according to some principles, robots can undertake automatic reasoning and calculation, and finally generate executable commands for robots automatically. Therefore, the key problem of designing the task-level automatic operating system for robots is to explore methods for acquiring operation commands and for realizing automatic programming through task-level language use.

Generally, operation commands are acquired through touch screen input, document interaction with other software, and natural language input. Choi et al. [3] developed the EL-E mobile robot which can automatically acquire and pass the target objects to users and the EL-E obtains the operation commands through the input using a laser pen and touch screen. By using the DXF (Drawing eXchange Format) automatic withdrawal method for a workpiece model, Xing et al. [4] analyzed operation information and then realized automatic programming for industrial robots based on VRML (Virtual Reality Modelling Language). Connell et al. [5] investigated the ELI robot and controlled its motion so as to get objects through natural language. This system can not only be used to execute pre-defined actions and recognize objects but also learn new nouns and verbs through speech training to construct a new semantic table. The system, designed by Fasola et al. [6], enables the mobile robot PR2 to fulfil automatic grabbing tasks under the control of non-professional users through the use of natural language.

Comparison of different input methods for operation commands reveals that if the natural language can be used as the input of operation command to realize automatic operation of robots, the robots can be more intelligent and easier to use. In view of indoor mobile robot service requirements, Matuszek et al. [7] studied a natural language parser to realize automatic programming of robots. This system can translate the natural language describing the indoor path into LISP-like robot control language with which the robots can arrive at the destination in unknown indoor environments by analyzing path commands. Based on the study of the natural Chinese language used to describing paths, Li et al. [8, 9] designed, and realized, visual navigation for mobile robots based on restricted natural language in indoor environment by establishing an intention map of navigation for mobile robots, moreover, they proposed the method for directly drawing movement paths of robots by using natural language describing paths. All of the above automatic operating systems are designed for mobile service robots, while in the field of traditional industrial robots, automatic operating systems based on natural language are rarely seen. Actually, if they can be controlled by natural language, industrial robots can be operated by non-professional staff without special programming experience, which is revolutionary with regard to the improved usability of industrial robots. In addition, the above systems are used for automatic operation when users give correct rules, but cannot be correctly operated when given wrong or invalid rules. This research proposes a "rule-scene" matching and interaction algorithm by integrating the advanced technologies into the design of the automatic operating system for industrial robots. These technologies include the 3D environmental perception of robots, human-robot-environment interaction, and automatic programming. By utilizing the algorithms, human-robot-environment interaction can be realized and the system can be used to guide users to give correct rules through speech when the rules are wrong or invalid. After obtaining the correct rules, automatic operation can be realized through the automatic programming and execution algorithm.

2 Structure of System

The designed automatic operating system for industrial robots mainly includes the 3D visual perception, the "rule-scene" matching and interaction, and the automatic programming and execution modules. The system structure is shown in Fig. 1. The 3D visual perception module provides high-quality semantic map information about the operating environment for the whole system. While the "rule-scene" matching and interaction module produces action of intention by combining rules obtained through human-robot interaction with real-time semantic map received through real-time 3D environmental perception. Furthermore, the automatic programming and execution module transforms action of intention with semantic information into executable commands for robots to control the manipulator to complete target actions.

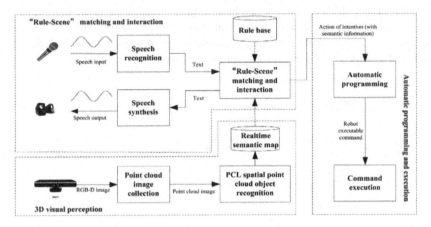

Fig. 1. Structure of the designed automatic operating system for industrial robot

2.1 3D Visual Perception

The 3D visual perception pipeline in this paper is shown in Fig. 2. It consists of three parts: (1) object modeling, (2) object recognition and pose estimation, and (3) semantic map file generation. In our process, the ISS (Intrinsic Shape Signatures) [10] key points are extracted from the scene and calculated the CSHOT (Color Signature of Histograms of OrienTations) [11] feature vector from key points; the candidate model is generated by the 3D feature matching based on the distance threshold; the transformation hypothesis is generated by RANSAC (RANdom SAmple Consensus) algorithm [12]; the solution of the hypothesis is verified by the iterative closest point algorithm, which generates a solution to maintain the consistency of the scene. Finally, the object's identification and geometric information (*object$_i$= category, name, color, shape, x, y, z, size*, which means object category, name, color, sharp, coordinate *x, y, z* and size respectively.) are written to the XML (eXtensible Markup Language) semantic map file [13].

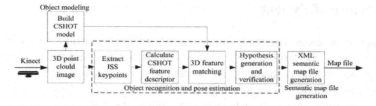

Fig. 2. Pipeline for 3D visual perception

2.2 "Rule-Scene" Matching and Interaction

Figure 3 shows the algorithm flow of the "rule-scene" matching and interaction algorithm. An open source speech recognition system (PocketSphinx) for the embedded platform is used for speech recognition in the system [14] and the multi-language and cross-platform open source speech synthesis system Ekho is employed for speech synthesis [15]. The whole matching and interaction process mainly includes three functional modules: rule acquisition, "rule-scene" matching, and guidance.

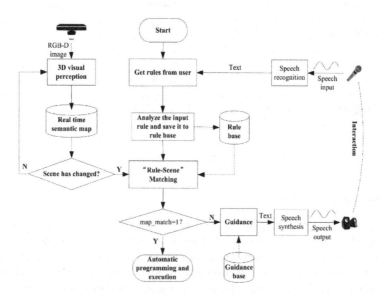

Fig. 3. Flowchart for the "rule-scene" matching and interaction algorithm

Rule Acquisition. Firstly, robots actively ask the users what they need to do. For example, robots ask "Hello, I am WUSTER, what can I do for you?" Then, the natural language input of users is transformed into text through speech recognition and then the attributes of rules can be obtained based on lexical analysis, syntactical analysis, and semantic analysis. Finally, the rule attributes are saved into the rule base.

A. Lexical Analysis. Since each different language has different lexical analysis processes, here we choose Chinese as an example to explain the process of lexical analysis. Chinese lexical analysis comprises Chinese word segmentation, part of speech tagging, syntactic dependency analysis, and word sense disambiguation. A word is the minimum unit of language that can be used independently. Differing from English words separated using spaces, Chinese words are written continuously with a large character set. Therefore, Chinese words have to be segmented to divide a complete sentence in natural language into independent words, so that computers can obtain the definite boundary of Chinese words and understand semantic information contained in the text. In our system, the NLPIR/ICTCLAS Chinese word segmentation system (2015 version) [16] developed by Dr. Zhang Pinghua (Chinese Academy of Sciences) is used for lexical analysis.

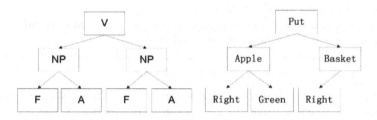

(a) The definition of the syntax tree structure (b) An example of the syntax tree structure

Fig. 4. An example of the syntax tree structure (Note: V, NP, F, and A represent the verb, noun phase, noun of locality, and adjective, respectively in (a))

B. Syntactical Analysis. Before carrying out syntax analysis, a syntax pre-processing stage is used: irrelevant modifiers in commands in natural language are eliminated. The pre-processed commands in natural language only comprise verbs, nouns, and modifiers relative to nouns. Syntax analysis uses the words in each lexical unit generated by the lexical analyzer to construct the intermediate representation of the syntax tree. This intermediate representation provides the syntax structure of lexical cell flow generated from lexical analysis. Each internal node in the tree stands for a word and its child node represents its attribute. In view of the characteristics of commands in natural language in the system, the verbs that represent actions can be used as the central word of the whole command, while noun phrases are attached to the verbs and shown to be child nodes of actions in the syntax tree: the nouns of locality and adjectives for modifying nouns are subjected to noun phrases and expressed as child nodes of noun phrases in the syntax tree (Fig. 4).

C. *Semantic Analysis.* Semantic analysis is carried out to check whether, or not, the generated syntax tree and semantic map are matched to examine whether, or not, commands in natural language are consistent with the semantic information defined by the language. The nodes in the syntax tress are matched with elements in the XML semantic map to obtain the operation sentence with semantic information, including designated actions and coordinates of target objects. The action requiring a robot to move an object from one place to another is defined as a composite action. The specific process is shown as follows: firstly, if the end-effector of robots is not in the same location as the target object, the end-effector is moved to this location and then the actions, such as capturing the object, moving to the target location, and putting down the object are executed. The two actions, grabbing and putting down, belong to sub-sets of the composite action and only the two designated simple actions need to be executed.

D. Definition of Rule Base. The rule base is defined by utilizing a topic tree structure to describe each term of information in rules and relationship of information. There are three nodes in the topic tree: a topic node, an intermediate node, and a leaf node (Fig. 5).

(1) Each topic node represents the root of a topic tree and indicates the category of the rules and relevant knowledge base.
(2) An intermediate node is a set of necessary attributes contained in the rules, that is, information required to make the rules effective.
(3) A leaf node is used to store sub-attributes of intermediate nodes and is utilized as the supplementary attributes of intermediate nodes.

Each node has a binary effective state descriptor. The effective state descriptor of nodes represents whether, or not, the information about each node is effective, that is, whether, or not, it is confirmed by its users. Each state set has a corresponding dialogue generating function and the set of these dialogue generating functions constitutes the guidance base. In different system states, different response inputs can be obtained by calling this function and each dialogue generating function only responds to its corresponding state sets: they do not affect each other in the design and modification.

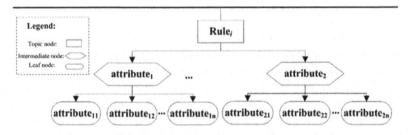

Fig. 5. Storage structure of rule base

According to the above definitions, while designing an object sorting system, the attribute of the rule base can be defined as a topic tree including the names of those objects to be sorted and the names of placement destination thereof (Fig. 6).

Where, *Obj_name*, *Obj_size*, *Obj_color*, and *Obj_location* represent the name, size, color, and location of the objects, respectively. *Des_name* and *Des_location* indicate the name and location of the placement destination of objects, separately.

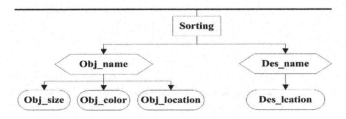

Fig. 6. Storage structure of rule base of object sorting system

"Rule-scene" matching. In "rule-scene" matching, the rules obtained by the system and real-time semantic map files of scenes generated in Sect. 2.1 are used as input to calculate whether, or not, rules match with scenes. In other words, the objects *expobj* referred to in the rules are sought in the scene semantic map and the matching formula of Eq. (1) is used:

$$map\_match = \begin{cases} 1, & if\ expobj = object_i \cap \text{expobj} \\ 0, & if\ expobj \neq object_i \cap \text{expobj} \end{cases} \tag{1}$$

Where, $object_i$ and *expobj* represent the object in a realtime scene and the object that is expected to be grasped in the rules, respectively.

When *map_match = 1*, it indicates that there are expected objects in the scenes. While *map_match = 0*, it shows that there are no expected object in the scenes, thus it is judged that the rules are invalid in the current environment. At this time, the guidance mode of the system is started to inform users of the object situations in the current scene and ask users to provide effective expectations.

Guidance. As described in the definition of the storage structure of rule base in Section **Rule acquisition**, each node has a corresponding dialogue generating function and the set of these dialogue generating functions constitutes the guidance base *GuidanceBase*. In accordance with the state set *node_state* of binary effective state descriptors of all nodes in the topic tree, the system calls the guidance solution *guidance_solution*. The definition of *GuidanceBase* is given in Eqs. (2) and (3):

$$GuidanceBase = (GC_1, GC_2, \ldots, GC_n) \tag{2}$$

$$GC_1 = (node\_state_i, guidance\_solution_i) \tag{3}$$

2.3   Automatic Programming and Command Execution

Figure 7 shows the automatic programming and execution process of our system. The action of intention with semantic information and realtime semantic map file are used as the input of automatic programming for the robot. Then a series of executable commands that can be understood by the robot are transformed through intention

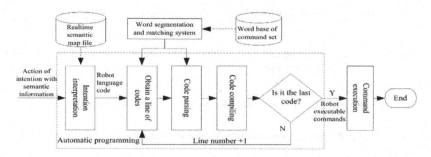

Fig. 7. Automatic programming and execution process

interpretation, code parsing, and code compiling. Finally, the command execution stage indicates that the robot executable commands are transferred to the rotation angle of each joint of the robot through the communication bus to control actual movements of the manipulator.

Automatic Programming. Intention interpretation means that the action of intention with semantic information is transformed into a series of program codes for the robot through robot programming language. In this paper, a set of robot language has been designed, which includes the command of *INIT, BEGIN, MOVEP, HANDON, HANDOFF, TIMER* etc. The pseudo-codes of intention interpretation are shown as **Algorithm 1**. Firstly, the operation rules are sought according to relevant attributes of objects in the scene to generate program templates. Then passing points are obtained in accordance with the coordinates of objects to be sorted and destination for placement, then, the tracks of each passing points are planned. Finally, the instantiated program is generated and the instantiated programs of all objects are collected to constitute robot language source codes. Figure 8 shows the program code with which the robot picks an object at point **A** and then places it at point **B**. It is assumed that the coordinates of points **A** and **B** are (*−0.080, 0.331, 1.350*) and (*0.486, 0.372, 1.230*).

Algorithm 1 Pseudo-codes of intention interpretation

1: **Input:** $SI \cap RSM$ //SI (Semantic Intention), RSM (Real time Semantic Map)
2: **Output:** $RLSC$ //$RLSC$ (Robot Language Source Codes)
3: **for** $(i = 0, i < N, i++)$ **do**   //The total number of objects in the scene is N
4:    **Search** operation rules $(rule_i, attribute_1, attribute_2)$ by the relevant attributes of $Object_i$ $(Object_i = \{category, name, color, shape, x, y, z, size\})$
5:    **Generate** the current operation program template
6:    $P_{pick} \leftarrow$ object coordinate of $attribute_1$
7:    $P_{place} \leftarrow$ object coordinate of $attribute_2$
8:    **Calculate** passing points $P_0, P_1, ..., P_{pick}, ..., P_{place}, ..., P_m..., P_0$ according to P_{pick} and P_{place} // P_0 indicates the location of the manipulator at initial status
9:    **Plan** trajectory of each path
10:    **Generate** instantiated program
11: **end for**
12: **Generate** $RLSC$

Command Execution. Command execution is the last step of our system, which using machine instruction to control the manipulator. The machine instructions are implemented one by one, until all the instructions are performed.

Next, we take the application of fruit sorting as an example to explain the whole process of rule-scene matching and interaction of our proposed algorithm. Assume that there are three red apples and two green apples on the platform and two baskets to the left and right of the robot are used to place the sorted fruit.

```
//Note: Manipulator is used to pick the object at point A (-0.080, 0.331,
1.350) and place it at point B (0.486, 0.372, 1.230)

    INIT          // initialize state of the manipulator
    BEGIN         // move the manipulator to the preparation state
    MOVEP P₁      // move to point P₁ (-0.080, 0.331, 1.450), which is 0.1 m
                     above point A in the direction of the z-axis
    HANDON        // turn on end-effector
    TIMER 1       // manipulator waits for 1 s in its current state
    MOVEL P₂      // (P₂) move to point A (P₂)
    HANDOFF       // turn off end-effector
    TIMER 1       // manipulator waits for 1 s in its current state
    MOVEL P₁      // return back to point P₁
    MOVEL P₃      // move to point P₃ (0.486, 0.372, 1.330), which is 0.1 m
                     above point B in the direction of the z-axis
    MOVEL P₄      //move to point B (P₄)
    HANDON        //turn on end-effector
    TIMER 1       // manipulator waits for 1 s in its current state
    MOVEL P₃      // move to point P₃
    BEGIN         // move the manipulator to the preparation state
```

Fig. 8. Robot language source code with which the robot picks an object at A and then places it at B

[**Robot**]: Hi, I am WUSTER, what can I do for you? (This sentence is pre-set in the system and is used to guide the user to give initial operation rules.)

[**User**]: Sort fruit. (The rule is unclear, because the user only provide an object (fruit) while the location where the objects are placed in is unclear. As shown in Fig. 6, only one in the two attributes of the intermediate node is matched, so the rule is incomplete. The robot reports the object situation in the scene to the user according to that real-time object information in the scene, so as to guide the user to give effective rules.)

[**Robot**]: Here are three red apples, two green apples, and two baskets to the left and right, please give the rules. (The robot informs the user of the object situation in the current scene to guide the user to present rules.)

[User]: Place all red apples in the left basket. (This rule is clear, because two attributes of the intermediate node are defined clearly and *map_match* = *1* is obtained by matching the scene and the expected objects, that is, this rule is effective and can be executed.)

[Robot]: Okay. (After response, the robot executes the command.)

[Robot]: Here are still two green apples, in which basket should they be placed? (After sorting red apples, the robot finds the rule of placing all red apples in the left basket does not match with the scene and *map_match* = *0* through matching the rule and the scene. The guidance mode is then started and the robot reports the object situations in the current scene and guides the user to define new rules.)

[User]: Put all green apples in the right basket. (This rule is clear, because there are clear definitions of two attributes of the intermediate node and *map_match* = *1* is obtained by matching the scene and the expected objects.)

[Robot]: Okay. (After response, the robot executes this task.)

[Robot]: Sorting is completed. (After sorting, the robot finds that there is no object to be sorted by matching the rule and the scene and then reports that the sorting process is completed.)

3   Experiment

In this section, an application scene for sorting fruit by using the Chinese speech control system is designed to verify effectiveness and practicability of the system. The main body of the manipulator used in the experiment is WUST-ARM modular manipulator [17] and each function module is run on an Ubuntu (Version 12.04) desktop computer with Intel $Core^{TM}$ i5 CPU 650 @ 3.20 GHz × 4 with 8 GB of RAM. Moreover, the Kinect is fixed in front of the WUST-ARM, as shown in Fig. 9. There are three species of fruit in the experimental platform.

Fig. 9. Experimental platform of WUST-ARM fruit-sorting system (Color figure online)

Three types of fruit to be sorted were placed on the production line, including red apples, green apples, and carambola: three baskets were put on the left, right, and back of the WUST-ARM. Users determined rules through interaction with the system in Chinese natural language. In this experiment, users, including five males and five females, were invited to interact individually with the robot 10 times, giving a total of 100 tests. Before testing, a large number of training and testing runs were conducted on the open source speech recognition system PocketSphinx and the speech synthesis system Ekho on a few, limited test-sets of natural language, which realized 100% accurate speech recognition and synthesis of test statements. Therefore, there is no error of speech recognition and synthesis in this test. The test data for fruit sorting are shown in Table 1. It can be seen from the table that the rule acquisition rate of the "scene-rule" matching algorithm is equivalent to the recognition rate of objects after eliminating errors of speech recognition. This proves the effectiveness and practicability of the automatic operating system.

Table 1. Experimental results of fruit sorting

User	Correct sorting rate	Correct recognition rate of objects
Male	94%	94%
Female	96%	96%

4 Conclusion

An automatic operating system for a robot, based on interaction with Chinese natural language, and integrating advanced technologies, such as: 3D visual perception, human-robot-environment interaction, and automatic programming was designed for use in the automatic operating system of an industrial manipulator. In the experiments, a fruit-sorting application was tested. The test results demonstrate that: (1) the "rule-scene" matching and interaction algorithm designed in this study is effective and correct, and can correctly guides users to provide correct rules. Moreover, the sorting accuracy of objects depends on correct recognition rate of objects. (2) The system enables the manipulator to sort objects automatically under the control of natural language.

For future works, our improvement efforts will focus on the research of the feature descriptor algorithm of 3D visual perception, and finally, to improve the accuracy of object recognition in our automatic operating system.

Acknowledgements. This work is supported by National Key R&D Program of China (Project No. 2017YFB1300400) and Natural Science Foundation of China (Project No. 61673304).

References

1. Liu, K., Li, S., Wang, B.: Research of the direct teaching system based on universal robot. Sci. Technol. Eng. **15**(28), 22–26 (2015)
2. Cai, Z.: Robotics, 2nd edn. Tsinghua University Press, Beijing (2009). (in Chinese)
3. Choi, Y., Chen, T., Jain, A., et al.: Hand it over or set it down: a user study of object delivery with an assistive mobile manipulator. In: 18th International Symposium on Robot and Human Interactive Communication (RO-MAN), pp. 736–743. IEEE Press, New York (2009)
4. Xing, J., Gan, Y., Dai, X.: Auto-programming system based on the workpiece model for industrial robot. Robot **39**(1), 111–118 (2017)
5. Connell, J., Marcheret, E., Pankanti, S., Kudoh, M., Nishiyama, R.: An extensible language interface for robot manipulation. In: Bach, J., Goertzel, B., Iklé, M. (eds.) AGI 2012. LNCS (LNAI), vol. 7716, pp. 21–30. Springer, Heidelberg (2012). https://doi.org/10.1007/978-3-642-35506-6_3
6. Fasola, J., Mataric, M.: Interpreting instruction sequences in spatial language discourse with pragmatics towards natural human-robot interaction. In: IEEE International Conference on Robotics and Automation (ICRA), pp. 2720–2727. IEEE Press, New York (2014)
7. Matuszek, C., Herbst, E., Zettlemoyer, L., Fox, D.: Learning to parse natural language commands to a robot control system. In: Desai, J., Dudek, G., Khatib, O., Kumar, V. (eds.) International Symposium on Experimental Robotics, vol. 8, pp. 403–415. Springer, Heidelberg (2013). https://doi.org/10.1007/978-3-319-00065-7_28
8. Li, X., Zhang, X., Dai, X.: A visual navigation method of mobile robot based on constrained natural language processing. Robot **33**(6), 724–749 (2011)
9. Li, X., Zhang, X.: A route instruction method using natural language processing for indoor intelligent robot navigation. Acta Automatica Sin. **40**(2), 289–305 (2014)
10. Alexandre, L.: 3D descriptors for object and category recognition: a comparative evaluation. In: Workshop on Color-Depth Camera Fusion in Robotics at the IEEE/RSJ International Conference on Intelligent Robots and Systems (IROS), pp. 1–6. IEEE Press, New York (2012)
11. Tombari, F., Salti, S., Di Stefano, L.: A combined texture-shape descriptor for enhanced 3D feature matching. In: 18th IEEE International Conference on Image Processing (ICIP), pp. 809–812. IEEE Press, New York (2011)
12. Rusu, R.: Semantic 3D object maps for everyday manipulation in human living environments. KI-Künstliche Intelligenz **24**(4), 345–348 (2010)
13. Lin, Y., Min, H., Zhou, H., Pei, F.: A human-robot-environment interactive reasoning mechanism for object sorting robot. In: IEEE Transactions on Cognitive and Developmental Systems, pp. 1–13 (2017)
14. Huggins-Daines, D., Kumar, M., Chan, A., et al.: Pocketsphinx: a free, real-time continuous speech recognition system for hand-held devices. In: IEEE International Conference on Acoustics, Speech and Signal Processing (ICASSP), pp. 185–188. IEEE Press, New York (2006)
15. Ekho Chinese Text-to-Speech Software Homepage. http://www.eguidedog.net/cn/ekho_cn.php. Accessed 16 June 2017
16. NLPIR/ICTCLAS Chinese word segmentation system. http://ictclas.nlpir.org/. Accessed 25 Mar 2017
17. Lin, Y., Min, H., Wei, H.: Inertial measurement unit-based iterative pose compensation algorithm for low-cost modular manipulator. Adv. Mech. Eng. **8**(1), 1–11 (2016)

An Adaptive Fuzzy Terminal Sliding Mode Control Methodology for Uncertain Nonlinear Second-Order Systems

Anh Tuan Vo[1], Hee-Jun Kang[2(✉)], and Tien Dung Le[3]

[1] Graduate School of Electrical Engineering, University of Ulsan,
Ulsan, South Korea
voanhtuan2204@gmail.com
[2] School of Electrical Engineering, University of Ulsan, Ulsan, South Korea
hjkang@ulsan.ac.kr
[3] The University of Danang - University of Science and Technology,
Danang, Vietnam
ltdung@dut.udn.vn

Abstract. This paper introduces a novel control strategy for uncertain non-linear second-order systems. Our strategy proposes a novel adaptive fuzzy sliding mode controller, which based on a combination of a new non-singular fast terminal sliding variable and a continuous control algorithm. In this paper, our main contribution is to contain benefits of non-singular fast terminal sliding variables such as fast convergence and no singularity drawback along with strong robustness in the frame of perturbation and uncertainty. In the suggested controller, a continuous control law is added to refuse the drawbacks of sliding mode control about consideration on upper bound value of perturbations and uncertainties. Unfortunately, it is remarkable difficult for a real system to identify those upper bound limits in advance. It is reminded that to deal with above concern, a fuzzy logic control algorithm with adaptive updating law can be used to approximate switching control law. Thanks to this technique, the perturbations and uncertainties can be eliminated without chattering behavior in control input. Accordingly, the strong robustness and the stability of the suggested system is then secured with high accuracy performance. The robustness topic of the suggested system is also completely proven by Lyapunov approach. In our numerical simulation, performances comparison among the suggested control strategy, a sliding mode controller, and a non-singular terminal sliding mode controller are specifically performed. Our simulation result demonstrates the effectiveness, practicality of suggested control strategy for the joint position tracking control of a 3-DOF PUMA560 robot.

Keywords: Uncertain nonlinear second-order systems
Non-singular fast terminal sliding mode · Fuzzy logic system
Adaptive control law

© Springer International Publishing AG, part of Springer Nature 2018
D.-S. Huang et al. (Eds.): ICIC 2018, LNCS 10954, pp. 123–135, 2018.
https://doi.org/10.1007/978-3-319-95930-6_13

1 Introduction

In recent decades, the rapid and vigorous development of science and technology has facilitated the improvement of product quality. A noticeable fact that the more developed product quality, the more complex system, and manufacturing. Accordingly, the exact identification of unknown components in the controlled system has the difficult challenges such as structural variation of the dynamic system, perturbations, and system uncertainty. Those challenges attract researchers to constantly study advanced control approaches to improve precision, reliability and product quality in the controlled system. To control uncertain nonlinear systems, several robust control methods have been successfully applied. It should be mentioned PID controller [1], neural network controller [2, 3], fuzzy logic controller [4], adaptive controller [5] sliding mode controller (SMC) [6–8]. Among those above controllers, it is well known that SMC has the merit characteristics to match perturbations and system uncertainties. However, the conventional SMC still has several limitations (e.g. a lager chattering behavior still exist in control input, inefficiently adapt to the rapid changes of perturbations or faults and cannot obtain a finite-time convergence).

To deal with those problems, several advanced methods have been proposed and applied a nonlinear sliding variable instead of a linear sliding variable in traditional SMC. Those control methodologies are well-known as the terminal sliding mode control (TSMC) [9–11].

In generally technical view, the TSMC provide a finite-time convergence but still exist the singularity problem along with slower convergence time than SMC when the state variables are far from the original point. To solve singularity problem, non-singular terminal sliding mode control (NTSMC) has been suggested and applied into nonlinear magnetic bearing system and robotic manipulators [12, 13]. Another problem is fast convergence which can be addressed by the fast terminal sliding mode (FTSMC). It has been proposed for controlling uncertain nonlinear second-order systems [14, 15]. Unfortunately, both NTSMC and FTSMC only deal with their own problem at all. Thus, to continuously solve both singularity and fast convergence, the non-singular fast terminal sliding mode control (NFTSMC) has been proposed [16, 17].

On the other hand, chattering phenomena is an unexpected matter in the real systems including all of above schemes (e.g. TSMC, FTSMC, NTSMC, NFTSMC) with a high-frequency switching control law. This chattering behavior can cause faults and damage of controlled system. Hence, several effective methodologies have been suggested to treat this matter, such as disturbance observer [18], boundary layer algorithm [19], fuzzy-SMC [20–22], high-order sliding mode [23, 24]. Here, those mentioned methods limitations are facing a trade-off between chattering elimination and the tracking positional accuracy and requiring an excessive magnitude of initial control input, sometimes. Among techniques to eliminate chattering behavior, fuzzy-SMC has strong advantages for designing a controller in which chattering behavior is eliminated without reducing the accuracy of the controlled system.

Therefore, the motivation of our paper is to synchronously solve drawbacks of both SMC and TSMC. The main objective of our paper is to propose a control methodology for uncertain nonlinear second-order systems as follows:

- Eliminates singularity drawback, provides fast convergence, low state error along with robustness against perturbation and system uncertain.
- Provides continuous control inputs with smooth and free chattering behavior.
- Eliminates essential knowledge of the upper limits for both perturbations and uncertainties.
- The stability and robustness topic of the suggested system has been completely proven by the Lyapunov approach.

The remainder of our study is organized as follows. The problem statements required for the proposed NFTSM variable and control are presented in Sect. 2. Section 3 presents the design procedure of the suggested control methodology. In Sect. 4, the suggested control methodology is applied to the joint position tracking control simulation for a 3-DOF PUMA560 robot. And its trajectory tracking performance is compared with those of the SMC [6] and the TSMC [11]. Finally, some concluding remarks are depicted in Sect. 5.

2 Problem Statement

The following general nonlinear second-order system is considered

$$
\begin{cases}
\dot{X}_1 = X_2 \\
\dot{X}_2 = H(X,t) + B(X,t)U + \Delta(X,t)
\end{cases}
\tag{1}
$$

where $X = [X_1,\ X_2]^T \in \mathbb{R}^n$ denotes the system state vector, $H(X,t) = H_n(X,t) + \delta H(X,t)$, $B(X,t) = B_n(X,t) + \delta B(X,t)$ in which $H(X,t) \in \mathbb{R}^n$, $B(X,t) \in \mathbb{R}^{n \times n}$ are the smooth nonlinear vector fields with $H(0) = 0$, $\delta H(X,t)$ represents structural variation of the dynamic system which is an uncertain term, $\delta B(X)$ is the input signal uncertainty. $\Delta(X,t)$ represent the perturbations and uncertainties, $U(t)$ are the actuation control inputs.

The following assumptions are assumed to the design procedure of the suggested control algorithm in the sequel.

Assumption 1. The inertia matrix $B(X,t)$ is invertible, positive definite and symmetric matrix for $\forall(X,t)$ bounding the following condition

$$
B_1 \leq B(X,t) \leq B_2
\tag{2}
$$

where B_1 and B_2 denote as positive coefficients.

Assumption 2. All perturbations and other uncertainties is a limited function satisfying the following constraint

$$
|L(X,t)| \leq \Upsilon
\tag{3}
$$

where Υ is unknown positive coefficient, $L(X,t)$ is termed as the lumped uncertainty and determined as $L(X,t) = \delta H(X,t) + \delta B(X,t) + \Delta(X,t)$.

Assumption 3. The reference trajectory vectors $X_d = \begin{bmatrix} X_d & \dot{X}_d \end{bmatrix}^T \in \mathbb{R}^n$ is a twice continuously differentiable function in terms of t.

The control goal of our article is that controlled variables X will reach the reference trajectory X_d under newly developed control scheme. Such the case that essential information of perturbations and uncertainties' upper limits are refused.

3   Design Procedure of Control Methodology

In this section, a new control strategy is developed for the system (1) and presented in the following main parts.

3.1   A Design Nonsingular Fast Terminal Sliding Variables

Let $\varepsilon = X - X_d$ as the tracking positional error, X_d indicated as reference trajectory values. So, the new NFTSM variables are designed as

$$s = \dot{\varepsilon} + \int_0^t \left(\mu_1 \varepsilon^{[\phi_1]} + \mu_2 \dot{\varepsilon}^{[\phi_2]} + \mu_3 \varepsilon + \mu_4 \, \text{sgn}(\dot{\varepsilon}) \right) d\varphi \tag{4}$$

where $s = \begin{bmatrix} s_1 & s_2 & \cdots & s_n \end{bmatrix}^T \in \mathbb{R}^{n \times 1}$ are the sliding variables, μ_1, μ_2, μ_3 and μ_4 are positive coefficients, ϕ_1, ϕ_2 satisfying the relation $0 < \phi_1 < 1$, $\phi_2 = \frac{2\phi_1}{1+\phi_1}$, and $\varepsilon^{[\phi]}$ are defined as (see [26])

$$\varepsilon^{[\phi]} = |\varepsilon|^\phi \, \text{sgn}[\varepsilon] \tag{5}$$

According to SMC theory, once the tracking positional error runs in sliding mode, the following constraint are approved [6]:

$$s = 0 \text{ and } \dot{s} = 0 \tag{6}$$

With Eqs. (4) and (6), it can be achieved that

$$\ddot{\varepsilon} + \mu_1 \varepsilon^{[\phi_1]} + \mu_2 \dot{\varepsilon}^{[\phi_2]} + \mu_3 \varepsilon + \mu_4 \, \text{sgn}(\dot{\varepsilon}) = 0 \tag{7}$$

Finally, the following sliding mode dynamics can be achieved as

$$\ddot{\varepsilon} = -\mu_1 \varepsilon^{[\phi_1]} - \mu_2 \dot{\varepsilon}^{[\phi_2]} - \mu_3 \varepsilon - \mu_4 \, \text{sgn}(\dot{\varepsilon}) \tag{8}$$

To prove the original point $\varepsilon = 0$ is globally asymptotically stable and to fulfill the proof of convergence, the following theorem establishes for this proof.

Theorem 3.1. Consider the dynamic system of Eq. (8). The original points $\varepsilon_i = 0 (i = 1, \cdots, n)$ are globally balanced points and the state variables of the system (8) converge to zero.

Proof. The following positive-definite Lyapunov functional is selected as

$$V_1 = \frac{\mu_1}{\phi_1 + 1} |\varepsilon|^{\varphi_1 + 1} + \frac{1}{2} \dot{\varepsilon}^2 + \frac{\mu_3}{2} \varepsilon^2 \qquad (9)$$

With Eq. (8) the time derivative of Eq. (9) is derived as

$$
\begin{aligned}
\dot{V}_1 &= \mu_1 |\varepsilon|^{\phi_1} \dot{\varepsilon} + \dot{\varepsilon}\ddot{\varepsilon} + \mu_3 \varepsilon \dot{\varepsilon} \\
&= \mu_1 |\varepsilon|^{\phi_1} \dot{\varepsilon} + \dot{\varepsilon}\left(-\mu_1 \varepsilon^{[\phi_1]} - \mu_2 \dot{\varepsilon}^{[\phi_2]} - \mu_3 \varepsilon - \mu_4 \mathrm{sgn}(\dot{\varepsilon})\right) + \mu_3 \varepsilon \dot{\varepsilon} \\
&= \mu_1 |\varepsilon|^{\phi_1} \dot{\varepsilon} - \mu_1 |\varepsilon|^{\varphi_1} \dot{\varepsilon} - \mu_2 \dot{\varepsilon}^{[\phi_2 + 1]} - \mu_3 \varepsilon \dot{\varepsilon} - \mu_4 |\dot{\varepsilon}| + \mu_3 \varepsilon \dot{\varepsilon} \\
&= -\mu_2 |\dot{\varepsilon}|^{[\phi_2 + 1]} - \mu_4 |\dot{\varepsilon}| \leq 0
\end{aligned}
\qquad (10)
$$

So, based on the LaSalle's invariant criterion [27], it can be proved that the original points $\varepsilon_i = 0 (i = 1, \cdots, n)$ are globally balanced points. The proof of Theorem 3.1 is done.

The suitable sliding variables of Eq. (4) has been designed. Next, to achieve the desired performance, the following control strategy is suggested for the system of Eq. (1).

3.2   Design Nonsingular Fast Terminal Sliding Mode Control

The NFTSMC algorithm is developed under the sliding mode approach as

$$U = U_{eq} + U_{sw} \qquad (11)$$

The equivalent control law U_{eq} holds the trajectory of the error state variables on the sliding variables. To obtain this result, it is necessary to get a derivative of sliding variables respect with time and to be $\dot{s} = 0$ for the nominal system without the existence of the perturbations and uncertainties.

The general nonlinear second-order system (1) is expressed in errors' state space as

$$\ddot{\varepsilon} = H_n(X, t) + B_n(X, t)U + L(X, t) - \ddot{X}_d(t) \qquad (12)$$

With Eq. (12), the time derivative of the sliding variables of Eq. (4) can be represented as

$$\dot{s} = H_n(X, t) + B_n(X, t)U + L(X, t) - \ddot{X}_d(t) + \mu_1 \varepsilon^{[\phi_1]} + \mu_2 \dot{\varepsilon}^{[\phi_2]} + \mu_3 \varepsilon + \mu_4 \mathrm{sgn}(\dot{\varepsilon}) \qquad (13)$$

The equivalent control law can be selected to satisfy the constraint condition $\dot{s} = 0$ as

$$U_{eq} = -B_n^{-1}(X, t)\left(H_n(X, t) - \ddot{X}_d(t) + \mu_1 \varepsilon^{[\phi_1]} + \mu_2 \dot{\varepsilon}^{[\phi_2]} + \mu_3 \varepsilon + \mu_4 \mathrm{sgn}(\dot{\varepsilon})\right) \qquad (14)$$

and the switching control law is constructed to deal with the lumped uncertainty as

$$U_{sw} = -B_n^{-1}(X,t)(\Gamma_1 s + \Gamma_2 \text{sgn}(s) + \Upsilon) \tag{15}$$

Then, the overall control law U obtains

$$U = U_{eq} + U_{sw} = -B_n^{-1}(X,t)\left(\begin{array}{c} H_n(X,t) - \ddot{X}_d(t) + \mu_1\varepsilon^{[\phi_1]} + \mu_2\dot{\varepsilon}^{[\phi_2]} \\ + \mu_3\varepsilon + \mu_4\text{sgn}(\dot{\varepsilon}) + \Gamma_1 s + \Gamma_2\text{sgn}(s) + \Upsilon \end{array}\right) \tag{16}$$

Theorem 3.2. For the dynamic system of Eq. (1) with the lumped system uncertainty satisfies the condition $|L_{max}(X,t)| < \Upsilon$ in which Υ is a positive coefficient, $L_{max}(X,t)$ is upper-bound of the lumped system uncertainty, if the suitable NFTSM variables have been chosen as Eq. (4), the actuation control input is designed as Eq. (16). Then, the sliding manifold motion guarantees to happen. It means that the error state variables reach sliding variables.

Proof. The following positive-definite Lyapunov functional is selected as

$$V_2 = \frac{1}{2}s^T s \tag{17}$$

Differentiating Eq. (17) with respect to time and combining with Eq. (13), we have

$$\dot{V}_2 = s^T\left(\begin{array}{c} H_n(X,t) + B_n(X,t)U + L(X,t) - \ddot{X}_d(t) \\ + \mu_1\varepsilon^{[\phi_1]} + \mu_2\dot{\varepsilon}^{[\phi_2]} + \mu_3\varepsilon + \mu_4\text{sgn}(\dot{\varepsilon}) \end{array}\right) \tag{18}$$

Substituting Eq. (16) into Eq. (18) results in

$$\begin{aligned} \dot{V}_2 &= s^T(-\Gamma_1 s - \Gamma_2\text{sgn}(s) - \Upsilon + L(X,t)) \\ &\le s^T(-\Gamma_1 s - \Gamma_2\text{sgn}(s) + L_{max}(X,t) - \Upsilon) \\ &\le -\Gamma_1 s^T s - \Gamma_2|s| \le 0 \end{aligned} \tag{19}$$

Accordingly, based on Lyapunov principle [27], it can be proven that the stability of the tracking error is secured under control laws of Eq. (16) despite whether there exist the perturbation and uncertainties. This completes the proof of Theorem 3.2.

Remark 3.1. For solving large perturbations and uncertainties, the switching control gains Γ in which $\Gamma = \max\{\Gamma_1, \Gamma_2\}$ in the control input of Eq. (16) must be large. That seriously lead to chattering behavior in the control input signal.

In the sequel, to solve chattering behavior matter, fuzzy logic control law $\hat{U}_{fls}$ with adaptive updating law is used to estimate switching control law $\Gamma_2\text{sgn}(s)$ in control input of Eq. (16).

3.3   Design Adaptive Fuzzy Nonsingular Fast Terminal Sliding Mode Control

The proposed controller (AFNFTSMC) is developed as same as the procedure of NFTSMC. However, the suggested system will use fuzzy logic control law $\hat{U}_{fls}$ with adaptive updating law is used to estimate switching control law $\Gamma_2 \mathrm{sgn}(s)$ instead of using a fixed gain Γ_2, adaptive updating law acts as an online tuning of the output weights in fuzzy logic system. Hence, the sliding motion assures to take place without the essential knowledge of perturbations and uncertainties' upper limits. The approximation of fuzzy logic system is given as follow

$$\hat{U}_{fls} = W^T h(s) \tag{20}$$

where W is termed as the parameter vector, $h(s)$ is the vector of fuzzy basic function. The column vector of the optimal parameter W^* is defined as follows

$$W^* = \arg \min_{W \in \Psi} \left[\sup_{s \in \mathbb{R}} \left| \hat{U}_{fls}(s|W) - U_{sw} \right| \right] \tag{21}$$

where Ψ is condition sets for W. Once W obtains the optimal parameter W^*, $\hat{U}_{fls}$ turns into

$$\hat{U}_{fls}(s|W^*) = W^* h(s) = \Gamma_2 \mathrm{sgn}(s) \tag{22}$$

Therefore, the overall actuation control input of Eq. (16) becomes

$$U = -B_n^{-1}(X,t) \left(\begin{array}{c} H_n(X,t) - \ddot{X}_d(t) + \mu_1 \varepsilon^{[\phi_1]} + \mu_2 \dot{\varepsilon}^{[\phi_2]} \\ + \mu_3 \varepsilon + \mu_4 \mathrm{sgn}(\dot{\varepsilon}) + \Gamma_1 s + \hat{U}_{fls} + \Upsilon \end{array} \right) \tag{23}$$

with adaptive updating law of fuzzy logic system $\hat{\tau}_{fls}$ is designed as follow

$$\dot{W} = I^* s h(s) \tag{24}$$

in which I^* is a positive coefficient.

Theorem 3.3. For the dynamic system of Eq. (1) with the lumped system uncertainty satisfies the condition $|L_{\max}(X,t)| < \Upsilon$ in which Υ is a positive coefficient, $L_{\max}(X,t)$ is upper-bound of the lumped system uncertainty, if the suitable NFTSM variables have been chosen as Eq. (4), the actuation control input is designed as Eq. (23) with adaptive updating law of fuzzy logic control law are designed as Eq. (24). Then, the sliding motion guarantees to happen. It means that the error state variables reach sliding variables despite whether there exist the perturbation and uncertainties.

Proof. The following positive-definite Lyapunov functional is selected as

$$V_3 = \frac{1}{2}s^T s + \frac{1}{2}\tilde{W}^T (I^*)^{-1}\tilde{W} \tag{25}$$

where $\tilde{W} = W - W^*$.

Differentiating Eq. (25) with respect to time results

$$\dot{V}_3 = s^T \dot{s} + \tilde{W}^T (I^*)^{-1}\dot{\tilde{W}} \tag{26}$$

Substituting Eqs. (13), (23) and (24) into Eq. (26) gives

$$\begin{aligned}
\dot{V}_2 &= s^T\left(-\Gamma_1 s - \Gamma_2 \mathrm{sgn}(s) - \Upsilon + L(X,t)\right) + s\tilde{W}^T h(s) \\
&= s^T\left(-\Gamma_1 s - W^T h(s) - \Upsilon + L(X,t)\right) + s\tilde{W}^T h(s) \\
&\le s^T\left(-\Gamma_1 s + W^{*T} h(s) - W^T h(s) - W^{*T} h(s)\right) + s\tilde{W}^T h(s) \\
&= s^T\left(-\Gamma_1 s - W^{*T} h(s)\right)
\end{aligned} \tag{27}$$

From result in Eq. (22) $W^{*T} h(s) = \Gamma_2 \mathrm{sgn}(s)$, hence,

$$\begin{aligned}
\dot{V}_2 &\le s^T\left(-\Gamma_1 s - \Gamma_2 \mathrm{sgn}(s)\right) \\
&\le -\Gamma_1 s^T s - \Gamma_2 |s| < 0
\end{aligned} \tag{28}$$

Accordingly, based on Lyapunov principle [27], it can be verified that the stability of the tracking error is guaranteed under control law of Eq. (23) despite whether there exist the perturbation and uncertainties. This completes the proof of Theorem 3.3.

4   Numerical Simulation Results

In this section, the suggested control strategy applies for the joint position tracking control of the robotic manipulators.

For an n-link rigid robotic manipulator, the corresponding dynamic equation can be given as (see [11])

$$M(\theta)\ddot{\theta} + C_m\left(\theta, \dot{\theta}\right)\dot{\theta} + G(\theta) + F_r\left(\dot{\theta}\right) + \tau_D = \tau \tag{29}$$

in which $\theta, \dot{\theta}, \ddot{\theta} \in \mathbb{R}^n$ are defined the system's state vector. $M(\theta) \in \mathbb{R}^{n \times n}$ is the inertia matrix, $C_m\left(\theta, \dot{\theta}\right) \in \mathbb{R}^{n \times 1}$ is defined the matrix resulting from Coriolis and centrifugal force, $G(\theta) \in \mathbb{R}^{n \times 1}$ is defined gravitational force term, $F_r\left(\dot{\theta}\right) \in \mathbb{R}^{n \times 1}$ is defined the friction matrix, $\tau \in \mathbb{R}^{n \times 1}$ is defined the torque produced by actuators, $\tau_D \in \mathbb{R}^{n \times 1}$ is defined a load disturbance matrix.

Equation (29) can be rewritten as

$$\ddot{\theta} = M^{-1}(\theta)\left[\tau - C_m\left(\theta,\dot{\theta}\right)\dot{\theta} - F_r\left(\dot{\theta}\right) - G(\theta) - \tau_D\right] \tag{30}$$

To simplify the analysis and design in next section, Eq. (30) can be given as

$$\ddot{\theta} = H\left(\theta,\dot{\theta}\right) + \Delta\left(\theta,\dot{\theta},t\right) + B(\theta)\tau \tag{31}$$

where $H\left(\theta,\dot{\theta}\right) = M^{-1}(\theta)\left[-C_m\left(\theta,\dot{\theta}\right)\dot{\theta} - G(\theta)\right]$, $B(\theta) = M^{-1}(\theta)$ and $\Delta\left(\theta,\dot{\theta},t\right) = M^{-1}(\theta)\left[-F_r\left(\dot{\theta}\right) - \tau_D\right]$.

Next, we denote $U = \tau$ as the control input, $X = [X_1, \ X_2]^T$ as the state vector in which X_1, X_2 are corresponding to $\theta, \dot{\theta} \in \mathbb{R}^{n \times 1}$. The robotic dynamic of Eq. (29) can be presented in the following state space form as

$$\begin{cases} \dot{X}_1 = X_2 \\ \dot{X}_2 = H(X,t) + \Delta(X,t) + B(X,t)U \end{cases} \tag{32}$$

in which $H(X,t) \in \mathbb{R}^n$, $B(X,t) \in \mathbb{R}^{n \times n}$ are the smooth nonlinear vector fields and $\Delta(x,t) \in \mathbb{R}^n$ presents the perturbations and uncertainties.

The proposed controller is directly applied to the robot system in Eq. (29) because of the dynamic model in Eq. (32) is exactly the formula of the general nonlinear second-order system in Eq. (1).

In this article, we perform a simulation example of the proposed control scheme, which is applied for trajectory tracking control simulation for first three joints of PUMA560 manipulator to exhibit its viability and effectiveness. The dynamic model with essential parameters of the manipulator was presented in [25]. All numerical simulation studies are performed in the Matlab/Simulink software with a fixed-step size 10^{-3} s.

3-DOF PUMA560 robot manipulator is investigated with the first three joints and the last three joints locked.

The Perturbations and uncertainties are assumed in these simulations to be as follows. The friction $F_r\left(\dot{\theta}\right)$ and perturbation τ_D are assumed to be

$$F_r\left(\dot{\theta}\right) + \tau_D = \begin{bmatrix} 2.1\dot{\theta}_1 + 2.02sign\left(3\dot{\theta}_1\right) + 7.2sin\left(\dot{\theta}_1\right) \\ 4.2\dot{\theta}_2 + 2.2sign\left(2\dot{\theta}_2\right) + 6.1sin\left(\dot{\theta}_2\right) \\ 1.1\dot{\theta}_3 + 1.15sign\left(2\dot{\theta}_3\right) + 4.15sin\left(\dot{\theta}_3\right) \end{bmatrix} \tag{33}$$

The reference joint trajectories for the position tracking are

$$\theta_d = [0.4 + \cos(t/5\pi) - 1, \quad -0.6 + \sin(t/5\pi + \pi/2), \quad 0.4 + \sin(t/5\pi + \pi/2) - 1]^T \tag{34}$$

The parameters for the NFTSM variables of Eq. (4) and the controlling input of Eq. (23) with the adaptive updating rule of fuzzy logic system in Eq. (24) are experimentally chosen as $\mu_1 = diag(15, 15, 15)$, $\mu_2 = diag(12.5, 12.4, 12.5)$, $\mu_3 = diag(64, 69, 64)$, $\mu_4 = diag(10, 10, 10)$, $\phi_1 = 1/2$, $\phi_2 = 2/3$. And $\Gamma_1 = diag(12, 12, 12)$, $I^* = diag(25, 25, 25)$, $\Upsilon = [0.1, \quad 0.1, \quad 0.1]^T$.

The trajectory tracking performances of the proposed control method are compared with the tracking performances of those SMC [6] and NTSMC [11] to present its effectiveness and feasibility. These control methods for the comparison are briefly explained as follows.

The SMC has the control input as

$$U = -B^{-1}(\theta)\left(H\left(\theta, \dot{\theta}\right) + \alpha\left(\dot{\theta} - \dot{\theta}_d\right) - \ddot{\theta}_d + g\mathrm{sgn}(s)\right) \tag{35}$$

in which the sliding variable is defined as follows $s = \dot{\varepsilon} + \alpha\varepsilon$.

The NTSM controller has the control input as

$$U = -B^{-1}(\theta)\left(H\left(\theta, \dot{\theta}\right) - \ddot{\theta}_d + \beta h/l(\dot{\varepsilon})^{2-\frac{l}{h}} + \kappa s + \left(l_g + \xi\right)\mathrm{sgn}(s)\right) \tag{36}$$

in which the sliding variable is defined as follow $s = \varepsilon + \beta^{-1}(\dot{\varepsilon})^{\frac{l}{h}}$.

The parameters of the controller in Eq. (35) were chosen as $\alpha = 2$ and $g = diag(10.4, 7.2, 9.5)$ to obtain the good performances.

The parameters of the controller in Eq. (36) were chosen as $l = 5$, $h = 3$, $\beta = 2$, $\kappa = diag(66.3, 117.6, 5.42)$, $l_g = 8$ and $\xi = 2.0$ to obtain the good performances.

The simulations were performed to compare the controllers in terms of their positional precisions, fast convergence and the existed chattering behaviors in their control inputs.

The tracking positional trajectories and tracking positional errors of the first three joints with three control methods are exhibited in Figs. 1 and 2, respectively. It can be seen trajectory tracking performances of three control method look like the similar good. However, fairly speaking, the tracking positional errors of the proposed control method are relatively smaller than those of the other control methods by the order of $10^{-6} \sim 10^{-7}$ rad and the convergence time of those positional errors is fast. The tracking positional errors of the other control methods are by the order of $10^{-4} \sim 10^{-5}$ rad with slower convergence time. Moreover, a special attraction is the comparison of the chattering behavior that exists in the control input and it is shown in Fig. 3. The chattering phenomena from the proposed NFTSM variable and control law were exhibited a lot less than those of the other control methods. The proposed control input system is efficiently smooth and chattering-free behavior.

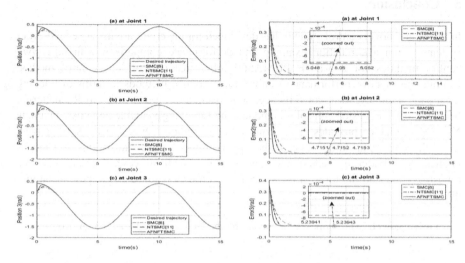

Fig. 1. Tracking positions: (a) at Joint 1, (b) at Joint 2, (c) at Joint 3

Fig. 2. Tracking errors: (a) at Joint 1, (b) at Joint 2, (c) at Joint 3.

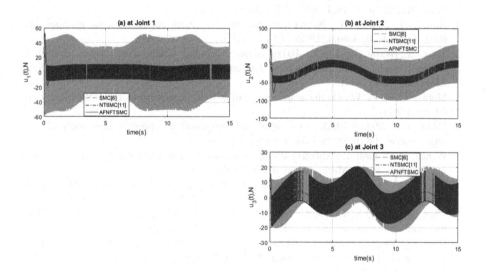

Fig. 3. Control input signals: (a) at Joint 1, (b) at Joint 2, (c) at Joint 3.

5 Conclusion

In this study, novel control strategy has developed for uncertain nonlinear second-order systems. From a simulation example and performance comparison with other two control methods for a 3-DOF PUMA560 robot manipulator, our control method exhibits the superior performances among three control methods in terms of tracking positional accuracy, fast convergence, low oscillation error, and small chattering behavior. The suggested control scheme has the following valuable benefits: (1) a new NFTSM variable provides fast convergence without singularity disadvantage, (2) without essential knowledge of the upper limits of the perturbations and uncertainties, (3) chattering behavior is impressively small in control input, (4) The robustness and stability topic of the system is rigorously secured by Lyapunov stability theory.

Acknowledgement. This work was supported by the University of Ulsan, Ulsan, Korea.

References

1. Su, Y., Muller, P.C., Zheng, C.: Global asymptotic saturated PID control for robot manipulators. IEEE Trans. Control Syst. Technol. **18**(6), 1280–1288 (2010)
2. Ge, S.S., Wang, C.: Adaptive neural control of uncertain MIMO nonlinear systems. IEEE Trans. Neural Netw. **15**(3), 674–692 (2004)
3. Kim, Y.H., Lewis, F.L.: Neural network output feedback control of robot manipulators. IEEE Trans. Rob. Autom. **15**(2), 301–309 (1999)
4. Tong, S., Wang, T., Li, Y.: Fuzzy adaptive actuator failure compensation control of uncertain stochastic nonlinear systems with unmodeled dynamics. IEEE Trans. Fuzzy Syst. **22**(3), 563–574 (2014)
5. Slotine, J.J.E., Li, W.: On the adaptive control of robot manipulators. Int. J. Rob. Res. **6**(3), 49–59 (1987)
6. Utkin, V.I.: Sliding Modes in Control and Optimization. Springer, Heidelberg (2013). https://doi.org/10.1007/978-3-642-84379-2
7. Zeinali, M., Notash, L.: Adaptive sliding mode control with uncertainty estimator for robot manipulators. Mech. Mach. Theory **45**(1), 80–90 (2010)
8. Sun, T., Pei, H., Pan, Y., Zhou, H., Zhang, C.: Neural network-based sliding mode adaptive control for robot manipulators. Neurocomputing **74**(14–15), 2377–2384 (2011)
9. Chen, M., Wu, Q.X., Cui, R.X.: Terminal sliding mode tracking control for a class of SISO uncertain nonlinear systems. ISA Trans. **52**(2), 198–206 (2013)
10. Chen, G., Song, Y., Guan, Y.: Terminal Sliding Mode-Based Consensus Tracking Control for Networked Uncertain Mechanical Systems on Digraphs. IEEE Trans. Neural Netw. Learn. Syst. (2016)
11. Yu, S., Yu, X., Shirinzadeh, B., Man, Z.: Continuous finite-time control for robotic manipulators with terminal sliding mode. Automatica **41**(11), 1957–1964 (2005)
12. Chen, S.Y., Lin, F.J.: Robust nonsingular terminal sliding-mode control for nonlinear magnetic bearing system. IEEE Trans. Control Syst. Technol. **19**(3), 636–643 (2011)
13. Feng, Y., Yu, X., Man, Z.: Non-singular terminal sliding mode control of rigid manipulators. Automatica **38**(12), 2159–2167 (2002)

14. Mobayen, S.: Fast terminal sliding mode controller design for nonlinear second-order systems with time-varying uncertainties. Complexity **21**(2), 239–244 (2015)
15. Yu, X., Zhihong, M.: Fast terminal sliding-mode control design for nonlinear dynamical systems. IEEE Trans. Circ. Syst. I Fundam. Theory Appl. **49**(2), 261–264 (2002)
16. Yang, L., Yang, J.: Nonsingular fast terminal sliding-mode control for nonlinear dynamical systems. Int. J. Robust Nonlinear Control **21**(16), 1865–1879 (2011)
17. He, Z., Liu, C., Zhan, Y., Li, H., Huang, X., Zhang, Z.: Nonsingular fast terminal sliding mode control with extended state observer and tracking differentiator for uncertain nonlinear systems. Math. Probl. Eng. **2014** (2014)
18. Zhang, J., Liu, X., Xia, Y., Zuo, Z., Wang, Y.: Disturbance observer-based integral sliding-mode control for systems with mismatched disturbances. IEEE Trans. Ind. Electron. **63**(11), 7040–7048 (2016)
19. Utkin, V.: Discussion aspects of high-order sliding mode control. IEEE Trans. Autom. Control **61**(3), 829–833 (2016)
20. Roopaei, M., Jahromi, M.Z.: Chattering-free fuzzy sliding mode control in MIMO uncertain systems. Nonlinear Anal. Theory Methods Appl. **71**(10), 4430–4437 (2009)
21. Nguyen, S.D., Vo, H.D., Seo, T.I.: Nonlinear adaptive control based on fuzzy sliding mode technique and fuzzy-based compensator. ISA Trans. **70**, 309–321 (2017)
22. Li, H., Wang, J., Wu, L., Lam, H. K., Gao, Y.: Optimal guaranteed cost sliding mode control of interval type-2 fuzzy time-delay systems. IEEE Trans. Fuzzy Syst. (2017)
23. Davila, J., Fridman, L., Levant, A.: Second-order sliding-mode observer for mechanical systems. IEEE Trans. Autom. Control **50**(11), 1785–1789 (2005)
24. Rubio-Astorga, G., Sánchez-Torres, J.D., Cañedo, J., Loukianov, A.G.: High-order sliding mode block control of single-phase induction motor. IEEE Trans. Control Syst. Technol. **22** (5), 1828–1836 (2014)
25. Armstrong, B., Khatib, O., Burdick, J.: The explicit dynamic model and inertial parameters of the PUMA 560 arm. In: Proceedings of the 1986 IEEE International Conference on Robotics and Automation, vol. 3, pp. 510–518. IEEE, April 1986
26. Polyakov, A., Fridman, L.: Stability notions and Lyapunov functions for sliding mode control systems. J. Franklin Inst. **351**(4), 1831–1865 (2014)
27. Slotine, J.J.E., Li, W.: Applied Nonlinear Control. Prentice, Englewood Cliffs (1991)

Robot Chain Based Self-organizing Search Method of Swarm Robotics

Yandong Luo[1], Jianwen Guo[1(✉)], Zhibin Zeng[2], Chengzhi Chen[1], Xiaoyan Li[1], and Jiapeng Wu[1]

[1] College of Mechanical Engineering, Dongguan University of Technology, Dongguan 523808, China
given_gjw@163.com
[2] Dongguan Hengli Mould Technology Development Limited Company, Dongguan 523460, China

Abstract. In this paper, we propose a self-organizing search method for swarm robotics based on the concept of robot chains. Through the local communication and distance measurement between robots, this method can establish robot chain in a self-organizing ay and connect starting position with search target to complete the search task. In particular, we consider consists in forming a path between two objects in decentralized communication and non-positioning condition. Finally, we demonstrate the feasibility and efficiency of our method using the physical experiments and the simulation experiments.

Keywords: Swarm robotics · Kilobots · Robot chains

1 Introduction

In recent years, swarm robotics have been widely concerned by theoretical research and practical applications. Swarm robotics originates from the nature inspired [1, 2] swarm intelligence method and the combination of multiple robots. The main objectives of swarm robotics research is to using simple rules and local interactions among individual robots resulting in desired collective swarm behavior. As for the system, it is no central coordination mechanism under this swarm behavior control. However, such swarm behavior has the characteristics of robustness, flexibility, scalability and so on [3]. Basing on these characteristics of collective behavior, swarm robots have become an important researching direction in many areas, such as exploration, surveillance, search and rescue.

In the research of swarm robotics search, there are many swarm robotics experiments mostly using centralized communication technology and absolute positioning. Although this communication technology and positioning mechanism have brought a lot of convenience to the research of group robot, there are still shortcomings. For example, even when simple messages are exchanged among robots, the communication bandwidth increases exponentially with the swarm scale if a centralized communication technique is used [4]. Thus bigger swarms require decentralized communication principles, like nearest-neighbor communication. With this kind of communication, the swarm robotics to finish search task is not as straightforward as with centralized

D.-S. Huang et al. (Eds.): ICIC 2018, LNCS 10954, pp. 136–146, 2018.
https://doi.org/10.1007/978-3-319-95930-6_14

communication. For example, if the distance between swarm robotics exceeds communication scope, this will affect the searching information which can't be updated in time and decrease search the searching efficiency.

In swarm robotics, inspiration is often taken from the foraging behavior of social insects [5, 6]. For example, ants of many species are foraging by laying pheromones in the environment, a chemical substance that attracts other ants. Inspired by the pheromone induction in the ant foraging behavior, Goss and Deneubourg [7] present the concept of robot chains. The concept of robot chains relies on the idea of locally manipulating the environment in order to attract other individuals and to form a global path. However, due to their lack of a substance such as pheromone, the robots constituting a chain need to serve as trail markers themselves. Meanwhile, in order to ensure the connectivity of the robot chains, the neighboring robots within a chain need sense each other by means of physical contact or communication.

Robot chain based searching method of swarm robotics has been put forward in this paper. Based on guarantee of communication network connectivity of swarm robotics, this method could establish the robot chains in a self-organizing way and connect starting position with search target to complete the searching task. Simulation and physical experiment certified the feasibility of this algorithm.

The remainder of this paper is organized as follows. In Sect. 2 we give a description of the considered problem and a short outline of our approach. In Sect. 3 we give a description of the control algorithm we used. In Sect. 4 we present the simulation and physical experimental results. Finally, in Sect. 5 we draw some conclusions and discuss possible future works.

2 Problem Description

The task that we have chosen as test-bed to analyze our control algorithm is illustrated in Fig. 1: a group of robots gathered in nest has to form a path between nest and target. And in this task, the robot has the following constraints:

1. The robot only has forward, leftward and rightward movement functions.
2. Distance between robots can be measured, perceived and communication can be possible only within communication scope.
3. The robot can't detect location information of another robot without the function of recording its own motion trail.
4. The robots have no a priori knowledge about the dimensions, or the position of any object within the environment.

The task that we have chosen as a test-bed to analyze our control algorithms is illustrated in Fig. 1. Explorer is a black robot with movement function, and its search task is to form a path between the target and the nest. At the beginning, a swarm of explorer gathers near the nest. Explorers continuously move. When certain conditions are met, they transfer to a fixed node. When one node perceives the target, the path connecting with the target is formed. All chains constitute a network that guarantees communication connectivity between swarm robotics.

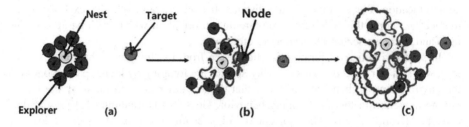

Fig. 1. The search process of the controller. We divide all the robots into four types, as described above figure.

3   Method

In the following section, we detail the controller, and then we give a global view of the controller by using the finite state automatons.

Controller is as Below:

1. Movement of explorer abides by edge-following [8]. node and nest broadcast their ID and status. After receiving this information, explorer will calculate the distance from the robot sending the message. If this distance exceeds one fixed value compared with the distance measured the last time (track radius, recorded as R), then it will turn left. When distance between explorer and nest or node is smaller than R, then it will turn right, or otherwise it will go straight as shown in Fig. 2.

Fig. 2. Red robot movement around black robot according to edge-following. (Color figure online)

2. If explorer finds that only one nest or node exists within communication scope, it means moving to the end of robot chain. Explorer will be transferred to a fixed node and constitute robot chain with itself being beacon and provide other Explorers with a new search direction.

3. If explorer detects the target and this target is detected among the swarm for the first time, explorer will stop movement and transmit the distance from the Target to each explorer through robot chain, and each explorer will enter "refractory period". The system will conduct the behavior of explorer transferring to node at the end detected with target in the "refractory period", while other explorers stop the above behavior until the distance from node to the target is smaller than the set value in the system.

At the time, target search is completed, and meanwhile, "refractory period" locking for all explorers is relieved, and new target search will be continued (Fig. 3).

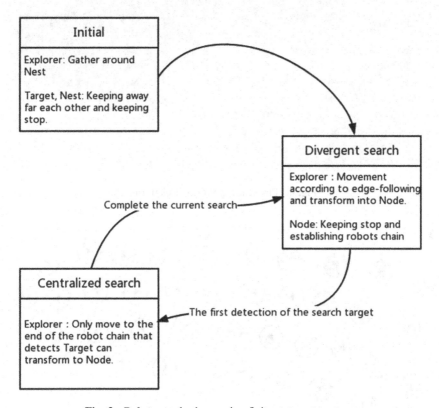

Fig. 3. Robot search phase using finite state automatons.

4   Experiment

In order to verify method effectiveness, simulation and physical experiment were respectively implemented. Kilobots [9, 10] platform was used in the physical experiment while simulation experiment was implemented on Kilombo [11] simulation platform. Kilombo is a simulation platform developed with kilobots swarm robotics experimental platform being the object, as shown in Figs. 4 and 5.

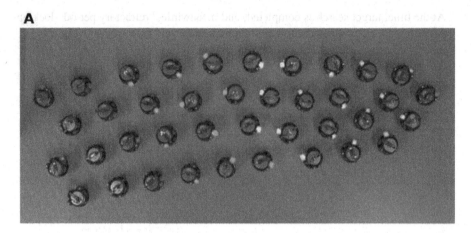

Fig. 4. Physical experimental platform.

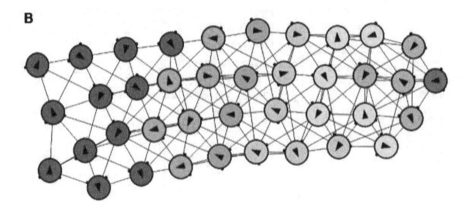

Fig. 5. Simulation experimental platform.

4.1 Simulation Experiment

The main objectives of experiments is to validate the effectiveness of the controller by single-target search and multi-target search experiment, and to analyze the performance of the controllers. In this session, we give the detailed setup of experiments.

Setup. A group of N simulated robots is placed within an unbounded arena for single target and multi-target search experiment. The nest is placed in the centrality of the arena, and the target is put at distance D (in mm). Other robots are initially gathered around the nest and the orientation are chosen randomly. The difficulty of experiment can be varied by changing the distance between nest and target. And for each experiment only if a robot chains establishes a connection to target which is composed of nodes that the experiment can be successful. In addition, the related parameters of the robots in the experiment are shown in Table 1.

Table 1. Simulation parameters

Parameters	Unit	Value
Maximum communication scope	Mm	100
Information transmission success rate	%	80
Error rate of distance measurement	%	2
Robot movement speed	mm/s	7

4.2   Result

We validate and analyze the controller through simulation experiments. In the following, we offer a detailed simulation process and results.

Effectiveness. We set the setups $N \in \{10, 14\}$ to validate the effectiveness of the controller in single-target and multi-target search experiments. Our experiments successfully finish single-target and multi-target search tasks. In Figs. 6 and 7, we respectively give the details of the process of experiments. As shown in Figs. 6 and 7, the search process is a continuous transformation that explorer transforms state into node to form the robot chains. These robot chains form a network topology, which will ensure timely communication and stable search of swarm robots.

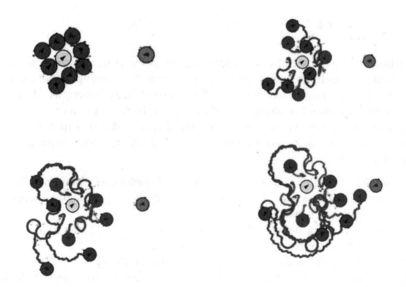

Fig. 6. Simulation experiment of single-target search.

Stability. In order to analyze the stability of controller, we investigated all setups (N, D), with $N = 9$, and $D \in \{150, 155, \cdots, 175, 180\}$. For each combination of the setups (N, D), we conducted 8 different target position trials which have same distance between nest and target. Figure 8 shows the completion times for all setups. The results are ordered by

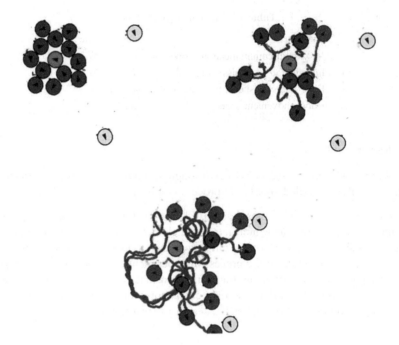

Fig. 7. Simulation experiment of multi-target search.

target distance. We can see that the completion time increases linearly with growing up target distance. This is not surprising, as search is finished by forming a robot chain, and the farther distance it is, the more time it will take to build the robot chains. Meanwhile, the exploring area grows quadratically with respect to the target distance.

In addition, we can get the time fluctuation of each experiment from Fig. 8. The reasons for the fluctuation and the difference of experimental time consumption are analyzed as follows:

1. In experiments, target is located at any location D millimeters away from the nest. In addition, the formation of a robot chain is random. Therefore, the existence of time fluctuation is certain.
2. Collision may occur between the explorers in the experiment, and as a result, the swarm robotics stay for a long time in one place.
3. We can't locate the position of the target. When explorer is converted to node and there is no other explorer nearby, the other explorer must to move through the edge-following algorithm to the robot chain that has been aware of the target.

Success Rate. In Fig. 9, we fit the relationship between success rate and target distance by curve fitting. The results are ordered by target distance. Obviously, when the target distance exceeds a certain range, the success rate will start to decrease. Because the number of robots is limited, so that the area of search and coverage is also limited. So we can increase the number of robots to expand the area of search and the success rate.

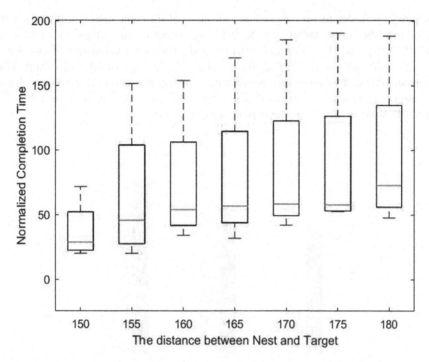

Fig. 8. The normalized completion time is shown and ordered by the target distance.

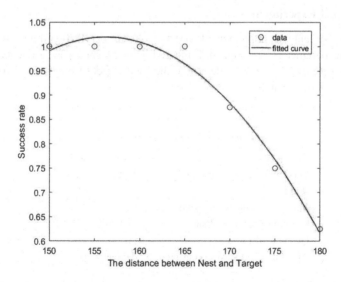

Fig. 9. Success rate is shown and ordered by the target distance.

Efficiency. In order to investigate the efficiency of different scale swarm robots to accomplish the same target search task, we selected the setups *(N, D)*, with N ∈ {9.12.15}, and D ∈ {170, 175, 180} for trials. For each combination of setups (N, D), we conducted 8 trails. In Fig. 10, we show the average of all trails result. The results are ordered by the distance between nest and target, and we can see that with the increase of robot scale, the time of search task is decreasing. This indicates that the search efficiency increases as the system scale increases.

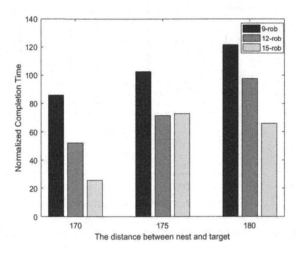

Fig. 10. The efficiency of different scale group robots to accomplish the same target search task.

4.3   Physical Experiment

Physical experiment was implemented on 0.4 × 0.6 m² acrylic board with 10 kilobots, and relevant parameters are seen in Table 2. Figure 11 shows experimental process. The robots gathering at the left side search the right target to finally form robot chain and complete target search.

Table 2. Experimental parameters

Parameter	Unit	Value
Maximum communication	mm	120
Radius	mm	60
Maximum movement speed	m/s	7
Area of experimental site	m * m	0.4 * 0.6
Quantity of robots	int	10
Angular speed in swerving process	degrees/s	0.64

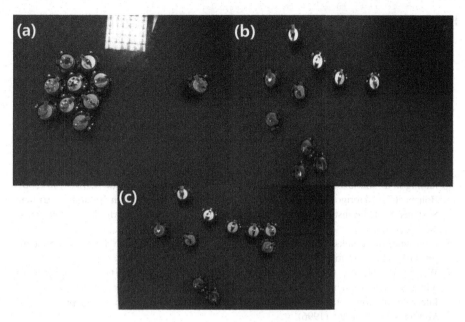

Fig. 11. Experimental screenshots of kilobots. Experiment snapshots from the initial situation (a), and a typical process when forming the chain (b) and a outcome when finishing the task.

5 Conclusion

The method proposed in this study can establish robot chain only through simple communication and distance measurement between robots in a self-organizing way so as to form the path connecting initial position and search target. Simulation experiment and physical experiment have verified effectiveness of the proposed method. In the future, the control method under more complicated environment will be studied. Meanwhile, network topology constituted by multiple robot chains will be investigated and optimized in order to optimize and reduce quantity of robots constituting robot chains.

Acknowledgments. The study was supported by the National Natural Science Foundation of China (Grant No. 61703102), the Natural Science Foundation of Guangdong (no. 2015A030310274, no. 2017A030313690, no. 2015A030310415, and no. 2015A030310315), the Dongguan Social Science and Technology Development Project (NO. 2013108101011, NO. 2017507140058, NO. 2017507140059), and Dongguan Industrial Science and Technology Development Project (NO. 2015222119), Department of Education of Guangdong China (Grant No: 2016KTSCX137, 2017KZDXM082). Scientific research foundation of advanced talents (innovation team), DGUT (No. KCYCXPT2016004), and Guangdong science and technology innovation foundation of university students (NO. pdjh2017b0492).

References

1. Giraldo, J.A., Quijano, N., Passino, K.M.: Honey bee social foraging algorithm for resource allocation. In: Kacprzyk, J., Pedrycz, W. (eds.) Springer Handbook of Computational Intelligence, pp. 1361–1376. Springer, Heidelberg (2015). https://doi.org/10.1007/978-3-662-43505-2_70
2. Liu, Y., Passino, K.M.: Biomimicry of social foraging bacteria for distributed optimization: models, principles, and emergent behaviors. J. Optim. Theor. Appl. **115**(3), 603–628 (2002)
3. Bayindir, L., Şahin, E.: A review of studies in swarm robotics. Turk. J. Electr. Eng. Comput. Sci. **15**(2), 115–147 (2007)
4. Schmickl, T., Möslinger, C., Thenius, R., et al.: Individual adaptation allows collective path-finding in a robotic swarm. Int. J. Factory Autom. Robot. Soft Comput. **4**, 102–108 (2007)
5. Bonabeau, E., Dorigo, M., Theraulaz, G.: Swarm Intelligence: From Natural to Artificial Systems. Santa Fe Institute Studies on the Sciences of Complexity. Oxford University Press, New York (2003)
6. Camazine, S., Franks, N.R., Sneyd, J., et al.: Self-organization in Biological Systems, pp. 110–118. Princeton University Press, New Jersey (2003)
7. Werger, B., Matarić, M.: Robotic food chains: externalization of state and program for minimal-agent foraging. In: From Animals to Animats 4, Proceedings of the 4th International Conference on Simulation of Adaptive Behavior (SAB 1996), pp. 625–634. MIT Press, Cambridge (1996)
8. Rubenstein, M., Cornejo, A., Nagpal, R.: Robotics. Programmable self-assembly in a thousand-robot swarm. Science **345**(6198), 795–799 (2014)
9. Tharin, J.: K-Team Mobile Robotics (2002–2015). http://ftp.kteam.com/kilobot/user_manual/Kilobot_UserManual.pdf. Accessed 12 Dec 2015
10. Rubenstein, M., Ahler, C., Hoff, N., Cabrera, A., Nagpal, R.: Kilobot: a low cost robot with scalable operations designed for collective behaviors. Robot. Auton. Syst. **62**(7), 966–975 (2014)
11. Jansson, F., Hartley, M., Hinsch, M., et al.: Kilombo: a Kilobot simulator to enable effective research in swarm robotics. Comput Sci. **43**(7), 793 (2015)

SOM-Based Multivariate Nonlinear Vector Time Series Model for Real-Time Electricity Price Forecasting

Ling Wang, ZhiYuan Chen, Tiehua Zhou[✉], Wenge Dong,
and Gongliang Hu

School of Information Engineering, Northeast Electric Power University,
Jilin City 132000, JiLin, China
smile2867ling@163.com, czy415@163.com,
thzhou55@163.com, wengedong@163.com,
hugongliang777@gmail.com

Abstract. Electricity price is the precondition and foundation of making decision and plan for the stakeholders in modern electric power market. The deregulation of power markets makes the market environment competitive in recent years. For business opportunities, many aggregators and retailers sprung up, which make electricity price appear fluctuation coupled with changing market structure and some influence factors. It undoubtedly increases the difficulty to analyze electricity price. Therefore, our proposed a novel method called multi-variable nonlinear vector time series (MNVTS) model which considering multidimensional influence factors in order to forecast real-time electricity prices. This forecasting model firstly clusters RTEP fluctuation process into the peak prices, abnormal prices, low prices and stable prices, then calculates and extracts the most influence factors. Our experiments show that our method has good prediction precision.

Keywords: Real time electricity price · Forecasting · Clustering
Multivariate time series

1 Introduction

With the repaid development of the global economy, higher requirements are put forward for departments of power system in the process of business model from monopoly to competition. The generation can be controlled by human beings, but the electricity can't be accumulated easily. Thus, demand generated by consumers determines the level of generating capacity.

Instead of being set by competent authorities as in the period of electric power regulation, the electricity prices are determined by the value of electricity commodity and influenced by various factors. If we forecast the electricity price accurately in advance, we will be in an advantageous position and gain more benefits in market competition. Therefore, electricity price forecasting is significant for setting up reasonable offers in the ultra-short term. Research on RTEP forecasting has therefore got in-depth development in recent years and is considered to be a valuable as well as

© Springer International Publishing AG, part of Springer Nature 2018
D.-S. Huang et al. (Eds.): ICIC 2018, LNCS 10954, pp. 147–159, 2018.
https://doi.org/10.1007/978-3-319-95930-6_15

challenging mission [1]. Accurate RTEP forecasting is very useful for generators and consumers [2] to determine offering and bidding strategies.

Some research studies have been done over the last years to probe the sophisticated problem of monthly electricity price forecasting via a multivariate time series analysis [3]. However, most of studies have postulated that the input data series are stationary [4]. RTEP series have been found to be non-stationary, as well as most of variables that may influence it. Therefore, it is essential to consider complex models that are able to deal with the nonlinear electricity price series in the case of multiple series are non-stationary. In order to hurdle the non-stationary time series problem, the neural network model-based and Auto-regressive Integrated Moving Average Model (ARIMA) approaches were applied on some research studies. However, the forecasting precision is greatly influenced by hidden layer nodes of neural networks and the ARIMA is not suitable to address RTEP problem since the price series has the complex characteristics of nonlinearity and high volatility.

The aim of this paper is to provide an accuracy model for RTEP forecasting. The raised method is considered for external nonlinear influential factors that relate to the electricity price volatility. In order to evaluate the forecasting performance, we conducted a comprehensive comparison of the predictive validity of the proposed method versus that of the wavelet transform and classical ARMA [5] as well as time series forecasting only [6].

The rest of the paper is organized as follows. Section 2 provides a brief review of the literature on electricity price forecasting models and methods. In Sect. 3, characteristic analysis of RTEP stochastic fluctuation process is presented and introduced a hybrid RTEP forecasting model based on improved traditional ARMA model. Section 4 describes the flowchart of the forecasting method. Experimental evaluation is implemented in Sect. 5. Section 6 contains conclusions.

2   Related Works

Various methods for forecasting electricity prices are reported in the literature: multivariate dynamic regression models and transfer function models, input/output hidden Markov models and wavelet models [7–10]. Time series forecasting methods was first used and played an important role in the field of economics. Zareipour et al. [11] established a model that the forecasting performance was not good because of the linear models could not be able to obtain non-stationary features of data. In order to compensate the disadvantage of single model, hybrid forecasting models by combining established method were proposed [12]. For hybrid models, Zhang et al. [13] propose a novel hybrid method based on wavelet transform, ARIMA, and least squares support vector machine.

One of modern intelligent methods, artificial neural network (ANN) is a common choice for such a numeric forecasting. Amjady et al. [14] put forward cascaded predictors that each predictor composes a neural network and an evolutionary algorithm. Catalao et al. [15] introduced a neural network approach to forecast next-week prices. Instead of ANN, Support Vector Machine (SVM) has become more and more popular as a data driven method. The main merit of SVM over ANN is that SVM can effectively solve the problem of data over fitting, local minimum and unpredictable outliers [16], but SVM is more suitable for small number of sample data and is difficult for multiple classification problems.

SOM [17], known as Kohonen map, is also a special type of ANN for clustering and visualization. The author [18] proposed an adaptive hybrid method based on SOM and SVM. Through SOM clustering, input data was partitioned into several clusters and then different SVMs were applied to each cluster to forecast the next day's electricity price curve. Hsu et al. [19] applied SOM to identify spatially homogeneous clusters of accurate groundwater level piezometers.

Because the electricity price is arranged chronologically discrete price, forecasting model of time series can be easily established and understand [20]. Contreras and Nogales [21, 22] separately adopted auto regressive, auto regressive integrated moving average to forecast the short-term electricity price and achieved good result. Garcia and Contreras [23] uses the generalized autoregressive conditional heteroskedasticity to describe electricity prices of heteroscedasticity and the results showed that the method has a good predictive accuracy.

3   Characteristic Analysis of RTEP Stochastic Fluctuation Process

3.1   Stochastic Fluctuation Process Division

It is important to consider factors for RTEP forecasting because electricity price shows volatility and randomness. Figure 1 shows the RTEP change curve that describes the average value of electricity price at half-hour intervals from the same day in four different seasons. How to identify the fluctuation of electricity price in a certain time period that related to some influencing factors is of vital importance to electricity price prediction.

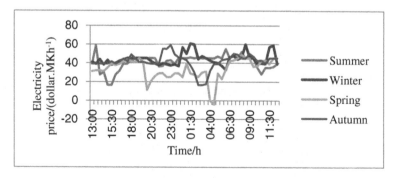

Fig. 1. RTEP change law in different seasons

It is shown from Fig. 1 that the RTEP change is great stochastic and fluctuating along with the emergence of abnormal electricity price. So the stochastic wave process of RTEP is defined as a change period from the maximal electricity price to the minimum price, then from the minimum to the maximum. The mathematical model is defined as formula (1):

$$v\{P_i\} = \begin{cases} P_1, & P_n \in \{P\} \\ P_m \in \{P_{max}\}, 1 < m < n \\ P_i \in \{P\}, & i = 1, \cdots, n \\ P_k \in \{P_{min}\}, & 1 < k < n \end{cases} \tag{1}$$

Where $v\{P_i\}$ is a RTEP fluctuation sequence. $\{P\}$ is a RTEP sequence for the whole period. $\{P_{max}\}$ is the local maxima value in RTEP sequence, which may contain abnormal spikes price. $\{P_{min}\}$ is a local minimum value that may also contain abnormal price such as negative value and zero electricity price. If the fluctuation is composed of n-point sequences, P_1 and P_n are the starting point and end point of the fluctuation.

3.2   Category Division of RTEP Stochastic Fluctuation Process

SOM Clustering Algorithm. SOM is an unsupervised learning network with the ability of self-organizing feature mapping. The SOM does not have implied layers, only including the input layer and the output layer. Input layer is a high-dimensional input vector, and the output layer is composed of a series of nodes which are organized on the low dimension grid to realize the transformation by the way of the extended-flutter order. Figure 2 shows the network topology of SOM.

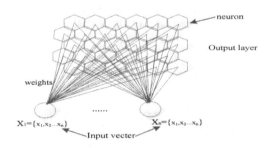

Fig. 2. SOM network topology.

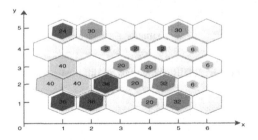

Fig. 3. Sample hits

Kalteh et al. [24] suggested that the hexagonal discrete grids are usually used to visualize for the SOM output layer. For the sake of complete the clustering of the input pattern dataset, a grid with 5 × 6 hexagons was applied to demonstrate the similar real time electricity price observations in the study field. Figure 3 above shows the several clusters based on SOM. Figure 4 shows the process of the two-level clustering of SOM and the different symbols different clusters.

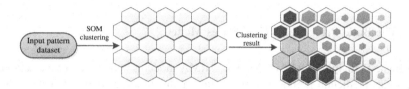

Fig. 4. The process of SOM.

The learning process of SOM algorithm is shown in Table 1.

Table 1. Learning steps of SOM algorithm.

learning steps of SOM algorithm
(1) Initialize weight vector $w_{ij}(0) = \left[w_{i1}(0), w_{i2}(0), \cdots, w_{ij}(0)\right]$.
(2) Provide input sample $\{P_i\}$ and Calculate European distance of weights vector using formula $\|P_i - N_k^*\| = min\{\|P_i - N_k\|$.
(3) Update neuron and weights according to the neighborhood function [24]. $w_{ij}(t) = w_{ij}(t-1) + \beta(t)h_{ki}(t)\left[P_i(t) - w_{ij}(t-1)\right]$.
(4) Learning rate function $\beta(t)$ and neighborhood function $h_{ki}(t)$ decay with iterative times.
(5) Update samples and repeat the step (2) until all the samples training are trained completely.

Where i is the number of neuron, j is the dimension of the weight vector. N_k^* is the most-matched node. $\beta(t)$ is the t step learning rate. $h_{ki}(t)$ is the neighborhood function of k.

Clustering Analysis of RTEP Stochastic Fluctuation. Due to the difference in the duration of stochastic fluctuations, the stochastic fluctuation of RTEP is the result of various influencing factors. According to the difference of electricity price amplitude and duration time, the state of RTEP stochastic fluctuation is divided into four categories. There are separately big fluctuation electricity price, small fluctuation electricity price, stable electricity price, abnormal electricity price. The abnormal RTEP contains the negative price and the peak electricity price that exceed the high price. For abnormal electricity price, it can be identified by the wave crest threshold. The threshold value of spike electricity price can take the average of the high electricity price in the whole time period. And threshold value of zero electricity price or negative electricity price can take the average of low electricity price. It is represented using the following formula:

$$THR_{spike} = \sum_{1}^{N} P_{max} \tag{2}$$

$$THR_{neg/zero} = \sum_{1}^{N} P_{min} \tag{3}$$

3.3 The Basic Principle of Time Series Model

The Basic Principle of ARMA Model. ARMA model is a method for non-stationary time series prediction proposed by Box and Jenkins in the early 70. Its general form [25] is:

$$\varphi(B)(1 - B)^d X_t = \theta(B)\varepsilon_t \tag{4}$$

Where X_t is a random time sequence on t moment. ε_t is a white noise process with a mean value of zero. B is defined as a delay operator, namely $BX_t = X_{t-1}$. d is a differential order. $\varphi(B)$ and $\theta(B)$ can also be defined as: $\varphi(B) = 1 - \varphi_1 B - \varphi_2 B^2 - \cdots - \varphi_p B^p$, $\theta(B) = 1 - \theta_1 B - \theta_2 B^2 - \cdots - \theta_q B^q$. Where p and q are the highest order with partial autocorrelation function and auto correlation function value.

The Basic Principle of ARIMA Model. The ARIMA model can be considered as the result of the difference operation combining the ARMA model. The purpose of the difference is to eliminate some fluctuations to stabilize the data because ARMA can only handle stationary data. For some data, the first-order difference may not be stable, such as RTEP data. So multiple differencing is required, but the significance of the data after that is difficult to explain. In most cases, the time series is also affected by many external factors, and it is not ideal to model the time series itself. Some literatures consider exogenous variable X on the basis of ARMA, and then establish ARMAX model considering exogenous variables. The expression [26] is described as follows:

$$Y_t = c + \sum_{i=1}^{p} \varphi_i Y_{t-i} + \sum_{i=0}^{q} \theta_i X_{t-i} + \sum_{j=1}^{m} \mu_j \varepsilon_{t-j} + \varepsilon_t \tag{5}$$

Where Y_{t-i} and X_{t-i} are lag. p and q are the highest order numbers with partial autocorrelation function values and autocorrelation function values that are significantly non-zero.

Multivariate Nonlinear Time Series Model. Although the ARMAX model of exogenous variables overcomes the traditional ARMA model that can't response to external factors, the exogenous variables considered are single and small quantities lacking of comprehensiveness and effectiveness. And this model is not ideal for large fluctuations in data. Based on the ARMAX model, the influence factor vectors are introduced in this paper to improve it. A multi-variable nonlinear vector time series

(MNVTS) model considering multidimensional influence factors was established. The expression is defined in Formula (6):

$$Y_t = c + \sum_{i=1}^{p} \xi_i Y_{t-i} + \sum_{j=1}^{k} \Phi_j \dot{Z}_{t-i} + \sum_{i=0}^{q} \Theta_i P_{F_{t-i}} \qquad (6)$$

Where E is the unit matrix. Φ_i and Θ_i are coefficient matrices. $P_{F_{t-i}}$ is the influence factor vector introduced. Of which, $\dot{Z}_{t-i} = \Phi_1 \dot{Z}_{t-1} + \cdots + \Phi_k \dot{Z}_{t-k} + a_t . a_t$ is a random vector of white noise.

4 RTEP Multivariable Nonlinear Time Series Modeling

The method proposed in the paper is firstly used to cluster and feature recognition by SOM. The original RTEP is transferred into several different categories of electricity price change state according to the impact on RTEP influence factors. Then, the proposed MNVTS model was used to predict the different RTEP categories. Finally, the forecasting results are obtained by linear combination of each forecasting result. The process of establishing the forecast model is shown in Fig. 5.

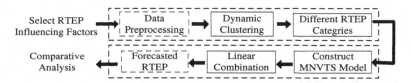

Fig. 5. Process of the prediction method

5 Example Evaluation

5.1 Predictive Accuracy Measurement

A data set was collected from Nyiso of New England [27] and PJM power market [28] separately. Due to the existence of abnormal electricity price, it is not appropriate to use the traditional average percent error to predict the result. For example, when there are zero electricity prices, the traditional formula loses its meaning. The Improved Absolute Percent Error (IAPE) is used to measure the error. The expression is defined as follows:

$$E_{av} = \frac{1}{n} \sum_{t=1}^{n} \frac{|p_t - \tilde{p}_t|}{\bar{p}_n} \qquad (7)$$

Where p_t is the real price. $\tilde{p}_t$ is the predicted electricity price, $\bar{p}_n$ is the average of p_t.

5.2 Results and Discussion

The results obtained by the proposed method were compared with the other two models. The first model is to consider the impact of various factors, and then it uses original ARMA model to predict each RTEP category (Model 1). The second model considers the RTEP factors, it directly only use multivariate vector time series model to predict RTEP (Model 2). The performance statistics of the proposed model and two different models are given in Table 2.

Graphical comparison of the actual values (blue line) to the forecasted values (green line) by the proposed method is presented in Figs. 6 and 7. The corresponding error results are shown in Tables 2 and 3. It is noteworthy that, due to space constraints, Tables 2 and 3 only give part of the prediction errors. As can be seen from Tables 2 and 3, the accuracy of the proposed method in this paper has been improved compared with the other two comparison models. However, the forecasting error is lower at about 1 o'clock than most other time period, the reason is that this period of time may be a lunch break with relatively less electricity consumption. At the period of 14:00–17:00 and 19:00–21:00, the RTEP forecasting error has gradually become larger. Since these two periods are at the peak of electricity use with large electricity changes, it is difficult for traditional models to forecast.

Table 2. Forecasting errors.

Time/min	Model 1 E_{av}/%	Model 2 E_{av}/%	Proposed model E_{av}/%
13:00	1.99	1.79	1.71
13:30	2.21	2.50	1.89
14:00	1.93	1.74	1.21
14:30	2.38	2.34	1.33
15:00	1.95	1.62	1.69
15:30	2.87	2.05	1.79
16:00	1.95	1.96	1.46
16:30	2.11	2.16	1.39
17:00	1.70	1.57	1.42
17:30	2.94	2.93	2.42
18:00	3.05	2.84	2.43
18:30	2.02	1.92	1.42
19:00	2.00	2.02	1.41
19:30	3.22	2.18	1.41
20:00	2.21	2.12	2.02
20:30	2.18	2.11	1.93
21:00	3.09	2.22	2.28

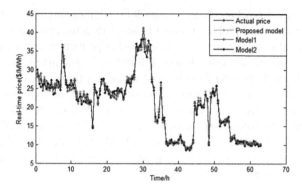

Fig. 6. Forecasting results from different models. (Color figure online)

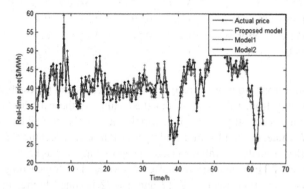

Fig. 7. Forecasting results from different models. (Color figure online)

Table 3. Forecasting errors from different models.

Time/min	Model 1 E_{av}/%	Model 2 E_{av}/%	Proposed model E_{av}/%
13:00	1.83	1.74	1.64
13:30	1.91	2.20	1.38
14:00	2.62	2.34	2.01
14:30	2.19	2.04	1.51
15:00	1.94	1.97	1.57
15:30	2.27	2.15	1.86
16:00	1.85	1.66	1.76
16:30	2.01	2.12	1.62
17:00	1.90	1.87	1.44
17:30	2.54	2.43	2.32
18:00	3.01	2.94	2.82
18:30	2.52	2.62	1.64
19:00	2.10	2.03	1.79
19:30	3.22	3.18	2.51
20:00	2.21	2.15	1.82
20:30	2.42	2.31	2.23

As you can see from Figs. 6 and 7, the RTEP forecasting curve shows that the prediction performance of proposed model is better than the other two models on the whole. There are also some points that have a higher prediction error, such as at about 26 h and 46 h in Fig. 6, at about 6 h and 30 h in Fig. 7. The mainly reason is that the RTEP appears lager fluctuation or influenced by other external factors.

In order to obtain a better idea of the proposed algorithm performance, the more complex datasets were used to verify the actual prediction effect of the proposed prediction model. The experiment compares the proposed model with different models: ARMA, ARMAX, VARMA and the proposed MNVTS model. In addition, posterior difference and correlation analysis were used to test the validity of four established prediction models, as shown in Table 4.

Table 4. Test results of RTEP forecasting model

Test items	ARMA	ARMAX	VARMA	Proposed
Large error rate L	0.12	0.09	0.13	0.09
Posterior error ratio P	0.45	0.41	0.44	0.39
Relevance R	0.75	0.71	0.71	0.78

The posterior error ratio is less than 0.5, which indicates that the established model is qualified. The results from Table 4 show that the L values of these models are all less than 0.2, and the P values are all less than 0.5, which validate the effectiveness of proposed models. From R value, the results predicted by RTEP model reflect the RTEP fluctuation more realistically. From the overall results, the proposed model has best prediction effect.

Table 5 shows RTEP forecast error of each model. The forecasting error of other contrast models at the time interval 7–12 and 18–22 is much higher than other time periods, because this period is at the peak power consumption during the day. Traditional time series can hardly have good prediction effect, so the prediction accuracy is low. The MNVTS model of considering stochastic fluctuation characteristics is obviously lower than the contrast model in the peak period. The reason is that the feature extraction of the SOM clustering model weakens the large RTEP fluctuation. The overall prediction error of proposed model reaches 2.11%, while VARMA is 2.71%. It is concluded that the RTEP forecasting results is significantly improved considering the fluctuation clustering analysis.

As shown in Fig. 8, the actual RTEP from Australia New South Wales in December 2015 and the forecasted electricity price curve are presented. The RTEP forecasting curve shows that the prediction performance of proposed model is better overall. However, before and after the periods 80, 551 and 620, the RTEP forecasting error is higher than that of other periods, and this period is just in the area where the RTEP fluctuates greatly. It is indicated that the accurate prediction results are difficult to obtain during sharply beating period, which also verifies the difficulty of RTEP forecasting. Through further in-depth research, it can be seen that RTEP levels of these points with higher prediction errors are significantly higher or lower than the changes of the day-ahead electricity prices. However, these predicted values are smaller before and after the periods 155 and 700 because of the fact that RTEP has changed considerably since the end of the holidays.

Table 5. Prediction error comparison.

时点	ARMA/%	ARMAX/%	VARMA/%	MNVTS/%
1	2.36	2.27	2.14	2.07
2	1.92	2.29	2.81	2.50
3	2.63	1.90	1.83	1.92
4	2.64	2.56	2.31	1.39
5	2.38	2.10	2.95	2.74
6	1.97	1.84	1.19	1.44
7	2.99	2.88	2.40	1.98
8	2.28	2.92	2.88	2.70
9	3.37	3.40	3.96	2.39
10	3.39	3.87	3.39	2.73
11	2.95	2.84	2.42	1.95
12	3.75	3.49	3.09	2.63
13	2.67	2.24	2.20	1.77
14	1.96	2.31	2.05	1.84
15	2.64	2.28	2.12	1.88
16	2.82	2.63	2.58	2.06
17	2.50	1.94	1.56	1.28
18	3.34	3.84	3.07	2.10
19	4.79	4.66	4.25	3.09
20	3.08	3.31	3.01	1.83
21	3.76	3.51	3.83	2.44
22	4.54	4.19	3.02	2.18
23	3.76	3.53	2.84	2.05
24	3.36	3.35	3.22	1.79
Average	2.99	2.92	2.71	2.11

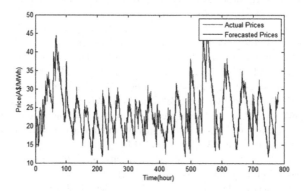

Fig. 8. RTEP forecasting results by proposed model.

6 Conclusion

A novel time series forecasting algorithm MNVTS has been developed to forecast RTEP in the New England, PJM power market and a more complicated datasets from Australia. The method presented overcomes the problem that the traditional time series model can only deal with the single variable and stationary sequence. The main contribution of the proposed methodology is the ability to forecast fluctuation RTEP by using simple series time models with input features selected with data mining techniques. The proposed methodology can be utilized by market participants in forecasting electricity prices.

Acknowledgments. This work was supported by the Project-sponsored by SRF for ROCS, SEM, by the Education Department Foundation of Jilin Province (No. 201698), and by the Science Research of Education Department of Jilin Province (No. JJKH20170108KJ).

References

1. Zareipour, H., Canizares, K.: The operation of Ontario's competitive electricity market. IEEE Trans. Power Syst. **22**(4), 1782–1793 (2007)
2. Sandhu, H.S., Fang, L., Guan, L.: Forecasting day–ahead price spikes for the Ontario electricity market. Electr. Power Syst. Res. **34**(5), 451–459 (2016)
3. Au, S.T., Ma, G.O.: Automatic forecasting of double seasonal time series with applications on mobility network traffic prediction 23(5), 99–11 (2010). web2.research.att.com
4. Du, S.H., Hou, Z.J., Jiang, C.W., Zhijian, H.: A new short–term grey forecasting procedure of spot price. J. Grey Syst. **14**(4), 351–358 (2012)
5. Li, N., ZhengFang, W.: Hybrid model based on wavelet transform and ARIMA for short–term electricity price forecasting. Appl. Res. Comput. **31**(3), 688–693 (2014)
6. Liao, R.H., Zhou, J.: Traffic forecasting model based on cloud–self–organizing neural network. J. Transp. Syst. Eng. Inf. Technol. **14**(4), 154–160 (2014)
7. Conejo, A.J., Contreras, R., Espinola, M.A.: Forecasting electricity prices for a day-ahead pool-based electric energy market. Int. J. Forecast. **21**(7), 435–462 (2005)
8. Zhang, J., Tan, Z.: Day-ahead electricity price forecasting by a new hybrid method. Comput. Ind. Eng. **63**(39), 695–701 (2006)
9. Nogales, F.J., Contreras, A., Conejo, R.: Forecasting next-day electricity prices by time series models. IEEE Trans. Power Syst. **17**(2), 342–348 (2005)
10. Gonzalez, A.M., Roque, J.: Modeling and forecasting electricity prices with input/output hidden Markov models. IEEE Trans. Power Syst. **20**(1), 13–24 (2009)
11. Zareipour, H., Cañizares, C.A.: Application of public domain market information to forecast Ontario's wholesale electricity prices. IEEE Trans. Power Syst. **21**(4), 1707–1716 (2006)
12. Yan, X., Chowdhury, N.A.: Mid–term electricity market clearing price forecasting utilizing hybrid support vector machine and auto–regressive moving average with external input. Int. J. Electr. Power Energ. Syst. **63**(5), 64–70 (2014)
13. Zhang, J., Tan, Z., Yang, S.: Day–ahead electricity price forecasting by a new hybrid method. Comput. Ind. Eng. **63**(3), 695–701 (2012)
14. Amjady, N., Keynia, F.: Day-ahead price forecasting of electricity markets by mutual information technique and cascaded neuro-evolutionary algorithm. IEEE Trans. Power Syst. **24**(1), 306–312 (2009)

15. Catalao, J., Mariano, S., Mendes, V.: Short-term electricity prices forecasting in a competitive market: a neural network approach. IEEE Trans. Power Syst. Res. **77**(10), 1297–1304 (2007)
16. Son, H., Kim, C.: Forecasting short–term electricity demand in residential sector based on support vector regression and fuzzy-rough feature selection with particle swarm optimization. Proc. Eng. **118**(2015), 1162–1168 (2012)
17. Kohonen, T.: The self-organizing map. Proc. IEEE **78**(9), 1464–1470 (1990)
18. Fan, S., Chen, I.: Short-term price forecasting based on an adaptive hybrid method. IEEE Trans. Power Syst. **21**(1), 392–401 (2010)
19. Hsu, J., Li, S.: Clustering spatial-temporal precipitation data using wavelet transform and self-organizing map neural network. Adv. Water Resour. **33**(5), 190–196 (2010)
20. Zhang, X., Wang, J.X.: Short-term electricity price forecasting based on price subsequences. Autom. Elect. Power Syst. **31**(3), 4–8 (2009)
21. Nogales, F., Contreras, J.: Forecasting next-day electricity prices by time series models. IEEE Trans. Power Syst. **17**(2), 342–348 (2002)
22. Contreras, J., Espinola, R.: ARIMA models to predict next-day electricity prices. IEEE Trans. Power Syst. **18**(3), 1014–1020 (2002)
23. Garcia, R., Contreras, J.: A GARCH forecasting model to predict day-ahead electricity prices. Journal **20**(2), 867–874 (2005)
24. Kalteh, A.M., Hjorth, P.: Review of the self–organizing map (SOM) approach in water resources: analysis, modeling and appliance. Environ. Model. Softw. **23**(5), 835–845 (2008)
25. Ming, Z., Zheng, Y.: A novel ARIMA approach on electricity price forecasting with the improvement of predicted error. Proc. CSEE **24**(12), 63–68 (2004)
26. XioaDong, N., Da, L.: Exogenous variable considered generalized autoregressive conditional heteroscedastic model for day–ahead electricity price forecasting. Power Syst. Technol. **22**(31), 44–48 (2007)
27. LNCS Homepage, /lncs. Accessed 18 June 2016. https://www.nyiso.com
28. https://www.pjm.com

A Situation Analysis Method for Specific Domain Based on Multi-source Data Fusion

Haijian Wang[1], Zhaohui Zhang[1,2,3(✉)], and Pengwei Wang[1,2,3]

[1] School of Computer Science and Technology, Dong Hua University,
Shanghai, China
zhzhang@dhu.edu.cn
[2] The Key Laboratory of Embedded System and Service Computing,
Ministry of Education, Tongji University, Shanghai, China
[3] Shanghai Engineering Research Center of Network Information Services,
Shanghai, China

Abstract. External Internet data, as a supplement of internal data, plays an important role to decision analysis of decision makers. However, the key point of this process is to solve the problem of inconsistency between multi-source heterogeneous data. In this paper, a situation analysis method based on multi-source data fusion is proposed to analyze the situation in a specific domain. The approach consists of three main steps. Firstly, Naive Bayes multi label classification algorithm is used in the process of text topic classification and quantization to overcome the structural inconsistency of multi-source data. Secondly, a time difference correlation analysis method is used to address the time inconsistency the two time series. Finally, Support vector machine regression algorithm (SVR) is used for situation assessment in related fields. In this study, the effectiveness of the model is verified by the free-trade zone (FTZ) platform shares data, Internet news text data, and Internet statistics. The experimental results show that the method has achieved good results in the situation estimation of the related indexes.

Keywords: Multi-source data fusion · Inconsistency
Naive Bayes multi label classification · Causal lag analysis · SVR
Situation estimation

1 Introduction

With the rapid development of economy, decision-makers and regulators in different fields hope to grasp the current or future development trend to reduce future uncertainty and lessen the risks, so that the decision-making goal can be smoothly achieved. The traditional methods are to analyze the development of the future situation through the statistical data within the Department. However, the statistical data from the statistical department always lag behind, so that the decision-makers cannot get the latest data in time. Internet data make up for the shortage of statistical data because of its strong timeliness. The Internet has a wide range of data and complex structure. How to handle these data and mine its valuable information has brought us a great challenge.

© Springer International Publishing AG, part of Springer Nature 2018
D.-S. Huang et al. (Eds.): ICIC 2018, LNCS 10954, pp. 160–171, 2018.
https://doi.org/10.1007/978-3-319-95930-6_16

The main step of multi-source heterogeneous data fusion studied in this paper is quantitative processing on unstructured text. The existing Text quantization methods can be divided into two groups. One group is the sentiment analysis to calculate the sentiment score of each text [5–7]. The other group is to extract the topics of the text to find the topics distribution of each text [8, 9]. However, compared to English language, the research on Chinese starts late, and Chinese language has distinct linguistic characteristics, so the effect of obtaining sentiment score of text is not good. The latter is often very poor because of the uncertainty of the number of topics and the uncertainty of the subject. Furthermore, the lag time difference between the two variables is not considered in process of the correlation analysis. The key contribution of this study is to try to propose a suitable method for situation analysis in specific fields. (1) The multi label classification technique based on Bayesian model is applied to divide text topics. (2) An improved time difference correlation analysis is proposed to solve the problem of the lag effect between the two variables. (3) It combines multi label classification, association analysis and regression prediction algorithm to carry out situation analysis. The method includes the main steps: quantitative text analysis, causal lag analysis, and trend prediction of related indicators.

The rest of the paper is organized as follows: Sect. 2 reviews the existing main approaches of the model. Section 3 introduces the proposed method. Section 4 illustrates and discusses the experimental results. Finally, the conclusion is drawn.

2 Related Work

At present, the multi label classification algorithms are basically divided into two types, one is to transform the problem into the traditional methods, and the other is to improve the existing methods. A classical method for the former is the Binary Relevance (BR) [1], It constructs training set for each tag, classifies each sample according to whether it belongs to this label or not, and then synthesizes multiple results. The drawback of the algorithm is that the computation complexity increases exponentially with the growing number of labels. The topic and label of the text are considered to be consistent in label correlation mixture model (LCMM) [2]. Considering the correlation between tags, texts can be classified by word similarity. The latter has multi label neighbor method based on improved KNN [3]. It proposes to estimate the value of each tag based on the maximum posterior probability in the probability distribution of each tag in the known K nearest neighbors. However, the algorithm does not take into account of the relevance between labels in the text classification.

Correlation analysis method is to measure the relationship between the data and the strength of the relationship. The existing main methods respectively are information entropy method and correlation coefficient method. Information entropy is a method to measure the correlation between the feature values of a text. Paninski [7] present some new results on the nonparametric estimation of entropy and mutual information. The correlation coefficient is a statistical index of the close relationship between the reaction variables. For example, the correlation coefficient is used to measure the normalized consistency in the signs of the remainders of two random vectors about their sample means in literature [11].

There are many prediction models in the prediction model of time series data. Bao et al. [13] put forward the fuzzy support vector machine regression algorithm, and applied it to the prediction of stock constituent index, and displayed the excellent algorithm in data processing. However, these methods are not suitable for the prediction of this data.

In terms of data fusion, Shroff et al. [10] describe the integration of Internet data and internal data construction framework using blackboard structure and local sensitive hash technology, and real-time alert the business status of enterprises. In this paper, Naive Bayes multi label classification algorithm is used to quantify news data. The trend development of relevant indicators of the field is predicted by integrating internal data and external Internet data.

3   Data Fusion and Situation Analysis Methods

3.1   The Fusion Method for External Internet Data and Internal Data

Data fusion, in nature, is to join, associate and combine data and information from all data sources in order to eliminate the inconsistency of multi-source data. The method of data fusion include two parts in this paper. The first step dealing with data fusion is to classify the unstructured news text by text classification algorithm to solve the inconsistency of data structure. The second step is to solve the inconsistency of the time causality among variables. Figure 1 shows an overview of the method. Firstly we collect the content and time of news related to the specific domain. Naive Bayes multi label classification algorithm is used to divide the news topic and quantify the topics according to the time series. Secondly, the time difference correlation method is used to study the time difference between variables. Multiple groups of time series will be obtained. Finally, these time series are used as inputs of SVR model to predict domain related indicators.

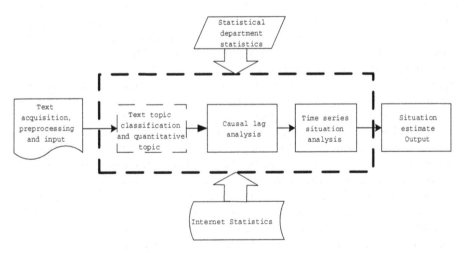

Fig. 1. The overview of the method

3.2   Text Topic Quantization for Data Structural Consistency

The most important part of text topic quantification is to determine the topics of text. In this paper, the topics are determined by the domain knowledge, and the Byes classification is used to divide the topics of each text.

We quantify the topics from the concepts of corpus, documents, topics, and words.

Definition 1. A corpus is a collection of words of N documents, which are recorded as $C = \{W_1, W_2, \ldots, W_N\}$, in which $W_i = d_i = \{w_{i1}, w_{i2}, \ldots, w_{im}\}$. Each document contains one or more topics, denoted as $d_i = \{tc_1, tc_2, \ldots, tc_k\}$. Each topic is made up of n words, so the topic jof the document i is expressed as $tc_{ij} = \{w_{i1}, w_{i2}, \ldots w_{in}\}$.

Naive Bayes classification [14] is a generation of model based on Bayes theorem and independent assumption of attribute. Formula (1) is used to calculate the conditional probability of each news appearing under the condition of class tc_i.

$$P(tc_i|d) = \frac{P(tc_i) \prod_j P(W^{(j)}=w^{(j)}|tc_i)}{\sum_i p(tc_i) \prod_j P(W^{(j)}=w^{(j)})} \tag{1}$$

From the formula (2), we can see that the denominator is the same to all categories, and therefore the main category of each news can be found according to formula (2).

$$Y = \arg\max_{tc_i} P(tc_i) \prod_j P(W^{(j)}=w^{(j)}|tc_i) \tag{2}$$

Considering the number of topics in each article, we can use formula (3) to find 0 or more times labels.

$$D = \frac{\text{MAX}_{tc_i} P(tc_i|d) - P(tc_j|d)}{\sum_{tc_i} P(tc_i|d)}. \tag{3}$$

Formula 3 is used to calculate the distance of posterior probabilities between other label and the main label. The tag corresponds to the sub label of the news when the distance D is less than a constant. Aforementioned method is used to train a classifier by the given training set. The classifier is used to extract topics of each document to solve the content consistency of the text data. Then the data were quantified based on the topics and time. The process is like formula (4).

$$\{date, \{w_{i1}, w_{i2}, \ldots, w_{im}\}\} \rightarrow \{date, \{P_{tc_1}, P_{tc_2}, \ldots, P_{tc_k}\}\} \tag{4}$$

$\{date, \{w_{i1}, w_{i2}, \ldots, w_{im}\}\}$ represents the time and word distribution of text, and $\{date, \{P_{tc_1}, P_{tc_2}, \ldots, P_{tc_k}\}\}$ represents the time and topic distribution of text. p_{tc_i} indicates whether the topic tc_i exists in the text. Date is the release time of the news.

Since the time of the statistical indexes is monthly statistics, the dimension of the sequence of topics should be reduced to keep the sequence of topics and the sequence of statistical indexes consistent in the time dimension. The process is like formula (5).

$$\{\{date, \{P_{tc_1}, P_{tc_2}, \dots, P_{tc_k}\}\}\}_N \rightarrow \{ym \in \{YM\}, \{P_{tc_1}, P_{tc_2}, \dots, P_{tc_k}\}\}_{YM} \quad (5)$$

$\{YM\}$ is the collection of all the months of the news date, and P_{tc_k} indicates whether the text belongs to the tc_k. The algorithm for text quantification of all texts is as follows.

Algorithm 1. Text topic quantization

Output: Set⟨document,documentDate⟩;   Set⟨time   slot⟩:{ym};   Set⟨topics⟩:

   {tc₁, tc₂, ..., tc$_k$}; Textclassifier : textClassfier(d);

Input: Time series theme value:topic[ym][tc$_k$]

1: // Initialization of topic matrix topic[ym][tc];

2: **for all** ym∈ {ym} **do**

3:      **for all** tc$_k$∈ Set⟨topics⟩ **do**

         topic[ym][tc$_k$] = 0;

4:      **end for**

5: **end for**

6: // Quantizing all kinds of topics in time order and putting values into the matrix

7: **for all** document∈ Set⟨document,documentDate⟩ **do**

8:      // Categorize each document to get the topic vector

9:      {P$_{tc1}$, P$_{tc2}$, ..., P$_{tck}$} ← textClassfier(document);

10:     **if** documentDate ∈ym **then**

11:         **for all** tc$_k$ ∈Set ⟨topics⟩ **do**

12:             topic[ym][tc$_K$] ← topic[ym][tc$_K$] + P$_{tck}$;

13:         **end for**

14:     **end if**

15: **end for**

16: **return** topic[ym][tc]

The Algorithm 1 is used to get the theme matrix topic[ym][tc] to complete the transformation of the text data to the numerical data, so that the multi-source data structure is consistent.

3.3   Causal Lag Analysis for Time Lag of External and Internal Data

Regression forecasting is the basis of the correlation principle of prediction and the factors that affect the forecast target are found. However there is a causal relationship of the first, the consistent and the lagging between the time series variables. For example, the state has issued a policy to control the operation of the macro-economy. It does not have an immediate impact on the economy, but has an impact after a period of lag. Thus the time difference correlation analysis (TTCA) is adopted to solve this problem in this paper. The relative index is chosen as the benchmark variable $y = \{y_1, y_2 \dots y_m\}$, and the number of news topics in the simultaneous segment and the Internet Statistics $x = \{x_1, x_2 \dots x_3\}$ is an alternative variable. We can use the formula (4) to get the time difference value and the corresponding correlation coefficient.

$$R_{x,y} = \max_l \left| \frac{\sum_{i=1}^{k}(x_{l+i} - \bar{x})(yi - \bar{y})}{\sqrt{\sum_{i=1}^{k}(x_{l+i} - \bar{x})^2 \sum_{i=1}^{k}(y_i - \bar{y})^2}} \right|, l = 0, \pm 1, \pm 2, \pm 3 \ldots \pm L; \quad (6)$$

In Formula (4), l indicates lead time and lag time. When l takes negative value, it indicates that x variable has a lagging effect on reference variables. It shows that X has a lagging effect on benchmark variables when l is positive. L is the maximum delay time for the lag effect. $R_{x,y}$ indicates the maximum correlation coefficient between benchmark variable and the alternative variable. $l_{R_{x,y}}$ is the value of l when the maximum correlation coefficient $R_{x,y}$ is obtained. The corresponding time difference value TD is obtained according to the maximum value, that is:

$$TD = l_{R_{x,y}} \tag{7}$$

Time difference correlation is often used to calculate the relationship between two time series. But an abnormal result is often obtained by the method. First of all, present the definition of this abnormal phenomenon is presented.

Definition 2. Pseudo correlation: A large correlation coefficient is obtained in the process of correlation analysis of two groups of time series. However, the correlation coefficient can not reflect the real causality between them, but just a coincidence in numbers. This relationship is called pseudo correlation.

For example, in Fig. 2, the relationship between operating income and enterprise litigation is a negative correlation through qualitative analysis, However, the maximum correlation coefficient calculated by TTCA is 0.58 when the time difference is d = 0, which is a positive correlation. The positive correlation s obviously a pseudo correlation. At the same time, it can be found that the correlation coefficient is −0.573 at d = −2, which has a negative correlation, and the absolute value distance is the most relevant coefficient. The negative correlation is a right result.

The news topics used in this paper have a clear impact on the related indicators, so the relevant factors of each topic can be determined by qualitative analysis to solve the problem of pseudo correlation, and the time difference correlation analysis is used to determine the time difference d between Time Series. The definition of the related factors is as follows:

Definition 3. Related factors (RF): Positive and negative correlation tendencies of each news theme and index, and its related factors can be determined according to the domain knowledge.

So the time difference correlation analysis can be improved to the time difference correlation analysis method based on the tendency factor (TTDCA), which can be expressed as formula (6).

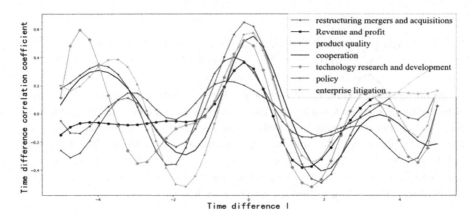

Fig. 2. The correlation of the time difference between news topic and Total income from FTZ (TDCA)

$$R_{x,y} = \max_l RF \; \frac{\sum_{i=1}^{k}(x_{l+i} - \bar{x})(yi - \bar{y})}{\sqrt{\sum_{i=1}^{k}(x_{l+i} - \bar{x})^2 \sum_{i=1}^{k}(y_i - \bar{y})^2}}, \; l = 0, \pm 1, \pm 2, \pm 3 \ldots \pm D; RF = \pm 1 \quad (8)$$

The algorithm for causal lag analysis is described as follows:

Algorithm 2. Causal lag analysis

Intput: Time series theme value:topic[ym][tc];Correlation Factor:RF;   Time series of related indexes:index[ym];

Output: Topic Set

1: //Initialize a set of sequence of topics and the corresponding time difference:
   Topic Set⟨topic[ym][tc], TD⟩ = {}

2: **for all** tc$_k$ **in** Set ⟨topics⟩ **do**

3:     //The maximum correlation coefficient calculated by the formula 5 to the subject and the related index

       r=R(RF,topic[tc$_k$],index)

4:     **if** r >0.3 **then**

5:         TD =l$_r$

6:         Topic Set.add(topic[tc$_k$],TD)

7:     **end if**

8: **end for**

We can get the correlation coefficient between each topic and related indicators, so that we can filter topics and find out the time difference d by above algorithm.

3.4   Situation Analysis of Related Indexes

Because of the short formation time of China's various fields, its statistical data are small and unstable. The traditional method based on empirical risk minimization cannot achieve good results. SVR is an application of support vector machine in solving regression problems. It adopts the idea of structural risk minimization, has better learning performance and generalization performance, and has the ability to deal with complex problems such as small samples and nonlinear problems. Moreover, it has achieved good results in many fields [5, 13]. Therefore, the result of Algorithm 2 are used to filter the reference topics and get its time difference TD, and the SVR model is used to make the regression prediction. The algorithm is as follows.

Algorithm 3. Causal lag analysis

Intput: topic Set; Internet Statistics set $\langle IS \rangle$;; $Time series of related indexes : index[ym]$

Output: predict

1: Initialize datum variables Set={};

2: //The topic sequence is adjusted according to the corresponding time difference topic set.d
and inputted as the reference variables of the model. ;

3: **for all** topic set **do**

datum variables Set.add(topic set.topic[t1-topic set.TD : $t2 - topic set.TD$])

4: **end for**

5: //Internet Statistics are inputted as the reference variables of the model.;

6: **for all** Internet Statistics set **do**

datum variables Set.add(Internet Statistic[t1-IS.TD : $t2 - IS.TD$]))

7: **end for**

8: //The predict value are calculated according to the SVR model ;

9: predict=SVR(datum variables Set[t1 :t2],index[t1 :t2]).predict(datum variables Set[t2+1])

---

4   Experiments and Analysis

4.1   Data Sources

The main purpose of this experiment is to study the correlation between Internet data and internal data, and make a trend analysis of the internal relevant indicators of FTZ by combining external data. Therefore, the data set mainly includes the private internal data and the external Internet data.

Internal data: Statistical data (structured data) within the FTZ. The shared data of the FTZ platform is a non-public data, so it will be presented in the form of normalization.

External data: More than 40000 relevant news text data (unstructured data) are obtained by means of web crawler based on the domain keyword search. Meanwhile, the relevant index data (structured data), numeric type, is captured on the relevant statistics department website, such as copper price, crude oil price, US dollar against RMB exchange rate, GDP, CPI, PPI and so on.

4.2   Experimental Results and Analysis

Experiment 1. According to the domain knowledge, the theme of financial news can be divided into: restructuring mergers and acquisitions, revenue and profit, technology research and development, cooperation, policy, enterprise litigation and others. In the experiment, 1500 news items were manually tagged, of which 1200 were used as training set and 300 were used as test set. The cross validation method was used to train the model. The average accuracy rate is 81.3%, which can be used as a good classifier.

Therefore, the classifier constructed by the above method divides 40026 documents on the subject, and the results are as follows.

Experiment 2. Text topics are quantified in time by using the result of the topic partition in Table 1. Then the correlation between the text topics and the benchmark variable is determined by the causal lag analysis.

Table 1. The number distribution of each topics

Restructuring merger and acquisition	Product quality	Revenues and profits
9736	6914	6378
Technology research and development	Policy	Enterprise litigation
6507	7452	6067

Figure 2 is obtained by the TDCA method and Fig. 3 is obtained by using the TTDCA method. As can be seen from Fig. 3, there is an interdependency between each topic and a certain index.

Most of the indicators get the maximum correlation when l is 0. However technology research and development and enterprise litigation have a lag effect on the index, and the lag time is 5 months and 2 months. This shows that the effect of the new product R&D will benefit after 5 months which means that the new product R&D enterprises will produce benefits after 5 months. The new product R&D exert an influence after 5 months. Business litigation will have a negative impact on the index after two months.

In the same way, the formula 4 can be used to calculate the correlation between the internet statistics and the reference variables (Fig. 4).

From Fig. 3, it can be found that the statistical data has a certain correlation with the related indexes, and there is a time difference of the advance or the lag effect between them.

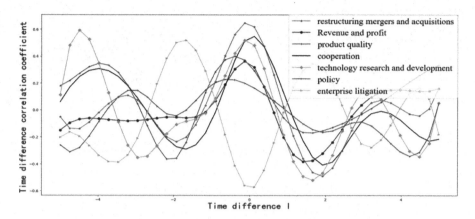

Fig. 3. The correlation of the time difference between news topic and Total income from FTZ (TTDCA)

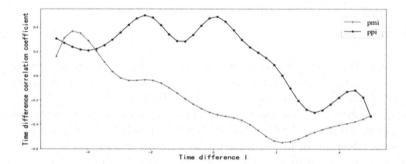

Fig. 4. The correlation of the time difference between Internet Statistics and Total income from FTZ

Experiment 3. This paper chooses the CPI, PPI, financial news, and the free trade zone platform data from 2014 to 2016 as experimental data. Multiple groups of time series will be obtained by means of the proposed method in this paper. We use TTCA and TTDCA to do the causal analysis and get multiple sets of time series. The time series are used as inputs of the SVR prediction model (TDCA − SVR and TTDCA − SVR) to predict Total income from FTZ (TIFTZ). At the same time, we use the ARIMA method predict the TIFTZ by only using internal data. Then the three methods are compared by the experiments.

Goodness of fit refers to the fitting degree of predict value and observation data. The statistic of goodness of fit is R^2. The maximum value of R^2 is 1. The closer the R^2 value is to 1, the better the fitting degree. From Fig. 5, Table 2, it is found that our method is better than the other two methods.

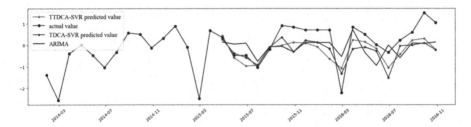

Fig. 5. Experimental prediction results of Total income from FTZ

Table 2. Performance evaluation index

Predict method	R^2
TDCA-SVR	0.758
TTDCA-SVR (our method)	0.760
ARIMA	0.484

5   Conclusion

In this paper, the above methods are used to deal with the problem of inconsistency of multi-source heterogeneous in the content, structure and time and make a situation analysis of related indicators in specific fields. There are three main steps in the method, including text topic quantization, causal lag analysis and situation analysis. The experimental results show that this method has a good effect on the analysis of the situation development of the FTZ, and it has been successfully applied to the wind control cloud service platform in FTZ. However, there is a lot of space to be raised in this way. In further work, we will try to use the method of unsupervised topic extraction and emotion calculation in the text quantization template so that this framework is applied in more fields. Furthermore, we can make comparisons with the significant methods.

Acknowledgement. This work was supported by National Natural Science Foundation of China (No. 61472004, 61602109), Shanghai Science and Technology Innovation Action Plan Project (No. 16511100903), and by The Key Laboratory of Embedded System and Service Computing of Tongji University of Ministry Education (2015).

References

1. Boutell, M.R., Luo, J., Shen, X., Brown, C.M.: Learning multi-label scene classification. Pattern Recogn. **37**(9), 1757–1771 (2004)
2. He, Z., Wu, J., Lv, P.: Label correlation mixture model for multi-label text categorization. In: Spoken Language Technology Workshop, pp. 83–88. IEEE, South Lake Tahoe (2015)
3. Zhang, M.L., Zhou, Z.H.: ML-KNN: a lazy learning approach to multi-label learning. Pattern Recogn. **40**(7), 2038–2048 (2007)

4. Sehgal, V., Song, C.: SOPS: stock prediction using web sentiment. In: IEEE International Conference on Data Mining Workshops, ICDM Workshops, pp. 21–26. IEEE, Omaha (2008)
5. Li, X., Xie, H., Chen, L., Wang, J., Deng, X.: News impact on stock price return via sentiment analysis. Knowl. Based Syst. **69**(1), 14–23 (2014)
6. Li, J., Xu, Z., Xu, H., Tang, L., Yu, L.: Forecasting oil price trends with sentiment of online news articles. Proc. Comput. Sci. **91**, 1081–1087 (2016)
7. Paninski, L.: Estimation of entropy and mutual information. Neural Comput. **15**(6), 1191–1253 (2014)
8. Xue, L., Xiong, Y., Zhu, Y., Wu, J., Chen, Z.: Stock Trend Prediction by Classifying Aggregative Web Topic-Opinion. In: Pei, J., Tseng, Vincent S., Cao, L., Motoda, H., Xu, G. (eds.) PAKDD 2013. LNCS, vol. 7819, pp. 173–184. Springer, Heidelberg (2013). https://doi.org/10.1007/978-3-642-37456-2_15
9. Dueñas-Fernández, R., Velásquez, J.D., L'Huillier, G.: Detecting trends on the web: a multidisciplinary approach. Inf. Fusion **20**, 129–135 (2014)
10. Shroff, G., Agarwal, P., Dey, L.: Enterprise information fusion for real-time business intelligence. In: Proceedings of the International Conference on Information Fusion, pp. 1–8. IEEE (2011)
11. Yang, J.H., Yang, M.S.: A control chart pattern recognition system using a statistical correlation coefficient method. Comput. Ind. Eng. **48**(2), 205–221 (2005)
12. Guo, Z.H., Wu, J., Lu, H.Y., Wang, J.Z.: A case study on a hybrid wind speed forecasting method using BP neural network. Knowl. Based Syst. **24**(7), 1048–1056 (2011)
13. Bao, Y.K., Liu, Z.T., Guo, L., Wang, W.: Forecasting stock composite index by fuzzy support vector machines regression. In: International Conference on Machine Learning and Cybernetics, vol. 6, pp. 3535–3540. IEEE (2005)
14. Mccallum, A.: A comparison of event models for Naive Bayes text classification. In: Proceedings of AAAI-1998 Workshop on Learning for Text Categorization, vol. 62, pp. 41–48 (1998)

Multi-scale DenseNet-Based Electricity Theft Detection

Bo Li[1], Kele Xu[2,3], Xiaoyan Cui[1(✉)], Yiheng Wang[4], Xinbo Ai[1],
and Yanbo Wang[5]

[1] Beijing University of Posts and Telecommunications, Beijing 100876, China
deepblue.lb@gmail.com
[2] School of Computer, National University of Defense Technology,
Changsha 410073, China
kelele.xu@gmail.com
[3] School of Information Communication, National University of Defense
Technology, Wuhan 430015, China
[4] The University of Melbourne, Parkville 3010, Australia
yihengwl@student.unimelb.edu.au
[5] China Minsheng Bank, Beijing 100031, China
wangyanbo@cmbc.com.cn

Abstract. Electricity theft detection issue has drawn lots of attention during last decades. Timely identification of the electricity theft in the power system is crucial for the safety and availability of the system. Although sustainable efforts have been made, the detection task remains challenging and falls short of accuracy and efficiency, especially with the increase of the data size. Recently, convolutional neural network-based methods have achieved better performance in comparison with traditional methods, which employ handcrafted features and shallow-architecture classifiers. In this paper, we present a novel approach for automatic detection by using a multi-scale dense connected convolution neural network (multi-scale DenseNet) in order to capture the long-term and short-term periodic features within the sequential data. We compare the proposed approaches with the classical algorithms, and the experimental results demonstrate that the multi-scale DenseNet approach can significantly improve the accuracy of the detection. Moreover, our method is scalable, enabling larger data processing while no handcrafted feature engineering is needed.

Keywords: Electricity theft detection · Convolutional neural network
DenseNet · Multi-scale

1 Introduction

In many countries, power utilities suffered a dramatic economic loss due to the electricity theft behavior, which has been a notorious problem in traditional power systems. Take the United States as an example, electricity theft makes power suppliers lose around six billion dollars every year [1]. In China, the electricity theft phenomenon is more common than before. Compared to previous mechanical meter, smart meter is

© Springer International Publishing AG, part of Springer Nature 2018
D.-S. Huang et al. (Eds.): ICIC 2018, LNCS 10954, pp. 172–182, 2018.
https://doi.org/10.1007/978-3-319-95930-6_17

more vulnerable as it turns to suffer more advanced attacks. Thus, timely identification of the electricity theft in the power system is crucial for its safety and availability.

There are two kinds of losses, which affect the management and the safety of power utilities [2]: (1) Technical loss caused by internal actions in the power system. They are mainly induced from electrical system components such as transmission lines and power transformers; (2) Non-Technical Losses (NTL), which were usually caused by external operation in the power system or electricity theft behaviors.

As NTL has been a major problem for most of energy companies, in this paper, we focus on the NTL detection leveraging the data of daily electricity consumption, with the aim to improve the accuracy of the electricity theft detection. Currently, NTL electricity theft detection methods can be divided into three groups:

1. State-based detection [3, 4]: by using improved sensors and other special devices to monitor user's behavior, the advanced equipment can improve the accuracy of electricity theft detection and reduce the false-positive rate. However, this method requires lots of equipment costs and software costs;
2. Game theory-based detection [5], a game between the electric utility and the thief is created. Although these methods cannot lead to an optimal solution, a low-cost and reasonable result may be achieved to relief energy theft. However, it is still a challenge to formulate the utility function of all players. Here the players include distributors, regulators and thieves. In addition, formulating potential strategies is also an issue;
3. Machine learning-based detection [6–10], by employing the machine learning algorithms to learn the user's electricity consumption behavior data achieved from sensors, a classifier can be trained to detect fraudulent users and recognize their behavior patterns. As the power enterprises have mass of user's historical electricity consumption data, this technique can make full use of them. The data mining strategy cannot only enhance the accuracy of detection to reduce the error rate, but also economize high software costs and equipment costs, avoid faults due to human intervention.

The machine learning-based approach has gained great interests as the detection accuracy has been improved consistently by using larger data. On the other hand, this kind of approach is privacy-preserving as only daily consumption is used to explore the theft pattern. Indeed, sustainable efforts have been made using different machine-learning approaches. However, all these classifiers relied on the handcrafted features, which are domain-specific and data-specific. For example, different features provide different performance, and the performance varies dramatically on different data by using the same feature sets.

Recently, deep learning methods have achieved dramatic success in different fields, such as, image classification [11], speech recognition [12]. Unlike previous efforts, which employ the handcrafted features and the shallow-architecture classifier, deep learning can be used to learn the feature in an automatic manner without the domain-specific constraints. On the other hand, with the dramatic increase of the data size, how to make fully use the big-data draws lots of attention, especially with the advance of the deep learning technique in the machine-learning field.

Here, we explore the use of convolutional neural network [11] for the detection task. Inspired by the statistical model approach, the periodicity of sequential data is of great importance for the classifier, and the series may have weekly, monthly, seasonal or annual periodicity. For the electricity theft detection, the pattern of electricity consumption is very salient for different user. Thus, efficient description the periodicity can be very helpful to improve the accuracy of electricity theft detection. Concretely, we propose to modify the multi-scale DenseNet, which can automatically capture the long-term and short-term periodic features of the sequential data. In summary, our contributions come with two folds:

- We propose to use DenseNet-based convolutional neural network to analyze the time series, with the aim to predict the probability of the theft behavior. Electricity consumption data has very strong periodicity, and convolutional neural networks can automatically capture some periodic features, different structures of networks can capture different kinds of periodic features, thus overcome the problems that classical machine learning model cannot get the periodic features;
- To capture more salient and abstract periodic features, we propose to modify the traditional dense block. Since the multi-scale dense block is given, the periodic features can be extracted within different time scales. Moreover, we suggest modifying the connection sequence of the layers in the dense block. On one hand, it greatly reduces the vibration of the training error, which makes the prediction more stable. On the other hand, it improves the accuracy significantly and provides superior performance.

Based on the aforementioned modifications, the accuracy of detection has been significantly improved. The rest of the paper is organized as follows. Section 2 gives the short summary on related work. Section 3 presents the proposed methods, while Sect. 4 provides the comparison experiment results using various algorithms. In the end, Sect. 5 draws conclusions of this paper.

2 Related Work

Sustainable efforts have been made to analyze the user's behavior patterns of electricity consumption, using the machine learning algorithm to establish a classification model to solve the problem. In summary, the classification methods include: statistical models [13], random forest [14], neural network [15], Support Vector Machines (SVM) [16]. In more detail, in [7], the author used SVM to detect electricity theft based on user's historical electricity consumption data over two years. This method used the past electricity consumption data. The problem is the longer-term data may be a noise for the model. In addition, it is of difficulties for the SVM classifier to learn the periodic features directly from the raw data. Moreover, the SVM is very time-consuming when using a large amount of data.

In [10], the author made an attempt to solve the problem by combining the decision tree and the SVM. This strategy improved the accuracy compared with the single SVM. However, this method is still hard to describe the periodic features. In [6], the author made use of statistical features, including the mean of power consumption, minimum of consumption, maximum of consumption, sum of consumption, standard deviation of power consumption, etc. Then, the k-means algorithm was employed to detect abnormal behavior. This kind of statistical features only depicts user's electricity consumption patterns in a certain aspect, which is not enough to improve the detection accuracy greatly. The handcrafted features replied on the domain knowledge, and the dimension of training data would be very large, which would induce a sharp increase in training time, Moreover, the features are task-dependent and data-dependent.

In [9], the author proposed to employ an ensemble algorithm to improve the accuracy of the model. Various machine learning algorithms are combined, including decision tree, SVM and optimum path forest, which improve the accuracy of the model by 2%–10%. However, the computing cost is dramatically increased. This is not applicable to the real case in the practical industrial world.

3 DenseNet-Based Electricity Theft Detection

In this section, we firstly, explore the traditional CNN to classify the electricity daily consumption time-series, then the DenseNet-based classification method is given. Moreover, the proposed multi-scale DenseNet is described. To make a quantitative comparison between CNN-based methods [17–20] and the traditional hand-crafted features-based shallow architecture classifiers, random forest and gradient boosting machines are also tested in our experiments, which can act as the benchmarks. And a short summary of the random forest and gradient boosting machines is given at the end of this part.

3.1 Convolutional Neural Network

The main difference between CNNs and traditional neural networks is that: CNNs have an automatic feature extractor, which consists of a convolution layer and a down-sampling layer (or pooling layer). Generally, a convolution layer includes a couple of feature maps, and each has some neurons. Normally, the parameters of the convolution kernel are initialized randomly (or using same specified initialization methods), and will be adjusted during the training step. The advantages achieved by the convolution kernel are that: the connections between each layer are reduced, and it decreases the risk of overfitting. Sub-sampling is also called pooling, and it usually includes mean sampling and maximum sampling methods [22]. The convolution layer and the sub-sampling layer lower the complexity of the model and substantially reduce the number of parameters [24]. In this paper, as our task can be viewed univariable time-series classification, 1D-CNN is employed in this paper.

3.2 DenseNet

DenseNet is a CNN architecture with dense connections, which provides dramatic performance improvement with comparison to previous CNN architectures. The DenseNet consists of many dense blocks, and each dense block connects to a transition layer except the last one which connects to the global pooling and the fully-connected layer directly. There is a direct connection between any two layers in each dense block. The input to each layer of the network is the union of the outputs from all the previous layers. In the network, each layer is connected with the previous layer directly in order to re-utilize features. At the same time, each layer is designed narrow to restrict the number of feature maps that can be learned and reduce the redundancy. Compared with the classical CNN, DenseNet not only performs better in image classification, but also has a higher utilization rate for the original data and less feature information loss. It reinforces feature propagation, supports feature re-utilization, solves vanishing gradient problems effectively and significantly reduce the number of parameters to economize the training time.

In the architecture, the transition Layer is a stacked series of nonlinear transformations, including Bach Normalization, ReLU and 1D-convolutional layer. Notably, the last Dense Block connects to a average pooling layer and a fully connected layer instead of a transition layer. Finally, the classification result is obtained by sigmoid function.

3.3 Multi-scale DenseNet

In the classical DenseNet, the size of the convolution kernel is very small (such as 3), thus it is difficult to get the middle-term and long-term periodical features, which are very important to improve the classification accuracy. Moreover, the kernel size is fixed in a small range. Different kinds of periodical features will be achieved if use a large range, and will help to get more abstract features after combine all the periodical features. Finally, if only modify the classical DenseNet in order to fit the one-dimensional data rather than change the whole structure, the outputs of the model will be unstable. Therefore, the traditional DenseNet architecture is not suitable for the electricity theft detection. To address aforementioned issues, two main modifications are proposed for the DenseNet architecture:

- The new multi-scale dense block is designed with a goal to capture the long-term and short-term periodic features within the sequential data. There are 24 different convolution layers in each multi-scale Dense Block, each layer has different length of convolution kernels. The length of the convolution kernel is longer in the upper layer and can generate more feature maps to extract long-term and medium-term user electricity features. In the lower layer, the convolution kernel receives all outputs in the previous layers and the original input. In consequence, it allows us to decrease the length of the convolution kernel and prompt the feature maps to grasp more short-term periodic features of electricity consumption. Since the network structure supports to re-utilize the feature maps, the multi-scale dense block concatenates all periodic features with different lengths into the transition layer for further manipulations. After this block, the model can extract a certain number of long and short-term user electricity behavior features from the original features and integrate them into the next layer.
- According to the Bach normalization-ReLU-Convolution layer connections in classical dense blocks, the outputs of 1D-DenseNet is unstable. In order to maintain the stability of the model and reduce the fluctuation of the outputs, the order of the

connections between each convolution part in dense blocks is modified into Convolution-Bach normalization-ReLU (as shown in Fig. 1.)

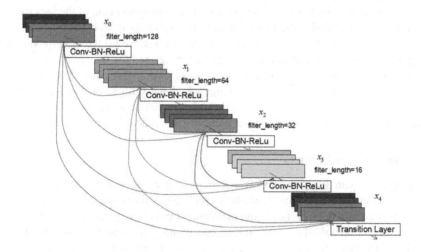

Fig. 1. Multi-scale dense block

3.4 Shallow Architecture Classifiers

In this part, a brief summary is conducted on two kinds widely-used shallow classifiers: Random Forest and Gradient Boosting Machines.

Random Forest. Random Forest (RF) is a decision tree-based ensemble algorithm. By constructing multiple fully-grown decision trees, the final forecast is obtained by voting for each of their single prediction. In a RF model, all decision trees are trained independently, and usually do not need to prune. Different decision trees may learn different rules from the data. If it is a classification problem, each decision tree will get a prediction according to the result of its learning, then all the decision trees will vote to get the final prediction. If it is a regression problem, the prediction of each tree is averaged to get the final prediction. The stochastic characteristics for RF can be summarized in two aspects: sampling by rows and sampling by columns (features). These two strategies cannot only offset some negative effects induced by outliers, but also prevent overfitting to those significant features. To sum up, RF performs well in accuracy and has strong robustness.

Gradient Boosting Machines. Gradient Boosting Machines (GBMs) [21] is a tree-based gradient boosting algorithm. By continuously fitting the residuals of the training samples, each new tree reduces the errors produced from the prediction of the previous tree. The strategy of reducing residuals greatly improves the prediction accuracy of the model. GBMs trains faster and more efficiently than many other algorithms, thus it is a popular machine learning algorithm.

4   Experimental Results

The data includes real electricity usage records of nearly 10,000 users collected in China. Each sample includes the daily electricity consumption data of the user over the past year. The data label is whether electricity theft behavior exits in the time-series. As we all know, the real data is often having missing values, we fill the missing values by using the Lagrange interpolation method [23]. Before feed the data into the model, Z-score standardization is employed to pre-process data. After using Z-score normalization, the data is transformed into the range which has 0 for mean and 1 for variance. On the one hand, standardization can reduce the impact of outliers on the model, and improve the performance, on the other hand, it can reduce the computation cost.

4.1   Handcrafted Features

In accordance with users' historical electricity consumption data, a series of handcrafted features are created for the shallow-architecture classifiers. In brief, the features can be divided into three parts: (1) the maximum, minimum, mean, variance, mean and median for the records during the recent one month, two months, three months, six months and a year; (2) the average electricity consumption for each month; (3) the maximum, minimum, variance, median, skew and divergence for the data generated in part two. These three parts of features will be used in Random Forest and Gradient Boosting Machines.

4.2   Performance Measures

Logloss and AUC are used to evaluate the performance of different models [25, 26]. Logloss is also called cross entropy, which is a commonly used loss function for the classification. A lower logloss indicates better model performance.

AUC is a measure of the quality of the model. The AUC is the area of the area covered by the ROC curve. The value of AUC is generally between 0.5 and 1. Obviously, a higher AUC value corresponds to better performances of the model.

4.3   Experimental Results

And the experimental whole flow chart is shown below. To ensure the stability and reliability of the results, all our experiments are conducted using 5 folds cross-validation. The whole framework of our experiments is given in Fig. 2.

Experiment 1: Random Forest. Firstly, the Random Forest method is tested on the data by using the aforementioned handcrafted features. By a large number of experiments, the ranges of parameters are shrunken within reasonable bounds. To sum up, the best prediction results for Random Forest are 0.289135 for logloss and 0.8430 for AUC.

Experiment 2: Gradient Boosting Machines. LightGBM model, which is an excellent implementation of GBM, are employed in our experiments. In order to find the best performance by using GBM model, grid search is used to find the best

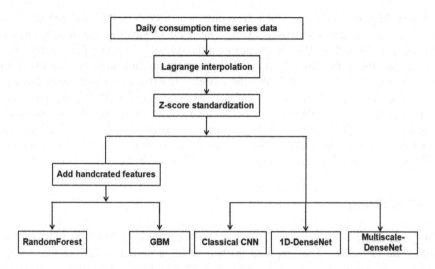

Fig. 2. Framework of electricity theft detection algorithms

parameters with small learning rate. The best performance using LightGBM is 0.287355 for logloss, and AUC is 0.8464.

Experiment 3: Classical Convolutional Neural Network. An user's daily electricity consumption information for a year can be considered as a time series and manipulated by 1D convolution. The best results of classical CNN is 0.2833 for logloss, and AUC is 0.85728. In accordance with these experiments, when the length of the convolution kernel is 7, the model learns some weekly electricity consumption features. Whereas, when the value increases to 14 or even higher, the model cannot catch longer periodic features as we expected. Besides, two fully connection layers with 64 and 32 kernels is good enough to get a better result than LightGBM, both increase and decrease the number of kernels will reduce the performance of the model.

Experiment 4: 1D-DenseNet. After modified the convolution kernel in DenseNet into 1D-convolution kernel and properly changed the structure of the network, the improved 1D-DenseNet was employed into the dataset and achieved better results. The best performance using 1D-DenseNet is 0.27880 for logloss, and AUC is 0.86314. Take one 1D-DenseNet for example, there are two dense blocks in this structure and the first dense block is composed of 6 identical parts. There are two convolution layers in each part. The first layer has 128 kernels and the length of the kernel is 1. The second layer has 32 kernels and the length of the kernel is 3. Setting the length of the kernel 3 to get short-term periodical features, compared with 1D-DenseNet-A, the deeper structure is employed to extract more abstract features. However, if the model has 3 Dense Blocks, it tends to overfitting and does not perform well on the test set.

Experiment 5: Multi-scale DenseNet. The proposed Multi-scale DenseNet is based on the two aspects of improvement on traditional dense block. The first dense block is composed of 6 parts with same structures, each part has 4 convolution layers and their length of kernels decrease in order. The output of a layer will be merged into next layer's input and generate new features. This kind of structure extracts features in different periods and completely obtains periodic rules of electricity behaviors. As we can see from these structures, the network that contains a single multi-scale dense block performs very bad. The reason is that the model only extract various periodic features. The lack of those advanced characteristics leads to underfitting. By adding the second multi-scale dense block, the network boosts a lot due to the further abstraction for the caught features. we can learn that the performance has greatly been improved, and the outputs are more stable than the 1D-DenseNet. In all the models, Multi-scale DenseNet performances best. The improved logloss is 0. 2585 while the AUC value is 0. 8670.

5 Conclusion

This paper presents a novel classification method for electricity theft detection using the multi-scale DenseNet. The proposed method builds a new deep architecture for the CNN to investigate the time-series data. The CNN architecture mainly employs the multi-scale dense block to capture the salient patterns of the time-series data at different time scales.

In our experiments, both RF and GBM do not perform well for our time-series classification task, and the handcrafted features are domain-specific. Compared to the shallow-architectures classifiers, the conventional convolutional neural networks and 1D-DenseNet provide better performance as they can extract periodic features from the raw data. The proposed multi-scale DenseNet remarkably enhances the model performance and enabling for timely identification problem. The key advantages of proposed multi-scale DenseNet are: (1) Multi-scale DenseNet can capture periodic information from the electricity records, in an end-to-end manner without handcrafted feature engineering. (2) The multi-scale dense block extracts seasonal trends in different periods, hence extracted features have more discriminative power, (3) Feature extraction and classification are unified in one CNN model, and their performances are mutually enhanced. Through the experiments, we demonstrate the proposed multi-scale DenseNet outperforms other models, and the AUC is improved from 0.8430 to 0.8670. We, therefore, believe that the proposed method can serve as a competitive tool of feature learning and classification of our time-series classification problem.

References

1. Mcdaniel, P., Mclaughlin, S.: Security and privacy challenges in the smart grid. IEEE Secur. Priv. 7, 75–77 (2009)
2. Navani, J.P., Sharma, N.K., Sapra, S.: Technical and non-technical losses in power system and its economic consequence in Indian economy. Int. J. Electr. Comput. Sci. Eng. 1(2), 757–761 (2012)
3. Lo, C.H., Ansari, N.: CONSUMER: a novel hybrid intrusion detection system for distribution networks in smart grid. IEEE Tran. Emer. Topic Comput. 1, 33–34 (2013)

4. Xiao, Z., Xiao, Y., Du, H.C.: Non-repudiation in neighborhood area networks for smart grid. Commun. Mag. IEEE. **51**, 18–26 (2015)
5. Cardenas, A.A., Amin, S., Schwartz, G., Dong, R.: A game theory model for electricity theft detection and privacy-aware control in AMI systems. In: 2012 50th Annual Allerton Conference on Communication, Control, and Computing (Allerton), pp. 1830–1837 (2015)
6. Angelos, E.W.S., Saavedra, O.R., Cortés, O.A.C., De Souza, A.N.: Detection and identification of abnormalities in customer consumptions in power distribution systems. IEEE Trans. Power Delivery **26**, 2436–2442 (2011)
7. Depuru, S.S.S.R., Wang, L., Devabhaktuni, V.: Support vector machine-based data classification for detection of electricity theft. In: Power Systems Conference and Exposition (PSCE), pp. 1–8 (2011)
8. Depuru, S.S.S.R., Wang, L., Devabhaktuni, V., Green, R.C.: High performance computing for detection of electricity theft. Int. J. Electr. Power Energ. Syst. **47**, 21–30 (2013)
9. Di, M., Decia, F., Molinelli, J., Fernández, A.: Improving electric fraud detection using class imbalance strategies. In: International Conference on Pattern Recognition Applications and Methods, vol. 3, pp. III-841–III-844 (2012)
10. Jindal, A., Dua, A., Kaur, K., Singh, M., Kumar, N., Mishra, S.: Decision tree and SVM-based data analytics for theft detection in smart grid. IEEE Trans. Ind. Inform. **12**, 1005–1016 (2016)
11. Krizhevsky, A., Hinton, G.E., Sutskever, I.: ImageNet classification with deep convolutional neural networks. In: Advances in Neural Information Processing Systems, vol. 25 (2012)
12. Hinton, G., Deng, L., Yu, D., Dahl, G.E., Mohamed, A., Jaitly, N., Senior, A., Vanhoucke, V., Nguyen, P., Sainath, T.N.: Deep neural networks for acoustic modeling in speech recognition: the shared views of four research groups. IEEE Sig. Process. Mag. **29**, 82–97 (2012)
13. Johnston, G.: Statistical Models and Methods for Lifetime Data, pp. 264–265. Wiley, New York (1982)
14. Svetnik, V., Liaw, A., Tong, C., Culberson, J.C., Sheridan, R.P., Feuston, B.P.: Random forest: a classification and regression tool for compound classification and QSAR modeling. J. Chem. Inf. Comput. Sci. **43**, 1947 (2003)
15. Haykin, S.: Neural Networks: A Comprehensive Foundation, pp. 71–80. Prentice Hall PTR, Upper Saddle River (1994)
16. Hearst, M.A., Dumais, S.T., Osman, E., Platt, J., Scholkopf, B.: Support vector machines. IEEE Int. Syst. Appl. **13**, 18–28 (1998)
17. Simonyan, K., Zisserman, A.: Very deep convolutional networks for large-scale image recognition. In: Proceedings of the IEEE Conference on Computer Vision and Pattern Recognition
18. Szegedy, C., Liu, W., Jia, Y., Sermanet, P., Reed, S., Anguelov, D., Erhan, D., Vanhoucke, V., Rabinovich, A.: Going deeper with convolutions. In: Proceedings of the IEEE Conference on Computer Vision and Pattern Recognition, pp. 1–9 (2014)
19. He, K., Zhang, X., Ren, S., Sun, J.: Deep residual learning for image recognition. In: Proceedings of the IEEE Conference on Computer Vision and Pattern Recognition, pp. 770–778 (2015)
20. Huang, G., Liu, Z., Van Der Maaten, L., Weinberger, K.Q.: Densely connected convolutional networks. In: Proceedings of the IEEE Conference on Computer Vision and Pattern Recognition (2016)
21. Friedman, J.H.: Greedy function approximation: a gradient boosting machine. Ann. Stat. **29**, 1189–1232 (2001)

22. Xu, K., Roussel, P., Csapo, T.G., Denby, B.: Convolutional neural network-based automatic classification of midsagittal tongue gestural targets using B-mode ultrasound images. J. Acoust. Soc. Am. **141**, EL531–EL537 (2017)

23. Berrut, J.P., Trefethen, L.N.: Barycentric lagrange interpolation. SIAM Rev. **46**, 501–517 (2004)

24. Xu, K., Feng, D., Mi, H.: Deep convolutional neural network-based early automated detection of diabetic retinopathy using fundus image. Molecules **22**, 2054 (2017)

25. Shore, J., Johnson, R.: Axiomatic derivation of the principle of maximum entropy and the principle of minimum cross-entropy. Inf. Theor. IEEE Trans. **26**, 26–37 (1980)

26. Huang, J., Ling, C.X.: Using AUC and accuracy in evaluating learning algorithms. IEEE Trans. Knowl. Data Eng. **17**, 299–310 (2005)

Resource Utilization Analysis of Alibaba Cloud

Li Deng[1,2], Yu-Lin Ren[1,2], Fei Xu[1,2], Heng He[1,2], and Chao Li[3(✉)]

[1] College of Computer Science and Technology, Wuhan University of Science and Technology, Wuhan 430065, China
[2] Hubei Province Key Laboratory of Intelligent Information Processing and Real-Time Industrial System, Wuhan 430065, China
[3] Department of Information Development and Management, Hubei University, Wuhan 430062, China
dengli@wust.edu.cn

Abstract. Currently, low resource utilization and high costs of cloud platform are becoming big challenges to cloud provider. However, due to confidentiality, few cloud platform providers are willing to publish resource utilization data of their cloud platform. This poses great difficulties in designing an effective cloud resource scheduler. Fortunately, Alibaba released its cloud resource usage data in September 2017. This paper analyzes Alibaba cloud trace data deeply from different aspects and reveals some important features of resource utilization. These features will help to design effective resource management approaches for cloud platform: (1) The maximum resource utilization of online services is closely related to their average utilization. (2) The longer a batch instance runs, the longer it may last. (3) The type of job that runs in a container can be estimated according to the amount of consumed resources and life time of this container. (4) Actual resources used by different batch jobs vary with time greatly and static resource allocation would make resource wasted seriously.

Keywords: Cloud platform · Online services · Batch jobs
Resource utilization ratio

1 Introduction

Nowadays, there are more and more computing nodes in cloud platform, such as Google Compute Engine, Amazon's EC2, Mesos [1], and OpenStack. Cloud platform provides flexibility and cost effectiveness for end-users and cloud operators. However, due to multi-tenant and diversity of load type, resource utilization rate of large-scale data centers is very low. Total CPU usage of thousands of servers in Twitter platform within one month is always lower than 20%. There are many researchers focusing on improving resource utilization of reserved resources in clusters [2]. Some pay attention to long-running services [3] or high-performance computing loads [4]. However, loads on real cloud computing platform are diversified and focusing on a certain type load

---

This work was supported by the National Natural Science Foundation of China (61602351).

does not effectively improve resource utilization of entire cloud platform. Moreover, due to the confidentiality of commercial application data, most of existing researches are based on simulation data [5] and lack of strong persuasiveness.

Recently, there are also some work based on real cloud trace. Reiss et al. [6] conducted detailed analysis of resource usage data in Google cloud platform, which uses Borg [7] as cluster manager. Reiss [6] reveals the heterogeneity of cloud computing and the dynamics of resource usage. Cortez et al. [8] also analyze loads in real cloud environment. But its cloud platform scheduler is based on virtual machines (VMs). The creation and destruction of virtual machines will incur greater costs, so resource utilization in these cloud environments may be lower than that in container-based environments [9]. The scheduler in Alibaba cloud platform is mainly based on Pouch, which is a rich container something like virtual machine. It has Init process and provides some system services. The research in [10] reveals that, resource distribution is unbalanced in Alibaba cloud. But it does not analyze the difference of resource usage for different types of workloads.

In this paper, our contributions are listed below: (1) This paper conducts trace analysis of Pouch container instead of virtual machine based on real cloud platform. (2) We summarize some characteristics of resource usage to help the design of effective resource management in cloud platform. (3) We also contrast the resource usage of Alibaba cloud trace and that of google trace.

2   Configuration and Workloads of Alibaba Cloud Platform

In September 2017, Alibaba released resource usage data of more than 1,300 servers for twelve consecutive hours. These data can be available at website https://github.com/Alibaba/clusterdata. There are seven trace files altogether (about 1 GB size). Two files are for resource usage description of computing nodes. Two files are for resource usage of batch loads, while the other two files are for online services. The last file is an explanatory document. Only two kinds of workloads, batch jobs and online services, are involved in the trace.

Each computing node in Alibaba cloud is configured with a certain amount of resources and its real resource usage is traced. Measurements of usage are taken in 60 s intervals. The average values over 300 s are recorded in trace files. From these files we know that, each batch job consists of at least one task and every task is composed of multiple instances. Each instance is configured with a fixed amount of resources (such as CPU, RAM). A batch instance is the minimum scheduling unit and runs on a specific computing node. The configuration and resource usage of each instance are recorded, such as start time, end time, the ·status of batch instance (failure or success), the maximum number of CPUs actually used, average number of CPUs used, the maximum memory usage (normalized value), average memory usage (normalized value). Similarly, configuration and running status of each online service are recorded, such as the amount of resource requested, actual resource usage, average CPU load, average cycles per instruction (CPI), cache miss.

For reasons of commercial confidentiality, the tracking data does not provide accurate information of workloads and resource configuration. Instead, *jobID*, *taskID*, and *containerID* are used to mark different types of workloads. These identifiers can be used to infer whether the application is a batch job or an online service. But the specific details of jobs can't be known. Also, memory size and disk capacities are respectively normalized based on the maximum values. The configuration of Alibaba cloud platform is shown in Table 1.

Table 1. The configuration of Alibaba cloud platform.

Number of machines	CPUs	Normalized memory
736	64	0.6899697150104833
365	64	0.6900007765381927
199	64	0.6900059534594777
7	64	0.9999585846297208
5	64	1
1	64	0.5747883933424792

From Table 1, it can be seen that the number of CPU cores of all nodes are the same. But there are multiple memory size. The memory size of all nodes exceeds the half of the maximum size. Among 1,313 machines, there were 39 nodes (about 3% of the whole) with soft errors during the tracking process. Two nodes had soft errors due to insufficient disks. But there was no any hard error happening. When a server has software or hardware errors, existing services and jobs can still be running on the node although new online services and batch jobs should not be dispatched to it. Disk capacity is also normalized. And almost all the nodes have the same disk capacity.

3   Characteristics of Resource Usage in Alibaba Cloud

In Alibaba cloud trace, only two kinds of workloads, batch jobs and online services, are involved. Each batch job contains multiple tasks. And every task can be divided into several instances. Each instance independently runs on servers. Online service is long-running service like mail service, news website and etc. These two kinds of workloads share same servers in Alibaba cloud. By default, resource utilization ratio means the ratio of the amount of resource actually used to what have been applied. Timestamp is relative to the start of trace period in seconds.

In this section, resource usage status of physical nodes is first discussed. Then, resource usage of online service and batch jobs is respectively studied. We summarize the characteristics of resource usage in Alibaba cloud in the end.

3.1   Resource Usage of Servers

There are 1,313 servers involved in the trace. These machines have the same number of CPU cores and nearly the same disk capacity. But they have different memory size. According to normalized value of memory size, almost 99% of machines have the same normalized memory size 0.69. Only 13 servers have different memory size, larger or less. All in all, Alibaba cloud is nearly homogeneous, which can make task scheduling easy to some degree. Online services and batch jobs share these servers at the same time. Resource usage status of a server is mainly based on resource demands of workloads residing on the server.

Figure 1 depicts resource usage of servers varying with time. At each timestamp, the minimum measurement (denoted as *min* in Fig. 1) and average resource usage (denoted as *ave* in Fig. 1) of all the servers are recorded. To accurately reflecting real resource usage, we use the 95th percentile measurement (denoted as *95th* in Fig. 1) rather than the maximum value. Also, average variance of all the servers is computed at each timestamp. Figure 1(a) shows CPU resource usage varying with time. It can be seen that during the entire tracking process, average CPU usage was between 10% and 45%, while the minimum CPU usage was almost close to 0. The 95th percentile CPU usage was fluctuant between 30% and 60%. Figure 1(b) lists dynamic memory resource usage. It is known that average memory usage rate fluctuated between 35% and 65%, while the 95th percentile memory usage was between 45% and 80%. Just like CPU, the minimum memory usage was near close to 0 all the time. There were big gaps both on CPU and memory usage among different servers during the tracking. And, memory usage was larger than CPU usage, which means that more CPUs were idle and much energy was wasted. Figure 1(c) shows average variance of resource usage at each timestamp. Average variance is enlarged 10,000 times to make results clearer. It is obvious that, memory usage fluctuates significantly among different servers at the same timestamp, while CPU usage fluctuation is relatively small.

To understand the distribution of servers on resource usage, cumulative distribution function (CDF) of CPU and memory is respectively figured out. For each server, average resource usage (denoted as *ave* in Fig. 2), the minimum (denoted as *min* in Fig. 2) and the 95th percentile measurement (denoted as *95th* in Fig. 2) during the whole tracking process are respectively computed first. Based on these values, the corresponding CDF is then figured out. Figure 2(a) depicts CDF of CPU usage ratio. It shows that, there are 90% of nodes whose average usage did not exceed 35% and the minimum of all the servers had less than 30%. The minimum usage of 90% of nodes did not exceed 15%. 90% of nodes' 95th percentile usage was less than 50%. That shows, CPU usage ratio of most nodes were very low. Figure 2(b) depicts CDF of memory usage ratio. It shows that, 90% of nodes' average utilization rate did not exceed 60% and 90% of nodes' minimum usage did not exceed 45%. 95% of nodes' 95th percentile memory usage was less than 70%.

In summary, memory utilization rate of servers in Alibaba cloud platform is higher than CPU usage. More CPU resource was idle and much energy was wasted. Effective CPU configuration methods should be designed for less energy. At the same time, memory usage had large gaps among different servers. It can be considered that assigning workloads with different resource bottlenecks on a same server would decrease resource contention.

3.2   Resource Usage of Online Services

Online services refer to long-running, latency-critical services such as web service and mail service. There were 11,102 online services during the tracking period and 10,910

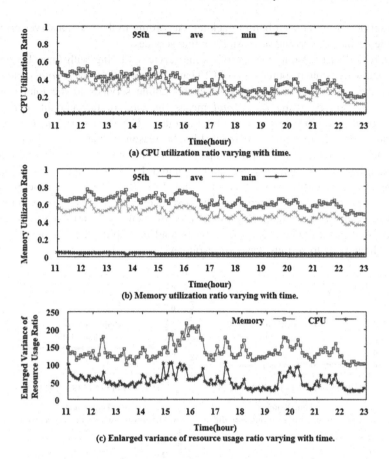

(a) CPU utilization ratio varying with time.

(b) Memory utilization ratio varying with time.

(c) Enlarged variance of resource usage ratio varying with time.

Fig. 1. Resource utilization ratio of servers varying with time.

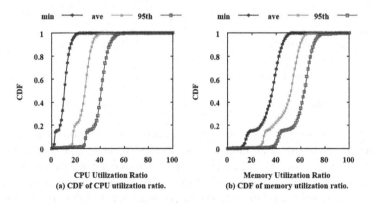

(a) CDF of CPU utilization ratio.

(b) CDF of memory utilization ratio.

Fig. 2. CDF of resource utilization ratio.

online services were created before tracking. According to the trace, a server hosted at least one online service and at most 19 online services.

Figure 3 depicts resource usage of online services varying with time. Figure 3(a) and (b) show dynamic resource ratio of CPU and memory. *Requested/Total* denotes the ratio of the amount of resource requested by online services to the total resource provided by servers, while *Used/Total* means the ratio of the amount of resource actually used by online service to the total resource. There are sharp changes in both subfigures about from hour 13 to hour 14. The reason is that, nearly half of measurements are missing at these timestamps. But it would not affect the overall trend of resource usage. These sharp changes can be ignored. It can be seen that, the amount of CPU requested by online services accounts for 70%–75% of the total amount, while only 6%–11% of total CPUs were actually used by online services. A large number of processor resource were idle. *Requested/Total* of memory is 80%–85%, while the actually used part occupies 32%–36%. Users tend to amplify their resource requirements which leads to serious waste of resources.

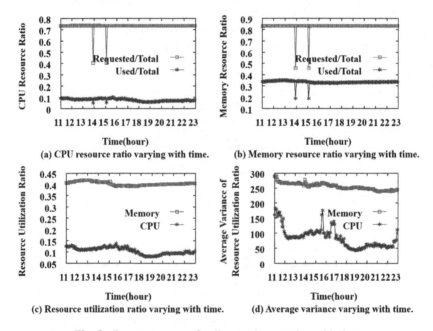

Fig. 3. Resource usage of online services varying with time.

Figure 3(c) depicts average resource utilization ratio of online services at each timestamp. Memory utilization ratio is about 40%–42%, while CPU usage is between 7% and 14%. Figure 3(d) shows average variance of resource utilization ratio. Variance is enlarged 10,000 times to make results clearer. It can be seen that, average variance of memory usage is between 230 and 300, larger than that of CPU. The difference of memory usage between multiple online services is larger. CPU usage fluctuation rate is relatively small. Although average memory usage rate is stable over time, different online services vary greatly in memory usage at the same timestamp and online services have polarization of memory usage. Average variance of CPU usage varies with time. Dynamic resource management would be suitable for processor resource.

Based on average utilization ratio (represented as *ave*) and the 95th percentile of CPU usage (denoted as *95th*) of each online service, Fig. 4 describes CDF of process utilization ratio. It can be shown that 80% of online services' average and 95th percentile usage are less than 20%. Most online services have low average CPU usage. Almost average CPU usage of all services is less than 40%. Although the 95th percentile CPU usage for most online services is basically very low, there are still nearly 5% of online services' 95th percentile CPU usage greater than 50%. This shows that when predicting the usage of online services, it is necessary to take some high CPU usage of online services into account. Combining prediction techniques with dynamic resource management would to some degree improve effective process usage.

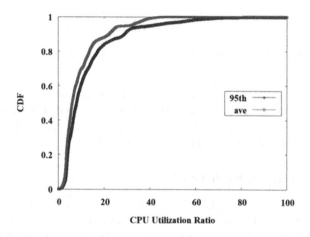

Fig. 4. CDF of CPU utilization ratio for online services.

According to average resource usage, online services are divided into two categories. The distribution of the maximum resource usage in different categories of online services is also analyzed. Figure 5 describes CDF of resource utilization ratio for classified online services. Figure 5(a) lists the distribution of memory resource usage. Online services are classified into two kinds (the set *more than 40%* and the set *less than 40%* in Fig. 5(a)) based on average memory utilization ratio. The threshold is 40%. Figure 5(b) lists the distribution of CPU usage. Online services are divided into two categories (the set *more than 30%* and the set *less than 30%* in Fig. 5(b)) according to the threshold 30% – average memory utilization ratio.

From Fig. 5 it can be seen that, for 50% of the set *more than 40%*, the maximum memory usage exceeds 60%, while the maximum is less than 40% for 90% of the set *less than 40%*. When average CPU usage exceeds 30%, 50% of the online services have a maximum CPU usage exceeding 50%. When average CPU usage is less than 30%, 99% of these online services have a maximum CPU usage of no more than 40%. It can be concluded that, if average usage of a container is low, its maximum usage rate

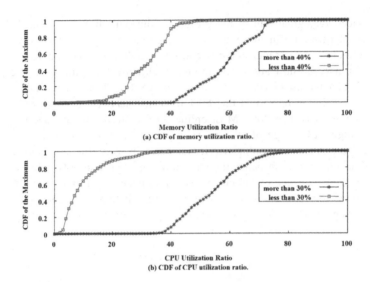

Fig. 5. CDF of resource utilization ratio for classified online services.

over the entire life cycle would be lower. Current resource allocation is static and the amount of allocated resource is determined based on predicted peak demand of applications in life cycle. The amount of resource for online services with low average utilization could be appropriately reduced.

3.3   Resource Usage of Batch Jobs

In the tracking period, there are 12,887 batch jobs, 80,177 batch tasks, and 11,777,514 batch instances. Some task may have 64,486 instances and every task at least has one instance. More than 2,487 batch instances are scheduled on a same node and one node hosts 10,798 instances at most according to the trace. For batch workloads, most memory data are missing in the trace. Therefore, only CPU usage is analyzed in the following part.

Figure 6 describes resource usage status, life time, and arrival rate of batch jobs. Figure 6(a) is based on the maximum number of CPUs actually used for each batch instance. 90% of batch instances actually use no more than 5 CPU, although there are still a small number of batch instances using 32 CPUs at most. Due to the fact that the number of CPUs allocated by Alibaba cloud scheduler is fixed, resource utilization is very low. A recommendation algorithm [2] can be used to test applications first and then to allocate appropriate resources. Thus, resource usage would be greatly improved.

Figure 6(b) shows the distribution of life cycle for batch instances. It can be seen that, for 60% of batch instances, life cycle is within 60 s, and for 90% of batch instances, life time is within 300 s. However, there are indeed some batch instances with long life time in the trace. The longest time is more than 3,000 s. But the number of such instances is very small. If a batch job or batch instance runs for a long time, it is likely that it will still run for a longer time. It can help to predict life time of batch jobs.

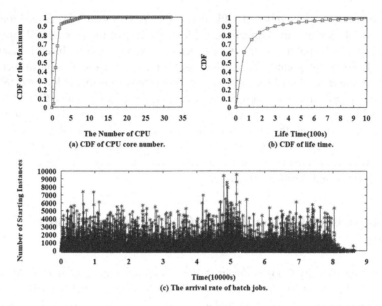

Fig. 6. Resource usage, lifetime, and arrival rate of batch jobs.

Figure 6(c) lists the number of starting instances at each timestamp. The number of batch instances scheduled per second varies greatly. The highest number is close to 10,000 instances at one timestamp. We can infer that the throughput of cloud scheduler must be very large. This feature is consistent with the analysis of azure [8]. Even if Alibaba's data only cover 12 h, this feature can still be seen.

3.4 Overall Characteristics of Resource Usage

According to above analysis, the characteristics of resource usage in Alibaba cloud platform can be summarized in the following: (1) If average resource usage rate of online service is relatively high, then it is likely to have a higher maximum CPU or memory usage rate. If average resource rate of online service is relatively low, then its maximum CPU or memory resource usage may be relatively low. (2) The longer a batch instance runs, the longer time it may need to complete. This is of great significance in predicting the complete time of a batch job. (3) If a container uses more CPU or memory than other containers, it is likely that the container is running batch jobs. If a container runs for a long time, the container is likely to run an online service. (4) Because resource usage of batch instances fluctuate widely, a fixed resource allocated will cause a lot of waste. If resource usage of batch instance can be predicted through application trial or other methods, resource utilization of cloud platform would be improved greatly. (5) From the view of the entire cloud platform or online service, the polarization of memory and CPU usage is significant.

Online services and batch jobs have different characteristics in resource usage. During the tracking period, the maximum number of CPU cores for online service is 12 and the maximum number of CPU cores used for batch instance is 32. If a container uses a relatively large number of CPU cores, it may be a batch instance. From the perspective of memory, the normalized value of the largest memory actually used by online service is 0.10506, and the largest memory actually used by batch instances is close to 1. From this point, if a container uses a relatively large amount of memory, then this container is likely to run a batch job. From the perspective of life time: 90% of the batch jobs' life time was within 5 min and online services were alive throughout the entire tracking period. So, if a container is running for a long time, it is likely to run an online service. Through the analysis of online services and batch jobs, if these characteristics can be used by some machine learning algorithms such as KNN [11] to classify workloads running in the containers, it would greatly improve cloud schedulers' efficiency and resources usage rate.

4   Comparison Between Alibaba Cloud Trace and Google Trace

The scheduler used by Google cluster is Borg based on regular container. Alibaba's scheduler is based on its own container – Pouch, which is a rich container and its internal application experience is similar to a virtual machine. Google has about 10 types of servers in its trace, as shown in Table 2. And Alibaba has published six types of servers, as shown in Table 1. Obviously, there is obvious heterogeneity in google cloud platform and Alibaba cloud platform is almost homogeneous.

Table 2. The normalized configuration of servers in google trace [6]

Number of machines	Normalized CPUs	Normalized memory
6732	0.50	0.50
3863	0.50	0.25
1001	0.50	0.75
795	1.00	1.00
126	0.25	0.25
52	0.50	0.12
5	0.50	0.03
5	0.50	0.97
3	1.00	0.50
1	0.50	0.06

Task submission rate published in google trace is up to more than 400 tasks one second, of which repeated tasks account for the majority. The number of instances submitted in Alibaba cloud is nearly 10,000. So, the throughput must be large enough when scheduler is designed. In addition, the number of re-submitted tasks in google trace reaches as many as 300 in one second sometimes, while the number of retried tasks in Alibaba trace is relatively small, which is less than 10. Production-level jobs in google trace use more CPU time than non-production-level jobs. However, CPU used by online services on Alibaba cloud platform is less than that of batch instances. This may be due to the large number of batch loads on Alibaba cloud platform. Compared to google tracking data analysis, this paper contributes to predicting the actual use of resources by workloads: When average resource usage rate of online services is relatively large, it may have a relatively large maximum resource usage rate. If a container uses more CPU or memory, it is very likely that a batch job runs in this container.

5 Conclusion

This paper analyzes Alibaba's cloud resource usage in detail and reveals some important characteristics from many aspects, such as resources usage used of online services, life time of batch instances, resource used by batch instance and the difference between batch instance and online service. These could help to design effective resource management for cloud platform by powerful analysis machine learning frameworks such as MLLib [12] and TensorFlow [13]. This paper also provides useful guidance for better integration of batch workload and online service to improve resource utilization in cloud platform.

References

1. Hindman, B., Konwinski, A., Zaharia, M., et al.: Mesos: a platform for fine-grained resource sharing in the data center. In: Proceedings of the 8th USENIX Conference on Networked Systems Design and Implementation, pp. 429–483. USENIX (2011)
2. Delimitrou, C., Kozyrakis, C.: Quasar: resource-efficient and QoS-aware cluster management. In: Proceedings of International Conference on Architectural Support for Programming Languages and Operating Systems, pp. 127–144. ACM (2014)
3. Poggi, N., Carrera, D., Gavalda, R., et al.: Characterization of workload and resource consumption for an online travel and booking site. In: Proceedings of IEEE International Symposium on Workload Characterization, pp. 1–10. IEEE (2010)
4. Zheng, Z., Yu, L., Tang, W., et al.: Co-analysis of RAS log and job log on Blue Gene/P. In: Proceedings of Parallel & Distributed Processing Symposium, pp. 840–851. IEEE (2011)
5. Li, C., Dai, B., Kuang, Z., et al.: Research on task scheduling with multiple constraints based on genetic algorithm in cloud computing environment. J. Chin. Comput. Syst. **38**(9), 1945–1949 (2017)
6. Reiss, C., Tumanov, A., Ganger, G.R., et al.: Heterogeneity and dynamicity of clouds at scale: google trace analysis. In: Proceedings of ACM Symposium on Cloud Computing, pp. 1–13. ACM (2012)
7. Verma, A., Pedrosa, L., Korupolu, M., et al.: Large-scale cluster management at Google with Borg. In: Proceedings of Tenth European Conference on Computer Systems. ACM (2015)
8. Cortez, E., Bonde, A., Muzio, A., et al.: Resource central: understanding and predicting workloads for improved resource management in large cloud platforms. In: Proceedings of Symposium on Operating Systems Principles, pp. 153–167 (2017)
9. Delimitrou, C., Kozyrakis, C.: HCloud: resource-efficient provisioning in shared cloud systems. In: Proceedings of International Conference on Architectural Support for Programming Languages and Operating Systems, pp. 473–488. ACM (2016)

10. Lu, C., Ye, K., Xu, G.: Imbalance in the cloud: an analysis on Alibaba cluster trace. In: Proceedings of IEEE International Conference on Big Data. IEEE (2017)
11. Yang, S., Zhang, Q.: Research on k-nearest neighbor text classification algorithm of approximation set of rough set. J. Chin. Comput. Syst. **38**(10), 2192–2196 (2017)
12. Meng, X., Bradley, J., Yavuz, B., et al.: MLlib: machine learning in apache spark. J. Mach. Learn. Res. **17**(1), 1235–1241 (2015)
13. Abadi, M., Barham, P., Chen, J., et al.: TensorFlow: a system for large-scale machine learning. In: Proceedings of the 12th USENIX Symposium on Operating System Design and Implementation, pp. 265–283. USENIX (2016)

Co-evolution Algorithm for Parameter Optimization of RBF Neural Networks for Rainfall-Runoff Forecasting

Jiansheng Wu[1,2(✉)]

[1] College of Mathematics and Computer, Guangxi Science and Technology
Normal University, Laibin 546199, Guangxi, People's Republic of China
Wjsh2002168@163.com
[2] School of Information Engineering, Wuhan University of Technology,
Wuhan 430070, Hubei, People's Republic of China

Abstract. In this paper, an effective co-evolution algorithm strategy is presented to optimize the parameters of Radial Basis Function Neural Networks by
incorporating the metropolis process of Simulated Annealing (SA) into the
movement mechanism and parallel processing of Particle swarm optimization
(PSO), namely HPSOSA, for rainfall-runoff forecasting model. Firstly, this
paper is constructed the co-evolutionary algorithm, the HPSOSA algorithm is
combined the advantage of fast computation of PSO, and the advantage of SA
searching in the direction of the global optimum solution, helping PSO jump out
of local optima, avoiding into the local optimal solution early and leading to a
good solution quality. Secondly, the co-evolution algorithm is used to optimize
the structures and parameters of RBF neural networks, namely HPSOSA-
RBFNN. Finally, The developed HPSOSA-RBFNN model is being applied for
daily rainfall-runoff forecasting in Liuzhou of Guangxi, China. The performance
of HPSOSA is compared to pure PSO in these Basis Function Neural Networks
design problems, showing the co-evolution algorithm strategy is more effective
global search ability and avoid falling into local solution. Experimental results
reveal that the predictions using the proposed approach are consistently better
than those obtained using the other methods presented in this study in terms of
the same measurements. Therefore, the HPSOSA-RBFNN model is a promising
alternative for rainfall-runoff forecasting tool.

Keywords: Co-evolution algorithm · Radial basis function neural networks
Simulated annealing algorithm · Particle swarm optimization
Rainfall-runoff forecasting

1 Introduction

The rainfall-runoff forecasting modeling plays an important role in optimizing and
planning water resources, managing reservoirs particularly during drought periods and
also has become a challenging task of research in hydrology recently, because accurate
and timely runoff prediction can avoid accidents, such as the life risk, economic losses,
etc. [1–3]. Developing a practical rainfall-runoff modeling has been a difficult subject in

© Springer International Publishing AG, part of Springer Nature 2018
D.-S. Huang et al. (Eds.): ICIC 2018, LNCS 10954, pp. 195–206, 2018.
https://doi.org/10.1007/978-3-319-95930-6_19

hydrology due to a variety of non-linear factors involved [3–5], such as, rainfall characteristics, watershed morphology, water level and soil moisture, etc. [6, 7]. Artificial Neural Network (ANN) techniques have been considered as the most effective modeling method for precipitation prediction than conventional statistical models, due to the ability of extracting essential characteristics relationships through empirical data [8, 9].

Radial Basis Function Neural Network (RBF-NN) has been successfully used in runoff prediction modeling, mainly because it is a universal approximations, so it can approximate any continuous function with any precision [10, 11]. In the application of the actual precipitation prediction, the effects of RBF-NN applications strongly depend upon the operator's experience for the lack of strict theoretical support. The technique of finding the suitable parameters of radial functions is very complex because inappropriate network is inefficient and sensitive to over-fitting. Such a model might be doing well in predicting past incidents, but unable to predict future events. Recently, many evolutionary algorithms have been proposed, such as Simulated Annealing (SA) [12] and Particle Swarm Optimization (PSO) [13], which has provided a more robust and efficient approach for solving complex real-world problems. By combining the two methods, the HPSOSA algorithm has the advantage of both fast calculation and searching in the direction of the global optimum solution, helping PSO jump out of local optima, avoiding into the local optimal solution early and leading to a good solution quality.

In order to overcome these drawbacks from RBFNN, it is necessary to find some effective approach to play the advantages of each algorithm, and avoid misleading in the local optimum. This paper propose a novel and specialized co-evolution strategy by incorporating SA into PSO to search for optimal or approximate optimal beginning connection weights and thresholds for the network, then using the RBFNN to adjust the final weights, namely HPSOSA-RBFNN. As such, this paper aims at identifying an optimal RBFNN's for rainfall-runoff forecasting model. The rest of this study is organized as follows. Section 2 describes the proposed HPSOSA-RBFNN, ideas and procedures. For further illustration, the method has been used to establish a prediction model for rainfall-runoff forecasting in Sect. 3 and conclusions are drawn in the final Section.

2 Radial Basis Function Neural Network

Radial Basis Function (RBF) networks were introduced into the neural network literature by Broomhead and Lowe [14]. The basic structure of a three layer RBF-NN is illustrated in Fig. 1. The network is generally composed of three layers: an input layer, a single hidden layer of nonlinear processing neurons, and an output layer. Suppose we are given training data $(x_t, y_t)_{t=1}^{N}$, where $x_t \in R^n$ is the input vector; $y_t \in R$ is the output value and n is the total number of data dimension. The output of the RBF-NN is calculated according to

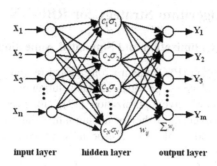

Fig. 1. The basic structure of a three layer RBF-NN

$$y_t = \sum_{i=1}^{N} \omega_{ti} \phi_i(x, c_i) = \sum_{i=1}^{N} \omega_{ti} \phi_i(\|x - c_i\|_2) \tag{1}$$

where $\phi_k(\cdot)$ is a radial basis function from R^n (set of all positive real numbers) to R, $\|\cdot\|$ denotes the Euclidean norm, ω_{ti} are the weights of the links that connects hidden neuron number i and output neuron number t in the output layer, N is the number of neurons in the hidden layer and c are the RBF centers in the input vector space[]. A normalized Gaussian function usually is used as the radial basis function as follows:

$$\phi_i(x, c_i) = \exp\left(-\|x - c_i\| / r_i^2\right) \tag{2}$$

where r_i denote the radius of the i-th node. It has been proved that when enough units are provided, RBF-NN can approximate any multivariate continuous function as much as desired. Thus, the main problem in RBF-NN design concerns establishing the number of hidden neurons to use and their centers and radii. Generally, the performance of RBF-NN depends mainly on the parameters of the Gauss function and the parameters of the connection weight. Among them, the parameters of the Gauss function are obtained by k-means clustering method and the parameters of the connection weight are obtained by least mean square or recursive least square [15].

The number of hidden, the location of the center, the value of the center location of center, the radii and the weights are very important for the performance of RBF-NN. When these values are not sufficient, the approximation offered by the net are sensitive to over-fitting or poor performances and these values interact with each other. Very small parameters will lead to over-fitting, and parameters large than necessary can give even worst results [16]. In RBF-NN actual runoff forecasting modelling application, establishing the number of neurons (that is, the number of centers and values related to them) is one of the most important tasks researchers have faced in this field. Thus, in order to obtain a good performance, network parameters need consider in coordination. In this paper a hybrid PSOSA algorithm is used to find the best components of the RBF-NN that approximate a function representing a rainfall-runoff time-series.

3 Co-evolution Algorithm Strategy for RBF-NN

3.1 Particle Swarm Optimization and Simulated Annealing

Particle swarm optimization (PSO) has a strong global optimization ability, which is more powerful, robust and able to provide accurate solution in the problem of nonlinear optimization and multi-objective optimization [13]. However, PSO has premature convergence, is easily trapped in the local optimum solution and is ineffective in balancing exploration and exploitation, especially in complex multi-peak search functions.

Simulated Annealing (SA) is a heuristic global optimization method inspired from the physical process of annealing of metals, which is based on the statistical thermo-dynamics to simulate the behavior of atomic arrangements in liquid or solid materials during the annealing process. A distinct feature of the SA algorithm is its integration with the Metropolis Monte Carlo procedure [12]. The Metropolis process in SA has a better ability of jumping away from a local optimum solution. However, SA requires very slow temperature variations, and therefore needs more calculation time.

3.2 Co-evolution Algorithm Strategy by Incorporating SA into PSO

To obtain the global or near global optimum solutions, this paper proposed a co-evolution algorithm strategy by integrating SA into PSO in order to combine the advantages of SA (ability to jump away from local optimum solutions and converge to the global optimum solution) and the advantages of PSO (fast calculation and easy mechanism). Firstly, particle swarms are randomly created by the PSO structure. Secondly, velocity and position update. The SA metropolis acceptance rule is incor-porated into the parallel processing of PSO at this stage. Thirdly, The rule determines whether to accept the new position or recalculate another candidate position according to the fitness function difference between the new and old positions. The conceptual diagram of HPSOSA is shown in Fig. 2.

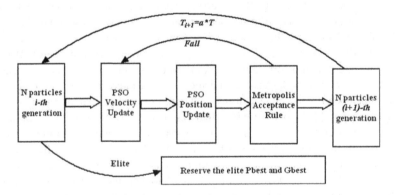

Fig. 2. The HPSOSA conceptual diagram

3.3   Chromosome Representations

In this paper co-evolution algorithm is used to establish the best components of the RBF-NN that approximate a function, including the number of hidden neurons, the centers, c_i, radii, r_i and weights. The hidden nodes are encoded as binary code string, 1 with connection to input and output nodes and 0 with no connection. The centers, radii and weights are encoded as float string, with string length $H = N * n + N + N * m$ (n is the number of input nodes, m is the number of output nodes, N is the number of hidden nodes). Each string corresponds to a chromosome, which consists of some gene sections, tabulated as follows: (Table 1)

Table 1. Schematic diagram of encoding chromosome. The number of hidden nodes is encoded in binary type, and other parts in real value.

Hidden nodes	c_1, r_1, ω_1	c_2, r_2, ω_2	......	c_H, r_H, ω_H
1, 0, 0, 1, 1	0.1,0.2,0.1	0.2, 0.4, 0.3	......	0.1, 0.4, 0.2

3.4   Fitness Function

Suppose that we have a training set $D = \{(x_t, y_t), t = 1, 2, \ldots, m\}$, where y_t is the output. The error function E for the network e_i output. Fitness function f for network e_i in the proposed method is defined by:

$$f_{fitness} = 1 / \left[1 + \frac{1}{m} \sum_{i=1}^{m} (e_i(x_j) - y_j)^2 \right] \tag{3}$$

3.5   The Establishment of Co-evolutionary RBF-NN

The proposed HPSOSA-RBFNN model, dynamically optimizing all values of RBF-NN's parameters through co-evolutionary process by SA into PSO, and all parameters of RBF-NN to construct optimized RBF model in order to proceed prediction. Figure 3 illustrates the algorithm process of the HPSOSA-RBFNN model. Details of our proposed HPSOSA-RBFNN are described as follows:

1. Particle initialization and PSO parameters setting: Generate initial particles comprised of the RBF-NN's parameters. The connection weights and thresholds are encoded as float string, randomly generated within $[-1, 1]$.
2. Input training data and calculate the fitness of all particles, which determine G_{best} and P_{best} by a simple comparison of their fitness values according to Eq. (3).
3. Repeat this step until the stopping criterion is satisfied.
   (1) Perform PSO operators. Update individual velocity and its position.
   (2) Stop condition. If the number of generation is equal to a given scale, then the best chromosomes are presented as a solution, otherwise go to the step 1 of the SA part. PSO will deliver its best individual to SA for further processing.

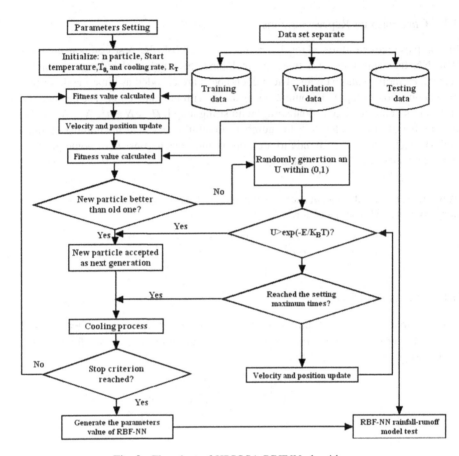

Fig. 3. Flowchart of HPSOSA-RBFNN algorithm.

(3) Perform SA operators. Evaluate Δffitness = ffitness(sk + 1)-ffitness (sk), and then randomly select a number R in [0,1]. If Δffitness > 0 meaning that, the new position is improved for increasing fitness function, then position sk + 1 is accepted according to the following criterion: exp(Δffitness/Ti) > R. Proceed to Step (e) when the velocity of all particles are determined, or return to Step (a) for those particles failing to be accepted, and generate new velocities using the same evaluation process. Too many failures (i.e., 100 in our study) for the same particle will force the last velocity to be accepted, in consideration of CPU time, thereby prevent entering into an endless loop.

(4) Renew each particle to the new position and modify G_{best} and P_{best} by simple comparison their fitness values. When the evolution process has achieved a satisfactory condition (or maximum evolution number is reached), proceed to Step 4, otherwise, modify the inertia weight, and annealing temperature $T_{i+1} = k * T_i$, let $k = k + 1$, k is set at 0.8 in this paper. If the predetermined temperature is reached, then stop the algorithm and the latest state is an approximate optimal solution. Otherwise, go to step 3.

4. Once the termination condition is met, output the final solution, obtain the number of hidden neurons, the centers, c_i, radii, r_i and weights. Input testing data and Output forecasting results by the HPSOSA-RBFNN.

4 Experiments Results and Analysis

To evaluate the purposed model performance, the obtained results compared with that other two model for rainfall-runoff prediction, namely RBFNN and pure PSO-RBFNN. RBFNN model was established by the trial-and-error approach following the previous studies [8], and PSO-RBFNN model bullied by the real code PSO to evolve initial the values of weights and thresholds for RBFNN, which the method has presented by Sedki et al. [10]. The parameter settings of the proposed HPSOSA-RBFNN model with other models are shown in Table 2.

Table 2. Results of the proposed HPSOSA-RBFNN model with other models.

Model	The description of parameter	Value
PSO	Number of generations	100
	Population size	20
	Crossover probability	0.80
	Mutation probability	0.05
SA	Initial temperature	1000
	Termination temperature	0.9
RBFNN	Architecture (input-hidden-output)	6-6-1
	Transfer function	Sigmoid-Sigmoid-Tanh

4.1 Study Area and Data

Daily flow as runoff data from Daqiao stream flow gauging stations, were obtained from the observation archives of Liuzhou Water Management Information System. The collected data were prepared for the period between 2006 and 2010 years, a total of 5 years with 1826 data points. The data were divided into two parts, the first fours years (2006–2008) samples with 1461 data points for model training and the remaining two years (2009 and 2010) samples with 730 data points for model testing.

4.2 Input Variables

In general, the causal variables involved in rainfall-runoff relations are those associated with rainfall, previous water level, evaporation, temperature, etc. Most studies applied rainfall and previous flow (or water level) as inputs variables [13]. According to the literature (see Ref. [13]), runoff at time steps $t - 1\left(X_{(t-1)}\right)$, $t - 2\left(X_{(t-2)}\right)$, $t - 3\left(X_{(t-3)}\right)$, $t - 4\left(X_{(t-4)}\right)$, $t - 5\left(X_{(t-5)}\right)$ and rainfall at time step $t - 1\left(R_{(t-1)}\right)$ are considered as inputs to the model by the stepwise regression method to eliminate low impact factors and choose the most influential factors.

4.3   Performance Evaluation

There four indexes are applied for performance evaluation of rainfall-runoff forecasting model in this paper, such as root mean square error (RMSE), mean absolute percentage error (MAPE), coefficient of efficiency (CE), and coefficient of efficiency for peak values (CE_{peak}), which can be found in many paper [8].

4.4   Application HPSOSA-RBFNN for Rainfall-Runoff

Figure 4 shows the curve of fitness in the learning stage for PSO-RBFNN approach arising from the iteration number. One can see that the maximum, average and the minimum fitness are unstable with increase of iteration 100. Figure 5 shows the curve of fitness in the learning stage for HPSOSA-RBFNN approach arising from the iteration number. One can see that the maximum, average and the minimum fitness are tending towards stability with increase of iteration 55. As can be seen from Fig. 5, with the increase in the number of training, fitness is quickly stabilized. The result show HPSOSA-RBFNN can avoid problems of premature convergence and escape from local optima.

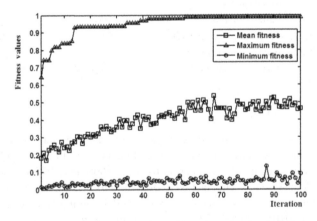

Fig. 4. Fitness function curve of PSO-RBFNN approach in the learning stage

4.5   Analysis of the Results

Figures 6, 7, and 8 show the observed and the forecasting by the proposed HPSOSA-RBFNN model with other two models at 730 testing data points. From the graphs, we can generally see that the results of HPSOSA-RBFNN model are closer to the corresponding observed runoff values than those of the RBFNN and PSO-RBFNN models. Especially, the HPSOSA-RBFNN model produced better forecasting for high runoff values, while other two models were slightly overestimated or underestimated.

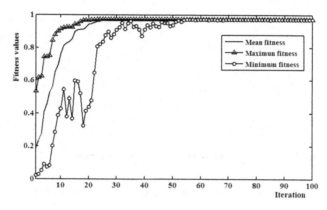

Fig. 5. Fitness function curve of HPSOSA-RBFNN approach in the learning stage

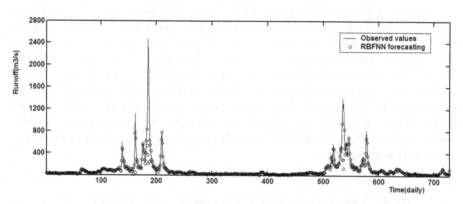

Fig. 6. Comparison between observed and predicted of RBFNN model.

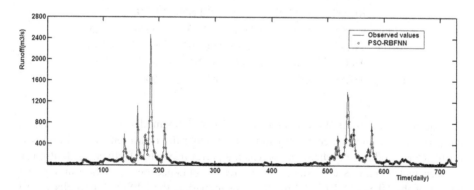

Fig. 7. Comparison between observed and predicted of PSO-RBFNN model.

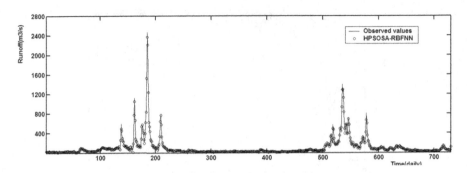

Fig. 8. Comparison between observed and predicted of HPSOSA-RBFNN models.

Table 3 illustrates the fitting and forecasting accuracy and efficiency of the model in terms of various evaluation indices for RBFNN, PSO-RBFNN and HPSOSA-RBFNN, respectively. As shown in Table 3, for the training data, the RMSE for RBFNN model is 0.3567, and for PSO-RBFNN model it is 0.2914; while for HPSOSA-RBFNN model, RMSE reaches 0.1066. Similarly, for the testing data, the RMSE for the RBFNN model is 0.8743, and for the PSO-RBFNN model it is 0.6520; while for the HPSOSA-RBFNN model, RMSE reaches 0.2133. Focusing on the MAPE indicator of the training and testing, the HPSOSA-RBFNN model is also less than the other two models. These results indicate that the errors of the HPSOSA-RBFNN are the smallest. In addition, the CE of HPSOSA-RBFNN is the maximum in all models. This also means that HPSOSA-RBFNN is capable to capture the average change tendency of the daily water level.

Table 3. Results of the proposed HPSOSA-RBFNN model with other models.

	Fitting (training)			Forecasting (testing)		
	HPSOSA RBFNN	PSO RBFNN	RBFNN	HPSOSA RBFNN	PSO RBFNN	RBFNN
RMSE	**0.1066**	0.2914	0.3567	**0.2133**	0.6520	0.8743
MAPE	**19.21%**	31.422%	45.86%	**14.22%**	25.66%	39.69%
CE	**0.9881**	0.8875	0.9340	**0.9915**	0.8732	0.7587
CE_{peak}	**0.9858**	0.8754	0.8265	**0.9972**	0.8014	0.7357

The empirical results show that the HPSOSA-RBFNN model are very promising in the runoff forecasting under the research where either the measurement of fitting performance is goodness or where the forecasting performance is effectiveness. It also can be seen that there was consistency in the results obtained between the training and testing of the HPSOSA-RBFNN model. To summarize, the HPSOSA-RBFNN model is superior to the other two models presented here in terms of RMSE, MAPE and CE for water level prediction under the same network input.

From the experiments presented in this paper, the HPSOSA algorithm also helps to improve the fine-tuning ability, the hill-climbing ability, the speed of convergence to the global optimum solutions or near global optimum solutions, and avoid trapping into local minimum than PSO did, thus, outperform the PSO-RBFNN model. For example, in Table 3, the CE_{peak} of the PSO-RBFNN model is 0.8754 in training samples; however the CE_{peak} of the HPSOSA-RBFNN model reaches 0.9858 which has obvious advantages over PSO-RBFNN models. Thus, it once again reveals that HPSOSA algorithm is much appropriate than PSO in parameter adjustments to achieve forecasting accuracy improvement by integrated into the RBFNN model.

The more important point in the runoff modeling which makes sure that our model is most promising when comparing with different models is the capability of the proposed model in estimating peak values. Where as the estimation of peak values is usually the most important part of the flood mitigation program for actual runoff application, as final step of the modeling process. From Table 3, the efficiency of HPSOSA-RBFNN model is 0.9858 in training data, and it is 0.9972 in testing data. There is an identical high capability of our models for prediction of peak flows in testing data. Therefore, not only the proposed model is appropriate in monitoring peak values, but also it can be considered as most powerful tool for runoff forecasting which is necessary in the water resources systems management, where it is directly in fluency by stream flow forecasting.

5   Conclusion

The rainfall-runoff system is one of the most active dynamic weather systems. In the design of RBFNN, a novel method based on the cooperative co-evolutionary and the PSOSA is proposed to model the runoff forecasting. The HPSOSA algorithm enables the solution to jump out of local optima, and decreases the vibration near the end of locating a solution by incorporating the metropolis acceptance criterion of SA into PSO randomly acceptance rule. The rule determines whether to accept the new position or recalculate another candidate position according to the fitness function difference between the new and old positions. This enables the solution to jump out of local optima, and decreases the vibration near the end of locating a solution. These characteristics improve the quality of the solution and increase the rate of convergence. The new hybrid HPSOSA-RBFNN approach is compared with the other two methods with a set of benchmark mathematical functions. According to the results obtained in this paper, we can draw the following conclusions that the HPSOSA-RBFNN model can be used as an alternative tool for actual runoff forecasting application to obtain better forecasting accuracy and improve the prediction quality further in view of empirical results.

Acknowledgment. This work was supported the Natural Science Foundation of Guangxi Province under Grant No. AD16450003, and by the Guangxi Education Department under Grant Nos. 2013YB281 and YB2014467, and Key Laboratory for Mixed and Missing Data Statistics of the Education Department of Guangxi Province under Grant No. GXMMSL201405.

References

1. Mehr, A.D., Nourani, V.: A Pareto-optimal moving average-multigene genetic programming model for rainfall-runoff modelling. Environ. Modell. Softw. **92**, 239–251 (2017)
2. Bartoletti, N., Casagli, F.: Marsili-Libelli, S., Nardi, A., Palandri, L.: Data-driven rainfall/runoff modelling based on a neuro-fuzzy inference system. In: Environmental Modelling & Software (2017)
3. Chang, T.K., Talei, A., Alaghmand, S.: Choice of rainfall inputs for event-based rainfall-runoff modeling in a catchment with multiple rainfall stations using data-driven techniques. J. Hydrol. **545**, 100–108 (2017)
4. Sinha, J.: A comparison of network types in artificial neural network-based rainfall-runoff modelling. Int. J. Appl. Res. Inf. Technol. Comput. **8**(1), 41–50 (2017)
5. Makungo, R., Odiyo, J.O., Ndiritu, J.G.: Rainfall-runoff modelling approach for ungauged catchments: a case study of Nzhelele river sub-quaternary catchment. Phys. Chem. Earth **35** (13–14), 45–62 (2010)
6. Wu, J.: An effective hybrid semi-parametric regression strategy for rainfall forecasting combining linear and nonlinear regression. Int. J. Appl. Evol. Comput. **2**(4), 50–65 (2011)
7. Makungo, R., Odiyo, J.O., Ndiritu, J.G., Mwaka, B.: Rainfall-runoff modelling approach for ungauged catchments: a case study of Nzhelele river sub-quaternary catchment. Phys. Chem. Earth **35**, 596–607 (2010)
8. Chadalawada, J., Havlicek, V., Babovic, V.: A genetic programming approach to system identification of rainfall-runoff models. Water Resour. Manag. **31**(12), 3975–3992 (2017)
9. Vahid, N., Özgür, K., Mehdi, K.: Two hybrid artificial intelligence approaches for modeling rainfall–runoff process. J. Hydrol. **402**, 41–59 (2011)
10. Qian, X., Huang, H., Chen, X., Huang, T.: Efficient construction of sparse radial basis function neural networks using L1-regularization. Neural Netw. **94**, 239–254 (2017)
11. Talei, A., Chua, L.H.C., Quek, C.: A novel application of a neuro-fuzzy computational technique in event-based rainfall-runoff modeling. Expert Syst. Appl. **37**, 7456–7468 (2010)
12. Eglese, R.W.: Simulated annealing: a tool for operation research. Eur. J. Oper. Res. **46**, 271–281 (1990)
13. Shi, X.H., Liang, Y.C., Lee, H.P., Lu, C., Wang, L.M.: An improved GA and a novel PSO-GA-based hybrid algorithm. Inf. Process. Lett. **93**(5), 255–261 (2005)
14. Chen, S., Cowan, C.F., Grant, P.M.: Orthogonal least squares learning algorithm for radial basis function networks. IEEE Trans. Neural Netw. **2**(2), 302–309 (1991)
15. Lu, J., Hu, H., Bai, Y.: Generalized radial basis function neural network based on an improved dynamic particle swarm optimization and AdaBoost algorithm. Neurocomputing **152**, 305–315 (2015)
16. Abedinia, O., Amjady, N.: Short-term load forecast of electrical power system by radial basis function neural network and new stochastic search algorithm. Int. Trans. Electr. Energ. Syst. **26**(7), 1511–1525 (2016)

Multicellular Gene Expression Programming-Based Hybrid Model for Precipitation Prediction Coupled with EMD

Hongya Li[1], Yuzhong Peng[1,2(✉)], Chuyan Deng[1], Yonghua Pan[1],
Daoqing Gong[1], and Hao Zhang[2]

[1] Key Laboratory of Scientific Computing and Intelligent Information
Processing in Universities of Guangxi, School of Computer and Information
Engineering, Guangxi Teachers Education University, Nanning 530001, China
jedison@163.com
[2] School of Computer Science, Fudan University, Shanghai 200433, China

Abstract. Accurate and timely precipitation prediction is very important to people's daily activities and production plans. However, the impact factors of meteorological precipitation are numerous and complex, making it difficult to predict, and the prediction effect by traditional methods is difficult to meet the public expectations. This work proposes to use Multicellular Gene Expression Programming (MC_GEP) algorithm for modeling the historical precipitation data series decomposed by Empirical Mode Decomposition (EMD). Then we design a novel Multicellular Gene Expression Programming based method coupled with Empirical Mode Decomposition, named as EMGEP2RP, for precipitation modeling and prediction. Using RMSE and MAE as evaluation indicators, simulation experiments were conducted on three different types of real precipitation data sets in different regions. The comparing results show that the EMGEP2RP algorithm significantly outperforms not only the existing Gene Expression Programming (GEP) algorithm, but also the Back Propagation and Support Vector Machine algorithms which are widely used in meteorological modeling and predictions.

Keywords: Gene Expression Programming · Empirical Mode Decomposition
Precipitation modeling · Precipitation prediction · Time series prediction

1 Introduction

Meteorological precipitation is an intricate process. Affected by such factors as land-sea distribution, atmospheric circulation, terrain and human activities, precipitation varies significantly in different regions and seasons and changes remarkably year to year. Thus there is non-determinacy in the volume of precipitation. Heavy precipitation in a region within a short term may easily bring about flood disaster and affect the national economy and the people's livelihood. Based on accurate and timely precipitation prediction, people can prepare well for daily activities and production plans and take corresponding measures of flood control and drought relief to reduce unnecessary loss. However, traditional prediction approaches for precipitation based on synoptic

© Springer International Publishing AG, part of Springer Nature 2018
D.-S. Huang et al. (Eds.): ICIC 2018, LNCS 10954, pp. 207–218, 2018.
https://doi.org/10.1007/978-3-319-95930-6_20

meteorology and statistics are deficient in prediction accuracy and relatively poor in prediction quality.

In recent years, many scholars at home and abroad have used data mining and intelligent computing theory and technology to study and improve precipitation prediction modeling methods, and obtained a number of outstanding results. These have improved the accuracy of precipitation prediction and forecasting to some extent, and won more time for disaster reduction and prevention. Therefore, using data mining and intelligent computing theory and technology to study and improve the quality of precipitation prediction has become a research focus in this field [1]. Among them, there are mainly researches on precipitation prediction based on artificial neural network methods, genetic computing and its fusion algorithms, support vector machines (SVM), Bayesian methods, decision tree methods, and association rules mining. For example, Geetha [2] constructed a backpropagation neural network model to predict precipitation using a dataset of humidity, dew point temperature and barometric pressure in an Indian location. The prediction accuracy was higher. Kisi [3] proposed a wavelet-SVM joint model for daily precipitation forecasting. The model first decomposed and reconstructed the precipitation time series with discrete wavelets, and then used the basic model component support vector regression machine for modeling and forecasting. The experimental result was better than that of the neural network prediction model. Chaoping [4] revised the forecast products provided by the mesoscale ensemble forecast model in the southwest region based on the Bayesian probability decision theory, which eliminated the false prediction to some extent. Mishra [5] designed a model based on Back Propagation Algorithm and Levenberg-Marquardt training function, using Regression Analysis, MSE and MRE as evaluation indicators, which has a good modelresult. Du [6] proposed a prediction model based on AVM and PSO, optimization of SVM parameters. The results showed that the SVM-PSO algorithm has significantly outperformed the direct prediction model and the SVM-PSO model is effective and useful in machine learning. Zainudin [7] studied Naive Bayes, Decision Tree, Neural Network and Support Vector Machine (SVM) and Random Forest for predicting precipitation in Malaysia, and the results showed that a set of random forest classifier can united beat a single classifier. Shamshirband [8] proposed an adaptive neuro fuzzy inference (ANFIS) and support vector regression (SVR) model for Serbia's precipitation forecasting, the probability of precipitation was estimated by using polynomial, linear and radial basis function (RBF) as the kernel function of SVR. Compared with SVR estimation, ANFIS method can improve the prediction accuracy and generalization ability. Geetha [9] used the decision tree to predict fog, rain, cyclones, thunderstorms and other weather phenomena, and more related properties can be used to predict related variables. Abbot and Marohasy [10] proposed a GP method for predicting precipitation. And compared the performance of GP and MCRP with 21 different data sets in Europe, the results showed that the performance of GP was better than that of MCRP.

In general, It is difficult to establish accurate models with traditional data mining and intelligent computing methods due to the multi-level non-stationary and nonlinear characteristics of meteorological data, thus reducing the accuracy of meteorological forecasts. Therefore, it is necessary to study new effective methods for meteorological data modeling and prediction.

This paper attempts to conduct a fusion modeling and prediction of precipitation data combining the MC_GEP with strong symbolic regression analysis ability and the EMD with strong non-stationary series decomposition ability. It proposes a precipitation prediction algorithm EMGEP2RP based on MC_GEP coupled with EMD. The simulation experiment results on three data sets of real precipitation data showed that the proposed algorithm has better performance.

The remaining of this paper include: (1) an overview of relevant theoretical basis of multicellular GEP algorithm and Empirical Mode Decomposition; (2) clearly describesEMGEP2RPalgorithm; (3) experiment and results analysis; (4) concludes the work followed by future directions.

2   Theoretical Basis Overview

2.1   Multicellular Gene Expression Programming

GEP is a new self-adaptive evolutionary algorithm that inherits GA fixed-length linear coding and GP tree structure [11]. GEP chromosome is represented by a fixed-length character string containing one or more genes. Each gene contains two parts: head and tail. The head can be composed of function set F and terminal set T, while the tail can only be composed of terminal set. The relationship between the head length H and the tail length T is: $T = H \times (N - 1) + 1$, where N is the maximum number of operations for all functions in the function set. Each gene corresponds to a K-expression (i.e., genotype) and an expression tree (i.e., phenotype), and the two can be converted to each other. That is, the K-expression can be obtained from the expression tree traversing from top to bottom and from left to right.

The MC_GEP algorithm, with introduction of homologous genes and cellular systems based on the previous GEP, is an improved GEP algorithm of complex individual consisting of a number of common genes (in order to distinguish from homologous genes, here the gene in the standard GEP algorithm is collectively referred to as the common gene) and homologous genes [12]. Each character in the terminal set of the homologous gene represents a gene. The operator at the head of the homologous gene is the symbol flexibly linking different genes to form a chromosome. The complex individual composed of multiple genes is generated. It greatly simplifies the process of building a powerful genotype/phenotype system. Figure 1 shows a multicellular GEP chromosomal genotype containing two common genes and one homologous gene (for convenience of understanding, the number in the 1st row of Fig. 1 is only used to indicate each gene position but no other meaning, the 2nd row is a genotype coding string of chromosomes, and the genes are separated by 2 vertical dashed lines). Head lengths of both its common gene and homologous gene are 4. The functions set F of the common gene and the homologous gene is both $\{+, -, *, /, \wedge\}$. The terminal set T of the common gene and the homologous gene is $\{?, a, b, c, d, e, f\}$ and $\{0, 1\}$ respectively. DC domain is the corresponding constants randomly generated from $\{A, B, C, D, E, F, G, H, I, J\}$.

```
0 1 2 3 4 5 6 7 8 9 0 |0 1 2 3 4 5 6 7 8 9 0 |0 1 2 3 4 5 6 7 8 9
+ - * c d a f c B e ? |^ / b d ? b ? e a b F|+ * 1 0 1 1 1 1 1 0
```

Fig. 1. Multicellular chromosome coding structure

Therefore, the chromosome can be decoded into mathematical expression (1):

$$((c - d) + a \times f + 1) \times (d \div ?)^b \tag{1}$$

2.2 Empirical Mode Decomposition (EMD)

EMD is a time-frequency analysis method for signal smoothing. Its core idea is the intrinsic mode function (IMF), that is to extract a frequency of physical meaning from unstable, nonlinear and complex signals by rule. The IMF highlights the local characteristics of the data. It satisfies two conditions. Firstly, in the data series $X(t)$, the total number M_e of maxima and minima must be equal to the number M_z of zero-crossings, or the difference is at most one, as in Eq. (2). Secondly, at any time t, the mean value of the upper envelope $U_j(t)$ and the lower envelope $V_j(t)$ consisting of local maxima and minima is zero, as in Eq. (3).

$$M_z - 1 \leq M_e \leq M_z + 1 \tag{2}$$

$$[U_j(t) + V_j(t)]/2 = 0 \quad t \subset [a, b] \tag{3}$$

Where [a, b] is a time interval. The basis function decomposed by EMD has no fixed expression, which is determined by the data series itself. The basic functions of different data series decompositions are different. The EMD decomposition process of the data series are described in reference [13].

3 EMGEP2RP Algorithm

3.1 Description of Precipitation Modeling and Prediction

Precipitation data is time series data formed by recording precipitation at certain time intervals. Precipitation prediction based on computing method is mainly to construct a model through historical precipitation data and use this model to predict the future precipitation. Its mathematical description is shown in Eq. (4):

$$\hat{y}(k) = f(y(k-1), y(k-2), y(k-3), \ldots, y(k-n)) \tag{4}$$

Where $\hat{y}(k)$ is the value at current time k, that is, the value that needs to be predicted, $y(k-1), y(k-2), y(k-3), \ldots y(k-n)$ are the values of n times before k, and f is a linear/nonlinear mapping function. It is in nature to perform regression analysis on the time series. The transformation rules of and relations among the given time series data are dug out to interpret them as mathematical models, and the past information is used to accurately predict future data. At any time k, if the values of the past n times exist, this model can be used to predict the value of the current moment k.

3.2   Basic Idea of EMGEP2RP Algorithm

The GEP algorithm has a powerful ability to express problems. The simple and flexible coding structure can not only encode any imaginary program, but also allow them to effectively search in the global search space. Its optimization ability is strong and the computational efficiency is high. Researches have shown that the GEP algorithm has a very strong ability to solve regression analysis problems [14–17]. The coding form of the MC_GEP algorithm is more flexible than the GEP, making it easier to extend the excellent structure, expand the search space, and increase the diversity of solutions. Therefore, research and application of MC_GEP for meteorological data modeling and forecasting will have good prospects.

The volume of precipitation is affected by many factors. As a result, the historical precipitation data are indeterminate and have nonlinear and non-stationary characteristics. So it is difficult to analyze the data characteristics and to obtain the correlation between the data, thus making it difficult to accurately predict precipitation. Empirical Mode Decomposition (EMD) can decompose complex signals into a number of stationary signal components and residuals that are relatively easy to characterize. This allows a more straightforward presentation of features and relationships between data signals on different scales. Furthermore, it can completely reconstruct the original signal (even if there is an error, it is very small). Therefore, using EMD to decompose the precipitation series can effectively filter and denoise the data. Then modeling and prediction can improve the prediction accuracy. Based on the above idea, the EMGEP2RP algorithm is designed to improve the effectiveness of the prediction performance.

The core ideas of EMGEP2RP are:

(1) EMD decomposition of precipitation time sequence. Via EMD, precipitation data is decomposed into several high frequency and low frequency stable signal components and residuals with descriptive features. The multi-component data signals that are difficult for observation and analysis can be decomposed into signal components that are easy to be observed and analyzed. Refer to Subsect. 2.2 for detailed decomposition process.
(2) Modeling for precipitation data. After data decomposition, it models for each component's training set. The current data values are predicted based on the first six data values of the current data. Refer to Subsect. 3.1 for detailed modeling method.
(3) MC_GEP function exploration. After modeling for each component, it explores algorithm via MC_GEP function. Under the guidance of chromosome's fitness function $\frac{1}{1+RMSE_i}$, through genetic operations of selection, cross and mutation, fitting expressions and fitting results for each component's training set are deduced.

(4) Computing the prediction value. Each component's prediction value (i.e. the value of corresponding year for test set) can be computed as per fitting expressions of each component's training set. As per the principle that each component's prediction value with the same year should be added, the final value of test data for each year (i.e. Prediction Value) can be computed. This process is also a remodeling for EMD. Figure 2 is a schematic diagram showing the basic logic of the algorithm:

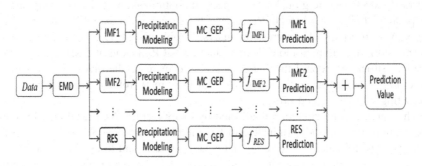

Fig. 2. The basic idea of the EMGEP2RP algorithm

3.3 The EMGEP2RP Algorithm Process

From the above ideas, the EMGEP2RP algorithm flow chart (Fig. 3).

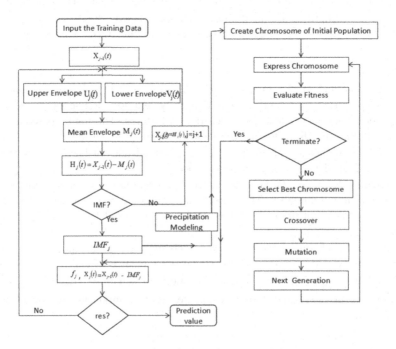

Fig. 3. The EMGEP2RP algorithm flow chart

4 Experiment and Result Analysis

4.1 Experimental Data and Parameter Setting

This paper presents modeling and prediction via three precipitation datasets in three different regions and climates to test the performance of the proposed algorithm. The datasets include Zhengzhou's annual precipitation (1951–2014, with a sample length of 64, hereinafter referred to as "Zz"), Nanning's precipitation in June (1951–2013, with a sample length of 63, hereinafter referred to as "Nn") and monthly precipitation in Australia's Melbourne during winter months (1911–2012, with a sample length of 102, hereinafter referred to as "Mb"). As shown in Fig. 4, for the Zz precipitation sample, its mean value is 634.828 mm, and its curve fluctuates greatly while its extremum ratio is 2.955 (that is, the maximum value is 2.955 times of the minimum value). For the Nn precipitation sample, its mean value is 223.381 mm, but its extremum ratio is 16.933, both the fluctuation amplitude and the extremum ratio fluctuate rather greatly. For the Mb precipitation sample, its mean value is 48.326 mm, though its extremum ratio is 8.391, its curve fluctuates relatively gently on the whole.

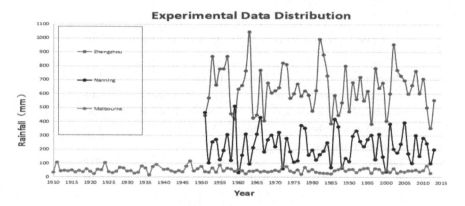

Fig. 4. Experimental datadistribution

85% of each data set above is taken as training sample and the rest 15% as test sample. This paper conducts comparative experiments on fitting modeling and prediction among the proposed EMGEP2RP algorithm, the GEP algorithm and the MC_GEP algorithm, as well as the BP neural network algorithm and SVR algorithm that are widely used in current precipitation prediction research. Wherein, the BP model is built with the BP toolbox in MATLAB, and the SVR built with the software package LIBSVM-3.22. The fitting and prediction performance of each algorithm is evaluated using root-mean-square error (RMSE) and mean absolute error (MAE) as indicators. The main parameters of the GEP related algorithm are shown in Table 1, where the terminal a, b, c, d, e, f respectively represent the previous 6 data $y(k-1)$, $y(k-2)$, $y(k-3)$, $y(k-4)$, $y(k-5)$, $y(k-6)$ of the current prediction data $\hat{y}(k)$ in the model. The parameters of BP neural network and SVR are shown in Table 2.

Table 1. **Main** parameters of GEP algorithm

Parameter	Algorithm		
	GEP	EDGEP	EMGEP2PR
Gene head length	8	8	8
Home gene head length	NULL	4	4
Terminal set	{a, b, c, d, e, f}	{?, a, b, c, d, e, f}	{?, a, b, c, d, e, f}
DC mutation rate	NULL	0.25	0.25
Gene mutation rate	0.3	0.3	0.3
Home gene mutation rate	NULL	0.2	0.2
Crossover rate	0.3		
Constant mutation	0.05		
Function set	{+ , -, *,/, sin, cos, sqrt}		
Constant array length	10		
Population size	100		
Select mode	Championships select		
Iteration	2000		
Fitness function	$\frac{1}{1+\text{RMSE}}$		

Table 2. Main parameters of BP algorithm and SVM algorithm

BP		SVM	
Parameter	Value	Parameter	Value
Number of hidden neurons	6	Kernel function	REF
Number of hidden neurons	3	P value of loss function	0.5
Optimization method	Trainlm	SVM_type	e-SVR
Activation function	transig	s	3
Epochs	10000	c	4096
lr	0.01	g	0.0625
goal	0.00001		

4.2   EMGEP2RP Experimental Result and Discussion

Firstly, GEP, MC_GEP, EMGEP2RP, BP and SVR algorithms are used to perform fitting modeling and prediction on the precipitation datasets of Zhengzhou, Nanning, and Melbourne, respectively. The evaluation indicator is RMSE, and the curves of the obtained results are shown in Figs. 5, 6 and 7, respectively. From these three figures, it can be clearly seen that result of each algorithm has a certain deviation from the actual precipitation data. But the EMGEP2RP algorithm's fitting effect on the precipitation data of Zhengzhou and Melbourne fluctuates with the real data values, which are relatively close to the real data curves. The other algorithms do not have such good fitting results for the precipitation data of these two cities. Each algorithm's fitting effect on Nanning's precipitation data is relatively poorer than that of Zhengzhou and Melbourne. But the fitting and prediction results of EMGEP2RP are still better than the other four algorithms, as shown in Fig. 6.

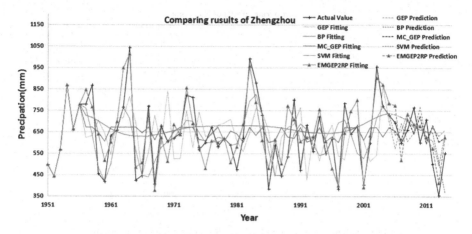

Fig. 5. The effect of precipitation fitting and prediction in Zz

In order to completely compare the performance of these algorithms, comparative experiments are done on modeling and prediction of three different precipitation datasets, using RMSE and MAE as evaluation indicators. And the compared results are shown in Table 3, which show that: (1) Regardless of whether the evaluation indicator is RMSE or MAE, the EMGEP2RP algorithm has the best fitting and prediction results on the three test datasets, followed by SVR, MC_GEP, BP, and GEP, respectively. This indicates that: A. The improvement of multicellular GEP on GEP is effective; B. The proposed method to use EMD for signal decomposition and make joint modeling prediction with MC_GEP is more promising. (2) In the experiment of Zhengzhou's precipitation dataset, the MAEs obtained by the EMGEP2RP fitting model are reduced by 49.374%, 41.913%, 35.129%, and 11.26%, respectively, compared with the MAEs obtained by GEP, BP, MC_GEP, and SVR fitting models. The MAEs of prediction are reduced by 49.761%, 40.782%, 34.743%, and 13.972%, respectively. (3) In the experiment of Nanning's precipitation dataset, the MAEs obtained by the EMGEP2RP fitting model are lowered by 39.501%, 36.331%, 29.463%, and 22.01%, respectively, compared with the MAEs obtained by GEP, BP, MC_GEP, and SVR fitting models. The MAEs of prediction are reduced by 40.54%, 37.387%, 29.645%, and 22.275%, respectively. (4) In the experiment of Melbourne's precipitation dataset, the MAE sobtained by the EMGEP2RP fitting model are reduced by 46.8%, 39.63%, 34.059%, and 5.402%, respectively, compared with the MAEs obtained by GEP, BP, MC_GEP, and SVR fitting models. The MAEs of prediction are reduced by 48.018%, 44.713%, 33.257%, and 14.352%, respectively. (5) The fitted MAEs of EMGEP2RP in Zhengzhou's precipitation dataset with the smallest extremum ratio is 11.725% of the mean values of the original series in the corresponding dataset, accordingly in Nanning's precipitation dataset with the largest extremum ratio is 23.981%, while in Melbourne's precipitation dataset is 17.684%. The predicted MAEs of EMGEP2RP in these three datasets are 12.579%, 26.541%, and 18.695% respectively of the mean values of the original series in the corresponding datasets. These show that the smaller the extremum ratio of the dataset is, the better the algorithm predicts.

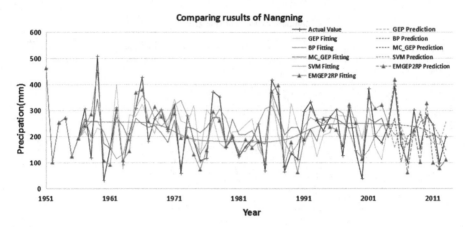

Fig. 6. The effect of precipitation fitting and prediction in Nn

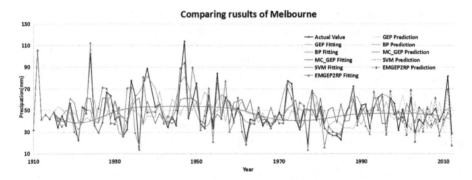

Fig. 7. The effect of precipitation fitting and prediction in Mb

Table 3. The experimental results of fitting and prediction models

Fitness		Algorithm									
		GEP		MC_GEP		**EMGEP2RP**		BP		SVM	
		Fitting	Test	Fitting	Test	**Fitting**	**Test**	Fitting	Test	Fitting	Test
RMSE	Zz	176.651	186.250	149.009	156.271	**89.431**	**92.096**	158.677	168.364	126.118	133.752
	Nn	106.780	118.287	85.800	97.933	**64.952**	**71.365**	92.205	102.251	80.763	89.005
	Mb	20.298	25.744	17.342	21.229	**12.602**	**15.114**	18.444	23.783	14.991	17.387
MAE	Zz	146.998	158.954	116.145	122.374	**74.434**	**79.857**	126.578	134.852	83.879	92.827
	Nn	88.177	99.712	75.628	84.271	**53.346**	**59.289**	83.786	94.691	68.401	76.280
	Mb	16.064	17.381	12.960	13.537	**8.546**	**9.035**	14.156	16.342	9.034	10.549

5   Conclusion

This paper presents a novel precipitation prediction algorithm, named as EMGEP2PR, based on multicellular gene expression programming coupled with Empirical Mode Decomposition. The real precipitation datasets in Zhengzhou, Nanning and Melbourne are used for simulated experiments. The results verify the effectiveness of the EMGEP2RP algorithm, comparing with not only the canonic GEP algorithm and MC_GEP algorithm, but also the BP neural network algorithm and SVR algorithm that are widely used in current precipitation prediction research. The experimental results show that the proposed algorithm using RMSE and MAE as evaluation indicators has better fitting and prediction performance. As the EMGEP2RP divides the data into many IMFs, each IMF needs to be modeled separately with increasing execution time of the algorithm, thus making it more time-intensive than other algorithms. However, in practical application, climate precipitation prediction mainly emphasizes the accuracy, and the time efficiency is almost negligible. It can be seen that EMGEP2RP will have good prospects for meteorological applications and can be also transferred to other time series predictions.

Acknowledgments. This work was supported in part by the National Natural Science Foundation of China Grant #61562008, #41575051, and the Natural Science Foundation of Guangxi Grant #2017GXNSFAA198228 and #2014GXNSFDA118037, and the grant of "Bagui Scholars" Program of Guangxi Zhuang Autonomous Region of China. Yuzhong Peng is the corresponding author.

References

1. Peng, Y., Wang, Q., Yuan, C., et al.: Review of research on data mining in application of meteorological forecasting. J. Arid Meteorol. **33**(1), 19–27 (2015)
2. Geetha, G., Selvaraj, R.S.: Prediction of monthly rainfall in chennal using back propagation neural network modeal. Int. J. Eng. Sci. Technol. **3**(1), 211–213 (2011)
3. Kisi, O., Cimen, M.: Precipitation forecasting by using wavelet-support vector machine conjunction model. Eng. Appl. Artif. Intell. **25**(4), 783–792 (2012)
4. Chen, C., Feng, H., Chen, J.: Application of Sichuan heavy rainfall ensemble prediction probability products based on Bayesian method. Meteorol. Monthly **36**(5), 32–39 (2010)
5. Mishra, N., Soni, H.K., Sharma, S., et al.: Development and analysis of artificial neural network models for rainfall prediction by using time-series data. Int. J. Intell. Syst. Appl. **10**(1), 16–23 (2018)
6. Du, J., Liu, Y., Yu, Y., et al.: A prediction of precipitation data based on support vector machine and particle swarm optimization (PSO-SVM) algorithms. Algorithms **10**(2), 57 (2017)
7. Zainudin, S., Jasim, D.S., Bakar, A.A.: Comparative analysis of data mining techniques for Malaysian rainfall prediction. Int. J. Adv. Sci. Eng. Inf. Technol. **6**(6), 1148 (2016)
8. Shamshirband, S., Petkovićet, G., et al.: Soft-computing methodologies for precipitation estimation: a case study. IEEE J. Sel. Top. Appl. Earth Observations Remote Sens. **8**(3), 1353–1358 (2015)

9. Geetha, A., Nasira, G.M.: Data mining for meteorological applications: decision trees for modeling rainfall prediction. In: IEEE International Conference on Computational Intelligence and Computing Research, pp. 1–4. IEEE (2015)

10. Abbot, J., Marohasy, J.: Application of artificial neural networks to forecasting monthly rainfall one year in advance for locations within the Murray Darling basin, Australia. Int. J. Sus. Dev. Plann. **12**(8), 1282–1298 (2017)

11. Ferreira, C.: Gene Expression Programming: Mathematical Modeling by an Artificial Intelligence, 2nd edn. Springer, Heidelberg (2006). https://doi.org/10.1007/3-540-32849-1

12. Peng, Y., Yuan, C., Chen, J., Wu, X., Wang, R.: Multicellular gene expression programming algorithm for function optimization. Control Theory Appl. **27**(11), 1585–1589 (2010)

13. Wang, X., Bi, G.-h., Tang, J.-R.: Composite forecasting model of sunspot time sequences based on EMD. Comput. Eng. **37**(24), 176–179 (2011)

14. Peng, Y.Z., Yuan, C.A., Qin, X., et al.: An improved gene expression programming approach for symbolic regression problems. Neurocomputing **137**(15), 293–301 (2014)

15. Zhong, J., Ong, Y.S., Cai, W.: Self-learning gene expression programming. IEEE Trans. Evol. Comput. **20**(1), 65–80 (2016)

16. Emamgolizadeh, S., Bateni, S.M., Shahsavani, D., et al.: Estimation of soil cation exchange capacity using genetic expression programming (GEP) and multivariate adaptive regression splines (MARS). J. Hydrol. **529**, 1590–1600 (2015)

17. Roushangar, K., Akhgar, S., Salmasi, F.: Estimating discharge coefficient of stepped spillways under nappe and skimming flow regime using data driven approaches. Flow Meas. Instrum. **59**, 79–87 (2018)

Data Preprocessing of Agricultural IoT Based on Time Series Analysis

Yajie Ma[1,2], Jin Jin[1,2(✉)], Qihui Huang[1,2], and Feng Dan[1,2]

[1] College of Information Science and Engineering, Wuhan University of Science
and Technology, Wuhan 430081, China
jinjin19920929@163.com
[2] Engineering Research Center for Metallurgical Automation
and Detecting Technology, Ministry of Education, Wuhan 430081, China

Abstract. Large-scale agricultural internet of things will generate a large amount of data every moment. After a certain period of time, the amount of data can reach hundreds of millions. It is very meaningful to analyze and mine agricultural big data and replace artificial experience with analysis results. However, the agricultural production environment is complex, and the raw data collected include a variety of anomalies, which can not be directly followed by analysis and mining. In this paper, a data preprocessing method based on time series analysis is proposed, which can quickly and efficiently obtain the prediction model, and can be used to fill and replace the abnormal data. On this basis, we add data preprocessing layer to the traditional three-layer Internet of things system (IoT), which is located between the application layer and the transmission layer, and designs a four layer of Agricultural IoT system. The system not only realizes the basic functions of data acquisition, transmission and storage, but also provides better data sources for subsequent analysis.

Keywords: Agricultural IoT · Time series analysis · Data preprocessing

1 Introduction

In recent years, China has paid more and more attention to the improvement and innovation of agricultural production management methods. The Outline of the 13th Five-Year Plan adopted in 2016 [1] pointed out that we must fully promote the construction of agricultural informatization and vigorously promote the application of the Internet of Things in agricultural production. According to the 13th Five-Year Plan proposed by the Central Government, Hubei Province plans to build a high-standard agricultural IoT demonstration area with an overall scale of more than 20,000 μ in each county (city, district) [2]. Large-scale agricultural internet of things will generate hundreds of millions of environmental data. We can analyze or excavate these data and find the influence of environmental parameters on crop growth. Using analysis results instead of artificial experience to guide production is of great significance. It is very meaningful to guide the production by analyzing results instead of artificial experience.

Data mining is the process of revealing meaningful relationships, trends, and models by analyzing large amounts of data [3]. Data preprocessing is an indispensable

© Springer International Publishing AG, part of Springer Nature 2018
D.-S. Huang et al. (Eds.): ICIC 2018, LNCS 10954, pp. 219–230, 2018.
https://doi.org/10.1007/978-3-319-95930-6_21

part of data mining. To enable data mining algorithms to get meaningful output, we must provide data that is not polluted, with low error rate and low repetition rate. However, the agricultural production environment such as cultivating farmland and greenhouses is very complicated. The noise generated by various reasons such as human, nature and hardware has a great impact on the collected environmental data. Because the raw data contaminated by noise can not carry out subsequent data analysis and data mining directly, we need to prepocess the raw data. Data preprocessing can effectively improve the quality of data mining and reduce data mining time.

Data preprocessing mainly solves two problems: data quality and data quantity [4]. Data quality problems include missing values, outliers, duplicate values and so on. Data quantity problems include too much data and too small data. Considering that the collected data is greatly influenced by noise, this paper focuses on data quality. Data cleaning is one of the steps of data preprocessing. Its main task is to deal with missing values, outliers, and delete duplicate data.

The processing of missing values is usually divided into three types: delete records, data interpolation, and no processing. The processing of outliers is usually treated as a missing value or deleted directly. Therefore, this paper regards outliers and outliers as similar. Common methods of data interpolation include (1) mean/median interpolation, (2) fixed-value interpolation, (3) nearest-neighbor interpolation. The data type determines the method of data analysis and processing. Agricultural environment data is collected at a certain frequency and is typical time series data. Time series data usually have a strong autocorrelation, that is, there is an influence between the sequence values. Dealing with timing using traditional missing-value processing will ignore the correlation between the sequence values, so we can't use the above method to handle the missing values of the timing.

In order to solve the missing values and abnormal values in the original data quickly and efficiently, the time series analysis method is used to model the original data, and a fitting model is obtained. We can use this model and some historical data to predict subsequent data and use prediction data to replace missing values and outliers. On this basis, we have added a data preprocessing layer to the hierarchical design of agricultural internet of things and designed a four-layer structure for agricultural internet of things.

The remainder of this paper is organized as follows. The second chapter introduces the four-layer structure of agricultural internet of things. The third chapter introduces the data source and the basic flow of time series modeling. The fourth chapter introduces the method of time series preprocessing. Section 5 introduces the experimental flow and results of the stationary timing modeling. Section 6 tests the model and gives the model prediction results. Section 7 summarizes the work of this paper.

2 Architecture of Agricultural IoT

Usually, the agricultural IoT can be divided into three levels: the sensor layer, the network layer and the application layer [5, 6]. Their main works are data acquisition, data transmission, data storage and application. Taking into account the importance of data preprocessing discussed above, a four-layer architecture is designed for the agricultural IoT, and a data preprocessing layer is added between the network layer and the application layer. The hierarchy of agricultural IoT is shown in Fig. 1:

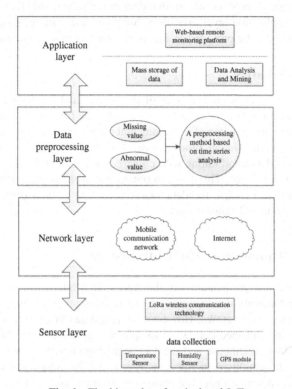

Fig. 1. The hierarchy of agricultural IoT

2.1 Sensor Layer

The sensor layer is the bottom layer of the system and mainly responsible for data collection. The sensor layer is composed of many wireless sensor terminals and a few sink nodes. Each terminal node is made up of several sensors such as humidity, Ardunio microcontroller, LoRa wireless communication module [7, 8] and GPS module. Terminal nodes transmit environmental information such as humidity, geographic location and time to the sink node through LoRa wireless communication technology.

2.2 Network Layer

The network layer is mainly responsible for data transmission and processing. On the one hand, the network layer performs data conversion on the data collected by the sensor layer, then uses the GPRS network and the Internet to transmit the data to a remote server. On the other hand, the user feedback information in the application layer is passed to the sensing layer through network layer.

2.3 Data Preprocessing Layer

The data preprocessing method based on time series analysis is embedded in this layer. We use this method to process the original data in the server and store it in a separate database. If the application layer user has subsequent data mining work requirements, the processed data can be provided to the user, which improves the user experience.

2.4 Application Layers

The application layer is the upper layer of the system and its main function is data storage and application. Users can log in to the web-based remote monitoring platform to query or obtain various types of environmental data in the server database according to different types and different time periods. The platform realizes visual data presentation, displays data trend online with polygonal graph, and provides data hints for peak value and mean value. The user can obtain two types of data: raw data, preprocessed data. Users can choose the data source according to their own needs for subsequent data analysis and mining work.

3 Data Description and Modeling Process

After the design and implementation of the Agricultural Internet of things, we carried out atmospheric humidity data acquisition experiments in greenhouse farms or cropland of many farms in downtown. However, we found that there were various kinds of abnormalities in the collected data. For example, we carried out five days of humidity data acquisition in a greenhouse and the data acquisition frequency was set to 10 min. We should get 720 humidity data, but actually only got 700 data, and there are obvious missing data and abnormal data. The above results show that agricultural greenhouse and other agricultural monitoring environment is complex. The environmental noise caused by various reasons has great influence on the collection data, and a lot of abnormal data are generated. In addition, data may be uploaded multiple times or may be missing due to hardware device failures.

The original environmental data with missing values and abnormity values can not be directly analyzed and mined. Therefore, we need to preprocess the original data reasonably and prepare for the subsequent data analysis and even data mining.

The humidity data collected in this paper is a typical time series. In order to process data quickly, efficiently, and at low cost, this chapter uses time series analysis methods to process humidity data [9]. The main flow of time series analysis is as follows: (1) time series preprocessing, (2) time series model identification, (3) time series model parameter estimation, (4) time series fitting and prediction. This paper will introduce the entire analysis process and experimental results according to the timing analysis process.

4 Time Series Preprocessing

After obtaining the real humidity time series, the first and key step is to preprocess the time series. Time series preprocessing includes stationarity detection and pure randomness detection. The purpose of these two detection is to classify the acquired time series. Different types of time series should be analyzed by different methods.

4.1 Stability Detection

Time series can be divided into stationary time series and non-stationary time series. Two kinds of time series are analyzed by different methods. Therefore, we should first determine whether the collected raw time series data is stable. stationary time series can be divided into strictly stationary and wide stationary according to the restriction conditions. In practical applications, most of the studies are wide stationary random sequences, so the stationary sequences usually refer to wide stationary sequences. If a random sequence $\{X_t\}$ meets the following conditions:

(1) The mean of a sequence is a constant that is independent of time t;
(2) The variance of a sequence is a constant that is independent of time t;
(3) The covariance of a sequence is a constant that is only related to the time interval k and is independent of time t;

It can be determined that the random sequence is stationary.

In this paper, we use the adftest function provided by the Matlab system identification toolbox to detect the stability of 700 humidity experimental data. The results showed that $h = 0$, indicating that the sequence was not stable. When the time series is non-stationary, one or more differential operations can be performed on the sequence. After we perform a first-order differential operation on the sequence, the result shows that $h = 1$ and the sequence is stationary.

4.2 Randomness Detection

Not all stationary sequences are worth modeling analysis. If a sequence is a purely random sequence, it means there is no correlation between the sequence values of the sequence. The past behavior has no influence on the future. Such a pure random sequence is of no analytical value. Therefore, after the stationarity detection, the sequence needs to be tested for pure randomness. The autocorrelation coefficient (ACF) of delay k-degree of a pure random sequence should fluctuate near zero. We use the autocorr (s) function provided by Matlab to get the autocorrelation coefficient of the sequence. The ACF of the humidity sequence is shown in Fig. 2, the ACF is still greater than 0 after the K-order delay. we can judge that the sequence is a non-pure random sequence.

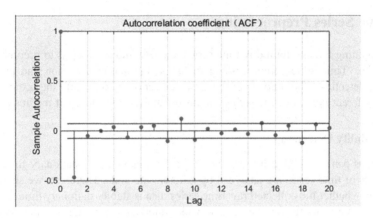

Fig. 2. Autocorrelation coefficient of stationary sequence

5 Modeling of Stationary Time Series

After stationarity detection and pure randomness detection, we can determine that the experimental sequence of this group is a stable and non-pure random time series, which means that the sequence values of the series contain relevant information and are worth modeling analysis [10]. For a stable non-pure random sequence, a linear model can usually be established to fit the development trajectory of the stationary sequence.

ARMA (autoregressive moving average) model can be used to construct linear polynomials. It is commonly used to fit stationary time series and has a wide range of applications. Liu et al. used the ARIMA model to predict the incidence of hand-foot-mouth disease in Sichuan, China, and provided support for HMFD prevention and control [11]. Anghelache et al. used the ARMA model to predict the future changes of GDP in Romania [12]. Compared with other linear fitting models, ARMA model has the advantages of simplicity, versatility, and flexibility. Therefore, this article uses the ARMA model to model the humidity sequence.

The ARMA model can be specifically divided into AR (autoregressive) models, MA (moving average) models, and ARMA models. The AR model and the MA model are special cases of the AMRA model. The general form of the ARMA model is

$$x_t - \varphi_1 x_{t-1} - \cdots - \varphi_p x_{t-p} = \varepsilon_t - \theta_1 \varepsilon_{t-1} - \cdots - \theta_q \varepsilon_{t-q} \qquad (1)$$

ε_t is a zero mean white noise sequence. This model is called the p-order autoregressive q-order moving average mixture model, which is denoted as ARMA(p, q) model. Specifically, when $p = 0$, it is called the pure moving average model, and it is denoted as MA(q). If $q = 0$, it is called pure autoregressive model, and it is denoted as AR(p). If $p = q = 0$, the model degenerates to ε_t.

This paper uses the ARMA model to model and analyze the sequence of the first 600 data in the 700 humidity data, and use the following data for model testing.

The basic steps of ARMA model estimation are as follows:

(1) Calculate the autocorrelation coefficient (ACF) and partial autocorrelation coefficient (PACF) values for this set of sequence values.
(2) Judging the ARMA(p,q) model of the appropriate order based on the autocorrelation coefficient and the trailing or truncated nature of the partial autocorrelation coefficient.
(3) Estimate unknown parameters in the ARMA(p,q) model.
(4) Use the obtained fitting model to fit the data and predict the future trend of the sequence.

5.1 ARMA Model Recognition

The correlation coefficient is a measure of the degree of interaction between two different random events, while the autocorrelation coefficient is a measure of the degree of interaction between an event at different times. For a stationary time series $\{X_t\}$, defined ρ_k as self correlation coefficient. The definition is like formula (2)

$$\rho_k = \frac{\gamma(k)}{\gamma(0)} \tag{2}$$

Where $\gamma(k)$ is K-order delayed autocovariance function of series $\{X_t\}$, $\gamma(0)$ is the variance of the series.

The autocorrelation coefficient actually incorporates the influence of other variables on X_t and X_{t-k}, and the partial correlation coefficient reflects the influence of X_{t-k} on X_t. Partial autocorrelation coefficient can be calculated based on autocorrelation coefficient, according to formula (3)

$$\phi_{kk} = \frac{\hat{D}_k}{\hat{D}} \tag{3}$$

Where

$$\hat{D}_k = \begin{pmatrix} 1 & \rho_1 & \cdots & \rho_{k-1} \\ \rho_1 & 1 & \cdots & \rho_{k-2} \\ \vdots & \vdots & & \vdots \\ \rho_{k-1} & \rho_{k-2} & \cdots & 1 \end{pmatrix} \qquad \hat{D}_k = \begin{pmatrix} 1 & \rho_1 & \cdots & \rho_1 \\ \rho_1 & 1 & \cdots & \rho_2 \\ \vdots & \vdots & & \vdots \\ \rho_{k-1} & \rho_{k-2} & \cdots & \rho_k \end{pmatrix} \tag{4}$$

In this paper, we use the autocorr (s) and parcorr (s) functions provided by Matlab to compute the autocorrelation function (ACF) and the partial autocorrelation function (PACF) with a confidence degree of 95%. The results of the ACF and PACF of the sequence are shown in Fig. 3.

Table 1 shows the model features of the AR, MA, and ARMA models. Truncation property means that the sequence of ACF or PACF abruptly decreases to zero near K when P is equal to P. The trailing property is that the ACF or PACF of the sequence is still greater than 0 after the K-order delay.

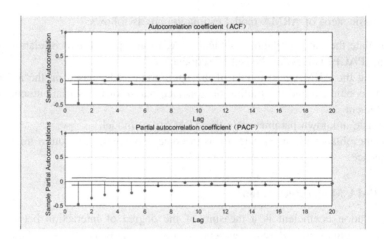

Fig. 3. ACF and PACF of the sequence

Table 1. Sequence characteristics of ARMA(p,q) Model

Model	ACF	PACF
AR(p)	Trailing	Truncation at k = p
MA(q)	Truncation at k = q	Trailing
ARMA(p,q)	Trailing	Trailing

It can be seen from the Fig. 3 that the ACF and PACF of the sequence all exhibit trailing properties. According to the conclusion of Table 1, we can confirm that the sequence is ARMA(p,q).

The p and q values of the AR(p) model and the MA(q) model can be judged from the partial autocorrelation and autocorrelation graph properties, respectively, while the p and q parameters of the ARMA(p,q) model cannot be determined directly. We need to build ARMA models from (1, 1), then increase the values of p and q, and find out a series of models. To find out an optimal model, This article uses the AIC criterion to judge. The AIC criterion is an ARMA(p,q) model identification method proposed by the Japanese scholar Chichi. Formula (5) is the calculation method of AIC.

$$AIC(p,q) = \log \sigma_k^2 + \frac{2(p+q+1)}{N} \tag{5}$$

Where σ_k^2 is the residual variance of the model. If $p = p0$, $q = q0$, the value of AIC (p,q) is the smallest, which means that the applicable fitting model is ARMA($p0$, $q0$). The program flow of ARMA model recognition shown in Fig. 4.

In this paper, the maximum order of the fitting model is 10. We increase the values of P and Q sequentially, and calculate the ARMA model of the experimental data and the value of each model AIC. The experimental results show that when $p = 2$, $q = 10$, the value of AIC is the smallest. Therefore, we can confirm that the model of this sequence is ARMA(2,10).

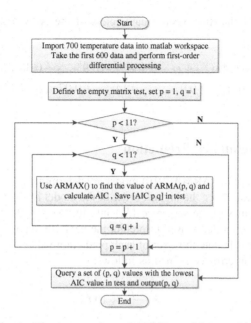

Fig. 4. The program flow of ARMA model recognition

5.2 Estimation of Model Parameters

It has been confirmed that the fitting model is ARMA(2,10), and the next step is to identify the unknown parameters in the model. The unknown parameters of the ARMA (2,10) model are $\varphi_1, \varphi_2, \theta_1, \ldots, \theta_{10}$.

In this paper, the least squares estimation method is used to estimate the parameters of the ARMA(2,10) model. The principle of least squares estimation is to calculate the sum of squares of residuals of a model. We find out a set of parameter values that minimize the sum of squares of the model residuals, which is the best parameter set of the model. The process flow of parameter estimation is shown in Fig. 5.

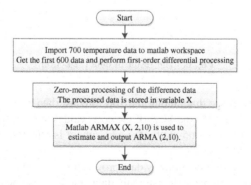

Fig. 5. The process flow of parameter estimation

The results show that the specific parameters of the ARMA(2,10) model are as follows:

$$x_t = -0.47x_{t-1} + 0.40x_{t-2} + \varepsilon_t - 0.46\varepsilon_{t-1} - 0.93\varepsilon_{t-2} + 0.38\varepsilon_{t-3} + 0.10\varepsilon_{t-4}$$
$$- 0.03\varepsilon_{t-5} + 0.015\varepsilon_{t-6} + 0.07\varepsilon_{t-7} - 0.14\varepsilon_{t-8} + 0.07\varepsilon_{t-9} - 0.05\varepsilon_{t-10} \tag{6}$$

6 Stationary Sequence Prediction

After the parameters of the ARMA(2,10) model are determined, the model can be used for data prediction test. Since the highest order of the model is 10, we need to input the first 10 sequence values of the sequence $\{X_t\}$ into the ARMA(2,10) model.

The ARMA(2,10) model is used to predict the next 100 data. We compare it with the original data and calculate the error of each prediction value. The result is shown in Fig. 6. The prediction error is defined as follows:

$$e = \frac{|s - p|}{s} \times 100\% \tag{7}$$

where s is the original data, p is the prediction data. After the experiment, the minimum error of 100 forecast data is 0.03%; the maximum error is 16.7%; the average error is 5.9%. The error distribution is shown in Fig. 7.

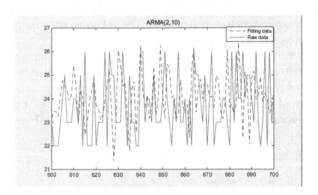

Fig. 6. The result of data prediction

From Fig. 7 we can see that among the 100 forecast data errors, nearly 50% of the error values are less than 0.05 and 80% of the error values are less than 0.1. The results show that the fitting model obtained by the experiment is effective and can predict the change trend of humidity series well.

In the subsequent humidity anomaly data processing, we can use the prediction data obtained by the model to replace the anomaly data, and can also be used to supplement the missing humidity data. The common data filling methods include (1) the fixed value

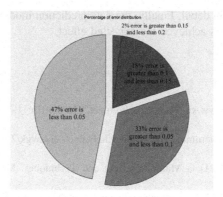

Fig. 7. The error distribution

(the mean value of all data), (2) the most recent value. To verify the effect of the prediction model, we randomly delete 10 of the 100 data. We use the ARMA(2,10) model and two methods mentioned above to fill the data and calculate the relative error between the filling data and the original data. The result of the test is shown in Fig. 8.

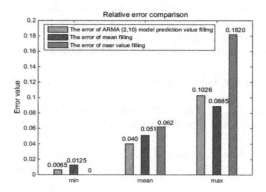

Fig. 8. Relative error comparison

The results show that the minimum of the error of the ARMA(2,10) model is the smallest of the three methods, and the mean of the error of the model is the smallest of the three methods. It shows that the prediction model obtained in this paper is superior to the common method in the data filling.

7   Conclusions

This paper expounds the reason and significance of using time series analysis to preprocess environmental data in Agricultural IoT. After introducing the structure of the four layer of Agricultural Internet of things, the whole process of time series analysis is introduced in detail. Finally, the data prediction model is tested. The results show that the model has achieved the expected effect.

References

1. Xinhuawang. http://www.xinhuanet.com/fortune/2016-09/01/c_129265606.htm. Accessed 1 Sept 2016
2. Hubei Provincial Government. http://www.hubei.gov.cn/zwgk/201703.htm. Accessed 23 Mar 2017
3. Han, J., Kamber, M.: Data Mining Concept and Techniques. Machinery Industry Press, Cambridge (2001)
4. Yan, R., Liu, Y., Gao, R.X.: Correlation dimension analysis: a non-linear time series analysis for data processing. Instrum. Meas. Mag. IEEE **13**(6), 19–25 (2010)
5. Shi, D.L.: Research and design of intelligent agriculture management system based on the internet of things. Appl. Mech. Mater. **687–691**, 1868–1871 (2014)
6. Fang, S., Xu, L.D., Zhu, Y., et al.: An integrated system for regional environmental monitoring and management based on internet of things. IEEE Trans. Industr. Inf. **10**(2), 1596–1605 (2014)
7. Wixted, A.J, Kinnaird, P., Larijani, H., et al.: Evaluation of LoRa and LoRaWAN for wireless sensor networks. Sensors. IEEE, pp. 1–3 (2017)
8. Silva, J.D.C., Rodrigues, J.J.P.C., Alberti, A.M., et al.: LoRaWAN—A low power WAN protocol for internet of things: a review and opportunities. In: International Multidisciplinary Conference on Computer and Energy Science. IEEE (2017)
9. Rojas, I., Pomares, H.: Time series analysis and forecasting. Contributions to Statistics (2016)
10. Shumway, R.H., Stoffer, D.S.: Time series analysis and its applications. Publ. Am. Stat. Assoc. **97**(458), 656–657 (2013)
11. Liu, L., Luan, R.S., Yin, F., et al.: Predicting the incidence of hand, foot and mouth disease in Sichuan province, China using the ARIMA model. Epidemiol. Infect. **144**(1), 144–151 (2016)
12. Anghelache, C., Grabara, J., Manole, A.: Using the dynamic model ARMA to forecast the macroeconomic evolution. Romanian Statistical Review Supplement (2016)

Fuzzy PID Controller for PCC Voltage Harmonic Compensation in Islanded Microgrid

Minh-Duc Pham and Hong-Hee Lee[(⊠)]

School of Electrical Engineering, University of Ulsan,
Ulsan 680-749, South Korea
minhducpham2009@gmail.com, hhlee@mail.ulsan.ac.kr

Abstract. In this paper, an intelligent control scheme based on fuzzy proportional-integral-derivative controller (FPIDC) is proposed for PCC voltage harmonic compensation in islanded microgrid. The proposed FPIDC method is composed of a closed-loop control of the virtual impedance at harmonic frequency to absorb the harmonic current from the nonlinear load, control harmonic sharing between distributed generators. As a result, the PCC voltage quality is improved with the total harmonic distortion (THD) significantly reduced. With the feedback of the PCC voltage and well-designed fuzzy controller, the uncertainty and unstable value of proportional-integral-derivative (PID) parameters are removed by adaptive tuning. Therefore, the PCC voltage quality is improved smoothly, and the system becomes more stable even load condition is changed. Compared with the traditional PID controller, the dynamic response and stability of the microgrid system are improved with the proposed FPIDC. The comparison and analysis of the proposed control with the conventional control are carried out to verify the superiority of the proposed method.

Keywords: Islanded microgrid · Fuzzy logic control
Voltage harmonic compensation · Secondary control

1 Introduction

Microgrids are becoming an important part to integrate power electronic interface converter and distributed energy storage system [1, 2]. Microgrid is composed of distributed generation (DG), energy storage system and renewable energy resources that can operate in both grid-connected and islanded mode [3].

Figure 1 shows the basic configuration of the microgrid. An indispensable function of microgrid is to achieve desired power management among DG units in islanding operation. When the static switch is open, the microgrid is isolated from the main grid and DGs have to operate independently. Therefore, each DG has to regulate not only the voltage but also the frequency to ensure the stability and autonomously power sharing feature of the microgrid.

© Springer International Publishing AG, part of Springer Nature 2018
D.-S. Huang et al. (Eds.): ICIC 2018, LNCS 10954, pp. 231–242, 2018.
https://doi.org/10.1007/978-3-319-95930-6_22

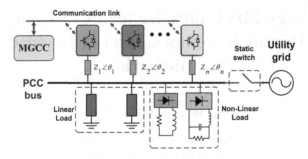

Fig. 1. Typical islanded microgrid configuration.

For this purpose, DG with the droop control is commonly used to maintain the constant voltage at the point of common (PCC) bus. In order to apply droop control concept, DGs are generally operated in the voltage control mode (VCM) [4]. In addition, the control scheme can be easily extended to a number of DGs in microgrid without voltage and current information. Meanwhile, because the droop control generally controls the fundamental component, any uncontrolled harmonic power can make microgrid power quality and stability worse; when nonlinear load is connected to the bus, a high harmonic current is induced, and the PCC bus voltage quality is markedly reduced.

Many harmonic compensation methods have been studied. In [5, 6], the harmonic compensation approaches are introduced based on virtual resistance at the harmonic frequency. In this control strategy, the resistive active power filter (R-APF) can simply view as the virtual harmonic resistor. However, in case of long transmission line, due to the increased inductive feeder impedance, this method is not effective to compensate harmonics. In [7], the negative virtual harmonic impedance was introduced to improve the voltage quality. However, this method requires exact value of the feeder impedance, which is hard to detect in practical applications.

In order to control indirectly the output impedance without detecting the feeder impedance, the concept for selective harmonic distortion (HD) compensation was presented in [8]. This concept is robust and the virtual harmonic impedance is adaptively tuned according to the error of the PCC harmonic distortion through the proportional-integral-derivative (PID) controllers. However, the performance of the controller is seriously dependent on PID parameters which are easily affected by many non-ideal system parameters such as noise and disturbance. Especially, the microgrid system is basically nonlinear due to many DGs connected to PCC bus. Therefore, the conventional PID controller used in [8] is not effective for the microgrid system because it reduces the efficiency of the compensator and decreases the stability of the whole microgrid system.

Therefore, it is important to remove the effect of the system parameter uncertainty. In this paper, we propose an enhanced fuzzy proportional-integral-derivative controller (FPIDC) to compensate the voltage harmonics in islanded microgrid. The FPIDC is developed by coordinating the advantages of the nonlinear control of FLC and the small steady-state error of PID controller. Even though the FPIDCs are already

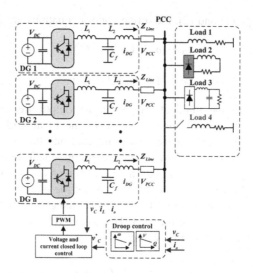

Fig. 2. Basic structure of parallel DGs in islanded microgrid

introduced in [9, 10], the application of FPIDC for voltage harmonic compensation in microgrid has not been presented up to now. Moreover, this paper develops a feedback compensation method for FPIDC to improve PCC voltage quality, power sharing performance and harmonic current sharing between DGs. The proposed compensator provides outstanding performance and guarantees system stability in spite of the load variation. The reliability of the proposed FPIDC for the islanded microgrid is verified by simulation using MATLAB/PLECS.

2 Islanded Microgrid with Parallel Inverters

2.1 Droop Control Theory

Figure 2 shows the basic structure of parallel DGs which are used in islanded microgrid. Droop controller [11] is used to control the output voltage and frequency of the inverter in order to autonomously share the power among DG units in islanding operation. From the active and reactive power, the phase angle and voltage magnitude differences between the inverter output voltage and PCC bus voltage can be expressed as

$$\delta \approx (XP)/(EV). \tag{1}$$

$$E - V \approx (XQ)/E. \tag{2}$$

where E and V are the amplitudes of inverter output voltage and PCC bus voltage, respectively, δ is the phase angle difference between E and V. R and X are the line resistance and line inductance of the feeder, respectively.

Based on the Eqs. (1) and (2), the droop control equations are defined as

$$\omega = \omega_0 - G_P(P - P^*),$$ (3)

$$E = E_0 - G_Q(Q - Q^*),$$ (4)

where ω_0 and E_0 are the nominal values of the output frequency and voltage, respectively, which are usually equal to those of the main grid. P and Q are the fundamental active power and reactive power injected into the grid from the inverter. G_P and G_Q are the frequency and amplitude droop coefficients, respectively. In islanded mode, because the inverter determines the power sharing without the support from the main grid, P^* and Q^* are set to zero. When the microgrid transferred to the islanded mode, the output power of DG systems is changed according to their droop characteristics to supply power to the load. The output voltage reference which is fed into the outer and inner loop can be written as

$$V^* = E \sin\left(\int \omega \, dt\right)$$ (5)

In (5), the reference of the conventional droop controls only the fundamental component. Therefore, when non-linear load is connected to the grid, the PCC voltage is distorted, and the system can be unstable with poor power quality.

2.2   Outer and Inner Loop Control

In order to control the output voltage, the following multi proportional-resonant (PR) voltage and current controllers are applied:

$$G_V(s) = K_{pV} + \sum_{i=1,3,5,7} \frac{2.K_{Vi}.\omega_{CV}.s}{s^2 + 2.\omega_{CV}.s + \omega_o^2},$$ (6)

$$G_I(s) = K_{pI},$$ (7)

where K_{pV} and K_{pI} are the proportional gains, K_{Vi} is the resonant gains of the PR controller. ω_{CV} is the cut-off frequency of the voltage control loop. Figure 3 shows the control diagram of outer and inner loop controller with PR controller; the double loop control is used to increase the phase margin, and to stabilize the system.

For fast dynamic response, the values of K_{pV}, K_{pI}, K_{Vi} and ω_{CV} of two control loop should be optimized. Typically, the current control loop bandwidth is selected 10 times higher than the voltage control loop to guarantee the system stability and phase margin [12].

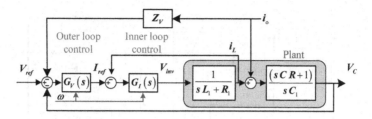

Fig. 3. Outer and inner loop control diagram.

3   Proposed Fuzzy Controller for Voltage Harmonic Compensation in Microgrid

3.1   Harmonic Virtual Impedance Control

From the close-loop transfer function of the inner loop shown in Fig. 3, the capacitor voltage (V_C) is obtained as following:

$$V_C = \frac{G_I G_V Z_C}{Z_C + Z_L + G_I + G_I G_V Z_C} V_{ref}(s) - \frac{Z_C(Z_L + G_I)}{Z_C + Z_L + G_I + G_I G_V Z_C} i_o(s) \quad (8)$$

$$= G_{control}(s) V_{ref}(s) - Z_O(s) i_o(s), \quad (9)$$

where $Z_L = sL_1 + R_1$ and $Z_C = (sCR + 1)/sC$.

In Eq. (9), the capacitor voltage (V_C) can be controlled by two elements, $G_{control}$ and Z_O: $G_{control}(s)$ is the voltage gain and $Z_O(s)$ is equivalent value of the inverter output impedance. Since only fundament sinusoidal signal is controlled when $V_{ref} = V_{droop}^*$ in (5), V_{ref} should be modify to control the harmonic component.

To compensate the PPC harmonic voltage, the harmonic virtual impedance is proposed by considering the harmonic voltage drop on the line impedance, and the new V_{ref} is defined as following:

$$V_{ref} = V_{droop}^* - V_{Vir\_har}^* = V_{droop}^* - V_{PCC\_h} \cdot K_{Vir}^h, \quad (10)$$

where $V_{Vir\_har}^*$ and K_{Vir}^h are the virtual harmonic impedance drop and the harmonic impedance factor, respectively. V_{droop}^* is the inverter droop voltage reference, which is calculated from the droop equation in (5) by replacing V^* with V_{droop}^*. V_{PCC} is measured by MGCC with the harmonic extraction method in [13], and the HD is calculated as

$$HD_h = \sqrt{\left(V_d^h\right)^2 + \left(V_q^h\right)^2} / \sqrt{\left(V_d^1\right)^2 + \left(V_q^1\right)^2} \quad (11)$$

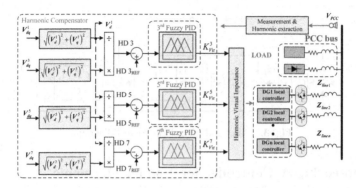

Fig. 4. The proposed FPIDC control diagram.

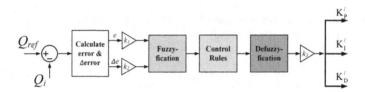

Fig. 5. Structure of the proposed FLC method

To compensate the harmonic voltage properly, the virtual harmonic impedance factor at hth frequency is adjusted by the external secondary loop control through PID controller:

$$K_{Vir}^h = K_{P\_vir}\left(HD_{ref} - HD_h\right) + K_{I\_vir} \int \left(HD_{ref} - HD_h\right) dt + K_{D\_vir}\ d\left(HD_{ref} - HD_h\right)/dt$$

where HD_{ref} is the demanded HD value to meet the PCC voltage requirement, and the HD_h is the current HD at hth frequency of the PCC. In this paper to keep the quality of the V_{PCC} as good as possible, the HD_{ref} at 3rd,5th, 7th harmonic requirement are set to be 1.0%.

3.2   Fuzzy Logic Controller Design

The FLC is widely used to solve problems under uncertainty and high nonlinearities. To improve the performance and dynamic response of the traditional FLC, there are some hybrid combination between the intelligent fuzzy controller and the traditional controllers like PD, PI, or PID [10, 14]. In this paper, the FPIDC is applied to increase the dynamic response and stability in islanded microgrid. Figure 4 shows the proposed FPIDC control block diagram. The general continuous-time PID is described as following:

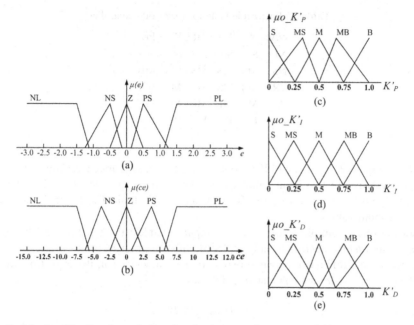

Fig. 6. Membership functions design for the input and output variable. (a) Input error e. (b) Input variable (Δe). (c) Output variable K'_p. (d) Output variable K'_I. (e) Output variable K'_D

$$G_{PID}(t) = K_P e(t) + K_I \int e(t)dt + K_D \dot{e}(t) \qquad (12)$$

To find three gains K_P, K_I, K_D of PID controller, it is better to define the range of $[K_{P\min}, K_{P\max}], [K_{I\min}, K_{I\max}], [K_{D\min}, K_{D\max}]$ firstly to implement the Fuzzy algorithm with less complexity and faster convergence to the steady state. It can be carried out using open loop tuning or Ziegler-Nichols method [15]. From Ziegler-Nichols method, K'_P, K'_I, and K'_D are defined as (13), and they are normalized in [0,1]:

$$\begin{cases} 0 \leq K'_P = \frac{K_P - K_{P\min}}{K_{P\max} - K_{P\min}} \leq 1 \\ 0 \leq K'_I = \frac{K_I - K_{I\min}}{K_{I\max} - K_{I\min}} \leq 1 \\ 0 \leq K'_D = \frac{K_D - K_{D\min}}{K_{D\max} - K_{D\min}} \leq 1 \end{cases} \qquad (13)$$

Then, the gains for the PID controller are obtained by modifying (13):

$$\begin{cases} K_P = (K_{P\max} - K_{P\min})K'_P + K_{P\min} \\ K_I = (K_{I\max} - K_{I\min})K'_I + K_{I\min} \\ K_D = (K_{D\max} - K_{D\min})K'_D + K_{D\min} \end{cases} \qquad (14)$$

Table 1. Fuzzy rule table for fuzzy PID controller

e/Δe	NL	NS	ZR	PS	PL
NL	S	S	S	S	S
NS	MS	MS	MS	MS	MB
ZR	MS	MS	M	MB	MB
PS	MS	MB	MB	MB	MB
PL	B	B	B	B	B

From Eqs. (12)–(14), the FPIDC control scheme is configurated as shown in Fig. 5, where it consists of two input variables and three output variables. It is clear that no saturation blocks are used because PID gains are already limited within their minimum and maximum values.

In the fuzzy controller, there are two input variables: the error (e) and the error change rate (de/dt). de/dt is usually replaced with the gradient of error (Δe) which is obtained by the difference between two sampling values. Then, the error (e) and the gradient of error (Δe) are calculated as follows:

$$e(k) = \left(HD_{ref} - HD_h(k) \right) \tag{15}$$

$$\Delta e(k) = e(k) - e(k-1) \tag{16}$$

where k and (k-1) represent the present and the previous sampling time, respectively. It is clear that the comprehensive ranges of fuzzy input and output variables, e(t) and Δe (t), symmetrically spread on both positive and negative sides. The design of the proposed FPIDC algorithm is divided into three main parts: fuzzification, fuzzy rule, and

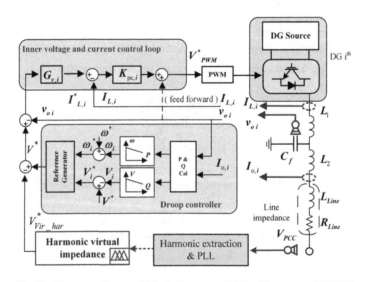

Fig. 7. The overall control block diagram of DG with proposed FPIDC

defuzzification. For reduced computation, the asymmetrical triangular membership functions (MFs) are selected, and five memberships are designed as shown in Fig. 6. Finally, the control rules are described in Table 1. The membership function and rules of FPIDC for 5th, 7th harmonic compensation are determined similarly with those of FPIDC for 3rd harmonic compensation except the range of PID parameters due to the same interaction effect of the PID controller. The overall control block diagram of DG with proposed FPIDC is shown in Fig. 7.

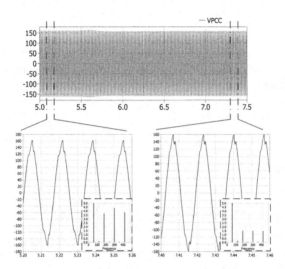

Fig. 8. The PCC harmonic voltage compensation with the proposed FPIDC

4   Simulation Results

In order to verify the effectiveness of the proposed FPIDC method, the microgrid configuration in Fig. 2 is simulated by using MATLAB/PLECS. The simulation parameters are given as follows: L_1 = 1.2 mH, L_2 = 1.2 mH, C = 20uF, G_P = 0.012, G_Q = 0.015, V_o = 110 V_{rms}/50 Hz, Z_{line1} = 0.1 + 0.39j Ω, Z_{line1} = 0.2 + 0.3j Ω, R_{load1} = R_{load4} = 20 Ω, L_{load1} = L_{load4} = 15 mH, $L_{nonlinear}$ = 0.64 mH, $R_{nonlinear}$ = 100 Ω, $C_{nonlinear}$ = 150 uF.

Figure 8 shows the performance of PCC harmonic compensation in islanded microgrid when the proposed FPIDC is applied at 5.5 s. In islanded mode, each DG uses the droop controller to share the active and reactive power with sinusoidal PCC bus voltage. However, because the droop controller regulates only fundamental component, the quality of the PCC bus is significantly reduced when the nonlinear load is connected to the grid; the HD_3, HD_5, and HD_7 are 3.749%, 4.41%, and 3.85%, respectively. When the controller is applied at 5.5 s, the harmonic current is shared between DGs, which reduces the HD to 1% as shown in Fig. 8.

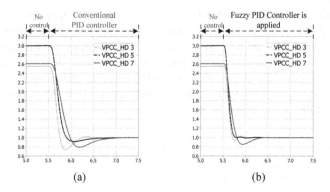

Fig. 9. (a) The dynamic response of the conventional PID controller (b) The dynamic response of the proposed FPIDC

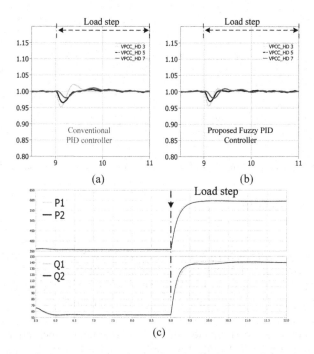

Fig. 10. The dynamic response comparison when the load step at 9.0 s. (a) The active and reactive power output of DG1, DG2. (b) The conventional PID controller. (b) The proposed FPIDC.

Figures 9(a) and (b) show the dynamic responses of the conventional PID controller and proposed FPIDC with the same initial PID parameters. As can be seen in Fig. 9, the HD of PCC is compensated to 1% for both controllers. But, the dynamic performance of the FPIDC is faster than that of the conventional PID controller.

In addition, the conventional PID shows more oscillation before steady state, and its transient performance is not smooth when the compensating algorithm starts compared with the FPIDC.

Figures 10 shows the dynamic responses when load 4 in Fig. 2 is connected to the grid at 9.0 s. The PCC HDs are shown in Figs. 10(a) and (b). Even through both PID controllers achieve the desired HD requirement, 1%, the dynamic performance with the FPIDC is faster approximately 5 s than that of the conventional PID controller because PID parameters are not selected properly in the new working state. As a result, it is clear that the proposed FPIDC has very good dynamic performance compared with the conventional PID controller. Figure 10(c) shows the active and reactive power sharing when load 4 is connected. As we can see, 2 DGs smoothly reach a new equilibrium operating point to keep the accurate active and reactive power sharing.

From simulated results, it is obvious that the performance of the FPIDC is much better than the conventional PID controller, we can say that the membership function and rules are designed correctly because the optimal PID parameters are adaptively tuned in spite of the microgrid condition variation.

5 Conclusion

This paper has presented an intelligence harmonic impedance control method based on the fuzzy logic controller to improve the dynamic performance of the PCC harmonic voltage compensation in islanded microgrid. In the proposed FPIDC, PID gains are adaptively tuned according to the load condition. With the proposed FPIDC, the harmonic virtual impedance is adjusted to share the harmonic current between DGs, and the HD of PCC is reduced to HDref value which is limited to be 1% in this paper. Furthermore, the microgrid operating system becomes more stable despite any load change due to exact PID gains tuning. The dynamic performance of the proposed FPIDC is evaluated by comparing with the conventional PID controller, and it is verified that the performance of the FPIDC is much better than that of the conventional PID controller.

Acknowledgment. This work was partly supported by the National Research Foundation of Korea Grant funded by the Korean Government (NRF-2015R1D1A1A09058166) and the Korea Institute of Energy Technology Evaluation and Planning (KETEP) and the Ministry of Trade, Industry & Energy (MOTIE) (No. 20174030201490).

References

1. He, J., Li, Y.W., Blaabjerg, F.: An enhanced islanding microgrid reactive power, imbalance power, and harmonic power sharing scheme. IEEE Trans. Power Electron. **30**(6), 3389–3401 (2015)
2. Mahmood, H., Michaelson, D., Jiang, J.: Accurate reactive power sharing in an islanded microgrid using adaptive virtual impedances. IEEE Trans. Power Electron. **30**(3), 1605–1617 (2015)

3. Olivares, D.E., et al.: Trends in microgrid control. IEEE Trans. Smart Grid **5**(4), 1905–1919 (2014)
4. Guerrero, J.M., Matas, J., De Vicuña, L.G., Castilla, M.: Wireless-control strategy for parallel operation of distributed generation inverters. IEEE Trans. Ind. Electron. **53**(5), 1461–1470 (2006)
5. Lee, T.L., Hu, S.H.: Discrete frequency-tuning active filter to suppress harmonic resonances of closed-loop distribution power systems. IEEE Trans. Power Electron. **26**(1), 137–148 (2011)
6. Lee, T.L., Li, J.C., Cheng, P.T.: Discrete frequency tuning active filter for power system harmonics. IEEE Trans. Power Electron. **24**(5), 1209–1217 (2009)
7. Sreekumar, P., Khadkikar, V.: A new virtual harmonic impedance scheme for harmonic power sharing in an islanded microgrid. IEEE Trans. Power Deliv. **31**(3), 936–945 (2016)
8. Moussa, H., Shahin, A., Martin, J.-P., Nahid-Mobarakeh, B., Pierfederici, S., Moubayed, N. N.: Harmonic power sharing with voltage distortion compensation of droop controlled islanded microgrids. IEEE Trans. Smart Grid, 1 (2017)
9. Li, Y., Tong, S., Li, T.: Hybrid fuzzy adaptive output feedback control design for uncertain MIMO nonlinear systems with time-varying delays and input saturation. IEEE Trans. Fuzzy Syst. **24**(4), 841–853 (2016)
10. Pham, Minh-Duc, Lee, Hong-Hee: Fuzzy PID controller for reactive power accuracy and circulating current suppression in islanded microgrid. In: Huang, De-Shuang, Hussain, Abir, Han, Kyungsook, Gromiha, M.Michael (eds.) ICIC 2017. LNCS (LNAI), vol. 10363, pp. 241–252. Springer, Cham (2017). https://doi.org/10.1007/978-3-319-63315-2_21
11. Guerrero, J.M., De Vicuña, L.G., Matas, J., Castilla, M.: Output impedance design of parallel-connected UPS inverters with wireless load-sharing control. IEEE Trans. Ind. Electron. **52**(4), 1126–1135 (2005)
12. Cha, H., Vu, T.-K., Kim, J.-E.: Design and control of proportional-resonant controller based photovoltaic power conditioning system. In: 2009 IEEE Energy Conversion Congress and Exposition, pp. 2198–2205 (2009)
13. Rodríguez, P., Luna, A., Candela, I., Mujal, R., Teodorescu, R., Blaabjerg, F.: Multiresonant frequency-locked loop for grid synchronization of power converters under distorted grid conditions. IEEE Trans. Ind. Electron. **58**(1), 127–138 (2011)
14. Sahu, R.K., Panda, S., Yegireddy, N.K.: A novel hybrid DEPS optimized fuzzy PI/PID controller for load frequency control of multi-area interconnected power systems. J. Process Control **24**(10), 1596–1608 (2014)
15. Gude, J.J., Kahoraho, E.: Modified Ziegler-Nichols method for fractional PI controllers. In: Proceedings of the 15th IEEE International Conference on Emerging Technologies and Factory Automation, ETFA 2010 (2010)

Supervised Learning Algorithm
for Multi-spike Liquid State Machines

Xianghong Lin[✉], Qian Li, and Dan Li

College of Computer Science and Engineering, Northwest Normal University,
Lanzhou 730070, China
linxh@nwnu.edu.cn

Abstract. The liquid state machines have been well applied for solving large-scale spatio-temporal pattern recognition problems. The current supervised learning algorithms for the liquid state machines of spiking neurons generally only adjust the synaptic weights in the output layer, the synaptic weights of input and hidden layers are generated in the process of network structure initialization and no longer change. That is to say, the hidden layer is a static network, which usually neglects the dynamic characteristics of the liquid state machines. Therefore, a new supervised learning algorithm for the liquid state machines of spiking neurons based on bidirectional modification is proposed, which not only adjusts the synaptic weights in the output layer, but also changes the synaptic weights in the input and hidden layers. The algorithm is successfully applied to the spike trains learning. The experimental results show that the proposed learning algorithm can effectively learn the spike trains pattern with different learning parameter.

Keywords: Spiking neural network · Liquid state machine
Bidirectional modification · Synaptic plasticity

1 Introduction

The Spiking Neural Network (SNN) is an important research direction in artificial intelligence. However, neurophysiological studies have shown that the efficient information processing in the nervous system can also be established at the precise timing with the real spiking neurons [1, 2]. A large number of studies have shown that the spiking neurons are the main medium for the information transmission through biological intelligence activities [3, 4]. In the networks of spiking neurons, the neural information is represented by the precisely timed spike trains. Compared with the traditional artificial neural networks based on spike rate coding, the SNNs overcome the problem that the computational power of neural networks made of threshold or sigmoidal units is not strong enough [5], which are more efficient computational models for spatio-temporal information processing [6].

Maass et al. [7, 8] proposed a new form of neural network model, named Liquid State Machine (LSM), which is capable of conducting universal computations [9]. This model consists of three main parts, an input component, a liquid component with recurrent structure, and a readout component [7]. In fact, the LSM is a special recurrent

© Springer International Publishing AG, part of Springer Nature 2018
D.-S. Huang et al. (Eds.): ICIC 2018, LNCS 10954, pp. 243–253, 2018.
https://doi.org/10.1007/978-3-319-95930-6_23

neural network and provides an efficient way to train the connection weights of network [10]. Throughout the training process, the linear output layer is usually trained only by standard linear regression method, and the part of the nonlinear recursive network remains unchanged [11]. The computational model of LSM based on SNN has gained increasing attention during the past decade, whose elements and structure are highly inspired from real biological nervous system. The multi-spike LSM has been applied in many ways. Sala et al. [12] proposed a framework which used the paradigm of reservoir computing to control the collaborative robot BAXTER. Zhang et al. [13] presented a bio-inspired digital LSM with spike-based learning hardware implementation. Furthermore, Jin et al. [14] leveraged the inherent redundancy of the targeted spiking neural networks to achieve both high performance for the application of speech recognition. Obada et al. [15] adopted LSMs to recognize the emotional state of an individual based on EEG data, and so on.

The LSM of spiking neurons is essentially a recurrent spiking neural network with a random synaptic weights and almost identical neural elements. The early forms of LSM were built without weight updating within the network. Some researches presented the unsupervised learning methods for multi-spike LSMs using the Spike Timing-Dependent Plasticity (STDP) rule [16]. Using the STDP and Widrow-Hoff rules [17], Ponulak and Kasinski [18] proposed a supervised learning algorithm for LSMs with the Remote Supervised Method (ReSuMe). However, the learning algorithm is suitable to train the single-layered SNNs, which can only adjust the output layer synaptic weights of LSMs. The synaptic weights of input and hidden layers are generated in the process of network structure initialization and no longer change. Although the ReSuMe method has improved the performance and accuracy for the spike train learning tasks, the dynamic characteristics of the LSMs is often neglected.

In this paper, we present a supervised learning algorithm for LSMs with spiking neurons using the bidirectional modification mechanism of presynaptic neuronal excitability [19]. The multi-spike learning algorithm can not only efficiently adjust the synaptic weights of output layer, but also update the synaptic weights of input and hidden layers in LSMs based on temporal coding. The algorithm is successfully applied to the spike train pattern learning problem. By analyzing the learning performance of the spike trains, we find that the proposed method can improve the learning capacity of the multi-spike liquid state machines.

2 The Architecture of Multi-spike Liquid State Machines

The liquid state machines are the other pioneer method of reservoir computing, developed simultaneously and independently from echo state networks. The LSM is a special recurrent network with a complex structure which contains three layers, including the input layer (or input component), hidden layer (or liquid component) and output layer (or readout component). In the LSMs, the neurons between the input layer and the hidden layer are feedforward fully connected, and the connectivity of the neurons between the hidden layer and the output layer are the same. In addition, the neurons in the hidden layer are recurrently connected with the given connectivity probability. In the particular learning task, the connection probability between the hidden neurons can be set according to the actual situation. The structural diagram of the LSMs is shown in Fig. 1.

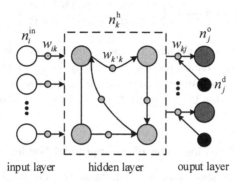

input layer   hidden layer   ouput layer

Fig. 1. The structure of a liquid state machine model

In the LSMs, three classes of neurons need to be considered in the learning process: the input neuron sets $N^{in}=\{n_1^{in}, n_2^{in}, \ldots, n_{N_I}^{in}\}$, the hidden neuron sets $N^h=\{n_1^h, n_2^h, \ldots, n_{N_H}^h\}$, and the output neuron sets $N^o=\{n_1^o, n_2^o, \ldots, n_{N_O}^o\}$, where N_I, N_H and N_O represent the number of neurons in the input layer, the hidden layer and the output layer, respectively. A hidden neuron n_k^h is randomly connected to an input neuron n_i^{in}, and the synaptic weight is expressed as w_{ik}. A hidden neuron is randomly connected to an output layer neuron n_j^o, the synaptic weight is expressed as w_{kj}. The corresponding synaptic weight between the hidden neurons is expressed as $w_{k'k}$. The learning variable quantity of hidden layer synaptic weight $w_{k'k}$ is solved by a learning algorithm based on bidirectional regulation.

3 Supervised Learning for Multi-spike Liquid State Machines

3.1 The Synaptic Learning Rule of the Output Layer

The change of synaptic weight w_{kj} in the output layer depends not only on the time difference of the spike firings between the presynaptic and postsynaptic neurons, but also on the desired spike train of the output neuron. The hidden layer spike trains, the output layer spike trains, and the desired spike trains are represented as $S^h(t)$, $S^o(t)$ and $S^d(t)$, respectively. By using the ReSuMe method [18], we can get the learning rule of synaptic weight w_{kj}. The adjustment rule of the excitatory synaptic weight between the hidden neuron k and the output neuron j is:

$$\Delta w_{kj} = \eta \int_0^\infty \left[S_j^d(t) - S_j^o(t) \right] \left[a + \int_0^\infty W(s) S_k^h(t-s) ds \right] dt \tag{1}$$

where $\eta > 0$ represents the learning rate, $a > 0$ is a non-Hebbian term, $W(s)$ is the learning window of the STDP, the learning rule of the STDP is as follows:

$$W(s) = \begin{cases} A_+ \exp(-s/\tau_+) \text{ if } s \geq 0 \\ -A_- \exp(s/\tau_-) \text{ if } s < 0 \end{cases} \tag{2}$$

where A_+ and A_- are the amplitudes, τ_+ and τ_- are the time constants of the learning window, $s = t_{post} - t_{pre}$ represents the time delay between the postsynaptic and presynaptic firings. The adjustment rule of the inhibitory synaptic weights between the hidden layer and the output layer is opposite to the updating rule of excitatory synapse weights.

3.2 The Synaptic Learning Rules of the Input and Hidden Layers

Biological experiments show that the high frequency electrical activity caused by transient synaptic plasticity, at the same time the neuron excitability change is induced by only comparing synaptic plasticity. In general, synaptic plasticity induces long-term potentiation (LTP) and long-term depression (LTD) respectively by high and low frequency stimulation. The LTP is induced by high frequency stimulation, the frequency of postsynaptic neurons will increase, which is caused by presynaptic input neurons. Related electrical activity induced synaptic plasticity, along with dendritic spatial integration of the bidirectional modification [19].

The neuronal properties other than synaptic efficiency are also altered by their electrical activity. The correlation between the presynaptic and postsynaptic neuron spike trains not only causes synaptic LTP and LTD, but also synaptic enhancement and inhibition will retrograde and rapidly propagate to the presynaptic neuron dendritic synapses, so bidirectional modification of presynaptic neurons in the overall excitability. The synaptic LTP regulates the average value of interval of neuronal spike in presynaptic neuron. In contrast, the synaptic LTD regulates the average value of inter-spike interval. Therefore, the application of synaptic plasticity of the bidirectional modification mechanism can be constructed hidden layer synaptic weights $w_{k'k}$, Fig. 2 shows the illustration of bidirectional modification for LSM architecture.

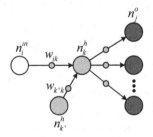

Fig. 2. Illustration of bidirectional modification for LSMs

In the training process, the learning rules of the synaptic weight in the hidden layers $\Delta w_{k'k}$ can be described by the following formula:

$$\Delta w_{k'k} = \begin{cases} \rho \frac{1}{N_O} \sum_{j=1}^{N_O} \Delta w_{kj} & \text{if } w_{k'k} \geq 0 \\ -\rho \frac{1}{N_O} \sum_{j=0}^{N_O} \Delta w_{kj} & \text{if } w_{k'k} < 0 \end{cases} \tag{3}$$

where $\rho > 0$ represents the backpropagation rate, N_O is the number of neuron in the output layer, $w_{k'k} \geq 0$ represents the excitatory synaptic, and $w_{k'k} < 0$ represents the inhibitory synaptic.

The learning rule of the excitatory and inhibitory synaptic weights Δw_{ik} between the input layer and the hidden layer is same as the hidden layer synaptic weights $w_{k'k}$:

$$\Delta w_{ik} = \begin{cases} \rho \frac{1}{N_O} \sum_{j=1}^{N_O} \Delta w_{kj} & \text{if } w_{ik} \geq 0 \\ -\rho \frac{1}{N_O} \sum_{j=0}^{N_O} \Delta w_{kj} & \text{if } w_{ik} < 0 \end{cases} \tag{4}$$

4 Learning Sequences of Spikes

4.1 The Error Function of the LSMs

The error function indicates the deviation between the actual output and the desired output. The goal of the spike trains learning is that the output neurons eventually fire the desired spike trains for a given input spike train by adjusting the synaptic weights. The key of this algorithm is the selection of the spike trains error function. In this experiment, the spike trains error function in the network is computed by the inner products of spike trains. S_o^a and S_o^d are the actual output spike train and desired output spike train respectively. The error function $E(t)$ is defined as:

$$E(t) = \frac{1}{2} \sum_{o=1}^{N_O} \left[f_{S_o^a}(t) - f_{S_o^d}(t) \right]^2 \tag{5}$$

The error function E in the time interval $\Gamma = [0, T]$ can be expressed as:

$$E = \int_0^T E(t)dt = \frac{1}{2} \sum_{o=1}^{N_O} \int_0^T \left[f_{S_o^a}(t) - f_{S_o^d}(t) \right]^2 dt$$

$$= \frac{1}{2} \sum_{o=1}^{N_O} \left[F(S_o^a, S_o^a) - 2F(S_o^a, S_o^d) + F(S_o^d, S_o^d) \right] \tag{6}$$

where $F(S_i, S_j)$ is the inner product of the spike trains of S_i and S_j. That is, the learning error of the spike neurons can be converted into the inner product of the two spike trains [20].

In the experiment, the neuron model in the network structure use the spike response model (SRM) [21]. In SRM, the time constant of the spike response function is $\tau = 5$ ms, the time constant of the refractory period function is $\tau_R = 50$ ms, the excitation threshold $\theta = 1.0$ and absolute refractory period $t_{ref} = 1.0$ ms, the range of synaptic weights between adjacent neurons is [–0.2, 0.2]. There are 20 input neurons, 100 hidden neurons and an output neuron. The maximum time of a single neuron is 200 ms, and the time step is 0.1 ms, while the other parameter settings remain the same. We set the learning rate $\eta = 0.0001$ and the backpropagation rate $\rho = 0.0001$. On each testing trial the learning algorithm is applied for a maximum of 600 learning epochs, the results are averaged over 100 trials. And each input and desired spike train is generated randomly according to the Poisson process with rates of 20 Hz and 50 Hz.

4.2 Analysis of Learning Process

First, we analyze the learning process of the learning algorithm and the change of synaptic weights. The learning effect is shown as follows. Figure 3(a) illustrates the complete learning process, which includes the initial output spike train Δ, the actual output spike trains $\cdot$ and the desired output spike train ∇ at some learning epochs during the learning process. From the graph, we can know that it is possible to derive the desired output spike train from the initial output spike train by about 32 steps. The evolution of the total network error in the time interval T is represented in Fig. 3(b). We note that the error function value decreases rapidly at the beginning of the learning and is reduced to 0 after 32 learning epochs. That is, the actual output spike train is equals to desired spike train. Figure 3(c) and (d) illustrates the synaptic weight changes before and after the training process. In this experiment, there are 200 synaptic weights in total. The synaptic weights in the learning process are generated as the uniform distribution in the interval [–0.2, 0.2].

4.3 Analysis of Learning and Backpropagation Rates

Figure 4 shows the learning results of spike trains with learning rates 0.00005, 0.00008, 0.0001, 0.0003 and 0.0005. Figure 4(a) shows the learning error of different learning rates. We can see that the learning error decreases at first, then increases when the learning rate reaches 0.0001, so the suitable learning rate is 0.0001. For example, when the learning rates are 0.00008, 0.0001 and 0.003, the corresponding errors are 1.1827, 0.8383 and 1.3087 respectively. Figure 4(b) shows the learning epochs when the error reaches the minimum value. When the learning rates are 0.00005 and 0.00008, the average learning epochs is larger, and the average learning epochs is relatively smaller when the learning rates are 0.0001, 0.003 and 0.005. For example, when the learning rates are 0.00008 and 0.0001, the average learning epochs are 264.32 and 188.26 respectively.

Figure 5 shows the learning results of spike trains with different backpropagation rates 0.00005, 0.00008, 0.0001, 0.0003 and 0.0005. Figure 5(a) shows the learning

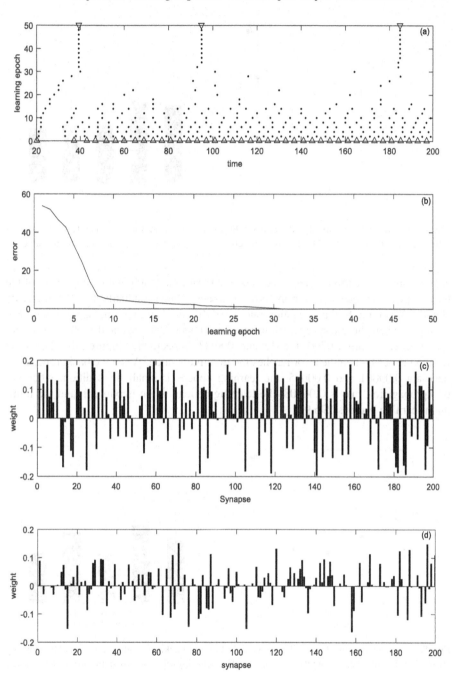

Fig. 3. The learning process of the spike train. (a) The complete learning process. (b) The evolution of the learning error. (c) Synaptic weight changes before training process. (d) Synaptic weight changes after training process.

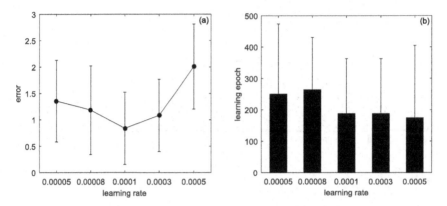

Fig. 4. The learning results of spike trains with different learning rates. (a) The learning error of different learning rates. (b) The learning epochs when the error reaches the minimum value.

error of different learning rates. As can be seen from the graph, with the increase of the backpropagation rates, the learning error becomes smaller gradually and then increases slightly. When the backpropagation rate is 0.0001, the learning error is minimum. For example, when the backpropagation rates are 0.00008, 0.0001 and 0.003, the corresponding errors are 1.0790, 0.8313 and 0.8987 respectively. Figure 5(b) shows the learning epochs when the error reaches the minimum value. When the backpropagation rates are 0.00001 and 0.00005, the learning epochs is small. In addition, when the backpropagation rate is 0.00008, the learning epochs is the maximum, and the minimum learning error is 296.91.

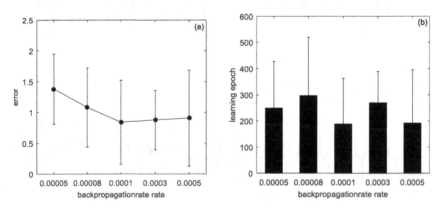

Fig. 5. The learning results of spike trains with different backpropagation rates. (a) The learning error of different backpropagation rates. (b) The learning epochs when the error reaches the minimum value.

4.4   Learning Performance of Different LSM Structures

Figure 6 shows the learning results of spike trains with different connection probabilities in hidden layer neurons. The connection probabilities increase from 0.2 to 0.8,
and increments constant is 0.1. Figure 6(a) shows the minimum error of different
connection probabilities. It can be seen from the graph that the error decreases at first,
then increases when the connection probabilities of input spike trains increase gradually. The learning error reaches the minimum value when the connection probability is
0.5. The learning error increases faster when the connection probabilities continue to
increase after 0.5. For example, when the connection probabilities are 0.5, 0.7 and 0.8,
the learning errors are 0.8305, 1.2420 and 2.0188 respectively. Figure 6(b) shows the
learning epochs when the error reaches the minimum value. When the connection
probabilities are 0.5, 0.6 and 0.8, the learning epochs is relatively small, and the
corresponding learning epochs are 268.31, 217.93 and 151.14 respectively. In addition,
the learning epochs is the maximum when the connection probability is 0.4, and the
average of epochs is 365.91 when the learning error is minimum.

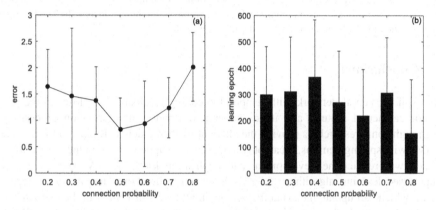

Fig. 6. The learning results of spike trains with different connection probabilities in LSMs
neurons. (a) The learning error of different connection probabilities. (b) The learning epochs
when the error reaches the minimum value.

Figure 7 shows the learning results of spike trains with different numbers of neurons in the hidden layers. The number of synaptic inputs increases from 20 to 200 with
an interval of 20. The number of synaptic inputs increases from 20 to 200 with an
interval of 20. Figure 7(a) shows the learning error different numbers of neurons in the
hidden layers. With the increase of the number of neurons in the LSMs, the learning
error decreases gradually. When the number of neurons in the LSMs increases from 80
to 100, the learning error decreases rapidly. Then the number of neurons in the LSMs
continues to increase, the learning error decreases slowly. For example, when the
number of neurons in the LSMs are 80, 100 and 200, the learning errors are 1.8950,
1.4391 and 0.8305 respectively. Figure 7(b) shows the learning epochs when the error
reaches the minimum value. When the number of neurons in the LSMs are 20, 80 and
120, the average learning epochs is smaller, and the corresponding average learning

epochs are 155.51, 143.01 and 164.12 respectively. In addition, when the number of neurons in the LSMs is 140, the average learning epoch is 338.61.

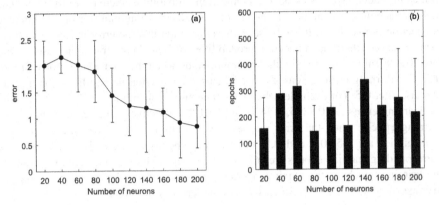

Fig. 7. The learning results of spike trains with different numbers of neurons in the hidden layers. (a) The learning error different numbers of neurons in the hidden layers. (b) The learning epochs when the learning error reaches the minimum value.

5 Conclusions

The spiking neural network with temporal encoding mechanism is closer to the real biological neural system, and it has better performance in information processing. Although some researchers provide the efficient method (such as ReSuMe) to train the LSMs with spiking neurons, it can only adjusts the synaptic weights in the output layer, which will restrict the development and application for LSMs. We present a new supervised learning algorithm of LSMs based on bidirectional modification mechanism, which can update the synaptic weights in all layers of LSMs. In the experiment, we demonstrate the learning ability of the proposed algorithm for training an LSM to reproduce the desired spike train patterns. The results show that the supervised learning algorithm of multi-spike LSMs can successfully learn the desired spatio-temporal spike patterns.

Acknowledgements. The work is supported by the National Natural Science Foundation of China under Grant No. 61762080, and the Medium and Small Scale Enterprises Technology Innovation Foundation of Gansu Province under Grant No. 17CX2JA038.

References

1. Bohte, S.M.: The evidence for neural information processing with precise spike-times: a survey. Nat. Comput. **3**(2), 195–206 (2004)
2. Quiroga, R.Q., Panzeri, S.: Principles of Neural Coding. CRC Press, Boca Raton (2013)
3. Izhikevich, E.M.: Which model to use for cortical spiking neurons? IEEE Trans. Neural Netw. **15**(5), 1063–1070 (2004)

4. Ostojic, S., Brunel, N.: From spiking neuron models to linear-nonlinear models. PLoS Comput. Biol. **7**(1), e1001056 (2011)
5. Maass, W.: Lower bounds for the computational power of networks of spiking neurons. Neural Comput. **8**(1), 1–40 (2014)
6. Ghosh-Dastidar, S., Adeli, H.: Spiking neural networks. Int. J. Neural Syst. **19**(4), 295–308 (2009)
7. Maass, W., Natschläger, T., Markram, H.: Real-time computing without stable states: a new framework for neural computation based on perturbations. Neural Comput. **14**(11), 2531–2560 (2002)
8. Maass, W.: Liquid state machines: motivation, theory, and applications. In: Computability in Context: Computation and Logic in the Real World, pp. 275–296. Imperial College Press, London (2011)
9. Rosselló, J.L., Alomar, M.L., Morro, A., et al.: High-density liquid-state machine circuitry for time-series forecasting. Int. J. Neural Syst. **26**(5), 1550036 (2016)
10. Lukoševičius, M., Jaeger, H.: Reservoir computing approaches to recurrent neural network training. Comput. Sci. Rev. **3**(3), 127–149 (2009)
11. Burgsteiner, H., Kröll, M., Leopold, A., et al.: Movement prediction from real-world images using a liquid state machine. Appl. Intell. **26**(2), 99–109 (2007)
12. Sala, D.A., Brusamarello, V.J., Azambuja, R.D., et al.: Positioning control on a collaborative robot by sensor fusion with liquid state machines. In: 2017 IEEE International Instrumentation and Measurement Technology Conference, pp. 1–6. IEEE, Turin, Italy (2017)
13. Zhang, Y., Li, P., Jin, Y., et al.: A digital liquid state machine with biologically inspired learning and its application to speech recognition. IEEE Trans. Neural Netw. Learn. Syst. **26**(11), 2635–2649 (2015)
14. Jin, Y., Li, P.: Performance and robustness of bio-inspired digital liquid state machines: a case study of speech recognition. Neurocomputing **226**, 145–160 (2017)
15. Zoubi, O.A., Awad, M., Kasabov, N.K.: Anytime multipurpose emotion recognition from EEG data using a liquid state machine based framework. Artif. Intell. Med. **86**, 1–8 (2018)
16. Xue, F., Guan, H., Li, X.: Improving liquid state machine with hybrid plasticity. In: Advanced Information Management, Communicates, Electronic and Automation Control Conference, pp. 1955–1959. IEEE, Xi'an, China (2017)
17. Kroese, B., van der Smagt, P.: An Introduction to Neural Networks, 8th edn. The University of Amsterdam, Amsterdam (1996)
18. Ponulak, F., Kasinski, A.: Supervised learning in spiking neural networks with ReSuMe: sequence learning, classification, and spike shifting. Neural Comput. **22**(2), 467–510 (2010)
19. Li, C.Y., Lu, J.T., Wu, C.P., et al.: Bidirectional modification of presynaptic neuronal excitability accompanying spike timing-dependent synaptic plasticity. Neuron **41**(2), 257–268 (2004)
20. Lin, X., Wang, X., Hao, Z.: Supervised learning in multilayer spiking neural networks with inner products of spike trains. Neurocomputing **237**, 59–70 (2017)
21. Gerstner, W., Kistler, W.M.: Spiking Neuron Models: Single Neurons, Populations, Plasticity. Cambridge University Press, New York (2002)

Cross-Media Correlation Analysis
with Semi-supervised Graph Regularization

Hong Zhang[1,2(✉)] and Tingting Qi[1]

[1] College of Computer Science and Technology,
Wuhan University of Science and Technology, Wuhan 430081, China
{zhanghong_wust,qitingting_wust}@163.com
[2] Hubei Province Key Laboratory of Intelligent Information Processing
and Real-Time Industrial System, Wuhan, China

Abstract. With the rapid development of multimedia data such as text, image, cross-media retrieval has become increasingly important, because users can retrieve the results with various types of media by submitting a query of any media type. The measure of relevance among different media is a basic problem. Existing methods usually only consider the original media instances (such as images, texts) but ignore their patches. In fact, cross-media patches can emphasize the important parts and improve the precision of cross-media correlation. What's more, existing cross-media retrieval methods often focus on modeling the pairwise correlation with the similarity matrix is a constant matrix, while the similarity matrix which is not a constant matrix can improve the accuracy. In this paper, we propose a novel algorithm for cross-media data, called cross-media correlation analysis with semi-supervised graph regularization (CMCA), which can not only take full advantage of both the media instances and their patches in one graph, but also explore the similarity matrix which can improve the correlation between data. CMCA explores the sparse and semi-supervised regularization for different media types, and integrates them into a unified optimization matter, which increases the performance of the algorithm. Comparing with the current state-of-the-art methods on two datasets (i.e., Wikipedia, XMedia), the comprehensive experimental results demonstrate the effectiveness of our proposed approach.

Keywords: Cross-media · The measure of relevance · Similarity matrix
Cross-media patches

1 Introduction

In some real applications, data are usually represented in diverse ways or collected from different domains. So, the data connected with the same basic content or object may exist in different modalities and show heterogeneous properties. For instance, when visiting the Palace Museum, we may record it by taking photos, posting a piece of microblog, or recording a video clip. These data show the same content in different forms. With the rapid development of such multimedia data, there is an immediate need for efficiently analyzing the data in different modalities. Although much attention has been paid to multimodal data analysis [1–4], it is more important to integrate multiple

© Springer International Publishing AG, part of Springer Nature 2018
D.-S. Huang et al. (Eds.): ICIC 2018, LNCS 10954, pp. 254–265, 2018.
https://doi.org/10.1007/978-3-319-95930-6_24

modalities to improve the learning performance. In this paper, we concentrate on cross-media retrieval, and its goal is to take one type of data as query to retrieve relevant data objects of another type. For example, a user can use an image to retrieve relevant textual descriptions or videos. Cross-media retrieval can make users to take any modality of content at hand as a query, and the retrieved results are rich in multiple modalities. Thus the results of cross-media retrieval are more comprehensive than that of traditional single-media retrieval.

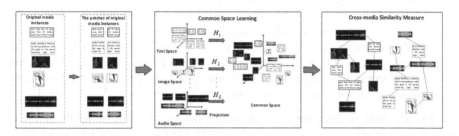

Fig. 1. Framework of our proposed method, which first divide the original media instances into patches, then projects the heterogeneous data of different modalities from their own feature spaces into a common space, finally computes the cross-media similarity based on the nearest neighbors among all the media.

As compared to separate analysis of single media, the additional information derivable from joint analysis of multiple media allows semantic correlations among multiple media to be explored in cross-media retrieval. On the one hand, users need such cross-media retrieval algorithm to retrieve results across various media [5]. On the other hand, because of the semantic gap between low level feature and human cognition, the similarity measure among objects of the same media type has always been a difficult issue. So, how to measure the content similarity among different media is much more challenging.

Some existing cross-media retrieval methods tend to concentrate on modeling the pairwise correlation and semantic information. On the one hand, pairwise correlation can provide accurate relationship among objects of different media types. On the other hand, semantic information focusing on high level abstraction of cross-media data, is consistent with human cognition. However, existing methods usually only consider the raw media data but ignore their patches. In fact, cross-media patches can intensify the important parts and make cross-media correlations more accurate. What's more, existing works in the literature often consider the pairwise correlation and semantic information without exploring the benefit of the similarity matrix with parameters. In fact, the similarity matrix which is not a constant matrix is more flexible, and can improve the accuracy.

To tackle above problems, in this paper, we proposes a novel joint learning framework (as shown in Fig. 1) for the cross-media retrieval problem, named cross-media correlation analysis with semi-supervised graph regularization (CMCA), which is not only able to make full use of both the media instances and their patches in one graph, but also explore a more flexible similarity matrix. It studies the semi-supervised and sparse regularization for different media types, and integrates them into a unified optimization problem.

In summary, our approach has several distinct advantages and contribution.

(1) CMCA can explore a more flexible similarity matrix, which can improve the correlation between data.
(2) CMCA considers both the media instances and their patches in one graph, and it makes cross-media correlations more precise.
(3) CMCA studies the semi-supervised and sparse regularization for different media types. It explores both the labeled and unlabeled data of all the different media, unlabeled instances of different media types increase the diversity of training data and boost the performance of our approach. And it learns sparse projection matrix for different media synchronously, different media can align each other and so, be robust for the noise in different media types.

The remainder of this paper is organized as follows. In Sect. 2, we briefly overview related work on the cross-media retrieval. Section 3 introduces the problem definition and the objective function of CMCA. Section 4 describes the optimization algorithm for our proposed approach. In Sect. 5, we report experimental results on two cross-media datasets. Finally, we conclude the paper in Sect. 6.

2   Related Works

In cross-media retrieval, exploring the semantic relationships shared across different modalities is the key challenge. Canonical correlation analysis (CCA) [6] is a representative solution to analyze the correlation between two multivariate random vectors. Through CCA, we could learn the subspace that maximizes the correlation between two groups of heterogeneous data. Kernel canonical correlation analysis (KCCA) has also been focused and been used for the fusion of text and image for spectral clustering [7]. To obtain the final heterogeneous similarity, Zhai et al. [8] proposed mining the semantic correlation between different media. In [8], the correlation is defined as must-link constraint and cannot-link constraint between two media types, respectively. While, the correlation propagation is attained by combination of image propagation and text propagation, respectively, and it means only one media type is considered in either propagation. Pereira et al. [9] prove the benefit of combining correlations and semantic information. The method compares three cross-media retrieval methods (correlation matching, semantic matching, and semantic correlation matching) and finds that combining the correlations and semantic information behaves the best. In the JRL [10] method, Zhai et al. jointly model the correlations and semantic information in a unified framework. However, the above methods only consider the original media instances but ignore their patches, and the similarity matrix usually is not flexible. In this paper, our proposed CMCA approach can employ one joint graph to simultaneously model all media types. On the one hand, it is able to make full use of both the media instances and their patches in one graph. On the other hand, CMCA can jointly model the correlations and semantic information in a unified framework with the similarity matrix is more flexible, and can improve the accuracy of retrieval.

3 Objective Function of CMCA

We first present the formulation of problem definition for cross-media retrieval. $D = \{x_i^1, x_i^2, \ldots, x_i^S\}_{i=1}^N$ denotes the N labeled multimedia documents, and each document contains data from S different modalities representing the same semantic. y_i represents the corresponding semantic label of x_i^p. The corresponding the label matrix is denoted as $Y = [y_1, y_2, \ldots, y_N]^T$. $X_p = [x_1^p, x_2^p, \ldots, x_N^p] \in R^{d_p \times N}, p = 1, 2, \ldots S$ denote the labeled data matrices from S modalities, respectively. $X_p^b = [x_1^p, x_2^p, \ldots, x_{N+E_p}^p] \in R^{d_p \times (N+E_p)}, p = 1, 2, \ldots S$ represent the matrices of both labeled and unlabeled data, and the pth modality has N labeled instances and E_p unlabeled samples embedded in the d_p dimensional space. $j^{(p)}$ denotes the number of patches of each data in pth media. $C_p^{(t)}$ denotes all the tth patches of each data in pth media.

The goal of our method is to learn the projection matrix for each modality of data, which can be used to project data from different modalities into a common space. Once all the media data are projected into the same cross-media feature space, we can measure the similarity between two media instances by using the existing methods such as the KNN classifier and the Euclidian distance and then achieve the goal of cross-media retrieval.

In existing cross-media retrieval methods [10, 11], correlation relationship between different types of media is widely used, i.e., if different media objects exist in the same document or are used jointly to represent for a given topic, so they should have the same semantic. In order to explore jointly the correlation and semantic information, we define the loss function that is proposed by [10] as follows

$$loss(H_1, \ldots, H_S) = \sum_{p=1}^{S} \sum_{q=1}^{S} \left\| X_{ap}^T H_p - X_{aq}^T H_q \right\|_F^2 + \sum_{p=1}^{S} \left\| X_p^T H_p - Y_p \right\|_F^2 \quad (1)$$

Here, $\|A\|_F$ denotes the Frobenius norm of matrix A, X_{ap} and X_{aq} denote two sets of media objects from pth media and qth media with the same labels.

Graph regularization has been widely used to protect the similarity between multimedia data [10, 12, 13]. The edge weights in the graph express the affinities of multimedia data. We define the weight matrix W of the graph as follows

$$W_{ij} = \begin{cases} \exp\left(-z_{ij}^{pq}/2\sigma^2\right), & \text{if } f_i^p \in N_k\left(f_j^q\right) \text{ or } f_j^q \in N_k(f_i^p). \\ 0, & \text{otherwise.} \end{cases} \quad (2)$$

where $f_i^p, p = 1, 2, \ldots, S$ is the projected object of x_i^p in the common space, z_{ij}^{pq} is the Euclidean distance between f_i^p and f_j^q, i.e., $z_{ij}^{pq} = \left\| f_i^p - f_j^q \right\|^2$, $\sigma = max\left(z_{ij}^{pq}\right)$, and $N_k(f_i^p)$ represents the set of k nearest neighbors of f_i^p.

The smooth function $\Omega(H_1, \ldots, H_S)$ denotes the smoothness of a projected feature vector f. The value $\Omega(H_1, \ldots, H_S)$ punishes big changes of the mapping function f between two objects. Based on the multimodal graph, we can define the smooth function as

$$\Omega(H_1, \ldots, H_S) = \frac{1}{2} \sum_{i=1}^{\hat{N}} \sum_{j=1}^{\hat{N}} W_{ij} \left\| f_i^p - f_j^q \right\|^2 = Tr\left(FLF^T\right) \tag{3}$$

where $\hat{N}$ is the number of the total instances from all modalities, $F = \left(F_1^T, \ldots, F_S^T\right) = \left(H_1^T X_1^b, \ldots, H_S^T X_S^b\right)$ represents for all modalities of projected data in the common space. $L = D - W$ is the Laplacian matrix. Based on the above definition, Eq. (3) can be transformed into

$$\Omega(H_1, \ldots, H_S) = \sum_{p=1}^{S} \sum_{q=1}^{S} Tr\left(H_p^T X_p^b L_{pq} \left(X_q^b\right)^T H_q\right) \tag{4}$$

What's more, to make full use of both the media instances and their patches, we extend the traditional framework as

$$\Omega(H_1, \ldots, H_S) = \sum_{p=1}^{S} \sum_{q=1}^{S} Tr\left(H_p^T K_p^b L_{pq} \left(K_q^b\right)^T H_q\right) \tag{5}$$

Here, $K_p = X_p + \sum_{t=1}^{j^{(p)}} C_p^{(t)}$.

In conclusion, we obtain the objective function as follows

$$\begin{aligned}
\min_{H_1, \ldots, H_S} \ &\alpha \sum_{p=1}^{S} \sum_{q=1}^{S} \left\| X_{ap}^T H_p - X_{aq}^T H_q \right\|_F^2 + \lambda_1 \sum_{p=1}^{S} \left\| H_p \right\|_{2,1} \\
&+ \lambda_2 \sum_{p=1}^{S} \sum_{q=1}^{S} Tr\left(H_p^T K_p^b L_{pq} \left(K_q^b\right)^T H_q\right) + \lambda_3 \sum_{p=1}^{S} \left\| X_p^T H_p - Y_p \right\|_F^2
\end{aligned} \tag{6}$$

4 Optimization Algorithm

In this subsection, we will introduce the optimization of the objective function (6).

Let $\varphi(H_p)$ denote the objective function in (6). Differentiating $\varphi(H_p)$ with respect to H_p and setting it to zero, we have the following equation

$$
\begin{aligned}
\frac{\partial \varphi}{\partial H_p} &= \alpha \sum_{q \neq p} X_{ap}\left(X_{ap}^T H_p - X_{aq}^T H_q\right) + \lambda_1 R_p H_p \\
&+ \lambda_2 \sum_{q \neq p} K_p^b L_{pq}\left(K_q^b\right)^T H_q + \lambda_2 K_p^b L_{pp}\left(K_p^b\right)^T H_p \\
&+ \lambda_3 X_p\left(X_p^T H_p - Y_p\right) = 0
\end{aligned}
\tag{7}
$$

Equation (7) can be rewritten as

$$
\begin{aligned}
&\left(\alpha \sum_{q \neq p} X_{ap} X_{ap}^T + \lambda_1 R_p + \lambda_2 K_p^b L_{pp}\left(K_p^b\right)^T + \lambda_3 X_p X_p^T\right) H_p \\
&= \left(\alpha \sum_{q \neq p} X_{ap} X_{aq}^T - \lambda_2 \sum_{q \neq p} K_p^b L_{pq}\left(K_q^b\right)^T\right) H_q + \lambda_3 X_p Y_p
\end{aligned}
\tag{8}
$$

Next, we propose an optimization method to minimize the objective function (6). The general procedure of this approach is that we initialize H_p as identity matrix first, then, in each iteration, we calculate $H_1^{t+1}, \ldots, H_p^{t+1}$ on the condition that $H_1^t, \ldots, H_p^t$ are given. Detailed steps will be explained in Algorithm 1.

In the process of optimization method, the iteration continues until convergence. In our real experiments, the iteration repeats about six rounds before convergence.

Until now, we have learned S projection matrices H_S for the primitive features of multiple media types, and using them we can project all of the raw data x_i^p into the common space $f_i^p = H_p^T x_i^p$. Then, the cross-media similarity measure is need to be explored in the constructed common space. The cross-media similarity is defined as the marginal probability. The probability expresses the semantic similarity of the two media data regardless of their media types. The marginal probability of f_i^p and f_j^q can be defined that is proposed by [10] as

$$
Sim\left(f_i^p, f_j^q\right) = P\left(y_i = y_j | f_i^p, f_j^q\right) = \sum_l p(y_i = l | f_i^p) p\left(y_j = l | f_j^q\right)
\tag{10}
$$

where $y_i(y_j)$ denotes the label of $f_i^p(f_j^q)$, $p(y_i = l|f_i^p)$ denotes the probability of f_i^p belonging to category l. $p(y_i = l|f_i^p)$ that is proposed by [10] is defined as

Algorithm 1. Cross-Media Correlation Analysis with Semi-Supervised Graph Regularization (CMCA)

Input:

The matrix of both labeled and unlabeled data $X_p^b \in R^{d_p \times (N + E_p)}$;

The matrix of labeled data $X_p \in R^{d_p \times N}$;

The matrix of labels $Y \in R^{N \times c}$.

Output:

The projection matrices $H_p \in R^{d_p \times c}, p = 1, 2, ..., S$.

Initialize $H_p^0, p = 1, 2, ..., S$ as identity matrix and set $t = 0$;

Repeat:

1. Compute the graph Laplacian matrix L^t according to $H_p^t \in R^{d_p \times c}, p = 1, 2, ..., S$;

2. By solving the linear system problem in Eq. (8), update H_p^{t+1} according to following equation:

$$
H_p^{t+1} = \left(\alpha \sum_{q \neq p} X_{ap} X_{ap}^T + \lambda_1 R_p + \lambda_2 K_p^b L_{pp} \left(K_p^b \right)^T + \lambda_3 X_p X_p^T \right)^{-1}
$$

$$
\left(\left(\alpha \sum_{q \neq p} X_{ap} X_{aq}^T - \lambda_2 \sum_{q \neq p} K_p^b L_{pq} \left(K_q^b \right)^T \right) H_q^t + \lambda_3 X_p Y_p \right)
$$
(9)

3. $t = t + 1$.

until Convergence

$$
p(y_i = l|f_i^p) = \frac{\sum_{f \in N_k(f_i^p) \wedge y = l} \sigma \left(\| f_i^p - f \|_2 \right)}{\sum_{f \in N_k(f_i^p)} \sigma \left(\| f_i^p - f \|_2 \right)}
$$
(11)

$N_k(f_i^p)$ denotes the k-nearest neighbors of media data f_i^p in training set, and y denotes the label of media object f. $\sigma(z) = (1 + \exp(-z))^{-1}$ is the sigmoid function.

5 Experimental Results

In this section, we will research the performance of our proposed method in cross-media retrieval. We will introduce the datasets, evaluation metrics, comparison methods, and parameter setting of the experiments. Then, we will describe the detailed comparison results of the proposed method with the other methods.

Datasets

The experiments will be conducted on the two real-world cross-media retrieval datasets, i.e., Wikipedia, and XMedia.

Wikipedia Dataset [14]: It is chosen from a collection of 2700 feature articles. These articles are divided into 29 classes, and the Wikipedia dataset is made up of the 10 most populated ones. The final dataset contains a total of 2866 documents and is randomly divided into a training set of 2173 documents and a testing set of 693 documents. They are all text-image pairs and labeled by a glossary of 10 semantic classes.

XMedia Dataset [10]: It is composed of 5000 texts, 5000 images, 500 videos, 1000 audios, and 500 3D models. All the media data are crawled from the Internet websites, including Wikipedia, Youtube, Flickr, freesound, findsound, 3D Warehouse, and Princeton 3D Model Search Engine. This dataset is separated into 20 categories with 600 media data per category, and is randomly divided into a training set of 9600 media data and a testing set of 2400 media data. We randomly select several instances of different media types from three categories of the XMedia dataset.

Evaluation Metrics

In the experiments, we use the mean average precision (MAP) [15] to evaluate the whole performance of the tested algorithms. To obtain MAP, we first calculate the average precision (AP) of a set of R retrieved documents by

$$AP = \frac{1}{T}\sum\nolimits_{r=1}^{R} P(r)\delta(r) \tag{12}$$

where T is the number of relevant files in the retrieved set, $P(r)$ represents the precision of the top r retrieved files. If the rth retrieved document is relevant $\delta(r) = 1$, otherwise $\delta(r) = 0$. The MAP is calculated by averaging the AP values in all queries in the query set. The larger the MAP, the better the performance.

Besides the MAP, precision-recall curve [15] is also used to evaluate the effectiveness of different methods.

Compared Methods

We compare our proposed CMCA algorithm with three different basic methods, which are summed up as follows.

(1) JGRHML [16]. It explores the heterogeneous metric, which can measure the content similarity between different media types.

(2) CMCP [17]. It learns to propagate the correlation between heterogeneous modalities, and can simultaneously treat with positive correlation and negative correlation between different media.

(3) JFSSL [13]. It is able to jointly cope with correlation measure and coupled feature selection in a joint learning framework.

Table 1. MAP comparison of different methods on the Wikipedia dataset

Dataset	Task	JGRHML	CMCP	JFSSL	CMCA
Wikipedia Dataset	Image→Text	0.250	0.388	0.493	**0.532**
	Text→Image	0.194	0.351	0.429	**0.459**
Average		0.222	0.370	0.461	**0.496**

Table 2. MAP comparison of different methods on the XMedia dataset

Dataset	Task	JGRHML	CMCP	JFSSL	CMCA
XMedia Dataset	Image→Text	0.458	0.727	0.876	**0.909**
	Image→Video	0.289	0.510	0.460	**0.555**
	Text→Image	0.367	0.723	0.882	**0.920**
	Text→Video	0.338	0.439	0.460	**0.581**
	Video→Image	0.251	0.485	0.436	**0.550**
	Video→Text	0.339	0.399	0.430	**0.562**
Average		0.340	0.547	0.591	**0.680**

Parameter Setting

The objective function in Eq. (6) mainly involves four parameters $\alpha, \lambda_1, \lambda_2, \lambda_3$. α is the parameter of the loss function of cross-media regularization. λ_1 is the weighting parameter of the $l_{2,1}$-norms, the weighting parameter of the multimodal graph

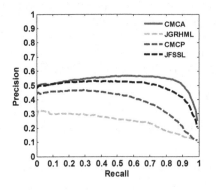
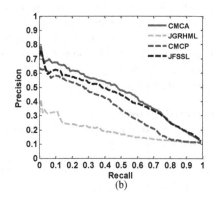

Fig. 2. Precision recall curves of cross-media retrieval on Wikipedia dataset. (a) Image→Text. (b) Text→Image.

regularization is denoted as λ_2, and λ_3 is the parameter of the label consistent term. We tune them from {0.001, 0.01, 0.1, 1, 10, 100, 1000} by cross validation. For the compared methods, we tune their parameters on the basis of the corresponding literature.

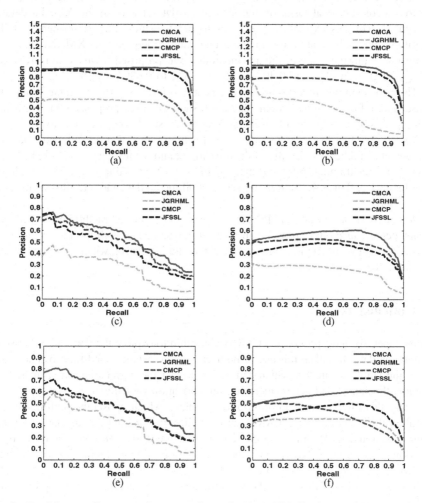

Fig. 3. Precision recall curves of cross-media retrieval on XMedia dataset. (a) Image→Text. (b) Text→Image. (c) Image→Video. (d) Video→Image. (e) Text→Video. (f) Video→Text.

Experimental Results

In this section, we compare our proposed CMCA with three state-of-the-art methods for cross-media retrieval. Note that each image feature is a 4,096-dimensional CNN (convolutional neural network) feature vector, and each text feature is a 3,000-dimensional bag of words vector in our experiment. Particularly, in the XMedia dataset, each audio is represented by a 29-dimensional MFCC feature vector, each video is represented by a 4,096-dimensional CNN feature vector and each 3D model is represented by the concatenated 4,700-dimensional vector of a set of Light Field descriptors. We conduct all methods in the experiment with same feature vectors for fair comparison.

Tables 1 and 2 show the efficiency of different methods ground on the MAP scores of cross-media retrieval tasks on the Wikipedia dataset and on the XMedia dataset respectively (In this experiment, in order to better achieve the performance of the algorithm, we only search for images, text, and video on the XMedia dataset.). Compared with the state-of-the-art methods, our proposed CMCA improves the average MAP from 0.461 to 0.496 and from 0.591 to 0.680, respectively.

The corresponding precision-recall (PR) curve is plotted in Fig. 2 on the Wikipedia dataset and in Fig. 3 on the XMedia dataset. It shows that our proposed CMCA obtains higher precision at most recall levels, surpassing those of compared methods.

We observe that MAP on XMedia dataset is higher than that on Wikipedia dataset. The reason could be that the ratio of the training set and the testing set is 3:1 (2173 text-image pairs for training, 693 text-image pairs to test) on Wikipedia dataset while on XMedia dataset that is 4:1 (a training set of 9600 media data and a testing set of 2400 media data).

As the MAP scores and the PR curves show, our proposed method has significant advantages because our approach can take full advantage of both the media instances and their patches, and explore a more flexible similarity matrix, which can improve the correlation between data.

6   Conclusions

In this paper, we have proposed a CMCA algorithm to explore the correlation between multimedia data. It makes full use of both the media instances and their patches, and learns the similarity matrix which can better protect the correlation between data. It explores the sparse and semi-supervised regularization for different media types, and integrates them into a unified optimization problem, which boosts the performance of the algorithm. Furthermore, CMCA also studies the semantic information of the original data and further improve the retrieval accuracy. We present an iterative optimization algorithm to solve the corresponding optimization problem. The experiments on two cross-media datasets show the effectiveness of our proposed approach, compared with the state-of-the-art methods. In the future, we intend to further optimize the algorithm to better express the similarity of cross-media data.

Acknowledgement. This research is supported by the National Natural Science Foundation of China (No. 61003127, No. 61373109).

References

1. Ranjan, V., Rasiwasia, N., Jawahar, C.V.: Multi-label cross-modal retrieval. In: IEEE International Conference on Computer Vision (ICCV), pp. 4094–4102 (2015)
2. Hua, Y., Wang, S., Liu, S., Cai, A., Huang, Q.: Cross-modal correlation learning by adaptive hierarchical semantic aggregation. IEEE Trans. Multimedia (TMM) 18(6), 1201–1216 (2016)
3. Peng, Y., Ngo, C.-W.: Clip-based similarity measure for query-dependent clip retrieval and video summarization. IEEE Trans. Circ. Syst. Video Technol. 16(5), 612–627 (2006)
4. Escalante, H.J., Hérnadez, C.A., Sucar, L.E., Montes, M.: Late fusion of heterogeneous methods for multimedia image retrieval. In: Proceedings of 1st ACM International Conference on Multimedia Information Retrieval, pp. 172–179 (2008)
5. Liu, J., Xu, C., Luo, H.: Cross-media retrieval: state-of-the-art and open issues. Int. J. Multimedia Intell. Secur. 1(1), 33–52 (2010)
6. Hotelling, H.: Relations between two sets of variates. Biometrika 28(3–4), 321–377 (1936)
7. Blaschko, M., Lampert, C.: Correlational spectral clustering. In: Proceedings of IEEE International Conference on Computer Vision Pattern Recognition, pp. 1–8, June 2008
8. Zhai, X., Peng, Y., Xiao, J.: Cross-modality correlation propagation for cross-media retrieval. In: Proceedings of International Conference on Acoustics, Speech, Signal Processing, pp. 2337–2340 (2012)
9. Pereira, J.C., et al.: On the role of correlation and abstraction in cross-modal multimedia retrieval. IEEE Trans. Pattern Anal. Mach. Intell. 36(3), 521–535 (2014)
10. Zhai, X., Peng, Y., Xiao, J.: Learning cross-media joint representation with sparse and semisupervised regularization. IEEE Trans. Circ. Syst. Video Technol. 24(6), 965–978 (2014)
11. Li, D., Dimitrova, N., Li, M., Sethi, I.: Multimedia content processing through cross-modal association. In: Proceedings of ACM International Conference on Multimedia, pp. 604–611 (2003)
12. Peng, Y., Zhai, X., Zhao, Y., Huang, X.: Semi-supervised cross-media feature learning with unified patch graph regularization. IEEE Trans. Circ. Syst. Video Technol. (TCSVT) 26(3), 583–596 (2016)
13. Wang, K., He, R., Wang, L., Wang, W., Tan, T.: Joint feature selection and subspace learning for cross-modal retrieval. IEEE Trans. Pattern Anal. Mach. Intell. 38(10), 2010–2023 (2016)
14. Rasiwasia, N., et al.: A new approach to cross-modal multimedia retrieval. In: Proceedings of ACM International Conference on Multimedia (ACM-MM), pp. 251–260 (2010)
15. Rasiwasia, N., Pereira, J.C., Coviello, E., Doyle, G., Lanckriet, G., Levy, R., Vasconcelos, N.: A new approach to cross-modal multimedia retrieval. In: Proceedings of 18th ACM International Conference on Multimedia, pp. 251–260 (2010)
16. Zhai, X., Peng, Y., Xiao, J.: Heterogeneous metric learning with joint graph regularization for cross-media retrieval. In: Proceedings of 27th AAAI Conference on Artificial Intelligence, pp. 1198–1204 (2013)
17. Zhai, X., Peng, Y., Xiao, J.: Cross-modality correlation propagation for cross-media retrieval. IEEE Int. Conf. Acoust. 22(10), 2337–2340 (2012)

An Improved Endpoint Detection Algorithm Based on Improved Spectral Subtraction with Multi-taper Spectrum and Energy-Zero Ratio

Tiantian Bao[1,3(✉)], Yaxin Li[1,3(✉)], Kena Xu[1,3(✉)],
Yonghao Wang[2,3(✉)], and Wei Hu[1,3(✉)]

[1] College of Computer Science and Technology, Wuhan University of Science
and Technology, Wuhan, China
xukena1992@foxmail.com
[2] Digital Media Technology Lab, Birmingham City University,
Birmingham, UK
[3] Hubei Province Key Laboratory of Intelligent Information Processing
and Real-Time Industrial System, Wuhan, China

Abstract. Endpoint detection plays a crucial role in speech recognition systems. An effective endpoint detection algorithm can not only reduce the processing time, but also can interfere with the noise of the silent segment. The traditional endpoint detection algorithms are mostly processed in a noise-free environment, so there will be problems such as weak noise immunity. In the problem of low SNR, this paper proposes an improved endpoint detection algorithm based on improved spectral subtraction with multi-taper spectrum and energy-zero ratio. The algorithm uses the improved spectral subtraction method of multi-window spectrum estimation to reduce the speech noise, and then combines the energy-zero ratio with endpoint detection. Experiments show that the proposed algorithm has better robustness under different SNR conditions.

Keywords: Endpoint detection · Energy-zero ratio
Multi-window spectrum estimation

1 Introduction

Speech signal endpoint detection [1] refers to: for a given period of continuous speech, the starting position and ending position of this speech are determined. Accurate speech endpoint detection not only improves the efficiency of the system, but also improves the recognition rate of the system. Traditional speech endpoint detection is conducted

This paper was supported by National Undergraduate Innovation Project with Granted No. 201710488004 and Fund of Hubei Province Key Laboratory of Intelligent Information Processing and Real-time Industrial System with Granted No. znxx2018MS03 and znxx2018QN07.

in a noiseless environment, but in real environment noise has been in existence. So in recent years, the endpoint detection in noisy environment has been studied. In a noisy environment the starting point and the ending point of the speech signal are detected from the speech signal, which can reduce the amount of data collected and remove the background noise and silence segments, so as to reduce the calculation amount and processing time of feature extraction. Finally, it can improve the accuracy of speech recognition. Therefore, accurately detecting the start and end positions of speech in a noisy environment is conducive to improving the performance of the speech system.

There are commonly endpoint detection methods which are energy threshold [2, 3], pitch detection [4], spectrum analysis [5, 6], and cepstrum analysis [1]. Among them, the double-threshold method [7–9] is adopted for the energy threshold, which is the classical and most widely used endpoint detection method. Many improved algorithms for the double threshold method are proposed. In [10], the end point detection method is combined with the energy value and the spectral entropy method, but as the noise becomes larger, the effect of the endpoint detection will be reduced. Therefore, this method has poor detection performance under high noise conditions. In [11], a method of endpoint detection based on weighted spectral entropy is proposed. The principle is that the spectral entropy can distinguish the speech signal and the noise signal by using the inherent characteristics of the spectrum. But the effect of this method to distinguish an unsatisfactory. In [12], the support vector machine (SVM) [13–15] is used and the speech feature vectors are used to detect the endpoint. This algorithm improves the high discrimination of endpoint detection, but endpoint detection for different noisy environments is not very good. This paper proposes an improved endpoint detection algorithm based on improved spectral subtraction with multi-taper spectrum [16] and energy-zero ratio, which not only improves the detection accuracy, but also has a good endpoint detection effect in response to different noise environments.

2 Related Work

The double-threshold method mainly uses the two parameters of energy and zero-crossing respectively to make corresponding judgments. When the signal-to-noise ratio decreases, its short-term energy and short-time zero-crossing rate are shown in Fig. 1 for noisy speech.

Figure 1 shows that in this speech, the short-term energy [17] of the speech interval is convex, but the curve of the short-term zero-crossing rate [18] is correspondingly depressed downward.It can be obtained from it that the short-term energy is large and the short-termzero-crossing ratio is small in the speechsegments. At the same time, when the short-term energy is small in the noise segments, the value of the short-term zero-crossing ratio is large. Therefore, in order to make the speech interval more prominent and the value of the noise interval become smaller. The short-term energy value can be divided by the value of the zero-crossing rate, so that the numerical difference between the speech interval and the noise interval becomes larger. It will be easy to detect the beginning and ending points of the voice.

The speech signal time is $x(m)$. The speech signal of windowing and framing division is $x_i(m)$. Its frame length is set to N. The energy of each frame is as follows:

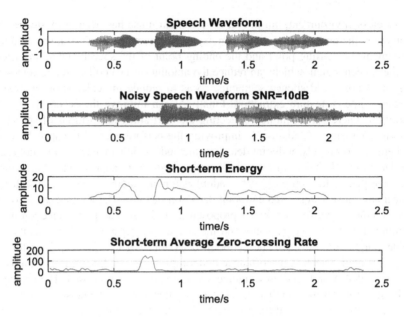

Fig. 1. Noisy speech waveforms

$$AMP_i = \sum_{m=1}^{N} x_i^2(\mathrm{m}) \tag{1}$$

Energy calculation relationship can be calculated.

$$LE_i = \log_{10}(1 + AMP_i/a) \tag{2}$$

where, AMP_i is signal energy per frame. a is a constant. When a large value is used for a, the amplitude of AMP_i varies greatly in LE_i, so the value of a can be appropriately selected based on the calculated AMP_i.

Short-term zero-crossing rate of each frame signal is as follows:

$$ZCR_i = \sum_{m=1}^{N} |sign[x_i(\mathrm{m})] - sign[x_i(\mathrm{m} - 1)]| \tag{3}$$

where,

$$sign[x_i(\mathrm{m})] = \begin{cases} 1 & |x_i(\mathrm{m})| \geq 0 \\ -1 & |x_i(\mathrm{m})| < 0 \end{cases} \tag{4}$$

Energy-zero ratio is calculated according to (2) and (3).

$$EZR_i = LE_i/(ZCR_i + b) \tag{5}$$

where b is a small constant. It prevents overflow when the value of ZCR_i is 0.

3 Proposed Approach

3.1 Overview of the Improved Endpoint Detection

Most speech signals contain noise. The traditional energy-zero ratio endpoint detection method has better detection effect when the signal to noise ratio SNR (Signal-Noise Ratio) is more than 10 dB. When the signal to noise ratio is reduced, sometimes some voice will be missed. The principle of improved zero ratio endpoint detection proposed isoperating noise reduction for noisy speech signals and then using energy-zero ratio endpoint detection method. This paper uses the improved spectral subtraction of multi-window spectrum estimation. The improved energy-zero ratio endpoint detection is shown in Fig. 2.

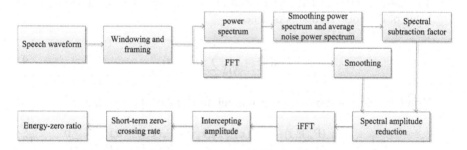

Fig. 2. Improved endpoint detection algorithm flow chart

3.2 Improved Energy-Zero Ratio Endpoint Detection Procedure

(1) The noisy speech is x(m). The speech after preprocessing is $x_i(m)$. There will be overlapping parts between adjacent voice frames.

(2) We can operate FFT transform for preprocessed speech signals. And then we can get amplitude spectrum $|Xi(k)|$ and phase spectrum $\theta_i(k)$. The adjacent speech frames are processed smoothly, and the average amplitude spectrum $|\overline{X}_i(k)|$ can be obtained.

$$|\overline{X}_i(k)| = \frac{1}{2M+1} \sum_{j=-M}^{M} |X_{i+j}(k)| \tag{6}$$

Taking the i frame speech signal as the center and taking the M frame speech signals before and after the center speech frame. A total of 2M + 1 frame signals are averaged. Normally, M = 1, that is, average processing is performed in the 3frame speech signal.

(3) Multi-window spectrum estimation processing is performed on the preprocessed speech signal $x_i(m)$ to calculate the power spectral density P(k, i).

$$P(k, i) = PM[x_i(m)] \tag{7}$$

where, i is the speech signal of i frame. k is kth spectral line. PM stands for multispectral power spectral density estimation.

(4) The smoothing power spectral density $P_y(k, i)$ is obtained by performing smoothing processing between adjacent speech frames on the estimated value obtained in the previous step.

$$P_y(k, i) = \frac{1}{2M+1} \sum_{j=-M}^{M} P(k, i+j) \tag{8}$$

(5) The noise section has NIS frames, and the average power spectral density $P_n(k)$ value of the noise section is obtained.

$$P_n(k) = \frac{1}{NIS} \sum_{i=1}^{NIS} P_y(k, i) \tag{9}$$

(6) Using the spectral subtraction relationship to calculate the gain factor.

$$g(k, i) = \begin{cases} (P_y(k, i) - \alpha P_n(k))/P_y(k, i) & P_y(k, i) - \alpha P_n(k) \geq 0 \\ \beta P_n(k)/P_y(k, i) & P_y(k, i) - \alpha P_n(k) < 0 \end{cases} \tag{10}$$

(7) The spectral amplitude-reduced amplitude spectrum can be obtained from the average amplitude spectrum obtained from the formula (6) and the formula (10).

$$|\widehat{X}_i(k)| = g(k, i) \times |\overline{X}_i(k)| \tag{11}$$

(8) Amplitude spectrum $|\widehat{X}_i(k)|$ and phase spectrum $\theta_i(k)$ are processed by the FFT. The purpose is to restore the frequency domain to the time domain, so that the reduced noise signal $\widehat{x}_i(m)$ can be obtained.

(9) Perform center-amplitude processing on the noise-reduced speech signal $\widehat{x}_i(m)$ to obtain the center-amplitude speech signal $\widetilde{x}_i(m)$.

(10) The zero-crossing rate ZCR_i of each frame signal is calculated for the center-amplitude speech signal $\widetilde{x}_i(m)$.

(11) Based on the calculated short-term energy and zero-crossing rate obtained by (2), the energy zero ratio EZR_i is obtained by (3).

4 Experimental Results and Discussion

4.1 Overview of the Improved Endpoint Detection

The test recordings are taken in an acoustically treated vocal booth with a DPA 4090 measurement microphone. Sampling frequency of the device is 44100. We choose 120 singers to sing /æ/ and /i:/ tone scalerange in B4. Each tone is spoken 3 times, and lasts for 3–5 s. So audio database contains 120 * 2 * 3 = 720 files.

We choose an /i:/speech file of a confirmed category as the experimental data and set the SNR to 0 dB. The SNR is the ratio between the effective signal and the noise in the speech signal. The smaller the value of the SNR, the greater the noise contained in the speech. This is the case where the SNR will be 0 dB, using the traditional energy-zero ratio endpoint detection method. The result is shown in Fig. 3. As can be seen from the figure, a segment of the tail is obviously not detected, and thus a lot of valid voice data will be missed. Endpoint detection accuracy is reduced.

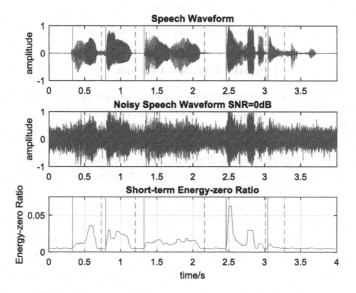

Fig. 3. The results of energy-zero ratio endpoint detection

As can be seen from Fig. 4, the improved energy-zero ratio endpoint detection method not only detects the speech signal detected using the traditional energy-zero ratio endpoint detection method, but also detects the speech signals that are not detected by the traditional endpoint detection method. It can also be well detected. Therefore, the improved algorithm can solve the problem that when the SNR is low, the accuracy of the endpoint detection will be reduced. Even if the SNR is low, all speech data segments containing the speech information can still be detected.

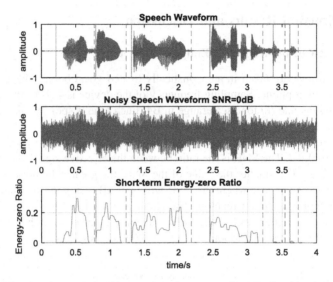

Fig. 4. The results of improved energy-zero ratio endpoint detection

4.2   Overview of the Improved Endpoint Detection

The improved energy-zero ratio endpoint detection algorithm can detect speech segments and noise segments when SNR = 0 dB. Then, judging the merits of an algorithm needs to be tested in different environments. For speech signal processing, the improved energy-zero ratioendpoint detection algorithm requires detection in different noise programs. Therefore, we can detect the speech segments under different SNRs. The following figures show the effect of endpoint detection when the SNR = 10 dB, the SNR = 5 dB, and SNR = −5 dB.

From the above Figs. 5, 6 and 7, we can see that when the SNR is 10 dB, the noise is relatively small. So we can observe from Fig. 5 that both the solid line and the dotted line are more accurate to handle the beginning and end of the speech segment. When SNR is 5 dB, the corresponding noise also increases. Observed from Fig. 6. The start and end points of most speech segments can be accurately marked. However, the solid line at the beginning of the marked voice at 2.5 s is not very accurate. When the SNR is −5 dB, that is, the noise is already quite large at this time. The improved energy-zero ratio endpoint detection result is shown in Fig. 7. The first half speech can be detected better. However, at the final tail, the gap between the end voice position and the end of the actual voice is slightly larger. All in all, the overall number of voice segment frames detected is more than 90% of the actual voice frame.

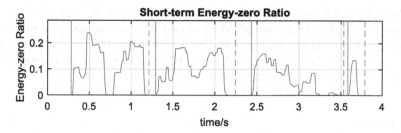

Fig. 5. SNR = 10 db

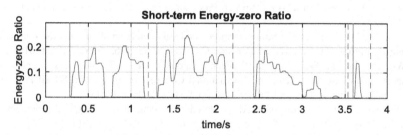

Fig. 6. SNR = 5 db

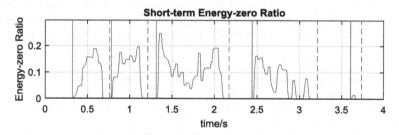

Fig. 7. SNR = − 5 db

5 Conclusion and Future Work

Energy-zero ratioendpoint detection can better distinguish the speech segments and the noise segments when the SNR is 10 dB. However, when the signal-to-noise ratio is reduced to 0 dB, some speech segments are not detected. At this time, the improved spectral subtraction of multi-taper spectrum estimation is used to perform noise reduction, so that the speech segments can be detected better. When Gauss white noise is increasing, the improved energy-zero ratio endpoint detection algorithm can still detect speech segments and noise segments accurately. The experiment shows that the algorithm has good robustness.

With the complexity of noise in the real environment, the endpoint detection methods also need to be diversified. The endpoint detection method proposed in this paper is applicable under Gaussian white noise. But in real life, there are other noise such as pink noise, factory noise, and so on. When adding these noises, we need to further improve the endpoint detection method.

References

1. Ying, G.S., Mitchell, C.D., Jamieson, L.H.: Endpoint detection of isolated utterances based on a modified Teager energy measurement. In: IEEE International Conference on Acoustics, Speech, and Signal Processing: Speech Processing, pp. 732–735. IEEE Computer Society (1993)
2. Cao, Y., Gao, J., Yang, G.: Study on speech endpoint detection algorithm based on wavelet energy entropy. In: Control and Decision Conference, pp. 3965–3969. IEEE (2016)
3. Wu, D., Tao, Z., Wu, Y., et al.: Speech endpoint detection in noisy environment using spectrogram boundary factor. In: International Congress on Image and Signal Processing, Biomedical Engineering and Informatics, pp. 964–968. IEEE (2017)
4. Zhang, C., Dong, M.: An improved speech endpoint detection based on adaptive sub-band selection spectral variance. In: Control Conference, pp. 5033–5037. IEEE (2016)
5. Junqua, J.C., Reaves, B., Mak, B.: A study of endpoint detection algorithms in adverse conditions: incidence on a DTW and HMM recognize. In: European Conference on Speech Communication and Technology, Eurospeech 1991, Genova, Italy, September. DBLP, pp. 757–762 (1991)
6. Shin, W.H., Lee, B.S., Lee, Y.K., et al.: Speech/non-speech classification using multiple features for robust endpoint detection. In: IEEE International Conference on Acoustics, Speech, and Signal Processing, ICASSP 2000, Proceedings, vol. 3, pp. 1399–1402. IEEE (2000)
7. Lynch, J.J., Josenhans, J., Crochiere, R.E.: Speech/Silence segmentation for real-time coding via rule based adaptive endpoint detection. In: IEEE International Conference on Acoustics, Speech, and Signal Processing, ICASSP, pp. 1348–1351. IEEE (1987)
8. Rui, H.U.G., Wei, X.D.: Endpoint detection of noisy speech based on cepstrum. Acta Electronica Sinica 28(10), 95–97 (2000)
9. Wilpon, J.G., Rabiner, L.R.: Application of hidden Markov models to automatic speech endpoint detection. Comput. Speech Lang. 2(3–4), 321–341 (1987)
10. Huang, L.S., Yang, C.H.: A novel approach to robust speech endpoint detection in car environments. In: IEEE ICASSP-2000, vol. 3, pp. 1751–1754 (2000)
11. Shen, J.L., Hung, J.W., Lee, L.S.: Robust entropy-based endpoint detection for speech recognition in noisy environments. In: International Conference on Spoken Language Processing, Incorporating the, Australian International Speech Science and Technology Conference, Sydney Convention Centre, Sydney, Australia, November–December. DBLP (1998)
12. Ramirez, J., Yelamos, P., Gorriz, J.M., et al.: SVM-based speech endpoint detection using contextual speech features. Electron. Lett. 42(7), 426–428 (2006)
13. Ganapathiraju, A., Hamaker, J., Picone, J.: Support vector machines for speech recognition. In: International Conference on Spoken Language Processing, Incorporating the, Australian International Speech Science and Technology Conference, Sydney Convention Centre, Sydney, Australia, November–December, pp. 2348–2355. DBLP (2002)

14. Matsumoto, M., Hori, J.: Classification of silent speech using support vector machine and relevance vector machine. Appl. Soft Comput. **20**(7), 95–102 (2014)
15. Lim, C., Chang, J.H.: Enhancing support vector machine-based speech/music classification using conditional maximum a posteriori criterion. IET Signal Process. **6**(4), 335–340 (2012)
16. Tseng, Y.H., Chiu, T.H., Lin, J.M., et al.: Linear precoding and adaptive multi-taper spectrum detector for cognitive radios. In: International Symposium on VlSI Design, Automation and Test, pp. 1–4. IEEE (2016)
17. Dai, Y.H., Chen, H.C., Qiao, D.J., et al.: Speech endpoint detection algorithm analysis based on short-term energy ratio. Commun. Technol. **42**(2), 181–183 (2009)
18. Kumar, S., Phadikar, S., Majumder, K.: Modified segmentation algorithm based on short term energy & zero crossing rate for Maithili speech signal. In: International Conference on Accessibility to Digital World, pp. 169–172. IEEE (2017)

Timing Algorithm and Application Based on EM and Small-Cap Stocks Risk Indicator

Xie Qi$^{(\boxtimes)}$ and Cheng Gengguo$^{(\boxtimes)}$

School of Information Science and Engineering,
Wuhan University of Science and Technology, Wuhan 430065, Hubei, China
85169023@qq.com

Abstract. In order to solve problems those are unsteadily to choose buy and sell points by using MACD and other popular classical technical indicators, this paper presents a new technical indicator and a timing algorithm based on it, based on the money flow characteristic of large-cap stocks and small-cap stocks of A-share market of China. Moreover, in order to overcome the problem of unstable returns resulting from generating stochastic initialization parameters, this paper improved the timing algorithm by using EM. With the GEM fund index data from November 2011 to September 2016, the results of our methods show that the return by using timing algorithm based on EM and small-cap stocks risk indicator is 177.23% better than those by using timing algorithm based on MACD and other popular classical technical indicators.

Keywords: Small-cap stocks · EM · Timing model · Returns model

1 Introduction

Quantitative investment refers to the trading method for obtaining stable income by means of the mathematical models and computer programs. In foreign countries, quantitative investment gradually developed since 1960s. In 1959, the Random-Walk theory argued that the stock volatility is random [1]. In 1970, Paul and Eugene propose the efficient market hypothesis, whose core idea is that all valuable information has been timely, accurately and adequately reflected in the stock price trend [2]. And Rose [3] proposed the Arbitrage Pricing Theory (APT) in 1976. In 2010, Kannan and Sekar made stock forecasts with data mining methods and got good results [4]. In China, with the development of capital market, quantitative investment has risen gradually in recent years. In 2015, the method of deep learning was successfully applied to event driven stock forecasts [5].

The core of quantitative investment is timing ability and stock selection ability of investors. The timing ability refers to the ability of investors to predict the overall trend of the market, and stock selection ability refers to the ability of investors to select winning stocks. In the field of quantitative investment, how to design an accurate timing algorithm which can forecast the top and bottom of stock market has become one of the most important research areas. In the last century, the classical investment

© Springer International Publishing AG, part of Springer Nature 2018
D.-S. Huang et al. (Eds.): ICIC 2018, LNCS 10954, pp. 276–286, 2018.
https://doi.org/10.1007/978-3-319-95930-6_26

analysis indexes, such as MACD, KDJ and RSI, have been proposed. The index of Moving Average Convergence Divergence (MACD) and its transaction method were put forward respectively in 1970 and 1985 by Gerald Appel [6, 7].

MACD and other classical indexes are widely used in actual investment, especially in the quantitative timing field. In 2008, Chong, Terence and Wing-Kam [8] studied in the London Stock Exchange Index FT30 with the indexes of MACD and RSI, which showed that its return is more than buy and that with held-to-maturity strategy. And Xue [9] used the timing algorithm based MACD to predict the domestic stock market in 2011. However, the MACD has two defects: hysteresis and poor robustness.

In view of the problem above, this paper proposed a new index and timing algorithm based on small-cap stocks, based on the capital flow characteristics between the small-cap and large-cap stocks and the sector rotation phenomenon in the domestic stock market. Moreover, in order to overcome the problem of unstable returns resulting from generating stochastic initialization parameters, this paper improved the timing algorithm with EM.

2 Timing Model

2.1 Small-Cap Stocks and Large-Cap Stocks

The obvious difference between small-cap and large-cap stocks, make scholars in China and abroad do in-depth research. In China, Wang [10] using statistical method and multifractal theory find that in the China's stock market, the large-cap stocks' risk is less than the small-cap stocks'. In 2014, Zhou [11] find that there is the rotation between the large-cap and small-cap in China A-stock market, that is to say, the return of the small-cap and large-cap stocks have the obvious phenomenon of positive-negative conversion and cycle. And the large- and small-cap rotation effect is very strong. In other countries, Melvyn [12], using CRSP stock and mutual fund data, find that there is a conversion between the growth stocks of small-capitalization and the value stocks of large-capitalization, and proves that the returns from the conversion is more than momentum model. In 2008, Cheol [13] shows that in the market as the leading of large-cap stock investment, the small-cap stock investment cannot replace the large-cap stock.

According to the domestic and foreign research results and the actual situation of the China A-stock market, we summarize the following 3 characteristics: (1) there is cycle conversion between small-cap and large-cap stocks, (2) short-term funds favor small-cap stocks and the risk of small-cap stocks is higher than large-cap stocks, (3) in the short term, the conversion between small-cap and large-cap stocks is driven by the capital.

Based on 3 characteristics above, we propose the following 2 hypotheses:

Hypothesis 1: there are turn-up or turn-down orderliness between small-cap and large-cap stocks trend.

Hypothesis 2: because the risk of small-cap stocks is higher than large-cap stocks, so when short-term risk comes, investors will prior sell small-cap stocks, which leads the descent speed of small-cap stocks higher than large-cap stocks. On the contrary, when short-term opportunities comes, the rising speed of small-cap stock is higher.

On the basis of the two hypotheses, stock market price will cause small-cap stocks to laggard play or fill down, whereas the rise or fall of small-cap stocks will drive stocks to laggard play or fill down. So there exists a mutual influence relationship between small-cap and large-cap stocks. We use the ratio of small-cap and large-cap stock market value to quantify this relationship.

Firstly, we conclude some expressions used in this paper for simplicity and clarity (Table 1).

Table 1. Simply expressions and their definition.

Expressions	Definition
Pb	The opening price of a stock at the buying time
Ps	The opening price of a stock at the selling time
Pl	The lowest price of a stock in a period of time
Po	The opening price for a period of time
MKT	The current market values of all the companies
S	The market values of small-cap stock
L	The market values of large-cap stock
N	The amount of small-cap or large-cap stocks
T	The amount of trading days for calculating
CRR	The cumulative return rate of multiple transactions over a period of time
PRR	The positive return rate for a period of time
MMD	The mean of max drawdowns for a period of time
MRR	The mean of return ratios for a period of time
MTD	The mean of trading days for a period of time

2.2 Small-Cap Stocks Risk Indicator (SSRI)

The small-cap stocks risk indicator is divided into two steps, calculating the large-cap stocks and small-cap stocks market value ratio (LSMVR) and generating SSRI based on LSMVR.

Calculating LSMVR: Every work day, after the stock market closed, all shares will be sorted in accordance with the ascending order of the stock market value. Then, the sum of the market value of first N stocks is the market value of small-cap stocks, as well as the market value of the large-cap stocks is calculated by the last N stocks.

Finally, the LSMVR is that the sum of N small-cap stock market value S_i divided by the sum of N large-cap stock market value L_i.

$$\text{LSMVR} = \frac{\sum_{i=1}^{N} S_i}{\sum_{i=1}^{N} L_i} \tag{1}$$

In addition, the daily stock market data needs cleaning. Because the suspension and new shares cannot reflect the true flow of funds, we need exclude these shares in the calculation of LSMVR.

Generating SSRI: Compared with the LSMVR of that day and the every last T trading day one-by-one, every time the current LSMVR is greater than before, the correct count C increases by 1, otherwise unchanged, repeated T times in total. Finally, the value obtained by the following formulation is SSRI.

So the SSRI ranges from 0 to 100. When SSRI is 100, there are three possibilities.

(a) Capital inflow of small-cap stocks is faster than large-cap stocks;
(b) Capital outflow speed of small-cap stocks is slower than large-cap stocks;
(c) Capital inflow speed of small-cap stocks is equal to outflow of large-cap stocks.
    So it's estimated that in these T days, the small-cap stocks appear at the top.

On the contrary, when SSRI is 0, the small-cap stocks reach the bottom for last T days.

2.3   The Timing Algorithm Based on SSIR (TA-SSIR)

This paper chooses the rate of return as objective function, formula is as follows:

$$\mathbf{f} = \frac{P_s - P_b}{P_b} \times 100\% \tag{2}$$

Assuming that the SSRI in the T trading day was 0, it indicates the bottom of small-cap stocks appears. Because SSRI is calculated after the market closed, the T + 1 day is the buying time. And the buying price includes two kinds. One is the stock's opening price of the T + 1 day, the other is the closing price of the T + 1 day.

On the other hand, when SSRI of the T trading day was 100, the small-cap stocks appears at the top, so as to sell at the T + 1 day. And the selling price will be the opening price or the closing price of the T + 1 day.

Algorithm 1. TA-SSRI

```
const:N,T
var: MKT, Po, Pl
```
begin:
```
step 1: loading and cleaning all the data of stock
market value over a period of time.
   step 2: calculating the SSRI with the empirical N and
T.
   step 3: finding all the buying and selling time by
the time sequence.
   step 4: calculating CRR, PRR, MMD, MRR, MTD by the
buying and selling time.
```
   return: CRR, PRR, MMD, MRR, MTD.
end.

---

3 Optimization Model of TA-SSRI

Seeing that the N and T for calculating the TA-SSRI are based on historical experience, it will cause uncertainty. Based on the considerations above, POBEM was introduced as a part of optimization algorithm to estimate the optimal parameters N and T. Then we can use the optimal N and T to re-calculate the SSRI.

SSRI involves two hidden parameters: N and T, which are interplay. For maximizing the return function, how to set the value of N and T is a question. And the parameters N and T of the maximum return are represented by the following formula.

$$N, T = \arg\max(f(N, T)) \tag{3}$$

3.1 Parameter Optimization Algorithm Based on Expectation Maximization (POBEM)

As a machine learning algorithm, the expectation maximization algorithm (EM) consists of two steps: expectation and maximization [14, 15]. In order to gain maximum expectation, the process is repeating these two steps until the objective function converges to a value.

Algorithm 2. POBEM

```
const: MKT, Po, Pl
var: max,N,T
begin:
step 1: estimating T to find the optimal return func-
tion using the fixed N.
step 2: estimating N to find the optimal return func-
tion using the N obtained from step 1.
step 3: 1repeating the iteration of the above two
steps until the maximum return converged.
return: optimal solution, optimal N, optimal T.
end.
```

Since the return function is usually not a completely convex function in practical application, the global optimal solution is not always found in the parameter optimization process. And it may be a local optimal solution.

3.2   Timing Algorithm Based on EM and Small-Cap Stocks Risk Indicator (TAEM)

TAEM is a method using the POBEM algorithm to estimate the value of N and T in TA-SSRI algorithm and finding the maximum objective function.

Algorithm 3. TAEM

```
const: MKT, Po, Pl
var: N, T
begin:
step 1: loading and cleaning all the data of A-shares
over a period of time.
step 2: calculating the SSRI using the initial N and
T.
step 3: searching the first time of buying and selling
from the time sequence, if found, jumping to the step 4,
otherwise to the end.
step 4: calculating the indexes, such as CRR, PRR,
MMD, by the buying and selling time.
step 5: finding the optimal N and T using POBEM.
step 6: jumping to the step 2 and recalculating the
indexes of SSRI using the optimal N and T.
return: CRR, PRR, MMD, MRR, MTD.
end.
```

---

4 Stock Analysis with SSRI

In order to verify the validity of the SSRI, this paper choose the GEM ETF index fund (code: 159915) as the experimental data for calculating the return rate. All the data is provided by the terminal of Wind financial advisory, and the software used in the whole process is Python 2.7.

4.1 Stock Analysis with TA-SSRI

In this part, the parameters N and T of TA-SSRI are respectively 30 and 600, and the data time is from December 9 in 2011 to September 30 in 2016.

By the calculation with the TA-SSRI, the cumulative trading is 17 times, including 16 times for positive return and 1 times for negative returns, which obtains the PRR 94.12%. And within nearly 5 years, the CRR, MRR and MD are respectively 710.216%, 12.812% and -5.074%. Moreover, the MTD is 29 days. The results are shown in Table 2

And if buying the stock 159915 on December 9 in 2011 and selling with the opening price until September 30 in 2016, the CRR is 159.37%. In addition, at the same buying time, when sold on June 5, 2015 which is the highest point in this period, the stock 159915 can obtain the CRR, 383.65%.

With the comparison of the CRR of the three methods above, it's obvious that the method of TA-SSRI can obtain much more return than the other two. So it shows that the TA-SSRI can get more return under certain conditions.

Table 2. The results of stock timing analysis with TA-SSRI.

No.	Buying time	Selling time	Buying price	Selling price	CRR (%)	MD (%)	TD
1	2011-12-13	2012-02-17	0.788	0.726	−7.9	−21.447	41
2	2012-03-29	2012-06-19	0.698	0.758	8.6	−2.579	53
3	2012-07-12	2012-08-23	0.723	0.751	3.9	−5.671	30
4	2012-11-09	2013-01-16	0.678	0.780	15.0	−11.947	45
5	2013-02-05	2013-02-25	0.767	0.833	8.6	0.000	9
6	2013-04-16	2013-05-14	0.827	0.990	19.7	0.000	17
7	2013-07-09	2013-07-19	1.026	1.124	9.6	0.000	8
8	2013-09-13	2013-10-15	1.210	1.372	13.4	0.000	15
9	2014-04-29	2014-06-09	1.225	1.315	7.3	−2.449	26
10	2014-11-12	2015-03-04	1.423	1.881	32.2	−1.616	73
11	2015-04-20	2015-05-14	2.355	3.030	28.7	−1.062	17
12	2015-07-02	2015-08-14	2.550	2.580	1.2	−26.000	31
13	2015-09-02	2015-10-16	1.646	2.307	40.2	0.000	25
14	2016-01-12	2016-02-24	2.011	2.102	4.5	−10.144	26
15	2016-03-07	2016-04-07	1.840	2.199	19.5	−2.174	22
16	2016-05-13	2016-06-28	1.925	2.095	8.8	−0.571	30
17	2016-08-01	2016-09-08	2.024	2.115	4.5	−0.593	28

4.2   Validation of POEM

In order to prove that the POBEM algorithm can converge to the extreme value or the maximum value. The same data is divided into two groups by the different time intervals that one is from December 9 in 2011 to September 30 in 2016, the other is from January 1 in 2014 to September 30 in 2016. The initial values of T and N are randomly generated in the interval [10, 100] and [10, 600], respectively. And each group uses the POBEM algorithm to calculate 30 times, then gets the statistical results. Under the same conditions, we can obtain the result with the enumeration method for comparison.

The results of the first group are shown in Table 3. In the whole process, we obtain the four convergence values of CRR: 758.01%, 730.68%, 718.09%, 706.53%, 758.01% and corresponding convergence times (CT) are 9, 6, 10 and 5. After 2 cycles, the CT is 25 in total, while the result of 3 cycles is 5 and all is 758.01%. With the same parameters, the enumeration method obtains 4751 different values by 53960 times, which the maximum value is 758.01%. And all the values is shown in Fig. 1.

Table 3. The results of the first group with POBEM.

CRR	CT	CT of 2 cycles	CT of 3 cycles	N	T
758.01%	9	4	5	28	32
730.68%	6	6	0	570	30
718.09%	10	10	0	427	31
706.53%	5	5	0	46	33

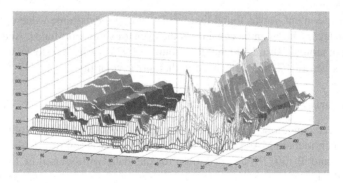

Fig. 1. The results of the first group with the enumeration method.

Similarly, for the second group, the convergence values of CRR are 435.01% and 419.6%, whose CT are respectively 22 and 8. After 2 cycles, the CT is 12 in total, while the result of 3 cycles is 18 and all is 435.01%. The result of the second group are shown in Table 4. The results with the enumeration method are shown in Fig. 2, which the maximum value calculated from the 53960 CRR is 435.01%.

Table 4. The results of the second group with POBEM.

CRR	CT	CT of 2 cycles	CT of 3 cycles	N	T
435.01%	22	4	18	28	32
419.6%	8	8	0	46	33

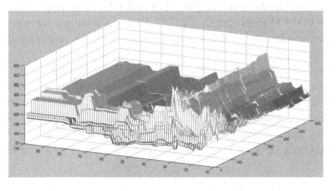

Fig. 2. The results of the second group with the enumeration method.

By the results of these two groups, it proves that the POBEM method can converge to the maximum value of the objective function. While the objective function is not a completely convex function, so the convergence values with the POBEM method may be the extreme values, not the maximum value. With respect to the performance of these two method, the time complexity of enumeration method is $O(n2)$, while POBEM's time complexity is $O(n)$. So, the POBEM method can greatly reduce the time complexity.

In our experiment, it was found that the T can find the extreme value or maximum in about 30 days, which can be considered the cycle of short-term funds. And the N has two ranges at around 30 and 500 that for a total of 2,935 stocks, the stocks whose CMV ascending is before 1% and 12% have higher stock volatility.

4.3   Comparison of TAEM and Other Common Methods

Under the same conditions, we make a comparison with the TAEM and other common methods which include MACD, KDJ, RSI and BIAS. Based on results above, the two groups of N and T, (30, 30) and (30, 600) for this data, will be used in TA-SSRI and TAEM method.

And the final results are shown in Table 5. It shows that all the CRR of TA-SSRI and the TAEM methods is more than 300%, which is much higher than other common methods. For the PRR of our methods, all is 100%, while that of the other methods is less than 72%. And the MRR of our methods is more than 11%, as well as the other is less than 8%. Despite the better MMD of MACD, that of our methods is nearly −5% which is better than the other three methods. Moreover, except the RSI method for 85 days, the MTD of the other methods is about 30 days. These results show that the return rates by using timing algorithm based on small-cap stocks risk indicator is

177.23% more than others by using timing algorithm based on MACD and other popular classical technical indicators.

Table 5. Comparison of TAEM and other common methods.

Methods	CRR	Trading times	PRR	MRR	MTD	MMD
TAEM (1)	312.472%	11	100%	11.425%	23	−4.799%
TAEM (2)	378.982%	8	100%	18.685%	32	−4.733%
MACD	166.041%	9	33.33%	7.969%	40	−2.163%
KDJ	128.161%	7	71.43%	3.56%	21	−10.539%
RSI	109.465%	2	50%	4.935%	85	−22.572%
BIAS	96.155%	11	63.64%	0.529%	26	−12.137%

5 Conclusion

According to the capital flow between the large-cap and small-cap stocks in China A-stock market, this paper proposes the SSRI, its timing algorithm and optimization algorithm for obtaining the excess returns. The results prove that our methods based on SSRI can obtain more benefits, and some conclusions are as follows:

- Compared with MACD, KDJ, RSI and BIAS, the method of SSRI can more accurately find small-cap stocks at the top and bottom, which can be used in actual investment.
- POBEM method solve the optimization problem of the initial N and T in TA-SSRI method. And the POBEM method can converge to the extreme value or the maximum value, which greatly reduces the time complexity compared with the enumeration method.

For further optimization, POBEM method can be improved to ensure obtaining the maximum value, rather the extreme value. And it needs to mend the SSRI for improving accuracy of top and bottom position of small-cap stocks, further reducing the retracement to expand returns.

Acknowledgments. This work was supported in part by the grants of Natural Science Foundation of China (61304129).

References

1. Osborne, M.F.M.: Brownian motion in the stock market. Oper. Res. **7**(2), 145–173 (1995)
2. Malkiel, B.G., Fama, E.F.: Efficient capital markets: a review of theory and empirical work. J. Finan. **25**(2), 383–417 (1970)
3. Ross, S.A.: The arbitrage theory of capital asset pricing. J. Econ. Theor. **13**(3), 341–360 (1976)
4. Kannan, K.S., Sekar, P.S., Sathik, M.M., et al.: Financial stock market forecast using data mining techniques. Lect. Not. Eng. Comput. Sci. **2180**(1) (2010)

5. Ding, X., Zhang, Y., Liu, T., et al.: Deep learning for event-driven stock prediction. In: International Conference on Artificial Intelligence, pp. 2327–2333. AAAI Press (2015)

6. Appel, G.: Technical Analysis: Power Tools for Active Investors. FT Press, Upper Saddle River (2005)

7. Appel, G.: The moving average convergence-divergence trading method: advanced version. Scientific Investment Systems (1985)

8. Ng, W.K.: Technical analysis and the London stock exchange: testing the MACD and RSI rules using the FT30. Appl. Econ. Lett. **15**(14), 1111–1114 (2008)

9. Xue, H.: Quantitative investment management and research based on the MACD and the other technical parameters. University of International Business and Economics (2011)

10. Wang, D., Suo, Y., Li, X.: A study on differences between large-cap-stock and small-cap-stock based on multifractal theory. Chin. J. Manag. **09**(7), 1025–1031 (2012)

11. Zhou, J.: The wheeled effect between big share and small share in Chinese A-share market, pp. 4–5. Southwestern University of Finance and Economics (2014)

12. Teo, M., Woo, S.J.: Style effects in the cross-section of stock returns. J. Finan. Econ. **74**(2), 367–398 (2004)

13. Eun, C.S., Huang, W., Lai, S.: International diversification with large- and small-cap stocks. J. Finan. Quant. Anal. **43**(2), 489–524 (2008)

14. Dempster, A.P., Laird, N.M., Rubin, D.B.: Maximum likelihood from incomplete data via the EM algorithm. J. Roy. Stat. Soc. **39**(1), 1–38 (1977)

15. Wu, C.F.J.: On the convergence properties of the EM algorithm. Ann. Stat. **11**(1), 95–103 (1983)

Preprocessing Technique for Cluster Editing via Integer Linear Programming

Luiz Henrique Nogueira Lorena[1]([✉]), Marcos Gonçalves Quiles[1],
André Carlos Ponce de Leon Ferreira de Carvalho[2],
and Luiz Antonio Nogueira Lorena[3]

[1] Federal University of São Paulo - UNIFESP, São José dos Campos, Brazil
luiz-lorena@hotmail.com, quiles@unifesp.br
[2] University of São Paulo - USP, São Carlos, Brazil
andre@icmc.usp.br
[3] National Institute for Space Research - INPE, São José dos Campos, Brazil
luizlorena54@gmail.com

Abstract. This paper addresses the Cluster Editing problem. The objective of this problem is to transform a graph into a disjoint union of cliques using a minimum number of edge modifications. This problem has been considered in the context of bioinformatics, document clustering, image segmentation, consensus clustering, qualitative data clustering among others. Here, we focus on the Integer Linear Programming (ILP) formulation of this problem. The ILP creates models with a large number of constraints. This limits the size of the problems that can be optimally solved. In order to overcome this limitation, this paper proposes a novel preprocessing technique to construct a reduced model that feasibly maintains the optimal solution set. In comparison to the original model, the reduced model preserves the optimal solution and achieves considerable computational time speed-up in the experiments performed on different datasets.

Keywords: Cluster Editing · Preprocessing technique · Unsupervised learning

1 Introduction

The Cluster Editing (CE) problem [5], also referred in the literature as Correlation Clustering [2], ask to transform an undirected graph G by minimizing the number of editions, i.e., insertions or deletions of edges, to create a vertex-disjoint union of cliques. Figure 1 shows a CE instance were three editions are made on the graph resulting on two disjoint cliques represented by the vertices set {A, B, C, D} and {E, F, G}.

This problem has been considered in the context of bioinformatics [4, 6], clustering documents [2], image segmentation [14], consensus clustering [1, 10], and qualitative data clustering [10, 11]. The CE problem is a NP-hard problem [2, 19], thus heuristics, approximations and data reduction methods were proposed in the literature [3, 9, 12, 18].

© Springer International Publishing AG, part of Springer Nature 2018
D.-S. Huang et al. (Eds.): ICIC 2018, LNCS 10954, pp. 287–297, 2018.
https://doi.org/10.1007/978-3-319-95930-6_27

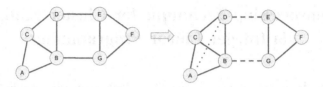

Fig. 1. Cluster Editing instance. Solid lines represent edges that are maintained, dashed lines are removed edges and dotted lines are edges that were inserted.

Grotschel and Wakabayashi [11] introduced an ILP model for the CE. They derived it from a mathematical analysis of the corresponding problem polytope, proposing several partition inequalities for this purpose. As there are an exponential number of these inequalities, they followed a cutting plane approach where the inequalities are added to the Linear Program only if a current fractional solution violates them.

The ILP creates models with a large number of constraints that limits the size of the problems that can be optimally solved. Therefore, here, we propose a preprocessing technique to construct a reduced model. In comparison to the original model, the reduced model preserves the optimal solution and achieves remarkable computational time speed up in the experiments performed on different datasets.

This work is organized as follows. In Sect. 2 the problem is defined and the integer linear programming formulation is presented. Section 3 presents the new preprocessing technique. The computational experiments are provided in Sect. 4. Finally, conclusions are drawn in Sect. 5.

2   Cluster Editing via ILP

The CE problem can be formulated as a maximization or minimization problem [7]. This paper considers a minimization version of graph clustering where the goal is to minimize the number of editions (edges deleted between clusters plus the number of edges inserted inside clusters).

The following notation is introduced to explain the proposed preprocessing technique. Given an undirected graph $G = (V, E)$ where V is the set of vertices and E is the set of edges. The edge weight values, represented by w_{ij}, are 1 if the edge exists in the graph and -1 for missing edges. The number of vertices in the graph is defined as $n = |V|$, while $m = |E|$ represent its number of edges.

The following ILP formulation can be used for cluster editing [6, 11]:

$$(CE_{ILP}) : \text{ Minimize } \sum_{e \in E} w(e) - \sum_{i<j} w_{ij} x_{ij}$$

$$\text{subject to } \quad x_{ij} + x_{jk} - x_{ik} \leq 1 \quad i < j < k \tag{1}$$

$$x_{ij} - x_{jk} + x_{ik} \leq 1 \quad i < j < k \tag{2}$$

$$-x_{ij} + x_{jk} + x_{ik} \leq 1 \quad i < j < k \tag{3}$$

$$x_{ij} \in \{0, 1\} \quad i, j \in [1..n]$$

where $x_{ij} = 1$ if i and j are part of the solution and 0 otherwise.

The edge editions are considered on (CE_{ILP}) model solution in the following way: an edge is inserted if $x_{ij} = 1$ for $w_{ij} = -1$ and is removed if $x_{ij} = 0$ for $w_{ij} = 1$. Therefore, the objective function returns the minimum number of editions.

Constraints (1–3) are called "transitivity constraints". Constraint 1, for instance, enforces that: if vertex i is in the same cluster as vertex j, and j is in the same cluster as k then i is in the same cluster as k.

3 Preprocessing Technique

The (CE_{ILP}) model has $O(n^2)$ variables and $O(n^3)$ constraints, which creates models with a large number of redundancies. Though not critical to solving the problem, these constraints affect the solver efficiency and might prohibit its usage.

Grotschel and Wakabayashi [11], in 1989, were the first to propose a strategy to deal with such redundancy. A cutting plane algorithm was created to identify violated constraints during the execution of a relaxed version of this problem. It was confirmed experimentally that, for many instances, the cutting plane ends with a small fraction of the transitivity constraints. This fact evidences a great redundancy of transitivity constraints in the original model (CE_{ILP}).

Other techniques, introduced in the context of clique partitioning problem [16] and modularity optimization in complex networks [8], try to identify redundancies in the ILP model by analyzing the graph representation of the clustering instance. Recently, Nguyen et al. [17] generalized the approach of Dinh et al. [8] to some constrained clustering problems. All those techniques were capable of reducing the size complexity of the transitivity constraints from $O(n^3)$ to $O(nm)$.

In the context of Cluster Editing, Bocker et al. [5, 6] introduced techniques that focus on reducing the problem size instead of improve the ILP model. They identify patterns that can be removed from the problem instance without changing the groups obtained in the optimal solution. The reduced problem is solved exactly by using the cutting plane proposed by Grotschel and Wakabayashi [11] and a fixed parameter branching algorithm. The experimental results shows that those techniques were capable of solving large graphs with 1000 vertices and several thousand edge modifications.

Here a new approach is introduced, focusing on the identification of the ILP model redundancy. It tries to identify redundancies by analyzing the graph representation of the clustering instance [8, 16]. Our approach is based on the identification of triangles formed by edges (graph edges and missing edges) that correspond to the transitivity constraints of a model (CE_{ILP}).

Figure 2 presents the edge weight distribution within triangles considered by the transitive constraints of the (CE_{ILP}) model. The analysis of such triangles helps to identify constraints that do not need to be considered in the model as they do not result in editions.

Transitivity constraints corresponding to triangles T1, T3 and T4 do not need to be considered because of the following reasons:

- T1: the optimization objective of (CE_{ILP}) tries to set all variables $x_{ij} = x_{jk} = x_{ik} = 1$ for $w_{ij} = w_{jk} = w_{ik} = 1$ since all transitivity constraints are satisfied.
- T2: the optimization objective of (CE_{ILP}) tries to set all variables $x_{ij} = x_{jk} = x_{ik} = 0$ for $w_{ij} = w_{jk} = w_{ik} = 0$ since all transitivity constraints are satisfied.
- T3: the optimization objective of (CE_{ILP}) tries to set the value 1 to the variable relative to the positive edge weight and 0 to the remaining variables. This satisfies all the transitivity constraints.

Constraints that deals with triangles of type T2 must be considered as the transitivity constraints are not satisfied when variables corresponding to positive edges are set to 1. To satisfy those constraints variables corresponding to the negative edge weights must be set to 1 which leads to one edge editing.

There is another possibility, removing an edge of the graph. Those circumstances are considered in Fig. 3. There are three possible variants of such triangle depending on vertex order:

- T2A: considering the optimization objective of (CE_{ILP}), the best possibility of editing (1 edition) that satisfy all transitivity constraints are the following:
  - $x_{ij} = x_{jk} = x_{ik} = 1$ (1 edge inserted)
  - $x_{ij} = 1, x_{jk} = x_{ik} = 0$ (1 edge removed)
  - $x_{jk} = 1, x_{ij} = x_{ik} = 0$ (1 edge removed)
- T2B: considering the optimization objective of (CE_{ILP}), the best possibility of editing (1 edition) that satisfy all transitivity constraints are the following:

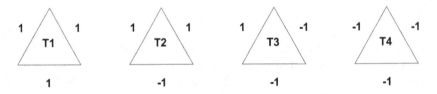

Fig. 2. Possible edge weights distribution.

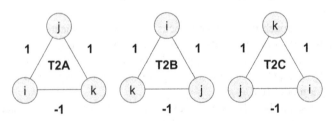

Fig. 3. Possible edge label distribution for triangles of type T2.

- $x_{ij} = x_{jk} = x_{ik} = 1$ (1 edge inserted)
- $x_{ij} = 1, x_{jk} = x_{ik} = 0$ (1 edge removed)
- $x_{ik} = 1, x_{ij} = x_{jk} = 0$ (1 edge removed)
- T2C: considering the optimization objective of (CE_{ILP}), the best possibility of editing (1 edition) that satisfy all transitivity constraints are the following:
  - $x_{ij} = x_{jk} = x_{ik} = 1$ (1 edge inserted)
  - $x_{ik} = 1, x_{ij} = x_{jk} = 0$ (1 edge removed)
  - $x_{jk} = 1, x_{ij} = x_{ik} = 0$ (1 edge removed)

T2 triangles are also known as conflict triples in literature and are the roots to some data reduction methods [5, 6]. In this paper, the information of the edge weight distribution within triangles is used directly in the model (CE_{ILP}) to identify what transitivity constraints should be considered while constructing the ILP model. The data is not reduced or modified, and only transitivity constraints corresponding to T2 triangles need to be taken into account. Hence, the following reduced model is proposed:

$$(CER_{ILP}) : \text{Minimize} \sum_{e \in E} w(e) - \sum_{i < j} w_{ij} x_{ij}$$

$$\text{subject to} \quad x_{ij} + x_{jk} - x_{ik} \leq 1 \quad i, j, k \in S1 \tag{4}$$

$$x_{ij} - x_{jk} + x_{ik} \leq 1 \quad\quad\quad i, j, k \in S2 \tag{5}$$

$$-x_{ij} + x_{jk} + x_{ik} \leq 1 \quad\quad\quad i, j, k \in S3 \tag{6}$$

$$x_{ij} \in \{0, 1\} \quad\quad\quad i, j \in [1..n]$$

where

$$S1 = \{i < j < k | w_{ij} = +1 \wedge w_{jk} = +1 \wedge w_{ik} = -1\}$$

$$S2 = \{i < j < k | w_{ij} = +1 \wedge w_{jk} = -1 \wedge w_{ik} = +1\}$$

$$S3 = \{i < j < k | w_{ij} = -1 \wedge w_{jk} = +1 \wedge w_{ik} = +1\}$$

The sets S1, S2, and S3 enforce that only constraints corresponding to T2 triangles must be considered while creating the model. This can be considered as a preprocessing technique that produces a model (CER_{ILP}) with a small number of constraints in comparison to the original model (CE_{ILP}). For instance, considering the example depicted in Fig. 1, the model (CE_{ILP}) has 105 constraints in contrast to the model (CER_{ILP}), which has only 11 constraints. Both models find the best number of editions, but the reduced model has 1.89 of speedup in computational times.

4   Experimental Results

The experiments and algorithms were coded in C++14 and executed on a computer with the following configuration: Intel Core i7-6770HQ (3,5 GHz) with 32 GB RAM running Windows 10 64-Bit. The commercial solver IBM ILOG CPLEX [13] 12.7.1 was used to solve the ILP models.

The following datasets were used to compare the performance and the quality of the solution of the proposed model (CER_{ILP}) against the original model (CE_{ILP}):

- *LFR benchmark networks.* Networks created with the benchmark developed by Lancichinetti-Fortunato-Radicchi (LFR) [15]. The following parameters were used: number of vertices $n = \{50, 100, 200\}$, the average degree was set to 5 and the maximum degree was set to 10. The default values were used for the remaining parameters. Networks with increasing values for the mixing parameter (μ) were used to blur the community distinction.
- *Random unweighted graphs.* Proposed by [6] in the following manner: given a number of nodes n and parameter k, uniformly selected an integer $i \in [1, n]$, which defines a cluster with i vertices. The process continues with the remaining $n \leftarrow n - 1$ vertices until $n \leq 5$ holds. In this case, all remaining vertices are assigned to the last cluster. Finally, an estimated value $k' \leq k$ is used to perform uniformly editions (add/remove edges). This dataset can be found online[1]. Datasets with sizes $n = \{100, 200, 300, 1000, 1500, 2000\}$ were selected for the experiments.

4.1   Experiments with LFR Networks

LFR benchmark networks [15] were created to evaluate the performance of the proposed preprocessing technique. Given a number of vertices n, one network with predefined clusters was created for each mixing parameter (μ). The amount of edges between clusters increase proportionally to μ. As a consequence, the clusters become more interconnected and the clustering problems more difficult.

Table 1 presents the results obtained for the (CE_{ILP}) and (CER_{ILP}) models on the LFR benchmark networks. Column n represents the number of vertices of the graph; μ is the mixing parameter; columns *Obj*, *#C* and *Time* are defined for both models and represent, respectively, the objective value, the number of constraints, and the computational time in seconds. Finally, column *%C* represents the percentage of constraints removed from the original model and S corresponds to the computational time speedup obtained by (CER_{ILP}).

It can be observed, based on Table 1, that (CER_{ILP}) achieves a better performance than (CE_{ILP}) while preserving the optimal number of editions for all the considered instances. This higher performance is due to a large number of redundant transitivity constraints disregarded by the preprocessing technique (above 99%). Consequently, the computational times are drastically improved providing speedups from 15 to 13755. All instances are solved by (CER_{ILP}) in less than 2 s.

---

[1] https://bio.informatik.uni-jena.de/data/.

Table 1. Results obtained by (CE_{ILP}) and (CER_{ILP}) on LFR benchmark networks.

n	μ	CE_{ILP}			CER_{ILP}				
		Obj	#C	Time	Obj	#C	Time	%C	S
50	0.1	50	58800	0.63	50	250	0.04	99.57	15.18
	0.2	65	58800	1.01	65	375	0.07	99.36	15.27
	0.3	71	58800	1.21	71	411	0.07	99.30	17.28
	0.4	80	58800	1.79	80	486	0.10	99.17	18.77
	0.5	78	58800	2.70	78	423	0.11	99.28	24.81
	0.6	74	58800	1.68	74	381	0.07	99.35	24.96
	0.7	88	58800	2.73	88	533	0.14	99.09	19.90
	0.8	87	58800	5.37	87	522	0.21	99.11	25.02
	0.9	89	58800	5.29	89	528	0.27	99.10	19.74
100	0.1	86	485100	6.23	86	428	0.05	99.91	118.44
	0.2	108	485100	6.52	108	627	0.05	99.87	126.95
	0.3	147	485100	11.73	147	893	0.15	99.82	75.86
	0.4	149	485100	16.01	149	890	0.11	99.82	143.08
	0.5	173	485100	46.76	173	1078	0.23	99.78	201.69
	0.6	183	485100	82.62	183	1153	0.63	99.76	131.68
	0.7	182	485100	74.43	182	1127	0.36	99.77	208.65
	0.8	193	485100	72.91	193	1260	0.27	99.74	267.89
	0.9	172	485100	49.43	172	1004	0.19	99.79	257.55
200	0.1	223	3940200	157.55	223	1118	0.18	99.97	865.92
	0.2	223	3940200	74.58	223	1270	0.11	99.97	651.88
	0.3	257	3940200	96.49	257	1563	0.11	99.96	911.65
	0.4	296	3940200	221.62	296	1688	0.17	99.96	1310.29
	0.5	331	3940200	558.39	331	2004	0.17	99.95	3460.32
	0.6	333	3940200	956.43	333	1925	0.20	99.95	4697.06
	0.7	342	3940200	5074.88	342	1908	0.37	99.95	13755.17
	0.8	376	3940200	10547.42	376	2333	1.73	99.94	6084.59
	0.9	481	3940200	5855.49	481	3388	2.03	99.91	2877.47

Problems with $n > 300$ vertices were not tested because the (CE_{ILP}) fails to solve them due to lack of memory. It is worth noting that the (CER_{ILP}) can solve problems with a higher number of vertices, based on the results presented in Table 1.

4.2   Experiments with Random Unweighted Graphs

Proposed by Bocker et al. [6], these datasets were generated by disturbing an ideal cluster graph using random edge insertions and deletions. Given a number of vertices n, 10 networks with predefined clusters were created for each corresponding k. The values of k were selected according to the following rule $k = c * n$ with $c = \{0.25, 0.5, 1, 1.25, 1.5, 1.75, 2\}$.

Datasets with $n = \{100, 200, 300\}$ were considered initially for the experiments because (CE_{ILP}) failed to solve instances with $n > 300$ due to lack of memory. Table 2 presents the results obtained for the (CE_{ILP}) and (CER_{ILP}) models on such networks. Each row represents the average result for the 10 datasets considering each pair (n, k). Column n represents the number of vertices of the graph; k is the upper limit for the number of editions; columns *Obj*, $\#C$ and *Time* are defined for both models and represent, respectively, the average objective value, average number of constraints, and computational times in seconds. Finally, column $\%C$ represents the average percentage of constraints removed from the original model and $\underline{S}$ corresponds to the average computational time speedup obtained by (CER_{ILP}).

Next, we run a set of experiments with larger datasets $n = \{1000, 1500, 2000\}$ to test the scalability of the (CER_{ILP}). The objective value cannot be compared to (CE_{ILP}), but the percentage of constraint elimination can be estimated.

Table 2. Results obtained by (CE_{ILP}) and (CER_{ILP}) on random unweighted graphs.

		CE_{ILP}			CER_{ILP}				
n	k	Obj	#C	Time	Obj	#C	Time	%C	S
100	25	25.0	485100	10.85	25.0	1745.8	0.03	99.64	346.66
	50	49.0	485100	7.03	49.0	3001.4	0.04	99.38	164.36
	75	74.2	485100	5.35	74.2	4409.1	0.07	99.09	84.59
	100	97.8	485100	5.14	97.8	6547.8	0.11	98.65	49.88
	125	121.4	485100	4.79	121.4	7726.2	0.09	98.41	52.59
	150	145.6	485100	4.92	145.6	8369.7	0.10	98.27	52.64
	175	168.2	485100	4.61	168.2	10108.1	0.10	97.92	47.19
	200	192.4	485100	4.62	192.4	11542.6	0.11	97.62	43.14
200	50	50.0	3940200	242.98	50.0	5889.6	0.08	99.85	3266.11
	100	99.6	3940200	141.69	99.6	13819.6	0.22	99.65	728.56
	150	149.6	3940200	64.90	149.6	17697.5	0.23	99.55	299.32
	200	198.0	3940200	51.93	198.0	24434.4	0.30	99.38	188.05
	250	248.0	3940200	41.75	248.0	32242.4	0.36	99.18	121.56
	300	295.2	3940200	41.19	295.2	36497.3	0.40	99.07	105.71
	350	343.0	3940200	41.09	343.0	37213.1	0.41	99.06	103.98
	400	391.4	3940200	41.41	391.4	46879.4	0.50	98.81	85.44
300	75	74.8	13365300	1621.70	74.8	12447.7	0.15	99.91	11386.81
	150	149.4	13365300	875.95	149.4	29056.6	0.42	99.78	2265.54
	225	224.0	13365300	430.22	224.0	45402.3	0.70	99.66	683.91
	300	298.2	13365300	262.33	298.2	57886.7	0.79	99.57	340.42
	375	371.5	13365300	203.77	371.5	78641.0	0.98	99.41	215.10
	450	445.2	13365300	188.92	445.2	80117.3	0.98	99.40	206.10
	525	518.8	13365300	186.77	518.8	103056.6	1.19	99.23	165.03
	600	593.2	13365300	195.56	593.2	126468.7	1.46	99.05	140.61

Table 3 presents the results obtained for the (CER_{ILP}). Each row represents the average result for the 10 datasets and each pair (n, k). Column n represents the number of vertices of the graph; k is the upper limit for the number of editions; for (CE_{ILP}) only the average number of constraints is presented ($\#C$). Columns Obj, $\#C$ and $\underline{Time}$ represent the average objective value, average number of constraints and computational times in seconds for the (CER_{ILP}), respectively. Finally, column $\%C$ highlights the average percentage of constraints removed from the original model.

Table 3. Results obtained by (CE_{ILP}) and (CER_{ILP}) on large random unweighted graphs.

		CE$_{ILP}$		CER$_{ILP}$		
n	k	#C	Obj	#C	Time	%C
1000	250	498501000	249.8	162941.7	9.29	99.97
	500	498501000	499.8	338462.5	14.56	99.93
	750	498501000	749.0	475923.0	17.91	99.90
	1000	498501000	998.0	652948.9	21.47	99.87
	1250	498501000	1247.5	900571.3	25.61	99.82
	1500	498501000	1494.6	1040558.8	24.14	99.79
	1750	498501000	1743.2	1235538.8	32.12	99.75
	2000	498501000	1990.0	1183563.7	41.57	99.76
1500	375	1684126500	374.8	363423.8	15.95	99.98
	750	1684126500	749.6	717436.6	39.40	99.96
	1125	1684126500	1123.6	1167055.5	73.22	99.93
	1500	1684126500	1494.6	1040558.4	24.14	99.79
	1875	1684126500	1872.8	1817375.0	89.43	99.89
	2250	1684126500	2245.2	2125008.0	150.52	99.87
	2625	1684126500	2618.6	2276352.3	227.35	99.86
	3000	1684126500	2991.2	3048967.0	386.80	99.82
2000	500	3994002000	500.0	739933.5	76.19	99.98
	1000	3994002000	998.8	1149414.0	77.29	99.97
	1500	3994002000	1498.4	2138914.5	220.12	99.95
	2000	3994002000	1998.0	2708588.2	340.46	99.93
	2500	3994002000	2497.2	3210726.6	403.91	99.92
	3000	3994002000	2994.4	4150590.4	346.08	99.90
	3500	3994002000	3495.8	4417540.9	742.18	99.89
	4000	3994002000	3993.0	5290777.8	1389.50	99.87

From Table 3, it can be verified that a large number of redundant transitivity constraints are disregarded by the preprocessing technique (above 99%). All instances are solved by (CER_{ILP}) in less than 23 min. This result shows the scalability of the proposed technique on the datasets proposed by Bocker et al. [6].

5 Conclusions

A novel preprocessing technique for the Clique Edition problem was proposed in this work. The experimental results showed that the reduced model (CER_{ILP}) provided a considerable reduction of transitivity constraints, preserved the optimal solution set, and improved the computational speedup compared to model (CE_{ILP}) for all considered instances.

This technique has the advantage of working as complementary to other techniques, like the cutting plane proposed by Grotschel and Wakabayashi [11]. The (CER_{ILP}) can provide means to increase the size of future instances that the Integer Linear Programming approach can execute.

Our preprocessing technique might also be combined with other approaches, such as the method for reducing the input graph proposed in [6]. Thus, the preprocessed reduced graph may speed up the solution of the ILP problem even further.

Finally, we expect that the results obtained in the ILP context can be used to guide the construction of better heuristic techniques for the unweighted Cluster Editing problem.

Acknowledgements. The authors thanks FAPESP (Grant No. 2011/18496-7), CNPq (Grant No. 310908/2015-9 and 301836/2014-0), CAPES and IBM for support.

References

1. Ailon, N., Charikar, M., Newman, A.: Aggregating inconsistent information. J. ACM **55**, 1–27 (2008). https://doi.org/10.1145/1411509.1411513
2. Bansal, N., Blum, A., Chawla, S.: Correlation clustering. Mach. Learn. **56**, 89–113 (2004). https://doi.org/10.1023/B:MACH.0000033116.57574.95
3. Bastos, L., Ochi, L.S., Protti, F., Subramanian, A., Martins, I.C., Pinheiro, R.G.S.: Efficient algorithms for cluster editing. J. Comb. Optim. **31**(1), 347–371 (2016). https://doi.org/10.1007/s10878-014-9756-7
4. Ben-Dor, A., Shamir, R., Yakhini, Z.: Clustering gene expression patterns. J. Comput. Biol. **6**(3–4), 281–297 (1999). https://doi.org/10.1089/106652799318274
5. Böcker, S., Baumbach, J.: Cluster editing. In: Bonizzoni, P., Brattka, V., Löwe, B. (eds.) CiE 2013. LNCS, vol. 7921, pp. 33–44. Springer, Heidelberg (2013). https://doi.org/10.1007/978-3-642-39053-1_5
6. Böcker, S., Briesemeister, S., Klau, G.W.: Exact algorithms for cluster editing: evaluation and experiments. Algorithmica **60**, 316–334 (2011). https://doi.org/10.1007/s00453-009-9339-7
7. Charikar, M., Guruswami, V., Wirth, A.: Clustering with qualitative information. J. Comput. Syst. Sci. **71**, 360–383 (2005). https://doi.org/10.1016/j.jcss.2004.10.012
8. Dinh, T.N., Thai, M.T.: Toward optimal community detection: from trees to general weighted networks. Internet Math. **11**, 181–200 (2014). https://doi.org/10.1080/15427951.2014.950875
9. Fellows, M., Langston, M., Rosamond, F., Shaw, P.: Efficient parameterized preprocessing for cluster editing. In: Csuhaj-Varjú, E., Ésik, Z. (eds.) FCT 2007. LNCS, vol. 4639, pp. 312–321. Springer, Heidelberg (2007). https://doi.org/10.1007/978-3-540-74240-1_27

10. Gionis, A., Mannila, H., Tsaparas, P.: Clustering aggregation. ACM Trans. Knowl. Discov. Data (TKDD) **1**(1), 4 (2007). https://doi.org/10.1145/1217299.1217303
11. Grötschel, M., Wakabayashi, Y.: A cutting plane algorithm for a clustering problem. Math. Program. **45**, 59–96 (1989). https://doi.org/10.1007/bf01589097
12. Guo, J.: A more effective linear kernelization for cluster editing. Theor. Comput. Sci. **410**, 718–726 (2009). https://doi.org/10.1016/j.tcs.2008.10.021
13. IBM: IBM ILOG CPLEX 12.7.1 (1987–2017)
14. Kim, S., Yoo, C.D., Nowozin, S., Kohli, P.: Image segmentation using higher-order correlation clustering. IEEE Trans. Pattern Anal. Mach. Intell. **36**, 1761–1774 (2014). https://doi.org/10.1109/tpami.2014.2303095
15. Lancichinetti, A., Fortunato, S., Radicchi, F.: Benchmark graphs for testing community detection algorithms. Phys. Rev. E Stat. Nonlin. Soft Matter Phys. **78**, 046110 (2008). https://doi.org/10.1103/physreve.78.046110
16. Miyauchi, A., Sukegawa, N.: Redundant constraints in the standard formulation for the clique partitioning problem. Optim. Lett. **9**, 199–207 (2014). https://doi.org/10.1007/s11590-014-0754-6
17. Nguyen, D.P., Minoux, M., Nguyen, V.H., Nguyen, T.H., Sirdey, R.: Improved compact formulations for a wide class of graph partitioning problems in sparse graphs. Discrete Optim. **25**, 175–188 (2017). https://doi.org/10.1016/j.disopt.2016.05.003
18. Protti, F., da Silva, M.D., Szwarcfiter, J.L.: Applying modular decomposition to parameterized cluster editing problems. Theory Comput. Syst. **44**, 91–104 (2007). https://doi.org/10.1007/s00224-007-9032-7
19. Shamir, R., Sharan, R., Tsur, D.: Cluster graph modification problems. In: Goos, G., Hartmanis, J., van Leeuwen, J., Kučera, L. (eds.) WG 2002. LNCS, vol. 2573, pp. 379–390. Springer, Heidelberg (2002). https://doi.org/10.1007/3-540-36379-3_33

A Bearing Fault Diagnosis Method Based on Autoencoder and Particle Swarm Optimization – Support Vector Machine

Duy-Tang Hoang[1] and Hee-Jun Kang[2(✉)]

[1] Graduate School of Electrical Engineering,
University of Ulsan, Ulsan, South Korea
hoang.duy.tang@gmail.com
[2] School of Electrical Engineering, University of Ulsan, Ulsan, South Korea
hjkang@ulsan.ac.kr

Abstract. Rolling element bearing is an important part of rotary machines. Bearing fault is a big issue because it can cause huge cost of time and money for fixing broken machines. Thus, early detecting fault of bearing is a critical task in machine health monitoring. This paper presents an automatic fault diagnosis of bearing based on the feature extraction using Wavelet Packet Analysis, feature selection using Autoencoder, and feature classification using Particle Swarm Optimization - Support Vector Machine. First, bearing vibration signals are decomposed at different depth levels by Wavelet Packet Analysis. Then the wavelet packet coefficients are used to compute the energy value of the corresponding wavelet packet node. After that, an Autoencoder is exploited to select the most sensitive features from the feature set. Finally, classification is done by using a Support Vector Machine classifier whose parameters are optimized by Particle Swarm Optimization. The effectiveness of the proposed intelligent fault diagnosis scheme is validated by experiments with bearing data of Case Western Reserve University bearing data center.

Keywords: Artificial neural network · Autoencoder · Bearing fault diagnosis
Particle Swarm Optimization · Support Vector Machine
Wavelet Packet Analysis

1 Introduction

Rolling Element Bearing (REB) is a critical component of rotary machine. Healthy operation of bearings is the necessary factor for rotary machines to operate efficiently. REBs are account for almost 45−55% of machine failures [1], which can cost a huge amount of time and money for maintaining. As a result, it is very important to early detect the existing faults in bearings. It will help to cut down costs necessary for emergency maintenance, replacement and delay production.

Due to the ease of measurement and the ability to provide a lot of dynamic information reflecting the condition of the mechanical systems, vibration signal-based fault diagnosis has been widely applied in machine health monitoring [1–4]. The abnormal vibration is the first clue of rotary component failure. Vibration analyzing can

© Springer International Publishing AG, part of Springer Nature 2018
D.-S. Huang et al. (Eds.): ICIC 2018, LNCS 10954, pp. 298–308, 2018.
https://doi.org/10.1007/978-3-319-95930-6_28

detect all types of faults, either localized or distributed. Furthermore, low-cost sensors, accurate results, simple setups, specific information on the damage location, and comparable rates of damage are other benefits of the vibration analysis method [5]. The condition monitoring of a rolling element bearing based on vibration signals can be considered as a pattern recognition problem which has been successfully applying intelligent diagnosis methods. Generally, a general intelligent diagnosis methodology includes three steps as follows: feature extraction, feature reduction, and feature classification.

Feature extraction is the first step mapping the original fault signals onto statistical parameters which reflect the working status of machines. Feature extraction requires expert knowledge, human labor, and signal processing techniques. To extract representative features indicating health conditions of bearings, vibration signals can be analyzed in time domain [2], frequency domain [3], or time-frequency domain [6, 7].

The output of feature extraction is a feature set which consists of features reflecting the characteristic of signals. Normally, the feature set has high dimensionality with a lot of features. The high dimensional feature set often reduces the classification accuracy of fault diagnosis system. Moreover, there is no guarantee that all features are equally usefully in reflecting the health status of bearings [8]. As a result, feature selection is often exploited to reduce the dimensionality of the feature set.

Two approaches are available to perform dimensionality reduction are feature selection and feature extraction. In feature selection approach, a subset of all the features is selected without the transformation. On the other hand, in feature extraction approach, a new feature set is created by the transformation of the existing features. The most well-known algorithm in this approach is principal component analysis (PCA) and its derivatives.

Recently, unsupervised learning autoencoder (AE) has been applied in signal-based fault diagnosis [9, 10]. Hongmei Liu et al. [9] used a stacked autoencoder which formed by stacking several AEs, to extract fault features. Three AEs are stacked together to extract fault features in frequency domain from a spectrogram of vibration signals. This approach obtained high performance but too complex and still missing features in time domain. Tao et al. [10] also used AE with many layers, but directly on raw vibration signal and can only extracted time domain feature.

In feature classification step, after the high sensitive feature set is determined, machine learning based classifiers are employed to detect the health condition of bearings. Among current classification algorithms, artificial neural network (ANN) has proved to be a powerful tool with high accuracy fault detection. However, ANN is not suitable for handling algorithm with few samples for training [11]. Moreover, ANN has some drawbacks, including generalization ability, and slow convergence [8]. Support Vector Machine (SVM) is a supervised machine learning algorithm which based on statistical learning theory. Compared to ANN, the advantages of SVM are (1) better generalization, and (2) does not require many samples for training.

Particle Swarm Optimization (PSO), inspired by the bird swarm behavior of preying on food, emerged as a powerful optimization technique. In PSO, the set of candidate solutions to the optimization problem is defined as a swarm of particles which may flow through the parameter space defining trajectories which are driven by their own and neighbors' best performances [12]. The concept and implementation of

PSO are simple so it is widely applied in many optimization problems, including optimization of SVM parameters. The parameter selection method based on PSO can not only ensure the learning ability of SVM but also improve the generalization ability and accordingly improve the integrated performance of SVM classification [13].

In this paper, a novel bearing fault diagnosis method is proposed. First, bearing vibration signals are analyzed in time-frequency domain by Wavelet Packet Analysis, the energy features from nodes of wavelet packets at different decomposition levels are calculated to characterize the health conditions of the bearing. After that, an AE is applied to reduce the dimensionality of the feature set, only select the most sensitive features. Finally, for feature classification, a SVM classifier whose parameters are optimized by PSO is exploited. To evaluate the effectiveness of the proposed scheme, experiments are carried out with bearing data from Case Western Reserve University (CWRU) bearing data center [14]. Moreover, the proposed method is also investigated with the signals under various low signal-to-noise ratio (SNR) conditions to show the effectiveness under noisy environment.

The remainder of this paper is organized as follows. Section 2 presents brief overviews of AE and PSO. Section 3 explains the proposed algorithm in detail. Section 4 described the experiments. Finally, in Sect. 5 we conclude the paper.

2 Background

2.1 Autoencoder

AE was first introduced by Rumelhart et al. [15] as an unsupervised machine learning algorithm. As shown in Fig. 1, an AE is a feed-forward neural network with three layers: input layer, hidden layer, and output layer where the input and output layer have the same size. The structure of an AE can be considered as an encoder followed by a decoder. The encoder includes the input layer and hidden layer, try to map the input vector to the hidden layer. In opposite, the decoder tries to reconstruct the input vector from the hidden layer vector. Taking an input vector x^i, the computation of AE includes two steps: encoding and decoding as follow:

$$a^i = f\left(W_e x^i + b_e\right) \tag{1}$$

$$\hat{x} = f\left(W_d a^i + b_d\right) \tag{2}$$

where W_e and b_e are respectively the weight matrix and the bias vector of the encoder; W_d and b_d are respectively the weight matrix and the bias vector of the decoder; $f(.)$ denotes the activate function.

With an input set includes m samples $\{x^i, i = 1 : m\}$, the AE will produce m output samples $\{\hat{x}^i, i = 1 : m\}$. Since the goal of the AE is to make the reconstructed vector $\hat{x}$ as close as possible to the input x, the cost function can be defined as:

$$J(W_e, W_d, b_e, b_d) = \frac{1}{2m} \sum_{i=1}^{m} (\hat{x}^i - x^i)^2 \tag{3}$$

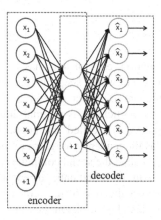

Fig. 1. Autoencoder

2.2 Particle Swarm Optimization

PSO introduced by Kennedy et al. [16] was inspired by the movement of organisms in a bird flock or fish school. The goal of PSO is finding out the solution for the optimization problem. Each candidate solution is called a particle and represents a point in a d-dimensional space, d is the variable number of the function need to be optimized. PSO performs searches using a population (swarm) of particles that are updated their characteristic after each step of movement. The principle algorithm of PSO can be explained as follows [12].

Assume that a swarm includes n particles $X : \{x_1, x_2, \ldots x_n\}$. Each particle is characterized by three properties:

- Position $x_i = \begin{bmatrix} x_i^1 & x_i^2 & \ldots & x_i^d \end{bmatrix}$
- Velocity $v_i = \begin{bmatrix} v_i^1 & v_i^2 & \ldots & v_i^d \end{bmatrix}$
- Personal best p_i the position of the best solution obtained so far by that particle

In each iteration, each particle moves to a new position, the movement in the searching space of a specific particle is computed by:

$$x_i(t+1) = x_i(t) + v_i(t+1) \tag{4}$$

where t and $t+1$ indicate two successive iterations. The velocity vectors govern the way particles move across the search space and are made of the contribution of three terms: the first one, defined the inertia or momentum prevents the particle from drastically changing direction, by keeping track of the previous flow direction; the second term, called the cognitive component, accounts for the tendency of particles to return to their own previously found best positions; the last one, named the social component, identifies the propensity of a particle to move towards the best position of the whole swarm (or of a local neighborhood of the particle, depending on whether a global or partial PSO is implemented). Based on these considerations, the velocity of particles is defined as:

$$v_i(t+1) = \omega v_i(t) + c_1 r_1 (p_i - x_i(t)) + c_2 r_2 (g - x_i(t)) \tag{5}$$

In the update velocity equation, each of term has its own distinct roles. The first term $\omega v_i(t)$ is the inertia component which keeps the particle move in the old direction. The value of the inertial coefficient ω is often chosen in the range $[0.8, 1.2]$, which can either dampen or accelerate the particle in its original direction.

The second term $c_1 r_1 (p_i - x_i(t))$, called the cognitive component, acts as the particle's memory, make it has the tendency to return to the position of its best fitness. The third term $c_2 r_2 (g - x_i(t))$, called the social component, causes the particle to move to the best position (g) the swarm has found until current step. The cognitive coefficient c_1 and social coefficient c_2 affect the size of the movement step of particles in each iteration. They are usually chosen close to 2. On the other hand, r_1 and r_2 are diagonal matrices of random numbers generated from a uniform distribution in $[0.1]$ cause the corresponding components to have a stochastic influence on the velocity update. Accordingly, the trajectories drawn by the particles are semi-random in nature, as they derive from the contribution of systematic attraction towards the personal and global best solutions and stochastic weighting of these two acceleration terms.

Consider the optimization problem:

$$\text{Given } f : R^D \to R$$
$$\text{Find } x_{opt} \ \lor \ f(x_{opt}) \geq f(x) \, \forall x \in R^D \tag{6}$$

The problem is solved by PSO algorithm described step by step as follows.

Step 1. Initialization: For each of the n particles:

1. Randomly initialize the position $x_i(0)$
2. Initialize the particle's best position by its initial position $p_i = x_i(0)$
3. Calculate the fitness of each particle $f(x_i(0))$
4. Initialize the global best by g by the position of particle which has the best fitness

Step 2. Until stop condition is met, repeat:

5. Update velocity by Eq. (2)
6. Update position by Eq. (1)
7. If $f(x_i(t+1)) \geq f(p_i)$, then $p_i = x_i(t+1)$
8. If $f(x_i(t+1)) \geq f(g)$, then $g = x_i(t+1)$

When the stop condition is satisfied, the iterative process stops, the best solution is presented by global best value g.

3   Proposed Bearing Fault Diagnosis Method

The proposed new bearing fault diagnosis scheme has 3 main steps: feature extraction, feature selection, and feature classification. Each step is described in detail as follows.

Step 1: Feature Extraction

At this step, vibration signals are decomposed into various levels by Wavelet Packet Analysis (WPA). All nodes at all levels are considered because it is difficult to declare definitively that those at a certain depth are better than those at another. Subsequently, we calculate wavelet packet energy for every node and use those as time-frequency features of the signal.

Consider a time-domain vibration signal $x(t)$ consists of S sample. By decomposing $x(t)$ into N levels, we obtain the result of the decomposition as follows. Energy of each node in the decomposition is computed by [17]:

$$E_j^n = \sqrt{\frac{1}{M}\sum_{t=1}^{M}\left(x_j^n(t)\right)^2} \tag{7}$$

where $M = \frac{S}{2^j}$ is the number of samples at the node x_j^n. Since the decomposition has $\sum_i^N 2^i$ nodes, we obtain the corresponding number of node energy features.

Step 2: Feature Selection

In this second step, an AE is exploited to reduce the dimensionality by mapping the energy features into high sensitive features. The energy feature set from step 1 is considered as the input of the AE. The training process using backpropagation with the cost function defined in (6). After training, the output of hidden layer of the SA becomes the input for classification step.

Step 3: Feature Classification

In this step, a SVM is employed to classify the selected features. The classification accuracy of the SVM is affected by three factors: kernel function, kernel parameters, and penalty parameter. PSO algorithm is used to find those parameters of the SVM. After finding out the SVM model with best parameters, the feature set is given into SVM to recognize the type of faults.

4 Experiment

4.1 Test-Bed

To evaluate the effectiveness of the proposed method, experiments are carried out with data of faulty bearing from the Case Western Reserve University (CWRU) bearing data center. This dataset is public and widely used to validate the effectiveness of bearing fault diagnosis algorithms. The test-bed shown in Fig. 2 includes a dynamometer (right), a 2 HP motor (left), and a torque transducer/encoder (center). The test-bed also consists of a control electronics but not shown in the figure. The motor shaft is supported by the test bearings. Single point faults were introduced to these bearings using electro-discharge machining with fault diameters of 7 mils, (1 mil = 0.001 inches). Vibration data are collected by using accelerometers, which are attached to the housing with magnetic bases. Accelerometers are placed at the 12 o'clock position at both the drive housing. Vibration signals are collected using a 16 channel DAT recorder, including three operating conditions: fault at ball, fault at inner race, and fault at outer race. These operating conditions are operated with bearings 6205-2RS JEM SKM, which are deep groove ball bearing type. All experiments are conducted for one load condition (2 HP load), where the rotation speed was 1797 revolutions per minute (rmp). Data were collected at 12 kHz sampling frequency from both drive end (DE) and fan end (FE). Figure 3 shows the vibration signals of three operating conditions.

Fig. 2. Bearing fault diagnosis test-bed

4.2 Vibration Signal Pre-processing

Vibration signals are from both fan end and drive end of the motor shaft at sample rate 12000 Hz. To have enough samples for the training process of the machine learning based classifier, at first, each vibration signal of each health condition is split into non-overlapping segments with the same length. For every condition, 100 samples are acquired, so totally we obtain 300 sample for three bearing health conditions.

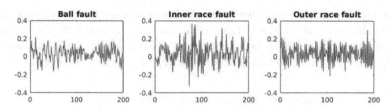

Fig. 3. Vibration signals

4.3 Feature Extraction and Dimensionality Reduction

In the next step, we compute the energy features using WPA. The mother wavelet is Daubechies 4, each segment of the signal is decomposed to 4 level. For each sample, we have $\sum_i^4 2^i = 30$ nodes corresponding to 30 energy features. The energy features are calculated from all 30 nodes of the decomposition tree.

AEs are exploited to extract most sensitive features from the energy feature set in other to reduce the dimension. The input size of every AE is equal to the number of energy features, while the size of the hidden layer can be varied. Our goal when using AEs is to compress the data feature as much as possible. So, we investigate AEs with various hidden layer size from 1 to 5 (neuron). Backpropagation is used to train the AEs. Finishing the training, the hidden vectors are used as compressed feature set.

4.4 PSO-SVM Classification

In our experiments, to classify features, SVM classifiers with RBF kernel are used. The classification performance highly depends on the parameters C, γ. To search the optimum values of those parameters, we exploited the PSO algorithm with parameters as follows.

- Number of particles in population: 6
- Searching dimension: 2 (including two values need to be optimized: C and γ)
- Searching space: $C \in [0, 100]$, $\gamma \in [0, 1]$
- Inertial coefficient, cognitive coefficient, and social coefficient are selected by the method of Clerc et al. [18] proposed as follows.

$$\chi = \frac{2\kappa}{\left| 2 - \phi - \sqrt{\phi^2 - 4\phi} \right|},$$

$$\omega = \chi, \ c_1 = \chi\phi_1, \ c_2 = \chi\phi_2, \ 0 < \kappa \le 1, \ \phi_1 + \phi_2 = \phi \ge 4 \tag{8}$$

By choosing $\kappa = 1$, $\phi_1 = \phi_2 = 2.05$, the cognitive coefficient and social coefficient are $c_1 = c_2 = 1.496$, the inertial coefficient $\omega = 0.73$.

In the last step – feature classification, we use k-fold ($k = 5$) cross validation for SVM classification with parameters found by the above PSO method. The feature sets are divided into k subsets; each subset is used for once for testing while the remain $k - 1$ subsets are used for training. The final classification accuracy is the average of k classification times. The results with different feature sets supplied by varying the size of AE hidden layers are shown in Table 1. We can see that even with the 2-feature set (AE with 2 neurons in hidden layer), the classification results still achieve satisfactory performance (99:05%). And from 3 neurons above, the classification accuracy is 100%.

Table 1. Classification results of different AE models

Size of hidden layer in AE	1	2	3	4	5
Classification accuracy (%)	93.39	99.05	100	100	100

4.5   Robustness Investigation

The proposed scheme can classify bearing fault efficiently with original vibration signal. However, in real industrial environments, the sensory signals are contaminated by noise [7]. Thus, now we analyze the robustness of the proposed scheme under low signal-to-noise ratio (SNR) condition. The additive Gaussian white noise (AWGN) with different standard variances are added to the original vibration signals to mimic the low SNR. The SNR is defined as follows:

where P_{signal} and P_{noise} are the power signal and noise respectively. Figure 4 shows the noised signal which made by adding the original signal with the AGWN.

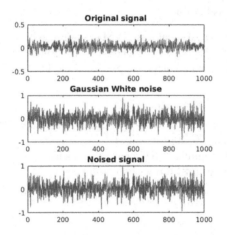

Fig. 4. A noisy signal with SNR $= -10$ dB

Table 2 shows the fault classification accuracy of the proposed scheme with SNR value varies in $[-2, -4, -6, -8, -10]$. Results show that even under very low SNR (-6 dB), the classifier is still capable to achieve absolute accuracy. Under the worst case (SNR $= -10$ dB), classification accuracy is 93.34% .

Table 3 shows the result of comparison between the proposed method with some other techniques mentioned in the publication [7] under the worst scenario

SNR $= -10$ dB. The comparison shows that the proposed scheme yields much superior classification accuracy robustness against noise compared to other methods.

Table 2. Classification results with different SNR values

SNR (dB)	-2	-4	-6	-8	-10
Classification accuracy (%)	100	100	100	98	94.34

Table 3. Classification results of different methods (SNR $= -10$ dB)

Scheme	Accuracy (%)
Malhi [3]	38.09
Lou [6]	33.33
Seker [19]	78.25
Li [20]	86
Yaqub [7]	91.23
Proposed method	94.34

5 Conclusion

In this paper, a feature reduction method by using AE was proposed and successfully applied in bearing fault diagnosis. At first, vibration signals are decomposed by WPA. Dual time-frequency domain features are extracted by computing the energy of every node in the decomposing tree. Unsupervised learning is applied to train simple AE with only 1 hidden layer to extract most sensitive features. Finally, features are classified by RBF kernel SVM with optimal parameters are optimized by PSO algorithm. Our proposed method can achieve very high accuracy and robustness even under very poor SNR condition. The effectiveness is also presented through comparisons with other existing fault diagnosis methods.

Acknowledgments. This research was supported by Basic Science Research Program through the National Research Foundation of Korea (NRF) funded by the Ministry of Education (NRF-2016R1D1A3B03930496).

References

1. Rai, A., Upadhyay, S.: A review on signal processing techniques utilized in the fault diagnosis of rolling element bearings. Tribol. Int. **96**, 289–306 (2016)
2. Samanta, B., Al-Balushi, K.: Artificial neural network based fault diagnostics of rolling element bearings using time-domain features. Mech. Syst. Sig. Process. **17**(2), 317–328 (2003)
3. Malhi, A., Gao, R.X.: Pca-based feature selection scheme for machine defect classification. IEEE Trans. Instrum. Measur. **53**(6), 1517–1525 (2004)

4. Yen, G.G., Lin, K.-C.: Wavelet packet feature extraction for vibration monitoring. IEEE Trans. Ind. Electron. **47**(3), 650–667 (2000)
5. Kharche, P.P., Kshirsagar, S.V.: Review of fault detection in rolling element bearing. Int. J. Innovative Res. Adv. Eng. **1**(5), 169–174 (2014)
6. Lou, X., Loparo, K.A.: Bearing fault diagnosis based on wavelet transform and fuzzy inference. Mech. Syst. Sig. Process. **18**(5), 1077–1095 (2004)
7. Yaqub, M.F., Gondal, I., Kamruzzaman, J.: Inchoate fault detection framework: Adaptive selection of wavelet nodes and cumulant orders. IEEE Trans. Instrum. Measur. **61**(3), 685–695 (2012)
8. Shen, C., Wang, D., Kong, F., Peter, W.T.: Fault diagnosis of rotating machinery based on the statistical parameters of wavelet packet paving and a generic support vector regressive classifier. Measurement **46**(4), 1551–1564 (2013)
9. Liu, Z., Cao, H., Chen, X., He, Z., Shen, Z.: Multi-fault classification based on wavelet SVM with PSO algorithm to analyze vibration signals from rolling element bearings. Neurocomputing **99**, 399–410 (2013)
10. Tao, S., Zhang, T., Yang, J., Wang, X., Lu, W.: Bearing fault diagnosis method based on stacked autoencoder and softmax regression. In: 2015 34th Chinese Control Conference (CCC), pp. 6331–6335. IEEE (2015)
11. Widodo, A., Yang, B.-S.: Support vector machine in machine condition monitoring and fault diagnosis. Mech. Syst. Sig. Process. **21**(6), 2560–2574 (2007)
12. Marini, F., Walczak, B.: Particle swarm optimization (PSO). a tutorial. Chemometr. Intell. Lab. Syst. **149**, 153–165 (2015)
13. Zhang, X., Guo, Y.: Optimization of SVM parameters based on PSO algorithm. In: Proceedings of Fifth International Conference on Natural Computation. ICNC 2009, vol. 1, pp. 536–539. IEEE (2009)
14. Loparo, K.A.: Bearing data center, Case Western Reserve University (2013)
15. Rumelhart, D.E., Hinton, G.E., Williams, R.J., et al.: Learning representations by back-propagating errors. Cogn. Model. **5**(3), 1 (1988)
16. Kennedy, J.: Particle swarm optimization. In: Encyclopedia of Machine Learning. Springer, Heidelberg, pp. 760–766 (2011)
17. Zarei, J., Poshtan, J.: Bearing fault detection using wavelet packet transform of induction motor stator current. Tribol. Int. **40**(5), 763–769 (2007)
18. Clerc, M., Kennedy, J.: The particle swarm explosion, stability, and convergence in a multidimensional complex space. IEEE Trans. Evol. Comput. **6**(1), 58–73 (2002)
19. Seker, S., Ayaz, E.: Feature extraction related to bearing damage in electric motors by wavelet analysis. J. Franklin Inst. **340**(2), 125–134 (2003)
20. Li, F., Meng, G., Ye, L., Chen, P.: Wavelet transform-based higher-order statistics for fault diagnosis in rolling element bearings. J. Vibr. Control **14**(11), 1691–1709 (2008)

Bayesian Probit Model with L^{α} and Elastic Net Regularization

Tao Li and Jinwen Ma$^{(\boxtimes)}$

Department of Information Science, School of Mathematical Sciences
and LMAM, Peking University, Beijing 100871, China
li_tao@pku.edu.cn, jwma@math.pku.edu.cn

Abstract. Most of the classification and regression models are established from the frequentist perspective. For certain models, the corresponding Bayesian versions have been developed. However, the Bayesian analysis of classification models has been rarely investigated yet, especially for penalized classification models. In this paper, we propose two probit models respectively with L^{α} regularization and elastic net regularization from a Bayesian perspective. It is demonstrated by the experiments on a real-world dataset that the proposed probit models can have certain advantages over the frequentist models.

Keywords: Bayesian classification · Probit model · L^{α} regularization
Elastic net

1 Introduction

Classification and regression are two basic tasks of supervised learning. Many models and algorithms have been established for classification and regression [1]. In frequentist statistical learning, a problem is generally formulated as optimizing an objective function (so-called loss), and the learning procedure is to minimize the loss via an optimization algorithm. For most classification and regression problems, the loss function is formulated as the negative logarithm likelihood.

However, minimizing the negative logarithm likelihood directly leads to the severe over-fitting problem occasionally. For preventing the over-fitting problem, many regularization techniques have been suggested, such as ridge regression [2], Lasso [3] and elastic net [4]. Besides, when the data points lay in a high dimensional space, certain regularization techniques can be implemented to produce sparse solutions, which means that feature selection can be embedded in the learning procedure. Intuitively, L^0 regularization helps to select a best subset of the features, but the corresponding problem is extremely difficult to optimize since it is a non-convex problem. Thus, L^1 regularization is often used as a convex surrogate for L^0 regularization. However, L^1 penalty tends to a result of shrinkage rather than a subset selection. Frank and Freidman [5] proposed L^{α} regularization with $\alpha \in (0, 1]$ which can be considered as a compromise between the shrinkage and subset selection. However, L^{α} regularization is also non-convex, thus it is not widely used and most related works pay attention to theoretical analysis of asymptotic properties [6, 7].

© Springer International Publishing AG, part of Springer Nature 2018
D.-S. Huang et al. (Eds.): ICIC 2018, LNCS 10954, pp. 309–321, 2018.
https://doi.org/10.1007/978-3-319-95930-6_29

Although most of the models and regularization terms are proposed in frequentist perspective, many Bayesian statisticians are devoted to explain the methods from the Bayesian perspective. For example, Polson [8] developed the theory of Bayesian SVM using data augmentation technique. But Lasso, elastic net and other penalized regression methods have been formulated as certain hierarchical models [9, 10]. Especially, Polson [11] developed a Bayesian version of L^{α} penalty for regression. The objective function is non-convex and thus difficult to optimize in the frequentist perspective, but in the Bayesian perspective, we can sample the parameters from posterior instead of optimizing the loss function.

Most related work emphasizes on penalized regression rather than penalized classification. In fact, penalized classification is very important and has various applications. For example, in microarray analysis, we hope to select related genes from thousands of gene expression levels with only less than one hundred observations. A logistic regression model with $L^{1/2}$ regularization was proposed for gene selection in cancer classification in [12], but the algorithm was based on the thresholding representation of $L^{1/2}$ [13], thus it is not so easy to extend the algorithm for general L^{α} regularization. An L^1 penalized logistic regression model for gene selection was proposed in [14], but it just estimated the parameters via optimization and the Bayesian analysis was only used for choosing the regularization parameter. Bae and Mallick [15] utilized a two-level hierarchical Bayesian model for gene selection, but only L^1 regularization and L^2 regularization were considered.

In this paper, we try to investigate the Bayesian elastic net regularization for probit model, which is far more general than L^1 and L^2 regularization. Actually, we propose two penalized classification models from the Bayesian perspective, and develop their learning algorithms by using Gibbs sampling [16]. We test these models on a real-world dataset and find out that their classification performances are satisfying.

The rest of this paper is organized as follows. In Sect. 2, we review some related works and make certain mathematical preparation. Two new classification models are proposed together with detailed derivations of full conditional distributions and Gibbs sampling algorithms in Sect. 3. In Sect. 4, we evaluate our models on a heart disease dataset and investigate how hyper-parameters affect the results. Finally, we give a brief conclusion and discussion in Sect. 5.

2 Related Works and Background

2.1 Notations

Suppose that we have n observations $x_i = \left(x_{i1}, x_{i2}, \cdots, x_{ip}\right), i = 1, 2, \cdots, n$ and corresponding class label $y_i = \pm 1$. Without loss of generality, we always assume that the data have been normalized to zero mean and unit variance. That is, for $\forall k = 1, 2, \cdots, p$ we have $\sum_{i=1}^{n} x_{ik} = 0$ and $\sum_{i=1}^{n} x_{ik}^2 = 1$. This can be achieved by minus the mean and divide the standard variance in each feature dimension. Let X be the design matrix and y be the response,

$$X = \begin{pmatrix} x_{11} & x_{12} & \cdots & x_{1p} \\ x_{21} & x_{22} & \cdots & x_{2p} \\ \vdots & \vdots & \ddots & \vdots \\ x_{n1} & x_{n2} & \cdots & x_{np} \end{pmatrix}, y = \begin{pmatrix} y_1 \\ y_2 \\ \vdots \\ y_n \end{pmatrix}$$

We consider the linear classification model, i.e., we want to estimate $\beta \in \mathbb{R}^p$ such that $y(x\beta) \geq 0$.

2.2 Probit Model

The probit model was proposed by Albert and Chib [17]. At first, we assume that $p(y = 1|x, \beta) = \Phi(x\beta)$, where Φ is the cumulative density function of standard normal distribution. Then the posterior is given as

$$p(\beta|X, y) \propto \prod_{i=1}^{n} \Phi(x_i\beta)^{\frac{1+y_i}{2}}(1 - \Phi(x_i\beta))^{\frac{1-y_i}{2}}\pi(\beta)$$

It seems difficult to sample β from the posterior directly. Using data augmentation technique, they introduce n independent latent variables $z = (z_1, z_2, \cdots, z_n)^T$ where $z_i \sim \mathcal{N}(x_i\beta, 1)$ and define

$$y = \begin{cases} +1, z \geq 0 \\ -1, z < 0 \end{cases}$$

We may assume $z = x\beta + \varepsilon$ with $\varepsilon \sim \mathcal{N}(0, 1)$, then

$$p(y = +1|x, \beta) = p(z \geq 0|x, \beta) = p(\varepsilon \geq -x\beta) = \Phi(x\beta)$$

This approach is similar to the well-known logistic regression model, the only difference is that logit link function is replaced by a standard normal cumulative density function (probit) here. In fact, the link function can be more complicated (for example, t-distributions) and they all belong to general linear model. Suppose that x, y and β are given, then the posterior probability of z is

$$z|x, y, \beta \sim \mathcal{N}(x\beta, 1)\mathbb{I}(yz \geq 0)$$

Here, $\mathbb{I}$ is the indicator function, taking value 1 when the condition in the brackets holds and 0 otherwise. Obviously, the posterior probability is a truncated normal distribution. We can sample from a truncated normal distribution using Rodriguez-Yam's Gibbs sampling method [18]. The probit model has been widely used in Bayesian classification, especially in bioinformatics such as gene selection [15, 19].

2.3 Bayesian L^α Penalized Regression

For regression problem, suppose $y = X\beta + \varepsilon$ where $\varepsilon \sim \mathcal{N}(0, \sigma^2 I_n)$, the bridge estimator β is the minimizer of

$$\frac{1}{2\sigma^2}\|y - X\beta\|_2^2 + v\sum_{j=1}^{p}|\beta_j|^\alpha$$

In the following discussion, σ^2 is always known. Without lost of generality, we assume $\sigma^2 = 1$ in this paper for simplicity. Let $\tau = v^{-1/\alpha}$, the Bayesian version of this problem can be expressed as estimating β from posterior given prior $\pi(\beta)$ and likelihood $\mathcal{L}(y, X|\beta)$

$$\pi(\beta|\alpha, \tau) \propto \prod_{j=1}^{p} \exp\left(-|\frac{\beta_j}{\tau}|^\alpha\right)$$

$$\mathcal{L}(y, X|\beta) \propto \exp(-\frac{1}{2}\|y - X\beta\|_2^2)$$

According to the theory of West [20], we can write the prior as a mixture of normal distributions, but the corresponding mixture distribution is too complicated. In fact, the mixture distribution is not known in closed form and we can only write it as an infinite series explicitly. Polson and Scott [11] suggested that the prior can be seen as a mixture of triangles (Barlett-Fejer kernels) as follows:

$$p(\beta_j|\tau, \omega_j, \alpha) \propto \frac{1}{\tau\omega_j^{1/\alpha}}[1 - |\frac{\beta}{\tau\omega_j^{1/\alpha}}|]_+$$
$$\omega_j|\alpha \sim \frac{1+\alpha}{2}\mathcal{G}(2 + \frac{1}{\alpha}, 1) + \frac{1-\alpha}{2}\mathcal{G}(1 + \frac{1}{\alpha}, 1)$$

Introduce latent variables $\Omega = (\omega_1, \omega_2, \cdots, \omega_p)$ and $u = (u_1, u_2, \cdots, u_p)$, the joint posterior of (β, Ω, u) is

$$\beta, \Omega, u|\alpha, y, X \propto \exp(-\frac{1}{2}\|y - X\beta\|_2^2)\prod_{j=1}^{p}p(\omega_j|\alpha)\mathbb{I}(|\beta_j| \leq b_j)\mathbb{I}(\omega_j \geq a_j)$$

Where $b_j = \tau^{-1}(1 - u_j)\omega_j^{1/\alpha}$ and $a_j = (\frac{|\beta_j/\tau|}{1-u_j})^\alpha$. Furthermore, the hyper parameter v may not be fixed. Suppose the prior for v is Gamma distribution $\mathcal{G}(c, d)$, then the posterior distribution of v is obviously $\mathcal{G}\left(c + \frac{p}{\alpha}, d + \sum_j|\beta_j|^\alpha\right)$.

The detailed derivation can be found in [11]. According to the full conditional distribution, we use the Gibbs sampling on variables (u, Ω, β) in the following way:

- Sample $u_j|\beta_j, \omega_j \sim \mathcal{U}\left(0, 1 - |\beta_j/\tau|\omega_j^{-1/\alpha}\right)$ where $\mathcal{U}$ denotes uniform distribution
- Sample ω_j from a mixture of truncated gamma distribution

$$\omega_j - a_j|\beta_j, u_j \sim \frac{1 - \alpha(1 - a_j)}{1 + a\alpha_j}\mathcal{G}(1, 1) + \frac{\alpha}{1 + a\alpha_j}\mathcal{G}(2, 1)$$

- Sample β from a truncated normal distribution

$$\beta \sim \mathcal{N}((X'X)^{-1}X'y, \left(X'X\right)^{-1}) \prod_{j=1}^{p} \mathbb{I}(|\beta_j| \le b_j)$$

2.4   Bayesian Elastic Net Regression

Consider the regression problem $y = X\beta + \varepsilon$ where $\varepsilon \sim \mathcal{N}(0, \sigma^2 I_n)$. The elastic net regularization was first proposed by Zou [4] and they proved many theoretical properties of this regularization. The loss function is defined as

$$\mathcal{L}(\lambda_1, \lambda_2, \beta) = \frac{1}{2\sigma^2} \| y - X\beta \|_2^2 + \lambda_1 \| \beta \|_1 + \lambda_2 \| \beta \|_2^2$$

We still assume $\sigma^2 = 1$. From the Bayesian perspective, this optimization problem is equivalent to maximum a posterior of β given prior $\pi(\beta)$ and likelihood $\mathcal{L}(y, X|\beta)$

$$\pi(\beta) \propto \exp\left(-\frac{1}{2}\left(\lambda_1 \| \beta \|_1 + \lambda_2 \| \beta \|_2^2\right)\right), \mathcal{L}(y, X|\beta) \propto \exp(-\frac{1}{2}\|y - X\beta\|_2^2)$$

Li and Lin [10] proved that $\pi(\beta)$ can be rewritten as a mixture of normal distributions in the following way

$$\pi(\beta) \propto \prod_{j=1}^{p} \int_{1}^{\infty} \sqrt{\frac{\omega_j}{1-\omega_j}} \exp\left(-\frac{\beta_j^2}{2}\left(\frac{\lambda_2 \omega_j}{\omega_j - 1}\right)\right) \omega_j^{-\frac{1}{2}} \exp\left(-\frac{\lambda_1^2 \omega_j}{8\lambda_2}\right) d\omega_j$$

We can see that for each dimension, the prior $\pi(\beta_j)$ is a mixture of normal distributions with a truncated Gamma as the mixture distribution. Formally, let's introduce latent variables $\Omega = (\omega_1, \omega_2, \cdots, \omega_p)$ with $\omega_j \sim \mathcal{G}\left(\frac{1}{2}, \frac{\lambda_1^2}{8\lambda_2}\right) \mathbb{I}(\omega_j \ge 1)$, then we have $\beta_j | \omega_j \sim \mathcal{N}\left(0, \frac{\omega_j - 1}{\lambda_2 \omega_j}\right)$.

Define $S = \lambda_2 \text{diag}\left(\frac{\omega_1 - 1}{\omega_1}, \frac{\omega_2 - 1}{\omega_2}, \cdots, \frac{\omega_p - 1}{\omega_p}\right)$, the full conditional distributions of β and ω_j are

$$\beta | \Omega, y, X \sim \mathcal{N}((X'X + S)^{-1}y, \left(X'X + S\right)^{-1})$$

$$\omega_j - 1 | \beta, y, X \sim \mathcal{GIG}\left(\frac{1}{2}, \frac{\lambda_1}{4\lambda_2}, \lambda_2 \beta_j^2\right)$$

where $\mathcal{GIG}(\lambda, \phi, \xi)$ denotes the generalized inverse Gaussian distribution.

3 Models and Algorithms

In this section, we propose two penalized classification models: Probit model with L^α regularization (L^α-Probit) and Probit model with elastic net regularization (EN-Probit). From the Bayesian perspective, the parameter estimation is achieved by sampling from the posterior. Since the original Metropolis-Hastings sampling algorithm is not efficient enough, we usually use the Gibbs sampling method instead. The key-point in these models is that the Gibbs sampling step for β is more complicated due to the difference of regression and classification and deeper hierarchical model. Therefore, we mainly concern about how to sample β from the full conditional distribution.

3.1 Probit Model with L^α Regularization

The L^α-Probit model combines probit classifier and L^α regularization. In the frequentist view, the loss function can be written as

$$\mathcal{L}(\beta) = \prod_{i=1}^{n} \Phi(x_i\beta)^{\frac{1+y_i}{2}}(1 - \Phi(x_i\beta))^{\frac{1-y_i}{2}}\pi(\beta) + \upsilon \sum_{j=1}^{p} |\beta_j|^\alpha$$

Obviously, this objective function is hard to optimize using existing algorithms due to the non-convexity. However, the estimation of β is easy via Gibbs sampling. Similar to the regression case, we define $\tau = \upsilon^{-1/\alpha}, b_j = \tau^{-1}(1 - u_j)\omega_j^{1/\alpha}$ and $a_j = (\frac{|\beta_j/\tau|}{1-u_j})^\alpha$. From the Bayesian perspective, the hierarchical model is as follows.

$$y = \begin{cases} +1, & z \geq 0 \\ -1, & z < 0 \end{cases}$$

$$z \sim \mathcal{N}(x\beta, 1)$$

$$\mathrm{p}(\beta_j|u_j, \omega_j, \upsilon) \propto \mathbb{I}(|\beta_j| \leq b_j)$$

$$\omega_j|\alpha, u_j \sim \left[\frac{1+\alpha}{2}\mathcal{G}\left(2+\frac{1}{\alpha}, 1\right) + \frac{1-\alpha}{2}\mathcal{G}\left(1+\frac{1}{\alpha}, 1\right)\right]\mathbb{I}(\omega_j \geq a_j)$$

$$u_j \sim \mathcal{U}(0, 1)$$

$$\upsilon \sim \mathcal{G}(c, d)$$

The full conditional distribution of β is similar to the regression case, and the only difference here is that the response y is replaced by the latent variable z. Therefore,

$$\beta \sim \mathcal{N}((X'X)^{-1}X'z, (X'X)^{-1}) \prod_{j=1}^{p} \mathbb{I}(|\beta_j| \leq b_j)$$

The detailed Gibbs sampling steps for L^α-Probit model can be found in Algorithm 1. We recommend Rodriguez-Yam's Gibbs sampling method [18] for sampling β from the multivariate truncated normal distribution. Note that we only collect the samples after the Markov chain turns to be stable.

It is noticeable that we need to calculate $(X'X)^{-1}$ in Algorithm 1, which may be intractable when dealing with high dimensional problems $(n<p)$ or the features are highly correlated. Therefore, we suggest use $(X'X+\varepsilon I_p)^{-1}$ with a small ε. According to Woodbury-Sherman-Morrison matrix identity, we have $(X'X+\varepsilon I_p)^{-1} = \epsilon^{-1}\left(I_p - X'(XX'+\varepsilon I_n)^{-1}X\right)$. This helps to accelerate the calculation of inverse when p is far larger than n.

Algorithm 1. Gibbs sampling for L^α-Probit

Input: $y \in \{0,1\}^n, X \in \mathbb{R}^{n\times p}, MAX\_ITER$
Output: $\hat{\beta} = (\hat{\beta}_1, \hat{\beta}_2, \cdots, \hat{\beta}_p)$
Variables: $\beta = (\beta_1, \beta_2, \cdots, \beta_p), z = (z_1, z_2, \cdots, z_n)$, $\Omega = (\omega_1, \omega_2, \cdots, \omega_p)$,
$u = (u_1, u_2, \cdots, u_p), v, \tau, a_j, b_j$
Initialize (z, u, Ω, β, v) via random guess
For $t \leftarrow 1$ to MAX_ITER **do**

$$v|\beta \sim \mathcal{G}(c + \tfrac{p}{\alpha}, d + \Sigma_j |\beta_j|^\alpha)$$

$$\tau = v^{-1/\alpha}, b_j = \tau^{-1}(1-u_j)\omega_j^{1/\alpha}, a_j = (\frac{|\beta_j/\tau|}{1-u_j})^\alpha$$

$$z_i|X,y,\beta \sim \mathcal{N}(x_i\beta, 1)\mathbb{I}(z_iy_i \geq 0)$$

$$u_j|\beta_j, \omega_j \sim \mathcal{U}(0, 1 - |\beta_j/\tau|\omega_j^{-1/\alpha})$$

$$\omega_j - a_j|\beta_j, u_j \sim \frac{1-\alpha(1-a_j)}{1+aa_j}\mathcal{G}(1,1) + \frac{\alpha}{1+aa_j}\mathcal{G}(2,1)$$

$$\beta \sim \mathcal{N}((X'X)^{-1}X'z, (X'X)^{-1})\prod_{j=1}^{p}\mathbb{I}(|\beta_j| \leq b_j)$$

Return $\hat{\beta} \leftarrow$ mean of all β after burn-in

3.2 Probit Model with Elastic Net Regularization

The EN-Probit model is a combination of probit classifier and the elastic net regularization. In the frequentist view, the loss function can be written as

$$\mathcal{L}(\beta) = \prod_{i=1}^{n} \Phi(x_i\beta)^{\frac{1+y_i}{2}}(1 - \Phi(x_i\beta))^{\frac{1-y_i}{2}}\pi(\beta) + \lambda_1 \parallel \beta \parallel_1 + \lambda_2 \parallel \beta \parallel_2^2$$

From the Bayesian perspective, the hierarchical mode is shown in the following.

$$y = \begin{cases} +1, & z \geq 0 \\ -1, & z < 0 \end{cases}$$

$$z \sim \mathcal{N}(x\beta, 1)$$

$$\beta_j | \omega_j \sim \mathcal{N}\left(0, \frac{\omega_j - 1}{\lambda_2 \omega_j}\right)$$

$$\omega_j \sim \mathcal{G}\left(\frac{1}{2}, \frac{\lambda_1^2}{8\lambda_2}\right) \mathbb{I}(\omega_j \geq 1)$$

Similar to the regression case, we define $S = \lambda_2 \text{diag}\left(\frac{\omega_1 - 1}{\omega_1}, \frac{\omega_2 - 1}{\omega_2}, \cdots, \frac{\omega_p - 1}{\omega_p}\right)$. The difference here is that response y is replaced by the latent variable z. Therefore, the full conditional posterior of β is

$$\beta | \Omega, y, X \sim \mathcal{N}((X'X + S)^{-1}z, \left(X'X + S\right)^{-1})$$

The Gibbs sampling steps for EN-Probit can be seen in Algorithm 2. Similar to the L^α-Probit model, we only collect the samples after the burn-in steps.

Algorithm 2 Gibbs sampling for EN-Probit

Input: $y \in \{0,1\}^n, X \in \mathbb{R}^{n \times p}, MAX\_ITER$
Output: $\hat{\beta} = (\hat{\beta}_1, \hat{\beta}_2, \cdots, \hat{\beta}_p)$
Variables: $\beta = (\beta_1, \beta_2, \cdots, \beta_p), z = (z_1, z_2, \cdots, z_p), \Omega = (\omega_1, \omega_2, \cdots, \omega_p), \lambda_1, \lambda_2$

Initialize (z, Ω, β) via random guess
For $t \leftarrow 1$ to MAX_ITER **do**

$$z_i | X, y, \beta \sim \mathcal{N}(x_i \beta, 1) \mathbb{I}(z_i y_i \geq 0)$$

$$\beta | \Omega, y, X \sim \mathcal{N}((X'X + S)^{-1}z, (X'X + S)^{-1})$$

$$\omega_j - 1 | \beta, y, X \sim \mathcal{GIG}(\frac{1}{2}, \frac{\lambda_1}{4\lambda_2}, \lambda_2 \beta_j^2)$$

Return $\hat{\beta} \leftarrow$ mean of all β after burn-in

---

4   Experiments

In this section, we evaluate the L^α-Probit model and EN-Probit model on a real-world dataset. First, we consider how hyper-parameters affect the results. Then we compare the performance with ordinary probit model, logistic regression model, decision tree

(CART) and SVM. Last, we plot the estimated β with error-bar to investigate the sparsity and robustness.

4.1 SPECTF Heart Dataset

SPECTF Heart dataset [21] consists of 267 instances with each describes diagnosing of cardiac Single Proton Emission Computed Tomography (SPECT) images of a patient. Every instance in this dataset is a 44-dimensional vector with continuous values and has been labelled as either normal (negative) or abnormal (positive). The training set contains 40 positive and 40 negative instances, while the testing set contains 172 positive and 15 negative instances. This dataset is similar to SPECT dataset, but the SPECT dataset only has 22 features taking integer values 0 or 1, while SPECTF dataset has 44 features taking continuous values. In fact, it's easy to achieve high accuracy on SPECT dataset using basic classification methods (e.g. logistic regression or SVM), but SPECTF is more difficult due to the flexibility and diversity of the features. In addition, in the training set we have a balanced number of instances labelled as positive or negative, but in the testing set more than 90% is labelled as positive. The imbalance in testing set also makes this classification task difficult.

4.2 The Effect of Hyper-parameters

First let's consider the L^α-Probit models. There are three hyper-parameters in this model: the power index α in L^α and (c, d) in the prior of $v \sim \mathcal{G}(c, d)$. For $\alpha \in \{0.25, 0.5, 0.75, 1.00\}$ and $(c, d) \in \{1, 2, 3\} \times \{1, 2, 3\}$, we train the L^α-Probit model with 20000 burn-in steps and collect 30000 samples after burn-in. The corresponding accuracy on testing set is shown in Table 1.

Table 1. Accuracy of L^α-Probit on SPECTF dataset under different α and (c, d)

$\alpha/(c, d)$	(1,1)	(1,2)	(1,3)	(2,1)	(2,2)	(2,3)	(3,1)	(3,2)	(3,3)
0.25	67.38%	62.03%	66.31%	66.31%	67.38%	65.78%	65.78%	64.71%	63.10%
0.50	66.31%	66.31%	65.78%	66.31%	67.38%	67.91%	66.31%	65.24%	66.85%
0.75	70.05%	70.59%	70.59%	71.12%	**72.19%**	71.12%	71.66%	70.59%	71.12%
1.00	69.52%	68.45%	70.05%	69.52%	69.52%	69.52%	69.52%	69.52%	69.52%

From this table, we can see that when we fix α, the change of (c, d) has relatively subtle influence on the results, and the choice of α seems to be vital. In this experiment, $\alpha = 0.75$ is a good choice. In general case, the choice of α can be decided by cross-validation. Note that it's also possible to view α as a random variable given certain prior distribution, and varying α according to the full conditional distribution in Gibbs sampling steps. However, sampling α in this model is complicated.

For the EN-Probit model, we mainly consider the effect of regularization coefficients λ_1 and λ_2. For $(\lambda_1, \lambda_2) \in \{0.1, 0.5, 1, 2, 5\} \times \{0.1, 0.5, 1, 2, 5\}$, we train the EN-Probit model with 20000 burn-in steps and collect 30000 samples after burn-in. The corresponding accuracy on testing set is shown in Table 2.

Table 2. Accuracy of EN-Probit on SPECTF dataset under different (λ_1, λ_2)

λ_1/λ_2	0.1	0.5	1.0	2.0	5.0
0.1	63.10%	64.71%	62.57%	63.64%	66.31%
0.5	64.71%	63.10%	64.71%	64.71%	67.38%
1.0	63.10%	63.64%	63.64%	64.17%	66.31%
2.0	64.71%	64.17%	63.10%	64.17%	67.91%
5.0	62.03%	63.64%	63.64%	65.24%	**68.98%**

We conclude that when regularization is weak (λ_1 and λ_2 are small), the accuracy is low, and the performance becomes better as we strengthen the regularization. This phenomena indicates the possibility of overfitting and the necessity of regularization.

4.3 Performance Comparison

We have already seen the accuracy of L^α-Probit model under best parameter setting is 72.19%, and the accuracy of EN-Probit model under best parameter setting is 68.98%. The accuracy is not high for such a binary classification problem. However, as we stated before, this task is difficult due to the imbalance in testing set. To further evaluate the performance, we train probit model, logistic regression model, decision tree (CART) and SVM on the training set, and the corresponding accuracy, precision, recall, F-value are reported in Table 3.

Table 3. Comparison of probit, logistic regression, CART, SVM, L^α-Probit and EN-Probit

Model	Probit	Logistic	CART	SVM	L^α-Probit	EN-Probit
Accuracy	64.71%	67.91%	63.10%	71.12%	**72.19%**	68.98%
Precision	92.74%	93.75%	94.02%	96.09%	**96.88%**	95.97%
Recall	66.86%	69.77%	63.93%	**72.09%**	**72.09%**	69.19%
F-value	0.7770	0.8000	0.7611	0.8238	**0.8267**	0.8041

According to Table 3 and Fig. 1, we can see that both L^α-Probit model and EN-Probit model outperforms original probit model, logistic regression model and decision tree. It's noticeable that Lα-Probit model gets a better result than SVM in terms of accuracy, precision and F-value. The accuracy of EN-Probit model is slightly lower than L^α-Probit model, but it still shows comparable results with SVM.

4.4 Sparsity and Robustness

In this part, we train L^α-Probit model with $\alpha = 0.75, c = 2, d = 2$ and EN-Probit model with $\lambda_1 = 5$, $\lambda_2 = 5$ on SPECTF dataset. Other experimental settings are same as before except that we estimate the standard variance of β during sampling. The estimated β with error-bar can be seen in Fig. 2. From Fig. 2 we can see that the estimations of both models have fairly low variance, which means the estimations are stable and robust. EN-Probit estimator seems to has a higher variance, but it's

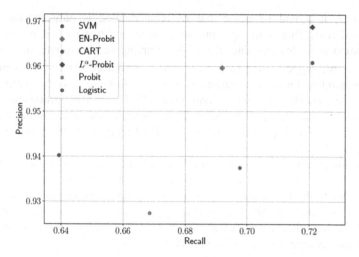

Fig. 1. Visualization of different models in terms of recall and precision

noticeable that the scales are quite different. Moreover, the results are sparse in the sense that some features have near zero coefficients, and EN-Probit estimator has better sparsity than L^α-Probit estimator. In summary, we claim that both L^α-Probit model and EN-Probit model are powerful.

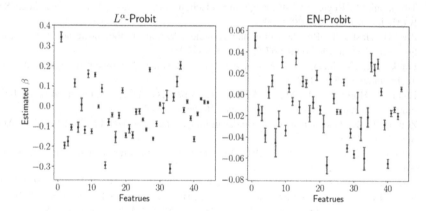

Fig. 2. Estimation of β with error-bar. Left: L^α-Probit model. Right: EN-Probit model

5   Conclusions and Discussions

Based on the review of probit model, L^α penalized regression and elastic net penalized regression from the Bayesian perspective, we have proposed two penalized classification models and derived the full conditional distribution and Gibbs sampling steps. The performances of the two proposed models are evaluated on SPECTF dataset and the experimental results are very satisfying.

However, it should be noted that L^α regularization is only suitable for low dimensional case. That is to say, suppose X is an $n \times p$ matrix, if $p > n$, then L^α regularization is not feasible since the Gibbs sampling steps involving the calculation of $(X'X)^{-1}$. We suggest using $(X'X + \varepsilon I)^{-1}$ as a surrogate for $(X'X)^{-1}$ to ensure the numerical stability. From the experiments, we see that the results of L^α regularization are better than elastic net regularization, especially in accuracy, while the sampling steps for multivariate truncated normal in L^α regularization models are time-consuming. In summary, L^α-Probit performs better than EN-Probit when it's feasible, but the training procedure needs a longer time.

In the future, other types of penalized classification models may be investigated. For example, one can extend Polson's Bayesian SVM in [8] to Bayesian L^α-SVM. However, due to the theoretical foundations of original SVM, this modification needs more theoretical analysis. Another challenging problem is how to use L^α regularization effectively in high-dimensional cases, which is vital in bioinformatics such as gene selection and genome-wide association analysis.

References

1. Friedman, J., Hastie, T., Tibshirani, R.: The Elements of Statistical Learning. Springer Series in Statistics, vol. 1. Springer, New York (2001)
2. Hoerl, A.E., Kennard, R.W.: Ridge regression: biased estimation for nonorthogonal problems. Technometrics **12**(1), 55–67 (1970)
3. Tibshirani, R.: Regression shrinkage and selection via the Lasso. J. Roy. Stat. Soc.: Ser. B (Methodol.) **58**, 267–288 (1996)
4. Zou, H., Hastie, T.: Regularization and variable selection via the elastic net. J. Roy. Stat. Soc. Ser. B (Stat. Methodol.) **67**(2), 301–320 (2005)
5. Frank, I.E., Friedman, J.H.: A statistical view of some chemometrics regression tools. Technometrics **35**(2), 109–135 (1993)
6. Huang, J., Horowitz, J.L., Ma, S.: Asymptotic properties of bridge estimators in sparse high-dimensional regression models. Ann. Stat. **36**, 587–613 (2008)
7. Zou, H., Li, R.: One-step sparse estimates in nonconcave penalized likelihood models. Ann. Stat. **36**(4), 1509 (2008)
8. Polson, N.G., Scott, S.L., et al.: Data augmentation for support vector machines. Bayesian Anal. **6**(1), 1–23 (2011)
9. Park, T., Casella, G.: The Bayesian Lasso. J. Am. Stat. Assoc. **103**(482), 681–686 (2008)
10. Li, Q., Lin, N., et al.: The Bayesian elastic net. Bayesian Anal. **5**(1), 151–170 (2010)
11. Polson, N.G., Scott, J.G., Windle, J.: The Bayesian bridge. J. Roy. Stat. Soc. Ser. B (Stat. Methodol.) **76**(4), 713–733 (2014)
12. Liang, Y., Liu, C., Luan, X.-Z., Leung, K.-S., Chan, T.-M., Xu, Z.B., Zhang, H.: Sparse logistic regression with a $L^{1/2}$ penalty for gene selection in cancer classification. BMC Bioinform. **14**, 198 (2013)
13. Xu, Z., Chang, X., Xu, F., Zhang, H.: $L^{1/2}$ regularization: a thresholding representation theory and a fast solver. IEEE Trans. Neural Netw. Learn. Syst. **23**(7), 1013–1027 (2012)
14. Cawley, G.C., Talbot, N.L.C.: Gene selection in cancer classification using sparse logistic regression with Bayesian regularization. Bioinformatics **22**(19), 2348–2355 (2006)
15. Bae, K., Mallick, B.K.: Gene selection using a two-level hierarchical Bayesian model. Bioinformatics **20**(18), 3423–3430 (2004)

16. Geman, S., Geman, D.: Stochastic relaxation, gibbs distributions, and the Bayesian restoration of images. IEEE Trans. Pattern Anal. Mach. Intell. **6**(6), 721–741 (1984)
17. Albert, J.H., Chib, S.: Bayesian analysis of binary and polychotomous response data. J. Am. Stat. Assoc. **88**(422), 669–679 (1993)
18. Rodriguez-Yam, G., Davis, R.A., Scharf, L.L.: Efficient gibbs sampling of truncated multivariate normal with application to constrained linear regression. Unpublished manuscript (2004)
19. Chang, S.-M., Chen, R.-B., Chi, Y.: Bayesian variable selections for probit models with componentwise gibbs samplers. Commun. Stat. Simul. Comput. **45**(8), 2752–2766 (2016)
20. West, M.: On scale mixtures of normal distributions. Biometrika **74**(3), 646–648 (1987)
21. Lichman, M.: UCI machine learning repository (2013)

Laplace Exponential Family PCA

Fangqi Li$^{(\boxtimes)}$ and Xudie Ren

School of Electronic Information and Electrical Engineering,
Shanghai Jiao Tong University, 800 Dongchuan Road, Min Hang District,
Shanghai 200240, People's Republic of China
solour_lfq@sjtu.edu.cn

Abstract. Considering numerous types of data, this paper discusses application of PCA to exponential family distributions. Reviewing the probabilistic basis of PCA, we propose a model using Laplace approximation, which was widely used in classification context, Laplace exponential family PCA (LePCA). The proposed approach provides a more probabilistic solution compared with numerous models before. Standard EM algorithm can be applied to this model, while only a degraded form of EM is applicable on previous exponential PCA models. LePCA absorbs probabilistic PCA, as well as the traditional PCA as its specialization by taking the Gaussian assumption for granted.

Keywords: Machine learning · Laplace approximation
Principal component analysis · EM algorithm

1 Introduction

Principal component analysis (PCA) [1] is a widely used algorithm for dimension-reduction. Intuitively, it looks for a low-dimensional subspace and represents the original data by their projection to a set of orthonormal basis in it. The projected data have a large variance, i.e. a relatively large amount of information is saved [2]. From a probabilistic perspective, we assume that there is a low-dimensional latent variable **z** subject to identity Gaussian. We further assume that one data term **x** is sampled from a linear Gaussian relationship. When the coefficient for covariance matrix is set to zero, the model degrades into a deterministic algorithm equal to the traditional PCA, the general latent variable approach is known as probabilistic principal component analysis (PPCA) [3]. PPCA is a special form of factor analysis, with traditional PCA as its specialization. It is the Gaussian linear hypothesis that gives rise to measurement of information by variance. But this hypothesis that data items are subject to Gaussian distribution is sometimes inappropriate. One generalization is replacing the Gaussian distribution with the more general exponential family distributions [4], which is known as exponential PCA (EPCA) [5].

Like all probabilistic models, PPCA needs to cope with the over-fitting caused by maximum likelihood estimation. It is naturally to introduce a-posteriori estimation of parameters on PPCA. Similarly, EPCA can also perform MAP as well. However, the prior distribution of the general exponential distribution is not normally distributed, which deprives the posterior estimation of a closed form. Hence the optimization of parameters for EPCA is usually done through an iterative approach. It is common to relate this procedure of optimization to expectation-maximum (EM) algorithm as long as they both aim at maximizing the likelihood for a model with latent variables.

Another problem with traditional PCA as well as PPCA is the dimensionality of latent variables. The dimensionality to which data are reduced to generally needs to be given beforehand, however, it is more reasonable to learn this parameter from data. Works have been done to solve the problem by applying Automatic Relevance Determination (ARD) [6] procedure on PPCA as well as EPCA, but in the second case, the conjugate nature of parameter subspace is lost.

In this paper, we concentrate on giving a closed form for PCA extended to exponential family distributions by making use of Laplace approximation as in [7], Chap. 8.4.1. Laplace exponential family PCA (LePCA). We will show how parameters can be estimated by using straightforward EM. The relationship between our method and other ones for parameter estimation in exponential PCA before is showed as well. We further illustrate that it is natural and foresightful to assume a Gaussian hypothesis for PPCA instead of other emission distributions. It is also showed that using Laplace approximation enables this model to generalize factor analysis (FA) [8] by introducing a variable Hessian.

The rest of paper is organized as follows: Sect. 2 reviews the generalization process of PCA. Section 3 introduces the proposed Laplace exponential family PCA. Experiment and illustration example are given in Sect. 4. Section 5 concludes the paper.

2 Related Work

2.1 Probabilistic PCA

The probabilistic perspective of principal component analysis was introduced in [3]. In PPCA, there is a latent variable z with fewer dimensionality $q < d$. Where d denotes the dimensionality of data x. After transformation, it forms the mean for the observed variables. The probabilistic graphical model for PPCA, together with its Bayesian version [9] is given in Fig. 1, N denotes the size of training set.

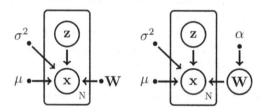

Fig. 1. PGM for PPCA and Bayesian PPCA

For ordinary PPCA, we use point estimation on μ, σ^2 and $\mathbf{W}$, where we have assumed a linear Gaussian relationship:

$$p(\mathbf{z}) = N(\mathbf{z}|\mathbf{0}, \mathbf{I}_q), \ p(\mathbf{x}|\mathbf{z}) = N(\mathbf{Wz} + \mu, \sigma^2 \mathbf{I}_d)$$

$$p(\mathbf{x}) = \int p(\mathbf{x}|\mathbf{z})p(\mathbf{z})d\mathbf{z} = N(\mu, \mathbf{WW}^T + \sigma^2 \mathbf{I}_d)$$

The ARD prior takes the form:

$$p(\mathbf{W}|\alpha) = \prod_{i=1}^{q} (\frac{\alpha_i}{2\pi})^{\frac{d}{2}} \exp\{-\frac{\alpha_i}{2} \mathbf{w}_i^T \mathbf{w}_i\}$$

Thanks to the linear Gaussian relationship, the parameter's update has a closed form given in [3].

PPCA provides a probabilistic approach of dimension reduction from d to q by identifying the most appropriate linear transformation introduced by loading matrix $\mathbf{W}$. Original data $\mathbf{x}$ is encoded by its correlated latent variable, which subjects to $p(\mathbf{z}|\mathbf{x})$. This posterior distribution has a closed form as long as the correlation is Gaussian.

One problem of PPCA is that q needs to be determined beforehand. In [9], an algorithm that can learn an appropriate q given data was proposed by introducing an ARD prior on loading matrix $\mathbf{W}$. A sparse solution can be obtained by learning hyper-parameter α.

2.2   Generalization of PCA to Exponential Family

The distribution of $\mathbf{x}$ introduced by PPCA is a Gaussian. However, there are numerous data with a non-Gaussian distribution. In order to applying the technique of PCA onto other data, a generalization of PCA to exponential family distributions was proposed [5]. It uses the same idea of looking for a subspace that reduces the dimension while preserves information to its best.

Under a Gaussian context, fitting the likelihood is equal to minimizing a loss function in a quadratic form, hence a least-square target. For general exponential family, the loss function is given by the Bregman distance. Thanks to the property of exponential family, this optimization task is convex with respect to two independent parameters. To estimate parameters, [5] gave an iterative solution during which $\mathbf{W}$ and $\mathbf{z}$ are optimized alternately. EPCA' s posterior form was studied by [10], where a MAP estimation for parameters are drawn through hybrid Monte Carlo given prior distribution for $\mathbf{W}$ and $\mathbf{z}$. The graphical model for EPCA is shown in Fig. 2.

And we have:

$$p(\mathbf{x}|\mathbf{z}, \mathbf{W}) = p_e(\mathbf{x}|\mathbf{Wz})$$

EPCA is different from PPCA in their ways of identifying conditional probability $p(\mathbf{x}|\mathbf{z}, \mathbf{W})$. PPCA uses a Gaussian distribution while EPCA uses a context-related exponential family distribution denoted by p_e.

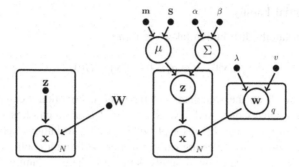

Fig. 2. PGM for EPCA and its Bayesian version, BXPCA

2.3 Simple Exponential Family PCA (SePCA)

As PPCA, EPCA faces the problem of finding the best dimensionality for latent variable, namely the number of principal components. One of the solutions has been proposed in [11], with a graphical model as Fig. 3.

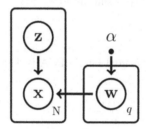

Fig. 3. PGM for SePCA

By using an ARD prior on W as in Bayesian PPCA, dimensionality is driven to its optimum by learning hyper-parameter α. It is noticeable that the PGM for SePCA is the same to that of Bayesian PPCA as shown in Fig. 1, after removing μ and σ^2.

On parameter estimation, SePCA uses an iterative method similar to that used in EPCA. It was interpreted in [11] that this iterative optimization on two set of parameters alternatively can be seen as a degenerated EM.

3 Proposed Model

In this section, we begin with reviewing some basic properties of exponential family distributions, followed by introducing the proposed model. By sticking to a PGM language, we list the theoretical improvements and some comparisons. On elaborating the method for parameter estimation, we elucidate the feature of proposed model further.

3.1   Exponential Family

An exponential family distribution takes the form:

$$p(\mathbf{x}|\boldsymbol{\theta}) = \exp\{\mathbf{x}^T\boldsymbol{\theta} + h(\mathbf{x}) - G(\boldsymbol{\theta})\}$$

Where we work on the sufficient statistics. The distribution's cumulant function $G(\boldsymbol{\theta})$ is always a convex one. And $h(\mathbf{x})$ is a scaling factor which often set to one. It is straightforward to recognize that a lot of widely-used distributions belongs to exponential family. Typical examples are Gaussian, Bernoulli, multinoulli, etc.

One significance for exponential family is that it is the only family with conjugate prior distributions, which can simplify the estimation as well as increase the interpretability.

The conjugate prior for an exponential family distribution takes the form:

$$p(\boldsymbol{\theta}|v,\lambda) = \frac{1}{Z(\lambda,v)}\exp\{\boldsymbol{\theta}^T\lambda - v \cdot G(\boldsymbol{\theta})\}$$

From the perspective of observation, this prior is tantamount to taking v prior data terms with their sufficient statistics sum up to λ.

3.2   Graphical Model

The structure of Laplace exponential family PCA is defined by Fig. 4.

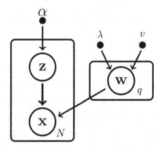

Fig. 4. PGM for LePCA

The conjugation on basis for the subspace is preserved by introducing conjugate prior on $\mathbf{W}$. However, as long as the dimensionality needs to be learned, an ARD prior is applied on $\mathbf{Z}$, denoting the collection of all latent variables. Unlike SePCA, the conjugate relationship saves more space for tuning parameters in a more intuitive and interpretive way. As it is pointed out in [12], the difference in status between $\mathbf{W}$ and $\mathbf{Z}$ is imposed. Traditional BPPCA obtains a sparse result by treating $\mathbf{Z}$ as latent variables and integrating out $\mathbf{W}$. It is also applicable to integrate out $\mathbf{Z}$, and yield a closed form of solution at length. In exponential family context, integrating out a ARD prior $\mathbf{Z}$ preserves sparsity and conjugation at the same time.

Formally, the prior for basis takes the form:

$$p(\mathbf{w}|v, \lambda) \propto \exp\{\mathbf{w}^T\lambda - v \cdot G(\mathbf{w})\}$$

Where $\mathbf{w}$ is one column of loading matrix $\mathbf{W}$. As for latent variables:

$$p(\mathbf{z}) = N((\mathbf{0}, 1)^T, \Sigma = \Lambda^{-1} = \text{diag}\{\alpha_1^{-1}, \alpha_2^{-1}, \ldots, \alpha_q^{-1}\})$$

By setting $\alpha_q^{-1} \approx 0$, the biased term is taken off, so the emission distribution is:

$$p(\mathbf{x}|\mathbf{W}, \mathbf{z}) = \exp\{\mathbf{x}^T\mathbf{W}\mathbf{z} - G(\mathbf{W}\mathbf{z})\}$$

Since this is a probabilistic model with latent variables, we naturally resort to EM on parameter estimation. In the E-step, estimate the conditional distribution on $\mathbf{z}$:

$$p(\mathbf{z}|\mathbf{W}, \mathbf{x}) = \frac{p(\mathbf{x}|\mathbf{W}, \mathbf{z})p(\mathbf{z})}{\int p(\mathbf{x}, \mathbf{z}'|\mathbf{W})d'} = \frac{\exp\{-\frac{1}{2}\mathbf{z}^T\Lambda\mathbf{z} + \mathbf{x}^T\mathbf{W}\mathbf{z} - G(\mathbf{W}\mathbf{z})\}}{c \cdot \int \exp\{-\frac{1}{2}\mathbf{z}'^T\Lambda\mathbf{z}' + \mathbf{x}^T\mathbf{W}\mathbf{z}' - G(\mathbf{W}\mathbf{z}')\}d\mathbf{z}'}$$

The difficulty within this approach is that on the denominator, latent variable $\mathbf{z}'$ can not be integrated out conveniently due to the variety in form of cumulant function G and the fraction takes an irregular form. However, it is still possible to estimate parameters with proper approximation. In previous works as EPCA or SePCA, an iterative and alternative approach is used. We are now to give another solution using approximation and indicate that the previous approaches are variants of the one we proposed.

3.3 On Parameter Estimation: Laplace Approximation

Interpretation between iterative optimization and EM within exponential PCA is given in [11]. It is also straightforward to elaborate the similarity by using point estimation to optimize joint likelihood instead of calculating the complete posterior distribution. Thus E-step is reduced to:

$$\mathbf{z}^{(t)} = \arg \min_{\mathbf{z}}\{-\log p(\mathbf{x}|\mathbf{W}^{(t-1)}, \mathbf{z}) - \log p(\mathbf{z})\}$$

Where it is assumed that the posterior distribution is peaked at its maximum. Using the literature of Bregman distance [13]:

$$B_F(\mathbf{p}||\mathbf{q}) = F(\mathbf{p}) - F(\mathbf{q}) - \nabla F(\mathbf{q})^T(\mathbf{p} - \mathbf{q})$$

The optimized target can be rewrite into:

$$B_F(\mathbf{x}||g(\mathbf{W}^{(t-1)}\mathbf{z})) + \frac{1}{2}\mathbf{z}^T\Lambda\mathbf{z}$$

Where we have $g = G'$ and $F = \int g^{-1}$. Once the context-related distribution is determined, these functions can be calculated analytically. This optimization of $\mathbf{z}$ is reduced into a standard convex optimization.

To estimate $\mathbf{W}$, applying the M-step:

$$\mathbf{W}^{(t)} = \arg\max_{\mathbf{W}}\{\int p(\mathbf{Z}|\mathbf{X}, \mathbf{W}^{(t-1)}) \log p(\mathbf{X}, \mathbf{Z}|\mathbf{W})d\mathbf{Z}\}$$

$$= \arg\max_{\mathbf{W}}\{\log p(\mathbf{X}, \mathbf{Z}^{(t)}|\mathbf{W})\}$$

$$= \arg\max_{\mathbf{W}}\{\log p(\mathbf{X}|\mathbf{Z}^{(t)}, \mathbf{W})\}$$

The cancellation of integral is due to the fact that the posterior distribution has been approximated as a Dirac function. The degenerated M-step is again a convex optimization.

To use the full EM algorithm, it is feasible to use Laplace approximation, which was firstly used in classification context to handle non-Gaussian likelihood, hence fit our issue. This equals using the second-order Taylor expansion of $f(\mathbf{z}) = G(\mathbf{Wz})$ to approximate the exact culumant function:

$$G(\mathbf{Wz}) = f(\mathbf{z})$$

$$= f(\mathbf{z}_0 + (\mathbf{z} - \mathbf{z}_0))$$

$$\approx f(\mathbf{z}_0) + \nabla f(\mathbf{z}_0)^{\mathrm{T}}(\mathbf{z} - \mathbf{z}_0) + \frac{1}{2}(\mathbf{z} - \mathbf{z}_0)^{\mathrm{T}}\nabla^2 f(\mathbf{z}_0)(\mathbf{z} - \mathbf{z}_0)$$

$$= G(\mathbf{Wz}_0) + \nabla G(\mathbf{Wz}_0)^{\mathrm{T}}\mathbf{W}(\mathbf{z} - \mathbf{z}_0) + \frac{1}{2}(\mathbf{z} - \mathbf{z}_0)^{\mathrm{T}}\mathbf{W}^{\mathrm{T}}\nabla^2 G(\mathbf{Wz}_0)\mathbf{W}(\mathbf{z} - \mathbf{z}_0)$$

$$= \frac{1}{2}\mathbf{z}^{\mathrm{T}}\mathbf{W}^{\mathrm{T}}\nabla^2 G(\mathbf{Wz}_0)\mathbf{Wz} + (\nabla G(\mathbf{Wz}_0)^{\mathrm{T}}\mathbf{W} - \mathbf{z}_0^{\mathrm{T}}\mathbf{W}^{\mathrm{T}}\nabla^2 G(\mathbf{Wz}_0)\mathbf{W})\mathbf{z} + c$$

$$= \frac{1}{2}\mathbf{z}^{\mathrm{T}}\mathbf{W}^{\mathrm{T}}\mathbf{H}\mathbf{Wz} + \boldsymbol{\alpha}^{\mathrm{T}}\mathbf{z} + c$$

Where we denote the Hessian matrix for G at $\mathbf{z}_0$ by $\mathbf{H}$ and absorb constant term into C and coefficient of linear term into $\boldsymbol{\alpha}$.

Laplace approximation addresses the marginal likelihood which is independence of $\mathbf{z}$, and we can now approximate the full posterior of $\mathbf{z}^{(t)}$ as $N(\boldsymbol{\mu}^{(t)}, \boldsymbol{\Sigma}^{(t)})$, where the Hessian is computed as $\mathbf{z}_0 = \mathbf{z}^{(t-1)}$ for better approximation:

$$\boldsymbol{\Sigma}^{(t)} = (\boldsymbol{\Lambda} + \mathbf{W}^{(t-1)\mathrm{T}}\mathbf{H}^{(t-1)}\mathbf{W}^{(t-1)})^{-1} \tag{1}$$

$$\boldsymbol{\mu}^{(t)} = \boldsymbol{\Sigma}^{(t)}(\mathbf{W}^{(t-1)\mathrm{T}}\mathbf{x} + \boldsymbol{\alpha}^{(t-1)}) \tag{2}$$

In M-step, we optimize $\mathbf{W}$ with respect to the expectation of log-likelihood under posterior distribution obtained before can be processed using Laplace approximation again:

$$\int \mathrm{p}(\mathbf{z}|\mathbf{W}^{(t-1)}, \mathbf{x})\{-\frac{1}{2}\mathbf{z}^{\mathrm{T}}\mathbf{\Lambda}\mathbf{z} + \mathbf{x}^{\mathrm{T}}\mathbf{W}\mathbf{z} - G(\mathbf{W}\mathbf{z})\}\mathrm{d}\mathbf{z}$$

$$\approx \int \mathrm{p}(\mathbf{z}|\mathbf{W}^{(t-1)}, \mathbf{x})\{-\frac{1}{2}\mathbf{z}^{\mathrm{T}}(\mathbf{\Lambda} + \mathbf{W}^{\mathrm{T}}\mathbf{H}^{(t-1)}\mathbf{W})\mathbf{z} + (\mathbf{x}^{\mathrm{T}}\mathbf{W} + \boldsymbol{\alpha}^{(t-1)\mathrm{T}})\mathbf{z} - C\}\mathrm{d}\mathbf{z}$$

Which can be solved analytically since it only composed of the expectation of first and second order moment of a Gaussian variable.

If the Hessian is a constant, calculation can be reduced sharply. A trivial case is Gaussian distribution with cumulant function $G(\theta) = \theta^2/2$, hence yields a constant Hessian. For general exponential family, the Hessian needs to be computed for times during iterations. Addressing the similarity between our model and FA, FA assumes $\mathrm{p}(\mathbf{x}|\mathbf{W}, \mathbf{z})$ to be normally distributed $N(\mathbf{W}\mathbf{z}, \mathbf{\Psi})$. Applying Laplace approximation on the emission distribution results in a variable $\mathbf{\Psi}^{(t)}$ (notice the similarity between Eq. (1) and the E-step for FA, and sufficient statistics we used through does not equal the one for Gaussian in factor analysis). The approximation should only be used during estimation so the overall distribution of data remains general (Table 1).

LePCA's parameter estimation can be concluded as follow:

Table 1. EM algorithm for LePCA

- **Initialization:** give $\mathbf{z}^{(0)}$ and $\mathbf{W}^{(0)}$ according to prior distribution; $t = 0$;
- **Iteration:** while not converge:
  - Compute $\mathbf{H}^{(t)} = \nabla^2 G(\mathbf{W}\mathbf{z})|_{\mathbf{z}^{(t)}}$;
  - E-step: Compute $\mathrm{p}(\mathbf{z}^{(t+1)})$ w.r.t $\mathbf{H}^{(t)}$, $\mathbf{W}^{(t)}$ using equation (1) and (2), then calculate expectation for $\mathbf{z}$ and $\mathbf{z}\mathbf{z}^{\mathrm{T}}$;
  - M-step: Optimize $\mathbf{W}^{(t+1)}$ w.r.t the expectations calculated before; similar to M-step for FA;
  - $t = t + 1$;

3.4 On Parameter Estimation: Evidence Framework

To learn the dimension of principal components q, we take $\mathbf{W}$ as granted and integrate out $\mathbf{z}$ to learn the hyper-parameter $\boldsymbol{\alpha}$. In PPCA, $\mathbf{z}$ can be integrated out easily using Gaussian properties. The difficulty within exponential family model is the same as the one arises in E-step. Hence it can be handled by using Laplace approximation as well. For latent variable $\mathbf{z}$, we collect all ith component in it from N data into $\mathbf{Z}_i$, using the evidence framework as linear regression, we have the re-estimation formula as in [9] or [14], Chap. 3.5:

$$\alpha_i := \frac{d - \alpha_i \text{tr}(\mathbf{H}^{-1})}{\mathbf{Z}_i^{\text{T}} \mathbf{Z}_i}$$

Where $\mathbf{H}$ is the Hessian for $\log p(\mathbf{Z}_i | \mathbf{X}, \mathbf{W})$ with respect to $\mathbf{Z}_i$, which consists of constant terms thanks to the Laplace approximation. In practice, it is possible to ignore the second term in numerator to simplify the re-estimation further. By evidence framework, the dimension with less support from the data will increase related component in α quickly. This ends in an decrease in the prior variance for that dimension. Since we have the prior to be a zero-mean Gaussian, this is tantamount to drive all components in that dimension to zero.

4   Experiment and Illustration

On synthetic data set used in [10] we test SePCA. The data set consists of $N = 120$ data that subject to Bernoulli distribution with $d = 16$. The "genuine" data is divided into three groups with consistent content. Each component in each data item is further flipped with probability 0.1 to form the training set. The data set is illustrated in Fig. 5, where vertical and horizontal axes represent dimension and data respectively.

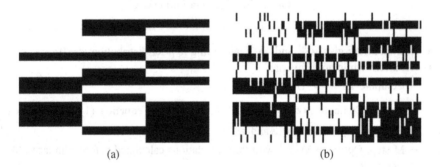

(a)   (b)

Fig. 5. Synthetic data set, the prototype **(a)** and the one added with noise **(b)**

Bernoulli distribution belongs to the exponential family. As for this data set, its cumlant function takes a concise form:

$$G(\boldsymbol{\theta}) = - \sum_{i=1}^{d} \log(1 - \sigma(\theta_i))$$

Where σ denotes sigmoid function. With this fact the gradient and Hessian (which is diagonal) can be easily computed.

Three typical phases in iteration are selected and demonstrated in Fig. 6:

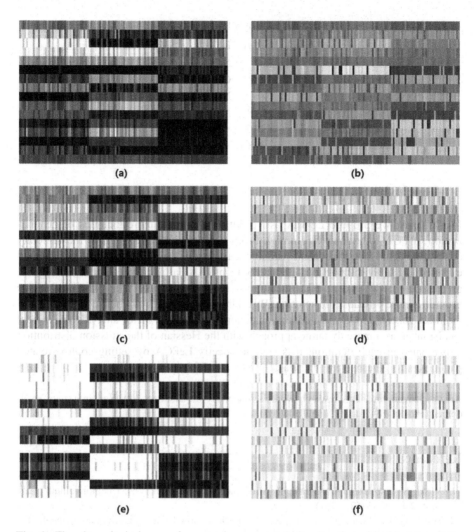

Fig. 6. The chronological mean for natural parameter during iterations (**a**), (**c**), (**e**), and the corresponding difference between the mean and the actual data set (**b**), (**d**), (**f**).

To quantify the effect, we calculate the log-likelihood log $p(\mathbf{X}|\mathbf{W}, \mathbf{Z})$ at the final stage, and reach a log-likelihood on training set and genuine data on -302.37 and -266.95 respectively. Together with features in model structure, we compare LePCA with other models thoroughly (Table 2):

It is noticeable that both SePCA and LePCA yield to relatively better effect on both training set and genuine data. EPCA suffers from over-fitting by using MLE. BXPCA handles with Bayesian approach but the number for free parameters are still too large (due to the unidentifiability of models). SePCA results in a better performance. However, its probabilistic characteristic is reduced by using MAP instead of a full EM. Allowing conjugate prior, LePCA increase the interpretability significantly. And it is also possible to apply a sequential learning process based on conjugation.

Table 2. Comparison between models (figures for some previous models are tested in [11])

	BPPCA	EPCA	BXPCA	SePCA	LePCA
Prior of $\mathbf{z}$	$N(\mathbf{0}, \mathbf{I})$	Flat	$N(\boldsymbol{\mu}, \boldsymbol{\Sigma})$	$N(\mathbf{0}, \mathbf{I})$	ARD
Prior of $\mathbf{w}$	ARD	Flat	Conjugate	ARD	Conjugate
Canceled	$\mathbf{w}$	No	No	$\mathbf{w}$	$\mathbf{z}$
Full EM	Applicable	MAP	MAP	MAP	Applicable
Train LL		−13.3	−202.7	−338.3	−302.4
Prototype LL		−808.1	−517.4	−231.0	−267.0

5 Conclusion

In this paper we propose LePCA as a solution to handle the difficulties of generalization of PCA to exponential family. From a probabilistic perspective, the difficulty arises from the fact that latent variable can hardly be integrated out for general exponential family distributions. Instead of using MAP as an approximation during EM as previous models, we use Laplace approximation to cancel latent variables, hence provide a more probabilistic solution. By addressing the similarity between LePCA and the most general model with continuous latent variables as factor analysis, our model can be taken as a more dynamic approach with the Hessian of the emission distribution re-estimated at each iteration. It is easy to specify LePCA by setting context-related parameters to give rise to other basic models as PPCA.

Acknowledgement. This research work is funded by the National Key Research and Development Project of China (2016YFB0801003).

References

1. Wold, S., Esbensen, K., Geladi, P.: Principal component analysis. Chemom. Intell. Lab. Syst. **2**(1–3), 37–52 (1987)
2. Hotelling, H.: Analysis of a complex of statistical variables into principal components. Br. J. Educ. Psychol. **24**(6), 417–520 (1933)
3. Tipping, M.E., Bishop, C.M.: Probabilistic principal component analysis. J. Roy. Stat. Soc. **61**(3), 611–622 (1999)
4. Guo, Y.: Supervised exponential family principal component analysis via convex optimization. In: International Conference on Neural Information Processing Systems Curran Associates Inc., pp. 569–576 (2008)
5. Collins, M., Dasgupta, S., Schapire, R.E.: A generalization of principal component analysis to the exponential family. In: International Conference on Neural Information Processing Systems: Natural and Synthetic, pp. 617–624. MIT Press (2001)
6. Mackay, D.J.C.: Probable networks and plausible predictions—a review of practical Bayesian methods for supervised neural networks. Netw. Comput. Neural Syst. **6**(3), 469–505 (1995)
7. Robert, C.: Machine Learning, A Probabilistic Perspective. MIT Press, Cambridge (2012)
8. Akaike, H.: Factor analysis and AIC. Psychometrika **52**(3), 317–332 (1987)

9. Bishop, C.M.: Bayesian PCA. In: Advances in Neural Information Processing Systems DBLP, pp. 382–388 (1999)
10. Mohamed, S., Heller, K.A., Ghahramani, Z.: Bayesian exponential family PCA. In: Conference on Neural Information Processing Systems, Vancouver, British Columbia, Canada, December, pp. 1089–1096. DBLP (2008)
11. Li, J., Tao, D.: Simple exponential family PCA. IEEE Trans. Neural Netw. Learn. Syst. **24** (3), 485–497 (2013)
12. Nakajima, S., Sugiyama, M., Babacan, D.: On Bayesian PCA: automatic dimensionality selection and analytic solution. In: International Conference on Machine Learning Omnipress, pp. 497–504 (2011)
13. Miller, F.P., et al.: Bregman Divergence. Alphascript Publishing, Beau Bassin (2010)
14. Bishop, C.M.: Pattern Recognition and Machine Learning. Information Science and Statistics. Springer-Verlag New York Inc., Secaucus (2006)

Active Framework by Sparsity Exploitation for Constructing a Training Set

Maozu Guo[1,2], Weining Wu[3], and Yang Liu[4(✉)]

[1] School of Electrical and Information Engineering, Beijing University of Civil Engineering and Architecture, Beijing 100044, People's Republic of China
guomaozu@bucea.edu.cn
[2] Beijing Key Laboratory for Research on Intelligent Processing Method of Building Big Data, Beijing University of Civil Engineering and Architecture, Beijing 100044, People's Republic of China
[3] College of Computer Science and Technology, Harbin Engineering University, Harbin 150001, People's Republic of China
wuweining@hrbeu.edu.cn
[4] School of Computer Science and Technology, Harbin Institute of Technology, Harbin 150001, People's Republic of China
yliu76@hit.edu.cn

Abstract. This paper addresses the problem of actively constructing a training set for the linear model with sparse structure. This problem usually occurs in the scenario that no nonlinear mappings give similar performance for large-scale learning data, but it has to train a linear model quickly. In this paper, an active framework is proposed to reduce the time expense further in constructing the training set. The training examples are iteratively selected by matching partial components and their weights given by the classifier in pairs, in order to exploit model's sparsity to precisely separate out more informative examples from others in a short time. The proposed framework is evaluated on a group of classification tasks, including the texts and images.

Keywords: Classification · Sparse models · Active framework

1 Introduction

Recently, linear classifiers with sparse structure have played a significant role in many large-scale learning tasks [1]. By using a loss function to evaluate the difference between the predictions and actual labels, the linear classifier can be obtained efficiently to give comparable performance for large data. Usually, the parameter vectors of linear models are optimized via L1-normalization with the aim of cutting costs in the training procedure. Thus, the classification model with sparse structure can provide the considerable accuracy in a relatively short time.

In the other parallel line of research, actively constructing a training set by optimized experimental design has been considered as an effective way of minimizing total labeling costs in obtaining a classifier. Then, it has been widely applied in various fields and achieves favorable performance [2]. Among all active strategies, uncertain-based sampling strategy is the most popular one of all, in which the training set is constructed by iteratively selecting these examples that the current model is confused. It has a strong theoretical basis [3] that the current hyperplane space is halved gradually. However, when the uncertain-based sampling strategy is applied in the large-scale learning tasks, it is time-consuming to estimate the uncertainty of individual examples, because it needs to compute the prediction of the classifier for every example. That is, these confused examples should be chosen for training the classifier within tolerated time.

Inspired by the linear model with sparse structure, we attempt to actively construct a training set by using partial components to estimate the uncertainty of individual examples. The objective is to minimize the time expense in selecting these confused examples. We deploy a sparse model on the pool of unlabeled data. And then, we estimate the uncertainty of all examples by matching partial components and their weights given by the current classifier in pairs. The most confused example of all is selected by gradually tightening the bounds of the uncertainty of unlabeled examples. In every round, the selected example is labeled and added into the training set.

2 Related Work

Until now, the literature of active learning on both theories and applications has been vast [4]. Traditionally, training examples can be evaluated in many ways, e.g. variance reduction [5], density information [6], uncertainty minimization [7] and so on. Among all existing strategies, uncertain-based sampling criterion, which the example locating closest to hyperplane boundary is selected for future rounds [3], has been considered as the most cost-effective one of all and widely used in most applications. When a linear model is used, the uncertainty of unlabeled examples can be achieved by their absolute values of outputs or scores provided by the model. Based on this core idea, uncertain-based criterion has been applied to other models recently. For example, active convolution unit [8], recurrent neural network [9] or other deep generative models [10].

Note that individual examples in the unlabeled pool are usually evaluated by their locations relative to the hyperplane boundary which is defined according to the scores of the classifier in the current round [2]. To our knowledge, this approach is the first to explore the benefits of sparse structure in order to help to reduce time expense in computing classifier's outputs. In this approach, some components are matched with their weights given by the model in pairs, and then unlabeled examples are evaluated by tightening the bounds of classifier's outputs. As a result, only the survived samples get chance to be selected. To show its details, we list the proposed framework and other popular ways of constructing the training set in Fig. 1.

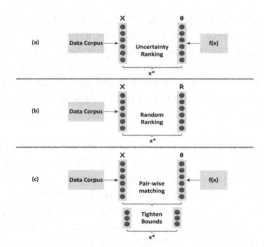

Fig. 1. Three different ways of constructing the training set for classification tasks. (a) Actively constructing the training set via uncertain-based sampling strategy; (b) Passively constructing the training set via random sampling; (c) The proposed active framework for constructing the training set. The x^* denotes the selected example for training the model $f(x)$ with the parameter vector θ. The X represents the set of unlabeled examples via feature extraction.

3   Active Framework

Suppose that we have a joint distribution $p(x, y)$ on the training set $T = \{(x_j, y_j)\}_{j=1}^{n}$ where the random variable $x \in X$ and its response variable $y \in Y$. A classification function $f_\theta : X \rightarrow Y$ is defined where y can be estimated by the $p(y \mid x; \theta)$. The objective is to iteratively select and label a group of examples from a pool of unlabeled examples $X_U = \{x_j\}_{j=n+1}^{n+m}, m >> n$. The most uncertain example of all [4] is selected as

$$x^* = \arg\max_{x \in X_U} 1 - p(y \mid x; \theta) \tag{1}$$

Usually, the parameter vector θ is obtained as an optimized solution of the classification function, and then $p(y \mid x; \theta)$ can be obtained by a sigmoid function according to the outputs of the classifier $\theta \cdot x + b$. In this work, we train a logistic regression model as the classifier while the proposed approach is not limited to linear models.

3.1   Uncertainty Estimation

When all examples are normalized, i.e. $b = 0$, the classification outputs only depend on θ and x. Suppose that there are a set of index items $I = \{i\}_{i=1}^{d}$ for the example x_j and the parameter vector θ, then we have

$$x_j = \left\{ x_j^i \right\}_{i=1}^d, \qquad \boldsymbol{\theta} = \left\{ \theta^i \right\}_{i=1}^d \tag{2}$$

The uncertainty of all examples in the unlabeled pool can be represented as $D = \left\{ D_j \right\}_{j=n+1}^{n+m}$ where $D_j = \left| \sum_{i=1}^d x_j^i \cdot \theta^i \right|$. Obviously, D is a set of m elements, in which every element is a sum of d pair-wise distances.

Proposition 1. D is a bounded set when x_j includes a group of non-negative components. The upper bound and lower bound of D_j can be obtained as

$$\begin{aligned}
D_j^{\min} &= \sum_{i=1}^d \theta^i \cdot x_j^i, \quad \forall \theta^i < 0 \\
D_j^{\max} &= \sum_{i=1}^d \theta^i \cdot x_j^i, \quad \forall \theta^i > 0
\end{aligned} \tag{3}$$

It is safe to assume that $x_j^i \geq 0, i = 1, 2, \dots, d$ in most learning tasks, such as text or image categorization, where x_j can be represented in forms of bag-of-words [8, 9], index vector [10] or other features. Then, the goal of uncertain-based sampling strategy is to find

$$D^* = \arg\min_{D_j \in D} D_j \quad s.t. \ D_j^{\min} \leq D_j \leq D_j^{\max}, \ 1 \leq j \leq m \tag{4}$$

3.2 Tightening Bounds

In every round of active learning, the uncertainty of individual examples are estimated by sequentially scanning θ_i. At beginning, a classification model is obtained on the current training set, then the upper bound $U_j^0 = D_j^{\max}$ and the lower bound $L_j^0 = D_j^{\min}$ for every example $x_j \in X_U$. The uncertainty estimation begins with two bounded sets $U^0 = \left\{ U_j^0 \right\}_{j=n+1}^{n+m}$ and $L^0 = \left\{ L_j^0 \right\}_{j=n+1}^{n+m}$. Similarly, let $U^i = \left\{ U_j^i \right\}_{j=n+1}^{n+m}$ and $L^i = \left\{ L_j^i \right\}_{j=n+1}^{n+m}$ be the bounds at the i_{th} position.

Proposition 2. For every x_j^i of the x_j, an identification function $I\left(x_j^i \right)$ is defined as follows:

$$\begin{aligned}
I\left(x_j^i \right) &= 0, \ x_j^i = 0 \\
I\left(x_j^i \right) &= 1, \ x_j^i > 0
\end{aligned} \tag{5}$$

Here, beginning with L_j^0 and U_j^0, the D_j in Eq. 4 can be obtained greedily by using the value θ_i at the position i to update the upper bound U_j^i and the lower bound L_j^i. In the sequential process, given the bounds U^{i-1} and L^{i-1} at the $i-1$ iteration and the i_{th} item of the current parameter vector θ, the bounds at the i_{th} iteration can be updated in Algorithm 1.

Algorithm 1: $Update\left(L^{i-1}, U^{i-1}, \theta^i\right)$

1. **Input:** $U^{i-1}, L^{i-1}, \theta^i$; **Output:** U^i, L^i
2.   If $I\left(x_j^i\right) \cdot \theta^i < 0$
3. $U^i \leftarrow U - Update\left(U^{i-1}, \theta^i, x_j^i\right)$
4.   Else
5. $L^i \leftarrow L - Update\left(L^{i-1}, \theta^i, x_j^i\right)$
6.   End
7. Return U^i, L^i

Given the i_{th} item of the parameter vector θ^i, both upper and lower bounds can be tightened according to θ^i in Algorithms 2 and 3.

Algorithm 2: $U - Update\left(U^{i-1}, \theta^i, x_j^i\right)$

1. **Input:** U^{i-1}, θ^i, x_j^i; **Output:** U^i
2. If $\theta^i < 0$
3. $U^i \leftarrow U^{i-1} - \theta^i \cdot x_j^i$
4. Elseif $\theta^i > 0$
5. $U^i \leftarrow U^{i-1} + \theta^i \cdot x_j^i$
6. End
7. Return U^i

Algorithm 3: $L - Update\left(L^{i-1}, \theta^i, x_j^i\right)$

1. **Input:** L^{i-1}, θ^i, x_j^i; **Output:** L^i
2. If $\theta^i < 0$
3. $L^i \leftarrow L^{i-1} + \theta^i \cdot x_j^i$
4. Elseif $\theta^i > 0$
5. $L^i \leftarrow L^{i-1} - \theta^i \cdot x_j^i$
6. End
7. Return L^i

Given lower and upper bounds returned by Algorithm 1, we formally define D^i in which every element is possible to become the minimum in Eq. 4 after being updated by θ^i. Firstly, we rank U_j^i in the descend order and obtain its index I^i. Then, given L^i and U^i, the valid uncertainty D^i can be obtained as

$$D^i = \left\{ U_j^i \mid \forall j \in I^{D^i}, L_j^i > \max_{j' \in I^i - I^{D^i}} U_{j'}^i \right\} \tag{6}$$

Where I^i denotes the index of current unlabeled pool and I^{D^i} is the index of top-ranking set D^i, $I^{D^i} \subseteq I^i$. D^i contains those unlabeled examples whose lower bounds are even higher than the remained examples in the unlabeled pool. It can be seen as the discriminative procedure by those scanned θ^i. When $\left|I^{D^i}\right|$ reaches the threshold, $D = D^i$ is used to find D^*. During the tightening steps, for individual examples, the values of lower bounds $L^i_j \in L^i$ increase while the values of upper bounds $U^i_j \in U^i$ decrease with respect to the number of observed θ^i. The number of elements contained in the valid distance D^i is reduced until it touches the stopping condition. At this point, we make $D = D^i$ be selected as the optimized solution D^* and the example x^* is selected according to D^*. Moreover, in order to speed up the tightening process, all elements $\{\theta^i\}_{i=1}^{d}$ can be sorted by their absolute values in a decending order. The main sampling process is given in Algorithm 4.

Algorithm 4: Actively constructing a training set for sparse models

---
Input: training set T, unlabeled set X_U

Output: the classification model

1. Obtain a classification model on the training set T
2. Sort the elements in the θ and obtain the indices set $I = \{i\}_{i=1}^{d}$
3. Compute lower bounds L^0 and upper bounds U^0
4. Given the threshold for I^{D^i}, t_{max}, $t = 0$, $i = 0$
5. **While** $t < t_{max}$, $t \leftarrow t+1$
6.   **While** $i < d$
7.   **While** $\left|D^i\right| > \left|I^{D^i}\right|$ **do** $i \leftarrow i+1$, $Update\left(L^{i-1}, U^{i-1}, \theta^i\right)$, Obtain D^i by Eq.6
8. $D = D^i$, Break
9.   Return x^* according to Eq. 4 and its label y^*
10.   Set $T' = T' \cup \{x^*, y^*\}$, $X_U^t = X_U^t - \{x^*\}$
11.   Retrain the classifier on the training set
12. Return the classifier

Runtime Analysis: We give some analysis of the runtime of the proposed active framework. In the worst case, there is $\Theta(m \cdot d \cdot T)$ to call the active sampling procedure. The keypoint of speeding up the sampling procedure is that we can choose partial components that should be exploited instead of utilizing all components in every active round. In real applications, the number of chosen components is very small, compared with the totality in the whole space, because the classification model with sparse structure only weights a small fraction of components in the whole space. The details of runtime are presented in experimental results in Sect. 4.

4 Experimental Results

We report the experimental results on two common benchmark datasets for classification tasks, i.e. 20-newsgroup dataset[1] for text classification and Caltech-256 image collection for object classification [11]. We compare three sampling criteria as follows: (1) Passive learning, where the next example-label pair is requested by random selection (in Fig. 1(b)); (2) Uncertain-based active learning, where the sampling criterion in Eq. 1 is used to find the most confused example of all (in Fig. 1(a)); (3) Proposed active framework (in Fig. 1(c)).

Text Classification: We firstly evaluate three methods on 15 classification tasks selected from 20-newsgroup dataset including 18774 examples. By using 1-vs-all framework to handle these multi-class problems, we obtain a group of binary classification tasks. For every classification task, all examples are obtained by extracting bag-of-words features from individual text files[1]. A random partition with 5 examples per class from the training partition is used as the initial dataset and the remained data is used as the unlabeled data pool. Then, the test partition is used as the test dataset for evaluate the performance of the classifier. All data is randomly divided for five times, and their average results of model's performance are reported for comparison. In every round, a logistic regression via L1-norm is implemented by Liblinear [1] with the default parameter settings and the threshold $|r| = 50$ is set for every task. We run every experiment for multiple times and report the experimental results.

Results: Firstly, we report the average precision of all methods in the tasks of text classification in Fig. 2(**Left**). In most tasks, when the number of sampling points increases, the average precision of the proposed method increases gradually and better results than other baselines can be obtained. It is worth noting that the average precision of the proposed method is lower than others at the early iterations, but it can overcome others after sampling enough points. The uncertain-based active framework seems even worse than the passive learning in parts, because the uncertainty is computed by using all components in a sparse space. However, the proposed method selects the most uncertain example of all by using partial dimensions in the whole space, which makes it be able to evaluate the information of individual examples more precisely. We also conclude similar results by showing the area-under-curve values of all methods in Fig. 2(**Right**).

Secondly, we compare the time expense (in Fig. 3(**Left**)) and example-to-boundary distances (in Fig. 3(**Right**)). In Fig. 3(**Left**), the passive learning needs least time of all for constructing the training set because of no needs to estimate the information of every example. Then, the proposed method needs less time to select informative examples than the uncertain-based sampling strategy which is considered as an efficient method to build the training set for the classification model. In Fig. 3(**Right**), it is worth noting that the training examples selected by the proposed framework are more close to the classification boundary than those selected by the passive learning. It is also the reason that the performance of the sparse model obtained by the proposed method is

---

[1] http://www.qwone.com/jason/20Newsgroups/.

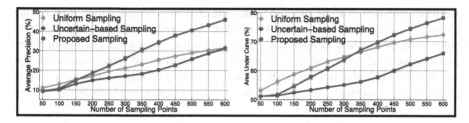

Fig. 2. The performance comparison on the 20-newsgroup dataset. (**Left**) The averaging precision. (**Right**) The values of area under curve.

better than that obtained by the passive learning. In most tasks, the example-to-boundary distance of the proposed approach is similar to that of the uncertain-based method, which is also shown in the following tasks of image classification. As shown in Fig. 2, the proposed method can obtain a more effective training set than the uncertain-based method in a short time. Thus, compared with other baselines, it is identified that the proposed method is more effective than other two baselines while needs less time to construct the training set.

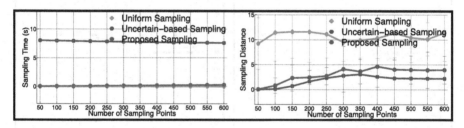

Fig. 3. The time expense (**Left**) and example-to-boundary distances (**Right**) of all methods on the 20-newsgroup database.

Image Classification: We perform our experiments by using Caltech-256 database including 256 ground-truth categories, and use the 1-vs-all formulation to handle these multi-class problems. Three kinds of descriptors in different dimensions ($d = 1024$, 2048 and 2659) [12, 13] are extracted from every image. We randomly select 25 examples per category (total 6400 examples) as the validation set. We also randomly select 5 examples per category (total 1280 examples) as the initial training set, and then remained examples are served as the unlabeled dataset for example selection. We learn a L1-normalized logistic regression as the classifier [1]. In every round, an unlabeled example is selected by using different sampling criteria. The threshold for updating bounds is set as 25 ($|I^{D^i}| = 25$). All experiments are averaged over multiple random training/test partitions.

Results: We show that the proposed sampling criterion can powerfully improve the classifier (average precision in Fig. 4(**Left**) and AUC values in Fig. 4(**Right**)). Compared with the random sampling, the proposed active framework can obtain better performance at all iterations. It is identified that the proposed active framework can obtain a more effective training set than passive learning. When the dimension size of every example increases, the performance of the classifier increases because more information is included.

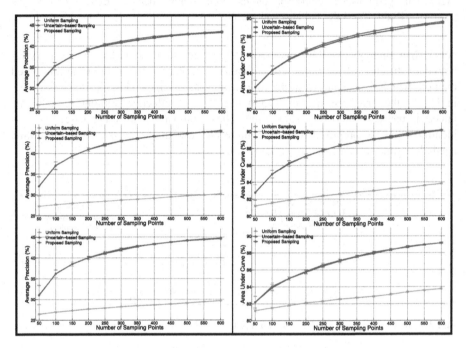

Fig. 4. The averaging precision (**Left**) and AUC values (**Right**) on Caltech-256 database. The values of $d = 1024$, $d = 2048$ and $d = 2659$ are listed from top to bottom.

Secondly, we show the time expense of three sampling criteria in Fig. 5(**Left**). Here, we only compute sampling time in every round, excluding time expense of training a classifier. Therefore, it only depends on the number of selected examples and independent with the number of iterations in the learning process. One unlabeled example is selected per round according to different sampling strategies, and the time expense in the selection procedure is reported. Note that the proposed method can be adapted to select multiple examples per round. Similarly, when multiple examples are selected in a round, the time expense will increase along with the number of selected examples. Random sampling needs least time of all because it does not need any computation or comparison in selecting an example. The traditional active learning with uncertain-based sampling criterion needs the most time expense of all. Moreover, the time expense increases with respect to the number of dimensions of unlabeled examples. The reason is that the distances are estimated by using all weights of the parameter vector and scanning all examples in the whole unlabeled pool.

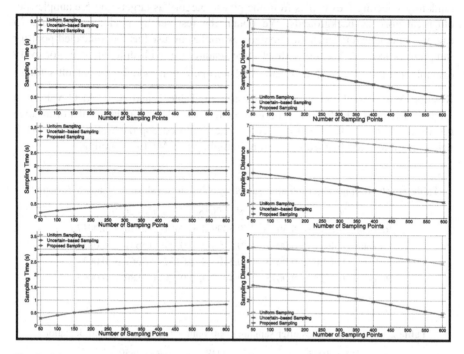

Fig. 5. The sampling time (**Left**) and example-to-hyperplane distances (**Right**) on Caltech-256 database. The values of $d = 1024$, $d = 2048$ and $d = 2659$ are listed from top to bottom.

Compared with the uncertain-based sampling criterion, the proposed sampling criterion can obtain nearly the same performance (in Fig. 4), while it needs less sampling time (in Fig. 5(**Left**)). We also compare three methods on example-to-hyperplane distances in Fig. 5(**Right**). The average distances of all sampling criteria decrease with respect to the number of sampling examples. The distance of random sampling is larger than other two sampling criteria because all unlabeled examples have the same probability to be selected. The uncertain-based sampling criterion and the proposed sampling criterion select the most close-to-boundary examples of all in every round, and then obtain smaller distances in the sampling process.

State-of-Art Comparison: We compare the effect of the proposed sampling strategy against a basic random sampling which is a standard choice in active learning. Then, we compare three sampling strategies recently applied on linear models for the tasks of classifying visual objects: (**a**) Sampling from the hyperplane zone F_0 [2] (F_0 zone) which is defined as that the classifier is maximally confused. The locations of unlabeled points are estimated by roughly summing the products of all components and their weights. (**b**) Maximum conflict-label equality sampling [2] (MCLE) using label equality conditions and maximum conflict to decide whether to sample a point from positive outer margin zone or hyperplane zone. **(c)** Proposed approach (Proposal) defined in this work. By using Caltech-256 dataset, we initialize the training set by randomly selecting 5 examples from the given category as targets and 5 examples per other category as backgrounds. We obtain the test set by randomly selecting 25 examples per category, then all remained examples are used as the unlabeled pool for sampling procedure. We also extract three feature sets with different sizes for every example, set the $|I^{D^i}| = 75$ and learn a L1-normalization model with default setting.

Results: We present the average precision in Table 1 and their AUC values in Table 2. It is observed that the proposed method can obtain similar results of average precision when $d = 1024$ and $d = 2659$. The proposed approach also obtains AUC values close to that obtained by F_0 zone sampling and MCLE sampling.

Table 1. The average precision after sampling 600 points. (Mean ± standard deviation)

	$d = 1024$	$d = 2048$	$d = 2659$
Random	$0.1972 \pm 5.3917e^{-4}$	0.3299 ± 0.0071	0.2920 ± 0.0017
F_0 zone	0.4651 ± 0.0047	0.5103 ± 0.0157	0.5661 ± 0.0031
MCLE	0.4636 ± 0.0053	0.5103 ± 0.0157	0.5661 ± 0.0031
Proposal	0.4627 ± 0.0075	0.4007 ± 0.0126	0.5880 ± 0.0065

Table 2. The AUC values after sampling 600 points. (Mean ± standard deviation)

	$d = 1024$	$d = 2048$	$d = 2659$
Random	$0.8003 \pm 9.6936e^{-4}$	0.7948 ± 0.0011	0.7699 ± 0.0024
F_0 zone	0.9035 ± 0.0021	0.9685 ± 0.0024	$0.9717 \pm 8.9525e^{-4}$
MCLE	0.9017 ± 0.0017	0.9685 ± 0.0024	$0.9717 \pm 8.9525e^{-4}$
Proposal	0.9004 ± 0.0019	0.9507 ± 0.0016	$0.9624 \pm 8.8640e^{-4}$

We also show sampling time in Table 3 and point-to-hyperplane distances of sampling points in Table 4. Both are computed by averaging results of all 600 sampling points during the whole procedure of active learning. The random method appears to be the fastest sampling method of all. As expected, the proposed sampling is faster than other active baselines. It is worth noting that MCLE sampling is slower than F_0 zone sampling among all three feature set because of more computation in evaluating the unlabeled points. It is identified that sampling time is proportionate to increased computation of point evaluation. Moreover, sampling time also increases with the size of dimensions of feature set among all three sampling strategies. Since our proposed sampling strategy uses partial components and weights to evaluate the unlabeled points, the size of dimension is much smaller than other baselines, and then its sampling time can be reduced. In Table 4, F_0 zone sampling can select the most confused example of all when $d = 2048$ and $d = 2659$.

Table 3. The sampling time expense averaged by all 600 points. (Mean ± standard deviation)

	$d = 1024$	$d = 2048$	$d = 2659$
Random	$0.0021 \pm 3.1067e^{-4}$	$0.0028 \pm 4.5752e^{-4}$	$0.0021 \pm 2.2538e^{-4}$
F_0 zone	0.3246 ± 0.0272	0.4059 ± 0.0495	0.4636 ± 0.0707
MCLE	0.6476 ± 0.0385	0.7224 ± 0.0729	1.0023 ± 0.0803
Proposal	0.2957 ± 0.0938	0.2632 ± 0.0924	0.2783 ± 0.0779

Table 4. The example-to-hyperplane distances of all 600 points. (Mean ± standard deviation)

	$d = 1024$	$d = 2048$	$d = 2659$
Random	5.8992 ± 1.4969	6.1009 ± 1.1968	6.0204 ± 1.4138
F_0 zone	2.1852 ± 1.3648	2.6066 ± 0.8046	2.3798 ± 1.0919
MCLE	2.2292 ± 1.2849	2.6066 ± 0.8046	2.3798 ± 1.0919
Proposal	2.1755 ± 1.2948	2.5197 ± 0.7876	2.4506 ± 1.0699

5 Discussions and Conclusions

In this paper, we focus on how to construct an effective training set. We propose an active scheme to construct the training set iteratively. We take advantage of the sparsity of the parameter of the classification model in order to select the most uncertain example of all in a short time. The proposed sampling criterion is compared on a group of tasks of text classification and image classification, and then the experimental results show the effectiveness of the proposed sampling criterion.

Acknowledgments. This work was supported by the Natural Science Foundation of China (Grant No. 61671188, 61571164, 61502122 and 61502117), the National Key Research and Development Plan Task of China (Grant No. 2016YFC0901902), and Natural Science Foundation of Heilongjiang Province QC2016084.

References

1. Fan, R.-E., Chang, K.-W., Hsieh, C.-J., Wang, X.-R., Lin, C.-J.: Liblinear: a library for large linear classification. J. Mach. Learn. Res. **9**, 1871–1874 (2008)
2. Gavves, E., Mensink, T., Tommasi, T., Snoek, C., Tuytelaars, T.: Active transfer learning with zero-shot priors: reusing past datasets for future tasks. In: ICCV, pp. 2731–2739. IEEE Press, New York (2015)
3. Tong, S., Koller, D.: Support vector machine active learning with applications to text classification. J. Mach. Learn. Res. **2**, 45–66 (2001)
4. Settles, B.: Active learning literature survey. Computer Sciences Technical report 1648, University of Wisconsin-Madison (2009)
5. Kapoor, A., Grauman, K., Urtasun, R., Darrell, T.: Gaussian processes for object categorization. Int. J. Comput. Vis. **88**, 169–188 (2010)
6. Huang, S.J., Jin, R., Zhou, Z.H.: Active learning by querying informative and representative examples. IEEE Trans. Pattern Anal. Mach. Intell. **36**(10), 1936–1949 (2014)

7. Lewis, D., Catlett, J.: Heterogeneous uncertainty sampling for supervised learning. In: ICML, pp. 148–156. Morgan Kaufmann, San Francisco (1994)
8. Jeon, Y., Kim, J.: Active convolution: learning the shape of convolution for image classification. In: CVPR, pp. 1846–1854. IEEE Press, New York (2017)
9. Huijser, M.W., Van Gemert, J.C.: Active decision boundary annotation with deep generative models. In: ICCV, pp. 5296–5305. IEEE Press, New York (2017)
10. Jayaraman, D., Grauman, K.: Look-ahead before you leap: end-to-end active recognition by forecasting the effect of motion. In: Leibe, B., Matas, J., Sebe, N., Welling, M. (eds.) ECCV 2016. LNCS, vol. 9909, pp. 489–505. Springer, Cham (2016). https://doi.org/10.1007/978-3-319-46454-1_30
11. Griffin, G., Holub, A., Perona, P.: Caltech-256 object category dataset. Computer Sciences Technical report 7694, California Institute of Technology (2007)
12. Torresani, L., Szummer, M., Fitzgibbon, A.: Efficient object category recognition using classemes. In: Daniilidis, K., Maragos, P., Paragios, N. (eds.) ECCV 2010. LNCS, vol. 6311, pp. 776–789. Springer, Heidelberg (2010). https://doi.org/10.1007/978-3-642-15549-9_56
13. Bergamo, A., Torresani, L., Fitzgibbon, A.: PiCoDes: learning a compact code for novel category recognition. In: NIPS, vol. 24, pp. 2088–2096 (2011)

Single Image Dehazing Based on Improved Dark Channel Prior and Unsharp Masking Algorithm

Liting Peng[(⊠)] and Bo Li

College of Computer Science and Technology, Hubei Province Key Laboratory
of Intelligent Information Processing and Real-Time Industrial System,
Wuhan University of Science and Technology, Wuhan, China
pltmsql1994@163.com

Abstract. In order to solve the problem of the "halo effect" and the bad color contrast after dehazing, a novel dehazing method based on the dark channel prior and the adaptive contrast enhancement algorithm is proposed. Using the hierarchical search method based on the quadratic tree space division to calculate the atmospheric light value, and then eliminate the "halo effect" caused by the guided filtering. By using the adaptive contrast enhancement algorithm based on unsharp masking algorithm to improve image information at the haze high concentration regional. Experimental results show that this algorithm can be more effective to dehaze and images after dehazing have a higher contrast.

Keywords: Dark channel prior · Guided filter · Single image
Unsharp masking

1 Introduction

Obtain outdoor natural landscape is essential for understanding the natural environment and performing visualization activities such as intelligent transportation systems, financial systems, object identification detection and other computer vision applications. However, in severe weather the visibility and contrast, the absorption and scattering of atmospheric particles, the color cast and the characteristic information are greatly reduced [1], and the effect of recognition and detection becomes worse. Therefore, in these cases it is necessary to use the haze removal technology.

Histogram equalization, wavelet transform, retinex algorithm and filtering are widely used in image removal. However, they do not take the reason or the process of image degradation into account, which will lead to incompleteness of dehazing effect or color distortion. In the single haze image processing method, most of the dehazing algorithms assume that the image satisfies the atmospheric scattering model, using an additional depth map or a separate projection pattern to enhance the visibility of the image. Tan's [2] haze removal method predicts scene albedo by maximizing local contrast, but the results tend to be supersaturated and appear to "halo effect". Fattal [3] assumes that the transmission and the surface model are statistically irrelevant to improve the image contrast, but the method needs to know a lot of color and brightness

© Springer International Publishing AG, part of Springer Nature 2018
D.-S. Huang et al. (Eds.): ICIC 2018, LNCS 10954, pp. 347–358, 2018.
https://doi.org/10.1007/978-3-319-95930-6_32

information in the haze scene. He et al. [4] have made a significant contribution in dark channel prior. By adding a prior constraint, the soft matting method is used to refine the initial projection graph and achieve the successful processing result. However, it will be paid expensive computational cost. according to this shortcoming, He et al. [5] are used the guided filter to replace the soft matting to speed up in 2010. But the computational complexity is still high, and there are problems such as poor processing effect in the sky area and the edge area. Yang et al. [6] found that haze mainly affects low-frequency component images, so wavelet transform is introduced. However, since the light of the scene is usually not as bright as the atmospheric light, the image after the haze looks very dark.

This paper puts forward some improvement measures to the shortcomings of the original dark channel prior: Firstly, in order to reduce the "halo effect", in the projection map calculation, the atmospheric light intensity A is calculated by the hierarchical search method based on the quadratic tree space division [7] instead of the original atmospheric light intensity; Secondly, the adaptive contrast enhancement algorithm based on unsharp masking technique [8] is used to solve the problem that after dark prior dehazing the color of the image become dim and in the high haze area local features is less. The above modified measure not only improve the image contrast, but also make the overall color of the image more natural and more depth.

2 Related Work

2.1 Atmospheric Scattering Model

In the haze weather conditions, a large number of particles suspended in the atmosphere have a strong scattering effect to the light. On the one hand, the reflected light on the surface of the object is attenuated by the scattering of the atmospheric particles. The attenuation of the light intensity directly leads to the weakening of the brightness of the image, and the forward scattering also causes the image blur resolution to decrease. On the other hand, natural light is involved in the imaging of the imaging sensor due to the backscattering of the atmospheric particles. This backscattering causes image saturation, contrast reduction and hue deviation [9].

In computer vision and computer imaging, the dark channel prior model is widely used to describe a fuzzy image information, the formula is as follows [2, 3, 10, 11]:

$$I(x) = J(x)t(x) + A(1 - t(x)) \tag{1}$$

$I(x)$ is the brightness of the haze image received by the observer at x pixel. $J(x)$ is dehazing images. A represents the atmospheric light intensity of the surrounding environment. $t(x) \in [0, 1]$ is the transmission map of the medium. Transform (1) can be obtained:

$$J(x) = \frac{I(x) - A}{t(x)} + A \tag{2}$$

From (2), the task of haze removal is to retrieve $J(x)$, A, $t(x)$ from $I(x)$. $J(x)t(x)$ is direct attenuation and shows how the scene emissivity decays with the medium. $A(1 - t(x))$ is called airlight results from previously scattered light and leads to the shift of the scene color. In general, a longer light propagation distance lead to acceleration of attenuation and scattering. So, transmission map $t(x)$ is as follows:

$$t(x) = e^{-\beta d(x)} \tag{3}$$

In (3), β is the scattering coefficient of the atmosphere and $d(x)$ is the scene depth.

The haze image model is shown in Fig. 1, it reveals the core idea of realizing haze removal from the perspective of image restoration, that is from the observed image brightness remove the atmospheric light participate part, while compensate the results of the attenuation light and can achieve restoring of the scene clarity. It means that in the RGB color space, vector $J(x)$, A, $I(x)$ from the geometric point of view is coplanar, the endpoint is collinear, the transmission $t(x)$ is the ratio of the length of the two lines [4]

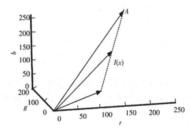

Fig. 1. Haze image formation model

$$t(x) = \left\| \frac{A(x) - I(x)}{A(x) - J(x)} \right\| \tag{4}$$

The method of using the haze image model to restore haze images to dehaze images can be divided into three categories: the first is based on the depth information method; the second is based on atmospheric light polarization characteristics of the dehaze algorithm; the third is a method based on dark channel prior. This paper uses the third method.

2.2   Haze Removal and Dark Channel Prior

The dark channel prior a statistical law of the foggy image, the previous dehaze method focuses on improving the contrast of the image, while He et al. [4] focus on the statistical characteristics of the foggy image by making a lots of experiments to collected foggy image, found that the objective existence of the statistical law, that is in most of the outdoor haze images in any local small pieces, there are always some (at least one) pixels where at least one color channel has very low intensity and close to zero, we call this the dark channel prior [4].

The low-channel values in real life are mainly caused by three factors: shadows, brightly colored objects or surfaces (such as green grassland, trees or red flowers) and darker objects or surfaces. We give the dark channel a mathematical definition, for any input image, the dark channel can be expressed as follows:

$$J^{dark}(x) = \min_{c \in (r,g,b)} (\min_{y=\Omega(x)} (J^c(y)))$$ (5)

In (5), $J(y)$ represents each channel of a color image, $\Omega(x)$ represents a window centered on pixel x. The expression of formula (5) is also very simple, First, the minimum value of each pixel RGB component is obtained, and a gray scale of the same size as the original image is stored, and then perform minimum filter to the gray scale. The radius of the filter is determined by the window size which expression is as follows: $WindowSize = 2 * Radius + 1$. Assuming that the atmospheric light intensity is known, the formula (1) can be written as:

$$\frac{I^c(y)}{A^c} = t(x)\frac{J^c(y)}{A^c} + (1 - t(x))$$ (6)

The superscript c is represented R/G/B, and in [4], the transfer rate is assumed to be a constant invariant in a local small patch and then described as the formula (5), Then the formula (6) is defined as:

$$\min_{c}(\min_{y \in \Omega(x)} (\frac{I^c(y)}{A^c})) = t'(x) \min_{c}(\min_{y \in \Omega(x)} (\frac{J^c(y)}{A^c})) + (1 - t(x))$$ (7)

From the dark channel prior principle we can see that the dark primary color J of the picture after dehazing is close to 0 and the atmospheric light intensity A has been given, and perform two minimum operations on both sides of Eq. (7),

$$J^{dark}(x) = \min_{c}(\min_{y \in \Omega(x)} \frac{J^c(y)}{A^c}) = 0$$ (8)

So, the transmission $t'(x)$ is represented by the following equation:

$$t'(x) = 1 - \min_{c}(\min_{y \in \Omega(x)} (\frac{I^c(y)}{A^c}))$$ (9)

This is the estimated value of the transmission. But in real life, even when the weather is fine, there will be some particles in the air, so look at the distant objects can still feel the impact of fog. In addition, the presence of fog can make people feel the depth of the scene. So when we do a fog experiment, it is necessary to deliberately preserve a certain degree of fog. Therefore, we introduce a factor between [0,1] in Eq. (9), then the formula (9) is modified to:

$$t'(x) = 1 - w \min_c (\min_{y \in \Omega(x)} (\frac{I^c(y)}{A^c})), 0 < w < 1 \tag{10}$$

Above is the calculation of transmission t. The next step is to introduce the law of atmospheric light A. It is assumed that the atmospheric light intensity is known. In practice, we can obtain the value from the haze image with the dark channel graph. The specific process of estimating the atmospheric light intensity is [4]: Firstly, from the dark channel map in accordance with the size of the brightness value to get the pixels of the brightness value of 0.1%; Secondly, find these pixels in the original image corresponding to the highest brightness of the pixel value as the atmospheric light value.

2.3 Guided Filter

The guided filter [5] derived from a local linear model computes the filtering output by considering the content of a guidance image, which can be the input image itself or another different image. Recently, guided filter is used to refine the transmission map. A local linear model between the guidance I and the filtering output q is the key assumption in the guided filter and can be defined as follows:

$$q_i = a_k I_i + b_k, \forall i \in w_k \tag{11}$$

Where i is the pixel index. Where (a_k, b_k) are some linear coefficients assumed to be constant in ω_k. Haze image I is used as the guided image, q is the filtering output. This model has been proven useful in image super-resolution [12], image matting [13], and dehazing [4]. To determine the linear coefficient (a_k, b_k), we seek a solution that minimizes the difference between the output q and the guided input p while maintaining the linear model (11). Specifically, we minimize the following cost function E in the window ω_k, E is defined as follows:

$$E(a_k, b_k) = \sum_{i \in \omega_k} ((a_k I_i + b_k - p_i)^2 + \varepsilon a_k) \tag{12}$$

Here, ε is a regularization parameter. Equation (12) is the linear ridge regression model (13) and (14) and its solution is given by

$$a_k = \frac{\frac{1}{|\omega|} \sum_{i \in \omega_k} I_i p_i - \mu_k \overline{p_k}}{\sigma_k^2 + \varepsilon} \tag{13}$$

$$b_k = \overline{p_k} - a_k \mu_k \tag{14}$$

(13) (14), μ_k and σ_k^2 are the mean and variance of I in ω_k, $|\omega|$ is the number of pixels in ω_k, and $\overline{p_k} = \frac{1}{|\omega|} \sum_{i \in \omega_k} p_i$ is the mean of p in ω_k. Having obtained the linear

coefficients (a_k, b_k), we can compute the filtering output q_i by (11). The value of q_i in (10) will be different if the pixel i is involved in many windows that contain i. One simple way is to use the average of all the possible values of q_i. Now we can get the filter result after computing (a_k, b_k) for all patches in the image. Then (10) can be rewritten as follows:

$$q_i = \frac{1}{|\omega|} \sum_{K:i\in\omega_k} (a_k I_i + b_k) \tag{15}$$

Noticing that $\sum_{k:i\in\omega_k} a_k = \sum_{k\in\omega_i} a_k$ due to the symmetry of the box window, we rewrite (11) by

$$q_i = \overline{a_i} I_i + \overline{b_i} \tag{16}$$

Where $\overline{a_i} = \frac{1}{|\omega|} \sum_{k\in\omega_i} a_k$ and $\overline{b_i} = \frac{1}{|\omega|} \sum_{k\in\omega_i} b_k$ are the average coefficients of all windows overlapping i. Eqs. (13), (14), and (16) are the definition of the guided filter.

3   The Improved Image Dehazing Method

3.1   Estimation of the Atmospheric Light Intensity

He's [4] dark channel prior using local patch minimum operation to calculate is bound to produce "halo effect". This is because if the pixel x is at the edge, the minimum value obtained by the above method is more likely to be in the position of the darker side near the edge, resulting in the final estimated value being smaller than the actual value. As shown in Fig. 2, since the pixels on the bright side near the edge have a lower minimum value than the entire bright area and have a higher minimum value than the entire dark area [14], so the overall edge of the projection is not well, the edge part of the final recovery image will appear light band effect, that is, "halo effect."

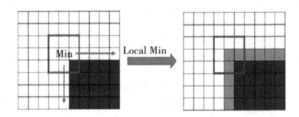

Fig. 2. Partial minimum operation

In order to solve the above problem, we propose a more effective method to improve the atmospheric light value. Atmospheric light A is usually used as the most bright color in the image as an estimate. Because a lot of haze usually leads to a bright (white) color, objects in this frame that are brighter than atmospheric light are usually selected, which can lead to an problem that the value should not be used as an atmospheric light is used as an estimate of atmospheric light, as shown in Fig. 3. In the picture the red box should be the correct estimate of the atmospheric light value A, while other three blue boxes are likely considered atmospheric light values. To estimate the atmospheric light more reliably, Kim et al. [7] exploited the fact that the variance of the pixels in those gray areas (the sky) is generally small.

Fig. 3. Error estimation of the atmospheric light value (Color figure online)

Based on this theory, Kim et al. [7] proposed a hierarchical search method based on quadratic tree space division. As shown in Fig. 4, we first divide the input image into four rectangular regions. Second, each sub-region is scored, and this score is calculated by subtracting the standard deviation of the pixels from the average of the pixels in the region. Use this method to select the region with the highest score and continue to divide it into smaller four sub-rectangles. We repeat this process until the selected area is less than a previously specified threshold. For example, the red part in Fig. 4 is the final selected area. We generally choose $\|(Ir(p), Ig(p), Ib(p)) - (255, 255, 255)\|$ to make the distance to minimize the color (including R/G/B three components) as the reference value of atmospheric light.

Fig. 4. Improved atmospheric light estimation method (Color figure online)

3.2   Contrast Enhancement Based on Unsharp Masking

In the dark channel prior we get the image has been dehazed, but after the process of the dark channel prior image's color is dim generally and image detail features are fewer in the high haze area, so using adaptive contrast enhancement algorithm based on unsharp masking technology to improve color contrast.

The unsharp masking technique was first used in photographic techniques to enhance the edge and detail of the image [15]. The optical method of operation is to superimpose the focused positives and defocused negatives which the result that the high-frequency components of the positive are enhanced and enhancing the contour, the defocused negative is equivalent to the "blur" template (mask), it is the opposite of the effect of sharpening, so the method is called the unsharp mask method. First, the original image is low-filtered to produce a passivation blurred image, second, the original image and the blurred image is reduced to retain the high-frequency compo-nents of the image, finally the high-frequency image with a parameter amplified and the original image superposition, which produces an image that enhances the edge [16]. The original image through the low-pass filter, because the high-frequency components are suppressed, so that the image is blurred, so the fuzzy image of high-frequency components are greatly weakened. The result of subtracting the original image from the blurred image will cause a lot of low-frequency components to be lost, and the high-frequency components are retained more completely. Therefore, the high-frequency components of the image with a parameter after amplification with the original image $f(x, y)$ after the superposition of high-frequency components to enhance the low-frequency components almost unaffected [17]. The process of unsharp masking:

Firstly, The passivation blurred image is obtained by low pass filtering of the original image;

Secondly, The original picture subtracts the passivated blurred image to get the image of the high frequency portion;

Then, The high-frequency part multiplied by a coefficient to get amplified high-frequency part;

Finally, The contrast enhanced image is by adding the original image to the high-frequency image after enlargement.

3.3   Recovering the Scene Radiance

The image dehaze algorithm process based on the dark channel prior principle and unsharp masking is as follows:

Step 1 Get dark channel map from the haze image;
Step 2 According to the dark channel prior principle to estimate the initial trans-mission and using the guided filter to obtain the final transmission;
Step 3 According to the dark channel prior principle and a hierarchical search method based on quadratic tree space division to estimate the correct atmospheric light value A_0;

Step 4 The scene radiance $J(x)$ recovered from $I(x)$. The final $J(x)$ is as follows:

$$J_0(x) = \frac{I(x) - A_0}{\max(t_0, t(x))} + A_0 \tag{17}$$

Step 5 The restored haze-free image color is dim, using unsharp masking to improve contrast about image.

4 Experiment Results

In order to verify the effectiveness of the proposed dehaze method, it is tested with multiple haze images and compared with the methods of Tan, Fattal, and He. All algorithms are implemented on MATLAB R2014a, and all the experiments are done on a PC with a processor of 3.50 GHZ and 4 GB RAM.

4.1 The Qualitative Comparison of Images

All of the dehaze algorithms have very good results for general outdoor images, but most of the existing dehaze algorithms are not sensitive to white, so this paper specifically deals with challenging images with white or gray areas more objectively compare the effects of these algorithms.

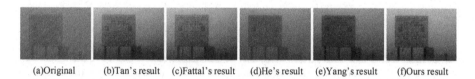

(a)Original   (b)Tan's result   (c)Fattal's result   (d)He's result   (e)Yang's result   (f)Ours result

Fig. 5. Dehazing effect comparison of brand

In Figs. 5, 6, 7, 8, 9 and 10 we can see that the results of Tan [2] are often supersaturated and appear "halo effect", Fattal [3] method in the scene near the skyline can only be partially removed, and after treatment the overall image is still a bit too full. The results of He [4] have better visual effects than the previous two methods, but the dark passages are poor when the scene brightness and atmospheric light are similar. Yang [6] looks better than He [4], but the overall picture after dehazing are still dim. Look at the method of this article, the image does not appear over saturation phenomenon, and in the sky and other bright areas can also get better processing results.

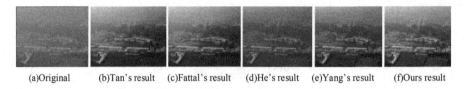

(a)Original   (b)Tan's result   (c)Fattal's result   (d)He's result   (e)Yang's result   (f)Ours result

Fig. 6. Defogging effect comparison of city 1

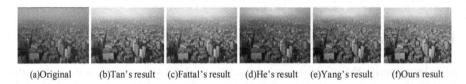

(a)Original   (b)Tan's result   (c)Fattal's result   (d)He's result   (e)Yang's result   (f)Ours result

Fig. 7. Defogging effect comparison of city 2

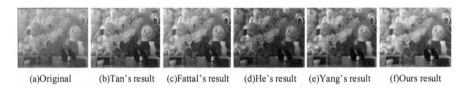

(a)Original   (b)Tan's result   (c)Fattal's result   (d)He's result   (e)Yang's result   (f)Ours result

Fig. 8. Defogging effect comparison of toys

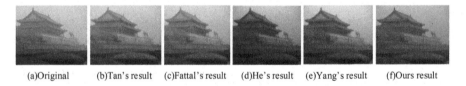

(a)Original   (b)Tan's result   (c)Fattal's result   (d)He's result   (e)Yang's result   (f)Ours result

Fig. 9. Defogging effect comparison of Tiananmen

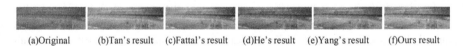

(a)Original   (b)Tan's result   (c)Fattal's result   (d)He's result   (e)Yang's result   (f)Ours result

Fig. 10. Defogging effect comparison of stadium

4.2   Quantitative Analysis

In order to compare the performance of each algorithm more intuitively, this paper will compare the contrast and information entropy of each algorithm result. The comparison results are shown in Table 1:

Table 1. The quantitative analysis of six images.

Image	Criterion	Original	Tan	Fattal	He	Yang	Our
Figure 5	Entropy	5.7754	7.412	7.523	7.6831	7.7015	**7.7832**
	Contrast	5.6976	8.0201	10.4229	12.7995	12.9421	**13.1301**
Figure 6	Entropy	6.1750	6.2174	6.3560	6.3961	6.2610	**6.9614**
	Contrast	27.5831	54.8575	70.3792	89.5614	92.1948	**96.2641**
Figure 7	Entropy	6.8101	7.4415	7.6356	7.9178	7.9214	**7.9595**
	Contrast	53.9254	68.0593	75.5398	82.7387	85.1397	**87.0599**
Figure 8	Entropy	7.0427	7.3874	7.4018	7.6196	7.7513	**7.7835**
	Contrast	43.0934	85.9620	113.4143	135.6250	139.8523	**142.8781**
Figure 9	Entropy	7.2641	7.2161	7.3903	7.4872	7.5237	**7.6419**
	Contrast	80.1589	92.1512	105.6297	115.9118	116.3764	**118.2729**
Figure 10	Entropy	6.5250	6.6048	6.7864	6.8423	7.0329	**7.0866**
	Contrast	32.2878	55.3728	63.8226	70.1816	70.9598	**72.5800**

From the Table 1, it be seen from the table for six test images, whether it's the experimental results of Contrast or Entropy, most of the images in this paper algorithm is in a state of a supreme, that is to say, for most of the images, the algorithm in this paper can achieve a good effect to dehazing.

5 Conclusions

In this paper, the improved dark channel prior principle and unsharp masking algorithm to deal with a single image, not only can retain simple and effective advantages of the dark channel prior, but also can remove the "halo effect" in dehazing process and solve the issue which after dehazing image color become dim. Experimental results show that the proposed algorithm improves the dehazing effect in high-haze areas, but there is a slight oversaturation in the brightly colored objects or surfaces in the image, which needs to be improved in the future.

References

1. Choi, L.K., You, J., Bovik, A.C.: Referenceless prediction of perceptual fog density and perceptual image defogging. IEEE Trans. Image Process. **24**(11), 3888–3901 (2015)
2. Tan, R.: Visibility in bad weather from a single image. In: Proceedings of the IEEE Conference on Computer Vision and Pattern Recognition, pp. 1–8 (2008)
3. Fattal, R.: Single image dehazing. In: Proceedings of the ACM SIGGRAPH, pp. 1–9 (2008)
4. He, K., Sun, J., Tang, X.: Single image haze removal using dark channel prior. In: Proceedings of IEEE Conference on Computer Vision and Pattern Recognition (2009)
5. He, K., Sun, J., Tang, X.: Guided image filtering. In: Proceedings of the European Conference on Computer Vision, vol. 6311, pp. 1–14 (2010)
6. Yang, Y.J., Fu, Z.Z., Li, X.Y., et al.: Improved single image dehazing using dark channel prior. J. Syst. Eng. Electron. **26**(5), 1070–1079 (2015)

7. Kim, J.H., Jang, W.D., Sim, J.Y., Kim, C.S.: Optimized contrast enhancement for real-time image and video dehazing. J. Vis. Commun. Image Represent. **24**(3), 410–425 (2013)
8. Peng, Y.-T., Cao, K.: Pamela: generalization of the dark channel prior for single image restoration. IEEE Trans. Image Process. **6**(27), 2856–2868 (2018)
9. Koschmieder, E.L.: Benard convection. Adv. Chem. Phys. **26**(177–212), 605 (1974)
10. Narasimhan, S.G., Nayar, S.K.: Chromatic framework for vision in bad weather. In: CVPR, pp. 598–605 (2000)
11. Narasimhan, S.G., Nayar, S.K.: Vision and the atmosphere. IJCV **48**, 233–254 (2002)
12. Zomet, A., Peleg, S.: Multi-sensor super resolution. In: Proceedings of IEEE Workshop Applications of Computer Vision (2002)
13. Levin, A., Lischinski, D., Weiss, Y.: A closed form solution to natural image matting. In: Proceedings of IEEE Conference on Computer Vision and Pattern Recognition (2006)
14. Polesel, A., Mathews, V.J., Ramponi, G.: Image enhancement via adaptive unsharp masking. IEEE Trans. Image Process. **9**(3), 505–510 (2000)
15. Draper, N., Smith, H.: Applied Regression Analysis, 2nd edn. Wiley, New York (1981)
16. Hastie, T., Tibshirani, R., Friedman, J.H.: The Elements of Statistical Learning. Springer, New York (2003). https://doi.org/10.1007/978-0-387-84858-7
17. Wang, K., Dunn, E., Tighe, J.: Combining semantic scene priors and haze removal for single image depth estimation. In: IEEE WACV (2014)

WT Model & Applications in Loan Platform Customer Default Prediction Based on Decision Tree Algorithms

Sulin Pang[1,2(✉)] and Jinmeng Yuan[1,2]

[1] Institute of Finance Engineering, School of Management, Jinan University,
Guangzhou 510632, China
pangsulin@jnu.edu.cn
[2] Guangdong Emergency Technology Research Center of Risk Evaluation
and Prewarning on Public Network Security, Guangzhou 510632, China

Abstract. Due to the huge losses caused by the bad credit customers, loan platforms attach great attention to testing and forecasting of bad loans. From perspectives of both loan customer type identification and loan default prediction, we initially constructed a WT early warning model for loan default client prewarning based on C5.0 decision tree, CART decision tree and CHAID decision tree in this paper. WT model is set with weighted calculating algorithms. Considering the data characteristics of loan platform, we designed a posteriori combination algorithm of three sub-models: C5.0, CART, CHAID, and performance test indicators: sensitivity, accuracy, warning rate, false alarm rate. In empirical research, we used the real loan transaction dataset of a bank in Taiwan to construct the WT model of the bank, and found that WT model overcomes the shortcomings of each sub-model respectively and achieves effective prewarning of customer default. The experimental results show that the alarm rate of test data set is 26.93% and the false alarm rate is 18.33% and the accuracy rate is 81.67% when applying the WT model. Loan platforms can acquire both high customer default prediction accuracy and high alarm rate by applying WT prewarning model. Both the research method and experiment results in this paper are meaningful to loan platform operation.

Keywords: WT model · Decision tree algorithms
Customer default prewarning · Loan platforms

1 Introduction

As of the end of June 2017, the loan balance of net loan industry increased to 1.044965 trillion China yuan (equals to 159456.3045 million dollars). In the first half of the year, the accumulated volume of net loan industry reached 1.39434 trillion yuan. According to the statistics, the active investors and active borrowers in net loan industry were 4.308 and 3.7353 million people rose by 3.82% and 15.96% respectively on a

---

Co-first author: Jinmeng Yuan

© Springer International Publishing AG, part of Springer Nature 2018
D.-S. Huang et al. (Eds.): ICIC 2018, LNCS 10954, pp. 359–371, 2018.
https://doi.org/10.1007/978-3-319-95930-6_33

month-on-month basis. Borrowers have been hitting monthly increases of over 10% since March. With the rise of internet finance, many relationships of borrowers and lenders are formatted, but this phenomenon is with limited information collection cost. It is necessary to control the risk of borrowing by predicting target borrowers in advance and keeping information. Accuracy of the prediction relates to the guarantee of investor's rights. This paper explores the prediction of bad credit borrowers.

At present, there are two kinds of methods to study the default forecast: the first is the traditional rule-based regulatory technology, which is based on the experience of experts or professionals to set regulatory conditions: alarm conditions. The alarm processing starts when a behavior is beyond the restrictions. Conditions are pre-set by platforms or banks, such as the overdue behavior appeared more than three times. The advantage of this method is that the risk is within control, and the platform or bank can choose a strict or lenient loan policy. However, the disadvantage is that the rules are not flexible and the regulatory effect cannot be absorbed and improved in time from the new forms of default. Another type of algorithm is technique "rules" with artificial intelligence self-learning methods. Customer information, including repayment records, is recorded and stored in the database for the system to learn and judge his/her behaviors habit. Judgment results also compares itself with the newly appeared behavior. Once an abnormal repayment appeared (deviation from the model predictions), system will alarm. Bache and Lichman (2013) found that artificial intelligence (AI) techniques such as neural networks, support vector machines and random forests are a good alternative to traditional linear statistical methods for building credit scoring models [1]. Tools based on the combination of artificial intelligence and big data are widely developed and applied, including decision tree, support vector machine, random forest and so on. However, with the flaws of being sensitive and over-fitting, the over-fitting of the previous data leads to the change of the model every time new data appears. Therefore, the integrated modeling technique has also been introduced into the field of credit scoring [2].

This paper proposes a new model by using three widely known models, C5.0, CART and CHAID, and forms a weight combination model for default early warning. The new model with a weighting mechanism is named WT model. Firstly, we discussed the process to execute customer default forecasting. Secondly, we describe the three models that appeared in our sub-model collection. Thirdly, we introduced a weighted combining mechanism used in WT model. Fourthly, we discuss an experimental research of WT model and analyzed our results.

2   Research Background

Machine learning methods are starting to be used in financial field research, and machine learning models are continually being trained on newly entered data [3], The machine learning approach to predict default can be divided into two categories. The first category is based on rules-based regulation. All customer information including repayment operations are recorded in the database and used to generate the model. Ge (2016) optimized the application of decision tree algorithm [4]. Imtiaz and Brimicombe (2017) studied credit scoring models using an interpolation technique and interpolation-free technique [5]. Bellalah et al. (2016) tested the predictive power of default probability from the structural model and explained the default behavior by incorporating the default probability as an explanatory variable into the unstructured model. But its limitation lies in its inability to overcome the default barrier of the model [6]. Xia et al. (2017) propose a cost-sensitive boosted tree loan evaluation model that introduces cost-sensitive learning and XGBoost to improve the ability to distinguish potential default borrowers [7]. Then a sequential integrated credit scoring model based on a gradient hoist variant (i.e. XGBoost) was proposed [8]. Abelian and Castellano (2017) extended the preliminary work on the selection of the best base classifier used in the collection on a credit dataset [9]. Sivakumar and Selvaraj (2018) applied various supervised classifiers such as DT, SVM, KNN, NB, NN and Improved DT to the data set [10]. Once the abnormal customer repayment behavior appeared (deviation from the normal model predictions), system will alarm.

This paper proposes a new model by combining three known models in the area of borrower credit forecasting. We propose the use of three decision tree models: C5.0, CART and CHAID. These are used to form a weight combination model to forecast customer default, which is WT model. Each sub-model is formed using the same set of customer information in parallel. When a new payment occurs, each sub-model outputs a judgment about whether the customer is good or bad. The result of WT model is jointly determined by the three sub-models. In the weighted combining mechanism, particular weights are allocated to each sub-model depending on the sub-models' performance on the test set. The structure of this paper is as follows. In Sect. 2, we introduced background researches about default forecasting. In Sect. 3, we discuss the proposed model in detail. In Sect. 4, we use a dataset of real credit card transactions from a bank in Taiwan (China) to evaluate the performance of WT model. We explain the evaluation criteria and then analyze the experimental results of WT model. Section 5 is Conclusion.

3 Structure of WT Model

This paper presents a comprehensive model for default customer warning, which contains the C5.0 decision tree, CART decision tree and CHAID decision tree and weight combination mechanism.

In this paper, three decision tree models C5.0, CART and CHAID are combined. C5.0 model, CART model and CHAID model are all one of the decision tree models. The decision tree, which can be understood as a set of if-then rules at each decision node and root node, and has actual attributes assigned to each branch. The training strategy is following principle of minimizing the loss function. New data from the root node followed by training down to the leaf node, and output test results. The difference between the CART algorithm and the C5.0 algorithm is that the attribute metrics used are exponents, the attribute value index is measured on the attribute, the left branch is required to meet the purity requirement, or the right branch is assigned. If the condition of stop splitting is met (the number of samples is less than the predetermined threshold, or the Gini index is less than the predetermined threshold), then the classification is stopped; otherwise, the above steps are recursively performed until the model stops branching, and the binary tree is less likely to generate data fragments.

WT model is proposed to conduct effective customer default forecasting through forming a weight combination mechanism. WT model is shown in Fig. 1. Our default prediction includes customer information, bank card payment information, and WT model. Training process is shown in Fig. 1. Figure 1 shows the data flow. Assume CSTi $(i = \{1, 2, \ldots \ldots, n\})$ stands for n customer, and each customer possess more than one bank card. Each card's history repayment records will be recorded and saved in the database. For example, CST1 possess three bank cards registered on the platform. His first bank card has three history bank transaction records. His second and third bank card has four records and three records respectively. The three models that constitute the sub-model collection are C5.0, CART and CHAID. These are three sub-models.

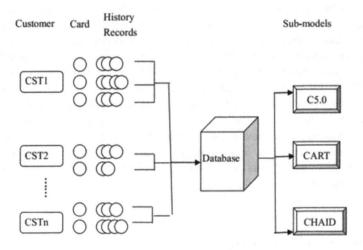

Fig. 1. Data processing flow

Repayment record information includes: customer number, transaction serial number, amount, date, time. The customer number is a unique number assigned to each customer. In each repayment, customer has a number, and the person's entire bank card belongs to the same customer number. The transaction serial number will be generated via a third party or other payment platform. Training of the model is carried out on each customer basis, and each customer corresponds to a WT model, as shown in Fig. 2. When a new repayment occurs, the model starts outputting the result of WT model using the weight mechanism.

Figure 2 shows the judgment process for CST1 when a new record occurs. The new record will update the database. Each sub-model will judge whether the customer is good or bad. WT model will produce judgment. "0" denotes good customer and "1" denotes bad customer.

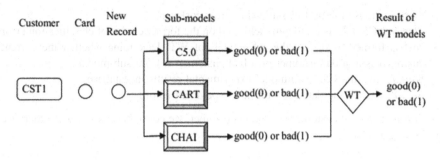

Fig. 2. Single WT model processing flow

4 Calculation of Sub-models

4.1 Sub-models

The goal of model development is to predict whether the borrower will default in the future. If the repayment occurred within a valid repayment period, the customer is considered a good customer; if the customer has no repayment in the meantime, the customer is regarded as a bad customer. The data source for this paper is a personal credit data set of a bank in Taiwan. A total of 30,000 customer samples are used Select the last 1500 customer as a test set, the front as a training set to execute model training. Microsoft Access is applied to build database, combining with both the characteristics of the data and the field includes the key information in the sample.

4.2 C5.0 Sub-model

All nodes in C5.0 decision tree have their corresponding attribute values, and the corresponding attribute meaning as we defined).

In this part, the algorism is as follows:

Step 1, user information is stored in the database, including user ID, name, gender, age, education level, marital status, emergency contact name, emergency contact number, credit limit, monthly repayment status, and customer type and so on. Among them, gender, age, education level, marital status, credit limit and monthly repayment status are used as decision tree classification attributes, and customer types are decision attributes, which are respectively defined.

Step 2, do the customer repayment behavior statistics according to the reality. And input the data to database.

Step 3, using C5.0 algorithm. The C5.0 classification index is calculated for each classification attribute, the current optimal classification index is selected as the classification node, and so on, until the customers in the classification node completely belong to one category, that is, all of customers in the node are good or bad customers [11].

Step 4, recognize classification error. Error 1 is to classify good customers as bad customers. Error 2 is to classify bad customers as good customers.

Step 5, choose the best classification error cost value.

Step 6, a C5.0 decision tree model based on the former choice of classification error cost is formed as one of the sub-models of WT. Determine whether the current customer is a good customer or a bad customer in C5.0 sub-model.

Step 7, save the C5.0 sub-model experimental results in database.

Step 8, Update the data set with new repayment record.

If it marked bad customer as a good customer, we consider it as *Error 1*, otherwise as *Error 2* (Table 1).

Table 1. Preliminary C5.0 decision tree model classification results

		Forecast category		Total	Correct rate	Error rate
		Good	Bad			
Actual	Good	21033	1163	22196	94.76%	(*Error 2*) 5.24%
category	Bad	4172	2132	6304	33.8%	(*Error 1*) 66.2%
Total		25205	3295	28500	81.28%	18.72%

As shown in Table 3, there are 21033 actually good customers who are judged to be good customers. 1163 actually good customers are classified to bad customers, which is marked as *Error 2*. 2132 actually bad customers are judged to be bad customers. There are 4172 actually bad customers who are judged to be good customers, that is marked as *Error 1* (Table 2).

Select the model of C(*Error 1*) = 4 is more appropriate, the total error rate does not exceed 25%, 33.30% is the lowest error rate, a better balance of the error rate requirements, although *Error 2* doesn't largely impact the degree of investor returns, we still hope to give more accurate prediction. Therefore, compared with the lowest and highest model, this paper is more inclined to take model 4. The prediction error rate does not exceed 25%. We use the calculating steps given above to generate C5.0 decision tree.

4.3   CART Sub-model

All nodes in CART decision tree have their corresponding attribute values, and the corresponding attribute meaning as we defined initially. In this part, the algorism is as follows:

Step 1, based on the initial definition of the property, the data set in the database is invoked.

Step 2, using CART algorithm. The CART classification index is calculated for each classification attribute, the current optimal classification index is selected as the classification node, and so on, until the customers in the classification node completely belong to one category, that is, all of customers in the node are good or bad customers [12].

Step 3, recognize classification error. Error 1 is to classify good customers as bad customers. Error 2 is to classify bad customers as good customers.

Table 2. C5.0 preliminary decision tree model classification error rate

	1	2	3	4	5
Error 2	5.24%	10.95%	15.64%	**22.01**	30.68%
Error 1	66.18%	48.26%	41.54%	**33.3%**	26.24%
Total	18.72%	19.20%	21.37%	**24.51%**	29.69%

Table 3. CART preliminary decision tree model classification error rate

	1	2	3	4	5
Error 2	5.09%	9.61%	10.52%	**19.52%**	36.77%
Error 1	69.56%	56.12%	44.47%	**40.47%**	27.93%
Total	19.35%	19.90%	23.7%	**24.45%**	34.82%

Table 4. C5.0 preliminary decision tree model classification error rate

	1	2	3	4	5
Error 2	5.78%	11.90%	**18.63%**	31.74%	37.41%
Error 1	66.90%	51.19%	**41.94%**	67.47%	61.49%
Total	19.30%	20.59%	**23.38%**	39.64%	42.73%

Step 4, choose the best classification error cost value.

Step 5, a CART decision tree model based on the former choice of classification error cost is formed as one of the sub-models of WT. Determine whether the current customer is a good customer or a bad customer in CART sub-model.

Step 6, save the CART sub-model experimental results in database.

Step 7, update the data set with new repayment record.

We observe the decision tree model indicators in order to pick the best value. Table 5 shows error rates under different classification error costs.

Table 5. WT model case

	Good performance	Bad performance
C5.0	9/10	5/7
CART	7/10	6/7
CHAID	7/10	7/7

Select the model of C(*Error 1*) = 4 is more appropriate, the total error rate does not exceed 25%, 40.47% is the lowest error rate, a better balance of the error rate requirements, although *Error 2* doesn't largely impact the degree of investor returns, we still hope to give more accurate prediction. Therefore, compared with the lowest *Error 1* and highest *Error 2* model 5, this paper is more inclined to take model 4 for prediction. The prediction error rate does not exceed 25%.

4.4   CHAID Sub-model

CHAID and CART belong to the same fork tree algorithm, the difference is that CART selects the best two partitions for each classification node, while CHAID calculates the chi-square test values for the variables at the nodes and divides them so as to be faster convergence, the formation of relatively bloated classification tree. Repeat the above process. Finally built CHAID decision tree. The result of CHIAD analysis is as follows (Table 4).

The customer category is set as the target variable and 11 categories attributes are set as classification variables. By running the CHAID algorithm, we summarize the results in Table 6.

Table 6. Decision tree model evaluation results

	AlarmRate	Sensitivity	Specificity	FalseRate
C5.0	29.87%	59.33%	78.51%	21.49%
CART	27.33%	57.53%	81.25%	18.75%
CHAID	26.40%	56.62%	82.19%	17.81%
WT	26.93%	57.23%	81.67%	18.33%

Select the model of C($Error\ 1$) = 3 is more appropriate, the total error rate does not exceed 25%, 23.38% is the lowest error rate, a better balance of the error rate requirements. This paper is more inclined to take model 3 for prediction. The prediction error rate does not exceed 25%.

4.5   WT Model

The following part describes the WT model construction process and evaluation methods.

4.5.1   Constructing WT Model

The WT model is a comprehensive model composed of C5.0, CART and CHAID. The three sub-models train and discriminate the same set of data set simultaneously. First, the model generates the original model by training the original data set. At this moment, if new data appears, the three sub-models train the new data at the same time to generate the discriminant results respectively, and the model results are calculated by the weight distribution algorithm.

In the WT model, the C5.0 model is treated as the main model, and the weights of the CART model and the CHAID model are modified to calculate the three weights for each customer/each WT model. One weight is calculated for a good customer, and the other weight is calculated for a bad customer. The weight model in this paper is based on the optimization model proposed by Kultur Y and Caglayan M U, and changes the combination ideas to improve the decision tree model.

Calculate the good customer's sub-model weight, as shown in formula (1):

$$W_{i,good} = \frac{P_{i,good}}{P_{C5.0,good} + P_{CART,good} + P_{CHAID,good}} \qquad (1)$$

Among them, $i = \{C5.0, CART, CHAID\}$. In good customer mode, $P_{i,good}$ is the probability of sub-model i predicts a good customer correctly, $W_{i,good}$ is a sub-model i calculates the weight of a good customer. That is $P_{C5.0,good}, P_{CART,good}, P_{CHAID,good}$ respectively represent the probability of predicting the good customer correctly with sub-model C5.0 and CART, CHAID.

Calculate the bad customer's model weight as shown in Eq. 2:

$$W_{i,bad} = \frac{P_{i,bad}}{P_{C5.0,bad} + P_{CART,bad} + P_{CHAID,bad}} \qquad (2)$$

Among them, $i = \{C5.0, CART, CHAID\}$. In good customer mode, $P_{i,bad}$ is the probability of sub-model i predicts a good customer correctly, $W_{i,bad}$ is a sub-model i calculates the weight of a good customer. That is $P_{C5.0,bad}, P_{CART,bad}, P_{CHAID,bad}$ respectively represent the probability of predicting the bad customer correctly with sub-model C5.0 and CART, CHAID. Formula (2) considers only bad customers, while formula (1) considers only good customers.

Example 1, assuming a platform default customer is a good customer; the platform database contains 10 good customers and 7 bad customers. The results of the sub-models are shown in the table below. The weight of the C5.0 model for good and bad customers is calculated as shown in Eqs. (3) and (4). Equation (3) is derived from Eq. (1) and Eq. (4) is derived from Eq. (2).

$$W_{i,good} = \frac{P_{i,good}}{P_{C5.0,good} + P_{CART,good} + P_{CHAID,good}} = \frac{9/10}{9/10 + 7/10 + 7/10} = \frac{9}{23} \qquad (3)$$

$P_{C5.0,good}$ is the prediction of good customer for C5.0 model, $W_{C5.0,good}$ is the good customer calculation weight of C5.0 model, $P_{CART,good}$ is the prediction of good customer of CART model and $P_{CHAID,good}$ is the good customer prediction of CHAID model. The bad customer's weighted calculation is as follows:

$$W_{i,bad} = \frac{P_{i,bad}}{P_{C5.0,bad} + P_{CART,bad} + P_{CHAID,bad}} = \frac{5/7}{5/7 + 6/7 + 7/7} = \frac{5}{18} \qquad (4)$$

$P_{C5.0,bad}$ is the C5.0 model bad client's forecast, $W_{C5.0,bad}$ is C5.0 model is bad customer computing weight, $P_{CART,bad}$ is the CART model's bad customer forecast, $P_{CHAID,bad}$ is the CHAID model bad customer's forecast. Similarly, the weight of the CART model is calculated according to Eqs. 3 and 4, changing only the numerator.

The WT model chooses C5.0 decision tree as the main model. Then, each customer is reevaluated using Eqs. (5) and (6), respectively. A single model output takes a value of 0 (good customer) or 1 (bad customer), and a WT model output value takes a continuous value between 0 and 1. Assuming that the platform chooses to trust customers to make good customer default for the sake of customer experience, the customer is regarded as a good customer when the integrated model result value is less than or equal to 0.5. When the result is greater than 0.5, the WT model considers the customer a bad customer.

$$R_{CST1} = R_{CST5.0,CST1} * W_{C5.0,good} + R_{CART,CST1} * W_{CART,good} + R_{CHAID,CST1} * W_{CHAID,good} \tag{5}$$

R_{CST1} is the WT model result for Customer 1, which is a good customer; $R_{C5.0,CST1}$ and $R_{CART,CST1}$ is the result of each model for Customer 1; $W_{C5.0,good}$ and $W_{CART,good}$ is the good customer weight for a single model. The bad customer's weighted result is as follows:

$$R_{CST2} = R_{CST5.0,CST2} * W_{C5.0,bad} + R_{CART,CST2} * W_{CART,bad} + R_{CHAID,CST2} * W_{CHAID,bad} \tag{6}$$

R_{CST2} is the WT model result for customer 2, which is a bad customer; $R_{C5.0,CST2}$ and $R_{CART,CST2}$ are the results of sub-models for customer 2; $W_{C5.0,bad}$, $W_{CART,bad}$ and $W_{CHAID,CST2}$ is the bad customer weight for the sub-models. The details of the WT process are shown in Fig. 3 below.

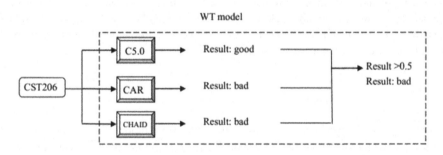

Fig. 3. Single WT model processing flow

The C5.0 model determines that the customer is a good customer, while the CART and CHAID models identify it as a bad customer. The weight of each sub-model can be calculated with formula 3, which is $\frac{9}{23}$, $\frac{7}{23}$, $\frac{7}{23}$ respectively. The result of the sub-model (0 stands for good and 1 stands for bad) is multiplied by the weight of sub-model. Because the result of the master model C5.0 model shows that CST206 is a good customer. Result values add up to more than 0.5, so the WT model determines that the customer CST206 is a bad customer.

4.5.2 WT Model Evaluation

Evaluation indicators include four indicators: sensitivity, accuracy, warning rate, false alarm rate. R_i is the true category of the customer, and D_i is the category of the customer that is identified by the decision tree model. Then four types of data will appear in each model: that is, the customer is decided to be good & the customer is actually good $R_{good}D_{good}$, The decision is a bad customer but is actually a good customer $R_{good}D_{bad}$, the customer is determined to be a good customer is actually a bad customer $R_{bad}D_{good}$, a customer is determined to be a bad customer, and a real bad customer $R_{bad}D_{bad}$.

AlarmRate, that is, the probability of being a bad customer among all customers:

$$AlarmRate = \frac{R_{good}D_{bad} + R_{bad}D_{bad}}{R_{good}D_{good} + R_{good}D_{bad} + R_{bad}D_{good} + R_{bad}D_{bad}} \qquad (7)$$

Sensitivity, that is, the probability of a bad customer being detected by the system and alarming:

$$Sensitivity = \frac{R_{bad}D_{bad}}{R_{bad}D_{good} + R_{bad}D_{bad}} \qquad (8)$$

Specificity, accuracy measures the probability that the system will not alarm and is a good customer:

$$Specificity = \frac{R_{good}D_{good}}{R_{good}D_{bad} + R_{good}D_{good}} \qquad (9)$$

FalseRate, that is, the probability of a good customer being alerted by the system:

$$FalseRate = \frac{R_{good}D_{bad}}{R_{good}D_{bad} + R_{good}D_{good}} = 1 - Specificity \qquad (10)$$

As can be seen from Table 6, C5.0 has a significantly higher sensitivity than the other models, with a significantly higher alarm rate. However, C5.0 false alarm rate is the highest. CHAID model has significantly higher accuracy than other models, but the sensitivity is too low. From the error rate column, it can be observed that in the WT model, the error rate of the master model C5.0 is significantly reduced, and the accuracy is improved. C5.0 can be regarded as a more pessimistic strategy choice, that C5.0 has a significant high alarm rate regardless of gains and losses. WT model shows high accuracy and high alarm rate.

5   Conclusions

In recent years, borrower escaping, defaults and other incidents occurred in large numbers. The top priority of the loan platform is to reduce losses. Models based on big data environments and artificial intelligence algorithms are also constantly being studied by scholars. This paper establishes a WT client type early warning model for multi-model combinations of borrower behavior, including C5.0, CART, CHAID model, and gives the weight calculation mechanism. The empirical results from a bank in Taiwan showed the ideal results. Under the action of WT model, the alarm rate was 26.93%, and the accuracy reached 81.67%. The loan platform can achieve the goal of high accuracy and high alarm rate through the WT model.

The WT model needs to be based on the customer's basic information and the repayment behavior information database. When new data appears, the model will be trained and do autonomous learning in turn. In the subsequent study, we plan to enrich the data inventory training model and set up an online shared credit system to integrate social network information and minimize the credit risk more effectively.

The main research content of this paper is to predict the customer's default behavior by constructing the weight combination mechanism of the three decision trees. In this paper, we propose the use of three decision tree model. These models are C5.0, CART, and CHAID. We initially invented a weight combination model, which is WT model, and evaluate the experimental results of our WT model. The main research method of this paper is to establish the mathematical model and test the model with the data in reality. The WT model can effectively predict the credit platform customer default event. Loan platforms can choose WT model for early default warning and acquire desired alarm rate.

Acknowledgement. The paper is supported by Natural Science Foundation of China (Grants No. 91646112); The Key Programs of Science and Technology Department of Guangdong Province (Grants No. 2016A020224001).

References

1. Bache, K., Lichman, M.: UCI Machine Learning Repository (2013)
2. Li, X.L., Zhong, Y.: An overview of personal credit scoring: techniques and future work. Int. J. Intell. Sci. **2**, 181–189 (2012)
3. Pawlak, Z.: Rough sets. Int. J. Inf. Comput. Sci. **11**(5), 341–356 (1982)
4. Ge, P., Peng, M.: The optimization and application of the decision tree algorithm in the classification prediction **47**, 936–939 (2016)
5. Imtiaz, S., Brimicombe, A.J.: A better comparison summary of credit scoring classification. Int. J. Adv. Comput. Sci. Appl. **8**, 1–4 (2017)
6. Bellalah, M., Zouari, S., Levyne, O.: The performance of hybrid models in the assessment of default risk. Econ. Model. **52**, 259–265 (2016)
7. Xia, Y., Liu, C., Liu, N.: Cost-sensitive boosted tree for loan evaluation in peer-to-peer lending. Electr. Commer. Res. Appl. **24**, 30–49 (2017)
8. Xia, Y.F., Liu, C.Z., Li, Y.Y., Liu, N.N.: A boosted decision tree approach using Bayesian hyper-parameter optimization for credit scoring. Expert Syst. Appl. **78**, 225–241 (2017)

9. Abelian, J., Castellano, J.G.: A comparative study on base classifiers in ensemble methods for credit scoring. Expert Syst. Appl. **73**, 1–10 (2017)
10. Sivakumar, S., Selvaraj, R.: Predictive modeling of students performance through the enhanced decision tree. Adv. Electr. Commun. Comput. **443**, 21–36 (2018)
11. Pai, P.-F., Chen, C.-T., Hung, Y.-M., Hung, W.-Z., Chang, Y.-C.: A group decision classifier with particle swarm optimization and decision tree for analyzing achievements in mathematics and science. Neural Comput. Appl. **25**(7–8), 2011–2023 (2014)
12. Luo, L., Zhang, X., Peng, H., Lv, W., Zhang, Y.: A new pruning method for decision tree based on structural risk of leaf node. Neural Comput. Appl. **22**(1), 17–26 (2013)

Study on Image Processing Based on Region Growing and Arc-Length Methods

Fei Jiao[1], Tianwen Huang[2(✉)], and Yichang Zhou[2]

[1] Zhaoqing University, Zhaoqing 526060, China
[2] Zhaoqing Meteorological Bureau, Zhaoqing 526040, China
huangtw_zq@126.com

Abstract. With the development of graphic and image processing technology, strong convective weather of Zhaoqing city subarea monitoring and early-warning system (ZQMES) is developed. It uses region growing and arc-length methods to process weather radar image. The specified intensity radar echo is extracted, and the map of Zhaoqing is drawn. The radar echo image and the map are overlapped. The principle of judging whether a strong echo existed in a certain town is judging whether a pixel point of the radar echo image existed in a polygon area of the Zhaoqing map. The system can judge automatically which areas was affected by strong convective echo by using improved arc-length method. As an example, the development of the system provides the experimental basis for the study on region growing and arc-length methods.

Keywords: Image processing · Region growing method · Arc-length method
Weather radar

1 Introduction

Doppler weather radar has enhanced monitoring, early-warning abilities, and weather forecast accuracy. However, in most Chinese meteorological departments, forecasters can only estimate the precipitation location and intensity visually based on their experience. Radar data are not completely utilized. With the development of computer technology and its application in meteorology field [1], a lot of system based on meteorological operation has been paid more and more attention [2–5].

Image processing has been widely applied to many fields of modem sciences, particularly those based on artificial intelligence [6, 7]. Image segmentation is one of the basic and key technologies in image processing. It can separate a target area of interest from the background. At present, the main methods of image segmentation include the threshold-based segmentation method [8], the region-based segmentation method [9] and the edge-based segmentation method [10]. The threshold-based segmentation method is generally used to deal with gray image, and the calculation is very large and the run time is long if it is used to deal with real-time color images. The edge-based segmentation method is suitable for images with obviously clear boundary, such as the segmentation of green leaves and red flowers. Considering that colorized radar images have real-time requirement and the boundary of the various echo color is not clear, the region-based segmentation may be a good choice. It can reduce the region

© Springer International Publishing AG, part of Springer Nature 2018
D.-S. Huang et al. (Eds.): ICIC 2018, LNCS 10954, pp. 372–377, 2018.
https://doi.org/10.1007/978-3-319-95930-6_34

segmentation and the target area can be completely partitioned in a complex environment [11]. The region growing method and arc-length method to process radar images are introduced in this paper to improve the monitoring and early-warning systems for strong convective weather in Zhaoqing. Moreover, the strong convective weather of Zhaoqing city subarea monitoring and early-warning system (ZQMES) is developed.

2 Image Segmentation

The dBZ is a physical quantity that represents the intensity of a radar echo. It is used to measure the intensity of rain, snow and the possibility of predicting the appearance of severe weather, such as hail and strong wind. Figure 1 shows an unprocessed image from the ZQMES. The signal of the strong echo has interference from weak echoes, echo noise, name texts and area boundary lines. This interference reduces the accuracy and efficiency of the prediction. Therefore, image segmentation is needed to extract the information for a strong echo. The traditional region-based method is used for separation, and it only reflects the gray distribution of the image without considering the spatial location of the pixels. So we read some documents and improved the region growing method [12, 13].

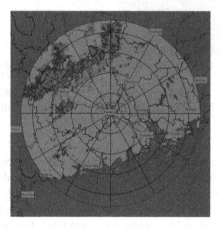

Fig. 1. Original distribution of radar echo (shading, dBZ) image in early-warning system on 20:06, May 11th, 2017 (Universal Time) over Zhaoqing.

Echo intensity of a radar image is described at different levels, which are represented by different colors. So it is easy to find seeds using the region growing method. The ZQMES used this method to extract the radar echoes at a specified intensity. The region growing method can directly detect homogeneous regions with similar pixel values in the image space. The algorithm used in programming is as follows:

(1) Scan the image sequence, that is, through the double loop traversal of the image, check the horizontal and vertical coordinates of every pixel, and find the first one which has not belong to the pixel setting to P_0;

(2) Pixel P_0 is the center, and its neighborhood pixel P is supposed to be traversed. If P is similar to P_0, and the growth criteria is met, then merge P with P_0. At the same time, P is pressed into the stack;

(3) Take a pixel from the stack and return it to step 2 as P_0;

(4) When the stack is empty, return to step 1.

Steps (1)–(4) are repeated until each point in the image has been tested. Refer to the unprocessed image as shown as Fig. 1, Fig. 2a shows the radar image after segmentation of the intensity echo. The strong echo whose intensity is above 35 dBZ is perfectly separated from the original image (Fig. 2a).

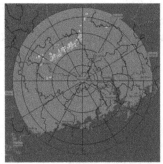 

(a) Segmented radar image   (b) Filtered radar image

Fig. 2. The distribution of radar echo (shading, dBZ) in (a) Segmented radar image, and (b) Filtered radar image. The strong echo whose intensity is above 35 dBZ is perfectly separated from the original image.

However, there are many meaningless echoes that compromise the accuracy of identification. Among these, super refraction echoes are particularly serious [14]. This system evaluates pixels using specific color recognition [15]. Figure 2b shows the filtered radar image according to above methods. Compared with Fig. 2a, extraneous information such as weak echoes, echo noise and texts, has been removed from the image. Removal of such information makes it possible for a computer to recognize colors automatically.

3 Judging Whether a Point Is in a Polygon Area

In order to monitor the distribution of the strong radar echo conveniently, the processed radar image and the Zhaoqing map should be overlaped. In fact, to judge whether a strong echo is in a certain town is also a process to judge whether a point is in a polygon area. The algorithm adopted by the ZQMES is based on arc-length method.

This method demands that the polygon should be a directed polygon. Generally, along the forward direction of the polygon and the left side of the polygon is called as inside of a region. Taking the measured point as the center circle, a unit circle is drawn. Then we make all the directed edges radial projection to the unit circle. The algebraic sum of the arc length on the unit circle is computed. If the algebraic sum is 0, the point is outside the polygon; if the algebraic sum is 2π, the point is in the polygon; if the algebraic sum is π, the point is on the edge of the polygon.

This method is improved and optimized here. Move the origin of the coordinate to the measured point P, then this new coordinate plane is divided into 4 quadrants. For each vertex of a polygon (P_i), its quadrant is only considered, and then access each vertex of the polygon (P_i) in the order of adjacency. As shown in Fig. 3, P_i and P_{i+1} are analyzed. If P_{i+1} is in the next quadrant of P_i, the arc length should be added by $\frac{\pi}{2}$; if P_{i+1} is in the last quadrant of P_i, the arc length should be subtracted by $\frac{\pi}{2}$; if P_{i+1} is in the opposite quadrant of P_i, cross product should be computed. $f = y_{i+1}x_i - x_{i+1}y_i$. If $f = 0$, the point is on the edge of the polygon. If $f < 0$, the sum of the arc length should be subtracted by π. If $f > 0$, the sum of the arc length should be added by π. In the end, algebraic sum is judged as in the case of the above.

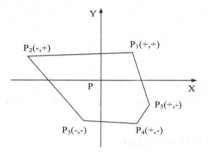

Fig. 3. The diagram of cumulative arc-length, P_1-P_5 are vertexes of the polygon, + and − denote positive and negative symbol of P_i in 4 quadrants, respectively.

More attention should be paid into the algorithm implementation. If one of the coordinates of a polygon vertex P is 0, its symbol is considered to be positive; if one point to be judged is coincident with the vertices of the polygon, it should be dealt with specially. Because it is a rare event, the point meeting the above conditions is considered in the polygon area when the system is developed. Based on this method, areas which are affected by strong echo are captured by the ZQMES.

4 Experimental Results

The ZQMES was implemented in Zhaoqing Meteorological Bureau since December, 2015. Figure 4a is the distribution of strong radar echo (a) on 20:06, May 11th, 2017 (Universal Time) over Zhaoqing. One hour later, the distribution of strong radar echo has changed. As Fig. 4b shows, the strong radar echo becomes scattered and gradually weakened. According to prompts of the system, forecasters make predictions based on experience, and the results are consistent with the facts. Affected by trough and shear line,on May 11th, 2017 (Universal Time), heavy rain and short-term strong wind occurred in most parts of Zhaoqing from north to south. There are 74 weather stations with accumulated rainfall of more than 10 mm and 21 stations with more than 20 mm. The maximum rainfall occurred in Huaicheng town is 35.5 mm. There are 31 weather stations which have recorded 6–8 level of short-term strong wind. The maximum wind speed occurred in the Huaiji county meteorological bureau is 19.8 meters per second (wind force 8).

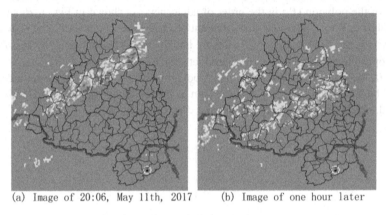

(a) Image of 20:06, May 11th, 2017   (b) Image of one hour later

Fig. 4. The distribution of strong radar echo (a) on 20:06, May 11th, 2017 (Universal Time) over Zhaoqing and the distribution of strong radar echo (b) one hour later.

5   Conclusions and Discussions

By using computer technologies to process images, the ZQMES is developed to improve the capability to monitor and early-warn strong convective weather of Zhaoqing. The Doppler weather radar images are processed, and the echo of a specified intensity level is acquired by using the improved image segmentation method for region growing. Then, the processed radar image and the map of all towns in Zhaoqing which is drawn by computer graphics technology are overlaped, so the system can judge which areas are affected by strong convective echo by using improved arc-length method. In summary, this system provides an intuitive, convenient and fast monitoring and early warning tool for forecasters.

Acknowledgments. The authors were supported by Science and technology innovation project of Zhaoqing (Grant No. 201624030904), Science and technology research project of Guangdong Meteorological Bureau (Grant No. 2016B51), Science and technology research project of Zhaoqing Meteorological Bureau (Grant No. 201609, No. 201708), Intelligent gridded forecasting team of Guangdong Meteorological Bureau (No. 201706).

References

1. Huang, T., Jiao, F.: Data transfer and extension for mining big meteorological data. In: Huang, D.-S., Bevilacqua, V., Premaratne, P., Gupta, P. (eds.) ICIC 2017. LNCS, vol. 10361, pp. 57–66. Springer, Cham (2017). https://doi.org/10.1007/978-3-319-63309-1_6

2. Hwang, Y.I., Paek, I., Yoon, K., Lee, W., Yoo, N., Nam, Y.: Application of wind data from automated weather stations to wind resources estimation in Korea. J. Mech. Sci. Technol. **24**, 2017–2023 (2010). https://doi.org/10.1007/s12206-010-0613-z

3. Wang, X., Jin, X.M., Chen, M.G., Zhang, K., Shen, D.: Topic mining over asynchronous text sequences. IEEE Trans. Knowl. Data Eng. **24**(1), 156–169 (2012)

4. Shen, P.Y., Luo, Y.Z.: Application of electronic sand table system in geological disasters meteorological service based on GIS. Meteorol. Environ. Res. **5**(2), 66–69 (2014)

5. Cremonini, R., Tiranti, D., Barbero, S.: The urban flooding early warning system of the greater turin (North-Western Italy) based on weather-radar observations. Eng. Geol. Soc. Territory **5**, 837–842 (2015)

6. Wan, C., Jin, X.M., Ding, G.G., Shen, D.: Gaussian cardinality restricted boltzmann machines. In: Twenty-ninth AAAI Conference on Artificial Intelligence, pp. 3031–3037 (2015)

7. Guo, Y.C., Ding, G.G., Jin, X.M., Wang, J.M.: Learning predictable and discriminative attributes for visual recognition. In: Proceedings of 29th AAAI Conf. on Artificial Intelligence (AAAI) (2015)

8. Otsu, N.: A threshold selection method from gray-level histograms. IEEE Trans. Syst. Man Cybern. **9**, 62–66 (1979). https://doi.org/10.1109/TSMC.1979.4310076

9. Schulz, D., Burgard, W., Fox, D., Cremers, A.B.: Tracking multiple moving targets with a mobile robot using particle filters and statistical data association. IEEE Int. Conf. Robot. Autom. **2**, 1665–1670 (2001)

10. Mallat, S., Zhong, S.: Characterization of signals from multiscale edges. IEEE Trans. Pattern Anal. Mach. Intell. **7**, 710–732 (1992)

11. Zhu, Q.D., Jing, L., Bi, R.: Exploration and improvement of Ostu threshold segmentation algorithm. In: 2010 8th World Congress on Intelligent Control and Automation (WCICA), pp. 6183–6188 (2010)

12. Giuliano, A., Michele, C.: Microarray image gridding with stochastic search based approaches. Image Vis. Comput. **25**(2), 155–163 (2007)

13. Shih, F.Y., Cheng, S.: Automatic seeded region growing for color image segmentation. Image Vis. Comput. **23**, 877–886 (2005)

14. Fu, Z., Yang, Y., Shu, C., Li, Y., Wu, H., Xu, J.: Improved single image dehazing using dark channel prior. Syst. Eng. Electron. Technol. **26**(5), 1070–1079 (2015)

15. Wang, X.Y., Bu, J.: A fast and robust image segmentation using FCM with spatial information. Digit. Signal Process. **20**(4), 1173–1182 (2010)

Separate Spaces Framework for Entities and Relations Embeddings in Knowledge Graph

Qinhua Huang[(✉)] and Weimin Ouyang

Department of Computer, Shanghai University of Political Science and Law,
Shanghai 201701, China
{hqh, oywm}@shupl.edu.cn

Abstract. Learning Entity and Relation Embeddings is to represent the entities and relations between them in knowledge graph. Recently a variety of models, starting from the work of TransE, to a series of following work, such as TransH, TransR, TransD, are proposed. These models take a relation as transition from head entity to tail entity in principle. The further researches noticed that relations and entities might be able to have different representation to be cast into real world relations. Thus it improved the embedding accuracy of embeddings. Based on these works, we further noticed that in some real world cases entities representation could be vary, especially when one entity in different positions of a relation. In this paper we proposed a new relation embedding by distinguishing head entity and tail entity. It is possible that an entity can have different semantic meanings in head and tail positions. Motivated by this fact, we defined two different entity spaces, head entities and tail entities, for each type of entities. With the training of the general embeddings learning model, we projected the two entity spaces into relation space to run the stochastic gradient descent (SGD) optimization. Experimental results show our method can have good performance in typical knowledge graphs tasks, such as link prediction.

Keywords: Relation embeddings · Head and tail entity · Knowledge graph

1 Introduction

Knowledge graphs is an important model in knowledge representation learning. Natural real world has high complexity when casting into digital model. One method is to find the entities and relations among them. Those entities and their relations are very rich in both structure and content. Knowledge graph is to encode structured information of entities and their rich relations. As this technique developed, currently a typical knowledge graph may contain millions of entities and billions of relational facts, but it is far from complete. The task of knowledge graph completion is to find and describe more relations in an existing knowledge graph. It aims at predicting relations between entities under supervision of the existing knowledge graph. The technique to find if there is a relationship between two entities is called link predication.

© Springer International Publishing AG, part of Springer Nature 2018
D.-S. Huang et al. (Eds.): ICIC 2018, LNCS 10954, pp. 378–388, 2018.
https://doi.org/10.1007/978-3-319-95930-6_35

Link predication is a popular research area with many important applications, including biology, social science, security, and medicine. Traditional approach of link prediction is not capable for knowledge graph completion. Recently, a promising approach for the task is embedding a knowledge graph into a continuous vector space while preserving certain information of the graph. Following this approach, many methods have been further explored, related details can be found in Tables 1 and 2.

Among these methods, TransE [1] and TransH are simple and effective, achieving the state-of-the-art prediction performance. TransE, inspired by the work of word embedding, learns vector embeddings for both entities and relationships. These vector embeddings are set in R^k. TransH is proposed to enable an entity having different representations when involved in various relations. Both TranH and TransE have the assumption that entities should be in the same space. TransR suggested in some situations, entity and relations should have different aspects due to the complexity of real world. Thus TransR proposed a method that the embeddings has different spaces for entities and relations.

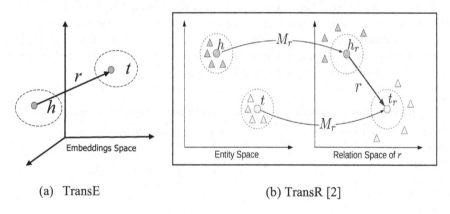

(a) TransE   (b) TransR [2]

Fig. 1. A simple comparison illustration between TransE and TransR

The basic idea of TransE and TransR can be naively illustrated in Fig. 1. In TransR, entities are looked as locations in a unique embeddings space. A relation is thought as the transition vector, which is the distance vector from head entity to tail entity in the same space. Based the complexity assumption, TransR goes further. It takes relations and entities as in different spaces. For each triple (h, r, t), entities in the entity space are first projected into r-relation space as h_r and t_r with operation Mr, and then $h_r + r \approx t_r$. The relation-specific projection can make the head/tail entities that actually hold the relation close with each other, and also get far away from those that do not hold the relation. Moreover, under a specific relation, head-tail entity pairs usually exhibit diverse patterns. It is insufficient to build only a single relation vector to perform all translations from head to tail entities [2].

This paper is organized as follows. First is the introduction part, we shortly give a intuitive description on models, from which our model is derived. In related work section, we list some important related embeddings models. We developed our method in section method and training. After presenting experiment results in experiment section, we conclude our method and propose the future work.

2   Related Works

In this section, we briefly listed the related models in knowledge graph embeddings. TransE firstly purposed projecting the entities into the same space, where the relation can be taken as a vector from head entity to tail entities. Formally, we have a triple (h, r, t), where $h, r, t \in \mathbb{R}^k$, h is the head entity vector, r is the relation vector and t is the tail entity vector. The TransE model represents the a relationship by a translation from head entity to tail entity, thus it holds $h + r \approx t$. By minimizing the score function $f(h, r, t) = \| h + r - t \|_2^2$, which means h + r is the closest to t in distance. This representation has very clear geometric meaning as it showed in Fig. 1.

TransH was proposed to address the issue of N-to-1, 1-to-N and N-to-N relations. It projected (h, r, t) onto a hyperplane of w_r, where w_r is the hyperplane normal vector of r. TransR noticed both TransE and TransH took the assumption that embeddings of entities and relations are represented in the same space $\mathbb{R}^k$. And relations and entities might have different semantic meaning. So TransE suggest project entities and relations onto different spaces in representation, respectively. The score function will be minimized by translating entity space into relation space.

Table 1. Entity and relation embedding models: embeddings and score functions

Model name	Embeddings	Score function s(h,r,t)
Neural Tensor Network (NTN)	$M_{r,1}, M_{r,2} \in \mathbb{R}^{k \times d},\ b_r \in \mathbb{R}^k$	$u_r^\top g\left(h^\top M_r t + M_{r,1} h + M_{r,2} t + b_r\right)$
Latent Factor Model (LFM) [10]		$h^\top M_r t$
Semantic Matching Energy (SME)	M_1, M_2, M_3, M_4 are weight matrices, $\otimes$ is the Hadamard product, b_1, b_2 are bias vectors	$((M_{1h}) \otimes (M_2 r) + b_1) \top ((M_3 t) \otimes (M_4 r) + b_2)$
TranE [1]	$h, r, t \in \mathbb{R}^k$	$\| h + r - t \|$
TransH [5]	$h, t \in \mathbb{R}^k, w_r, d_r \in \mathbb{R}^k$	$\| \left(h - w_r^\top h w_r\right) + d_r - \left(t - w_r^\top t w_r\right) \|$
TransD	$\{h, h_p \in \mathbb{R}^k\}$ for entity h, $\{t, t_p \in \mathbb{R}^k\}$, for entity t, $\{r, r_p \in \mathbb{R}^d\}$ for relation r	$\| \left(h + h_p^\top h r_p\right) + r - \left(t + t_p^\top t r_p\right) \|$
TransR [2]	$h, t \in \mathbb{R}^k, r \in \mathbb{R}^d, M_r \in \mathbb{R}^{k \times d},$ M_r is a projection matrix	$\| M_r h + r - M_r t \|$

There are some other models like Unstructured Model, which is a simplified TransE. It suppose that all r = 0; Structured Embedding, it adopted L_1 as its distance measure since it has two relation-specific matrices for head and tail entities; Neural Tensor Network (NTN), which has some complexity that makes it only suit for small knowledge graphs. For the convenience of comparison, we listed the embeddings and score functions of some models in Table 1.

In Table 2, the constraints of each models are presented. As we should point out that with the models developed, the embeddings and constraints actually become more complicated. One thing is sure that if the model is more complicated, the computation cost goes higher. This problem should be carefully considered in algorithm design.

Table 2. Entity and relation embedding models: constraints

Model name	Constraints
TranE	$h, r, t \in \mathbb{R}^k$
TransH	$h, t \in \mathbb{R}^k, w_r, d_r \in \mathbb{R}^k$
TransD	$\{h, h_p \in \mathbb{R}^k\}$ for entity h, $\{t, t_p \in \mathbb{R}^k\}$, for entity t, $\{r, r_p \in \mathbb{R}^d\}$ for relation r.
TransR	$h, t \in \mathbb{R}^k, r \in \mathbb{R}^d, M_r \in \mathbb{R}^{k \times d}$, M_r is a projection matrix.

3 Method and Training

3.1 Our Model

TransE takes entities and relations as same objects [1]. TransR [2] realized that entities and relations might be possible in different formations. Thus it will remove the dimension regulations and constraints of the same size rule for entities and relations. As a further consideration, we proposed a generalization method that maybe the heads and

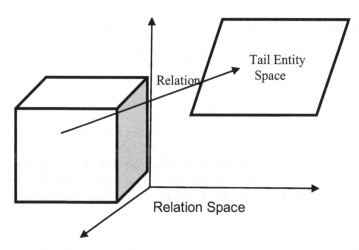

Fig. 2. A simple illustration of entity and relation spaces in our Embeddings model, where the space dimensions k, l and d might not be the same between any two spaces.

tails are different objects too. According to our idea, there are three spaces, head entity space, tail entity space and relation spaces, but these three spaces have different sizes. Figure 2 illustrates the basic idea of our method. For a relation in Fig. 2, the relations has a dimension size of 3, and head entity has size of 3 too, while the entity is 2 in dimension size. Our representation can contain both TransE and TransR, in which they will be a kind of special case. Following this idea, we formally give the definitions.

For each triple (h, r, t), $h \in \mathbb{R}^k, t \in \mathbb{R}^l$, while $r \in \mathbb{R}^d$. Here k, l and d can be different. For the sake of calculation, like in TransR and TransE, we need to project head entities and tail entities into relation space. The projected vectors of head entities and tail entities are defined as

$$h_r = hM_{hr}, t_r = tM_{tr}$$

Where $M_{hr} \in \mathbb{R}^{k \times d}, M_{tr} \in \mathbb{R}^{l \times d}$ are transition matrix.
Routinely the score function is thus defined as

$$f_r(h, t) = \| h_r + r - t_r \|_2^2$$

And there are also constraints on the norms of embeddings h, r, t and the transition matrix. As it showed below

$$\forall h, r, t, \quad \| h \|_2 \leq 1, \| r \|_2 \leq 1, \| t \|_2 \leq 1, \| hM_{hr} \|_2 \leq 1, \| tM_{hr} \|_2 \leq 1$$

With this representation, if $k = l = d$, it becomes TransE; if $k = l \neq d$, it becomes TransR. In some cases, one entity might have different semantic meanings in head position and tail position for a relation. To address this problem, we set k, l and d could not be necessarily identical. So in addition to k, l and d relations mentioned, there is also possible that we have $k \neq l, k \neq d, l \neq d$. Follow the naming tradition, we name this unified model as TransUF.

3.2 Training

The margin-based score function is defined as follow equation.

$$L = \sum_{(h,r,t) \in S} \sum_{(h',r,t') \in S'} \max(0, f_r(h, t) + \gamma - f_r(h', t'))$$

Where max(x,y) aims to get the maximum between x and y, γ is the margin, S is the set of correct triples and S' is the set of incorrect triples.

The optimization method is using stochastic gradient descent (SGD). And the initialization process of entities and relations is done by following the operations in TransR to avoid the overfitting, which actually is the entities embeddings result of TransE.

We implemented and trained our model using multiple thread framework [4, 7].

Algorithm 1. Training

Initialization: p threads, embeddings of each entity and relation

for epoch = 1; epoch < num epoches; epoch + + do

    for each thread do

        i=1

        repeat

            sample a golden triple (h, r, t)

            construct a corrupted triple (h', r, t')

            if $s(h, r, t) + \gamma - s(h0, r, t0) > 0$ then

                calculate the gradient of the embeddings of h, t, h', t', r

                subtract the gradient from the corresponding embeddings

            end if

            normalize h, t, h', t', r according to the constraints

            i++

        until i== number of triples a thread handles in an epoch

    end for

end for

save the embeddings into files

---

4 Experiments and Result

In this section firstly we describe our dataset. And then give the experiment result. With the convention of knowledge graph, we use link prediction to test the performance with other algorithms.

4.1 Data Sets and Experiment Setting

In this paper, we evaluate our methods with two typical knowledge graphs, built with Freebase [8] and WordNet [9]. These two datasets are chosen of the same with TransE [1], TransR [2] and other embedding models. The statics of the datasets are listed in Table 3.

These two datasets have a little difference in semantic meaning layer. We briefly

Table 3. Datasets statics

Dataset	#Rel	#Ent	#Train	#Valid	#Test
FB15 k	1,345	14,951	483,142	50,000	59,071
WN18	18	40,943	141,442	5,000	50,000

give a short introduction here. Freebase provides general facts of the world. For example, the triple (Steve Jobs, founded, Apple Inc.) builds a relation of founded between the name entity Steve Jobs and the organization entity Apple Inc. In WordNet, each entity is a synset consisting of several words, corresponding to a distinct word sense. Relationships are defined between synsets indicating their lexical relations, such as hypernym, hyponym, meronym and holonym.

4.2   Link Prediction

Link prediction aims to predict the missing h or t for a relation fact triple (h, r, t), used in [1–3]. In this task, for each position of missing entity, the system is asked to rank a set of candidate entities from the knowledge graph, instead of only giving one best result. As set up in [1–3], we conduct experiments using the data sets FB15 K and WN18.

In testing phase, for each test triple (h, r, t), we replace the head/tail entity by all entities in the knowledge graph, and rank these entities in descending order of similarity scores calculated by score function. Following TransR [2] and ParaGraphE [3], we use 4 measures as our evaluation metric. Each measure has 2 settings, raw and filtered. The meanings of each measure are showed in Table 4.

Generally, if link predictor can have lower mean rank or higher Hits@1, Hits@10,

Table 4. The meaning of each measure

Measure	Meaning
mr	The mean rank of correct entities
mrr	The value of mean reciprocal rank
hits@1	The proportion of ranks list at first
hits@10	The proportion of correct entities in top-10 ranked entities
Raw	The metrics calculated on all corrupted triples
Filter	The metrics calculated on corrupted triples without those already existing in knowledge graph

we think it had better performance. In fact, a corrupted triple may also exist in knowledge graphs, which should be also considered as correct. However, the above evaluation may under-estimate those systems that rank these corrupted but correct triples high. Hence, before ranking we may filter out these corrupted triples which have appeared in knowledge graph. We name the first evaluation setting as "Raw" and the latter one as "Filter".

We describe the experiment running parameters here. We implemented our algorithm based on the framework of ParaGraphE library [3]. The learning rate λ is set to be fixed to 0.001, the margin γ set 4. The thread number, which is defined by ParaGraphE, is set to 10 constantly. In the comparison tests with TransR and TransE, the dimensions of head entity, tail entity, relation are all set to 50. And the dimension of TransE test is set to 50. Also the head entity and relation dimensions in TransR are set to 50 too. And the batch size is set to 1000. For both datasets tests, WN18 and FB15K, they have the same configuration. The SGD method is employed to do optimization. All training processes go through 500 rounds.

Table 5. Experiment results on FB15k.

Method	Raw				Filter			
	mr	mrr	hits@1	hits@10	mr	mrr	hits@1	hits@10
TransE	241.813	0.14999	0.058582	0.339168	125.908	0.232222	0.115175	0.460818
TransR	314.001	0.142759	0.0684769	0.295881	185.407	0.215418	0.124333	0.39841
OurMethod	455.258	0.129644	0.0642193	0.264555	309.258	0.204106	0.122412	0.369403

Table 6. Experiment results on WN18.

Method	Raw				Filter			
	mr	mrr	hits@1	hits@10	mr	mrr	hits@1	hits@10
TransE	635.344	0.292859	0.0369	0.7577	621.89	0.382625	0.067	0.8861
TransR	649.733	0.273091	0.0303	0.7034	637.062	0.35026	0.0488	0.8150
OurMethod	740.927	0.285938	0.0342	0.7249	785.136	0.28574	0.0107	0.3825

Evaluation results on both WN18 and FB15K are shown in Tables 5 and 6. From the table we find generally these three method has approximate performance. For example, in FB15K we achieved better mrr. On other measure our results are little bit behind, but not with significant difference.

Table 7. Tests results with varied tail entity dimension, raw settings

Tail entity dimension	mr	hits@10	mrr	hits@1
50	314.001	0.295881	0.142759	0.0684769
60	487.077	0.257013	0.124818	0.0611383
70	477.613	0.25835	0.127649	0.0627888
80	488.437	0.258502	0.125599	0.0612568

Tables 7 and 8 showed the test results when we fixed the dimension of head entity and relation. These two fixed dimensions are set to 80, 50 respectively. To observe the affection of varied tail entity dimensions, we do the tests on dataset FB15K and let tail entity dimension to be 50, 60, 70, 80. It indicates that the performance is better at dimension 50. While on the other settings the performance has some fluctuation. As it should be pointed out here, setting a different tail entity dimension is to address the problem of one entity might have different semantic meaning in different relations.

Table 8. Tests results with varied tail entity dimension, filtered settings.

Tail entity dimension	f_mr	f_hits@10	f_mrr	f_hits@1
50	185.407	0.39841	0.215418	0.124333
60	341.201	0.359711	0.201627	0.123335
70	331.831	0.359059	0.201579	0.122488
80	342.431	0.361201	0.200859	0.121913

4.3 Epoch Loss Rate

The epoch loss rate indicated the efficiency of training. Figure 3 showed the epoch loss vs epochs comparison tests. We can observe when the test was running across 500 epoches, our method performs better. The red curve (our method) goes down more quickly than TransR and TransE, while the latter two curves performed very similar to each other. This indicates that our method converges more fast, which will be sure to improve the learning efficiency.

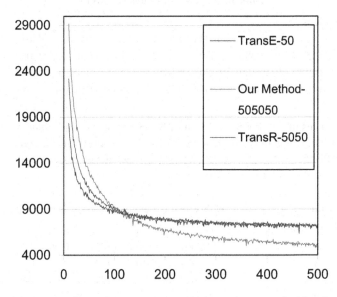

Fig. 3. Epoch loss vs epoches, when come across 100th epoch, our method has faster triple loss changing rate. (Color figure online)

Figure 4 showed epoch loss vs. epoches curve when we increase the tail entity dimension from 50 to 80, meanwhile keep the dimension of relations and head entity constant. The dimension size of head entities and relations are set to 80, 50 respectively. We can observe that when the tail entities dimension decreases, the epoch loss goes down a little bit, but not very significant. This might due to the addition of size of dimension is not too much significant compare to the original size. The four curves have similar trends and very close to each other.

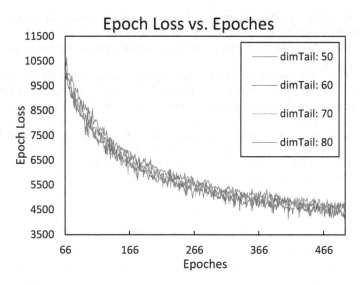

Fig. 4. Epoch loss vs. epochs, when the tail entities dimension varied from 50 to 80.

5 Conclusion

In this paper we proposed a new unified framework method using separate spaces for entity and relation embedding. Motivated by the TransR model, we go further to free the dimension constraint on tail entity representation. Based on the implementation framework, we test our algorithm on freebase dataset. The experiment results showed that our model can have similar algorithmic performance and smaller epoch loss in training. Meanwhile the operation of taking an entity as different vector representation by its locations in a relation can have potential benefits in semantic sense. And this representation is more flexible.

As our further work, we are going to investigate the semantic benefits of our representation. Also it's worth of keeping on improving the optimization method in the training process.

References

1. Bordes, A., Usunier, N., Garcia-Duran, A., Weston, J., Yakhnenko, O.: Translating embeddings for modeling multi-relational data. In: NIPS, pp. 2787–2795 (2013)
2. Yankai, L., Zhiyuan, L., Maosong, S., Yang, L., Xuan, Z.: Learning entity and relation embeddings for knowledge graph completion. In: AAAI, pp. 2181–2187 (2015)
3. Niu, X.-F., Li, W.-J.: ParaGraphE: a library for parallel knowledge graph embedding. arXiv: 1703.05614v3 (2017)
4. Recht, B., Re, C., Wright, S., Niu. F.: Hogwild: a lock-free approach to parallelizing stochastic gradient descent. In: NIPS, pp. 693–701 (2011)
5. Wang, Z., Zhang, J., Feng, J., Chen, Z.: Knowledge graph embedding by translating on hyperplanes. In: AAAI, pp. 1112–1119 (2014)

6. Xiao, H., Huang, M., Yu, H., Zhu, X.: From one point to a manifold: knowledge graph embedding for precise link prediction. In: IJCAI, pp. 1315–1321 (2016)

7. Zhao, S.-Y., Zhang, G.-D., Li, W.-J.: Lock-free optimization for nonconvex problems. In: AAAI, pp. 2935–2941 (2017)

8. Miller, G.A.: Wordnet: a lexical database for english. Commun. ACM **38**(11), 39–41 (1995)

9. Bollacker, K., Evans, C., Paritosh, P., Sturge, T., Taylor, J.: Freebase: a collaboratively created graph database for structuring human knowledge. In: Proceedings of KDD, pp. 1247–1250 (2008)

10. Jenatton, R., Roux, N.L., Bordes, A., Obozinski, G.R.: A latent factor model for highly multi-relational data. In: Proceedings of NIPS, pp. 3167–3175 (2012)

11. Ji, G., He, S., Xu, L., Liu, K., Zhao, J.: Knowledge graph embedding via dynamic mapping matrix. In: ACL, pp. 687–696 (2015)

Knowledge Reasoning Based on 2n-Valued Temporal Logic

Yuanxiu Liao, Jingli Wu, and Xudong Luo[✉]

Guangxi Key Lab of Multi-Source Information Mining and Security,
Guangxi Regional Multi-Source Information Integration and Intelligent
Processing, Collaborative Innovation Center, Faculty of Computer Science
and Information Engineering, Guangxi Normal University,
Guilin, Guangxi 541004, China
luoxd@mailbox.gxnu.edu.cn

Abstract. Knowledge representation and reasoning is one of the most important topics in the field of artificial intelligence. Classical 2-valued logic is often
used for representation and reasoning of precise knowledge. However, in
practical applications, intelligent systems need to understand and handle
knowledge involving time and fuzzy concepts, which classical 2-valued logic
cannot do well. To this end, in this paper we propose a 2n-valued temporal logic
for representing and reasoning of temporal, fuzzy knowledge. Moreover, we
discuss the characteristics of the logic, including some of theorems that hold in
the classical logic as well as some unique to the logical proposed. Finally, we
give an example of the impact of seasonal cooling on influenza to show how our
logic can be used in practice.

Keywords: Temporal logic · Knowledge representation and reasoning
2n-valued logic · Generalised effective formula

1 Introduction

Time and fuzziness are two important concepts that human beings cannot avoid.
Intelligent systems that interact with human beings must have the ability to learn,
understand, model, and apply knowledge concerning time and fuzziness to accomplish
given complex tasks. As the intelligence of an intelligent system is realised through
computer programs, formal methods are needed to model the robot's knowledge and
actions. Classical propositional logic and predicate logic provide a formal method for
knowledge representation and reasoning. However, they cannot handle knowledge
concerning time and fuzziness.

To this end, in this paper we propose a 2n-valued temporal logic for representation
and reasoning of temporal, fuzzy knowledge. We also discuss some properties of our
logic, which are the same as those in the classical logic, as well as some special ones of
our logical properties. In addition, we give an example, which uses the proposed logic
to describe that the seasonal cold weather caused the occurrence of influenza related
knowledge, and predict of influenza.

© Springer International Publishing AG, part of Springer Nature 2018
D.-S. Huang et al. (Eds.): ICIC 2018, LNCS 10954, pp. 389–399, 2018.
https://doi.org/10.1007/978-3-319-95930-6_36

The rest of this paper is organised as follows. Section 2 discusses the state of art on this topic. Section 3 presents the syntax and semantics of our fuzzy temporal logic. Section 4 reveals some properties of our logic. Section 5 illustrates how our logic can be used to solve a practical problem. Finally, Sect. 6 concludes the paper.

2 Related Work

In order to deal with knowledge concerning time, Arthur [1] proposed temporal logic. Afterwards, people carried out extensive, in-depth research on temporal logic, and widely used this kind of logic in knowledge representation and reasoning, program verification, and robot action planning.

In the aspect of knowledge representation and reasoning, Kreutzmann et al. [5] used temporal logic to handle the specification, identification, and interpretation. They treated the problem of recognition as a model checking problem, and provided an abstract process description of the robot's intelligent operation in a dynamic environment. For example, in a warehouse logistics scenario a robotic observer can anticipate a dangerous situation by identifying a process and inferring its own safe area. And Xu et al. [8] proposed a method for temporal logic reasoning based on prior information (i.e., provided a new method to derive a temporal logic formula that models the action of a robot arm). These formulas are defined by a set of parameters, which values are determined by minimising a cost function that balances the matching of the formula with the trajectory and the similarity between its predicates and a priori known predicates. The algorithm is tested by using the experimental trajectory generated by PHANToM Omni robot. Gol et al. [9] studied the problem of prediction and control based on temporal logic model, and proposed an optimal control strategy for discrete-time linear systems, which satisfies the temporal logic specifications of a set of linear predicates over their state variables. The specification is a syntactically common linear temporal logic formula that can be satisfied within a limited time.

In the aspect of program verification, in order to solve the problem of autonomous vehicles driving safely in the uncertain environment, Jha et al. [4] proposed a kind of temporal logic probability expansion, called opportunistic constrained temporal logic (C2TL), which is used to designate the correctness in case of uncertainty of perception demand. C2TL extends temporal logic by using opportunity constraints as predicates in formulas that allow modelling of perceived uncertainty while preserving the simplicity of its reasoning. And Favara et al. [6] discuss the application of temporal logic to safety monitoring and model-based risk monitoring. They extended the previously introduced safety monitoring framework and used temporal logic elements to improve accident prevention and dynamic risk Assessment. Dixon et al. [10] studied the application of the logic of temporal knowledge in the specification and verification of security protocols, proved that the typical hypotheses related to the desirability of the authentication protocol, and presented a case where a solution based on theorem prover is used to describe security protocol in temporal knowledge logic.

In the aspect of robot action planning, Koo et al. [2] proposed a multi-robot action planning framework based on temporal logic specifications. It is used to coordinate robot networks for formal requirements specification in temporal logic. Gillani et al. [7]

studied physics-based action planning with temporal logic norms, and proposed a method of linear temporal logic programming, which uses ontologies to enhance task description and reasoning. Their approach enhances the autonomy of robots and empowers them to make plans in discrete and continuous spaces to find the actions required to accomplish complex tasks. Levonevskiy et al. [3] proposed an integration model of a multi-level company service based on the functional temporal context of action-tense logical intelligence, and discussed the architecture of multi-modal information and navigation cloud systems. They used the integrated corporate information infrastructure to develop enterprise intelligence space, and used available resources to achieve the common goal of users.

In addition to knowledge concerning the concept of time, intelligent robots also need to deal with knowledge concerning the concept of fuzziness. As a result, researchers have further expanded the temporal logic and introduced various forms of fuzzy temporal logic. Mukherjee et al. [11] integrated the concept of real-time interval temporal logic (such as metrological interval temporal logic) with fuzzy logic to obtain a fuzzy real-time temporal logic. In their logic, the concept of fuzzy time is used to define the development of synchronisation specification. And to evaluate the control quality of time as the key factor of an embedded control system, they also proposed a method of computing the quality of specification satisfaction for a given implementation by using certain fuzzy membership functions.

Shi [12] constructed a quantitative method of linear temporal logic (LTL). Based on some kind of probability measure of Kripke structure, the satisfaction of LTL formula is defined as a quantitative concept extension of the classic case in model verification. In the work, Shi also introduced the concept of similarity between LTL formulas. Specifically, Shi applied a pseudo-metric into all LTL formula sets so that the LTL logical metric space can be constructed. Sotudeh et al. [13] proposed a fuzzy Kripke model and a verification method of fuzzy tree logic. They use fuzzy temporal logic as well as model abstraction and approximation to deal with the uncertainty in the model checking. Selvi et al. [14] proposed a new energy-efficient routing algorithm based on data collection and clustering, which uses fuzzy temporal logic to form clusters and perform cluster based routing. The cluster uses two cluster heads: one for data collection and the other for routing. Their experiments show that the fuzzy clustering algorithm based on time constraint can stretch the life cycle of high network, reduce energy consumption, and improve the quality of service (i.e., increasing packet delivery rate and reducing delay). Frigeri et al. [15] proposed a time frame, called fuzzy temporal logic, to deal with fuzziness in time. The framework defines a set of modes of fuzzy tense that can be defined by selecting a particular semantic of the conjunction. They proved that the semantics are reliable. To characterise inaccurate temporal knowledge in classical description logic, Ho et al. [16] proposed a kind of fuzzy temporal description logic and its formal grammar and semantics, and proved that the logic captures the lack of precise modelling and reasoning temporal information.

3 Logic System

This section presents a 2n-valued temporal logic (2nvTL), which is used to handle knowledge representation and reasoning with temporal and fuzzy concepts.

3.1 Syntax

Initial symbol

1. Proposition symbol $p, q, r, \cdots$, are used to represent the atomic proposition.
2. Propositional conjunctions $\neg, \wedge, \vee \rightarrow$, and $\leftrightarrow$ are used to construct compound formulas.
3. Tense operators G, F, H, and P are used to represent the concept of tense. Specifically, G is called a strong future operator, which means "will happen all the time in future"; F is called a weak future operator, which means "will happen once in future"; H is called strong past operator, which means "happened all the time in past"; and P is called a weak past operator, which means "happened once in past".

Formula rules

1. Proposition symbols are formulas.
2. If A is a formula, $\neg A$ is also a formula.
3. If A and B are formulas, $A \wedge B$, $A \vee B$, $A \rightarrow B$ and $A \leftrightarrow B$ are also formulas.
4. If A is a formula, GA, FA, HA, PA are also formulas.
5. All formulas are obtained through the above four steps.

Reasoning rules MP (Modus Ponens): If A and $A \rightarrow B$ then B.

2nvTL minimal system FT_0 reasoning rules is only one (i.e., the above separation rules).

3.2 Semantics

Definition 1. Let True $= \{\alpha_1, \cdots, \alpha_n\}$, where $\alpha_1, \cdots, \alpha_n$ represent different true degree; and False $= \{\beta_1, \cdots, \beta_n\}$, where $\beta_1, \cdots, \beta_n$ represent different false degrees; and $\alpha_1 > \cdots > \alpha_n > \beta_n > \cdots > \beta_1$. Then the set of truth values of a propositional formula in 2nvTL is defined as follows:

$$TF = True \cup False \qquad (1)$$

For example, it can be defined that α_1 stands for "absolutely true", α_2 for "extremely true", α_3 for "very true", and α_4 for "true"; and β_1 stands for "absolute false", β_2 for "extremely false", β_3 for "very false", and β_4 for "false".

Definition 2. The temporal structure of 2nvTL is a tuple (T, R), where T is a nonempty set of moments and R is a binary relation on T. For any $t_1, t_2 \in T$, $t_1 R t_2$ means that t_1 is the past of t_2 and t_2 is the future of t_1.

The rules for calculating the truth-value of any logical formula are as follows:

Definition 3. Let Q be the set of atomic propositions in 2nvTL. $V : Q \times T \rightarrow TF$ is a mapping from the set of atomic propositional formulas to the truth value set. Then the assignment of a formula in 2nvTL to structure (T, R) is recursively defined as follows: for any two formulas A and B, and for any moment $t \in T$,

1. the assignment of atomic proposition A at t is the value of mapping V at (A, t);
2. $V(\neg A, t) = \begin{cases} \beta_i \text{ if } V(A, t) = \alpha_i \\ \alpha_i \text{ if } V(A, t) = \beta_i \end{cases}$, where $i \in \{1, \cdots, n\}$;
3. $V(A \wedge B, t) = \min\{V(A, t), V(B, t)\}$;
4. $V(A \vee B, t) = \max\{V(A, t), V(B, t)\}$;
5. $V(A \rightarrow B, t) = \max\{V(\neg A, t), V(B, t)\}$;
6. $V(A \leftrightarrow B, t) = \min\{V(A \rightarrow B, t), V(B \rightarrow A, t)\}$;
7. $V(GA, t) = \min\{V(A, t')|\forall t' \in T(tRt')\}$;
8. $V(FA, t) = \max\{V(A, t')|\forall t' \in T(tRt')\}$
9. $V(HA, t) = \min\{V(A, t')|\forall t' \in T(t'Rt)\}$;
10. $V(PA, t) = \max\{V(A, t')|\forall t' \in T(t'Rt)\}$.

Definition 4. A model of 2nvTL is a tuple (T, R, V), where (T, R) is a structure and V is an assignment on (T, R).

Definition 5. Let (T, R, V) be a model of 2nvTL, A be a formula, and t be any element of T.

1. A is true at t if and only if $V(A, t) \in$ True, denoted as $(T, R, V) \vDash_t A$.;
2. A is satisfiable on (T, R, V) if and only if there exists $t \in T$ such that $(T, R, V) \vDash_t A$.;
3. A is valid on (T, R, V) if and only if $(T, R, V) \vDash_t A$. for any $t \in T$, denoted as $(T, R, V) \vDash A$.;
4. A is called the efficient if and only if A is valid on any model, denoted as $\vDash A$; and
5. A is called contradictory if and only if for any model (T, R, V), A is unsatisfiable.

4 Properties

When people conduct logical reasoning, they often use some logical properties. In this section, we will reveal that in our multi-valued logic 2nvTL, some classical logic properties still hold, while the other classical logic properties can appear only in weaker forms.

Theorem 1. $V(\neg\neg A, t) = V(A, t)$ for any model (T, R, V), any formula A and any time point $t \in T$.

Proof: If $V(A, t) = \alpha_i$, then $V(\neg A, t) = \beta_i$ and thus $V(\neg\neg A, t) = \alpha_i$. □

Theorem 1 states that the double law of the formula holds in 2nvTL.

Theorem 2. For any model (T, R, V), any formulas A and B, and any $t \in T$, we have:

1. $V(\neg(A \wedge B), t) = V(\neg A \vee \neg B, t)$
2. $V(\neg(A \vee B), t) = V(\neg A \wedge \neg B, t)$

Proof:

(1) Without loss of generality, suppose $V(A, t) = \alpha_i$, $V(B, t) = \alpha_j$ and $\alpha_i \gg \alpha_j$. On the one hand, since

$$V(A \wedge B, t) = \min\{V(A, t), V(B, t)\}$$
$$= \min\{\alpha_i, \alpha_j\}$$
$$= \alpha_j.$$

we know $V(\neg(A \wedge B), t) = \beta_j$. On the other hand,

$$V(\neg A \vee \neg B, t) = \max\{V(\neg A, t), V \neg B, t)\}$$
$$= \max\{\beta_i, \beta_j\}$$
$$= \beta_j.$$

Therefore, we have $V(\neg(A \wedge B), t) = V(\neg A \vee \neg B, t)$.

(2) Similarly, we can prove item 2.   □

Theorem 2 states that De Morgan law holds in 2nvTL.

Theorem 3. For any model (T, R, V), any of the formulas A, and any $t \in T$, if $V(A, t) = \alpha_i$ or $V(A, t) = \beta_i$ then

1. $V(A \vee \neg A, t) = \alpha_i$;
2. $V(A \wedge \neg A, t) = \beta_i$.

The proof of Theorem 3 is straightforward, so it is omitted here. We all know that in the multi-valued logic, the law of middle row and the contradiction do not hold. However, in 2nvTL we have $V(A \vee \neg A, t) \in \text{True}$, $V(A \wedge \neg A, t) \in \text{False}$. for any formula A. So, Theorem 3 states that in the case of "generalised truth" and "generalised false", formula $A \vee \neg A$ is valid in 2nvTL and formula $A \wedge \neg A$ is contradictory in 2nvTL.

Theorem 4. For any model (T, R, V), any formulas A and B, and any $t \in T$,

$$V(A \rightarrow B, t) = V(\neg B \rightarrow \neg A, t).$$

Proof: Without loss of generality, we assume $V(A, t) = \alpha_i$, $V(B, t) = \alpha_j$ and $\alpha_i \gg \alpha_j$. On the one hand,

$$V(A \rightarrow B, t) = \max\{V(\neg A, t), V(B, t)\}$$
$$= \max\{\beta_i, \alpha_j\}$$
$$= \alpha_j;$$

on the other hand,

$$V(\neg B \rightarrow \neg A, t) = \max\{V(\neg\neg B, t), V(\neg A, t)\}$$
$$= \max\{V(B, t), V(\neg A, t)\}$$
$$= \{\alpha_j, \beta_i\}$$
$$= \alpha_j.$$

Therefore, the theorem holds.   □

Theorem 5. In 2nvTL, the following formulas are valid:

1. $A \wedge B \rightarrow A$;
2. $A \rightarrow A \vee B$;
3. $(A \rightarrow B) \wedge A \rightarrow B$;
4. $(A \rightarrow B) \wedge (B \rightarrow C) \rightarrow (A \rightarrow C)$.

Proof: The proofs of items 1 and 2 are straightforward, so we omit them here.

The proof of item 3 is as follows. Assume that (T, R, V) is an arbitrary model, A and B are arbitrary formulas, t is an arbitrary element in T. Let $V(A, t) = \alpha_i$, and $V(B, t) = \alpha_j$, and $\alpha_i \gg \alpha_j$. Then by Definition 4, we have

$$V((A \rightarrow B) \wedge A \rightarrow B, t)$$
$$= \max\{V(\neg((A \rightarrow B) \wedge A), t), V(B, t)\}$$
$$= \max\{V(\neg(A \rightarrow B) \vee \neg A, t), \alpha_j\}$$
$$= \max\{\max\{V(\neg(A \rightarrow B), t), V(\neg A, t)\}, \alpha_j\}$$
$$= \max\{\max\{V(\neg(A \rightarrow B), t), \beta_i\}, \alpha_j\}.$$

And since $V(A \rightarrow B, t) = \max\{V(\neg A, t), V(B, t)\} = \max\{\beta_i, \alpha_j\} = \alpha_j$, we have $V(\neg(A \rightarrow B), t) = \beta_j$. Therefore, we have

$$V((A \rightarrow B) \wedge A \rightarrow B, t) = \max\{\max\{V(\neg(A \rightarrow B), t), \beta_i\}, \alpha_j\} = \alpha_j.$$

Then by the arbitrariness of (T, R, V), A, B and t, we have $(A \rightarrow B) \wedge A \rightarrow B$. Similarly, we can prove item 4.   □

Theorem 5 shows that some tautologies in the classical propositional logic are also established in the sense of "generalised truth" and "generalised falsehood".

5   Example

This section gives a 2nvTL application example. Formulas in 2nvTL are used to represent the relationship between influenza and the cooling of the weather, and reason the likelihood and magnitude of the influenza before it gets cool.

5.1   Background Knowledge

According to a city meteorological data and the number of influenza that occurred in the monthly data, we have the following knowledge:

1. The impact of the season on the influenza: Most of influenza happen in winter (December-February), less in spring (March-May), much less in autumn (September-November), and the least in summer (June-August).
2. The impact of cooling on influenza: Let D denote the decreasing range of daily average temperature. Then the probability of influenza is:
   (a) extremely high in the case of $20\ ^\circ C \le D$;
   (b) very high in the case of $15\ ^\circ C \le D < 20\ ^\circ C$;
   (c) high in the case of $10\ ^\circ C \le D < 15\ ^\circ C$; and
   (d) low in the case of $5\ ^\circ C \le D < 10^\circ C$.
3. According to the monthly incidence rate, denoted M, of urban population, the influenza can be divided into 5 grades:
   (e) an extremely large epidemic situation when $1\% \le M$;
   (f) major epidemic situation when $0.5\% \le M < 1\%$;
   (g) a larger epidemic situation when $0.2\% \le M < 0.5\%$;
   (h) a general epidemic situation when $0.1\% \le M < 0.2\%$; and
   (i) a slight epidemic situation when $M < 0.1\%$.

5.2   Correlation Among Temperature Dropping, Season and Influenza

According to the background knowledge, the correlation among temperature dropping, season, and influenza can be described as follows:

1. in the case of winter temperature dropping and $20\ ^\circ C \le D$, an extremely large epidemic situation of influenza will occur;
2. in the case of winter cooling and $15\ ^\circ C \le D < 20\ ^\circ C$, a major epidemic situation of influenza will occur;
3. in the case of winter temperature dropping and $10\ ^\circ C \le D < 15\ ^\circ C$, a great epidemic situation of influenza will occur;
4. in the case of winter temperature dropping and $5\ ^\circ C \le D < 10\ ^\circ C$, a general epidemic situation will occur;
5. in the case of spring temperature dropping and $20\ ^\circ C \le D$, a major epidemic situation of influenza will occur;
6. in the case of spring temperature dropping and $15\ ^\circ C \le D < 20\ ^\circ C$, a great epidemic situation of influenza will occur;
7. in the case of spring temperature dropping and $10\ ^\circ C \le D < 15\ ^\circ C$, a general epidemic situation of influenza will occur; and
8. in the case of spring temperature dropping and $5\ ^\circ C \le D < 10^\circ C$, a slight epidemic situation of influenza will occur.

Due to space limitations, only the data of winter and spring are listed above.

5.3   2nvTL Formulas of Relevant Knowledge and Reasoning

Let SpTD denote spring temperature dropping, and WiTD denote winter temperature dropping. Then SpTD → EIH means that if in spring temperature drops then an influenza will occur; and WiTD → EIH means that if in winter temperature drops then an influenza will occur.

Suppose that a weather forecast is as follows: from 25 December 2017 to 31 December 2017, the average daily temperature dropping is 16 °C; and 1 January 2018 to 5 January 2018 temperature will gradually pick up and the average daily temperature is 12 °C. Formally, for the model of 2nvTL (T, R, V), we have:

- $T = \{25.12.2017, \cdots, 5.1.2018\}$
- R is defined as follows: For any $t_1, t_2 \in T$, R if and only if R is the date after t_1.
- The assignment of V to the formula is as follows:

$$V(\text{WiTD} \to \text{EIH}, t) = \begin{cases} \alpha_3 \text{ if } t \in \{25.12.2017, \cdots, 31.12.2017\} \\ \alpha_4 \text{ if } t \in \{01.01.2018, \cdots, 05.01.2018\} \end{cases}$$

$$V(\text{WiTD}, t) = \alpha_2 \text{ if } t \in \{25.12.2017, \cdots, 05.01.2018\}$$

Thus, the forecast of influenza outbreak from 25 December 2017 to 31 December 2017 is inferred as follows. Since $V(\text{WiTD}, t) = \alpha_2$ and $V(\text{WiTD} \to \text{EIH}, t) = \alpha_3$, by Definition 3 we have

$$V(\text{WiTD} \to \text{EIH}, t) = \max\{V(\neg\text{WiTD}, t), V(\text{EIH}, t)\}$$
$$= \max\{\beta_2, V(\text{EIH}, t)\}$$
$$= \alpha_3;$$

and because $\beta_2 < \alpha_3$, we have $V(\text{EIH}, t) = \alpha_3$ Further, by Definition 3, we have

$$V(\text{GEIH}, t) = \min\{V(\text{EIH}, t') | \forall t' \in T(tRt')\}$$
$$= \alpha_3.$$

That is, we can predict that between 25 December 2017 and 31 December 2017 a major influenza will happen.

The reasoning of predicting the outbreak of influenza from 25 Dec 2017 to 5 January 2018 is as follows. It can be seen from the above reasoning that

$$\forall t \in \{25.12.2017, \cdots, 31.12.2017\}, \; V(\text{EIH}, t) = \alpha_3$$

holds true. For any $t \in \{1.1.2018, \cdots, 5.1.2018\}$, similar to the above proof, we can prove $V(\text{EIH}, t) = \alpha_4$. Because $\forall t \in \{25.12.2017, \cdots, 31.12.2017\}$,, $V(\text{EIH}, t) = \alpha_3$ and $\forall t \in \{1.1.2018, \cdots, 5.1.2018\}$, $V(\text{EIH}, t) = \alpha_4$, by Definition 3, we have

$$V(GEIH, t) = min\{V(EIH, t')|\forall t' \in T(tRt')\}$$
$$= min\{\alpha_3, \alpha_4\}$$
$$= \alpha_4.$$

That is, we can predicte that between 25 December 2017 and 5 January 2018 a big influenza will happen.

6 Conclusion

In order to carry out formal reasoning according to the knowledge concerning time and fuzziness, this paper extends the classical 2-valued temporal logic to a 2n-valued temporal logic (2nvTL), which has truth-values of n different degrees of "truth" and n different degrees of "falsehood". We also prove that this logic keeps some properties of the classical 2-valued logic, such as double negation law and De Morgan law. Moreover, we reveal the property of *generalised efficiency* for some formulas of 2nvTL in the sense of "generalised truth" and "generalised falsehood". Finally, our logic is used to predict the occurrence of influenza caused by seasonal temperature dropping. Due to the inaccuracy of weather forecast, it is impossible to predict the scale of influenza, and so it is difficult to use classic 2-valued temporal logic to represent the relevant knowledge. Therefore, our temporal logic 2nvTL has important theoretical significance and application prospect.

Acknowledgments. This work was supported by the National Natural Science Foundation of China (Nos. 61662007, 61762015, and 61762016), and Guangxi Key Lab of Multi-Source Information Mining and Security (No. 18-A-01-02).

References

1. Arthur, N.P.: Time and Modality. Oxford University Press, Oxford (1957)
2. Koo, T.J., et al.: A framework for multi-robot motion planning from temporal logic specifications. Sci. China Inf. Sci. **55**(7), 1675–1692 (2012)
3. Levonevskiy, D., Vatamaniuk, I., Saveliev, A.: Integration of corporate electronic services into a smart space using temporal logic of actions. In: Ronzhin, A., Rigoll, G., Meshcheryakov, R. (eds.) ICR 2017. LNCS (LNAI), vol. 10459, pp. 134–143. Springer, Cham (2017). https://doi.org/10.1007/978-3-319-66471-2_15
4. Jha, S., Raman, V., Sadigh, D., Seshia, S.A.: Safe autonomy under perception uncertainty using chance-constrained temporal logic. J. Autom. Reasoning **60**, 43 (2018)
5. Kreutzmann, A., et al.: Temporal logic for process specification and recognition. Intel. Serv. Robot. **6**(1), 5–18 (2013)
6. Favara, F.M., Saleh, J.H.: Application of temporal logic for safety supervisory control and model-based hazard monitoring. Reliab. Eng. Syst. Saf. **169**, 166–178 (2018)
7. Gillani, M., Akbari, A., Rosell, J.: Physics-based motion planning with temporal logic specifications. IFAC-PapersOnLine **50–1**, 8993–8999 (2017)
8. Xu, Z., Belta, C., Julius, A.: Temporal logic inference with prior information: an application to robot arm movements. IFAC-Papers OnLine **48**(27), 141–146 (2015)

9. Gol, E.A., Lazar, M., Belta, C.: Temporal logic model predictive control. Automatica **56**, 78–85 (2015)
10. Dixon, C., et al.: Temporal logics of knowledge and their applications in security. Electron. Notes Theoret. Comput. Sci. **186**, 27–42 (2007)
11. Mukherjee, S., Dasgupta, P.: A fuzzy real-time temporal logic. Int. J. Approximate Reasoning **54**, 1452–1470 (2013)
12. Shi, H.-X.: A quantitative approach for linear temporal logic. In: Fan, T.-H., Chen, S.-L., Wang, S.-M., Li, Y.-M. (eds.) Quantitative Logic and Soft Computing 2016. AISC, vol. 510, pp. 49–57. Springer, Cham (2017). https://doi.org/10.1007/978-3-319-46206-6_6
13. Sotudeh, G., Movaghar, A.: Abstraction and approximation in fuzzy temporal logics and models. Formal Aspects Comput. **27**, 309–334 (2015)
14. Selvi, M., et al.: Fuzzy temporal approach for energy efficient routing in WSN. In: Proceedings of the International Conference on Informatics and Analytics, p. 117 (2016)
15. Frigeri, A., Pasquale, L., Spoletini, P.: Fuzzy time in linear temporal logic. ACM Trans. Comput. Logic, **15**(4), Article 30 (2014)
16. Ho, L.T.T., Arch-int, S., Arch-int, N.: Introducing fuzzy temporal description logic. In: Proceedings of the 3rd International Conference on Industrial and Business Engineering, pp. 77–80 (2017)

Developing a Corporate Chatbot
for a Customer Engagement Program:
A Roadmap

Fernanda Castro[1], Patricia Tedesco[1(✉)], Havana Alves[1],
Jonysberg P. Quintino[1], Juliana Steffen[2], Frederico Oliveira[2],
Rafael Soares[2], Andre L. M. Santos[1], and Fabio Q. B. da Silva[1]

[1] Centro de Informática, Universidade Federal de Pernambuco,
Recife, Pernambuco, Brazil
{mfcc, pcart, hdaa, jpq, alms, fabio}@cin.ufpe.br
[2] Samsung SIDI Institute, Campinas, São Paulo, Brazil
{j.steffen, f.oliveira, rf.soares}@samsung.com

Abstract. There has been a steady increase in the popularization of chatbots solutions in recent years, in different areas, where the interface with users are replaced by software trained to answer like a human. This paper presents the stages, from conception to the publication, of a text based chatbot that is able to provide instant replies to users questions from an engagement program of a Brazilian company. Even though there is no consensus about the best development practices of such solutions, since each application domain brings along its own characteristics and requirements, this paper lists the activities, challenges faced and decisions made to build a chatbot solution.

Keywords: Chatbot development · ChatScript · Ontology

1 Introduction

The development of systems that interact with people through Natural Language has been a topic of interest in Computing since the 1950's. At first, these conversational agents, called chatbots, aimed to simulate the conversation between two human beings [1]. Such systems used text and were restricted to competitions and entertainment. ELIZA[1] is a representative of the first generation of chatbots, and was widely popular at the time of deployment.

However, with the evolution of AI (Artificial Intelligence) and the need for new approaches for human computer interaction, chatbots have become a communication channel capable of providing automated services in a personalized way [1]. Chatbots have since then been applied to different areas, including sales and customer support. Their presentation format has also changed: many conversational agents possess voice processing, embodiment and can act over the web as well as in modules integrated to other systems such as Facebook Messenger, WhatsApp or Skype. The success of

---

[1] http://www-ai.ijs.si/eliza-cgi-bin/eliza_script.

© Springer International Publishing AG, part of Springer Nature 2018
D.-S. Huang et al. (Eds.): ICIC 2018, LNCS 10954, pp. 400–412, 2018.
https://doi.org/10.1007/978-3-319-95930-6_37

virtual assistants such as Siri[2] and Cortana[3] reinforce the potential usage of conversational agents in customer services.

Chatbot development is still an open topic, and there are plenty of specific techniques and approaches for various domains [2, 3]. Different cases have different challenges, such as how to ensure Bot cover with respect to linguistic variations and regionalisms or how to provide easier user interactions and better experience.

This paper presents the steps executed to develop a text-based Chatbot by SIDI (Samsung Institute) and UFPE (Federal University of Pernambuco), both located in Brazil, which goal is to support users of a training and engagement program. In the program, users must attend training sessions and answer quizzes so they can accumulate points according to their performance. The chatbot will support users about the program rules, how to use the system and how to solve problems, replacing the customer support service previously adopted. The Chatbot was developed with the adoption of a proprietary ontology and the usage of ChatScript.

This paper is organized as follows: Section 2 discusses the popularization of Chatbots over time, and presents their main features and evolution. The focus is mainly on corporate and client service chatbots, as well as on the main questions related to their development (e.g. interaction style and security). Section 3 presents the planning of the Chatbot developed. t describes the Chatbot's main requirements, development scenario, knowledge base creation, comparative study on technologies and architecture. Section 4 discusses Chatbot implementation and details the main challenges found. Section 5 presents a performance analysis of the Bot and Sect. 6 brings the conclusions, main contributions and opportunities for further work.

2 Chatbots and Their Applications

The first Chatbot generation was motivated by Alan Turing's 'Imitation Game' [1]. According to Turing's game, a system would be considered intelligent if it could successfully impersonate a human being in interactions with humans. From then on, programs able to interpret text inputs and provide adequate answers were developed.

Chatbots were initially developed for entertainment, focusing on games and specific domains. With the appearance of the Loebner Prize, chatbots began to be compared with respect to techniques and performance [3]. In fact, the Loebner Prize has been one of the main incentives for the use and development of chatbots.

While the first chatbot generation used simple pattern matching, oftentimes using whole sentences, the current generation uses modular scalable architectures, markup languages and machine learning [1, 4]. Furthermore, the conception of a chatbot must always be consistent with its purposes and applicability. In fact, the study conducted by Baron [5] points out that people might not be very comfortable with systems that are too realistic: bots need to be designed according to their goal, which can be quite

---

[2] https://www.apple.com/ios/siri/.

[3] https://www.microsoft.com/en-us/windows/cortana.

simple. In the following sections, this paper discusses the application of chatbots in the corporate environments and the associated development challenges.

2.1 Chatbots, Enterprises and Client Service

Besides providing a friendly interface between users and companies, chatbots allow corporations to mine data obtained in conversations with users with respect to information about service quality and client sentiment (i.e. their feelings towards the service provided). Furthermore, the adoption of chatbots avoids employee overload, since customer service employees need to be always readily available. By filtering calls, automating part of the processes and resolving some of the problems found, the chatbot helps diminishing user waiting time and enables employees to focus on demands with a higher priority [6].

The usage of chatbots in corporations is not restricted to the relationship with customers – it can also be profitable within the organization [7]. Delloite[4] highlights the use of chatbots in human resources, business to business applications, and IT support. Lastly, one can employ conversational agents to provide training in activities that demand interaction [8], i.e., developing sales pitches.

It is also possible to find companies that become specialized in providing platforms, services and frameworks for chatbot developments, as well as configurable bots ready for use. Such resources, known as third-party agents, allow customers to customize their bots without need of chatbot development expertise [6]. Microsoft Bot Framework[5] and Amazon Lex[6] are well known examples of third-party agents.

When deciding whether to choose between a third-party agent and a proprietary agent (first-party), one should consider the following aspects: by using a third-party, the usage of the resulting chatbot must be compliant with the terms of use and privacy policy of the service provider. It is possible that developers need to integrate the bot to other development platforms, and this might hinder scalability. First-party agents demand investments in research and development, and may also incur in maintenance and evolution costs. For this project, ChatScript was the platform chosen for development. This is because it is an open source project, not connected to any profit organizations, it has shown to be flexible and easily integrated to other systems. In the next section, development challenges are discussed in detail.

2.2 Development Challenges

Some challenges related to commercial chatbots are still under discussion. The first relates to how to design interactions that maximize user satisfaction. Chatbots that deal with consumers do not necessarily need to be too realistic, but should be concerned with providing quick, effective customer service [5]. In the case of Customer Service

---

[4] https://www2.deloitte.com/insights/us/en/focus/signals-for-strategists/era-of-conversational-user-interface-chatbots.html.

[5] https://dev.botframework.com/.

[6] https://aws.amazon.com/pt/lex/.

Applications and FAQ (Frequently Asked Questions), text-based chatbots are common, since this is a quick and easy communication channel, that can be used via mobile devices anywhere [7]. In this light, a few good practices for commercial chatbots have been proposed. Chatbot Magazine, pinpoints, for instance, the importance of providing shortcuts and ready-made answers, to prevent users from wasting time typing[7]. Duijst [9] argues that, in the case of customer service chatbots, simple conversations are considered to be more effective.

Navigating in the conversational environment is also a concern, and an intuitive interface, with visual resources, such as interactive cards, buttons and hyperlinks is recommended [7]. Another challenge is to correctly classify each service, identifying what can be solved by the bot and what must be sent to human support. In that sense, the use of ontologies, can greatly improve bot performance (in terms of precision and assertiveness), since they reduce the search time on the Bot's knowledge base.

Security must also be taken into consideration, especially when bots use external channels to communicate with the customer. This is due to the care with information confidentiality (both from the user and the company) [10]. Lastly, companies must consider storing dialog logging and analysis, through reports or mining techniques, since these datasets might provide useful insights to improve customer support [7].

3   Chatbot Development Planning and Architecture

During conception, it was necessary to elicit the necessary and desirable functionalities. Then, the development scope was divided into sprints, beginning with the necessary requirements and then tackling the desirable ones. Firstly, it was decided that the bot's purpose was to provide information and services for users of a training and engagement program. The program consists of an application, installed on an electronic device given to partners, who receive training on sales and technical information. Contents are made available on a weekly basis to the users, that must undertake quizzes to test their knowledge. Assessments are graded, and users are ranked according to their perfor-mances. When they obtain a high classification in the ranking, users are eligible for prizes and advantages. In the previous application, customer support was performed by a human assistant. Customer service was carried out through text messages. Through this channel, it was possible to answer questions about the program rules, request support to perform small operations on the application and register complaints. Common questions included: "What should I do if my phone gets lost or stolen?", "What should I do if I miss a training session?"

Even though users could make their requests on a 24/7 basis, the answer to those was still dependent on employee responsiveness. This generated a delay on the responses, as well as an accrual of requests. Many of them consisted on the repetition of simple questions, which caused a work overload for system responsiveness.

---

[7] https://chatbotsmagazine.com/essential-tips-to-keep-in-mind-when-building-a-chatbot-d289a00d1
7d8.

Thus the chatbot aimed to answer questions and performs some other tasks. The chatbot solution should be hybrid, so that cases where an audit is required would be forwarded to the human assistant (employee in charge of customer support). Thus, there would be a reduction in response time and an increase in users' satisfaction. In order to exploit chatbot potential in acquiring information[8], we have also chosen to use the bot in other scenarios by launching small questionnaires whenever they showed an unsatisfactory performance in assessments as well as when they disliked the contents presented to them. Thus, it is also possible assign a new functionality to the chatbot, namely, feedback collection.

With these goals in mind, one of the requirements elicited was to keep conversation via text messages, abiding by the patterns to which users and customer support employees were used to. Moreover, since the chatbot would have to handle sensitive questions, such as problems, unsatisfactory outcomes, and bad user performance, the usage of emphatic messages [8] became fundamental.

Another requirement was the creation of mechanisms to facilitate the communication between bot and users, such as date masks and interactive cards with the main conversation topics. It was also considered that, in case the bot does not understand the messages (user's input not mapped into any regular expression), the bot should help the user to rephrase their question or to choose a topic from the interactive cards. In case the non-understanding persists in this second attempt, the chatbot forwards the query to the customer support employee. This behavior aims at avoiding frustrations and a long waiting time for problem solution. Particular characteristics of the planning phase will be discussed in the next subsections.

3.1 The Scenario

This work aimed for the development of a corporate customer support chatbot, able to interact with users about a restricted domain. Moreover, since the bot aimed to solve customer questions, clarifying questions about the program and its rules, we were also able to design the most adequate answers to questions.

In order to plan bot interactions, the log of users' interactions with customer services was analyzed. However, the previous base was quite limited and not classified with respect to the interactions. Thus, we chose to manually label interactions, based on the existing documentation.

Another characteristic of the Project was that only open third-party codes (or codes that allowed commercial use) could be used. Once we were able to delineate a scenario, we were able to collect information and structure the chatbot knowledge base, through the development of an ontology, described in the next section.

3.2 The Ontology

Ontologies can be defined as an 'explicit shared conceptualization of a knowledge domain' represented in a format that can be read by both humans and machines,

---

[8] https://chatbotsmagazine.com/replacing-surveys-with-chatbot-conversations-8295614af367.

detailing its key concepts and their relationships [11, 12]. The adoption of ontologies fosters knowledge management, performance, scalability and reuse. According to Abdul-Kader and Wood [13], a complex chatbot is as good as its knowledge base.

Since this bot interacts with users about a predefined set of topics, often related to each other, an ontology was chosen for this project. This allowed contexts to be updated modularly, keeping consistency. For instance, the loss of a participant's smartphone and the replacement of the missing training contents, are two related contexts. If these two questions are presented in sequence, there is a priority of answers to be followed. If the only question received is, e.g. loss of a smartphone, then the answer provided might be different.

The biggest challenge in developing ontologies is to choose a development methodology [11], since the same domain can be represented differently. In this project, we have used On-To-Knowledge [14], which has four stages: (1) kick-off, which encompasses the analysis of requirements, bases and other information sources with the goal of developing an initial taxonomy; (2) refinement, where classes and subclasses are refined and formalized with the help of domain experts; (3) evaluation, where the formalization is tested, revised and amplified; and (4) maintenance, where the ontology is expanded and modified, if and when needed.

The resulting ontology was a structure containing classes, subclasses, relationships and instances. Each class represents a context within the domain, whereas subclasses represent specific topics within contexts, and were generated through the observation of questions made by users and logged in the system's initial base. For instance, the class Ranking/Campaign represents the user's' performance in the engagement program. This class possesses three subclasses: program rules, ranking classification and contestation.

Relationships indicate contexts that either share information or are complementary (e.g. classes Theft/Loss and Replacement). Instances represent question-answer pairs (e.g. whenever users report that they had their smartphone stolen, the bot answers with the procedure for forwarding the appropriate police report). Once this stage was completed, a development platform was chosen. To this end, a comparative study of technologies, described in Sect. 3.3, was carried out.

3.3 A Comparative Study of Technologies

In the design stage, a comparative analysis of the main platforms for chatbot development [2, 3, 15] was carried out, in order to decide which would be more adequate. Platform choice considered the given context, problem constraints, desired functionalities and integration with the ontology.

IBM Watson[9] is a powerful platform based on rules and machine learning, that can be used by applying for membership plans. A web interface allows the training stage to be executed remotely, as well as the extraction of bot metrics. Watson is robust in its usage of machine learning for speech comprehension. However, integrating the bot with external applications might be complex, and costly [10].

---

[9] https://www.ibm.com/watson/.

LUIS[10] is a natural language comprehension framework, offered by Microsoft. It has a free membership plan, given that the bot does not exceed the allowed number of monthly transactions. It can be integrated to other Microsoft frameworks for Chatbot development. It learns and modifies its responses according to previous interactions.

Dialogflow[11], a tool acquired by Google, is a Machine Learning based platform. It can be easily integrated to ontologies, supporting classes and relationships. It can be trained through a web interface, although this channel is not very intuitive [10].

WIT.AI[12] is a free service for chatbot creation, focusing on Facebook, using its chat service as main usage channel. It also uses machine learning techniques. In 2017, the company announced Project discontinuation.

ChatScript[13] is a NLP and dialog management framework, used both in chatbot development and in Language processing services. It is an open source application based on rule matching. It has been used in scenarios with specific domains, besides allowing the integration of ontologies.

The use of Watson was discarded due to commercial constraints in its license and terms of use. LUIS was also discarded due to the limit on monthly transactions, and the fact that this paper's target domain did not allow message variation through learning, since messages sent to users should be precise and strictly abide by program rules. Dialogflow was also not considered since this project's data volume was too small for its needs. Lastly, WIT.AI was discarded due to having been discontinued and to demand extra coding to function on external chat tools.

ChatScript was chosen, due to being open source and also due to the ease of integration with other technologies. Moreover, it allows precise answers to be coded in the scripts, with no variation possibilities, an ideal choice for the context of this bot. The resulting bot architecture is described in detail in the next section.

3.4   System Architecture

Commercial chatbots normally present a transactional architecture [16]. The architecture proposed in this project is based on the retrieval model, compiled by Pavel [16] and used by Almeida [15]. In this model, the chatbot decodes the user message and subject context to search for the most adequate answer in a knowledge base. This model is known for its high predictability, since it is possible to predict all possible answers. Some architectures consider previous conversations and learning, while others only identify the current conversation topic [13]. Here, only the current context is taken into consideration. When the user sends a message, the Mobile Client module, sends requests to the Chat module at the Server. In its turn, the Chat module activates the Pre-processing Module to treat user entries, correcting small mistakes and understand meaning (this module uses Hunspell for grammar correction and a NLP submodule for syntactic and semantic analysis.

---

[10] https://www.luis.ai/.

[11] https://api.ai/.

[12] https://wit.ai/.

[13] https://github.com/bwilcox-1234/ChatScript.

After the treated message is returned, the Chat Module sends it to ChatScript, that performs pattern matching according to the regular expressions coded in its scripts, organized according to the ontology's classes and subclasses. ChatScript will return the appropriate text answer as well as the associated metadata (for instance, the indication of some service to be performed). This response is then saved on the database and returned to the Mobile Client module, allowing the user to see it in its screen. Whenever the application identifies a service request, the request is performed by the Server and returned to Mobile Client.

It is also possible to execute requests through the Admin Client module, where admin users can request reports through the web interface. Such requests are treated by the Server's Reports module, that retrieves log data from the database and organizes them according to the type of report requested (e.g. number of times that each conversation topic was tackled, number of tickets/questions solved, number of messages sent to the human support, etc.).

The implementation of this architecture is described in Sect. 4.

4   Implementation

Firstly, an analysis and categorization of users' questions and the answers provided by the human assistant was carried out. These interactions were stored in the database. After that, contents were structured. In the sequel, classes and subclasses were represented through regular expressions, so that users' questions were classified correctly. The answers to be provided by the chatbot were also coded in ChatScript. A Natural Language Process service was integrated to process users' entries, eliminating linguistic variations and typing mistakes. This structure was integrated to the software using JSON (JavaScript Object Notation)[14] for data encapsulation.

Other areas also presented challenges. Regarding testing, we had to plan and systematize automatic tests to assess the chatbot coverage of word variation and their synonyms, sentence formation and typos. This process is discussed in Sect. 5.

4.1   Regular Expressions and ChatScript

ChatScript is a pattern matching based technology [2]. Klüwer [17] argues that to provide coverage for a sentence in its different forms, it would be necessary to create as many patterns as the syntactic variations. Hence, it was decided to classify user's inputs through regular expressions. These are concise formalizations to identify strings [18]. They allow us to translate a sentence and its variations, generating a pattern that can be extracted from similar sentences.

ChatScript does not provide support for regular expressions, although, with some adjustments, they can be partially supported [15]. Thus, the regular expressions in this project were created using only classical logic operators, indicators of optional strings, substring indicators and commutation indicators in the order of substrings.

---

[14] https://www.json.org/.

Resulting expressions were organized in order to translate the ontology classes and subclasses, so that a given regular expression allowed the simultaneous matching of several domain instances. Each regular expression then guided the bot to a ChatScript rule, that contained the answers and options to be returned to the user. For instance, if a user asked 'Where can I find Program Rules?' or 'How can I access the rules' the same expression is captured, thus triggering a specific ChatScript rule that returns a message to the user specifying in which menu the rules can be found.

The main challenges related to the manipulation of regular expressions in Chat-Script were rule overlapping and precedence [17]. ChatScript verifies rules sequentially. Since regular expressions target domains wider than string matching, it is possible that part of the scope of the rules overlaps (for instance, imagine an expression that nominates cases where the user wishes to send the device for repairing and another that deals with devices that have not been returned from repairing on time). This means that one must code first more specific expressions.

In ChatScript, different topics are stored in different script files. Upon receiving an input, the system searches all scripts simultaneously. Whenever an overlap is identified, through manual testing or unexpected bot behavior, it is possible to: (a) refine regular expressions to solve conflicts; or (b) use ChatScript variables and functions to better identify the conversation topic. In order to better identify conversation topics, user input must be preprocessed, to exclude input errors and variations and to determine what the user intention really is. This preprocessing is described in detail in Sect. 4.2.

4.2   NLP

Chatbots need to handle varied inputs, sometimes unexpected and containing errors. In other words, a chatbot must be able to infer what is the user true intention [19]. In this particular case, it was important to count with a tolerance mechanism, since the mobile chat is an informal channel and thus very susceptible to typos, abbreviations, slangs or regionalisms. For instance, if we only used the matching in ChatScript, inputs "to die" and "to dye" would represent different concepts. However either of them could be a typo. Thus, it is necessary to pre-process the input, to identify contexts and intended meanings [4].

Text processing in NLP is done by using 5 modules: morphological analysis, lexical analysis, syntactic analysis, semantics and pragmatics [20]. In morphological analysis, punctuation marks are discarded and words are segmented in radicals, prefix and suffix. This means that derived words, such as plural words will be reduced to their canonical form to facilitate pattern matching. In lexical analysis, the words in a form close to the canonical one will be analysed – this allows typos to be corrected.

Syntactic analysis aims to verify the role of each word in a sentence, as well as its relationship to the others. Semantic analysis explores word meanings within the context of the sentence, whereas pragmatic analysis considers the speech sense related to the context. It is in semantic analysis that the issue of homonyms is solved.

In this project, NLP was performed with two free license tools: Hunspell[15] and CoGrOO[16]. Thus, user's text inputs are preprocessed by the NLP module before pattern matching, going first to Hunspell and then to CoGrOO. After this preprocessing, ChatScript receives a list with the words in their canonical forms and the closest lexical pairs, as well as an indication of the role of each word in the sentence. Since the bot needed to operate together with the existing customer engagement program, its integration to the Android and Web modules is described in Sect. 4.3.

4.3 Integration to the Android and Web Modules

One of the chatbot's requirements was that it should operate integrated to the already existing customer engagement program. This had two consequences: we needed to integrate the bot to the Android module, that handled message exchange with users and to the Server system, whose metrics are presented to program management.

Since ChatScript is script-based and does not possess any preexisting integration mechanisms, we had to choose a new technology to make the communication between modules. The solution was to use JSON, that allows the exchange of data objects in text form. Thus, requests and ChatScript results may be encapsulated and sent to the server, that transmits them, encapsulated as well, to either the web or mobile modules. Thus, it was possible to enhance the bot interaction power. For instance, when it is noted that a user is not being understood, the chatbot may send a Request, through JSON to the server, requesting that the Android Module shows a series of cards in the chat area, that inform the user of the main available subjects and services. The server retrieves the information pertaining to each card, encapsulate them in a JSON and return to the mobile mode.

Whenever the bot cannot understand the user's intentions in two consecutive attempts, it provides support through cards. The user can scroll them to explore the main themes discussed by the bot. When users do not agree with any cards, they can send messages directly to human support. If the user opts to click on one of these cards, the Mobile Client module will then send this topic to the Server, which will again communicate via JSON with the ChatScript, triggering the corresponding rule. This would not be possible if only the pattern matching scripts were used. Communication with the other modules is transparent to ChatScript. After decoding a JSON message, ChatScript will treat it as user input. This facilitates module reuse and maintenance. As the bot's functionalities were being implemented, their performance was tested (as described in Sect. 5).

---

[15] http://hunspell.github.io/.

[16] http://cogroo.sourceforge.net/.

5 Performance Analysis

Radziwill and Benton [21] have presented an overview of quality features used to assess the performance of conversational agents, that can be divided into three areas: efficiency, effectiveness and satisfaction. Efficiency is related to the bot performance, its response time and robustness in face of unexpected events. Effectiveness concerns the bot's ability to fulfill its domain requirements. Satisfaction is related to the ability of creating empathy, maintaining ethics during dialog and possess accessibility resources. The analysis described here only considered efficiency and effectiveness.

In order to test the large number of answer options, services and functionalities, we used both manual and automated tests. Automated testing was done with SoapUI[17], to verify whether the bot responded adequately to valid input. At this stage, real user questions were presented to bot, who answered accurately to all questions presented. All of the bot's functionalities (answering frequent questions, interacting with users to find out why they did not perform well in their quizzes and management report generation) were found to be compliant with the specification. After that, manual testing was done to check interface presentation and interactive elements. The Chatbot was also able to cover 100% of the previous customer support database.

Bot efficiency and robustness were thus assessed. A suite of tests to evaluate bot behavior in face of unexpected input was also designed. Tests were carried out to evaluate bot behavior in face of environment failure. Robustness with respect to spelling mistakes depends on Hunspell. The mechanism that forwards questions to human support in case of two consecutive failures, dealt with messages that could not be corrected properly. Thus, we could attend to the bot efficiency in dealing with adverse situations without interrupting service.

6 Conclusions and Further Work

Although chatbots have become increasingly popular, there are still many development challenges that must be considered carefully in order to guarantee a satisfactory user experience.

This paper described the design and development of a chatbot that aimed to interact with users in a customer engagement program. This bot functions as a way of answering questions, solving common problems and acting as a support center. Moreover, since the customer engagement program provides training and assessment, the chatbot also collects information about users that faced difficulties.

The research encompassed system design and scenario description, ontology conception, architecture, implementation and preliminary tests. In this paper, we have discussed the main development challenges, discussing development platforms and frameworks, justifying the adoption of ChatScript. This work has also contributed with the joint application of NLP and regular expressions to ChatScript. Preliminary testing has shown that the bot fulfils the requirements of speed and domain coverage.

---

[17] https://www.soapui.org/.

In the near future we intend to: (a) complete the testing process, studying other metrics; (b) carry out a field validation of the bot, with real users, collecting feedback from them; (c) implement context-aware mechanisms, that help the bot to adjust its answers to users; (d) adopt data mining techniques to infer user trends and refine the chatbot's functioning.

Acknowledgements. The results presented in this paper have been developed as part of a collaborative project between Samsung Institute for Development of Informatics (Samsung/SIDI) and the Centre of Informatics at the Federal University of Pernambuco (CIn/UFPE), financed by Samsung Eletronica da Amazonia Ltda., under the auspices of the Brazilian Federal Law of Informatics no. 8248/91.dentify applicable sponsor/s here. The authors would like to thank the support received from the Samsung/SIDI team. Professor Fabio Q. B. da Silva holds a research grant from the Brazilian National Research Council (CNPq), process #314523/2009-0.

References

1. Barros, F., Tedesco, P.: Agentes Inteligentes Conversacionais: Conceitos Básicos e Desenvolvimento. In: 35º JAI - Jornada de Atualização em Informática, pp. 169–218 (2016)
2. Shaikh, A., Phalke, G., Patil, P., Bhosale, S., Raghatwan, J.: A survey on chatbot conversational systems. Int. J. Eng. Sci. (2016)
3. Bradesko, L., Mladenic, D.: A survey of chabot systems through a loebner prize competition. In: Proceedings of Slovenian Language Technologies Society Eighth Conference of Language Technologies, pp. 34–37 (2012)
4. Dingli, A., Scerri, D.: Building a hybrid: chatterbot – dialog system. In: International Conference on Text, Speech and Dialogue, pp. 145–152 (2013)
5. Baron, N.: Shall we talk? conversing with humans and robots. Inf. Soc. **31**(3), 257–264 (2015)
6. Cui, L., Huang, S., Wei, F., Tan, C., Duan, C., Zhou, M.: SuperAgent: a customer service chatbot for e-commerce websites. In: Proceedings of ACL 2017, System Demonstrations, pp. 97–102 (2017)
7. AIBUSINESS. https://aibusiness.com/inbenta-ceo-interview-nlp/. Accessed 28 Feb 2018
8. Kenny, P., Hartholt, A., Gratch, J., Swartout, W., Traum, D., Marsella, S., Piepol, D.: Building interactive virtual humans for training environments. In: Proceedings of I/ITSEC, vol. 174 (2007)
9. Duijst, D.: Can we improve the User Experience of Chatbots with Personalisation? M.Sc Thesis, University of Amsterdam (2017)
10. ACUVATE. https://issuu.com/acuvate/docs/a_guide_to_choosing_an_enterprise_b. Accessed 28 Feb 2018
11. Morais, E., Ambrósio, A.: Ontologias: conceitos, usos, tipos, metodologias, ferramentas e linguagens. In: Technical Report - INF_001/07, December 2007
12. Borst, W.: Construction of engineering ontologies for knowledge sharing and reuse. Ph.D thesis, University of Twente, Netherlands (1997)
13. Abdul-Kader, A., Wood, J.: Survey on chatbot design techniques in speech conversation systems. (IJACSA) Int. J. Adv. Comput. Sci. Appl. **6**(7), 72–80 (2015)
14. Staab, S., Schnurr, H., Studer, R., Sure, Y.: Knowledge processes and ontologies. IEEE Intel. Syst. **16**(1), 26–34 (2001)
15. Almeida, O.: Beck: Um Chatbot Baseado na Terapia Cognitivo-Comportamental para Apoiar Adolescentes com Depressão. Master Thesis, UFPE, Brazil (2017)

16. Pavel, S.: Chatbot Architecture. http://pavel.surmenok.com/2016/09/11/chatbot-architecture/. Accessed 28 Feb 2018
17. Klüwer, T.: RMRSBot – using linguistic information to enrich a chatbot. In: Ruttkay, Z., Kipp, M., Nijholt, A., Vilhjálmsson, H.H. (eds.) IVA 2009. LNCS (LNAI), vol. 5773, pp. 515–516. Springer, Heidelberg (2009). https://doi.org/10.1007/978-3-642-04380-2_69
18. The IEEE and The Open Group Base Specifications. http://pubs.opengroup.org/onlinepubs/009695399/basedefs/xbd_chap09.html. Accessed 28 Feb 2018
19. Mateas, M., Stern, A.: Natural language understanding in façade: surface-text processing. In: Göbel, S., Spierling, U., Hoffmann, A., Iurgel, I., Schneider, O., Dechau, J., Feix, A. (eds.) TIDSE 2004. LNCS, vol. 3105, pp. 3–13. Springer, Heidelberg (2004). https://doi.org/10.1007/978-3-540-27797-2_2
20. Kumar, E.: Natural Language Processing, 1st edn. I. K. International Pvt Ltd, New Delhi (2011)
21. Radziwill, N., Benton, M.: Evaluating quality of chatbots and intelligent conversational agents. Softw. Qual. Prof. **19**(3), 25–36 (2017)

Application of Non-uniform Sampling in Compressed Sensing for Speech Signal

Changqing Zhang[(✉)], Gang Min, Huan Ma, and Xian Li

College of Information and Communication,
National University of Defense Technology, Xi'an 710106, China
zhangcq1108@163.com

Abstract. Currently, the most widely used Gaussian random observations in compressed sensing require that signals must be discrete, and the signal waveform must be known before observation, which greatly restricts the application of compressive sensing in speech. In response to this problem, this paper draws on the advantages of non-uniform sampling, constructs a non-uniform observation matrix, directly extracts the data from the signal waveform as observations, and gives a corresponding new method of reconstruction. The theoretical analysis and simulation results show that non-uniform observation can directly apply compressed sensing to analog speech signal processing, and the corresponding reconstruction method effectively enriches the means of compressive perception reconstruction.

Keywords: Compressed sensing · Non-uniform sampling · Speech signal

1 Introduction

1.1 A Subsection Sample

With the continuous development of information technology and the rapid growth of people's demand for information, traditional data compression technologies and information processing methods are facing more and more challenges. Its main features: the signal before transmission and storage should always go through sampling and compression; in order to ensure the integrity of the information, we usually collected more data, meanwhile, in order to ensure the efficient, before transmission large amounts of data should be compressed off, which causes a large waste of resources. These compression processes are often complicated, and increases the burden on the collection end, which runs counter to the requirements of simplifying the information collecting equipment under the background of big data processing such as wireless sensor network.

Sponsored by National Natural Science Foundation *(No. 61701535 and No. 61072125)*, Shaanxi Natural Science Foundation *(No. 2017JQ6033)*.

Compressed Sensing (CS) [1, 2], proposed by Donohoo in 2006, combine sampling and compression together, which provide a high effective sampling method for sparse signals. At present, the theory has achieved good application in image and video processing [3]. However, the application of CS in speech signal is relatively less. The reason is that the effect of sparse representation of speech is not satisfactory, on the other hand, there are still some obstacles in the practical application of CS in one-dimensional analog signals. This article mainly discusses and studies the practical application of CS in speech signal.

2 Compressed Sensing and Non-uniform Sampling

2.1 Compression Perception and Its Mathematical Model

CS pointed out that if a signal is sparse (or sparse on a transform base), it can be observed with a reduced-dimensional observation matrix and the original signal can be accurately reconstructed from the observations by an optimization algorithm. Obviously, CS theory contains three key parts: sparse representation, reduced-dimensional observation and optimized reconstruction. Subsequent paragraphs, however, are indented.

Let a signal to be $x = [x(1), x(2), \cdots, x(N)]^T \in R^{N \times 1}$, and $\Psi \in R^{N \times N}$ is a certain orthogonal base for sparse transform, then the signal x can be expressed as $\theta = \Psi x = [\theta(1), \theta(2), \cdots, x(N)]^T \in R^{N \times 1}$ under the base Ψ. If $||\theta||_0 = K < N$, the sparsity of θ can be measured as K, and this progress is called sparse representation of x. According to reference [1], this is the basic condition of using CS to process a signal, and if the sparsity of signal x is K, x can be expressed completely through M measures of x, while the number M depends on N, K and the method of measuring, in general there exists N > M > K, and this progress is called measuring or observation in CS, which can be expressed briefly as follow:

$$\alpha = \Phi^{M \times N} x = \Phi^{M \times N} \Psi^{-1} \theta \in R^{M \times 1} \tag{1}$$

At the receiving end, the original signal x can be reconstructed exactly from vector α with the information of Φ, Ψ and K through a reconstruction algorithm. More information can refer to [9].

Obviously from the mathematical model, CS combines the two processes of compression and sampling through formula (1), which greatly simplifies the process of information acquisition. Currently, one of the main factors restricting the practical application of CS is the design of observation matrix.

2.2 Non-uniform Sampling Theory

Non-uniform sampling is an information acquisition method with respect to uniform sampling. This method can solve the problem of spectrum aliasing under low sampling rate for uniform sampling, and thus can realize efficient sampling of information.

Let $f(t)$ be the finite band-pass continuous time signal, $F(j\omega)$ be the continuous spectrum, $f_n(n) = f(t = t_n)$ is the non-uniformly sampled signal and $F_n(jw)$ be the continuous spectrum of the non-uniformly sampled signal. When using non-uniform sampling, if the sampling time is random and independent, and all random sampling time in the signal duration obey uniform distribution, set its probability density function p_t, then:

$$p_t = \begin{cases} \frac{1}{T_d}, & t \in [0, T_d] \\ 0, & \text{others} \end{cases} \tag{2}$$

The expectation of the discrete-time Fourier transform of a non-uniformly sampled signal sequence is:

$$\begin{aligned}
E\left\{\sum_{k=1}^{N} f(t_k)e^{-j\omega t_k}\right\} &= \sum_{k=1}^{N} E\{f(t_k)e^{-j\omega t_k}\} \\
&= \sum_{k=1}^{N} \int_{-\infty}^{+\infty} f(t_k)e^{-j\omega t_k}p(t_k)dt_k \\
&= \frac{1}{T_d}\sum_{k=1}^{N} \int_{0}^{T_d} f(t_k)e^{-j\omega t_k}dt_k \\
&= \frac{N}{T_d}\int_{0}^{T_d} f(t)e^{-j\omega t}dt = \frac{N}{T_d}F(j\omega) \\
&= \frac{1}{\mu}F(j\omega)
\end{aligned} \tag{3}$$

In the formula above, $\mu = T_d/N$ is the average sampling interval. It can be seen that the expectation of Fourier transform of the non-uniformly sampled signal is the frequency spectrum of the original signal, but there is a constant proportional coefficient in the amplitude. There is no periodic extension in the uniform sampling spectrum, so the generation of aliasing is effectively avoided.

In fact, the aliasing signal is not completely suppressed, but its energy is evenly distributed randomly over the whole frequency domain by non-uniform sampling. The sampling frequency varies for non-uniform sampling, but the real spectrum of the signal does not change regardless of the sampling frequency. Since the aliasing signal is related to the sampling frequency and varies with the sampling frequency, through the spectrum analysis of the data obtained we find that the real spectrum is enhanced every time and the aliasing signals are randomly distributed to the entire frequency band. If set the sampling time interval completely random, you can reduce the aliasing effect to a minimum [4, 5]. In this way, as long as the original signal spectral components can be identified from the spectral noise caused by non-uniform sampling, the number of samples of the signal can be greatly reduced so as to achieve efficient data processing from the beginning. From this point of view, it has many similarities with the CS observation. Reference to non-uniform sampling is expected to provide a new breakthrough for the CS observation method for studying speech signals. In [6], researchers studied that Before CS observation, if signal is random sampled, it can effectively reduce the hardware requirements of the signal acquisition node, but it still does not solve that the signal waveform should be known before observation.

3 Compressed Sensing Based on Non-uniform Sampling

3.1 The Structure of Non-uniform Observation

CS as an information processing method, the advantages embodied in the observation mode, that is, the unity of sampling and compression. In [2], it is pointed out that in order to represent the signal with as few observations as possible, the observation matrix requires no correlation with the sparse transform base. Taking into account that the observation should have a lower correlation with the vast majority of sparse base, Gaussian random observations matrix became the first choice. Although this observation matrix has been widely used in image processing and other fields, it still has some shortcomings [7], especially when dealing with one-dimensional analog signals like speech. It can be seen from (1) that each value in the observation vector is obtained by inner product of the corresponding row vector in the observation matrix and the signal, that is, all information of the signal waveform needs to be known before observation, and the observed signal must be discrete signal. This is not suitable for processing speech, on the one hand, acquiring all or part of the signal before observing is bound to increase the delay, which is not desirable in real-time communication; On the other hand, the signal processed by CS is generally discrete, so before processing by CS, A/D sampling is needed, if do like that, the significance of the application of CS is lost in a large extent.

In order to solve the problem above, this paper draws on the advantages of non-uniform sampling [8, 9], and improves the application of CS in speech by constructing non-uniform observations. In order to facilitate to compare with Gaussian random observation, a non-uniform observation matrix is designed as shown in Fig. 1. The black areas in the figure represent the sampling locations and the sampling locations are random, but the following conditions should be met: (1) each row is considered as a sampling, each sampling only preserves a time-domain sampling value completely at sampling time; (2) each time it samples only one value, in order to avoid repetition, each value is sampled up to once; (3) there is a definite sequence between the samples, which stipulates that the time corresponding to the ith row than the $(i + 1)^{th}$ sample value should be earlier. Figure 2 shows the observation process of non-uniform observations and Gaussian random observations, where (a) is the sparse representation of the signal, (b) is the Gaussian random observation, from that we find that the observation vector is the result of the inner product of the observation matrix and the entire signal, (c) shows the process of signal observation in the non-uniform observation matrix, it can be seen that the observed value is a direct sampling of the signal value corresponding to the observation position, and each value of the Gaussian random observation is a "holographic" perception of the whole signal. In fact, if we decompose the signals into sparse forms, the non-uniform observations are inherently holographic sensing of signal sparse representation. Because very narrow pulses in the time domain have a wide range of perception in the frequency domain, which is directly related to Gaussian random observations in the time domain perception is consistent, only the perception matrix (CS operator A^{CS}) is different.

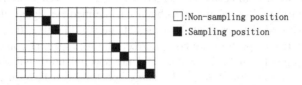

□:Non-sampling position
■:Sampling position

Fig. 1. Non-uniform observation matrix demonstration.

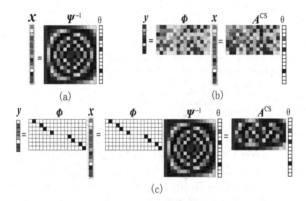

Fig. 2. Non-uniform observation and random observation.

In non-uniform observation, observation vector is actually the random extraction of the original signal, and each observation value has a certain time information, that is, the observation value depends only on the value of the signal at the corresponding time, without the need to get inner product of the observation matrix and the entire signal, so that non-uniform observation can be simplified to non-uniform sampling, and is no longer limited to discrete signals, which will provide convenience of using CS for speech signal.

3.2   Reconstruction Algorithm Based on Non-uniform Observation

CS simplifies the process of signal acquisition by increasing the complexity of the reconstruction algorithm. Since it needs to reconstruct high-dimensional signals from low-dimensional observation vectors, the common method is using reconstruction algorithm, such as Orthogonal Matching Pursuit (OMP) and Basis Pursuit (BP), through sparse representation as a bridge to realize reconstruction.

The non-uniform observation vector is different from other observation vectors. Each observation value of the non-uniform observation vector directly contains explicit time information. Therefore, this paper proposes a reconstruction algorithm based on the characteristics of non-uniform observation. Since non-uniform observation is essentially non-uniform sampling, the spectrum of the signal can be estimated by formula (3), as long as the spectrum of the original signal is not completely submerged (the spectrum will be completely submerged when the observed value is not enough), using part spectrum of signal we can obtain an intermediate signal, then filling the signal of unobserved positions with intermediate signal, like that, the approximation signal will get. If the spectrum found is correct, the real spectral components in the approximation signal will be intensified and the spectral noise will decrease. As a result, the previously submerged small energy spectrum can be visualized to better approximate the original signal, In order to achieve signal reconstruction, the process shown in Fig. 3.

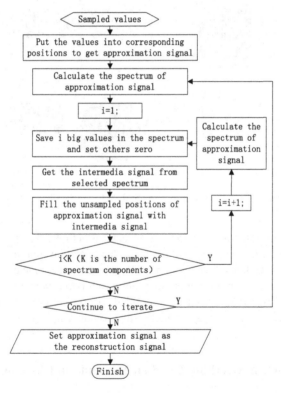

Fig. 3. Non-uniform observation matrix demonstration.

4   Simulation Experiments

The following verifies the performance of CS based on non-uniform observation through simulation experiments.

Experiment 1: Comparison of reconstruction for sinusoidal signals.

Set f_1 = 400 Hz, f_2 = 2 kHz, f_3 = 4 kHz and sample signal f(t) = sin($2\pi f_1 t$) + sin ($2\pi f_2 t$) + sin($2\pi f_3 t$) with f_s = 16 kHz to get the experimental signal f[n]. In this experiment, the length of f[n] is limited to 20 ms, which is 320 samples. The original signal is shown in Fig. 4(a), when the compression ratio (the ratio of the original signal length to the length of the observation vector, denoted as CR) is 6, the signal is observed by a Gaussian random matrix and a non-uniform observation matrix, the result is shown in Fig. 4(c) and (e), and the reconstructed signal is got separately using OMP from the two observation vectors, the results are shown in (d) and (f). The signal-to-noise ratio (SNR) of the reconstructed signal under random observation reaches 47.52 dB, while the SNR of the reconstructed signal under non-uniform observed reconstruction is 43.86 dB. Obviously, the two observation effects are close, which also confirms that the non-uniform observation is essentially consistent with the Gaussian random observations. In this experiment, we rebuild the signal from observation vector, which is show in Fig. 4(e), flowing the process shown in Fig. 3, and the results are shown in (g) ~ (m). With the increasing the number of approximations, the real spectrum of the signal is strengthened, while the spectrum noise decreases, and the reconstructed signal approximates the original signal gradually. However, unlike reconstruction under OMP, this method can only reduce the spectral noise and can not be eliminated, which can be seen in the band 5–7 kHz of Fig. 4(m). Figure 4(n) is the relationship between the number of approximations and the SNR. It can be seen that the quality of reconstructions increases with the number of approximations at the beginning, but tends to be constant after reaching a certain reconstructed mass. It can be seen from the Fig. 4(n) that the reconstruction method proposed in this paper can obtain a higher reconstruction signal quality (in this experiment, the signal to noise ratio can reach more than 60 dB). In order to illustrate the application effect of this method in the actual signal, it is verified by Experiment 2.

Experiment 2: Comparison of compression and reconstruction of speech signals.

We selected a speech signal from the voice library of the Chinese Academy of Sciences Institute of Automation as the experimental object. The sampling rate of the signal was 16 kHz, and the signal was overlapped and framed with a fixed frame length of 512 points, the overlap length was 128 points. To obtain the observed values of speech frames using non-uniform observation matrix in the condition that the CR (compression ratio) = 3 and CR = 4, respectively, the speech frames are reconstructed using the procedure in Fig. 3, and the number of large coefficients in each frame is set to 20, the number of iterations is set to 7, and the complete reconstructed speech is obtained by frame splicing, which is called the new method. The experimental results are compared with the results of compression perception using random observation and OMP reconstruction (we called the traditional method). In order to facilitate the analysis, the speech spectrogram of the reconstructed speech is drawn, the MOS score

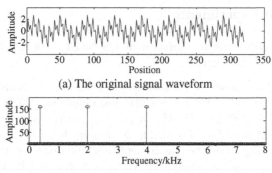

(a) The original signal waveform

(b) The amplitude spectrum of the original signal

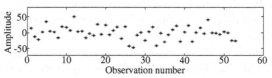

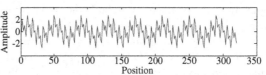

(c) The observation vector under Gaussian random observation

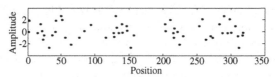

(d) The reconstruction signal from OMP under Gaussian random observation

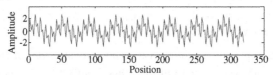

(e) The observation vector under Non-uniform observation

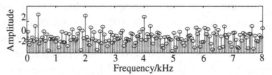

(f) The reconstruction signal from OMP under non-uniform observation

(g) The amplitude spectrum of the non-uniform observed signal

Fig. 4. The compression and reconstruction of sinusoidal signal.

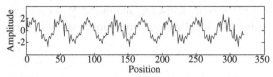

(h) The reconstructed signal for approximated first time

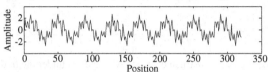

(i) The amplitude spectrum of the reconstructed signal for approximated first time

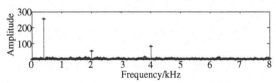

(j) The reconstructed signal for approximated 21st times

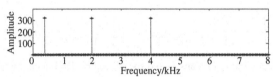

(k) The amplitude spectrum of the reconstructed signal for approximated 21st times

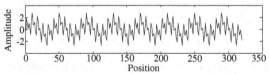

(l) The reconstructed signal for approximated 22^{ed} times

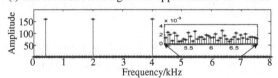

(k) The amplitude spectrum of the reconstructed signal for approximated 22^{ed} times

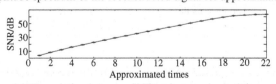

(n) The relation between SNR and approximated times

Fig. 4. (*continued*)

is obtained using the VQT scoring software, and the SNR is calculated. The results of the new method and the traditional method are compared, which are shown in Fig. 5 and Table 1.

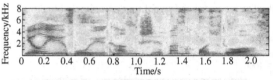

(a) Speech spectrogram of the original speech

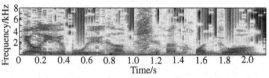

(b) Speech spectrogram of reconstructed speech when CR=3 under the traditional method

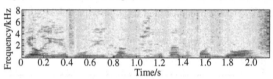

(c) Speech spectrogram of reconstructed speech when CR=3 under the new method

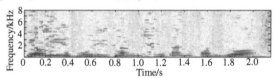

(d) Speech spectrogram of reconstructed speech when CR=4 under the new method

Fig. 5. The comparison of reconstructed speech.

Table 1. Table captions should be placed above the tables.

Indicator	CR	MOS	SNR/dB
The traditional method	3	2.89	13.09
The new method	3	2.86	8.95
	4	2.62	7.94

Figure 5(b) and (c) are the speech spectrogram of the reconstructed speech using the traditional method and the new method when the compression ratio is 3, in Fig. 5(c), it can be seen that the speech basically retains the spectral structure of the original speech (the most low-band and high-energy spectrum are retained), through subjective auditory experiments, we found that the reconstructed speech, which shown in Fig. 5(c), also completely preserves the original speech information, but there is obvious background noise. The white noise covering the entire frequency band is mainly caused by the noise caused by non-uniform sampling in the frequency domain and the reconstruction process can only suppress the noise but cannot completely eliminate the noise. After filtering, the noise is obviously suppressed, the reconstructed speech quality is similar to the reconstructed speech shown in Fig. 5(b) when we compare the MOS scores in Table 1. Figure 5(d) is the speech spectrogram of reconstructed speech using the new method when the compression ratio is 4, Comparing with the original speech spectrogram and analyzing Table 1, we can see that when the compression ratio increases, the speech quality of the new method will be significantly reduced. This is mainly due to the fact that too few observations cause a large amount of spectrum to be submerged by noise and cannot be recovered properly.

Comparing the compression performance of the actual speech signal with the traditional method and the new method, we find that CS and non-uniform sampling are essentially the same, both by reconstructing the original signal from a small number of observations of the sparse signal, but the CS perceives directly in the time domain, while uniform sampling is perceives the sparse domain through several time domain samples (in this article the sparse domain is the frequency domain). From the perspective of reconstruction, the traditional method is more efficient than the new method, but the new method can obtain a higher quality of the signal at the expense of a certain reconstruction efficiency, and the non-uniform observation matrix has more practical advantages than the traditional Gaussian random matrix.

5 Conclusion

In this paper, we start from the problems of application of CS in speech signal, non-uniform sampling theory is used for reference and non-uniform observation and reconstruction method are studied. Through theoretical analysis, the consistency and difference between CS and non-uniform sampling are revealed. Non-uniform observation matrix and signal reconstruction method based on non-uniform observation are constructed. Simulation results verify the effectiveness of the proposed method. It is worth mentioning that although this paper gives the application of CS in one-dimensional analog signals, there are still many problems worth further study. The first is the choice of sampling time. The ideal non-uniform sampling requires time completely random. Although the reference [9] had improved the problem of randomness of sampling positions by additive random sampling, but for a specific speech signal, each speech segment contains only a limited number of information, and theoretically there are a minimum number of sampling points to characterize the segment of speech, but the sampling position must not be randomly placed [10]. The placement rules for the sampling points have yet to be studied. The second is the reconstruction problem. The default sparse basis for non-uniform observations in this paper is the frequency domain, although the reference [11] gives wavelet non-uniform sampling method, but these transform domain coefficients all have a definite frequency attribute. Therefore, the study of non-uniform sampling problems that do not have frequency properties in sparse domains is expected to enrich CS reconstruction methods.

References

1. Guangming, S., Danhua, L., Dahua, G., et al.: Advances in theory and application of compressed sensing. Acta Electronica Sin. **37**(5), 1070–1081 (2009)
2. Donoho, D.L.: Compressed sensing. IEEE Trans. Inf. Theory **52**(4), 1289–1306 (2006)
3. Lu, X.: Research on MIMO Radar Imaging for Sparse Distributed Target. University of Science and Technology of China, pp. 87–88 (2017)
4. Jinchao, L., Zhaoxiang, D., Yanjun, J., et al.: Spectrum analysis and weak signal detection based on nonuniform sampling. J. Data Acquis. Process. **27**(3), 320–326 (2012)
5. Qian Hui, Yu., Lun, Z.H.: Performance analysis of parameters sparse signal nonuniform sampling. Comput. Digit. Eng. **39**(7), 24–26 (2011)
6. Kai, Yu., Yuanshi, L., Zhi, W., et al.: New method for acoustic signal collection based on compressed sampling. Chin. J. Sci. Instrum. **33**(1), 105–112 (2012)
7. Wenbiao, T., Guosheng, R., Haibo, Z., et al.: Non-uniform information acquisition and reconstruction within compressed sensing framework. J. Jilin Univ. (Eng. Technol. Edn.) **44** (4), 1209–1214 (2014)
8. Dongliang, G., Tiejun, Z., Xianhua, D.: Methods of signal frequency, amplitude and phase measurement based on non-uniform sampling. Syst. Eng. Electron. **34**(4), 662–665 (2012)
9. Qing, L.: The Research of Compressed Sampling at Sensor Network Nodes. Tianjin University of Technology, pp. 15–17 (2014)
10. Zhou, J., Shi, Z., Hu, L., et al.: Radar target one dimensional high resolution imaging based on sparse and non-uniform samplings in frequency domain. Acta Electronica Sin. **40**(5), 926–934 (2012)
11. Anming, W., Shu, W., Mingxin, C.: Study on spectrum of nonuniform sampling signals based on wavelet transform. J. Electron. Inf. Technol. **27**(3), 427–430 (2005)

GridWall: A Novel Condensed Representation of Frequent Itemsets

Weidong Tian[✉], Jianqiang Mei[✉], Hongjuan Zhou[✉],
and Zhongqiu Zhao[✉]

School of Computer and Information, Hefei University of Technology,
Hefei 230601, Anhui, China
Wdtian@hfut.edu.com

Abstract. A complete set of frequent itemset can be extremely and unexpectedly large due to redundancy when the given minimum support is low or when the transactional database is dense. To solve the problem, various concise representation strategies have been previously proposed, among which, some works undeniably well. Take Max Frequent Itemset, it has reached a high rate condensing frequent itemsets. But those existed models may consume too many resources which makes them not be suitable in some scenarios where restraints on time complex are strict but itemsets' support is not necessary at all. For this very kind of scenarios, this paper proposes a novel concept of frequent itemset border - Grid Wall, formed by positive border together with negative border, which tells whether an itemset is frequent, recovers all frequent itemsets but ignores their support. Grid-Wall founds on bi-partition and employs divide-and-conquer strategy, which make it fast and goes one step further than MFI on concise representation of frequent itemsets.

Keywords: Condensed representation · Itemset mining
Frequent itemset border · Grid Wall

1 Introduction

FI (Frequent Itemset) Mining is one of the most important subjects in field of Data Mining [1]. However, those itemsets are often huge in quantity which makes it even impossible holding the complete set of them [2, 3]. Mining so many FIs is not only a time consuming task, but also a space costing one, it then defeats the primary purpose of Data Mining as a consequence [4]. FIs could be reduced via increasing the given minimum support threshold but some significant itemsets can be filtered out. To address it, a new concept, namely MFI (Max Frequent Itemset), is proposed by Bayardo, as well as a special mining algorithm known as MaxMiner [5, 6]. Literally, a MFI itself and all its subsets are frequent while all its supersets are not [7, 8].

Apriori [9] is an algorithm to find all sets of itemsets that have support no less than specified minimum support. It is characterized as a width-first search algorithm using the well-known and interesting apriori property. Apriori scans database several times to generate all candidates and then the complete set of frequent itemset.

© Springer International Publishing AG, part of Springer Nature 2018
D.-S. Huang et al. (Eds.): ICIC 2018, LNCS 10954, pp. 425–433, 2018.
https://doi.org/10.1007/978-3-319-95930-6_39

Pasquier et al. proposed a concept named Frequent Closed Itemset and an algorithm called Pascal to min those itemsets [10], but still required multiple database scanning.

Han et al. introduced a data structure, namely FP-Tree, to hold and condense all necessary informations in memory [11]. FP-Tree only needs two times scanning database which makes it a milestone in the history of Data Mining.

Frequent Generator is proposed in [12, 13]. Set of frequent generators is non-redundant which makes it more appropriate than frequent closed itemsets on classification in some cases. But generators alone are not adequate for representing the complete set of frequent itemsets, it usually works together with negative borders.

Max frequent itemset firstly proposed by Bayardo in 1998 [6], all these itemsets together consist a border which splits itemsets into two parts, frequent part and non-frequent part. Besides, max frequent itemsets themselves are frequent, too, we call the complete set of them Positive Border.

In some scenarios, marketing as a typical case, recovering all frequent itemsets and telling frequency of an itemset need to be fast while itemsets' support is not necessary at all. This kind of scenarios magnifies shortage of some existed models which are focusing on lossless representation. Take the resolution in [14, 15], when it judges the frequency of a specified itemset, time cost increases exponentially.

In this paper, we present a new concept, namely Grid Wall, of concise representation of FI and a special algorithm, Grid-Wall, to mine it; the output implicitly and concisely represents all FIs. Grid-Wall sorts intermediate and final results in order, founds on bi-partition and employs divide-and-conquer strategy which makes it fast condensing FIs and telling whether a specified itemset is frequent with worst time complex $O(mn)$ while m and n separately being max length and size of grid wall. Grid-Wall shows a good performance on some data-sets, even just half size of MFI.

2 Preliminaries

Let $I = \{a_1, a_2, \ldots, a_n\}$ be a set of items and $D = \{t_1, t_2, \ldots, t_n\}$ be a transaction database, where t_i ($i \in [1, N]$) is a transaction, $t_i \subseteq I$ and t_i is called Itemset.

Definition 1 (Support): *The support of an itemset l in D is defined as* $support(l) = |\{t | t \in D \text{ and } l \in t\}| / |D|$. *It is a user specified threshold ranges from 0 to 1.*

Definition 2 (Frequent Itemset): *Itemset l is a frequent itemset when* $support(l) \geq min\_sup$, *denoted as FI.*

With given $min\_sup = 2/7$, itemset $\{A\}$ is frequent but not any more when $min\_sup = 5/7$. There is a dotted curve in Fig. 1 splits frequent itemsets apart from non-frequent one Itemsets on the upper side are all frequent, those on the other side, however, are all not.

Definition 3 (Max Frequent Itemset): *FI l is a MFI when none of its super sets is frequent. Iconically, the complete set of MFI is called Positive Border of FI, denoted as PB.*

Property 1: *PB can implicitly and concisely represent all FIs.*

From the definition above, we know that once we get a MFI, we can say all its subsets and the itemset itself are frequent while none of its super sets is. This is similar to apriori properties.

Table 1.
Example

Tid	Transaction
1	A
2	AC
3	ABC
4	BC
5	ABD
6	BCD
7	DE

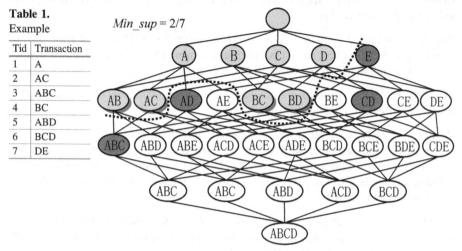

Fig. 1. Grid structure of *D*

Again the database shows in Table 1. In Fig. 1 with suggestion that *min_sup* = 2/7, itemsets in light gray are frequent, and those FIs with shadows are MFIs. Figure 1 clearly shows that all FIs are subsets of MFIs, take itemset {*D*}; it is subset of {*BD*}, for the latter one is frequent; the former is frequent, too. One step further, PB = {{*AB*}, {*AC*}, {*BC*}, {*BD*}} can stand for all FIs.

3   Grid Wall Representation

Definition 4 (Negative Border): *Itemset l itself is not frequent while all its subsets are, l is named Minimum Non-Frequent Itemset, denoted as MNFI. The complete set of MNFIs is Negative Border, NB for short.*

Property 2: *MNFI and its super sets are all non-frequent.*
One thing needs to be clear that subsets of MNFI cannot always be MFIs. In Fig. 1, Itemsets in light black are MNFIs, and {{*ABC*}, {*AD*}, {*CD*}, {*E*}} becomes negative border.

Definition 5 (Minimum Cluster): *Set of elements with relationships between each other is called Cluster. A cluster C has one and only one itemset NS ∈ NB, (C − NS)⊆PB, |C| ≥ 3 ,NS is called Cluster Kernel, denoted as CK, (C − NS) is named as Remote Area of Cluster, RAC for short.*

$\{\{ABC\}, \{AB\}, \{AC\}, \{BC\}\}$ is a minimum cluster in Fig. 1 with $\{ABC\}$ be its kernel and the other 3 be RAC. But $\{\{E\}, \emptyset\}$ is not a minimum cluster because of its size.

Lemma 1: *Compared against PB, replace RAC with CK can certainly and undoubtedly condense FIs.*

Proof: According to Definition 5, a minimum cluster has 3 elements at least, among which one is MNFI, others are MFIs. After replacement, PB decreases at least 1 itemset but still has full ability to represent all FIs concisely.

Property 3: *All RAC itemsets are same in length within a minimum cluster, say n, they contain n-1 items in common pairwise.*

Property 4: *Only those MFIs with same length might be replaced.*

Property 3 indicates that all RAC itemsets are direct sub sets of the corresponded CK, in other words, they are on the same level in grid structure. So if existed MFIs with different length, they will never belong to same minimum cluster. Thus, they can never be replaced by an MNFI.

Definition 6 (Grid Wall): *Find those MFIs play RAC role in the same time and replace them with corresponded CK, the new set obtained is called Grid Wall, denoted as GW and the itemsets are called Grid Brick, GB for short.*

Theorem 1: *Grid-Wall can certainly condense FIs and $|GW| \leq |PB|$ is correct for sure.*

Proof: It is proved the same way as proving Lemma 1.

Actually, this theorem is one of the guide rules of our job.

4 Grid-Wall Algorithm

4.1 Strategies Employed

Grid-Wall has two steps to mine final Grid Wall on framework level. First is to export intermediate result and second is to extract the final Grid Wall from intermediate result. We use some strategies to make condensing and recovering FIs more convenient.

The first one is a variant of divide-and-conquer strategy. Property 4 tells us that each time we find minimum clusters, we only focus on the set of MFIs with same length. Grid-Wall finds the complete set of MFIs and sorts them by their length. It starts with the section where MFIs only contains 1 item, extracts corresponding Grid Wall, then move to the next section of MFIs containing two items, and the same way down until none being not took into consideration.

The second is bi-partition strategy which mainly works for frequency checking. As intermediate result and final Grid Wall are all sorted in order, bi-partition can be used to locate region specified by itemsets' length.

4.2 Base Algorithms

Grid-Wall organizes intermediate result into a suitable form for search-space pruning. One of the 2 things to be done is to sort items one MFI by one, this is to make super and sub set checking more convenient. Sorting in as simple as alphabetical order can do a great favor.

The other thing is to sort them by their length and this is to help prune search-space. Property 4 implies that MFIs with same length should stand closely together rather than loosely, for this avoids searching the whole MFIs list from head to tail and only need to find those probabilities section by section. Divide and conquer strategy tells us that search can be faster in this way.

These 2 things together help improve performance of frequency checking. It is known to all that set l_1 can be sub set of l_2 only when $|l_1| \leq |l_2|$. This allows super and sub set checking to start from an appropriate position rather than from the very beginning of MFIs list when do frequency review.

As items lists are ordered, the algorithm just compare the two items in same position at the beginning, if they are equal, then go to next position synchronically, and if not, the position of child remains where it was while the position of parent goes to the next. At last, if parent contains all items of child, it is one of child's super sets.

Algorithm 1 only works during the first step of the two. Section 4.1 says only those itemsets whose length is no less than child's needs to be checked, since MFIs is sorted by their length, we start operation from the suitable position to prune some search-space. Bi-partition is used in line 7 to locate the position.

Algorithm 1. Algorithms for frequency checking

Input: MFIs : the complete set of MFIs with a certain length;

Itemset: the itemset to be checked.

Output: a Boolean value indicates whether an itemset is frequent or not

Flag ← ,false;

Position ← position of a MFI which has same length with itemset;

WHILE MFIs[Position] has same length with itemset DO

BEGIN

Flag ← IsSupersetOf(MFIs[Position], itemset);

IF Flag THEN exit (TRUE);

END;

FOR i ← Position TO MFIs.size DO

BEGIN

Flag ← IsSupersetOf(MFIs[i], itemset);

IF Flag THEN exit (TRUE);

END;

4.3   Grid-Wall

The above pruning strategies are just one part of all. Flag is used to skip those replaced MFIs before during search operation, and this can skip redundant search on one MFI on the one hand, on the other hand, it avoids useless replacements. For example, $\{AB\}$, $\{BC\}$ and $\{BD\}$ are 3 MFIs and $\{AB\}$ and $\{BC\}$ is replaced by $\{ABC\}$, while $\{BCD\}$ doesn't replace $\{BC\}$ and $\{BD\}$. The latter one doesn't help reduce Grid-Wall's size; In addition, $\{BD\}$ should not be replaced twice logically.

On the foundation of our preliminary work, main part of Grid-Wall algorithm becomes much easier.

Algorithm 2. GridWall Algorithm

Input: D: a transaction database; support: the given minimum support threshold.

Output: Grid-Wall retrieved.

ASet ← find and sort all MFIs from D with threshold support by their length;

FOR_EACH itemset1 which has not been replaced in ASet

BEGIN

    FOR_EACH the following itemset2 qualified in ASet

    BEGIN

        parent ← the super set of itemset1 and itemset2;

        IF all sub sets of parent are frequent THEN

            replace itemset1 and itemset2 with it;

    END;

END

ASet ← find k-MFIs;

In Algorithm 2, Flag is used in line 7 and 9 to mark whether the corresponding itemset is replaced. In line 9, "qualified" means itemset2 comes after itemset1 in MFIs list, its size equals itemset1's and it is not been replaced yet.

4.4   Frequency Checking for Specified Itemset

A specified itemset l, as there being no need to recover the l's support, Grid-Wall does a very simple job to complete the task. Generally, GW is split into two parts, those in one part are smaller in length that that specified itemset while those in the other part are larger. Grid-Wall uses itemsets in former part to say the l is non-frequent, if failed, then goes to the other part to check if l is frequent.

As for recovering all FIs, we use old simple solution for now. For those MFIs in GW, their entire sub sets and themselves are recursively given, and for those MNFIs in GW, they are not frequent but all their sub sets are.

5 Experimental Results

To examine the correctness and effectiveness, a whole system is implemented on the basis of this paper. Data-sets selected in different scales ranges from 20 to 28000 from UCI [3] and we even downloaded partial of a real supermarket transaction database to run our system on.

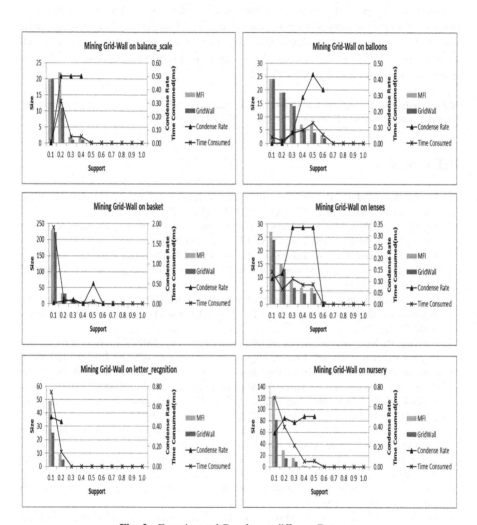

Fig. 2. Experimental Results on different Data-sets

In figures below, take Fig. 2, there are data not shown under some support values, this is because those support threshold is given too high to find any FI at all. In these scenarios, we assume no time consumed.

6   Conclusion

In this paper, we present a novel condensed representation model for FIs, describe it in details and implement and test it on variant data-sets. From the results above, we know that Grid-Wall works well on dense data-sets, on other data-sets, it performs not so well as dense ones. Next stage, we could lower restraints on minimum cluster and import other scenarios of concise representation.

Acknowledgement. The work described in this paper was fully supported by a grant from the National Natural Science Foundation of China (No. 61672203).

References

1. Moens, S., Aksehirli, E., Goethals, B.: Frequent itemset mining for big data. In: IEEE International Conference on Big Data, pp. 111–118 (2013)
2. Liu, G., Li, J., Wong, L.: A new concise representation of frequent itemsets using generators and a positive border. Knowl. Inf. Syst. **17**(1), 35–56 (2008)
3. Tseng, V.S., Wu, C., Fournier-Viger, P., Yu, P.S.: Efficient algorithms for mining the concise and lossless representation of high utility itemsets. IEEE Trans. Knowl. Data Eng. **27**(3), 726–739 (2015)
4. Liu, G., Li, J., Wong, L., et al.: Positive borders or negative borders: how to make lossless generator-based representations concise. In: SIAM International Conference on Data Mining – SDM (2006)
5. Hui-ling, P., Yun-xing, S.: A new FP-tree-based algorithm MMFI for mining the maximal frequent itemsets. IEEE Int. Conf. Comput. Sci. Autom. Eng. (CSAE) **2**, 61–65 (2012)
6. Bayardo Jr., R.J.: Efficiently mining long patterns from databases. In: ACM-SIGMOD International Conference on Management of Data, vol. 27, no. 2, pp. 85–93 (1998)
7. Burdick, D., Calimlim, M., Flannick, J., et al.: MAFIA: a maximal frequent itemset algorithm. IEEE Trans. Knowl. Data Eng. **17**(11), 1490–1504 (2005)
8. Lee, G., Yun, U.: Analysis of recent maximal frequent pattern mining approaches. In: International Conference on Computer Science & Its applications, pp. 873–877 (2016)
9. Agrawal, R., Srikant, R.: Fast algorithms for mining association rules. In: Proceedings of the 20th International Conference on Very Large Data Bases (VLDB 1994), pp. 487–499 (1994)
10. Bastide, Y., Taouil, R., et al.: Mining frequent patterns with counting inference. SIGKDD Explor. **2**(2), 66–75 (2000)
11. Pei, J., Han, J., Mao, R.: CLOSET: an efficient algorithm for mining frequent closed itemsets. In: Proceedings of the ACM SIGMOD International Conference on Management of Data (2000)

12. Bastide, Y., Pasquier, N., Taouil, R., Stumme, G., Lakhal, L.: Mining minimal non-redundant association rules using frequent closed itemsets. In: Lloyd, J., et al. (eds.) CL 2000. LNCS (LNAI), vol. 1861, pp. 972–986. Springer, Heidelberg (2000). https://doi.org/10.1007/3-540-44957-4_65
13. Kumar, A., Upadhyay, A.: An efficient algorithm to mine non redundant top K association rules. Int. J. Emerg. Trends Sci. Technol. **03**(01), 3491–3500 (2016)
14. Hamrouni, T., Denden, I., et al.: A new concise representation of frequent patterns through disjunctive search space. In: International Conference on Concept Lattices and their Applications (CLA 2007) (2007)
15. Hamrouni, T., Ben Yahia, S., Mephu Nguifo, E.: Towards faster mining of disjunction-based concise representations of frequent patterns. Int. J. Artif. Intell. Tools **23**(23), 315–335 (2014)

Lie Speech Time-Series Modeling Based on Dynamic Sparse Bayesian Network

Yan Zhou[1,2(✉)], Heming Zhao[2], and Li Shang[1]

[1] School of Electronics and Information Engineering,
Suzhou Vocational University, Suzhou 215104, Jiangsu, China
zhyan@jssvc.edu.cn
[2] School of Electronics and Information Engineering, Suzhou University,
Suzhou 215100, Jiangsu, China

Abstract. The lie detection from speech is difficult to be realized because it relates to many factors such as emotion, cognitive, willpower will and so on. Lying psychological state can dynamically change, and it influences the speech features obviously. However, the traditional modeling method does not fully consider the dynamic factors of speech signal, not mention to the psychological characteristics. Aiming at this problem, this paper presents a lie speech time-series modeling method based on Dynamic Sparse Bayesian Network (DSBN). This method analyzes the topology of the proposed DSBN model to achieve the probability dependence relationship of the state variables. Then the association relationship and the time series characteristic of the corresponding features can be calculated easily. The simulation experiments show that the established lying state time-series model have achieved a satisfied detection rate. The average correct detection rate has reached 76%. Therefore, the proposed time-series model is effectively, and it also provides a novel time-series modeling method for the psychological calculation.

Keywords: Lie speech · Time-series modeling · DSBN · Detection rate

1 Introduction

Lie detection from speech is a complicated psychological calculation problem [1, 2]. When someone lies, the physiological state such as blood pressure, heart rate, or the psychological state such as emotion, cognitive, willpower will all can be changed corresponding. As a result, these changes can also significantly affect the characteristics of speech signal [3, 4]. Under normal circumstances, the liar shows hesitated or fear of what he has said. For example, the speaking rate may be a little slower, the tone may be not very sure, and many pauses. Furthermore, due to the liar is nervous when telling a lie, the tone will also changes, such as the sudden loud sound, the high-pitched sound and so on. In addition, the liar may also be accompanied with repetition, or making perfect expression deliberately, etc. However, the lying psychological state can convert from normal to lying state over time, or the reverse process. And the corresponding speech signal will be reflected a variety of changes in response. Therefore, the lying state is a gradient and reciprocating process, and has significant time dynamic

© Springer International Publishing AG, part of Springer Nature 2018
D.-S. Huang et al. (Eds.): ICIC 2018, LNCS 10954, pp. 434–443, 2018.
https://doi.org/10.1007/978-3-319-95930-6_40

characteristic. So in order to mining the changes of speech characteristics over time, the detection lie from speech must consider the time-series modeling.

The Sparse Bayesian Networks (SBN) [5–7] is designed based on Bayesian Networks (BN). It is through the way of adding sparse constraint condition to the BN, which can control the number of kernel function in the algorithm. At the same time, the SBN structure and parameters can be obtained. As we know, the lying detection from speech is very complex. It is a learning process from the speech flow step by step, which also involves the abstract mapping relationship of each speech feature set. That is to say, it needs to analyze the feature parameters changed over time and also need to research their relationships [8, 9]. Then the psychological state and the state transition probability can be calculated. Through this way, the lying state can be correctly judged. Considering of the significant time-series characteristic of lying state, a time-series model of DSBN is proposed in this paper. It is efficient for analyzing the lie speech signal with time series characteristics. The proposed DSBN is a kind of model which can learn the change rule of feature variables over time, and the transfer relationship between them can also be learned. Because DSBN has the ability of expressing multilayer knowledge and processing time series data, it is suitable for modeling the correlation relationship between the feature variables.

The proposed DSBN model mainly processes the lie speech signal in this paper. As to detect the lie state, there are eight kind of acoustic features selected to be the input. They are the peak frequency, pitch frequency, short-term energy, roughness, fluctuation strength, the first order Mel Cepstrum coefficient, the second order Mel Cepstrum coefficient, the third order Mel Cepstrum coefficient respectively. However, in order to successfully determine the lie psychological state, the change rule of speech characteristics over time must be research, and the DSBN time-series model can complete the time-series modeling requirements.

2   Dynamic Sparse Bayesian Network

DSBN is an extension of SBN on the time series. SBN is a model which can only reflect the relationship between the variables of a state for one moment [10–12]. But this cannot meet the requirements of analyzing the time series change problem, for example the lie detection problem. However, DSBN can observe dynamic state because it can extends to deal with the issue of a series of variables that change over time. The dynamic change of SBN model is shown in Fig. 1.

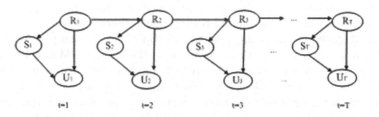

Fig. 1. The dynamic change process of sparse Bayesian Network.

From Fig. 1, it can be seen that DSBN has the significant time series characteristics. The probability model of the dependent relationship between the features parameters can be established depend on the DSBN structure. The established probability model can reflect the transfer probability relation between variables, furthermore, the variables change over time and the probability distribution are shown as following:

The variable Y_t contains three characteristic parameters $\{R_t, S_t, U_t\}$, at any time t, the state of S_t is determined by R_t, the state of U_t is determined by R_t and S_t. So the joint distribution of Y_t can be expressed as this:

$$P(Y_t) = P(R_t, S_t, U_t) = P(R_t)P(S_t|R_t)P(U_t|R_t, S_t) \tag{1}$$

However, the conditional probability distribution can be expressed as following:

$$
\begin{aligned}
P(U_t|R_t) &= \frac{P(U_t, R_t)}{P(R_t)} = \frac{\sum_m P(U_t, R_t, M_t = m)}{P(R_t)} \\
&= \frac{\sum_m P(R_t)P(M_t = m|R_t)P(U_t|R_t, M_t = m)}{P(R_t)} \\
&= \sum_m P(M_t = m|R_t)P(U_t|R_t, M_t = m)
\end{aligned}
\tag{2}
$$

Between the moment $t - 1$ and t, the state transition of R_t in the variable Y_t has produced, so the transition probability of variable Y_t is expressed as following:

$$P(Y_t|Y_{t-1}) = P(R_t|R_{t-1}) \tag{3}$$

From the analysis of above, due to the DSBN model estimate the changes of variables through observing the state, so the model parameters can reflect the feature changes over time. So the DSBN model can solve the dynamic problem.

3   Lie Detection from Speech Based on DSBN

The research of lying detection is a typical time series analysis process. In a length of speech, the lie information and normal information exists at the same time. As the change of time, lie state and the normal state can transform mutually. Therefore, the status of lie is a reciprocating process. From a physiological standpoint, the liar may feel nervous or relax from time to time in the process of telling a lie. Consequently, the speech signal feature may be changed corresponding. When analyzing the lie speech, the selected length of time or the moment will impact the different psychology state. Considering the complex of lie state, the traditional HMM and GMM modeling method all cannot express the state. This is because of that they do not calculate the significant time series characteristics sufficiently. DSBN model proposed in this paper is the development of the traditional Bayesian model, which can effectively learn the probability of the dependent relationship between speech variables and the rule of variables change over time. It shows the state change in the form of probability transfer diagram, and it is suitable for the analysis of correlation between the variables. The process of lie detection from speech based on DSBN model is shown as following:

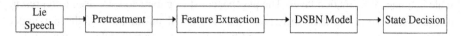

Fig. 2. The detection process for the lying psychological state.

The Hidden Markov chain is selected for calculating the probability transfer of hidden variables. DSBN can not only express the psychological states change as the speech features change, but also can fusion the whole information and the local details of the lie speech signal.

In this model, there is a variety of information contained in the unit time t. It not only includes the overall psychological affection information of the speaker, but also contains the local details such as psychological cognitive and psychological will. Therefore, combined with the whole information and local details of speech signal, this paper established multi-level information fusion model to implement the lie detection. The features in the DSBN structure are shown in Fig. 3. From this structure, dynamic probability of each layer is the first-order Hidden Markov process. The probability of future is only associated with the current moment, and nothing to do with the past, that is, X_t rely on X_{t-1}. Here, assuming that the topology structure of constructing network cannot be shifted over time.

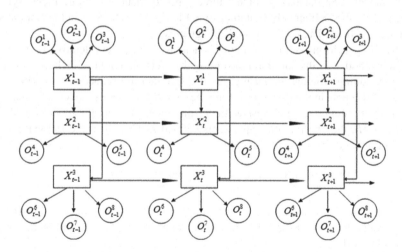

Fig. 3. The structure of time-series model for the lie speech features.

In order to capture the relationship of speech lie features changed over time, the DSBN time-series model can be expressed in Fig. 3.

In Fig. 3, the three layers of the model contain three state variables X_1, X_2, X_3 and eight observed variables are $o^1 \sim o^8$. State variable X^1 is used to express the whole dynamic information of psychological states that is the large-scale language information layer. The prosodic features are used to represent it. State variable X^2 is used to express the details which are caused by the subjective feeling of liar that is the middle-scale language information layer. The psychoacoustics parameters are used to represent it. State variable X^3 is used to express the will detail information of the liar under different psychological physiology condition that is the small-scale language information layer. The MFCC series parameters are used to represent it.

Therefore, the parameter settings are as following, the prosodic features include the peak frequency, pitch frequency, short-term energy are used for expressing the observation variable o^1, o^2, o^3 respectively. The psychoacoustics parameters of Roughness R and fluctuation strength F are used for expressing the observation variables o^4, o^5 respectively. The first order Mel Cepstrum coefficient, the second order Mel Cepstrum coefficient, the third order Mel Cepstrum coefficient are used for expressing the observation variables o^6, o^7, o^8. The proposed model can successfully express the time series characteristics of speech, and also can fuse the global and local information of lie speech signal.

4 Experiment Results and Analysis

4.1 Experimental Data

Suzhou University Lie Database (SULD) is adopted for testing the effectiveness of DSBN model. The database contents three types of data, the Inducement style Lie Speech (ILS), the Deliberately Imitating style Lie Speech (DILS), and the Natural style Lie Speech (NLS). The total size is 725 segments.

The lie speech time-series modeling is actually a reasoning iterative process. The specific method is as this, at first, performing the pretreatment task, including endpoint detection, adding window, denoising, and PCA dimension reduction, etc. Then using the part of ILS data in the database to train the DSBN time-series model, in which takes 70% as the training data, and 30% as the testing data. Training the model for M liars, defined as M_i, the observation sequence is defined as $O_{1:T}$, then the following calculation formula can be used for judging.

$$\gamma = \arg \max_i P(M_i|O_{1:T}) = \arg \max_i P(O_{1:T}|M_i)/P(O_{1:T}) \qquad (4)$$

In the equation above, $P(M_i)$ is the prior probability of model M_i, getting the average value for $1/N$, if the parameter of model M_i is Θ_i, then the following formula can be obtained.

$$P(O_{1:T}|M_i) = P(O_{1:T}|\Theta_i) \qquad (5)$$

Under the condition of a given observation sequence, then the Eq. (6) can be determined.

$$P(O_{1:T}) = 1 \tag{6}$$

Then combined with Eq. (4), the equation above can be expressed as following, which can realize the task of lie state identification.

$$\gamma = \arg \max_i P(M_i|O_{1:T}) = \arg \max_i P(O_{1:T}|M_i)/N \tag{7}$$

4.2 Experiment Results and Analysis

(1) Contrast of lie detection results

In the experiment of testing the performance of DSBN lie detection model, the ILS data is used for training the model, then using the corresponding data to identify the lie state. According to this method, the performance of the proposed model in this paper can be proved. First of all, randomly select 60 people from ILS, with 30male samples and 30 female samples. Here, taking the 70% samples use for the training signals and the 30% use for the test signals.

In order to objectively compare the performance of the proposed model with other models, this paper selects the Multiple Hidden Markov Model (MHMM) and SVM for comparison.

Tables 1 and 2 shows the lie detection results of the DSBN model, MHMM model and SVM model respectively.

(2) Robustness analysis of the DSBN model

In order to verify the robustness of the proposed time series model, the DILS data is used in this experiment. Since the DILS data is not recorded in a quiet environment, so the data may carry a lot of noise.

In order to measure the noise robustness of DSBN model, the experiment is designed as following. The training samples are randomly selected from the DILS, which are the denoising speech signals of 30 male and 30 female respectively. The testing samples adopt the denoising signal and original signal of the corresponding 60 people. However, this experiment mainly concerns the influence of the noise to the proposed model, so the contrast of detection effect for the denoising signal and original signal is the main purpose. Figures 4, 5 and 6 show the results of this experiment.

4.3 Experiment Analysis

There are two parts in the lie speech time-series modeling experiment based on DSBN. First of all, employing the MHMM and SVM model for comparison, the performance of the proposed time-series model is tested. Secondly, the noise robustness of the model is tested to examine the noise resistance property. The specific analysis for each experiment is as following.

(1)

The analysis of detection rate

The comparison results of correct detection rate in Tables 1 and 2 are obviously indicate that, DSBN time-series model proposed in this paper is better than the MHMM model and SVM model. The male sample testing and female sample testing results based on DSBN time-series model show that, the average correct detection rate is 78.38% for male and 76.22% for female. They are higher than the average detection rate of MHMM model for 10.36% and 8.85% respectively. However, they are incredibly higher than the average detection rate of SVM model for 20.60%and 13.74%

Table 1. Detection rate of different models (%) (For male).

Detection model	Testers														
	1	2	3	4	5	6	7	8	9	10	11	12	13	14	15
DSBN	75.4	78.2	80.4	75.3	80.9	78.5	71.7	81.4	82.3	75.5	73.7	76.4	81.4	80.4	76.4
MHMM	66.3	69.2	69.4	65.7	70.4	60.6	70.3	70.3	77.4	60.3	68.6	68.1	69.3	66.5	67.4
SVM	61.2	53.2	58.3	56.3	58	57.9	61.4	59.4	63.4	49.3	57.3	56.3	49.2	54.3	61.3
Detection model	Testers														
	16	17	18	19	20	21	22	23	24	25	26	27	28	29	30
DSBN	75.2	83.2	72.5	76.3	83	78.9	76.4	84.7	80.4	79.3	81.2	80.2	73.6	76.3	82.3
MHMM	66.3	69.2	69.4	65.7	70.4	60.6	70.3	71.3	77.4	60.3	68.6	68.1	59.3	66.5	77.4
SVM	60.4	53.2	58.4	54.3	46.9	54.5	60.7	55.4	72.3	57.4	63.7	56.4	51.4	60.4	71.4

Table 2. Detection rate of different models (%) (For female).

Detection model	Testers														
	1	2	3	4	5	6	7	8	9	10	11	12	13	14	15
DSBN	75.2	72.2	68.3	76.4	78.9	77.1	76.3	79.3	70.5	69.3	77.5	70.3	73.4	76.8	80.3
MHMM	67.5	68.4	70.8	65.3	61.5	70.5	63.3	61.5	77.2	65.3	66.8	64.1	62.3	63.5	78.3
SVM	61.2	64.7	68.4	54.3	55.9	66.3	57.4	55.6	70.3	61.4	62.7	58.4	59.4	61.4	71.4
Detection model	Testers														
	16	17	18	19	20	21	22	23	24	25	26	27	28	29	30
DSBN	76.5	83.2	76.3	82.7	78.6	77.9	76.3	79.8	80.4	69.3	77.4	82.6	73.2	80.1	70.6
MHMM	64.3	75.9	69.4	75.7	70.4	64.6	67.3	67.3	77.4	67.8	67.6	73.1	69.3	66.7	64.5
SVM	57.4	48.2	64.8	67.3	68.9	60.5	58.7	62.6	70.3	62.4	61.5	69.4	55.4	67.9	70.4

respectively.

For the male testing samples, the correct detection rate of 13 testers are more than 80% based on the DSBN model, in which the number 23 tester got the highest recognition rate of 84.7%. Moreover, the detection rate of number 10 tester is improved 15.2% from using the MHMM model to the DSBN model.

In addition, for the female testing samples, the correct detection rate of 6 female testers is more than 80% based on the DSBN model, in which the number 17 tester got the highest recognition rate of 83.2%. Moreover, the detection rate of number 8 tester is improved 17.8% from using the MHMM model to the DSBN model.

In general, among these three kinds of models, the detection rate of the proposed lying time-series model is the highest, and the detection rate of the SVM model is the lowest. The results indicate that the time series characteristic of the detection model is quite important. Since the proposed DSBN model adequately consider the temporal characteristics, so it obtained the best detection result.

(2) The analysis of robustness performance

In the experiment of analyzing the robustness performance of DSBN model, the original speech signal and the denoising signal are tested respectively. The experiment results in Figs. 5 and 6 indicate that, the proposed DSBN model has achieved the outstanding result.

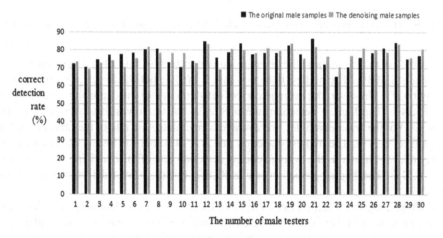

Fig. 4. The noise robustness testing results of DSBN model (male lie speech samples).

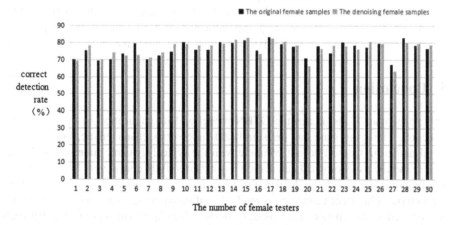

Fig. 5. The noise robustness testing results of DSBN model (female lie speech samples).

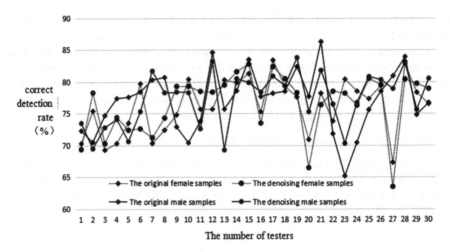

Fig. 6. The noise robustness contrast for male and female samples based on the DSBN model.

For the male samples, the average correct detection rate of the original speech signal is 77.03%. The highest detection rate is 86.3%, but the lowest detection rate is only 65.2%. However, the average correct detection rate of the denoising speech signal is 77.34%. The highest detection rate is 83.8%, but the lowest detection rate is only 69.3%. For the female samples, the average correct detection rate of the original speech signal is 76.27%. The highest detection rate is 83.4%, but the lowest detection rate is only 67.3%. However, the average correct detection rate of the denoising speech signal is 76.51%. The highest detection rate is 82.8%, but the lowest detection rate is only 63.5%.

Comparing the detection rate of original speech signal and denoising speech signal, the detection difference is not so obviously either for male or female. This implies that the DSBN lying state time-series model proposed in this paper has good noise robustness. This is because of that it extracts the characteristic components of major information that can represent the speech signal. However, in the process of speech signal sparse decomposition, the Gaussian white noise cannot be effectively decomposed and separated. This reduced the influence of Gaussian white noise to the detection model. Therefore, the DSBN lying state time-series model is a robust model.

5   Conclusion

This paper mainly researches the DSBN model and its application in the lying state detection. The lying state has the significant time series characteristic, but the traditional modeling method do not combine with the typical dynamic time series characteristic and the psychological calculation has not been fully considered. So they are not suitable for dealing with the detection of lie from speech. Considering of this problem, the method of lie speech time-series model based on dynamic sparse bayesian network is proposed in this paper. This method studies the dependent relationship between psychological state variables, and the variables change over time through analyzing the network topology. Taking this approach, the relationship between features which is corresponding the variables can be calculated and then the relationship of the time series also can be found. In addition, the model training adopted a variety of features, it not only can obtain overall lie speech signal characteristic information, but also be able to capture the details information. Moreover, the lie state can be deeply expressed with different time scales of feature extraction method. The experiment results proved that the lying state time-series model established in this paper can efficiently implement the lie speech detection. The average correct detection rate of male is 78.38%, and 76.22% for female. Thus it can be seen that the method proposed in this paper is an effective time-series model for lie detection, and it also provides a new research method for psychological calculation.

Acknowledgments. The authors acknowledge the National Natural Science Foundation of China (Grant: 61372146, 61373098), the Youth Natural Science Foundation of Jiangsu Province of China (Grant: BK20160361), the QingLan project of colleges and universities in Jiangsu province, the Professional Leader Advanced Research Project Foundation of Higher Vocational College of Jiangsu Province (Grant: 2017GRFX046).

References

1. Perez, G., Luis, A., Caballero, M., et al.: Multimodal emotion recognition with evolutionary computation for human-robot interaction. Expert Syst. Appl. **66**, 42–61 (2016)
2. Billard, A.: On the mechanical, cognitive and sociable facets of human compliance and their robotic counterparts. Robot. Auton. **88**, 157–164 (2017)
3. Marie, J.C.: Vocal fatigue induced by prolonged oral reading: analysis and detection. Comput. Speech Lang. Comput. Speech Lang. **28**(2), 453–466 (2014)
4. Schuller, B., Steidl, S., Batliner, A., et al.: Medium-term speaker state-A review on intoxication, sleepiness and the first challenge. Comput. Speech Lang. **28**(2), 346–374 (2014)
5. Raczynski, S.A., Vincent, E., Sagayama, S.: Dynamic bayesian networks for symbolic polyphonic pitch modeling. IEEE Trans. Audio Speech Lang. Process. **21**(9), 1830–1840 (2013)
6. Yang, Q., Xue, D.: The gait recognition based on the double scale dynamic bayesian network and multiple information fusion. J. Electron. Inf. **34**(5), 1148–1153 (2012)
7. Codecasa, D., Stella, F.: Classification and clustering with continuous time Bayesian network models. J. Intell. Inf. Syst. **45**(2), 187–220 (2015)
8. Lim, C., Chang, J.: Efficient implementation techniques of an SVM-based speech/music classifier in SMV. Multimedia Tools Appl. **74**(15), 5375–5400 (2015)
9. Han, W., Zhang, X.: The third lecture of the classical deep learning network model and the training methods. Mil. Commun. Technol. **37**(1), 90–97 (2016)
10. Zhang, X., Chen, F., Gao, J.: Sparse bayesian and its application in time series prediction. Control Decis. Making **21**(5), 585–588 (2006)
11. Chien, J.T., Ku, Y.C.: Bayesian recurrent neural network for language modeling. IEEE Trans. Neural Netw. Learn. Syst. **27**(2), 361–374 (2016)
12. Zhiyong, W., Cai, L.: The audio and video double modal speaker recognition based on dynamic bayesian network. Res. Dev. Comput. **43**(3), 470–475 (2006)

A Computationally Grounded Model
of Emotional BDI-Agents

Yun Su[1,2], Bin Hu[1(✉)], Yongqiang Dai[1], and Juan Rao[1]

[1] School of Information Science and Engineering,
Lanzhou University, Lanzhou, China
bh@lzu.edu.cn
[2] College of Computer Science and Engineering,
Northwest Normal University, Lanzhou, China

Abstract. This paper extends BDI (belief, desire, and intention) logic by incorporating well-being emotion modalities (joy and distress) based on Ortony, Clore, and Collins's (OCC) theory and obtain the emotional BDI logic called BDIE (belief, desire, intention and emotions) logic. We propose a new computational model of emotion triggers for BDI agents, called the interpreted observation-based BDIE system model (or BDIE model for short). The key point of this BDIE model is to express agent's emotions, such as joy and distress, as a set of runs (computing paths), which is exactly a system in the interpreted system model, a well-known agent-model due to Halpern et al. We present a sound and complete proof system with respect to our BDIE model and specify a simplified auction scenario to illustrate the construction of the BDIE model and the specification of multi-agent systems involving agents' emotional states using BDIE logic.

Keywords: Emotional model · BDI logic · OCC theory · Agents

1 Introduction

Computational models which concerned with the integration of emotional states into agents have gained significant attraction [1]. When building emotional agents, BDI (belief, desire, and intention) logic [2] has been proved to be one of the best options one can select, which offers expressive frameworks to represent agents' mental attitudes such as beliefs and desires and to reconstruct on their basis the cognitive layer of emotion [3, 4].

The interpreted system model [5], one of the two main semantic approaches to formalizing agent systems via BDI modal logic, offers a very natural interpretation, in terms of the states of computer processes, to S5 epistemic logic. The salient point of this approach is that we are able to associate the system with a computer program and formulas can be understood as properties of program computations. In this sense, the interpreted system model is computationally grounded [6, 7]. Thus, it is interesting to extend the interpreted system model to capture an agent's mental states, such as belief, desire, intention, and emotional states as well, such as joy and distress.

© Springer International Publishing AG, part of Springer Nature 2018
D.-S. Huang et al. (Eds.): ICIC 2018, LNCS 10954, pp. 444–453, 2018.
https://doi.org/10.1007/978-3-319-95930-6_41

This paper aims at extending BDI logic by incorporating well-being emotion modalities (joy and distress) and presenting a computationally grounded model of belief, desire, intention, and emotions, called the interpreted observation-based BDIE system model (or the BDIE model for short), by extending the interpreted system model. The key point of this BDIE model is to express agent's emotions, such as joy and distress, as a set of runs (computing paths), which is exactly a system in the interpreted system model. Our BDIE model can represent and formalize two kinds of emotions, i.e., joy and distress, their properties, and the elicited progress by the agents' cognitive mental states (belief, desire, and intention).

We ground our emotional extension of the BDI model in the OCC (named after the authors Ortony, Clore, and Collins) model [8] which is a psychological appraisal model of emotions. It considers that an emotion is the result of an evaluation process called appraisal. For example, people may experience joy or distress when they evaluate the consequences of an event according to their goals. This paper focuses on modelling joy and distress emotions, which these emotions seem to be the most important [9, 10] and they are also simpler to formalize since they require few variables.

The paper is structured as follows: related research is discussed followed by consideration of the BDIE model and BDIE logic constructed in our work. A simplified auction scenario illustrating the construction of the model is addressed. The paper closes with concluding observations and consideration of future work.

2   Related Work

There are no less than 150 psychological and philosophical emotion theories that have been proposed [11]. A theory widely used by computer scientists is the OCC emotion theory. The reasons are that it provides a clear classification of a broad range of emotion types, it lists concise descriptions of the conditions that elicit emotions, and for this it uses concepts that are well studied and relatively straightforward to formalize [12, 13].

Some rigorous formalizations which extend the logical theories of agents to incorporate emotions based on OCC theory have been proposed. Meyer et al. [14] use KARO, the agent logic based on the dynamic logic augmented with BDI-type modal operators to formalize the belief-desire preconditions of four basic emotions (happiness, sadness, anger, and fear) proposed in Oatley and Johnson-Laird's theory of emotion. Emotional-BDI logic presented by Pereira et al. [15] illustrates that agents' behavior is guided not only by beliefs, desires and intentions, but also by the role of emotions in reasoning and decision-making.

A more closely related work by Adam et al. [12] is devoted to a formalization of emotions in BDI terms, using an extended version of the Cohen-Levesque logic by integrating OCC's appraisal variables. More recently, Steunebrink et al. [13] use KARO to formalize the cognitive-motivational preconditions of the emotions considered in the OCC theory. Later, Gluz and Jaques [16] work with a probabilistic extension of the BDI model based on affective probabilistic logic (AfPL) to calculate and represent the intensity of the event-based emotions of the OCC model.

The work mentioned above only uses modal operators (such as B, D, and I) to provide the logical definitions of emotions, rather than extending emotions to the semantic model which will make it possible to use model checking tools to verify formally emotional attributes. There is the challenge of evaluating whether an emotional model works as intended [17]. Besides, to keep their logic simple, the work mentioned above lacks some temporal expressivity like *Until*. Finally, the semantics of models mentioned above are defined over the Kripke semantics in terms of possible worlds and accessibility relations, however, they are still not very clear how to obtain concrete emotional agent models with the belief, desire, and intention accessibility relations from specific agent programs.

The present paper, on the contrary, aims at presenting a computationally grounded model of belief, desire, intention, and emotions, called the BDIE model, by extending the interpreted system model. The key point of this BDIE model is to express agent's emotions, such as joy and distress, as a set of runs (computing paths), which is exactly a system in the interpreted system model. Our BDIE model can represent and formalize two kinds of emotions, i.e., joy and distress, their properties, and the elicited progress by the agents' cognitive mental states (belief, desire, and intention).

3 Computationally Grounded BDIE Model

In this section we introduce the interpreted system model, the interpreted BDI system model, and the interpreted BDIE system model. Here we follow the presentation given in [18], where agents have local observation.

3.1 Interpreted System Model

Interpreted systems are a well-known formalism for the multi-agent systems. They provide a convenient means for reasoning about time and agents' mental states. We assume a system composed of multiple agents $A = \{1, 2, \ldots, n\}$ in an environment e.

We represent *the system's state* or *the global state* as a tuple $(s_e, s_1, s_2, \ldots, s_n)$ where s_e is the environment's local state, and for each $i, i \in A$, s_i is the agent i's local state.

Let L_e be a set of possible local states of the environment and L_i be a set of possible local states for the agent i, $i \in A$. We take $G \subseteq L_e \times L_1 \times \cdots \times L_n$ to be the set of *reachable global states* of the system. A *run r* over G is a function from the time domain–the natural numbers in our case–to G. Thus, the run over G can be identified with a sequence of global states in G.

We refer to a pair (r, m) consisting of a run r and time m as a *point*. Given the point (r, m), we denote the first component of the tuple $r(m)$ by $r_e(m)$ which is the local states of the environment, and for each $i, i \in A$, the $(i + 1)$'th component of the tuple $r(m)$ by $r_i(m)$. Thus, $r_i(m)$ is the local state of the agent i in the run r at the "time" m.

For every agent $i, i \in A$, let $(r, u) \sim_i (r', v)$ denotes that $r_i(u) = r'_i(v)$. Intuitively, $(r, u) \sim_i (r', v)$ means that (r, u) and (r', v) are indistinguishable to the agent i.

The idea of the interpreted system semantics is that the run represents one possible computation of a system and the system may have a number of possible runs, so we say the *system* is a set of runs.

Assume that we have a set Φ of primitive propositions, which describes basic facts about the system. *An interpreted system $I = (\mathcal{K}, \pi)$ is a pair*, where $\mathcal{K}$ is a set of runs over a set of global states and π is a valuation function, which gives the set of primitive propositions true at each point in $\mathcal{K}$.

3.2 Interpreted BDI-System Model

Given a set G of global states and a system $\mathcal{K}$ over G, an agent's mental state, i.e. belief, desire, and intention, over the system $\mathcal{K}$ is a tuple $\mathcal{M} = \; <\mathcal{B}, \mathcal{D}, \mathcal{I}>$, where $\mathcal{B}, \mathcal{D}$, and $\mathcal{I}$ are systems (sets of runs over G). Thus, the runs in $\mathcal{B}_i$ are the possible computing paths from the viewpoint of the agent i; the runs in $\mathcal{D}_i$ are the computing paths that the agent i desires; and the runs in $\mathcal{I}_i$ are computing paths with the agent i's intentional choices of the possible actions.

We assume that $\mathcal{I} \subseteq \mathcal{K}$ and $\mathcal{D} \subseteq \mathcal{B} \subseteq \mathcal{K}$ because it is reasonable to assume that every desired computing path is possible one. Nevertheless, we need not assume that $\mathcal{I} \subseteq \mathcal{D}$ or even $\mathcal{I} \subseteq \mathcal{B}$ because an agent's intention may fail to achieve its goal and the real computing path may be beyond the agent's belief even though the agent has chosen and completed an intentional series of actions.

Then, we define *a BDI-system $\mathcal{S}$ is a structure* $\mathcal{S} = \; <\mathcal{K}, \mathcal{M}_1, \cdot \cdot \cdot, \mathcal{M}_n>$, where $\mathcal{K}$ is a system and for each $i, i \in \{1, \ldots, n\}$, $\mathcal{M}_i$ is the agent i's mental state over $\mathcal{K}$.

Assume that we have a set Φ of primitive propositions which describe basic facts about agents and their environment. *An interpreted BDI$-$system $I = (\mathcal{S}, \pi)$ consists of* a pair, where $\mathcal{S}$ is a *BDI*-system and π is a valuation function, which gives the set of primitive propositions true at each point in G.

3.3 Interpreted BDIE System Model

In order to extend the interpreted BDI-system model by incorporating well-being emotion modalities (i.e. joy and distress), the OCC theory of emotion described by Ortony, Clore and Collins is first introduced briefly in this section. Then, we address ourselves to the construction of a new computational model of emotion triggers for BDI agents, called the interpreted observation-based BDIE system model (or the BDIE model for short).

The OCC Theory of Emotion. Ortony, Clore and Collins propose a cognitive appraisal theory of emotion, known as OCC, which explains the eliciting condition of 22 emotion types grouped in six classes. The OCC theory is structured as a three-branch typology, corresponding to three kinds of stimuli: consequences of events, actions of agents, and aspects of objects.

In this paper, we focus on the well-being emotions (i.e., joy and distress). This class of emotions arises when an individual appraises an event that has just occurred while only focusing on the desirability of its consequences for herself/himself. That is, an agent experiences joy when she/he is pleased about a desirable event, and an agent experiences distress when she/he is displeased about an undesirable event. Consider an example where an agent feels pleased when she/he wins the auction because this is desirable for him, and an agent is in distress when she/he loses the auction because the consequence is undesirable for him.

The Interpreted BDIE System Model. The key point of this BDIE model is to express agent's emotions, such as joy and distress, as a set of runs (computing paths), which is exactly a system in the interpreted system model.

Given a set G of global states and a system $\mathcal{K}$ over G, an agent i's mental states (believe, desire, intention) and emotional states (joy and distress) over the system $\mathcal{K}$ are defined as a tuple $\mathcal{M}_i = \ <\mathcal{B}_i, \mathcal{D}_i, \mathcal{I}_i, \mathcal{JOY}_i, \mathcal{DISTRESS}_i>$ where $\mathcal{JOY}_i$ and $\mathcal{DISTRESS}_i$ are systems (sets of runs over G). The runs in $\mathcal{JOY}_i$ are computing paths that the agent i feels pleased; and the runs in $\mathcal{DISTRESS}_i$ are computing paths that the agent i feels displeased. According to the OCC theory, it is reasonable to assume that $\mathcal{JOY}_i = \mathcal{D}_i \cap \mathcal{I}_i$ and $\mathcal{DISTRESS}_i = \ \sim \mathcal{D}_i \cap \mathcal{I}_i$, that is, the agent i feels joy because the desirable consequences (or goals) is achieved after her/his intentional choices of the possible actions, similarly, she/he feels distress because the undesirable consequences happen after her/his intentional choices of the possible actions.

Then, *a BDIE system S* is defined as a structure $S = \ <\mathcal{K}, \mathcal{M}_1, \cdots, \mathcal{M}_n>$, where $\mathcal{K}$ is a system and for each i, $\mathcal{M}_i$ is the agent i's mental states (believe, desire, and intention) and emotional states (joy and distress) over $\mathcal{K}$.

Assume that we have a set Φ of primitive propositions which describe basic facts about agents and their environment. *An interpreted BDIE system $I = (S, \pi)$* consists of a pair, where S is a BDIE system and π is a valuation function, which gives the set of primitive propositions true at each point in G.

4   Computationally Grounded BDIE Logic

In this section, we introduce the multimodal logic of belief, desire, intention and emotions, called BDIE logic, where changes and computations of agents' beliefs, desires, and intentions are based on agents' observations (i.e. local states), and changes and computations of the agents' emotions (i.e. joy and distress) are based on agents' belief, desire, and intention as OCC theory described.

4.1   BDIE Logic Syntax

The syntactic primitives of our logic of emotions are as follows: a nonempty finite set of agents $A = \{1, 2, \ldots, n\}$, and a nonempty finite set of atomic propositions $\Phi = \{p_1, p_2, \ldots, p_m\}$. The variables i denotes the agents number; and p denotes propositional letters (propositional atoms). The language of BDIE logic is defined by the following BNF (Backus Naur Form) notations:

$$\varphi ::= \perp \mid p \mid \neg \varphi \mid \varphi \wedge \varphi \mid \bigcirc \varphi \mid \varphi \, \mathbf{U} \, \varphi \mid$$
$$B_i \varphi \mid D_i \varphi \mid I_i \varphi \mid$$
$$Joy_i \varphi \mid Distress_i \varphi$$

where p rangers over Φ, i ranges over A. The classical boolean connectives $\vee$ (disjunction), $\rightarrow$ (material implication), $\leftrightarrow$ (material equivalence), and $\top$ (tautology) are defined from $\neg$ (negation), $\wedge$ (conjunction), and $\perp$ (contradiction) in the usual manner. BDIE logic is the modal logic augmented with the future-time connectives $\bigcirc$ (next) and $\mathbf{U}$ (until), modal operators B_i, D_i, I_i, and emotional modal operators Joy_i and

Distress$_i$ for each agent *i*. Linear-time temporal logic (LTL) operators F and G can be defined as follows.

$$F\varphi \stackrel{def}{=\joinrel=} \top \ \mathbf{U}\varphi$$

$$G\ \varphi \stackrel{def}{=\joinrel=} \neg F\neg\varphi$$

Informally, $B_i\varphi$ means "the agent *i* believes that φ is true". Belief is understood as subjective knowledge, alias truth in all worlds that are possible for the agent. $D_i\varphi$ indicates "φ is desirable for the agent *i*". In our view, every goal is about something that is desirable. Thus, if a consequence of an event is a goal, then this consequence is desirable. $I_i\varphi$ denotes that "φ holds under the assumption that the agent *i* acts based on his intention". $Joy_i\varphi$ means "the agent *i* feels joy for φ". *Distress$_i\varphi$* means "the agent *i* feels distress for φ".

4.2 BDIE Logic Semantics

We now proceed to interpret BDIE logic formulas in terms of the interpreted BDIE system. Given an interpreted BDIE system $I = (\mathcal{S}, \pi)$, suppose that $\mathcal{S} = <\mathcal{K}, \mathcal{M}_1, \ldots, \mathcal{M}_n>$ and for each *i*, $i \in \{1, \ldots, n\}$, $\mathcal{M}_i = <\mathcal{B}_i, \mathcal{D}_i, \mathcal{I}_i, \mathcal{JOY}_i, \mathcal{DISTRESS}_i>$. Let *r* be a run in $\mathcal{K}$ and *u* be a natural number, in the following, we inductively define the satisfaction relation $\models_{BDIE}$ between a formula φ and a pair of the interpreted BDIE system *I* and a point (r, u).

- $(I, r, u) \models_{BDIE} B_i\varphi$ iff $(I, r', v) \models_{BDIE} \varphi$ for all (r', v)
  such that $r' \in \mathcal{B}_i$ and $(r, u) \sim_i (r', v)$;
- $(I, r, u) \models_{BDIE} D_i\varphi$ iff $(I, r', v) \models_{BDIE} \varphi$ for all (r', v)
  such that $r' \in \mathcal{D}_i$ and $(r, u) \sim_i (r', v)$;
- $(I, r, u) \models_{BDIE} I_i\varphi$ iff $(I, r', v) \models_{BDIE} \varphi$ for all (r', v)
  such that $r' \in \mathcal{I}_i$ and $(r, u) \sim_i (r', v)$;
- $(I, r, u) \models_{BDIE} Joy_i\varphi$ iff $(I, r', v) \models_{BDIE} \varphi$ for all (r', v)
  such that $r' \in \mathcal{JOY}_i$ and $(r, u) \sim_i (r', v)$;
- $(I, r, u) \models_{BDIE} Distress_i\varphi$ iff $(I, r', v) \models_{BDIE} \varphi$ for all (r', v)
  such that $r' \in \mathcal{DISTRESS}_i$ and $(r, u) \sim_i (r', v)$;

The semantics of atomic formulas *p* or formulas of the form $\neg\varphi$, $\varphi \wedge \varphi'$, $\bigcirc \varphi$ or $\varphi \mathbf{U}\varphi'$ can be dealt with in the usual manner. We use $\models_{BDIE} \varphi$ to denote that φ is valid in every interpreted BDIE system. According to our definition, $D_i\varphi$ is true iff φ is true along those runs that are desirable to the agent *i* and consistent with the agent *i*'s observations. Thus, $D_i\varphi$ intuitively means that the agent *i*'s goal implies that formula φ holds, $Joy_i\varphi$ is true iff φ is true along those runs that the agent *i* is pleased, and *Distress$_i\varphi$* is true iff φ is true along those runs that the agent *i* is displeased.

We use $(r, u) \sim_i^{spr} (r', v)$ to denote $u = v$ and, for every $j \leq u$, $r_i(j) = r_i'(j)$ (here *spr* stands for synchronous systems with perfect recall). For those agents with perfect recall and a global clock, we may use $\sim_i^{spr}$ instead of $\sim_i$ to interpret those formals with modalities B_i, D_i I_i, Joy_i and $Distress_i$ and get an alternative satisfaction relationship $\models_{BDIE}^{spr}$.

Proposition 1. *The following axioms are valid with respect to both* $\models_{BDIE}$ *and* $\models_{BDIE}^{spr}$:

- $\Delta_i (\varphi \Rightarrow \psi) \Rightarrow (\Delta_i \varphi \Rightarrow \Delta_i \psi)$
  where Δ stands for B, D, I, Joy or *Distress*.
- Relationship between belief, and intention
  $B_i \varphi \Rightarrow D_i \varphi$
- Relationship between desire, intention, joy, and distress
  $Joy_i \varphi \Leftrightarrow D_i \varphi \wedge I_i \varphi$

  $Distress_i \varphi \Leftrightarrow \neg D_i \varphi \wedge I_i \varphi$
- Temporal operators
  $\bigcirc(\varphi \Rightarrow \psi) \Rightarrow (\bigcirc\varphi \Rightarrow \bigcirc\psi)$

  $\bigcirc(\neg \varphi) \Rightarrow \neg\bigcirc\varphi$

  $\varphi \mathbf{U} \psi \Leftrightarrow \psi \vee (\varphi \wedge \bigcirc(\varphi \mathbf{U} \psi))$

Proposition 2. *The following axioms are valid with respect to* $\models_{BDIE}^{spr}$: $\Delta\bigcirc\varphi \Rightarrow \bigcirc\Delta\varphi$, *where Δ stands for any modality of B_i, D_i, I_i, Joy_i, and $Distress_i$.*

The formula $D_i\bigcirc\varphi \Rightarrow \bigcirc D_i\varphi$ says that if the agent i's current goal implies φ holds at the next point in time, then at the next point in time her goal will imply φ, that is, the agent i persists on her goal. And the formula $Joy_i\bigcirc\varphi \Rightarrow \bigcirc Joy_i\varphi$ says that if the agent i's current emotional state is joy for φ holding at the next point in time, then at the next point in time her emotional state is still joy for φ.

4.3   BDIE Logic Proof System

We now discuss a proof system, called the BDIE proof system, for those agents with the perfect recall and a global clock. The proof system contains the axioms of propositional calculus plus those in Propositions 1 and 2. It is closed under the propositional inference rules plus: $\frac{\vdash \varphi}{\vdash Joy_i\varphi}$ and $\frac{\vdash \varphi}{\vdash Distress_i\varphi}$ for every agent i.

Theorem 3. *The BDIE proof system for agents with perfect recall and a global clock is sound and complete with respect to interpreted BDIE systems with satisfaction relation* $\models_{BDIE}^{spr}$.

Proof: By Proposition 1 and the soundness and completeness of *BDI* proof system proposed by Su et al. [18].

5 A Case Study

In this section, we specify a simplified auction scenario to illustrate the construction of the BDIE model and BDIE logic specifications.

Let us consider the scenario of two agents ($ag1$ and $ag2$) participating in an auction. There are four global states: s_0, s_1, s_2, and s_3. In the state s_0, each agent bids, and in the states s_1, s_2, and s_3, a winner is announced. Specifically, in s_1, the winner is $ag1$, in s_2, the winner is $ag2$, and in s_3, the winner cannot be decided because $ag1$ and $ag2$ bid the same price. Moreover, each agent has initial belief about what he bids, and each one desires to win.

Now we define a BDIE system $\mathcal{S} = <\mathcal{K}, \mathcal{M}_1, \mathcal{M}_2>$, where

- $\mathcal{K}$ is the set of those runs r such that, for every natural number m, for each j, $j \in \{1,2,3\}$, if $r(m) = s_j$, then $r(m + 1) = s_j$, which means that if the winner is announced, then it will keep so.
- $\mathcal{M}_i = <\mathcal{B}_i, \mathcal{D}_i, \mathcal{I}_i, \mathcal{JOY}_i, \mathcal{DISTRESS}_i>$, for each i, $i \in \{1,2\}$.
- $\mathcal{B}_i$ is the set of those runs $r_i \in \mathcal{K}$ with $r_i(0) = s_0$. This means that each agent believes in how much he can bid. Notice that belief is just the information state of the agent, and there is no guarantee that the agent will win the auction.
- $\mathcal{D}_i$ is a subset of $\mathcal{B}_i$ such that, for each run $r_i \in \mathcal{D}_i$ there is a number m with $r_i(m) = s_i$. This means that each agent desires to win the auction.
- $\mathcal{I}_i$ is a subset of $\mathcal{K}$ such that, for each run $r_i \in \mathcal{I}_i$ and every natural number m, if $r_i(m) = s_0$, then $r_i(m + 1) = s_j$ ($j \in \{1,2,3\}$). This indicates each agent will bid immediately.
- $\mathcal{JOY}_i = \mathcal{D}_i \cap \mathcal{I}_i$, that is, $\mathcal{JOY}_i$ is the set of runs r_i such that for every natural number m, if $r_i(m) = s_0$, then $r_i(m + 1) = s_i$. This means that the agent i feels joy if he wins the auction after bidding.
- $\mathcal{DISTRESS}_i = {\sim}\mathcal{D}_i \cap \mathcal{I}_i$, that is, $\mathcal{DISTRESS}_i$ is the set of runs r_i such that for every natural number m, if $r_i(m) = s_0$, then $r_i(m + 1) = s_j$ ($j \in \{1,2,3\}$ and $j \neq i$).. This means that the agent i feels distress if he does not win the auction after bidding.

We may take $\{ag1.bid, ag2.bid, winner1, winner2, draw\}$ as the set Φ of primitive propositions which describe basic facts about the agents and their environment. Clearly, we may naturally define the valuation function π as follows.

$$\pi(s_0)(ag1.bid, ag2.bid, winner1, winner2, draw) = (1,1,0,0,0)$$
$$\pi(s_1)(ag1.bid, ag2.bid, winner1, winner2, draw) = (1,1,1,0,0)$$
$$\pi(s_2)(ag1.bid, ag2.bid, winner1, winner2, draw) = (1,1,0,1,0)$$
$$\pi(s_3)(ag1.bid, ag2.bid, winner1, winner2, draw) = (1,1,0,0,1)$$

Thus, we get the interpreted BDIE system $I = (\mathcal{S}, \pi)$. Furthermore, the agent has the ability to express emotions based on our formal description of emotional trigger. Now we can formalize these emotional attributes in the auction scenario.

1. **F** (**Joy**$_{ag1}$ (*winner1*)) indicates that eventually the agent *ag1* will be in joy for winning the auction.
2. **F** (**Distress**$_{ag2}$ (¬*winner2*)) indicates that eventually the agent *ag2* will be in distress for losing the auction.

6 Conclusions

In this paper we have addressed ourselves to a formalization of the appraisal process of event-based emotions based on the OCC theory in BDIE logic, which extends BDI logic by incorporating joy and distress emotion modalities, obtains the emotional BDI logic called BDIE logic, and proposes a new computational grounded model of emotion triggers for BDI agents, called the interpreted observation-based BDIE system model. With these constructs, an agent program can be built so that cognitive agents have the ability to automatically compute two kinds of emotions, that is, joy and distress, during runs.

This cross-disciplinary work brings an interesting contribution. The domain-independent model we designed is based on BDI logic which is already used in a great number of agent architectures. Our model is thus ready to be implemented in any BDI agent, whatever its application may be, thus facilitating the development of intelligent virtual agents with affective abilities.

We are aware that the proposed formalization is just a first step in an extension of the logical BDI framework for incorporating emotions. As for future work, we plan to further investigate the formalization of the other emotions of the OCC model and explore how to verify the emotion attributes by the model checking tools.

Acknowledgments. This work is supported by the National Basic Research Program of China (973 Program, No. 2014CB744600), the National Natural Science Foundation of China (No. 61632014, No. 61210010), the Program of Beijing Municipal Science & Technology Commission (No. Z171100000117005), the Program of International S&T Cooperation of MOST (No. 2013DFA11140), and the Northwest Normal University Foundation (NWNU-LKQN-14-5).

References

1. Reisenzein, R., Hudlicka, E., Dastani, M., Gratch, J., Hindriks, K., Lorini, E., Meyer, J.-J.C.: Computational modeling of emotion: toward improving the inter-and intradisciplinary exchange. IEEE Trans. Affect. Comput. **4**(3), 246–266 (2013)
2. Rao, A.S., Georgeff, M.P.: The semantics of intention maintenance for rational agents. In: Proceedings of the 14th International Joint Conference on Artificial Intelligence (IJCAI 1995), pp. 704–710. IJCAI, Melbourne, Australia (1995)
3. Meyer, J.-J.C., van der Hoek, W., van Linder, B.: A logical approach to the dynamics of commitments. Artif. Intell. **113**(1–2), 1–40 (1999)

4. Hu, X., Bai, K., Cheng, J., Deng, J., Guo, Y., Hu, B., Wang, F.: MeDJ: multidimensional emotion-aware music delivery for adolescent. In: Proceedings of the 26th International Conference on World Wide Web Companion (WWW 2017), pp. 793–794. ACM, Perth (2017)
5. Fagin, R., Halpern, J.Y., Moses, Y., Vardi, M.: Reasoning About Knowledge. MIT Press, Cambridge (2004)
6. Wooldridge, M.: Computationally grounded theories of agency. In: The 4th International Conference on MultiAgent Systems Proceedings (ICMAS 2000), pp. 13–20. IEEE, Washington, DC (2000)
7. Jin, L., Li, S., Hu, B.: RNN models for dynamic matrix inversion: a control-theoretical perspective. IEEE Trans. Industr. Inf. **14**(1), 189–199 (2018)
8. Ortony, A., Clore, G.L., Collins, A.: The Cognitive Structure of Emotions. Cambridge University Press, Cambridge (1990)
9. Bagozzi, R.P., Dholakia, U.M., Basuroy, S.: How effortful decisions get enacted: the motivating role of decision processes, desires, and anticipated emotions. J. Behav. Decis. Making **16**(4), 273–295 (2003)
10. Chen, J., Hu, B., Moore, P., Zhang, X., Ma, X.: Electroencephalogram-based emotion assessment system using ontology and data mining techniques. Appl. Soft Comput. **30**, 663–674 (2015)
11. Strongman, K.T.: The Psychology of Emotion. From Everyday Life to Theory. Wiley (2003)
12. Adam, C., Herzig, A., Longin, D.: A logical formalization of the OCC theory of emotions. Synthese **168**(2), 201–248 (2009)
13. Steunebrink, B.R., Dastani, M., Meyer, J.-J.C.: A formal model of emotion triggers: an approach for BDI agents. Synthese **185**(1), 83–129 (2012)
14. Meyer, J.-J.C.: Reasoning about emotional agents. Int. J. Intell. Syst. **21**(6), 601–619 (2006)
15. Pereira, D., Oliveira, E., Moreira, N.: Formal modelling of emotions in BDI agents. In: Sadri, F., Satoh, K. (eds.) CLIMA 2007. LNCS (LNAI), vol. 5056, pp. 62–81. Springer, Heidelberg (2008). https://doi.org/10.1007/978-3-540-88833-8_4
16. Gluz, J., Jaques, P.A.: A probabilistic formalization of the appraisal for the OCC event-based emotions. J. Artif. Intell. Res. **58**, 627–664 (2017)
17. Broekens, J., Bosse, T., Marsella, S.C.: Challenges in computational modeling of affective processes. IEEE Trans. Affect. Comput. **4**(3), 242–245 (2013)
18. Su, K., Sattar, A., Wang, K., Luo, X., Governatori, G., Padmanabhan, V.: Observation-based model for BDI-agents. In: AAAI Conference on Artificial Intelligence (AAAI 2005), pp. 190–195. AAAI, Pittsburgh (2005)

Frequent Sequence Pattern Mining
with Differential Privacy

Fengli Zhou[⊠] and Xiaoli Lin

Faculty of Information Technology, Wuhan College of Foreign Language
and Foreign Affairs, Wuhan 430083, China
thinkview@163.com, aneya@163.com

Abstract. Focusing on the issue that releasing frequent sequence patterns and
the corresponding true supports may reveal the individuals' privacy when the
data set contains sensitive information, a Differential Private Frequent Sequence
Mining (DPFSM) algorithm was proposed. Downward closure property was
used to generate a candidate set of sequence patterns, smart truncating based
technique was used to sample frequent patterns in the candidate set, and geometric mechanism was utilized to perturb the true supports of each sampled
pattern. In addition, to improve the usability of the results, a threshold modification method was proposed to reduce truncation error and propagation error in
mining process. The theoretical analysis show that the proposed method is ε-
differentially private. The experimental results demonstrate that the proposed
method has lower False Negative Rate(FNR) and Relative Support Error
(RSE) than that of the comparison algorithm named PFS2, thus effectively
improving the accuracy of mining results.

Keywords: Frequent sequence mining · Differential Privacy (DP)
Privacy protection · Geometric mechanism · Data mining

1 Introduction

Frequent Sequence Mining (FSM) is an important issue in data mining research fields,
its purpose is to find out the patterns frequently appearing in datasets with time or other
sequences, which are the basis of many other data mining tasks like association rule,
correlation analysis, classification and clustering. But frequent sequence pattern itself
and its corresponding support counts are likely to reveal private information.

Traditional privacy protection methods were based on k-anonymity [1] and partitioning mostly, these methods needed to make assumptions about the attack models
and the attacker's background knowledge in advance. Differential Privacy (DP) was a
strong privacy protection model proposed by Dwork [2] in 2006 based on data distortion. In recent years, it has become a hot research area in privacy protection. Literature [3, 4] proposed two kinds of frequent sequence pattern mining methods in non-
interactive framework, which focused on the release of sequence databases, so the
mining results were less useful. Literature [5, 6] proposed frequent sequence pattern
mining method in interactive framework, which focused on continuous sequence
patterns mining and generalized frequent sequence mining respectively. Although the

© Springer International Publishing AG, part of Springer Nature 2018
D.-S. Huang et al. (Eds.): ICIC 2018, LNCS 10954, pp. 454–466, 2018.
https://doi.org/10.1007/978-3-319-95930-6_42

differentially private frequent mining algorithm (PFS^2) based on the differential privacy proposed in literature [6] could solve privacy leakage problem existing in the process of frequent sequence pattern mining. But this algorithm doesn't consider the discrepancy of different candidate sequences' importance when sequence database was reconstructing.

To solve the above problems, this paper proposes a frequent sequence mining algorithm based on differential private (DP-FSM). The algorithm mainly improves the accuracy of mining results from two aspects. First, a heuristic method is proposed to design the scoring function which can distinguish the importance degree of different candidate sequences based on their discrepancy in candidate set. Secondly, a threshold modification strategy is proposed for the error problems of mining frequent sequence patterns.

2 Related Work

2.1 Frequent Item-Sets Mining

In recent years, frequent item-sets mining with differential privacy have made some significant progress. Bhaskar et al. have proposed two top-k frequent item-sets mining strategies that could satisfy differential privacy [7]. These two methods used Laplace and exponent mechanism respectively. Li et al. had proposed PrivBasis algorithm which can be used for high-dimensional data [8]. The algorithm combined θ-base with mapping technology to achieve differential privacy protection under the premise of ensuring computational performance. Zhang Xiaojian had proposed a method that used post processing technology to refine the noise count and the output results could satisfy the consistency requirements [9]. Zeng et al. had proposed a greedy method Smart-Trunc based on transaction truncation technique to improve the results' usability [10]. The differences between item-sets and sequence have made the above algorithm unsuitable for mining frequent sequence patterns.

Chen et al. have proposed a context-free classification tree [11] and combined top-down tree partitioning approach to publish datasets. Lee et al. have proposed a method to publish frequent item-sets privately by using prefix tree [12]. These two algorithms were frequent item-sets mining methods in non-interactive framework.

2.2 Frequent Sequence Mining

Chen et al. have proposed a method based on prefix tree to publish trajectory datasets privately [3] which could be used for frequent sequence mining. In addition, they have also proposed a method to publish sequence datasets based on *n-gram* model of variable-length in literature [4], this method mainly used the model to extract necessary information of sequence database and used exploring tree to reduce the amount of added noise. Bonomi et al. have proposed a two-stage mining strategy for frequent continuous sequence and could satisfy differential privacy protection [5]. It used a prefix tree to find all candidate sequences and then used local transformation technology of database to refine the candidate sequences' support. Xu et al. have proposed an algorithm PFS^2 which focused on generalized frequent sequence mining and ignored whether the items in sequence were continuous [6]. This algorithm used candidate pruning technique based on sampling to reduce the number of candidate sequences.

This study is more similar to algorithm PFS^2, but PFS^2 had treated all candidate sequences as equivalent in sequences' reconstruction. This paper has proposed an algorithm to distinguish the importance of different candidate sequences and put a threshold modification strategy to improve the accuracy of mining results.

3   Problem Definition

3.1   Differential Privacy

If two datasets D and D' differ by one record at most, then D and D' are called Neighboring Datasets. This paper has used symbol $D \sim D'$ to express that these two datasets are neighbors.

Definition 1. ε-differential privacy [2]. Given two neighboring datasets D and D' and a random algorithm A that *Range(A)* represents data range of A. If A satisfies the following inequality (1) for all output results $O(O \in Range\ (A))$ on datasets D and D', which called algorithm A satisfies ε-differential privacy.

$$\Pr(A(D) \in O) \leq \exp(\varepsilon)\Pr(A(D') \in O) \tag{1}$$

While parameter ε is called privacy budget, which can control the level of privacy protection. The smaller ε represents higher level of privacy protection.

Definition 2. Global sensitivity. With function $f: D \to \mathbb{R}^d$, the input is dataset D and the output is d-dimensional real number vector. For any neighboring datasets D and D', the global sensitivity of function f is as Eq. (2).

$$GS_f = \max_{D \sim D'} \|f(D) - f(D')\|_1 \tag{2}$$

While $\|\cdot\|_1$ indicates the distance of L_1.

The Laplace mechanism was a method of implementing differential privacy protection that proposed by Dwork et al. in 2006 [13]. A formal description of the mechanism is as follows.

Given a function $f: D \to \mathbb{R}^d$, if output result of random algorithm A satisfies Eq. (3), A satisfies ε-differential privacy.

$$A(D) = f(D) + Lap(\Delta f / \varepsilon)^d \tag{3}$$

For the case where the function's return value is an integer, Ghosh et al. have proposed a special case of Laplacian mechanism - geometric mechanism [14]. The noise added by this mechanism obeys bilateral geometric distribution that probability density function is in Eq. (4). This paper also uses geometric mechanism to add noise to the support for frequent sequences.

$$\Pr(\delta = x) \sim \exp(-\varepsilon|x|) \tag{4}$$

Theorem 1. Geometric mechanism. Suppose $f: D \to R^d$ is function which output is integer and sensitivity is Δf. If the output of a random algorithm A satisfies Eq. (5), A satisfies ε-differential privacy.

$$A(D) = f(D) + G(\Delta f/\varepsilon)^d \tag{5}$$

Generally, a complex privacy protection problem is solved usually by using differential protection algorithm many times. In order to ensure that the entire process satisfies ε-differential privacy, the entire budget needs to be reasonably allocated to each sub-process. At this time, a very important property of differential privacy protection algorithm - sequence composition will be used.

Property 1. Sequence composition. Suppose that there have algorithms $A_1, A_2, ..., A_n$ and each algorithm A_i provides ε_i-differential privacy, the sequence $\{A_1, A_2, ..., A_n\}$ can provide $\sum_{i=1}^{n} \varepsilon_i$- differential privacy to the same dataset D.

3.2   Frequent Sequential Pattern Mining

The alphabet is setting as $I = \{i_1, i_2, \cdots, i_{|I|}\}$ and a sequence S is an ordered arrangement of some items in alphabet. $S = s_1 s_2 \cdots s_{|S|}$ can be used to denote a sequence with length $|S|$ that is also called $|S|$-sequence. A sequence database D is a collection of input sequences and each input sequence represents a person's record. Before giving the definition of support, the inclusion relation of sequence is defined as follows.

Definition 3. Inclusion relation. Given two sequences $S = s_1 s_2 \cdots s_{|S|}$ and $T = t_1 t_2 \cdots t_{|T|}$. If there is an integer $w_1 < w_2 < \cdots w_{|S|}$ that can hold $s_1 = T_{w_1}, s_2 = T_{w_2}, \cdots s_{|S|} = T_{w_{|S|}}$, S is said to be contained in T and expressed as $S \subseteq T$. Where S is a subsequence of T and T is a super sequence of S.

Definition 4. Support. Given a sequence pattern S and a sequence database D, the support of T in D is defined as trace number of S that is included in D. That is $Sup(T) = |\{o|o \in D \bigwedge S \subseteq o\}|$.

Given a semantic trajectory database D, privacy parameter ε and a threshold σ_k, this paper has designed DP-FSM algorithm to mine all frequent sequence patterns that support are greater than σ_k and calculated the support for each frequent sequence. The algorithm needs to satisfy ε-differential privacy and has higher data availability.

4   DP-FSM Algorithm

In order to satisfy differential privacy, FSM has used down-closed attributes [15] to generate all candidate sequences and add perturbation noise to each candidate sequence's support, then frequent sequence will be selected according to noise support.

The amount of added noise which depends on the sensitivity of mining frequent sequence function needs to be reduced for accurately estimating which candidate sequences are frequent. According to the definition of sensitivity, the sensitivity for calculating k-sequence support is $\Delta k = \min\{|C_k|, |C_l^k|\}$, where $|C_k|$ represents the size of the candidate sequence. It can be seen that the function sensitivity is positively related to the sequence length. For this reason, the DP-FSM algorithm reduces function sensitivity by constraining sequence length to improve the availability of mining sequence patterns.

4.1   Algorithm Description

Algorithm1. DP-FSM Algorithm.
   Input: Sequence Database D, Percentage η, Threshold σ_k, Privacy Budget ε and Alphabet I.
   Output: Frequent sequences and their noise counts.
   1) $\langle l_1, l_2 \cdots, l_n \rangle = $ Estimate_Distribution(D, ε_1)

   2) Calculating the smallest integer l_{max} to make $\sum_{i=1}^{l_{max}} l_i \geq \eta$.

   3) $l_f = $ Estimate_Frequent_Maxlength$(D, \varepsilon_2, \sigma_k)$
   4) $FS = $ Mining_FrequentSequence$(D, l_{max}, \sigma_k, \varepsilon_3)$
   5) Outputting frequent sequence patterns and their noise count Perturb(FS, ε_3).

DP-FSM consists of three stages: preprocessing, mining and disturbance.
   Preprocessing stage (line (1) to (3)): Estimating the maximum length constraint l_{max} and the maximum length of frequent sequence l_f.
   l_{max} is calculated as follows: Calculating cardinality $|D|$ of series data set D firstly, then incrementally calculating the number of input sequences a_j for each j-length from the beginning of sequence length 1 until the value of $p = \left(\sum_{j=1}^{l_i} a_j \right) / |D|$ is η at least.
Geometric noise $G(\varepsilon_{11})$ needs to be added when calculating $|D|$ and $G(\varepsilon_{12})$ needs to be added to calculate a_j due to privacy requirements.
   l_f is calculated as follows: Calculating maximum support for i-sequence from 1 to l_{max}, then adding noise $G(\varepsilon_2/\log(l_{max}))$ to it, and selecting the integer i which support is greater than threshold σ_k.
   Mining Stage (Line (4)): The algorithm finds frequent sequences privately with ascending sequence lengths.
   Disturbance Stage (Line (5)): Adding the noise of geometric mechanism to the support of selected frequently frequent sequences.

4.2   Mining Frequent Sequence Pattern

Algorithm2. DP-Mining Frequent Sequence.
   Input: Maximum Length of Frequent Sequence l_f, Length Constraint of Truncated Sequence l_{max}, Privacy Budget ε and Threshold σ_k, and Alphabet I.
   Output: Frequent sequence collection FS.
   1) For $(k = 1; k \le l_f; k + +)$
   2)   if $k == 1$ then
$$D' \leftarrow D; C_k \leftarrow I$$
       Else
   3) $C_k \leftarrow \text{Generate\_CandidatSet}(S_{k-1});$
   4) $D' = \text{Truncate\_Database}(D, C_k, l_{max})$
   5) $S_k \leftarrow \text{Choose\_Candidate\_Seed}(D', C_k, \varepsilon', \sigma_k);$
   6) $FS_k = \text{Discover\_Frequent\_Sequence}(D', C_k, \varepsilon', \sigma_k);$
   7) $FS = FS + FS_k;$
   8) Return FS;

Algorithm 2 describes the process of mining frequent sequence patterns in privacy. The core idea is that use noise support degree of candidate sequences in truncated sequence database D' to determine whether the candidate sequence is frequent or not. At the same time, the algorithm modifies the threshold from two aspects in order to improve the accuracy of mining results: (1) The algorithm corrects the decision threshold σ_k to $\sigma_k - avg(\theta') + \theta'$, where $avg(\theta')$ is the sequence's average support in original database which is estimated by its noise support θ'. (2) The threshold that is used to judge whether a sequence generates has be fixed to $\sigma_k - max(\theta') + \theta'$, where $max(\theta')$ is the sequence's maximum support in the original database according to its noise support θ'.

(1) Truncated sequence database. This step can convert sequence database to satisfy the length constraints. So a method of shrinking sequence is proposed, which can contract the length of shrinking sequence by deleting irrelative item, compressing continuous patterns and reconstructing intelligent sequence.

    (a) Deleting irrelative item.

        Given a sequence, when an item is not included in any candidate sequence, it has no contribution to the support of any sequence pattern in candidate set. Such item is called irrelative item, which will not cause any loss frequency when it has been deleted from the sequence.

        Based on the down-closed property, if an item is an irrelative item for a candidate k-sequence, it is also an irrelative item for candidate sequence which length is greater than k.

    (b) Compressing continuous patterns.

        A sequence may contain a sequential pattern that occurs continuously. The estimates of k-sequence can compress j continuous sequence patterns into consecutive k sequence patterns without causing any loss of frequency information, that is because k items in the candidate k-sequence come from k patterns at most.

(c) Intelligent sequence reconstruction.

Deleting irrelative item and compressing continuous pattern effectively shrink the sequence without causing frequency (support) loss. But some sequences still violate the sequence length constraint after they had been processed by these two method. Intuitively, only subsequences of the frequent sequence need to be preserved when a sequence is shrunk, because the infrequent subsequences do not contribute to the support of frequent sequences have no contribution to the support of frequent sequences. Thus a heuristic method is proposed to predict whether a certain candidate sequence is frequent for finding frequent sequence patterns in privacy. If all subsequences of a candidate sequence are sufficiently frequent, the candidate sequence is likely to be a frequent sequence pattern. Based on this observation, a frequency score is assigned to each i-th candidate sequence ($i \geq 2$).

Definition 5. Frequency Score. Given a collection $GS = \{S_1, S_2, \cdots, S_d\}$ of frequency $(i-1)$-sequence, the frequency score of an i-sequence X is shown in Eq. (6).

$$fs(X) = \sum_{S_i \subset GS \wedge S_i \subset X} S_i \cdot sup' \tag{6}$$

Where $S_i \cdot sup'$ represents the noise support of sequence S_i.

It is necessary to reserve a sequence with high frequency score when it is shrunk, so the problem can be expressed as follows:

Finding the optimal l_{max}-sequence $S_{l_{max}}$ in the given set of candidate sequences $X = \{X_1, X_2, \cdots, X_d\}$, which can make this sequence has the highest coverage score. Where coverage score is $cs(S_{l_{max}}) = \sum_{S_i \subset S_{l_{max}} \wedge S_i \subset X} fs(S_i)$.

But the problem of finding the optimal l_{max}-sequence is NP-hard, a greedy algorithm is proposed to shrink the sequence. The idea is as follows: iteratively adding candidate sequences that are included in the input sequences with high frequency scores for each sequence until the reconstructed sequence reaches the maximum cardinality. Because the frequency score of a candidate sequence is not static, when some subsequences have been added to the reconstruction sequence, in order to include the candidate sequence in the reconstructed sequence, the number of items in subsequences that needs to be added into the candidate sequence must be less than the others. Therefore, the frequency score of affected candidate sequence should be updated.

The updating strategy is as follows: Suppose a candidate sequence X_i is added to the reconstructed trajectory sequence and the frequency score of set X needs to be updated. For each remainder of candidate sequence $X_j \subset \{X_1, \cdots, X_{i-1}, X_{i+1}, \cdots X_d\}$, the score for each item is $a_j = fs(X_j)/i$. Assume that the number of items that are included in the largest subsequence of reconstructed sequence is β_j, the frequency score of candidate sequence is updated to $fs(X_j) = fs(X_j) + \alpha_j * \beta_j$.

(2) Threshold modification. A new support estimation method which is inspired by the "double standard" mechanism [10] has been proposed to make up the loss of frequency information that generated by truncation and propagation errors. This method estimates the true support information of the sequence in original data set according to the noise support degree of the sequence in converted transaction database, it uses the average support to judge whether the sequence is frequent and the maximum support to determine whether the sequence will be used to generate candidate sequence. Specifically, the method mainly consists of the following two steps.

Step 1: Given the noise support θ' of sequence X, estimating the true support θ^{real} of sequence X in transformed dataset. According to the "Double Standard" theorem, Eq. (7) can be get.

$$\Pr\left(\theta^{\text{real}}|\theta'\right) \approx e^{-\varepsilon\left(\theta^{\text{real}}-\theta'\right)} \tag{7}$$

Step 2: Estimating the true support degree θ of sequence in the original dataset according to θ^{real}. The information loss that is generated by random interception method can be used to analyze the information loss generated by the truncated sequence database. Intuitively, given a sequence record t with length l which will be truncated to sequence record t', if t does not contain sequence S, then transaction t' must not contain sequence S. Thus truncated sequence record that contains sequence S can be used to calculate its support. For a sequence record with length l, a transaction t with length $|t|$ which contains i-sequence S is truncated to sequence record t' and the probability can be calculated by Eq. (8).

$$Pr = C_{l-i}^{|t|-i}/C_l^{|t|} \tag{8}$$

Considering the support degree in the transformed dataset for sequence S as a random variable, its expectation can be obtained by Eq. (9).

$$E\left(\theta^{\text{real}}\right) = \theta * \left(\sum_{j=1}^{l} \partial_j + \sum_{j=l+1}^{|t|} \partial_j \frac{c_{l-i}^{j-i}}{c_l^j}\right) \tag{9}$$

It's similar to the method of calculating the maximum and average support in the "Double Standard", if θ^{real} has been given, the average and maximum support for the sequence in original database can be get by Eqs. (10) and (11).

$$avg\left(\theta^{\text{real}}\right) = \frac{\theta^{\text{real}}}{\left(\sum_{j=1}^{l} \partial_j + \sum_{j=l+1}^{|t|} \partial_j \frac{c_{l-i}^{j-i}}{c_l^j}\right)} \tag{10}$$

$$\max(\theta'') = \begin{cases} \dfrac{\theta'' - \ln \rho + \sqrt{\ln \rho * (\ln \rho - 2\theta'')}}{\sum\limits_{j=1}^{l} \partial_j + \sum\limits_{j=l+1}^{|r|} \partial_j \dfrac{c_i^{j-i}}{c_i^j}} & \ln \rho \leq 2\theta'' \\[2em] avg(\theta'') & \ln \rho > 2\theta'' \end{cases} \qquad (11)$$

Combining the conclusions of first step and second step and giving the noise support θ' for the sequence in transformed dataset, the average and maximum support for the sequence in original dataset can be estimated by Eqs. (12) and (13).

$$avg(\theta') = \sum_{i=0}^{+\infty} \mathrm{pr}(i|\theta')avg(\theta') \qquad (12)$$

$$\max(\theta') = \sum_{i=0}^{+\infty} \mathrm{pr}(i|\theta')\max(\theta') \qquad (13)$$

In summary, threshold σ_k can be modified to $\sigma_k - avg\left(\theta'\right) + \theta'$ when the sequence is judged whether is frequent or not; and when determining whether a sequence is used to generate a candidate sequence, it is modified to $\sigma_k - \max\left(\theta'\right) + \theta'$.

5 Experimental Results and Analysis

The performance of DP-FSM algorithm is evaluated through experiments. All code is implemented in C++ programming language. The experiment compares DP-FSM with the most advanced algorithm PFS[2], which mines frequent sequences in privacy. Since the algorithm involves randomization, each algorithm runs 15 times and the average value will be extracted as the result to publics.

5.1 Experiment Settings

Two real sequence datasets named BIBLE [16] and MSNBC are used in the experiment. Table 1 shows the parameters of each dataset.

Table 1. Experiment dataset parameters

Dataset	Sequences number	Maximum length	Average length
MSNBC	989 818	14 795	5.70
BIBLE	36 369	100	21.64

The experiment has used the following two availability metrics to measure the performance of DP-FSM. Assume that $FSP(D)$ represents a real set of frequent sequence patterns in dataset D, and $FSP(\tilde{D})$ represents a set of frequent sequence patterns that have been mined by DP-FSM algorithm.

(1) False Negative Rate (FNR). Measuring the proportion that the patterns in $FSP(\tilde{D})$ are not in $FSP(D)$, which means the ratio of patterns that does not appear in output result to total patterns. The formula is in Eq. (14).

$$FNR = 1 - \frac{|FSP(D) \cap FSP(\tilde{D})|}{|FSP(D)|} \tag{14}$$

(2) Relative Support Error (RSE). It can measure the average error ratio of one pattern's support in $FSP(\tilde{D})$ which is relative to support threshold σ_k. The formula is as follows.

$$RSE = \frac{\sum\limits_{X \in FSP(D)} Sup(X) - \sum\limits_{\tilde{X} \in FSP(\tilde{D})} Sup(\tilde{X})}{\sigma_k |FSP(D)|} \tag{15}$$

Where $\sum\limits_{X \in FSP(D)} Sup(X)$ represents the support summation for all patterns in $FSP(D)$ and $\sum\limits_{\tilde{X} \in FSP(\tilde{D})} Sup(\tilde{X})$ represents the support summation for all patterns in $FSP(\tilde{D})$. The smaller the RSE, the higher the availability of the algorithm.

5.2 Experimental Results

Performance Comparison of DP-FSM and PFS^2 Algorithms

Figure 1(a) analyzes the effect with threshold parameter σ_k of dataset MSNBC on *FNR*. It can be seen from the experimental results that *FNR* of PFS^2 algorithm shows an increasing trend with the increase of σ_k, which means that the mining results' accuracy of PFS^2 algorithm is reduced and the performance variates. But *FNR* of DP-FSM

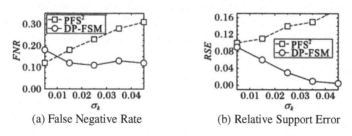

(a) False Negative Rate   (b) Relative Support Error

Fig. 1. Utility of algorithms on dataset MSNBC

algorithm generally shows a relatively stable trend and the performance is better than PFS^2 algorithm in terms of *FNR*.

Figure 1(b) analyzes the effect with threshold parameter σ_k of dataset MSNBC on the *RSE* of two algorithms. From the experimental results, it can be seen that *RSE* of DP-FSM algorithm does not change much with the increase of σ_k, while *RSE* of PFS^2 algorithm shows an increasing trend. This phenomenon is due to the requirement of threshold which sets by frequent sequence is higher with σ_k increases. PFS^2 does not consider the difference of candidate sequences when the sequence reconstructs, that has led to a large number of infrequent sequences in result set; and because of the low support of infrequent sequences, it can be obtained that the *RSE* of the PFS^2 will increase with the increase of σ_k according to Eq. (15).

Figure 2 shows the performance results of DP-FSM and PFS^2 on dataset BIBLE. It can be seen that this two algorithms' performance trend in BIBLE is similar to that on dataset MSNBC.

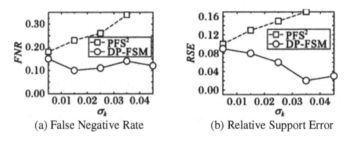

(a) False Negative Rate   (b) Relative Support Error

Fig. 2. Utility of algorithms on dataset BIBLE

Effect of Threshold Correction Strategy on Performance of DP-FSM Algorithm

In order to research the influence of threshold correction strategy on the DP-FSM algorithm's accuracy, the algorithm for mining frequent sequence patterns without threshold correction strategy is called DP-RS. Figure 3 shows the effect of threshold correction strategy on the algorithm's performance with σ_k changes. It can be seen that the performance of DP-FSM algorithm with threshold correction is better than DP-RS, and the performance difference between these two algorithms is greater with σ_k increases. Since σ_k increases, the requirements for frequent sequences are more strict, the existence of errors in DP-RS will cause many infrequent patterns existing in result set, which leads to an increase in *FNR*; and most infrequent patterns have low support, so the *RSE* of DP-RS is also large.

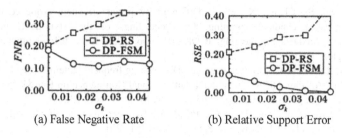

(a) False Negative Rate   (b) Relative Support Error

Fig. 3. Influence of threshold modification on utility

6 Conclusion

In order to solve the problem of privacy leakage in the process of frequent sequential pattern mining, this paper proposes an algorithm named DP-FSM that can satisfy the ε-difference privacy. It mines all frequent sequence patterns from the candidate set first, and then uses geometric mechanism to add geometric noise disturbance to the selected patterns' support. In addition, in order to improve the availability of mining results, this paper proposes a threshold correction strategy to reduce the truncation and propagation errors in mining process. Finally, the experiments verify that this algorithm can achieve higher data availability under the premise of satisfying differential privacy. The next work will study how to allocate privacy budget reasonably and make the accuracy of mining results reaches the highest value.

Acknowledgment. This work was supported in part by Research Project of Hubei Provincial Department of Education (No. B2017590).

References

1. Sweeney, L.: k-Anonymity: a model for protecting privacy. Int. J. Uncertainty Fuzziness Knowl.-Based Syst. **10**(5), 557–570 (2002)
2. Dwork, C.: Differential privacy: a survey of results. In: Agrawal, M., Du, D., Duan, Z., Li, A. (eds.) TAMC 2008. LNCS, vol. 4978, pp. 1–19. Springer, Heidelberg (2008). https://doi.org/10.1007/978-3-540-79228-4_1
3. Chen, R., Fung, B.C.M., Desai, B.C., et al.: Differentially private transit data publication: a case study on the Montreal transportation system. In: Proceedings of the 18th ACM SIGKDD International Conference on Knowledge Discovery and Data Mining, KDD 2012, pp. 213–221. ACM, New York (2012)
4. Chen, R., Acs, G., Castelluccia, C.: Differentially private sequential data publication via variable-length n-grams. In: Proceedings of the 7th ACM CCS Conference on Computer and Communications Security, CCS 2012, pp. 638–649. ACM, New York (2012)
5. Bonomi, L., Xiong, L.: A two-phase algorithm for mining sequential patterns with differential privacy. In: Proceedings of the 22nd ACM International Conference on Conference on Information & Knowledge Management, CIKM 2013, pp. 269–278. ACM, New York (2013)

6. Xu, S., Su, S., Cheng, X., et al.: Differentially private frequent sequence mining via sampling-based candidate pruning. In: Proceedings of the 31st IEEE International Conference on Data Engineering, ICDE 2015, pp. 1035–1046. IEEE Computer Society, Washington, DC (2015)
7. Bhaskar, R., Laxman, S., Smith, A., et al.: Discovering frequent patterns in sensitive data. In: Proceedings of the 16th ACM SIGKDD International Conference on Knowledge Discovery and Data Mining, KDD 2010, pp. 503–512. ACM, New York (2010)
8. Li, N., Qardaji, W., Su, D., et al.: PrivBasis: frequent itemset mining with differential privacy. Proc. VLDB Endowment **5**(11), 1340–1351 (2012)
9. Zhang, X.J., Wang, M., Meng, X.F.: An accurate method for mining top-k frequent pattern under differential privacy. J. Comput. Res. Dev. **51**(1), 104–114 (2014)
10. Zeng, C., Naughton, J.F., Cai, J.-Y.: On differentially private frequent itemset mining. Proc. VLDB Endowment **6**(1), 25–36 (2012)
11. Chen, R., Mohammed, N., Fung, B.C.M., et al.: Publishing set valued data via differential privacy. Proc. VLDB Endowment **4**(11), 1087–1098 (2011)
12. Lee, J., Cliftonc, W.: Top-k frequent itemsets via differentially private FP-trees. In: Proceedings of the 20th ACM SIGKDD International Conference on Knowledge Discovery and Data Mining, KDD 2014, pp. 930–940. ACM, New York (2014)
13. Dwork, C., McSherry, F., Nissim, K., Smith, A.: Calibrating noise to sensitivity in private data analysis. In: Halevi, S., Rabin, T. (eds.) TCC 2006. LNCS, vol. 3876, pp. 265–284. Springer, Heidelberg (2006). https://doi.org/10.1007/11681878_14
14. Ghosh, A., Roughgarden, T., Sundararajan, M.: Universally utility-maximizing privacy mechanisms. In: Proceedings of the 41th ACM STOC Annual Symposium on Theory of Computing, STOC 2009, pp. 351–360. ACM, New York (2009)
15. Agrawal, R., Srikant, R.: Fast algorithms for mining association rules. In: Proceedings of the 20th Conference of Very Large Data Bases, VLDB 1994, pp. 487–499. Morgan Kaufmann Publishers, San Francisco, CA (1994)
16. Zhang, C., Han, J., Shou, L., et al.: Splitter: mining fine-grained sequential patterns in semantic trajectories. Proc. VLDB Endowment **7**(9), 769–780 (2014)

A New Asymmetric User Similarity Model Based on Rational Inference for Collaborative Filtering to Alleviate Cold Start Problem

Dan Wang$^{(\boxtimes)}$ and Chengliang Wang$^{(\boxtimes)}$

School of Big Data and Software Engineering, Chongqing University,
Chongqing, China
danwang0815@qq.com, wcl@cqu.edu.cn

Abstract. For user-based collaborative filtering, the similarity methods used to calculate the target user's neighbors are very important. More similar neighbors lead to better recommendations and more accurate results. There are a lot of similarity methods till now, but there is still a room for improvement, especially when the data is sparse. It is well known that sparse data can easily lead to cold start problems. The performances of most traditional methods are disappointing under cold start conditions. In order to get a better performance under the cold start conditions, we proposed a new similarity method based on the idea that users with similar interests in the past will show similar tastes in the future. While considering similarities between items and rational inferences, the proposed method focuses on how to utilize more ratings information. At the same time, in order to reduce the time spent on calculations and reduce the impact of excessive ratings information, we have limited the range of items neighbors through experiments. Besides, the proportion of co-rate items to personally rated items is different from each user, base on which the asymmetric factor is considered. Experiments on the dataset MovieLens prove that the proposed method outperforms state-of-the-art methods.

Keywords: Asymmetric · Rational inference · Cold start
Collaborative filtering

1 Introduction

Recommender systems (RS) typically produce a list of recommendations through different ways. The main task of RSs is to contact users and information. On the one hand, the RSs can help users discover information that is useful to them. On the other hand, the RSs can allow information to be displayed to users who are interested in them. RSs have been used in many fields such as e-commerce [1, 2], large-scale retailing, music [3, 4], books [5–8], news [9], e-learning [10], etc.

Recommender methods commonly used in the RSs include collaborative filtering (CF), content-based recommendation, knowledge-based recommendation and hybrid recommendation methods. The fundamental of CF is that if users have the same interests in the past, they will have similar tastes in the future. Content-based recommendation is based on the availability of item or user descriptions and a profile that

© Springer International Publishing AG, part of Springer Nature 2018
D.-S. Huang et al. (Eds.): ICIC 2018, LNCS 10954, pp. 467–478, 2018.
https://doi.org/10.1007/978-3-319-95930-6_43

assigns importance to these characteristics. Knowledge-based recommendation typically makes use of additional information about the available items. The hybrid recommendation method combines two or more methods mentioned above to get the advantages of all of them [11].

CF can be divided into memory-based CF and model-based CF. Memory-based CF can be classified into user-based CF and item-based CF. User-based CF needs the following steps: given a rating database and the active user as input, identify the neighbors of the active user who had similar tastes in the past. And then a prediction for every item that the active user has not rated is calculated. The model-based CF processes the raw data off-line to form a model. At runtime, only the "learned" models are needed to make predictions. Model-based CF methods include matrix factorization, association rules mining and recommendation methods based on probability analysis.

There are a lot of research directions for RSs, including the traditional directions: recommender systems foundation, K-nearest neighbor algorithm, cold-start problem, similarity models, and RSs evaluation. Of course, there are some newer research directions, such as: social filtering, social tags, privacy issues in the recommendation systems, how to recommend and explain to a group of users [12]. This paper focuses on how to mitigate cold start problem.

The objective of our study is to propose a new method that performs well under cold start conditions. This method considers utilizing other possible ratings, asymmetry influence between users and some rational inferences in CF. The main contributions of this paper are summarized as follows:

- We propose the idea that users with similar preferences in the past will show similar preferences in the future.
- We extend a traditional CF method based on rational inferences and combine the extended method with distance-based item similarity to recommend more accurate items to users.
- We regard the proportion of users' co-rate items to personal rated item as an asymmetric factor to represent different impact on users.

The rest of the paper is organized as follows. In Sect. 2, the related work on CF, cold start problems and some asymmetric factors are reviewed. In Sect. 3, a new similar model is proposed. In Sect. 4, the experiments are described and the results are reported and analyzed. Finally, the conclusions are shown in Sect. 5.

2   Related Work

2.1   Traditional Similarity Models

In order to find neighbors for users or items, a lot of similarity measures have been proposed. Cosine similarity (COS), Pearson correlation coefficient (PCC), Jaccard and the mean square displacement (MSD) are the most commonly used similarity methods. Some improvements based on the four methods was proposed, including adjusted cosine similarity, sigmoid function based Pearson correlation coefficient (SPCC), Constrained Pearson correlation coefficient (CPCC) and JMSD.

Besides, Yang et al. [13] thought traditional methods were sometimes over confident. Those methods always disregarded some useful information and implied untrustworthy inferences. They proposed a method based on the fact that any two users may have some common interest genres as well as different ones. This method outperformed other traditional measures.

2.2 Cold Start Problem

In fact, users generally rate a few items, resulting in a very sparse rating matrix. Sparse data can easily lead to cold start problem. The cold start problem could lead to the loss of new users due to lack of accuracy. To solve this problem, the most direct approach is to utilize the user's additional information, such as age, gender, educational level, to help find users' neighbors. [14, 15] utilized the neural network to find more accurate result. [16–18] employed the transfer learning technology to get knowledge from other related domains. [19, 20] combined social networks with geographic location information to obtain more accurate predictions. These methods tried to take advantage of third-party information. However, the third-party information is always hard to be accessed by the system in most situations.

Ahn et al. [21] proposed a new heuristic similarity measure called PIP to improve the recommendation performance under cold start conditions. Different from methods mentioned previously, the PIP method only used the users' rating information. The method was divided into three parts: proximity, impact and popularity. The proximity factor only considered the distance between two users. The impact factor thought about how strongly an item is preferred or disliked by users. The popularity factor thought over the distance between the pair of ratings and the average ratings.

But the PIP only considered the absolute ratings and the local context, ignoring the proportion of common ratings and the global preference. Liu et al. [22] proposed a new similarity model called NHSM, which combined the local context information of user ratings with the global preference of user behavior, showing the superiority in recommended performance. They improved the PIP to PSS (Proximity-Significance-Singularity). Proximity denoted the distance between ratings. Significance considered the distance between rating pair and the median rating. The singularity represented the distance between the rating pair and other ratings.

In addition, Yong Wang et al. [23] put forward a new hybrid model to improve the prediction accuracy under cold start conditions. When calculating similarity, the previously mentioned methods only took into account ratings of co-rate items. However, this new method considered almost all ratings for all items. They used the Kullback–Leibler divergence method to calculate the similarity between items. And then they adjusted the PSS method to calculate the similarity between two ratings. They also considered the user's preference and an asymmetric factor. The user's preference factor was from NHSM method. The results of their experiments proved their proposal.

2.3 Asymmetric Collaborative Filtering

Most traditional CF methods are symmetric. But recently, some asymmetric methods are proposed. A recommender system based on asymmetric user similarity was exhibited in [24]. Parivash et al. [25] put forward an asymmetric similarity, which was defined as the proportion of co-rate items, normalized by number of rated items by active user. This measure could be combined with PCC, COS and MSD, leading to lower errors. Liu et al. [26] proposed an asymmetric factor which combined the proportion of co-rate items in the active user's rated item set and the proportion of co-rate items in the union of the two users' rated items. The authors combined this factor with traditional COS and MSD methods. Bin et al..[27] proposed a novel adaptive bidirectional asymmetric similarity measurement, which learned the asymmetric similarities at the same time through matrix factorization.

3 The Proposed Model

Assume user u and user v buy two different books, but these two books both describe the RSs. And there is no co-rate item for the two users. When using traditional CF methods, the similarity between the two users is 0. But this is unreasonable. Obviously, both users are interested in RSs, they may purchase other books about RSs in the future.

According to the above analysis, we will extend the basic of CF. If users show similar interests in the past, they will show similar preferences in the future.

Based on this idea, when calculating the similarity between two users, we need to consider all items rated by the two users, rather than only consider the co-rate items. First of all, the similarity between items should be taken into account. The similarity between two items is calculated with the Euclidean distance method, utilizing all ratings of two items. The formula is shown below.

$$S_{item}(i,j) = \frac{1}{1 + \sqrt{\sum_{u \in U}(r_{ui} - r_{uj})^2}} \tag{1}$$

u denotes user u. U denotes the user sets. i and j denote different items.

Another basic element is derived from [13] to obtain more accurate results, which considers the rational inferences in CF. In order to combine the rational inferences, we will adjust this formula as a basic element. The original formula is used to calculate the similarity between two users. Here, we need to adjust the formula to utilize more ratings. The formula is shown below.

$$S_{basic}(u,v) = \frac{UV_3 - |UV_1 - UV_2|}{\sqrt{\sum_{i \in I_{uv}}(r_{ui} - r_{med})^2}\sqrt{\sum_{j \in I_{uv}}(r_{vi} - r_{med})^2}} \tag{2}$$

r_{ui} and r_{vj} represent rating of user u on item i and rating of user v on item j. And I_{uv} represents the union of the movies sets of user u and user v. r_{med} denotes the median of the rating range. The definitions of UV_1, UV_2 and UV_3 respectively represent the private interest genres of user u, the private interest genres of user v and the common interest genres of user u and v.

$$UV_1 = \sum\nolimits_{i,j \in I_{uv}, r_{ui} - rv_j > p_1} S_{item}(i,j) \times \left| (r_{ui} - r_{med}) \times (r_{vj} - r_{med}) \right| \tag{3}$$

$$UV_2 = \sum\nolimits_{i,j \in I_{uv}, r_{vj} - r_{ui} > p_1} S_{item}(i,j) \times \left| (r_{ui} - r_{med}) \times (r_{vj} - r_{med}) \right| \tag{4}$$

$$UV_3 = \sum\nolimits_{i,j \in I_{uv}, |r_{ui} - r_{vj}| \le p_1} S_{item}(i,j) \times \left| (r_{ui} - r_{med}) \times (r_{vj} - r_{med}) \right| \tag{5}$$

According to definition in [13], the value of p_1 is $\frac{R_{max}}{2}$. R_{max} represents the max value of possible ratings.

However, utilizing all the ratings information would take a lot of time to calculate similarity. The ratings of dissimilar items would result in inaccurate results. When it comes to an exactly item, the number of the exactly item' neighbors should be limited to get more accurate results. When j is not the neighbor of item i, the similarity between them will be regard as 0.

Besides, in the traditional similarity calculation method, $sim(u, v) = sim(v, u)$. But for user u who has rated a lot of items and user v who has rated a few items, I_v may be a subset of I_u. For example, two users' ratings are shown in Table 1.

Table 1. Rating examples.

User	Item1	Item2	Item3	Item4	Item5
u	2	3	–	–	–
v	2	3	5	4	4

In this case, the impact of user u on user v is obviously different from the impact of user v on user u. The asymmetric factor should be considered. The most intuitive way is utilizing the proportion of co-rate items between user u and user v.

$$S_{asymmetric}(u, v) = \frac{|I_u \cap I_v|}{I_u} \tag{6}$$

But the value of the above formula would be 0 if there is no co-rate item between two users. Here we use the sigmoid function to adjust this factor.

$$S_{asymmetric}(u, v) = \frac{1}{1 + e^{-\frac{|I_u \cap I_v|}{I_u}}} \tag{7}$$

The formula considering all factors is as follow.

$$sim(u, v) = S_{asymmetric}(u, v) \times S_{basic}(u, v) \tag{8}$$

4 Experiments

In this section, we present experiments to prove our prediction. We also implement the COS, PCC MSD, Jaccard, JMSD, the method based on rational inference (RISM) [13], PIP [21] and NHSM [22].

4.1 Dataset

In this paper, we use the MovieLens-100 K to prove our prediction. There are 100000 ratings from 943 users on 1682 movies and each user has rated at least 20 movies.

4.2 Evaluation Metrics

In general, when a website provides a recommendation service, it generally gives the user a personalized recommendation list. Such prediction accuracy is generally measured by precision and recall [28]. Recall and precision respectively indicate the proportion of correctly recommended items in the test set and recommender list.

$$precision = \frac{|R \cap T|}{|R|} \tag{9}$$

$$recall = \frac{|R \cap T|}{|T|} \tag{10}$$

T denotes the test set. R.denotes the recommended list given by the recommendation system.

The x-axis of all figures in this paper represents the number of active users' neighbors.

4.3 The Selection of the Number of Similar Items

First of all, we need to select the number of items' neighbors. We set this value to 20, 40, 60 and 80, and we will determine the value through experiments.

Figure 1 shows the values of recall and precision when choosing different number of items' neighbors. The values of recall are very close to each other in different situations. But when the number is 20, 40 or 60, the values of recall are slightly higher. As for precision, when the number of users' neighbors is smaller, the values of precision are similar. But when the number of users' neighbors is larger, the values of precision for 20 items' neighbors are higher. Considering recall and precision comprehensively, the number of the neighbors of item is 20.

4.4 Experiments with Full Ratings

TPR represents the proposed method in this paper. Figure 2 shows the values of recall and precision with full ratings. The values of recall of the proposed method is higher than most other methods. The values of recall of our method is close to the values of

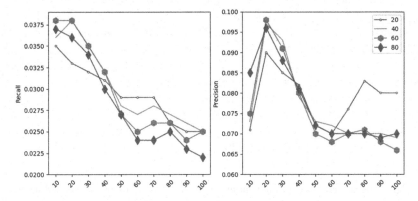

Fig. 1. Recall, precision with different item neighbors

RISM, sometimes the values of the proposed method even higher than the values of RISM. At the same time, the precision is higher than most methods. The recall and precision show that the recommender lists can match users' requirements. The results show that our approach makes full use of advantages of RISM.

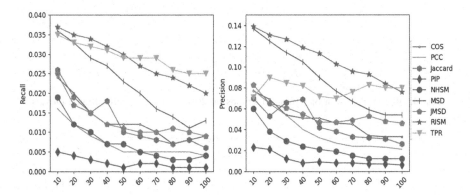

Fig. 2. Recall, precision with different methods

But how does this method perform under the cold start situations? In next subsection, we will use an artificial way to build cold start datasets and experiment on the constructed datasets (Figs 5 and 6).

4.5 Cold Start Tests

First of all, we need to manually simulate cold start dataset. The method is from [21]. In the full rating dataset, each user has rated at least 20 movies. Now, we will try to adjust the dataset to construct the cold start conditions. The percentage of cold start users is set to 20% and 40%. And the number of the rated movies for cold start users is set to 5, 10 or 15.

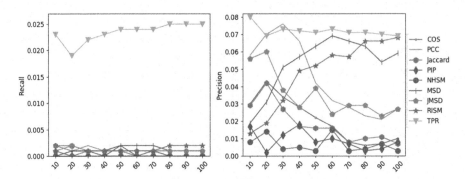

Fig. 3. Recall, precision of different methods on 20% cold users with 5 rated movies

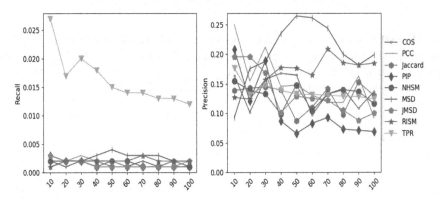

Fig. 4. Recall, precision of different methods on 20% cold users with 10 rated movies

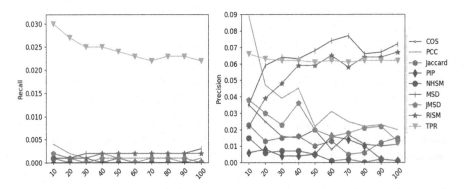

Fig. 5. Recall, precision of different methods on 20% cold users with 15 rated movies

Figure 3 shows the values of recall and precision with 20% cold users and each cold start user rates 5 movies. The values of recall of other methods is very small, even close to 0 in most cases. The results mean that items in recommended lists of these

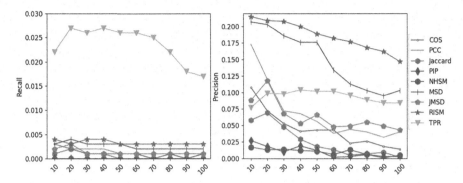

Fig. 6. Recall, precision of different methods on 40% cold users with 5 rated movies

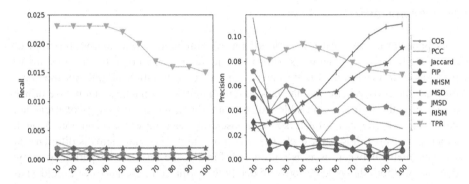

Fig. 7. Recall, precision of different methods on 40% cold users with 10 rated movies

methods hardly appear in the test sets. Due to lack of co-rate items, the similarity between users easily becomes 0, leading to dissimilar neighbors for cold start users. Finally, the items recommended for cold start users could not represent their true preferences. Because of making full use of the possible ratings, it is easier for the proposed method to find co-rate similar items. Our method can find more accurate neighbors for the target user and show a clear advantage under the cold start conditions. Finally, the values of recall of our method is much higher than others. As for precision, Fig. 3 shows that our method performs better than other methods.

Figure 4 through Fig. 8 show the comparisons of recall and precision for different cold-start scenarios, respectively. The values of recall of our method are obviously higher than others under different cold start conditions, which means that our method can recommend proper items under cold start conditions. And the values of precision keep stable. Although the values of our method are not highest, but the values are still higher than most other methods. These results prove the superiority of our method (Fig. 7).

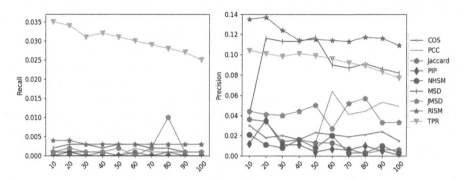

Fig. 8. Recall, precision of different methods on 40% cold users with 15 rated movies

5 Conclusions

This paper proposes an asymmetric similarity method for collaborative filtering, of which the basic idea is that users with similar preferences in the past will show similar preferences in the future. Taking into account that the traditional similarity measures only use the co-rate items and the lack of common items in most situations, we utilize as much of the ratings as possible. At the same time, too many ratings of irrelevant items will cause negative effects, so we choose the proper number of neighbors through experiments. In addition, for each user, the proportion of co-rate items is different, so the asymmetry between users is considered. Finally, the results of experiments show that our method shows superiority under both normal circumstances and cold start conditions.

Future directions for improvement will focus on how to improve the operating efficiency of the method should also be considered. More experiments and adjustments need to be implemented to ensure the reliability of the method.

References

1. Putra, A.A., Mahendra, R., Budi, I., Munajat, Q.: Two-steps graph-based collaborative filtering using user and item similarities: case study of E-commerce recommender systems. In: International Conference on Data and Software Engineering, pp. 1–6 (2017)
2. Aditya, P.H., Budi, I., Munajat, Q.: A comparative analysis of memory-based and model-based collaborative filtering on the implementation of recommender system for E-commerce in Indonesia: a case study PT X. In: International Conference on Advanced Computer Science and Information Systems, pp. 303–308. IEEE, Malang (2016)
3. Lee, S.K., Cho, Y.H., Kim, S.H.: Collaborative filtering with ordinal scale-based implicit ratings for mobile music recommendations. Inf. Sci. **180**(11), 2142–2155 (2010)
4. Naser, I., Pagare, R., Wathap, N.K., Pingale, V.: Hybrid music recommendation system: enhanced collaborative filtering using context and interest based approach. In: Annual IEEE India Conference, pp. 1–11. IEEE, Pune (2014)
5. Song, C.: Application of an improved collaborative filtering method on recommending books in college libraries. Libr. Inf. Serv. (2016)

6. Ng, Y.-K.: Recommending books for children based on the collaborative and content-based filtering approaches. In: Gervasi, O., et al. (eds.) ICCSA 2016. LNCS, vol. 9789, pp. 302–317. Springer, Cham (2016). https://doi.org/10.1007/978-3-319-42089-9_22

7. Dong, Y., Liu, S., Chai, J.: Research of hybrid collaborative filtering algorithm based on news recommendation. In: 9th International Congress on Image and Signal Processing, BioMedical Engineering and Informatics, pp. 898–902. IEEE, Datong (2016)

8. Mathew, P., Kuriakose, B., Hegde, V.: Book recommendation system through content based and collaborative filtering method. In: 2016 International Conference on Data Mining and Advanced Computing, pp. 47–52. IEEE, Ernakulam (2016)

9. Saranya, K.G., Sadasivam, G.S.: Personalized news article recommendation with novelty using collaborative filtering based rough set theory. Mob. Netw. Appl. **22**(1), 1–11 (2017)

10. Do, P., Nguyen, K., Vu, T.N., Dung, T.N., Le, T.D.: Integrating knowledge-based reasoning algorithms and collaborative filtering into e-learning material recommendation system. In: Dang, T.K., et al. (eds.) FDSE 2017. LNCS, vol. 10646, pp. 419–432. Springer, Cham (2017). https://doi.org/10.1007/978-3-319-70004-5_30

11. Jannach, D., Zanker, M., Felfernig, A., Friedrich, G.: Recommender Systems: An Introduction. Cambridge University Press, New York (2010)

12. Bobadilla, J., Ortega, F., Hernando, A.: Recommender systems survey. Knowl.-Based Syst. **46**(1), 109–132 (2013)

13. Yang, J.M., Li, K.F.: Recommendation based on rational inferences in collaborative filtering. Knowl.-Based Syst. **22**(1), 105–114 (2009)

14. Wei, J., He, J., Chen, K., Zhou, Y., Tang, Z.: Collaborative filtering and deep learning based recommendation system for cold start items. Expert Syst. Appl. **69**, 29–39 (2017)

15. Bobadilla, J., Ortega, F., Hernando, A., Bernal, J.: A collaborative filtering approach to mitigate the new user cold start problem. Knowl.-Based Syst. **26**, 225–238 (2012)

16. Grolman, E., Bar, A., Shapira, B., Rokach, L., Dayan, A.: Utilizing transfer learning for in-domain collaborative filtering. Know.-Based Syst. **107**(C), 70–82 (2016)

17. Li, B., Zhu, X., Li, R., Zhang, C.: Rating knowledge sharing in cross-domain collaborative filtering. IEEE Trans. Cybern. **45**(5), 1068–1082 (2015)

18. Pan, W., Liu, M., Ming, Z.: Transfer learning for heterogeneous one-class collaborative filtering. IEEE Intell. Syst. **31**(4), 43–49 (2016)

19. Nguyen, V.D., Sriboonchitta, S., Huynh, V.N.: Using community preference for overcoming sparsity and cold-start problems in collaborative filtering system offering soft ratings. Electron. Commer. Res. Appl. **26**, 101–108 (2017)

20. Chen, Z., Shen, L., Li, F., You, D.: Your neighbors alleviate cold-start: on geographical neighborhood influence to collaborative web service QoS prediction. Knowl.-Based Syst. **138**, 188–201 (2017)

21. Ahn, H.J.: A new similarity measure for collaborative filtering to alleviate the new user cold-starting problem. Inf. Sci. **178**(1), 37–51 (2008)

22. Liu, H., Hu, Z., Mian, A., Tian, H., Zhu, X.: A new user similarity model to improve the accuracy of collaborative filtering. Knowl.-Based Syst. **56**(3), 156–166 (2014)

23. Wang, Y., Deng, J., Gao, J., Zhang, P.: A hybrid user similarity model for collaborative filtering. Inf. Sci. **418–419**, 102–118 (2017)

24. Millan, M., Trujillo, M., Ortiz, E.: A collaborative recommender system based on asymmetric user similarity. In: Yin, H., Tino, P., Corchado, E., Byrne, W., Yao, X. (eds.) IDEAL 2007. LNCS, vol. 4881, pp. 663–672. Springer, Heidelberg (2007). https://doi.org/10.1007/978-3-540-77226-2_67

25. Pirasteh, P., Jung, Jason J., Hwang, D.: An asymmetric weighting schema for collaborative filtering. In: Camacho, D., Kim, S.-W., Trawiński, B. (eds.) New Trends in Computational Collective Intelligence. SCI, vol. 572, pp. 77–82. Springer, Cham (2015). https://doi.org/10.1007/978-3-319-10774-5_7

26. Liu, Z., Shihua, O.U., Hang, S.: Collaborative filtering recommendation algorithm based on asymmetric weighted user similarity. J. Chin. Comput. Syst. **38**(4), 721–725 (2017)

27. Cao, B.: Learning bidirectional asymmetric similarity for collaborative filtering via matrix factorization. Data Min. Knowl. Disc. **22**(3), 393–418 (2011)

28. Linden, G.: What is a good recommendation algorithm? https://cacm.acm.org/blogs/blog-cacm/22925-what-is-a-good-recommendation-algorithm/fulltext

Research on a New Automatic Generation Algorithm of Concept Map Based on Text Clustering and Association Rules Mining

Zengzhen Shao[1,2(✉)], Yancong Li[2], Xiao Wang[2], Xuechen Zhao[1], and Yanhui Guo[1]

[1] School of Data Science and Computer Science,
Shandong Women's University, Jinan 250002, China
shaozengzhen@163.com
[2] School of Information Science and Engineering,
Shandong Normal University, Jinan 250014, China

Abstract. As an important teaching tool of visualization, the concept map has become a hot spot in the field of smart education. The traditional concept map generation algorithm is hard to guarantee the construction process and quality because of the huge amount of work and the great influence of the expert experience. A TC-ARM algorithm for automatic generation of hybrid concept map based on text clustering and association rules mining is proposed. This algorithm takes full account of the attributes of the relationship between concepts, uses text clustering technology to replace the relationship between artificial mining concepts and test questions, combines association rules mining methods to generate the concept maps, and introduces consistency of answer record parameter to improve the efficiency of concept map generation. The experimental results show that the TC-ARM algorithm can automatically and rapidly construct the concept map, which not only reduces the impact of outside experts, but also dynamically adjusts the concept map based on the basic data. The concept map generated by the TC-ARM algorithm expresses the relationship between the concepts and the degree of closeness through the relationship pairs and relationship strength, and can clearly show the structural relationship between concepts, provide instructional optimization guidance for knowledge visualization.

Keywords: Concept map · Automatic generation · Text clustering
Association rules mining · Smart education

1 Introduction

With the continuous promotion of modern education information, smart education [1] has become the focus of attention of researchers at home and abroad. In order to improve the intellectualized level of education, a variety of educational technologies have been ubiquitously used [2]. Among them, the concept map [3] has become the research hotspot in the field of smart education with the advantages of strong relation and distinct hierarchy. The concept map was first proposed by Dr. Novak [4] at Cornell

© Springer International Publishing AG, part of Springer Nature 2018
D.-S. Huang et al. (Eds.): ICIC 2018, LNCS 10954, pp. 479–490, 2018.
https://doi.org/10.1007/978-3-319-95930-6_44

University in the United States in 1984 to show the relationship between concepts through an intuitive and close to natural language graphical representation. It mainly takes Novak's [5] network-shaped conceptual structure as a standard manifestation, using nodes to represent concepts, using directed edges to represent the connection between concepts and using prepositional labels to represent the dependencies between concepts relationship.

Scholars both at home and abroad have conducted various degrees of research [6, 7] on concept maps and have applied them to a wide range of disciplines, such as teaching diagnosis [8], knowledge building [9] and clinical nursing [10], and achieved some results. However, the early concept map building mainly relied on expert experience, which not only took a long time, but also its accuracy was influenced by the experience of experts. In recent years, automatic generation algorithms of concept map relying on data mining technology have been continuously proposed. Bai et al. [11] used the association rules mining to analyze the inherent relations between concepts based on fuzzy rules, but the relationship between concepts and problems was still classified by experts manually. Haiyan et al. [12] proposed a multi-layered literature topic model TMDOM based on thematic maps. They constructed the domain knowledge structure by multi-level topic association and automatically generated concept maps by clustering documents many times. However, the concepts need to be obtained manually. Huang et al. [13] proposed an algorithm to automatically generate concept maps under simulated datasets. They calculated the correlation between concepts through Apriori [14] algorithm, but the concepts originated from the existing data and did not achieve automatic acquisition. Atapattu et al. [15] extracted concept maps from lecture slides and the suitability of auto-generated concept maps as a pedagogical tool. However, they only considered the content of the slides, and did not pay enough attention to the test data of students. Santos et al. [16] used Natural Language Processing (NLP) techniques to automate the extraction of concepts, but they did not consider enough the connection between knowledge points.

In this paper, we propose a TC-ARM (Text Clustering-Association Rules Mining) algorithm based on text clustering and association rules mining, which is based on the test data of students. The TC-ARM algorithm uses text clustering to abstract test questions into concept clusters to reduce the inadequacy of constructing conceptual prior knowledge through artificial means. At the same time, it combines student answer records and introduces association rules mining to automatically generate concept maps. The experimental results show that the TC-ARM algorithm can automatically generate a concept map under the premise of reducing the labor intensity.

2   TC-ARM Concept Map Automatic Generation Algorithm

The TC-ARM algorithm consists of two stages: question text clustering and association rules mining of concept clusters. The basic idea of TC-ARM algorithm is as follows: The text clustering technique is used to cluster test questions with similar text features. Test questions are automatically divided into clusters to replace the manual classification process. As shown in Fig. 1, each clustered cluster abstracted into concept, and the category of clusters to which test questions belong is abstracted into the correspondence between the concepts and test questions. Through the process of concept clusters association rules mining phase, combined with answering records, we generate frequent item sets of test questions, and map the relationships between test questions in the text clustering stage to the degree of concept relevance, and finally generate concept maps.

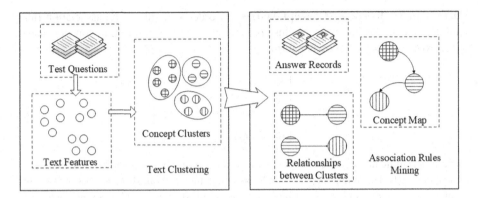

Fig. 1. TC-ARM algorithm schematic diagram

2.1 Question Text Clustering

Text clustering [17] is an unsupervised data mining method that can group semi-structured or unstructured text object data. The k-means [18] algorithm is a classic algorithm in the clustering analysis stage. It has many advantages such as low time complexity and can effectively handle large data sets. It is in accordance with the algorithm requirements for generating the concept map in this paper.

Using the text clustering algorithm, the questions can be clustered into abstract concepts instead of the artificial classification process to reduce the labor intensity.

Definition 1: Question Text Features. Using the VSM (Vector Space Model) [19], the text can be mapped into points in the vector space, and then expressed as a form that can be analyzed and processed by the clustering algorithm. When the text is mapped to a vector, it needs to be processed for word segmentation, stop word filtering, and weight calculation. The test questions are expressed as $Q = (Q_1, Q_2, \cdots, Q_m)$, where m is the number of questions. The text of the test question is processed to form a vector in space as a textual feature of a single question, denoted as $W = (W_{i1}, W_{i3}, \cdots, W_{ij}, \cdots, W_{ir})$, where W_{ij} is the weight of the feature item j in the i-th test text, and r is the dimension of the feature vector. Text weight calculation using the TF-IDF [20] method during the construction of the test questions text feature:

$$Wij = TFij \times IDFj \tag{1}$$

The TF_{ij} denotes the word frequency of feature item j in the text of the i-th test item; IDF_j denotes the number of feature items j appearing in the text, which is called reverse text frequency, i.e., word frequency is proportional to weight, and reverse text frequency is inversely proportional to weight.

Definition 2: Concept Cluster. The test questions are digitized into multidimensional vectors after text feature extraction. Using the k-means [18] algorithm, the digitized text features can be clustered. The division of test questions formed by clustering is defined as the concept cluster, which is expressed as $C = (C_1, C_2, \cdots C_x, \cdots, C_k)$, where k is the number of concept clusters.

Definition 3: Question - Concept Matrix. In order to facilitate the association rules mining calculation, the concept clusters are converted into a question-concept matrix QC, expresses as the following form:

$$QC = \begin{bmatrix} qc_{11} & qc_{12} & \cdots & qc_{1k} \\ qc_{21} & qc_{22} & \cdots & qc_{2k} \\ \vdots & \vdots & \ddots & \vdots \\ qc_{m1} & qc_{m2} & \cdots & qc_{mk} \end{bmatrix}$$

where qc_{ix} indicates whether the current test question Q_i belongs to the concept cluster C_x, $qc_{ix} \in \{0,1\}$. When $qc_{ix} = 1$ indicates that the test question Q_i belongs to the concept cluster C_x, and $qc_{ix} = 0$ indicates that the test question Q_i does not belong to the concept cluster C_x.

2.2 Concept Clusters Association Rules Mining

Association rules were first proposed by Agrawal et al. [21] in 1993 to find interesting associations hidden in data sets. The relationships mined by association rules mining can use the form of association rules or frequent item sets, which are expressed as $A \rightarrow B$. A and B are disjoint item sets, where $A \cap B = \emptyset$, A is called the predecessor of the rule, and B is called the post-item of the rule. Apriori [14] algorithm is used to mine the associations between the concept clusters.

Through the association rules mining method and the combination of the answer records data, the links between the concept clusters can be found and the concept map is finally generated [22].

Definition 4: Grade Matrix. Before mining associations between concept clusters, the student's answer records R need to be digitized into a grade matrix G, expresses as follows:

$$G = \begin{bmatrix} g_{11} & g_{12} & \cdots & g_{1n} \\ g_{21} & g_{22} & \cdots & g_{21} \\ \vdots & \vdots & \ddots & \vdots \\ g_{m1} & g_{m1} & \cdots & g_{mn} \end{bmatrix}$$

where $g_{iy} \in \{0,1\}$, $g_{iy} = 1$ means that student S_y correctly answered the question Q_i, and $g_{iy} = 0$ means that student S_y erroneously answered the question Q_i, and n is the total number of students.

Definition 5: Consistency of Answer Records. In order to remove unnecessary associations between test questions, we introduce in consistency of answer records. Consistency of answer records is the XNOR value between every two rows of grades in the grade matrix G, which means the number of students who all answered correctly or all answered incorrectly in every two questions. When the XNOR value is 0, it means that the student has answered one of the two questions correctly. Otherwise, it indicates that the student has all answered correctly or all answered incorrectly in the two questions. So we have

$$\text{Count}(Q_a, Q_b) = \sum\nolimits_{k=1}^{n} \left(g_{ap} \odot g_{bp} \right) \tag{2}$$

Use $\text{Count}(Q_a, Q_b)$ to indicate the consistency of answer records between question Q_a and question Q_b, and define the threshold as Min_{count} for the minimum consistency of answer records. When the value of the consistency of answer records is less than its threshold value, it indicates that the number of students who all answered correctly or all answered wrongly between the two questions is relatively rare. The relationship between the two questions is weak and will not be considered in the subsequent calculation.

Through the screening of the consistency of answer records and its threshold, the number of test questions to be calculated can be reduced, and the efficiency of generating concept map can be improved.

Definition 6: Relationship Strength of Test Questions. The correlation between two test questions is defined as the relationship strength of the test questions. That is, the confidence of the test questions. So we have

$$\text{Conf}(Q_a \rightarrow Q_b) = \frac{\text{Sup}(Q_a, Q_b)}{Q_a} \tag{3}$$

Among them, $\text{Sup}(Q_a, Q_b)$ indicates the support of questions Q_a and Q_b. In order to remove unnecessary associations between test questions, the minimum relationship strength Min_{Conf} is set as the threshold for the strength of the test questions. The greater the relationship strength of the test questions, the stronger the relationship between the two test questions and the greater the possibility of an association. Conversely, the weaker the relationship between the two test questions, the less likely they are related.

Definition 7: Relationship Strength of Concept Clusters. The relationship strength of concept clusters expresses the correlation between two concept clusters and can be embodied as the corresponding relationship between two concept clusters. Combining the question-concept matrix QC, use the following formula to map the relationship strength of test questions to the relationship strength of concept clusters:

$$\text{Rev}(C_u, C_v)_{Q_a \rightarrow Q_b} = qc_{au} \times qc_{bv} \times \text{Conf}(Q_a \rightarrow Q_b) \tag{4}$$

The greater the relationship strength of concept clusters, the stronger the relationship between the two concept clusters and the greater the likelihood of association. Conversely, the weaker the relationship between the two concept clusters is, the less likely it is to have associations.

The pseudocode and flow chart (Fig. 2) of the TC-ARM concept map generation algorithm are as follows:

Algorithm TC-ARM

Input Q, R, k, 〖Min〗_count, 〖Min〗_Conf
$m \leftarrow$ the number of Q
While $i \leq m$
   Segmenting question Q_i word
   Remove stop words
While $i \leq m$
   Construct question text feature W_ij with (1)
Select k question text features as initial centroids
Repeat
   Assign each question text feature to the nearest centroid, forming k clusters
   Recalculate the centroid of each cluster
Until Clusters do not change or reach the maximum number of iterations
Get clusters C and question-concept matrix QC
Grade matrix $G \leftarrow$ student answer records R
While $i \leq m$
   Calculate the consistency of answer records Count(Q_a,Q_b) between every two questions with (2)
     If Count(Q_a,Q_b)$\leq$ 〖Min〗_count
      Delete the relationship between Q_a and Q_b
While $i \leq m$ Calculate the consistency of answer records Count(Q_a,Q_b) between every two questions with (2)
     If Count(Q_a,Q_b)$\leq$ 〖Min〗_count
      Delete the relationship between Q_a and Q_b
While $i \leq m$
   Calculate the relationship strength of test questions Conf(Q_a→Q_b) with (3)
     If Conf(Q_a→Q_b) $\leq$ 〖Min〗_Conf
      Delete the relationship between Q_a and Q_b
While $x \leq k$
   Calculate the relationship strength of concept clusters 〖Rev(C_u,C_v)〗_(Q_a→Q_b) with (4)
   Repeat
     If there is more than one relationship between two concept clusters
      Retain relationship pair with the greatest relationship strength
   Until Complete deal with the data all concept clusters relationships
   Represented graphically
   Output Concept map

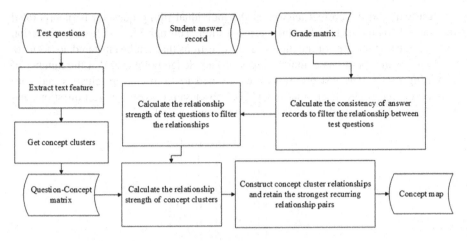

Fig. 2. TC-ARM algorithm flow diagram

2.3 Algorithm Complexity Analysis

TC-ARM concept map automatic generation algorithm mainly includes two stages. The k-means algorithm is mainly used in the question text clustering stage. The time complexity of the k-means algorithm is $O(tkm)$, where t is the number of iterations, k is the number of concept clusters, and m is the total number of question texts. The Apriori algorithm is used in the concept clusters association rules mining stage. In this paper, only the frequent 2-itemsets in the concept cluster are considered, so the time complexity is not more than $O(n^2)$.

3 Experiment and Result Analysis

3.1 Experimental Data Preparation

In order to verify the feasibility and effectiveness of the TC-ARM algorithm, this paper selects 298,070 authentic answer records from 3,635 college students in a large-scale examination of Computer Culture Foundation as the experimental data set. There are a total of 82 questions involving storage device, computer language, virus concept, hex conversion and operating system.

The experimental operating environment is the Windows 7 operating system. The programming language is C#, and the software development environment is Microsoft Visual Studio 2015 and SQL Server 2008.

3.2 Experimental Process and Results Analysis

The questions include four types: single selection, multiple selection, judgment, and individual filling. In the preprocessing section of the question text clustering stage, combine the options for single selection questions and multiple selection questions with the topic text as an entire test question, the title text of the judgment question is used as the test text, and the correct answer of the individual filling question is incorporated into the title as the text of the entire test question. Through the text feature extraction method, the text features of the digitized test questions that can be clustered are formed.

Because the questions contain 5 types of knowledge points, so that the number of clustering $k = 5$. The test questions are clustered to form concept clusters and converted into question-concept matrix QC. The abscissa represents the test questions and the ordinate represents the concept clusters, shown as follows:

$$QC = \begin{bmatrix} 1 & 0 & 0 & 0 & 0 \\ 0 & 0 & 0 & 1 & 0 \\ 0 & 0 & 0 & 1 & 0 \\ 0 & 0 & 0 & 1 & 0 \\ 0 & 1 & 0 & 0 & 0 \\ 0 & 0 & 0 & 0 & 1 \\ 1 & 0 & 0 & 0 & 0 \\ \vdots & \vdots & \vdots & \vdots & \vdots \\ 0 & 0 & 0 & 1 & 0 \end{bmatrix}.$$

fore screening the relationships between test questions, the student answer record is preprocessed as the grade matrix G, the abscissa indicates the test questions, and the ordinate indicates the students, shown as follows:

$$G = \begin{bmatrix} 0 & 0 & 0 & 0 & 0 & 0 & 0 & 0 & 0 & 0 & 1 & 0 & \cdots & 1 \\ 0 & 0 & 1 & 0 & 0 & 0 & 1 & 0 & 1 & 0 & 1 & 1 & \cdots & 0 \\ 0 & 1 & 0 & 0 & 0 & 0 & 0 & 0 & 0 & 0 & 1 & 0 & \cdots & 0 \\ 1 & 1 & 1 & 0 & 0 & 0 & 1 & 0 & 0 & 0 & 1 & 0 & \cdots & 1 \\ 0 & 0 & 0 & 1 & 0 & 0 & 0 & 1 & 1 & 0 & 0 & 1 & \cdots & 0 \\ 0 & 1 & 0 & 1 & 0 & 0 & 0 & 1 & 1 & 0 & 1 & 1 & \cdots & 0 \\ 1 & 0 & 1 & 1 & 1 & 1 & 0 & 0 & 0 & 1 & 0 & 1 & \cdots & 0 \\ 0 & 1 & 0 & 0 & 0 & 0 & 0 & 0 & 1 & 0 & 1 & 0 & \cdots & 0 \\ 0 & 1 & 0 & 1 & 1 & 0 & 1 & 1 & 1 & 1 & 0 & 0 & \cdots & 1 \\ 0 & 1 & 0 & 0 & 1 & 1 & 1 & 0 & 1 & 1 & 0 & 1 & \cdots & 0 \\ \vdots & \vdots & \vdots & \vdots & \vdots & \vdots & \vdots & \vdots & \vdots & \vdots & \vdots & \vdots & \ddots & \vdots \\ 0 & 1 & 0 & 0 & 0 & 0 & 1 & 0 & 0 & 0 & 1 & 1 & \cdots & 0 \end{bmatrix}$$

When setting the consistency of answer records threshold $Min_{count} = n * 50\%$, a total of 938 relationship pairs of test questions were deleted, and the relationships between the remaining 2383 pairs of test questions (Fig. 3a) were continuously calculated. The threshold of the strength of test questions determines the number of pairs of test questions to be screened. In order to avoid that the threshold is too large, which leads to the number of concept cluster relationship pairs is too small, and the threshold is too small, which leads to the complexity of the concept clusters, assume that the threshold of the strength relationship of test questions given by the expert [23] is $m = 0.85$, and the relationships between the test questions are further screened, and 1065 relationship pairs of test questions (Fig. 3b) are retained.

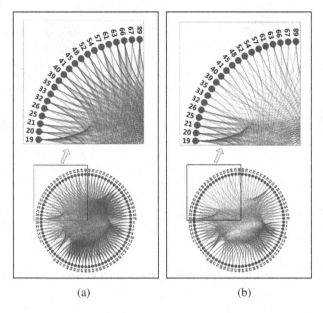

(a)   (b)

Fig. 3. Relationships between test questions after filtering

Combined with the question-concept matrix QC to calculate the strength of the concept cluster relationships, and map the question relationships to the concept cluster relationships. Relationships between the processed concept clusters are shown in Table 1, and they are graphically represented as the concept map as shown in Fig. 4a.

Table 1. Relationship between conceptual clusters

Order	Relationship	Strength
1	$C_1 \rightarrow C_5$	0.98
2	$C_1 \rightarrow C_3$	0.87
3	$C_2 \rightarrow C_1$	1
4	$C_2 \rightarrow C_5$	0.98
5	$C_2 \rightarrow C_3$	0.97
6	$C_3 \rightarrow C_5$	1
7	$C_3 \rightarrow C_4$	0.83
8	$C_4 \rightarrow C_5$	0.99

In order to verify that the conclusion of the concept map generated by the TC-ARM algorithm is correct, after the expert analysis of the content of each concept cluster, the corresponding relationships between concept clusters and knowledge points are shown in Table 2. In order to further verify the quality of the concept map automatically generated by the TC-ARM algorithm, we obtain the matrix QC by classifying the questions into knowledge points by experts, and combine the concept clusters

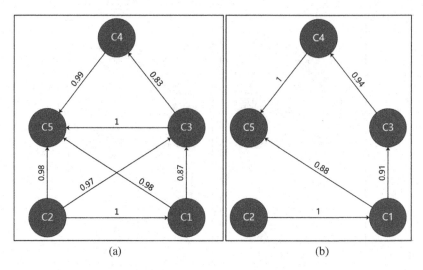

Fig. 4. The concept map generated by the TC-ARM algorithm and the concept map generated by expert assistance

Table 2. Relationships between concept clusters and knowledge points

Order	Concept cluster	Knowledge point
1	C_1	Storage Device
2	C_2	Computer Languages
3	C_3	Virus Concept
4	C_4	Hex Conversion
5	C_5	Operating System

association rules mining in the TC-ARM algorithm to generate concept maps with the help of experts. In order to facilitate comparison, the knowledge points are converted into a unified form according to Table 2, as shown in Fig. 4b. Comparing the concept map generated by the TC-ARM algorithm (Fig. 4a) with the expert-assisted concept map (Fig. 4b), the number of concept relationships in Fig. 4a is more, and the concept relationships in Fig. 4b is relatively concise, but the directions of the concept relationships are basically the same. Compare the relationship between the knowledge points with the authoritative literature on the computer culture foundation [24], the conclusions are basically the same. And the entire generation process is less than 1 min, which is far less than the time when the experts participate in the classification and then generate the concept map. It proves that the TC-ARM algorithm can reduce the artificial time cost and has reference value.

4 Conclusions

Aiming at the shortage of artificial dependencies and low efficiency existing in current concept map generation algorithms, this paper proposes a concept map automatic generation algorithm TC-ARM based on text clustering and association rules mining. Compared with previous researches, the TC-ARM algorithm has the following characteristics: (1) Low dependence on expert experience and good stability; (2) The concept map can be dynamically trimmed according to the basic data such as the consistency of answer records threshold, and the flexibility is high. The concept map generated by TC-ARM algorithm shows the direction and degree of relationship between concepts through relational pairs and relational strength. It shows the structure between concepts, and provides optimal guidance for teaching as a knowledge visualization tool. However, some parameters need to be given by the experts in the process of generating the concept map, such as the number of clusters, and the threshold of the relationship strength of test questions. Later we will consider using a deep learning algorithm to improve the text clustering process to generate higher quality concept maps. At the same time, collecting more information in the student's learning process and introducing students' learning psychology is our future research direction.

References

1. Uskov, V.L., Howlett, R.J., Jain, L.C.: Smart Education and e-Learning 2016. Springer, Berlin (2016)
2. Yang, X., Yu, S.: The architecture and key support technologies of smart education. China Educational Technology (2015) (in Chinese). (J. N. University)
3. Markham, K.M., Mintzes, J.J., Jones, M.G.: The concept map as a research and evaluation tool: further evidence of validity. J. Res. Sci. Teach. **31**(1), 91–101 (2010)
4. Novak, J.D., Gowin, D.B.: Learning How to Learn. Cambridge University Press, Cambridge (1984)
5. Novak, J.D.: Learning, creating, and using knowledge: concept maps as facilitative tools in schools and corporations. Concept Mapp. **56**(4), 392 (2010)
6. Maisonnasse, L., Chevallet, J.P., Berrut, C.: Incomplete and fuzzy conceptual graphs to automatically index medical reports. In: Kedad, Z., Lammari, N., Métais, E., Meziane, F., Rezgui, Y. (eds.) NLDB 2007. LNCS, vol. 4592, pp. 240–251. Springer, Heidelberg (2007). https://doi.org/10.1007/978-3-540-73351-5_21
7. Thomopoulos, R., Buche, P., Haemmerlé, O.: Different kinds of comparisons between fuzzy conceptual graphs. In: Ganter, B., de Moor, A., Lex, W. (eds.) ICCS-ConceptStruct 2003. LNCS (LNAI), vol. 2746, pp. 54–68. Springer, Heidelberg (2003). https://doi.org/10.1007/978-3-540-45091-7_4
8. Lee, C.H., Lee, G.G., Leu, Y.: Application of automatically constructed concept map of learning to conceptual diagnosis of e-learning. Expert Syst. Appl. **36**(2), 1675–1684 (2009)
9. Zhang, H.P., Zhou, N., Chen, Y.Y.: Research on application of concept map in knowledge organization. Inf. Sci. (2007)
10. Xiu-Mei, Q.I., Zhu, N.Q., Luo, Y.: Research of concept map and its application in nursing care. Nurs. J. Chin. Peoples Liberation Army (2006)
11. Bai, S.M., Chen, S.M.: Automatically constructing concept maps based on fuzzy rules for adapting learning systems. Expert Syst. Appl. **35**(1–2), 41–49 (2008)

12. Haiyan, T.: Study of Document Organization Method Based on Topic Map. Dalian University of Technology (2006) (in Chinese)
13. Huang, X., Yang, K., Lawrence, V.B.: An efficient data mining approach to concept map generation for adaptive learning. In: Perner, P. (ed.) ICDM 2015. LNCS (LNAI), vol. 9165, pp. 247–260. Springer, Cham (2015). https://doi.org/10.1007/978-3-319-20910-4_18
14. Toivonen, H.: Apriori algorithm. In: Encyclopedia of Machine Learning, pp. 39–40 (2017)
15. Atapattu, T., Falkner, K., Falkner, N.: A comprehensive text analysis of lecture slides to generate concept maps. Comput. Educ. **115**, 96–113 (2017)
16. Santos, V.: Concept Maps Construction Using Natural Language Processing to Support Studies Selection (2018)
17. Steinbach, M., Karypis, G., Kumar, V.: A Comparison of Document Clustering Techniques (2000)
18. Macqueen, J.: Some methods for classification and analysis of multivariate observations. In: Proceedings of, Berkeley Symposium on Mathematical Statistics and Probability, pp. 281–297 (1967)
19. Qingyun, Y.: Research of VSM-Based Chinese Text Clustering Algorithms. Shanghai Jiao Tong University (2008) (in Chinese)
20. Roelleke, T., Wang, J.: TF-IDF uncovered: a study of theories and probabilities. International ACM Sigir Conference on Research & Development in Information Retrieval ACM, pp. 435–442 (2008)
21. Agrawal, R., Imieliński, T., Swami, A.: Mining association rules between sets of items in large databases. ACM Sigmod Record **22**(2), 207–216 (1993)
22. Chen, S.M., Sue, P.J.: Constructing concept maps for adaptive learning systems based on data mining techniques. Expert Syst. Appl. **40**(7), 2746–2755 (2013)
23. Chen, S.M., Bai, S.M.: Using data mining techniques to automatically construct concept maps for adaptive learning systems. Expert Syst. Appl. **37**(6), 4496–4503 (2010)
24. Jian, Ma., Chengbing, T., Li, D.: Fundamentals of Computer Culture. Chemical Industry Press, Beijing (2010). (in Chinese)

Mandarin Prosody Prediction Based on Attention Mechanism and Multi-model Ensemble

Kun Xie$^{(\boxtimes)}$ and Wei Pan$^{(\boxtimes)}$

Fujian Key Laboratory of Brain-Inspired Computing Technique
and Applications, School of Information Science and Engineering,
Xiamen University, Xiamen, China
xierhacker@stu.xmu.edu.cn, wpan@xmu.edu.cn

Abstract. Prosodic boundary prediction is very important and challenging in the speech synthesis task, the result of prosodic prediction directly determines the quality of speech synthesis. In this paper, we proposed a prosodic boundary prediction method based on "encoding-decoding" frame while using an effective position attention mechanism to further improve performance. Finally, we investigate the use of Random Forest and Gradient Boosting Decision Tree to explore the potential of combined multiple models. The experimental results show that compared with the current best method of prosodic structure (Bi-LSTM), the proposed method presented a good result with F1-Score in terms of prosodic words, prosodic phrases, intonation phrases; the subjective experiment also shows that the proposed method can improve the quality and naturalness of synthesized speech.

Keywords: Speech synthesis · Prosodic boundaries prediction
Attention mechanisms · Bi-LSTM · Model ensemble

1 Introduction

Speech synthesis is a very important field in the study of speech interaction, it's main purpose is to convert normal language text into speech. Generally, we use naturalness and intelligibility [1] to judge the quality of the Speech synthesis system. In recent years, the performance of speech synthesis is getting better and better, but the naturalness of synthesized speech still has a certain gap compared with the real speech. One of the most important reason is that the performance of the existing prosody prediction model is not good enough.

Prosody prediction can be treated as a sequence labeling or sequence to sequence problem, we want to learn a function $f : X \rightarrow Y$ that maps an input sequence x to the corresponding label sequence y. In fact, we add labels to indicate the boundary of each word so we can restore the prosodic structure through this boundary information. Therefore, the essence of the prosody prediction problem is a boundary judgment problem at different prosodic levels. In mandarin speech synthesis systems, a typical hierarchical prosodic structure can significantly improve the accuracy of prosody

© Springer International Publishing AG, part of Springer Nature 2018
D.-S. Huang et al. (Eds.): ICIC 2018, LNCS 10954, pp. 491–502, 2018.
https://doi.org/10.1007/978-3-319-95930-6_45

prediction to distinguish different levels of pauses between words in speech. Normally, the prosodic boundaries are often classified into prosodic word (PW), prosodic phrase (PPH) and intonational phrase (IPH) [2].

In this work, we investigate how prosody prediction can benefit from the strong modeling capacity of sequence to sequence models. we also investigate the use of multi-model ensemble to explore the performance of combined multiple models at the decision level. The rest of this paper is organized as follows: Sect. 2 offers a brief overview of the existing research work that is related to this research; in Sects. 3, 4 the proposed approach is described in detail; experimental results are given in Sect. 5 to demonstrate the feasibility and performance of the proposed method; and, finally, a brief conclusion and future works are presented in Sect. 6.

2   Related Works

2.1   Traditional Methods

Traditional methods including rules-based methods and statistical machine learning based methods. Most of the early methods are rules-based methods, the idea of which is to use empirical rules to map the syntactic structure to the level of the prosodic structure [3]. With the development of statistical machine learning, many machine learning methods are gradually applied to the prediction of prosodic structures, such as decision trees [2] and conditional random fields (CRF) [4].

It is worth mentioning that the CRF model is one of the best models that we still using now, it's a very simple but surprisingly effective tagging model. therefore, in comparison with the traditional methods, we will focus on the CRF model.

The core idea of CRF is not complicated, for sequence labeling tasks, it is often beneficial to explicitly consider the correlations between adjacent labels [28]. Correlations between adjacent labels can be modeled as a transition matrix $T \in \mathbb{R}^{C \times C}$. Given a sentence $S = (c_1, c_2, \ldots, c_L)$, we have corresponding scores $E \in \mathbb{R}^{L \times C}$. For a label sequence $y = (y_1, y_2, \ldots, y_L)$, we define it's unnormalized score to be

$$s(S, y) = \sum_{i=1}^{L} E_{i, y_i} + \sum_{i=1}^{L-1} T_{y_i, y_{i+1}} \tag{1}$$

then we can takes the form of linear chain CRF [29], the probability of the label sequence is defined as

$$P(y|S) = \frac{e^{s(S, y)}}{\sum_{y' \in Y} e^{s(S, y')}} \tag{2}$$

where Y is the set of all valid label sequences. Then the loss of the proposed model is defined as the negative log-likelihood of the ground-truth label sequence y^*,

$$L(S, y^*) = -\log P(y^*|S) \tag{3}$$

During training, the loss function is minimized by back propagation. During test, Viterbi algorithm is applied to quickly find the label sequence with maximum probability.

2.2 Deep Neural Networks Based Method

As a sequence labeling problem, prosody prediction can be expressed like other sequence labeling problems as Eq. (4),

$$\hat{y} = \arg\max_y P(y|x) \tag{4}$$

means that we want to find the best label sequence y given an input sequence x.

In practical terms, we will considering all available information from the input and the emitted output sequences, it's turned to find the best parameter set θ that maximizes the likelihood which can be described by the following expression

$$\arg\max_\theta \prod_{t=1}^{T} P(y_t|y_1^{t-1}, x; \theta) \tag{5}$$

where y_1^{t-1} represents the predicted output sequence prior to time step t.

With the rapid development of deep neural networks in recent years, related technologies and models have been applied to many fields [5–7], those methods also been applied to sequence modeling problems [8, 9]. Vadapalli et al. [10] proposed a prosodic structure prediction model based on RNN and added word vectors as semantic features. Experimental results show that RNN-based methods can greatly improve the performance comparing by traditional machine learning methods (such as conditional random fields), and meanwhile, the vector feature of words can well adapt to the model of cyclic neural network. Ding et al. [11] from another perspective, the word (rather than the traditional grammatical word) as the unit of prediction of prosodic structure, use of vector as the RNN input to replace other traditional features directly. The advantage of this approach is it does not rely on the precision of other text analysis.

Recently, encoder-decoder neural network frames have been successfully applied in many sequence learning problems such as machine translation [8] and speech recognition [12]. The framework is firstly introduced in [8, 13] and the encoder and decoder are two separate RNNs. The main idea behind the encoder-decoder frame is to encode input sequence x into a dense vector c. This vector encodes information of the whole source sequence and then use this vector to generate corresponding output sequence, which can be expressed as follows:

$$P(y) = \prod_{t=1}^{T} P(y_t|y_1^{t-1}, c) \tag{6}$$

The attention mechanism introduced in [14] enables the encoder-decoder archi-tecture to learn to align and decode simultaneously.

3   Proposed Approach

3.1   Basic Encoder-Decoder Framework

The prosody prediction task is to generate the boundary labeling sequence y from the word sequence x, Let x_i to represent a word and 0 or 1 represent the Prosodic boundary. Considering the ability to better model long-term dependencies, we use LSTM [15] as the basic recurrent network unit. Furthermore, bidirectional LSTM as the encoder module and unidirectional LSTM as the decoder module. Based on this, the main Framework is illustrated in Fig. 1.

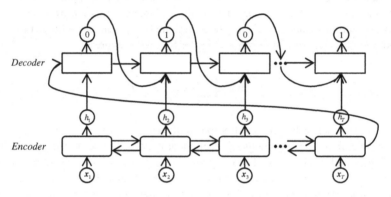

Fig. 1. Encoder-decoder model with aligned input

The encoder reads the word sequence forward and backward. The forward pro-cesses and the backward processes read the word sequence in its original order and reverse order, at the same time, generates forward hidden states and backward hidden states at each time step. The final encoder hidden state at each time step is a con-catenation of the forward states and backward states. The last state of the encoder carries information of the entire input word sequence and we use it and the initial decoder hidden state. The decoder reads the hidden states sequence forward and generate boundary labeling sequence. At each decoding step, the decoder state is calculated as a function of the previous decoder state, the previous emitted label, and the aligned encoder hidden state.

3.2   Encoder-Decoder Framework with Attention

The attention mechanism is proposed mainly to solve the problem of losing hidden states information [8, 13]. Under the encoder-decoder frame, the hidden states carry information of the whole input sequence, but along the forward and backward prop-agation, information may gradually lose. Thus, instead of only utilizing the hidden state at each step, the use of context vector c gives us any additional supporting information, especially those require longer term dependencies that is not being fully captured by the hidden state. Motivated by the use of attention mechanism in encoder decoder frames, we propose the attention-based model for prosody prediction which main framework is illustrated in Fig. 2.

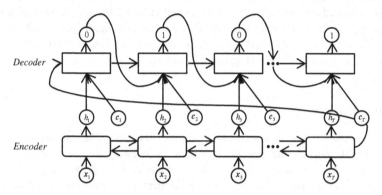

Fig. 2. Encoder-decoder model with aligned input and attention

In the proposed model, we still use the same basic encoder-decoder frame as above, encoder reads the word sequence forward and backward and generate hidden states. The only difference is that we add a context vector c_i at each time step. where the context vector c_i is calculated as a weighted average of the encoder hidden states $h = (h_1, h_2, \ldots, h_T)$. the context vector can be regarded as information that when model focusing on a few hidden states. we reuse the pre-computed hidden states h of the encoder to produce intent class distribution.

There are many studies on the mechanism of attention [16, 17] and which have achieved very good results. We adopt positional attention [18] to calculate attention weights which is suitable for prosody prediction task.

We can describe that progress as follows:

$$e(\boldsymbol{h}_j, \boldsymbol{p}_j) = \boldsymbol{V}^T \tanh(\boldsymbol{W}_H \boldsymbol{h}_j + \boldsymbol{W}_P \boldsymbol{p}_j + \boldsymbol{b}) \tag{7}$$

And,

$$\alpha_j = \frac{\exp(e(\boldsymbol{h}_j, \boldsymbol{p}_j))}{\sum\limits_{k=1}^{K} \exp(e(\boldsymbol{h}_k, \boldsymbol{p}_k))} \tag{8}$$

Finally, the context vector can be expressed by Eq. (9),

$$\boldsymbol{c}_i = \sum\limits_{j=1}^{T} \alpha_j \boldsymbol{h}_j \tag{9}$$

4 Multi-model Ensemble

Ensemble learning achieve a better generalization performance than a single learner by constructing and combining multiple learners. According to the generation of individual learners, the current ensemble learning methods can be broadly divided into two categories, the first is Boosting, it is a strong dependency between the individual learners, serialization methods must be generated in series; the second is Bagging, there is no strong dependencies between individual learner and can be generated in parallel. We use Random Forest(RF) and Gradient Boost Decision Tree (GBDT), the two most representative Boosting and Bagging algorithms as the main method of model ensemble.

4.1 Random Forest

Random Forest, RF [19] is an extended variant of Bagging. The training algorithm for random forests applies the general technique of bagging to tree learners. Given a training set $X = \{x_1, x_1, \ldots, x_n\}$ with responses $Y = \{y_1, y_1, \ldots, y_n\}$, bagging repeatedly selects a random sample with replacement of the training set and fits trees to these samples. After training, predictions for unseen samples x' can be made by averaging the predictions from all the individual regression trees on x':

$$\hat{f} = \frac{1}{T} \sum_{t=1}^{T} f_t(x') \tag{10}$$

or by taking the majority vote in the case of classification trees. Random forest is simple and easy to implement, and leads to better model performance because it decreases the variance of the model, without increasing the bias. This means that while the predictions of a single tree are highly sensitive to noise in its training set, the average of many trees is not, as long as the trees are not correlated. Simply training many trees on a single training set would give strongly correlated trees.

4.2 Gradient Boosting Decision Tree

Gradient Boosting Decision Tree, GBDT [20, 21] is a typical method belonging to Boosting. In the GBDT iteration, suppose the strong learner we obtained in the previous iteration is $f_{t-1}(x)$ and the loss function is $L(y, f_{t-1}(x))$. The goal of our current iteration is to find a CART [22] $T_t(x)$ minimizes the loss,

$$L(y, f_t(x)) = L(y, f_{t-1}(x)) + T_t(x) \tag{11}$$

A decision tree partitions the space of all joint predictor variant into disjoint regions R_j, $j = 1, 2, \ldots, J$. as represented by the terminal nodes of the tree. A constant γ_j is assigned to each such region. Then GBDT is a sum of such trees, where M is the numbers of trees in GBDT,

$$f_M(x) = \sum_{m=1}^{M} T_m(x) \tag{12}$$

GBDT is complicated than RF and can't be generated in parallel, but it may leads to a better performance in some tasks.

4.3 Structure

We adopt 4 classifier: CRF, BLSTM, Basic encoder-decoder frame(aligned model), Attention model as the basic classifier. The framework of multi-model ensemble is shown in Fig. 3.

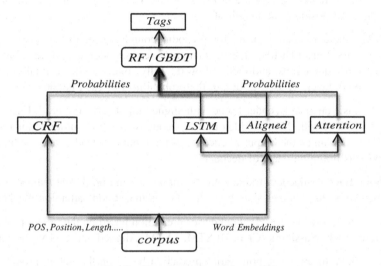

Fig. 3. Flowchart of model ensemble

Firstly, the CRF uses linguistic class features (POS, Position information etc.) and the LSTM, Aligned mode, Attention model use embedding features, then the probability of Breaks (PW, PPH and IPH) can be obtained by these four single classifiers; Next, the output probabilities, together with the important features are consisting of the inputs for model fusion module. During model fusion, two different methods, RF and GBDT are trained and employed to make the final prediction.

5 Experiments

5.1 Settings

Considering that the word is often used as the ideographic unit in Mandarin, which carries more semantic and boundary information than isolated character, we decide use word as basic input units in our experiments.

Datasets. We evaluate our model on a speech synthesis corpus which contains 66150 sentences. The whole corpus is partitioned into training, validation and test set for all experiments according to 7:2:1. Word segmentation and POS tagging are also annotated by expert annotator.

Word Embeddings. Word embeddings [23, 24] lead to a significant improvements over the linguistic features which don't take into account the distributional behavior of words [25]. And this issue has been addressed by word embedding which encodes a word as a low-dimensional vector. We used 8G Mandarin text corpus which word vocabulary size is 393255, The word embedding dimension is set as 128, the context window size is set as 5 during training, and we also use both hierarchical softmax and 5-word negative sampling.

CRF. Same as traditional labeling tasks, we train CRF basically on linguistic features like POS, word position, the length of word, etc.

Bi-LSTM. Standard method used for prosodic boundaries prediction based on Bi-LSTM. We use 2-layer Bi-LSTM as the main time series modeling part. each layer with 256x2 LSTM hidden units, and then followed a fully connected layer, finally, a binary output softmax layers is used to output the probabilities of each boundary.

Aligned. Basic encoder-decoder frame with aligned input show in Fig. 2. Encoder is a 1-layer Bi-LSTM with 256x2 hidden units. Decoder is a 1-layer unidirectional LSTM with 256 hidden units. the output of encoder as the input of decoder, and the final state of encoder as the initial state of decoder.

Attention. Encoder-decoder frame with attention show in Fig. 3. The main structure is the same as LSTM_Aligned, the only difference is that it adds attention mechanism.

RF. Random forest for prosodic boundaries prediction based on the output probability of the four single classifiers (CRF, BLSTM, LSTM_Aligned and LSTM_Attention).

GBDT. GBDT for prosodic boundaries prediction based on the output probability of the four single classifiers (CRF, BLSTM, LSTM_Aligned and LSTM_Attention).

All networks mentioned above are trained with Adam [26], the initial learning rate been set to 0.003 and reset the learning rate by exponential decay (decay rate set to 0.02) for every epoch. The batch size is 20.

5.2 Main Result

Table 1 gives the performances of all six methods described in 5.1, We report our results in terms of F1-Score, which is defined as the harmonic mean of precision and recall. We analyze the results below.

First, we compare standard statistical model CRF with the standard neural model LSTM. In Table 1, we can see clearly that the Bi-LSTM model perform well in all the hierarchies. Many successful research and experiences also shown that LSTM-based model performs better than statistical methods for complex annotation tasks.

Then, we compare Bi-LSTM, Aligned model and Attention model. These methods are all based on LSTM, the differences is the structure and whether the attention

Table 1. Performance of F1-Score

	CRF	Bi-LSTM	Aligned	Attention	RF	GBDT
PW	95.02	95.61	96.52	96.76	97.13	**97.22**
PPH	82.04	82.25	83.82	83.86	84.84	**84.91**
IPH	78.80	80.42	83.49	83.62	83.86	**83.89**

mechanism been applied. We can infer from the Table 1 that the aligned encoder-decoder structure perform better than basic Bi-LSTM, we suspect that compared to the use of LSTM to do classification, the prosody prediction task is more biased towards the seq2seq architecture. Not to our surprise, the model with attention perform best in all of them, and, the higher the prosodic level (such as IPH), the more obvious the effect of improvement. The probable reason for this is that for IPH, a longer span of contextual information is needed to make the correct prediction. This is the same with IPH labeling, the same need to consider the entire sentence information to give the final annotation results. With respect to aligned based encoder-decoder, attention-based encoder-decoder is precisely this predictive-enabling contextual information.

Finally, we can see from Table 1 that the integrated model (RF and GBDT) achieves superior performance than all the single models and GBDT shows an absolute increase of around 5% than CRF for IPH prediction. It can be explained by ensemble learning can learn the advantages of all the individual models, prevent over-fitting effectively and improve the generalization performance. In the comparison between GBDT and RF, GBDT performed slightly better, in fact the performance of the two are very close, all showed the ability of ensemble learning should have.

Considering that the linguistic class features can be also applied into the deep learning models, if we combine the word embeddings and all the linguistic feature together we can get a new word embeddings. Specifically, we concatenate the word POS tag, position, length, cumulative length and original word embeddings by last dimension and use this new word embeddings to train and test all the models, the result shown in Table 2.

Table 2. Performance of F1-Score on muilti-features

	CRF	Bi-LSTM	Aligned	Attention	RF	GBDT
PW	95.02	95.75	97.03	97.11	97.25	**97.26**
PPH	82.04	82.31	84.21	84.31	85.01	**85.06**
IPH	78.80	80.46	83.63	83.70	83.92	**83.96**

By comparing Tables 1 and 2, we can see a slight improvement in the performance of each model, and the convergence of multiple features is helpful for the final performance improvement. To evaluate the effects of four single classifiers for the best ensemble method GBDT. We further calculate the contribution of each feature by Gini importance [27], which is used as a general indicator of feature importance. Take IPH prediction for example, the degree of the top five features contributions for IPH prediction on word-based GBDT are listed in Table 3.

This means the output from the LSTM based method are playing the dominant role rather than CRF. The previous results in Table 1, we can see that the Attention-based approach can lead to more accurate results, and it does contribute the most to all methods in this ensemble method.

Table 3. Model contribution in GBDT

Methods	Contributions
CRF	12.96%
Bi-LSTM	22.43%
Aligned	31.77%
Attention	**32.84%**

6 Conclusions

In this paper, we explored strategies in utilizing explicit alignment information in the attention-based encoder-decoder neural network models for Mandarin prosodic boundaries prediction, and we also use ensemble learning to further enhance the generalization performance of this model. Our results show that attention-based model is more suitable than basic LSTM approach, compared to a single model, the use of multi-model ensemble can bring very large performance improvements. Meanwhile, the model fusion results indicate that the dependency of results on BLSTM is greater than CRF, and the features generated from feature ranking module can further boost the performance of prosodic boundaries prediction. In the future, We have two paths to go, the first is try to deepen the encoders and decoders to see if it can learn more abstract concepts. The second is to change the type of encoder and decoder, for example, we can use CNN to be the encoder or the decoder.

References

1. Taylor, P.: Text-to-Speech Synthesis. Cambridge University Press, Cambridge (2009)
2. Chu, M., Qian, Y.: Locating boundaries for prosodic constituents in unrestricted Mandarin texts. Int. J. Comput. Linguist. Chin. Lang. Process. **6**(1), 61–82 (2001). Special Issue on Natural Language Processing Researches in MSRA
3. Truckenbrodt, H.: Phonological phrases–their relation to syntax, focus, and prominence. Massachusetts Institute of Technology (1995)
4. Levow, G.A.: Automatic prosodic labeling with conditional random fields and rich acoustic features. In: Proceedings of the Third International Joint Conference on Natural Language Processing, vol. I (2008)
5. Geng, Y., Liang, R.Z., Li, W., Wang, J., Liang, G., Xu, C., Wang, J.Y.: Learning convolutional neural network to maximize pos@top performance measure. In: ESANN 2017 - Proceedings, pp. 589–594 (2016)

6. Geng, Y., Zhang, G., Li, W., Gu, Y., Liang, R.Z., Liang, G., Wang, J., Wu, Y., Patil, N., Wang, J.Y.: A novel image tag completion method based on convolutional neural transformation. In: International Conference on Artificial Neural Networks, pp. 539–546 (2017)

7. Zhang, G., Liang, G., Li, W., Fang, J., Wang, J., Geng, Y., Wang, J.Y.: Learning convolutional ranking-score function by query preference regularization. In: International Conference on Intelligent Data Engineering and Automated Learning, pp. 1–8 (2017)

8. Sutskever, I., Vinyals, O., Le, Q.V.: Sequence to sequence learning with neural networks. In: Advances in Neural Information Processing Systems, pp. 3104–3112 (2014)

9. Mikolov, T., Kombrink, S., Burget, L., et al.: Extensions of recurrent neural network language model. In: 2011 IEEE International Conference on Acoustics, Speech and Signal Processing (ICASSP), pp. 5528–5531. IEEE (2011)

10. Vadapalli, A., Prahallad, K.: Learning continuous-valued word representations for phrase break prediction. In: Fifteenth Annual Conference of the International Speech Communication Association (2014)

11. Ding, C., Xie, L., Yan, J., et al.: Automatic prosody prediction for Chinese speech synthesis using BLSTM-RNN and embedding features. In: 2015 IEEE Workshop on Automatic Speech Recognition and Understanding (ASRU), pp. 98–102. IEEE (2015)

12. Chan, W., Jaitly, N., Le, Q., et al.: Listen, attend and spell: a neural network for large vocabulary conversational speech recognition. In: 2016 IEEE International Conference on Acoustics, Speech and Signal Processing (ICASSP), pp. 4960–4964. IEEE (2016)

13. Cho, K., Van Merriënboer, B., Gulcehre, C., et al.: Learning phrase representations using RNN encoder-decoder for statistical machine translation. arXiv preprint arXiv:1406.1078 (2014)

14. Bahdanau, D., Cho, K., Bengio, Y.: Neural machine translation by jointly learning to align and translate. arXiv preprint arXiv:1409.0473 (2014)

15. Hochreiter, S., Schmidhuber, J.: Long short-term memory. Neural Comput. 9(8), 1735–1780 (1997)

16. Li, H., Min, M.R., Ge, Y., et al.: A context-aware attention network for interactive question answering. In: Proceedings of the 23rd ACM SIGKDD International Conference on Knowledge Discovery and Data Mining, pp. 927–935. ACM (2017)

17. Wang, X., Yu, L., Ren, K., et al.: Dynamic attention deep model for article recommendation by learning human editors' demonstration. In: Proceedings of the 23rd ACM SIGKDD International Conference on Knowledge Discovery and Data Mining, pp. 2051–2059. ACM (2017)

18. Chen, Q., Hu, Q., Huang, J.X., et al.: Enhancing recurrent neural networks with positional attention for question answering. In: Proceedings of the 40th International ACM SIGIR Conference on Research and Development in Information Retrieval, pp. 993–996. ACM (2017)

19. Breiman, L.: Random forests. Mach. Learn. 45(1), 5–32 (2001)

20. Friedman, J.H.: Greedy function approximation: a gradient boosting machine. Ann. Stat. 29, 1189–1232 (2001)

21. Friedman, J.H.: Stochastic gradient boosting. Comput. Stat. Data Anal. 38(4), 367–378 (2002)

22. Breiman, L., Friedman, J., Stone, C.J., et al.: Classification and Regression Trees. CRC Press, Boca Raton (1984)

23. Mikolov, T., Sutskever, I., Chen, K., et al.: Distributed representations of words and phrases and their compositionality. In: Advances in Neural Information Processing Systems, pp. 3111–3119 (2013)

24. Pennington, J., Socher, R., Manning, C.: Glove: global vectors for word representation. In: Proceedings of the 2014 Conference on Empirical Methods in Natural Language Processing (EMNLP), pp. 1532–1543 (2014)
25. Watts, O., Yamagishi, J., King, S.: Unsupervised continuous-valued word features for phrase-break prediction without a part-of-speech tagger. In: Twelfth Annual Conference of the International Speech Communication Association (2011)
26. Kingma, D.P., Ba, J.: Adam: A method for stochastic optimization. arXiv preprint arXiv: 1412.6980 (2014)
27. Menze, B.H., Kelm, B.M., Masuch, R., et al.: A comparison of random forest and its Gini importance with standard chemometric methods for the feature selection and classification of spectral data. BMC Bioinf. 10(1), 213 (2009)
28. Collobert, R., Weston, J., Bottou, L., et al.: Natural language processing (almost) from scratch. J. Mach. Learn. Res. 12, 2493–2537 (2011)
29. Lafferty, J., McCallum, A., Pereira, F.C.N.: Conditional random fields: probabilistic models for segmenting and labeling sequence data (2001)

Chinese Text Detection Using Deep Learning Model and Synthetic Data

Wei-wei Gao[1], Jun Zhang[1(✉)], Peng Chen[2], Bing Wang[3], and Yi Xia[1]

[1] School of Electrical Engineering and Automation, Anhui University,
Hefei 230601, Anhui, China
wwwzhangjun@163.com
[2] Institute of Health Sciences, Anhui University, Hefei 230601, Anhui, China
[3] School of Electrical and Information Engineering,
Anhui University of Technology, Ma Anshan 243032, China

Abstract. Detection of text in natural scene images is very challenging, and it is not completely solved. In this work we propose a fast and reliable algorithm to generate synthetic data of Chinese characters in images. The proposed algorithm make the text content cover the background in a natural way. To validate the proposed method effective, another dataset are generated by ordinary fusion method. Two dataset are used to train Faster-RCNN network. And the experimental result shows that the dataset are generated by proposed method achieve a better performance of detection than the normal way.

Keywords: Synthetic data · Text detection · Faster R-CNN

1 Introduction

Since ancient times, text, as one of the most important inventions of mankind, has played a significant role in human life. In everyday life, words are present through the universe, they can be found on documents, signposts, screens, advertising board, billboard and other objects such as cars or smart phones. Therefore, automatically detecting and reading text from real-time scene images is obviously an important and interesting part of the system, which can be used for several challenging tasks such as image-based machine translation, automatic car or video indexing.

While the detection and recognition of text within scanned documents is well studied and that have many Optical Character Recognition (OCR) systems which perform pretty well. However, spotting and reading texts by machine in natural scenes are quite difficult tasks. The main challenges in scene text detection and recognition can be roughly divide into three types: diversity of scene text, in document there are often with regular font, single color, almost same size and uniform arrangement, but in the natural scenes we may bear entirely different fonts, colors, scales and orientations, even in the same images. They also have much more complexity of background, inevitably there are various interference factors, for instance: noise, blur, distortion, low-resolution, non-uniform illumination and partial occlusion, and all of these may raise the rate of failures in text detection and recognition.

© Springer International Publishing AG, part of Springer Nature 2018
D.-S. Huang et al. (Eds.): ICIC 2018, LNCS 10954, pp. 503–512, 2018.
https://doi.org/10.1007/978-3-319-95930-6_46

To effectively detect and read natural scene text, the existed research works [2–4] developed end-to-end scene text recognition systems that generally consist of complicated two pipelines where the first is word detection that will generates a large set of word bounding box candidates. Previous work uses sliding window methods or region grouping methods are very successfully for detection. Next, these candidate detections are waiting for recognition.

In this paper we hit it from different perspective, inspired by [5] we would use a method which will generate synthetic images of Chinese characters that naturally blends text in existing natural scenes, and we will use depth learning methods and segmentation techniques that can be reach-me-down to align the characters with the geometry of the background image and are limited by the scene boundaries. And then as a contrast we generated another dataset that we regard the images which contain only Chinese character as foreground and the scene images as background and then we just fuse them. Next, we will use them to train the end-to-end network the Faster R-CNN.

2 Related Work

The synthetic data not only provides details of ground truth annotations, but also provides a cheap and scalable alternative to manually annotating images. They were often used diffusely to learn CNN models Wang et al. [6] and Jaderberg et al. [7] used synthetic data to train word-image recognition networks. Dosovitskiy et al. [8] used floating chair renderings to train dense optical flow regression networks. Detailed synthetic data has also been used to learn generative models — Dosovitskiy et al. [9] trained inverted CNN models to render images of chairs, while Yildirim et al. [10] used deep CNN features trained on synthetic face renderings to regress pose parameters from face images.

Traditional methods of text recognition are based on the sequential character classification of sliding windows [11] or components, and word prediction is then performed by grouping the borders in a bottom-up manner. Sliding window classifiers include Wang et al. [12] random forest [13] and CNN model. Recent work, such as [14] using over-segmentation, guided by supervised classifiers, generated candidate proposals that were later classified as characters or false positives. For example, Photo-OCR [15] used binarization and sliding window classifiers to generate candidate character regions to perform character recognition through string search driven by category scores, and then sort them using a 100000 word dictionary. [8] Used the CNN's convolutional property to generate a feature map for the classifier to score the dictionary words.

Our text detection network draws primarily on Faster R-CNN [1]. While the others proposed methods of using deep networks for locating class-specific or class-agnostic bounding boxes [16–19] in over-feat way, a fully-connected(FC) layer trained to predict the box coordinates for the localization task that assume a single object, bout R-CNN framework [20] combination of region proposals and features, the R-CNN has three steps-(1) generating object proposals, (2) Extracting CNN feature maps for each of proposal. (3) Filtering the proposals through class specifics.

3 Process of Generating Synthetic Data

3.1 Image Segmentation

In this paper we use ready-made deep learning, segmentation technology to arrange words into natural scene pictures, which can naturally integrate words into natural scene pictures.

In a real natural scene image, text rarely crosses two different areas. Therefore, we shall first divides the natural background image just searched, according to local color and texture, and divides it into continuous regions, and embeds the text in one of the regions.

In real images text tends to be contained in well defined regions. We approximate this constraint by requiring text to be contained in regions, in order to obtain the regions we first need use the gPb-UCM [21] algorithm to segment pictures.

First of all, we need the global probability of boundary which calculate the probability of each pixel as a boundary, which is the weight of the pixel. And then we need to convert the above gPb results to multiple closed regions. In the end, converted the above-mentioned regions set to hierarchical tree.

The basic building block of the Pb Contour Detector is to calculate the directional gradient signal G from the intensity image I(x, y). This calculation is performed by placing the disk at a position (x, y) which is divided into two half-discs at an angular diameter. For each half-disc, we analyze the histogram of the n intensity value of the pixels it covers. The gradient value G at the position (x, y) is defined by the χ^2 distance between the two half-disk histograms u and d:

$$\chi^2_{(u,d)} = \frac{1}{2} \sum_n \frac{(g(n) - h(n))^2}{g(n) - h(n)} \tag{1}$$

Then, we apply second-order filtering to enhance the local maximum and smooth multiple detection peaks in the orthogonal direction θ. We show the gPb signal:

$$gPb(x, y, \theta) = \sum_m \sum_n \alpha_{n,m} G_{m,\mu}(x, y, \theta) + \lambda \cdot sPb(x, y, \theta) \tag{2}$$

We subsequently rescale *gpb* using a sigmoid to match a probabilistic interpretation. Where the weights $\alpha_{n,m}$ and λ are learned by gradient ascent on the F-measure using the BSDS training images, and μ represent the multiple diameter, and θ represent the diameter's orientation.

3.2 Depth Estimation of Image

Text in natural scenes is generally on the surface of the background image, for the same effect, the text should be "placed" according to local surface normal, specifically, we need first obtains a pixel-level depth image through the CNN model proposed by [22]. Then use RANSAC [23] to fit the plane perpendicular to the normal vector. Then the text can be placed on the plane, so that the text is more naturally integrated into the background image.

The normal is estimated by first predicting one automatically dense depth map using [22] CNN as region split on top, then use a flat facet RANSAC [23].

When placing multiple texts in the same region, check that if there is an overlap between the text and the text. That is, a text should not be overlaid on the previous text. Note that not all segmented regions are suitable for embedding text, Such region should not be too small, the aspect ratio of the region shall not be too disparity. In addition, the regions have too much texture is not suitable for embedded text, the number of textures here, is measured by the strength of the third derivative of the RGB image.

In the processing here, we use the CNN model of [22] to estimate the depth image of a natural image. However, one mistake here we often made is that we often use the RGBD data to do it directly, ignoring the CNN depth estimation process. Here we use CNN to estimate a depth of image instead of using the data of RGBD directly.

Here in this, we will show some details about the proposed deep convolution model. Let i be the image, $x = [x_1, \ldots, x_n]^T$ be a vector of successive depth value corresponding to all n super-pixels in I similar to the traditional CRF, we model the conditional probability distribution of data with the following density function:

$$pr(x|i) = \frac{1}{z(i)} \exp\{-E(x, i)\} \tag{3}$$

In which E is the energy function, Z (i) is a partition function, respectively defined as:

$$E(x, i) = \sum_{a \in M} (x_a - y_a)^2 + \sum_{(a,b \in s)} \frac{1}{2} G_{ab}(x_a - x_b)^2 \tag{4}$$

$$Y(i) = \int_x \exp\{-E(x, i)\} dx \tag{5}$$

Where,

$$G_{ab} = \sum_{n=1}^{N} \alpha_n S_{ab}^{(n)} \tag{6}$$

In which y is the regression depth parameterized by θ (namely, y is the abbreviation of $Y(\theta)$, $\alpha = [\alpha_1, \ldots \alpha_k]$) the pairwise parameter, $S^{(n)}$ the n-th similarity matrix and the number of N pairwise entries were considered. To guarantee Y(i) is holonomic we require $\alpha_n > 0$. Our goal is to learn together $Y(\theta)$ and α here (Fig. 1).

Fig. 1. Here are some examples of the picture's depth. The up is the raw images which, and the down is the RGB predicted depth using the CNN model.

4   Fast Text Detection Network

This section we would like to use Faster R-CNN to train and test the data we have. Faster R-CNN is a deep learning based end-to-end object detection framework that does not require double counting. These advantages make R-CNN work faster and achieve higher target recognition accuracy. You can train a faster R-CNN network for object detection and return the object's position. In feature extraction, faster R-CNNs use ImageNet's classification network as a front-end network. The faster R-CNN provides three different network models with different levels of convolutional layers. By entering an image into a faster R-CNN network, you can get a feature map from multiple convolution layers in your front network. Faster R-CNN design an RPN (Regional Proposal Network) layer instead of a full connection, The RPN network is a fully convolutional network [13], which uses back-propagation and SGD (Stochastic Gradient Degradation) [14] to train the data end-to-end (Fig. 2).

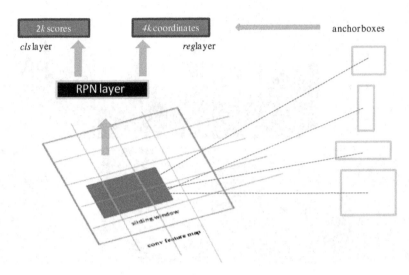

Fig. 2. Region Proposal Network (RPN).

5   Experiments and Discussions

In this work we use two datasets to train network respectively, one is fusion data another one is synthetic data to compare these two performance. All these images were crawled from Google Image Search through queries related to different objects/scenes and indoor/outdoor and natural/artificial locales (Fig. 3).

Fig. 3. The above shows partial synthetic data and its label.

As for the fusion data we searched the pictures from Baidu and from the novel websites we download the text Grimm's Fairy Tales (Fig. 4 and Tables 1 and 2).

At this work we use Approximate joint training (end-2-end) rather than Alternate training (alt-opt) as the end2end use of memory is smaller, and training will be faster, at the same time the accuracy of almost. Next we will check the loss of both (Figs. 5 and 6).

Fig. 4. The above shows partial fusion data.

Table 1. The two different data's train results, there is three different depth of networks were used and they respectively are ZF, VGG_CNN_M_1024 and VGG16. decimal fraction represent the mAP(mean Average Precision) to evaluate the experimental results, as we could figure it out that the deeper network the better performance, Synthetic data training performance better than Fusin data as well.

Model	Synthetic data	Fusion data
ZF	0.812	0.723
VGG_CNN_M_1024	0.831	0.768
VGG 16	0.874	0.809

Table 2. Test data all from Scene pictures and use pre-trained module by synthetic data and fusion data to apply the test.

Data	Average test accuracy
Synthetic data	0.713
Fusion data	0.634

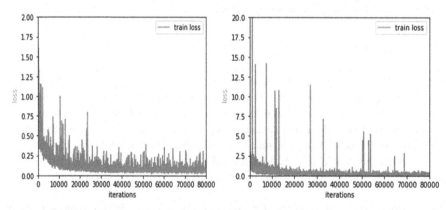

Fig. 5. *Left*: the loss curve of our Synthetic data. *Right*: the loss curve of our Fusion data. And both of them were trained by end2end. Even though the start loss of both is different it all converged in the end.

Fig. 6. Partial raw (before processed) tested result. These results are raw results and if the probability of word boxes is equal or greater than 0.5 then the boxes of word would be showed. And we obtained the overall accuracy by average every single word's accuracy.

6   Conclusion

In this work we have already trained the Faster R-CNN network use two different data, our Synthetic data overlay synthetic text to existing background images in a natural way, accounting for the local 3 dimensional scene geometry. Our Fusion data just fuse the text and background pictures together, so perhaps a little bit of not in a very natural way. And our Fusion data may trained as well but when it comes to test phase it performances not much well. And we trained the network all in end2end way since it would take up smaller memory and faster shared the same accuracy as well.

Acknowledge. This work is supported by Anhui Provincial Natural Science Foundation (grant number 1608085MF136).

References

1. Ren, S., He, K., Girshick, R., et al.: Faster R-CNN: towards real-time object detection with region proposal networks. In: International Conference on Neural Information Processing Systems, pp. 91–99. MIT Press (2015)
2. Geng, Y., Liang, R.Z., Li, W., et al.: Learning convolutional neural network to maximize Pos@Top performance measure (2016)
3. Geng, Y., et al.: A novel image tag completion method based on convolutional neural transformation. In: Lintas, A., Rovetta, S., Verschure, P.F.M.J., Villa, A.E.P. (eds.) ICANN 2017. LNCS, vol. 10614, pp. 539–546. Springer, Cham (2017). https://doi.org/10.1007/978-3-319-68612-7_61
4. Zhang, G., et al.: Learning convolutional ranking-score function by query preference regularization. In: Yin, H., et al. (eds.) IDEAL 2017. LNCS, vol. 10585, pp. 1–8. Springer, Cham (2017). https://doi.org/10.1007/978-3-319-68935-7_1
5. Gupta, A., Vedaldi, A., Zisserman, A.: Synthetic data for text localization in natural images, 2315–2324 (2016)
6. Wang, T., Wu, D.J., Coates, A., Ng, A.Y.: End-to-end text recognition with convolutional neural networks. In: Proceedings ICPR, pp. 3304–3308 (2012)
7. Jaderberg, M., Simonyan, K., Vedaldi, A., Zisserman, A.: Synthetic data and artificial neural networks for natural scene text recognition. In: Workshop on Deep Learning, NIPS (2014)
8. Dosovitskiy, A., Fischery, P., Ilg, E., et al.: FlowNet: learning optical flow with convolutional networks. In: IEEE International Conference on Computer Vision, pp. 2758–2766. IEEE Computer Society (2015)
9. Dosovitskiy, A., Brox, T.: Inverting visual representations with convolutional networks, pp. 4829–4837 (2015)
10. Yildirim, I., Kulkarni, T., Freiwald, W., et al.: Efficient analysis-by-synthesis in vision: a computational framework, behavioral tests, and comparison with neural representations. In: Conference of the Cognitive Science Society (2015)
11. Jaderberg, M., Vedaldi, A., Zisserman, A.: Deep features for text spotting. In: Fleet, D., Pajdla, T., Schiele, B., Tuytelaars, T. (eds.) ECCV 2014. LNCS, vol. 8692, pp. 512–528. Springer, Cham (2014). https://doi.org/10.1007/978-3-319-10593-2_34
12. Ozuysal, O.M., Fua, P., Lepetit, V.: Fast keypoint recognition in ten lines of code. In: IEEE Computer Society Conference on Computer Vision and Pattern Recognition, pp. 1–8. DBLP (2007)
13. Wang, K., Babenko, B., Belongie, S.: End-to-end scene text recognition. In: IEEE International Conference on Computer Vision, pp. 1457–1464. IEEE (2012)
14. Alsharif, O., Pineau, J.: End-to-end text recognition with hybrid HMM maxout model. Comput. Sci. (2013)
15. Bissacco, A., Cummins, M., Netzer, Y., et al.: PhotoOCR: reading text in uncontrolled conditions. In: IEEE International Conference on Computer Vision, pp. 785–792. IEEE (2014)
16. Szegedy, C., Toshev, A., Erhan, D.: Deep neural networks for object detection. Adv. Neural. Inf. Process. Syst. **26**, 2553–2561 (2013)
17. Sermanet, P., Eigen, D., Zhang, X., Mathieu, M., Fergus, R., LeCun, Y.: OverFeat: integrated recognition, localization and detection using convolutional networks. In: ICLR (2014)
18. Erhan, D., Szegedy, C., Toshev, A., et al.: Scalable object detection using deep neural networks. **3**(4), 2155–2162 (2013)

19. Szegedy, C., Reed, S., Erhan, D., et al.: Scalable, high-quality object detection. Comput. Sci. (2014)
20. Girshick, R., Donahue, J., Darrell, T., Malik, J.: Rich feature hierarchies for accurate object detection and semantic segmentation. In: CVPR (2014)
21. Arbelaez, P., Maire, M., Fowlkes, C., Malik, J.: Contour detection and hierarchical image segmentation. IEEE PAMI **33**, 898–916 (2011)
22. Liu, C.S., Lin, G.: Deep convolutional neural fields for depth estimation from a single image. In: Proceedings CVPR (2015)
23. Fischler, M.A., Bolles, R.C.: Random sample consensus: A paradigm for model fitting with applications to image analysis and automated cartography. Comm. ACM **24**(6), 381–395 (1981)

Learning Misclassification Costs for Imbalanced Datasets, Application in Gene Expression Data Classification

Huijuan Lu[1], Yige Xu[1], Minchao Ye[1(✉)], Ke Yan[1], Qun Jin[2], and Zhigang Gao[3]

[1] College of Information Engineering, China Jiliang University,
258 Xueyuan Street, Hangzhou 310018, China
yeminchao@cjlu.edu.cn
[2] Faculty of Human Sciences, Waseda University, Tokorozawa 359-1192, Japan
[3] College of Computer Science, Hangzhou Dianzi University,
Hangzhou 310018, China

Abstract. Cost-sensitive algorithms have been widely used to solve imbalanced classification problem. However, the misclassification costs are usually determined empirically, leading to uncertain performance. Hence an effective method is desired to automatically calculate the optimal cost weights. Targeting at the highest weighted classification accuracy (WCA), we propose two approaches to search for the optimal cost weights, including grid searching and function fitting. In experiments, we classify imbalanced gene expression data using extreme learning machine to test the cost weights obtained by the two approaches. Comprehensive experimental results show that the function fitting is more efficient which can well find the optimal cost weights with acceptable WCA.

Keywords: Cost-sensitive · Misclassification cost · Correct classification rate
Parameter fitting

1 Introduction

Classification of gene expression data reveals tremendous information in various application fields of biomedical research, such as cancer diagnosis, prognosis and predictions [1]. The classification of gene expression data is a cost-sensitive problem [2]. For imbalanced datasets, traditional classification algorithms with the correct classification rates (CCR) may bias towards the majority classes. To solve this problem, cost sensitive learning (CSL) can be introduced.

In this paper, we utilize weighted classification accuracy (WCA) as the measurement of cost-sensitive classification performance. The cost weights that lead to optimal classification performance are learned by grid searching. Function fitting is adopted to find the optimal cost weights. Experimental results show that function fitting is an effective way to find the optimal cost weights, targeting at high WCA.

© Springer International Publishing AG, part of Springer Nature 2018
D.-S. Huang et al. (Eds.): ICIC 2018, LNCS 10954, pp. 513–519, 2018.
https://doi.org/10.1007/978-3-319-95930-6_47

2 Related Works

CSL remains as one of the most active topics in the field of machine learning. Many works have studied on CSL and embedded the misclassification costs into various classifiers, such as the decision trees (DTs), support vector machines (SVMs) and extreme learning machines (ELMs). Lu et al. [3] made use of the cost-sensitive DTs as base classifiers and constructed a cost-sensitive rotational forest. The experiments performed on gene expression datasets showed that the cost-sensitive rotational forest not only guaranteed the classification accuracy but also reduced the misclassification costs.

Cao et al. [4] proposed to embed evaluation measures into the objective function to improve the performance of a cost-sensitive support vector machine (CS-SVM). It successfully assigned different misclassification costs to different label sets for reducing the overall misclassification cost. Zheng et al. [5] formally applied cost-sensitivity to extreme learning machine (ELM). Yan et al. [6, 7] extended Zheng et al.'s work and introduced a cost-sensitive dissimilar ELM (CS-D-ELM). The CS-ELM algorithms guarantee the classification accuracy and reduce the misclassification cost. Incremental results show that the CS-ELM has better performance in terms of accuracy, cost, efficiency and robustness over other existing classifiers.

3 Backgrounds

3.1 Classical Definition of Cost Matrix

Considering the binary classification problem, the confusion matrix shows four types of classification results according to the prediction values, namely, true positive (TP), false positive (FP), false negative (FN) and true negative (TN). The CSL seeks the overall minimum cost by introducing sensitive costs. This work focuses on the mis-classification cost. We define the minority class as positive (P), the majority class as negative (N), and construct the cost matrix C as follows:

Table 1. Cost matrix

Actual	Predicted	
	P	N
P	C_{00}	C_{01}
N	C_{10}	C_{11}

In Table 1, C_{00} and C_{11} are the cost of correct classification, which are set to 0. C_{01} and C_{10} are the cost of error classifications.

3.2   Correct Classification Rates Versus Weighted Classification Accuracy

For classical machine learning problems, the classification accuracy always refers to the correct classification rate [8], which is the proportion of all correct classified samples:

$$OA = \frac{TP + TN}{TP + FN + TN + FP} \times 100\% \tag{1}$$

However, for imbalanced datasets where the numbers of positive and negative samples differ significantly, the CCR might be misleading. Embedding a weight w_i into the i-th class, enforcing $w_1 + w_2 = 1$, we get the WCA as:

$$WCA = w_1 \cdot \frac{TP}{TP + FN} + w_2 \cdot \frac{TN}{TN + FP} \tag{2}$$

Equation (2) can be easily extended to multi-classification problems:

$$WCA_n = \sum_{i=1}^{n} w_i \frac{CM_i}{M_i}, \sum_{i=1}^{n} w_i = 1 \tag{3}$$

where n denotes the number of sample classes, M_i ($i = 1, 2,..., n$) denotes the number of the i-th class sample, and CM_i ($i = 1, 2,..., n$) denotes the number of the i-th class samples that are classified correctly. Since the WCA is more accurate describing the classification accuracy, we use the WCA to evaluate the classification performance of cost-sensitive classifiers.

4   Optimal Cost Weights Searching

From the University of California Irvine (UCI) standard classification dataset, we choose Leukemia, Colon, Prostate, Lung and Ovarian gene datasets for cost weights searching and further test, which are listed in Table 2.

Table 2. Specifications of datasets

Dataset	Sample number	Feature dimension	Number of classes
Leukemia	34	7130	2
Colon	62	2000	2
Prostate	136	12600	2
Lung	181	12533	2
Ovarian	253	15154	2

4.1   Optimal Cost Weights Searching by Grid Searching Algorithm

The optimal weights are searched by an adaptive algorithm using grid searching. There are two crucial factors to consider: the sample importance w and sample categorical distribution p. p is the proportion between the number of positive class and negative class in test set. It is necessary to study the relationship between the three factors, namely, w, p and WCA, where WCA is the fitness value for the grid searching. The grid searching can be described as follows:

1. Set the searching region as M, searching step size as T, and initial position as P_0;
2. Calculate the fitness of the current position, record the position P_{max} that has the best fitness f_{max} (f_{max} = WCA);
3. Update current location, $P = P+T$;
4. If the current fitness value is greater than f_{max}, update f_{max} and P_{max};
5. Return f_{max} and P_{max}.

Cost-sensitive extreme learning machine (CS-ELM) is a kind of ELM, which attaches a cost matrix on output layer. In this research, we set the number of hidden neurons at 10. Seven different gene expression datasets are used to obtain the classification results with CS-ELM as the classifier. CS-ELM minimizes the conditional risk by embedding misclassification cost in ELM.

$$\arg\min R(i|x) = \arg\min \sum_j P(j|x) \cdot C(i,j) \qquad (4)$$

where $R(i|x)$ is the conditional risk that sample x assigned to the class i, and $P(j|x)$ is the conditional probability that x belongs to j, $C(i, j)$ is the risk of misclassifying j to class i, and $i, j \in \{c_1, c_2, ..., c_m\}$, m is the number of classification categories.

The sample distribution p, the optimal weight $w_c = C_{01}/C_{10}$ and the highest fitness value of each dataset are plotted in Fig. 1.

Fig. 1. Optimal weights for different data set

4.2   Optimal Cost Weights Searching by Function Fitting

Here we use w and p as independent variables, and propose a function fitting as:

$$w_c = f(w, p) \tag{5}$$

where $w_c = C_{01}/C_{10}$, $w = w_1/(w_1 + w_2)$ and p represents the proportion of positive and negative classes. We set C_{10} to 1 to reduce the complexity of calculation, i.e., $f_c = C_{01}$.

We use an automatic fitting software named 1STOPT to do the function fitting [9]. In 1STOPT, Levenberg-Marquardt and Universal Global Optimization are used to fit functions. We compared 500 functions with different types, and selected the best function with the highest correlation coefficient:

$$
\begin{aligned}
w_{c1} &= f_1(w, p) \\
&= \frac{a_1 + a_2 \cdot w + a_3 \cdot w^2 + a_4 \cdot w^3 + a_5 \cdot a_{12} \cdot \ln p + a_6 \cdot (a_{12} \cdot \ln p)^2}{1 + a_7 \cdot w + a_8 \cdot w^2 + a_9 \cdot a_{12} \cdot \ln p + a_{10} \cdot (a_{12} \cdot \ln p)^2 + a_{11} \cdot (a_{12} \cdot \ln p)^3}
\end{aligned} \tag{6}
$$

where $a_1 = 1.323, a_2 = -2.278, a_3 = 3.047, a_4 = -1.286, a_5 = -1.746, a_6 = 0.998$, $a_7 = -0.400, a_8 = 0.369, a_9 = -2.606, a_{10} = 2.544, a_{11} = -0.818, a_{12} = 0.482$. The correlation coefficient R of f is 0.96346. Figure 2 shows the comparison of the three-dimensional interpolation of optimal weights and fitting function.

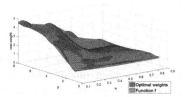

Fig. 2. The values of function w_c compared with the optimal weights

5   Classification Experiments

To compare the cost weights obtained from the grid searching and function fitting, we tested WCAs with four different datasets, namely, Ovarian, Prostate, Lung1 and Lung2. The majority over minority class proportion of datasets are 1.68, 2.5, 5 and 8 respectively. All WCAs are computed using ELM as the base classifier. For each dataset, we plot the weight variance with different values of w in Fig. 3, which shows that more unbalanced the dataset is, the higher degree of fitness we can get; and the cost weights obtained from function fitting are closer to the optimal weights.

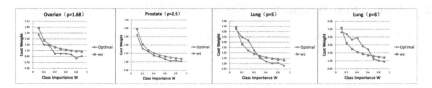

Fig. 3. Cost weight comparison using Ovarian dataset ($p = 1.68$), Prostate dataset ($p = 2.5$), Lung1 dataset ($p = 5$) and Lung2 dataset ($p = 8$).

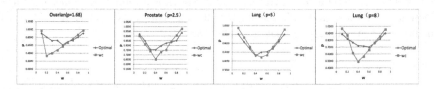

Fig. 4. WCA comparison with Ovarian dataset ($p = 1.68$), Prostate dataset ($p = 2.5$), Lung1 dataset ($p = 5$) and Lung2 dataset ($p = 8$).

We also illustrate the comparison of WCAs against different w values in Fig. 4. In Fig. 4, we can see that the WCAs of the fitting function is lower than the optimal accuracy when w is less than 0.5. The reason is that the fitting degree of the cost weights in this range is lower. Moreover, it can be seen from Fig. 4 that the WCAs of the fitting functions approach to the optimal accuracy with the increment of p.

6   Conclusion

In this paper, we propose two approaches to calculate optimal cost weights for gene expression data, including grid searching function fitting. They enrich the ways of calculating the cost weights for imbalanced data classification problems. In general, the function fitting approach is more efficient than the grid searching algorithm. The experimental results still show that the function fitting approach can well find the optimal cost weights for imbalanced gene expression datasets.

Acknowledgments. This study is supported by National Natural Science Foundation of China (Nos. 61272315, 61402417, 61602431 and 61701468), Zhejiang Provincial Natural Science Foundation (Nos. Y1110342, LY15F020037) and International Cooperation Project of Zhejiang Provincial Science and Technology Department (No. 2017C34003).

References

1. Golub, T.R., Slonim, D.K., Tamayo, P., et al.: Molecular classification of cancer: class discovery and class prediction by gene expression monitoring. Science **286**(5439), 531–537 (1999)
2. Yan, K., Ma, L.L., Dai, Y.T., et al.: Cost-sensitive and sequential feature selection for chiller fault detection and diagnosis. Int. J. Refrig. **86**, 401–409 (2018)
3. Lu, H.J., Yang, L., Yan, K., et al.: A cost-sensitive rotation forest algorithm for gene expression data classification. Neurocomputing **228**, 270–276 (2017)
4. Cao, P., Zhao, D., Zaiane, O.: An optimized cost-sensitive SVM for imbalanced data learning. In: Pei, J., Tseng, V.S., Cao, L., Motoda, H., Xu, G. (eds.) PAKDD 2013. LNCS, vol. 7819, pp. 280–292. Springer, Berlin (2013). https://doi.org/10.1007/978-3-642-37456-2_24
5. Zheng, E., Zhang, C., Liu, X., Lu, H., Sun, J.: Cost-sensitive extreme learning machine. In: Motoda, H., et al. (eds.) ADMA 2013. LNCS (LNAI), vol. 8347, pp. 478–488. Springer, Heidelberg (2013). https://doi.org/10.1007/978-3-642-53917-6_43
6. Liu, Y., Lu, H., Yan, K., et al.: Applying cost-sensitive extreme learning machine and dissimilarity integration to gene expression data classification. Comput. Intell. Neurosci. **2016** (2016). Article ID 8056253
7. Lu, H.J., Chen, J.Y., Yan, K., et al.: A hybrid feature selection algorithm for gene expression data classification. Neurocomputing **256**, 56–62 (2017)
8. Yan, K., Ji, Z.W., Shen, W.: Online fault detection methods for chillers combining extended Kalman filter and recursive one-class SVM. Neurocomputing **228**, 205–212 (2017)
9. Cheng, X.Y., Chai, F.X., et al.: 1stOpt and global optimization platform—comparison and case study. In: Proceedings of the 4th IEEE International Conference on Computer Science and Information Technology, Chengdu, China, pp. 18–21 (2011)

BIN1 rs744373 Variant Is Significantly Associated with Alzheimer's Disease in Caucasian but Not East Asian Populations

Zhifa Han[1], Tao Wang[1], Rui Tian[1], Wenyang Zhou[1],
Pingping Wang[1], Peng Ren[1], Jian Zong[1], Yang Hu[1],
Shuilin Jin[2(✉)], and Qinghua Jiang[1(✉)]

[1] School of Life Science and Technology,
Harbin Institute of Technology, Harbin, China
qhjiang@hit.edu.cn
[2] Department of Mathematics, Harbin Institute of Technology, Harbin, China
jinsl@hit.edu.cn

Abstract. Genome-wide Association studies (GWAS) and candidate gene studies have identified the association between BIN1 rs744373 variant and Alzheimer's disease (AD) in Caucasian populations. Recently, a number of studies investigated the association between BIN1 rs744373 and AD in Asian populations. However, both positive and negative results have been identified. We consider that the relatively small sample sizes may lead to the lower statistical power. Here, we selected 71,168 samples (22,395 AD cases and 48,773 controls) from 19 articles containing 38 studies and reinvestigated this association using meta-analysis method. We observed a significant genetic heterogeneity and identified a significant association between rs744373 polymorphism with AD in pooled populations ($P = 5 \times 10^{-07}$, OR = 1.12, and 95% confidence interval (CI) 1.07–1.17). In subgroup analysis, we not only identified significant genetic heterogeneity in East Asian but also in Caucasian. The subgroup meta-analysis only revealed significant association between rs744373 polymorphism with AD in the Caucasian populations ($P = 3.38 \times 10^{-08}$, OR = 1.16, 95% CI 1.10–1.22), but not in the East Asian populations ($P = 0.393$, OR = 1.057, and 95% CI 0.95–1.15). The regression analysis suggested no significant publication bias. In summary, this large-scale meta-analysis highlighted the significant association between rs744373 polymorphism and AD in Caucasian populations but not in the East Asian populations.

Keywords: BIN1 · rs744373 polymorphism · Alzheimer's disease
East Asian · Caucasian

Z. Han and T. Wang—These authors contributed equally to this work.

D.-S. Huang et al. (Eds.): ICIC 2018, LNCS 10954, pp. 520–526, 2018.
https://doi.org/10.1007/978-3-319-95930-6_48

1   Introduction

Alzheimer's disease (AD) is the most common neurodegenerative disease in elderly and causes 50–75% dementia types [1–3]. Large-scale genome-wide association studies (GWASs) have identified more than twenty genetic risk factors for AD in the populations of Caucasian ancestry [4, 5]. Among these loci, a single nucleotide polymorphism (SNP) rs744373, located upstream of BIN1, was one of the most significantly associated variant with AD identified by early GWAS of Caucasian (OR = 1.15, $P = 1.6 \times 10^{-11}$) [6].

Recently, the genetic association of the SNP rs744373 with AD has also been extensively investigated in East Asian populations [7]. However, in addition to the positive results, there are many other studies had also obtained negative results. For example, Tan et al. analyzed 1224 individuals (612 AD cases and 612 controls) in Chinese population, and they did not report any significant association using allele test ($P = 0.217$) and genotype test ($P = 0.547, 0.263$ and 0.397 for dominant, recessive and additive logistic genetic models) [8]. We believe that the relatively small sample size and the genetic heterogeneity of AD susceptibility loci among different populations may be important factors in the untrustworthiness of the inconsistent results in East Asian populations. In this study, we aimed to collect more studies and samples than before and to obtain more statistically significant results by performing genetic heterogeneity test and meta-analysis of the rs744373 polymorphism in the Caucasians, East Asians, and pooled populations.

2   Materials and Methods

2.1   Literature Selection and Data Extraction

In order to find available association studies, we mainly searched the PubMed database (https://www.ncbi.nlm.nih.gov/pubmed) and AlzGene database (http://www.alzgene. org/) with the Keywords "Alzheimer's disease", "Bridging Integrator 1" or "BIN1". We selected the studies which can provide sufficient data such as the OR value, 95% CI, the P value and so on.

From the selected studies, we firstly extracted the basic information containing the name of the first author, the numbers of AD cases and controls, etc. And then, the OR with 95% CI and other important information were obtained. Before the main analysis, we simply used t-test to investigate whether there were differences of the OR values and MAF values between the Caucasian populations and East Asian populations.

2.2   Method Description

We investigated the association of rs744373 polymorphism with AD risk in this meta-analysis primarily using the additive genetic model, which can be described as C allele versus T allele in this study [9]. The Dominant model (CC + CT versus TT) and Recessive model (CC versus CT + TT) also were used with the data containing exact genotype numbers.

We respectively tested genetic heterogeneity in East Asians, Caucasians, and pooled Populations by using Cochran's Q test. Statistics I^2 also be used to measure the genetic heterogeneity. For meta-analysis, if the P value of Cochran's Q test was less than 0.10, and I^2 value was greater than 50%, we selected the random effect model (DerSimonian–Laird) to calculate the overall OR; otherwise we selected the fixed effect model (Mantel–Haenszel). The signification of overall OR was measured by Z-test.

We evaluated the potential publication bias using funnel plot [4]. Begg's test and Egger's test was used to evaluate the asymmetry of the funnel plot [4]. The significant level was 0.01. All statistical tests above were performed using the program R (http://www.r-project.org/).

3 Results

3.1 Data Description

We obtained 186 articles by searching PubMed and AlzGene databases. Through a strict screening process as described in Fig. 1, 19 articles containing 37 studies were picked out for this meta-analysis. These 37 studies contained 22,395 AD cases and 48,773 control samples.

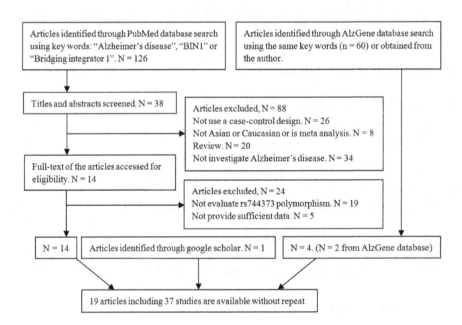

Fig. 1. Flow diagram of article inclusion or exclusion

There are 11 studies belong to East Asian populations. And the other studies belong to Caucasians. By using t-test to compare the MAF values between Caucasians and East Asians, we found a significant result with $t = 5.89$ and $P = 1.53 \times 10^{-6}$. However, the result of comparison of OR values didn't indicate a very significant distinction between the two populations ($t = 1.75$ and $P = 0.11$).

3.2 Meta Analysis

Based on the significant genetic heterogeneity test results in Caucasians ($P = 0.001, I^2 = 52.3\%$), East Asians ($P = 0.001$, $I^2 = 65.1\%$) and pooled populations ($P = 1.03 \times 10^{-5}$, $I^2 = 57.2\%$), we used random effect model to calculate the overall OR values. Meta-analysis results indicated significant association between rs744373 and AD in Caucasians with $P = 3.38 \times 10^{-08}$, OR $= 1.16$, 95% CI 1.10–1.22, and in pooled populations with P $= 5 \times 10^{-07}$, OR $= 1.12$, and 95% CI 1.07–1.17. However, we did not find any significant result in East Asians with $P = 0.393$, OR $= 1.057$, and 95% CI 0.95–1.15. The detailed results and forest diagram are described in Table 1 and Fig. 2.

Table 1. The genetic heterogeneity test and meta-analysis results of rs744373 polymorphism

Populations	Comparisons	I^2	OR	95% CI	P for meta
East Asian	C vs T	0.665	1.03	0.92–1.16	0.611
	CC + CT vs TT	0.500	1.06	0.93–1.21	0.391
	CC vs CT + TT	0.556	1.03	0.81–1.31	0.806
Caucasian	C vs T	0.402	1.17	1.12–1.22	1.35E−12
	CC + CT vs TT	0.149	1.20	1.14–1.27	5.99E−11
	CC vs CT + TT	0.356	1.26	1.14–1.39	0.00001
East Asian vs Caucasian	C vs T	0.541	1.12	1.06–1.19	0.000179
	CC + CT vs TT	0.330	1.17	1.12–1.23	3.95E−11
	CC vs CT + TT	0.499	1.19	1.10–1.29	1.35E−05

The funnel plot is a symmetrical inverted funnel that suggests no significant publication bias (Begg's test, $P = 0.471$; Egger's test, $P = 0.428$).

We further investigated the association of rs744373 polymorphism with AD risk using dominant model and recessive model. A total of 33,184 samples (12,717 AD cases and 20,467 controls) that contain genotype data were included. We obtained significant or insignificant results as same as the additive model. The detail information are described in Table 1.

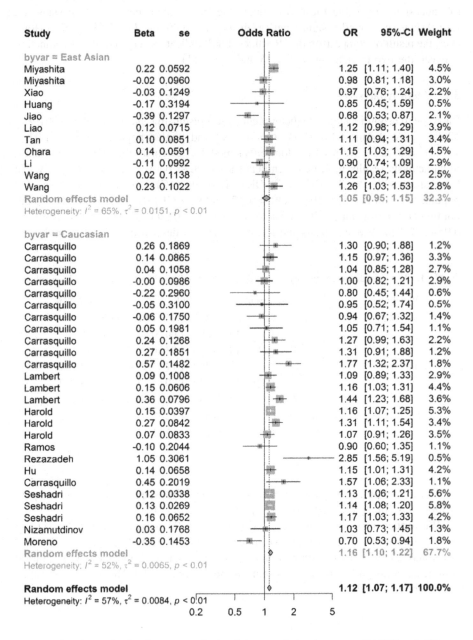

Fig. 2. Forest plot for the meta-analysis of the association between rs744373 and AD

4 Discussion

BIN1 have diverse functions in endocytosis, trafficking, immune response, calcium homoeostasis and apoptosis [10]. Genome-wide Association Studies showed that SNPs located in upstream of the coding region of BIN1 are strongly associated with AD [10]. SNP rs744373 located in this region was significantly associated with the expression of BIN1 in brain tissue by analyzing the GTEx datasets [3]. Chapuis et al. found that BIN1 affects tau pathology and suppress its neurotoxicity in AD using biological experiments [10]. To some extent, investigating the association between rs744373 polymorphism and AD risk is helpful to understand the role of BIN1 in AD pathogenesis.

However, taking into account the relatively small sample sizes, the results of the association studies between rs744373 polymorphism and AD risk in East Asians were always inconsistent. In order to obtain results with greater statistical power, we collected 33 studies involving 22,395 AD cases and 48,773 controls for analysis. To the best of our knowledge, this is the largest sample size by far.

In this large-scale meta-analysis, whatever the additive model or other models, the results always highlighted the significant association between rs744373 polymorphism and AD in Caucasian populations but not in the East Asian populations.

Our samples of East Asian ancestry mainly come from Chinese, Japanese and Koreans populations. Considering these samples may be can't represent the East Asian populations completely, we believe that a large sample size Genome-wide Association Study of East Asian populations is very necessary.

Acknowledgements. This work was supported by the National Science and Technology Major Project (2016YFC1202302), the National Nature Science Foundation of China (61571152), the Fundamental Research Funds for the Central Universities (AUGA5710001716), Natural Science Foundation of Heilongjiang Province (F2015006) and the National High-tech R&D Program of China (863 Program) (2015AA020101).

References

1. Liu, G., Yao, L., Liu, J., Jiang, Y., Ma, G., Genetic, Environmental Risk for Alzheimer's disease C, Chen, Z., Zhao, B., Li, K.: Cardiovascular disease contributes to Alzheimer's disease: evidence from large-scale genome-wide association studies. Neurobiol. Aging **35** (4), 786–792 (2014)

2. Liu, G., Jiang, Y., Wang, P., Feng, R., Jiang, N., Chen, X., Song, H., Chen, Z.: Cell adhesion molecules contribute to Alzheimer's disease: multiple pathway analyses of two genome-wide association studies. J. Neurochem. **120**(1), 190–198 (2012)

3. Zhu, R., Liu, X., He, Z.: The bridging integrator 1 gene polymorphism rs744373 and the risk of Alzheimer's disease in Caucasian and Asian populations: an updated meta-analysis. Mol. Neurobiol. **54**(2), 1419–1428 (2017)

4. Liu, G., Wang, H., Liu, J., Li, J., Li, H., Ma, G., Jiang, Y., Chen, Z., Zhao, B., Li, K.: The CLU gene rs11136000 variant is significantly associated with Alzheimer's disease in Caucasian and Asian populations. NeuroMol. Med. **16**(1), 52–60 (2014)

5. Harold, D., Abraham, R., Hollingworth, P., Sims, R., Gerrish, A., Hamshere, M.L., Pahwa, J.S., Moskvina, V., Dowzell, K., Williams, A., et al.: Genome-wide association study identifies variants at CLU and PICALM associated with Alzheimer's disease. Nat. Genet. **41** (10), 1088–1093 (2009)

6. Liu, G., Zhang, S., Cai, Z., Li, Y., Cui, L., Ma, G., Jiang, Y., Zhang, L., Feng, R., Liao, M., et al.: BIN1 gene rs744373 polymorphism contributes to Alzheimer's disease in East Asian population. Neurosci. Lett. **544**, 47–51 (2013)

7. Hu, X., Pickering, E., Liu, Y.C., Hall, S., Fournier, H., Katz, E., Dechairo, B., John, S., Van Eerdewegh, P., Soares, H., et al.: Meta-analysis for genome-wide association study identifies multiple variants at the BIN1 locus associated with late-onset Alzheimer's disease. PLoS ONE **6**(2), e16616 (2011)

8. Tan, L., Yu, J.T., Zhang, W., Wu, Z.C., Zhang, Q., Liu, Q.Y., Wang, W., Wang, H.F., Ma, X.Y., Cui, W.Z.: Association of GWAS-linked loci with late-onset Alzheimer's disease in a northern Han Chinese population. Alzheimer's Dement. J. Alzheimer's Assoc. **9**(5), 546–553 (2013)

9. Li, X., Shen, N., Zhang, S., Liu, J., Jiang, Q., Liao, M., Feng, R., Zhang, L., Wang, G., Ma, G., et al.: CD33 rs3865444 polymorphism contributes to Alzheimer's disease susceptibility in Chinese, European, and North American populations. Mol. Neurobiol. **52**(1), 414–421 (2015)

10. Cuyvers, E., Sleegers, K.: Genetic variations underlying Alzheimer's disease: evidence from genome-wide association studies and beyond. Lancet Neurol. **15**(8), 857–868 (2016)

Using Weighted Extreme Learning Machine Combined with Scale-Invariant Feature Transform to Predict Protein-Protein Interactions from Protein Evolutionary Information

Jianqiang Li[1], Xiaofeng Shi[1], Zhuhong You[2(✉)],
Zhuangzhuang Chen[1], Qiuzhen Lin[1], and Min Fang[1]

[1] Shenzhen University, Nanhai Road 3688, Shenzhen, China
{lijq,qiuzhlin}@szu.edu.cn, sxf7758258@sina.com,
chenzhuangzh@qq.com, gracefangcs@163.com
[2] No. 40, Beijing South Road, Urumqi, China
zhuhongyou@ms.xjb.ac.cn

Abstract. Protein-Protein Interactions (PPIs) play an irreplaceable role in biological activities of organisms. Although many high-throughput methods are used to identify PPIs from different kinds of organisms, they have some shortcomings, such as high cost and time-consuming. To solve the above problems, computational methods are developed to predict PPIs. Thus, in this paper, we present a method to predict PPIs using protein sequences. First, protein sequences are transformed into Position Weight Matrix (PWM), in which Scale-Invariant Feature Transform (SIFT) algorithm is used to extract features. Then Principal Component Analysis (PCA) is applied to reduce the dimension of features. At last, Weighted Extreme Learning Machine (WELM) classifier is employed to predict PPIs and a series of evaluation results are obtained. In our method, since SIFT and WELM are used to extract features and classify respectively, we called the proposed method SIFT-WELM. When applying the proposed method on three well-known PPIs datasets of Y east, Human and Helicobacter.pylori, the average accuracies of our method using five-fold cross validation are obtained as high as 94.83%, 97.60% and 83.64%, respectively. In order to evaluate the proposed approach properly, we compare it with Support Vector Machine (SVM) classifier in different aspects.

Keywords: Protein-protein interactions · Scale-invariant feature transform
Weighted extreme learning machine

1 Introduction

Protein-Protein Interactions (PPIs) get involved in many fundamental cellular functions, and the research on PPIs helps us to understand the molecular mechanisms of biological processes and to propose some new methods in practical medical field. So it is necessary and urgent to carry out the study of PPIs.

© Springer International Publishing AG, part of Springer Nature 2018
D.-S. Huang et al. (Eds.): ICIC 2018, LNCS 10954, pp. 527–532, 2018.
https://doi.org/10.1007/978-3-319-95930-6_49

Nowadays, a large amount of high-throughput methods have been developed to predict PPIs, such as yeast two-hybrid (Y2H) screening methods [1, 2], immunopre-cipitation [3], and protein chips [4]. However, there are some shortcomings in these experiments, such as high cost and time-consuming. Moreover, these methods yield high false positives and false negatives, which result in difficulties to predict unknown PPIs by experimental methods.

In addition, there are many biological databases, such as BIND [5], DIP [6] and MINT [7]. Protein sequences occupy an overwhelming advantage in quantity in these databases, so in order to efficiently utilize these sequence data, it is necessary to develop computational methods to predict PPIs from protein sequences. In general, sequence-based computational methods have two main parts: feature extraction and sample classification [8–10].

In first part, Scale-Invariant Feature Transform (SIFT) [11, 12] is applied to extract features from Position Weight Matrix (PWM) [13]. In order to reduce the effect of noise and shorten training time, Principal Component Analysis (PCA) is used to reduce the dimension of features.

In second part, Weighted Extreme Learning Machine (WELM) [14, 15] is used to identify protein pairs' interacting or non-interacting based on SIFT features. WELM only needs to set two parameters, which is fast to get the best parameters. Moreover, WELM gets better performance in generalization.

In this paper, a novel computational method based on SIFT algorithm and WELM is proposed to predict protein-protein interactions, which helps to insight into the molecular mechanisms of cells and explain the causes of some disease, and it may propose some new treatment methods in practical medical field.

2   Materials and Methods

2.1   Datasets

In our experiment, we collect Yeast dataset from DIP [6]. After removing protein pairs whose sequence length less than 50 and filtering out protein pairs whose sequence identity bigger than 40%, we get 5594 positive protein pairs, and we construct 5594 negative sample according to the results in [16].

To demonstrate the generality of our approach, we collect 3899 protein pairs as positive dataset by removing sequence identity bigger than 25%, and we construct 4262 negative protein pairs according to the work in [17]. In addition, Helicobacter. pylori dataset consists of 1458 positive protein sequence and 1458 negative protein sequence according to the result of Martin et al. [18].

2.2   Scale-Invariant Feature Transform

Scale-Invariant Feature Transform (SIFT) is an algorithm widely used in the field of computer vision, which can be applied to extract local features from images. SIFT was firstly introduced by Lowe in [11], which was summarized and perfected in [12]. SIFT algorithm can be applied in different fields, such as face recognition, 3D modeling and template matching because of its robustness to rotation, scaling, viewpoint and so on. In this paper, SIFT is used to extract features.

2.3   Weighted Extreme Learning Machine

Extreme Learning Machine (ELM) [14] is a single hidden layer feed-forward neural network (SLFN) algorithm, which is simple in theory but effective in practice. ELM just needs to set the hidden nodes in network before the use, and ELM produces the unique optimal result, so it gets fast in learning and achieves better performance in generalization. Weighted ELM (WELM) is proposed to process the data with imbalanced class distribution [15], which can maintain the advantages of original ELM, and extend to cost-sensitive learning according to user's needs.

2.4   Evaluation Criteria

In order to evaluate the performance of our method, we use the following evaluation criteria: accuracy, sensitivity, precision and Matthews correlation coefficient (MCC). They are calculated as

$$Accuracy = \frac{TN + TP}{TN + TP + FN + FP} \tag{1}$$

$$Sensitivity = \frac{TP}{TP + FN} \tag{2}$$

$$Precision = \frac{TP}{TP + FP} \tag{3}$$

$$MCC = \frac{TP \times TN - FP \times FN}{\sqrt{(TP + FN) \times (TP + FP) \times (TN + FP) \times (TN + FN)}} \tag{4}$$

where true positive (TP) stands for the number of true interacting pairs that predicted correctly; true negative (TN) represents the number of true non-interacting pairs that predicted correctly; false positive (FP) is the number of true non-interacting pairs that predicted incorrectly and false negative (FN) is the number of true interacting pairs that predicted to be non-interacting pairs falsely.

3   Results and Discussion

3.1   Evaluation of the Proposed Method

In our experiment, we set the same parameters for three datasets—Yeast, Human and H.pylori, which are classified by WELM. Here, $L = 10000$ and $C = 2^5$, where L means the number of hidden neurons and C represents the trade-off constant [15]. Five-fold cross validation is employed to evaluate the performance of our method, which can avoid over-fitting problem of our model and evaluate the stability of our model [19]. Results of our method are shown in Tables 1, 2 and 3.

Table 1. Five-fold cross validation results of our method applied on *Yeast* dataset.

Test Sets	Accu.(%)	Prec.(%)	Sen.(%)	Mcc.(%)	Auc.(%)
1	95.36	94.89	96.21	90.71	97.92
2	95.29	95.07	95.80	90.58	98.54
3	93.86	91.04	96.56	87.86	98.41
4	94.51	93.49	95.69	89.04	97.42
5	95.11	95.11	94.99	90.22	97.82
Average	94.83 ± 0.64	93.92 ± 1.74	95.85 ± 0.59	89.68 ± 1.21	98.02 ± 0.46

Table 2. Five-fold cross validation results of our method applied on *Human* dataset.

Test Sets	Accu.(%)	Prec.(%)	Sen.(%)	Mcc.(%)	Auc.(%)
1	97.06	95.06	99.05	99.05	99.48
2	97.61	96.01	99.25	95.28	99.63
3	98.17	97.15	99.03	96.34	99.60
4	96.97	95.40	98.48	93.99	99.33
5	98.17	97.47	98.90	96.34	99.59
Average	97.60 ± 0.57	97.60 ± 0.57	98.94 ± 0.29	95.23 ± 1.12	99.53 ± 0.12

Table 3. Five-fold cross validation results of our method applied on *H. pylori* dataset.

Test Sets	Accu.(%)	Prec.(%)	Sen.(%)	Mcc.(%)	Auc.(%)
1	85.02	89.29	78.13	70.35	90.74
2	82.02	87.20	77.30	64.64	89.32
3	86.14	88.28	83.70	72.40	90.06
4	83.52	84.17	80.16	66.92	87.93
5	81.48	86.18	76.26	63.51	89.88
Average	83.64 ± 1.97	87.02 ± 1.98	79.11 ± 2.94	67.56 ± 3.76	89.59 ± 1.06

From above tables, we can refer that WELM classifier combining with SIFT descriptors can predict PPIs effectively, and the low standard deviations of the results indicate that our approach is robust. The excellent results of our method lie in the following reasons: (1) When compared to sequence dataset, the corresponding PWM matrix can retain more prior information. (2) The SIFT descriptors extracted from datasets retain abundant information of protein pairs and have strong ability to resist noise. (3) WELM is faster than traditional neural network algorithm in training while guaranteeing the learning accuracy.

3.2 Comparison with SVM-Based Method

To further evaluate our method, we compare results of the proposed approach with the widely used SVM classifier LIBSVM, which is developed by professor Chih-Jen Lin of National Taiwan University [20]. From Table 4, we notice that WELM achieves better performance than SVM when proposing classification on Yeast, Human and H. pylori datasets. Thus we can conclude that WELM is superior to SVM.

Table 4. Performance comparison between the SIFT+WELM and the SVM prediction models

Dataset	Classifier	Accu.(%)	Prec.(%)	Sen.(%)	Mcc.(%)	Time (s)
Yeast	WELM	94.83 ± 0.64	93.92 ± 1.74	95.85 ± 0.59	89.68 ± 1.21	113.9 ± 1.6
	SVM	91.27 ± 1.06	90.39 ± 1.17	92.05 ± 0.55	82.55 ± 2.11	1033.2 ± 1.6
Human	WELM	97.60 ± 0.57	96.22 ± 1.06	98.94 ± 0.29	95.23 ± 1.12	56.7 ± 1.2
	SVM	96.55 ± 0.71	96.15 ± 1.49	97.12 ± 0.44	93.11 ± 1.41	403.7 ± 4.2
H.pylori	WELM	83.64 ± 1.97	87.02 ± 1.98	79.11 ± 2.94	67.56 ± 3.76	5.4 ± 0.3
	SVM	80.49 ± 1.40	77.79 ± 2.60	82.30 ± 2.72	61.11 ± 2.73	17.2 ± 0.1

4 Conclusions

The use of computational methods to predict PPIs is becoming more and more important because of its low cost and high efficiency when compared to the experimental methods. In this paper, we propose a novel prediction model by using scale-invariant feature transform and weighted extreme learning machine to predict PPIs. When compared to SVM-based methods, our method can increase the accuracy and shorten the training time greatly. The experimental results indicate that the proposed method is efficient, feasible and robust.

Acknowledgement. This work is supported in part by the Natural Science Foundation of SZU under Grant CYZZ20160304165036893 and Grant 2016048, in part by the National Natural Science Foundation of China under Grant U1713212, Grant 61572330, and Grant 61602319, and in part by the Technology Planning Project from Guangdong Province, China, under Grant 2014B010118005.

References

1. Gavin, A.-C., Bsche, M., Krause, R.: Functional organization of the yeast proteome by systematic analysis of protein complexes. Nature **415**(6868), 141–147 (2002)
2. Ito, T., Chiba, T., Ozawa, R., Yoshida, M., Hattori, M., Sakaki, Y.: A comprehensive two-hybrid analysis to explore the yeast protein interactome. Proc. Natl. Acad. Sci. USA **98**(8), 4569–4574 (2001)
3. Ho, Y., Gruhler, A., Heilbut, A.: Systematic identification of protein complexes in Saccharomyces cerevisiae by mass spectrometry. Nature **415**(6868), 180–183 (2002)
4. Snyder, M., Zhu, H., Bertone, P., Bidlingmaier, S.M., Bilgin, M., Casamayor, A.J., Gerstein, M., Jansen, R., Lan, N.: Global analysis of protein activities using proteome chips, p. 2101 (2004)

5. Alfarano, C., Andrade, C.E., Anthony, K., Bahroos, N., Bajec, M., Bantoft, K., Betel, D., Bobechko, B., Boutilier, K., Burgess, E.: The biomolecular interaction network database and related tools 2005 update. Nucleic Acids Res. **33**(Database issue), 418–424 (2005)
6. Salwinski, L., Miller, C.S., Smith, A.J., Pettit, F.K., Bowie, J.U., Eisenberg, D.: The database of interacting proteins: 2004 update. Nucleic Acids Res. **32**(1), D449 (2004)
7. Licata, L., Briganti, L., Peluso, D., Perfetto, L., Iannuccelli, M., Galeota, E., Sacco, F., Palma, A., Nardozza, A.P., Santonico, E.: MINT, the molecular interaction database: 2012 update. Nucleic Acids Res. **35**(Database issue), 572–574 (2012)
8. You, Z.H., Lei, Y.K., Gui, J., Huang, D.S., Zhou, X.: Using manifold embedding for assessing and predicting protein interactions from high-throughput experimental data. Bioinformatics **26**(21), 2744 (2010)
9. You, Z.H., Lei, Y.K., Zhu, L., Xia, J., Wang, B.: Prediction of protein-protein interactions from amino acid sequences with ensemble extreme learning machines and principal component analysis. BMC Bioinf. **14**(S8), 1–11 (2013)
10. Huang, Y.A., You, Z.H., Gao, X., Wong, L., Wang, L.: Using weighted sparse representation model combined with discrete cosine transformation to predict protein-protein interactions from protein sequence. BioMed Res. Int. **2015**, 1–10 (2015)
11. Lowe, D.G.: Object recognition from local scale-invariant features. In: The Proceedings of the Seventh IEEE International Conference on Computer Vision, p. 1150 (2002)
12. Lowe, D.G.: Distinctive image features from scale-invariant keypoints. Int. J. Comput. Vis. **60**(2), 91–110 (2004)
13. Stormo, G.D., Schneider, T.D., Gold, L., Ehrenfeucht, A.: Use of the 'Perceptron' algorithm to distinguish translational initiation sites in E. coli. U.S. Department of Commerce, National Bureau of Standards: for sale by the Superintendent of Documents, U.S. Government Printing Office (1982)
14. Huang, G.B., Zhu, Q.Y., Siew, C.K.: Extreme learning machine: a new learning scheme of feedforward neural networks. In: 2004 Proceedings of the IEEE International Joint Conference on Neural Networks, vol. 2, pp. 985–990 (2005)
15. Zong, W., Huang, G.B., Chen, Y.: Weighted extreme learning machine for imbalance learning. Neurocomputing **101**(3), 229–242 (2013)
16. Guo, Y., Yu, L., Wen, Z., Li, M.: Using support vector machine combined with auto covariance to predict protein-protein interactions from protein sequences. Nucleic Acids Res. **36**(9), 3025–3030 (2008)
17. You, Z.H., Yu, J.Z., Zhu, L., Li, S., Wen, Z.K.: A MapReduce based parallel SVM for large-scale predicting protein-protein interactions. Neurocomputing **145**(18), 37–43 (2014)
18. Martin, S., Roe, D., Faulon, J.L.: Predicting protein-protein interactions using signature products. Curr. Opin. Struct. Biol. **21**(2), 218 (2005)
19. Jiao, Y., Du, P.: Performance measures in evaluating machine learning based bioinformatics predictors for classifications. Quant. Biol. **4**, 1–11 (2016)
20. Lin, C.H., Liu, J.C., Ho, C.H.: Anomaly detection using LibSVM training tools. In: International Conference on Information Security and Assurance, pp. 166–171 (2008)

Characterizing and Discriminating Individual Steady State of Disease-Associated Pathway

Shaoyan Sun[1(✉)], Xiangtian Yu[2], Fengnan Sun[3], Ying Tang[2],
Juan Zhao[2], and Tao Zeng[2(✉)]

[1] School of Mathematics and Statistics Science, Ludong University,
Yantai 264025, China
Sunsy_2014@163.com
[2] Key Laboratory of Systems Biology, Institute of Biochemistry and Cell
Biology, Chinese Academy Science, Shanghai 200031, China
zengtao@sibs.ac.cn
[3] Medical Laboratory, Yantaishan Hospital, Yantai 264001, China
renasun@gmail.com

Abstract. Recently, individual heterogeneity is becoming a hot topic with the development of precision medicine. It is still a challenge to characterize the intrinsic regulatory convergence along with temporal gene expression change corresponding to different individuals. Considering the similar functions will be more suitable than the same genes to find consistent function rather than chaotic genes, we propose a computational framework (ABP: Attractor analysis of Boolean network of Pathway) to recognize the key pathways associated with phenotype change, which uses the network attractor to represent the steady pathway states corresponding to the final biological sate of individuals. By analyzing temporal gene expressions, ABP has shown its ability to recognize key pathways and infer the potential consensus functional cascade among pathways, and especially group individuals corresponding to disease state well.

Keywords: Pathway activity · Boolean network · Temporal gene expression
Personalized medicine · Steady state

1 Introduction

As known, the general gene-set or gene module can be used to find combinatory biomarkers or signatures corresponding to phenotypic change of a biological system, e.g. disease occurrence and progression [1]. However, the interpretability of such ab initio predictions is usually far from the requirement in biological study [2]. By contrast, the pathway centered analysis will provide a tradeoff between the discovery of interpretable biological functions charactering phenotypes and the detection of some new pathway elements.

Currently, many pathways have been carefully curated to supply more creditable functional details [3]. Actually, pathway centered approaches have been widely applied in many bioinformatics and biomedicine studies. For example, the pathway enrichment analysis or well-known gene set enrichment analysis (GSEA) [4] can identify

© Springer International Publishing AG, part of Springer Nature 2018
D.-S. Huang et al. (Eds.): ICIC 2018, LNCS 10954, pp. 533–538, 2018.
https://doi.org/10.1007/978-3-319-95930-6_50

dysregulated pathway by qualitatively measuring the changed status of a pathway. In the follow-up studies, many pathway-level aggregation methods have been proposed to study the phenotypic signatures (e.g. biomarkers) on the pathway activity level rather than gene expression level [5]. Besides, there are also many improvement based on the application of pathway analysis [6].

By contrast, there are few pathway studies in a temporal manner [7]. It is still lack of approaches to characterize the interactive behaviors among multiple pathways and their states corresponding to phenotypes. Thus, in this work, we propose a computational framework (ABP: Attractor analysis of Boolean network of Pathway) to study the dynamical process of pathway activities by the Boolean network model and use the network attractor to represent the steady pathway states of a biological system corresponding to control and case condition respectively. In dry-experiment validation, ABP has shown its ability on recognizing key pathways associated with phenotype change; inferring the potential consensus functional cascade during dynamical process; and especially grouping pathway states corresponding to disease states well.

2 Methods

The key approaches applied in ABP will be introduced in details in bellows:

(1) Collecting the normalized (temporal) gene expression profiles;
(2) Obtaining the pathway activity profiles by Pathway scoring approach, i.e. GSVA [8] for each sample with expression data and KEGG pathway data;
(3) Selecting key pathways by differential significance test, significance threshold is 0.05;
(4) Extracting the temporal pathway activity profile for each individual, so that, the size of activity profile is determined by the number of key pathways (should be less than the number of time points), the number of individuals and the number of time points;
(5) Building the Boolean network for each individual based on corresponding temporal pathway activity profile by BoolNet method [9], where binarize Time Series adopted "scan Statistic" method, window Size set from 0.1 to 0.2, sign.level set from 0.05 to 0.15; and reconstruct Network used "best fit" method and other default parameters.
(6) Extracting the attractors of each Boolean network to represent the steady pathway state of each individual, which can be used to discriminate different individual phenotypes by general hierarchical clustering (HCL) or SVM (radial kernel).
(7) Evaluating the performance of pathway activity features by attractor-based individual clustering with index Acc [10], and the robustness of pathway states as indicators with the AUC measurements of SVM under 80% vs. 20% random cross-validation.

3   Results and Discussion

To validate the efficiency of ABP, we mainly used the gene expression data from a serially sampled challenge study [11] named as Rhinovirus UVA data, which consists of a total of 16 subjects that were inoculated with live human rhinovirus (HRV). Each subject was sampled 15 times in all, once before the viral inoculum and 14 times after inoculation. Each subject was designated as a symptomatic subject (Sx) or an asymptomatic subject (Asx) according to a modified Jackson score [11].

After pathway selection based on the pathway activities across multiple samples, the number of deferentially activated pathways accesses maximum at 12th time point (i.e. 48 h after inoculation), and there are 13 pathways at 12th time point which would suggest a critical point between two groups of subjects. These pathways will be used as the indices to indicate the activity tendency of pathways in each subject across time points, and the final state of pathways will be determined by the steady sate of a subject's corresponding Boolean network among pathways.

Rhinovirus (RV) leads the majority of common colds. As shown in the detected key pathways (Table 1) by our methods, many signaling pathways would be productive entry pathways of human rhinoviruses, e.g. airway inflammation in rhinovirus infection, and some transmembrane protein needed for the replication of virus. Besides, some special biological processes and functions will also be associated with virus infection, e.g. HRV infection, as discovered in Table 1.

Table 1. The key pathways detected in HRV dataset

Metagene ID	Kegg ID	Pathway ID
Gene 1	hsa00062	Fatty acid elongation
Gene 2	hsa00260	Glycine, serine and threonine metabolism
Gene 3	hsa00630	Glyoxylate and dicarboxylate metabolism
Gene 4	hsa03320	PPAR signaling pathway
Gene 5	hsa04115	p53 signaling pathway
Gene 6	hsa04725	Cholinergic synapse
Gene 7	hsa04750	Inflammatory mediator regulation of TRP channels
Gene 8	hsa04920	Adipocytokine signaling pathway
Gene 9	hsa04972	Pancreatic secretion
Gene 10	hsa05143	African trypanosomiasis
Gene 11	hsa05146	Amoebiasis
Gene 12	hsa05160	Hepatitis C
Gene 13	hsa05164	Influenza A

For each individual, the topological structure of reconstructed Boolean network is shown in Fig. 1(A). These structures can't classify the Sx and Asx groups directly, although there are a few edges in the network have group-specific existence. For example, the edge Gene12 -> Gene13, i.e. the potential regulation association between "Hepatitis C" and "Influenza A" pathways, have multiple observations in the networks corresponding to Sx individuals (5 in 10) rather than Asx individuals (0 in 6). Thus the interaction between "Hepatitis C" and "Influenza A" pathways or related genes/proteins would also play important roles in HRV infection.

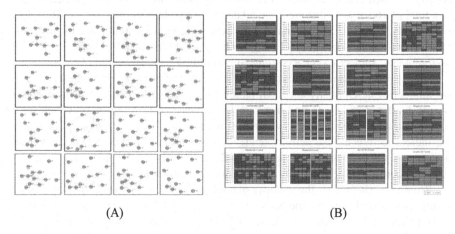

(A)   (B)

Fig. 1. The individual pathway features. (A) The topological structure of Boolean network corresponding to each subject. The Sx individual is labeled with black box; the Asx individual is labeled with grey box. (B) The attractor states of Boolean network corresponding to each subject. The Sx individual is labeled with black box; the Asx individual is labeled with grey box.

Similar to above comparison on the topological structures of whole Boolean network corresponding to each individual, we also compare the final states of each individual represented by the attractors of networks as shown in Fig. 1(B). Obviously, there are several common pathways would be in active states for Sx and Asx subjects, e.g. Hepatitis C and Influenza A, which would indicate the shared molecule mechanism underlying virus infection. Meanwhile, through the combination of all pathways' states, our simple clustering based on pathway activities can still discriminate the Sx and Asx with accuracy above than 0.8, and be more efficient than that based on structure, as shown in Fig. 2(A).

Next, the AUC of SVM with cross-validation has also been calculated for evaluating the robustness of the attractor based model rather than simple classification. The results in Fig. 2(B) show again the model based on pathways' states tend to robustly have similar or higher AUC than the model based on structure. These results support again the merit of system biology study on dynamical biological process, like virus infection, which would not only provide distinguishing features to characterize the difference of phenotypes, but also recover the interpretable mechanisms behind phenotypic change on individuals.

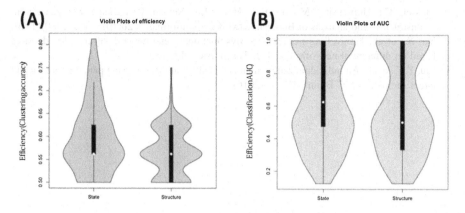

Fig. 2. The efficiency for discriminating Sx and Asx individuals by the state and structure of Boolean network respectively. (A) Statistic of Acc under different parameters. (B) Statistic of AUC under different parameters.

4   Conclusion

Indeed, by analyzing time-course gene expression datasets of HRV infections, our method has shown its ability and efficiency on selecting key pathways to explain phenotype change in a state-transition manner. How to build attractor-focused disease prediction model based on these pathways and network will be future direction.

References

1. Creixell, P., et al.: Pathway and network analysis of cancer genomes. Nat. Methods **12**(7), 615–621 (2015)
2. Alcaraz, N., et al.: De novo pathway-based biomarker identification. Nucleic Acids Res. **45**(16), e151 (2017)
3. Cerami, E.G., et al.: Pathway commons, a web resource for biological pathway data. Nucleic Acids Res. **39**(Database issue), D685–D690 (2011)
4. Subramanian, A., et al.: Gene set enrichment analysis: a knowledge-based approach for interpreting genome-wide expression profiles. Proc. Natl. Acad. Sci. USA **102**(43), 15545–15550 (2005)
5. Ihnatova, I., Budinska, E.: ToPASeq: an R package for topology-based pathway analysis of microarray and RNA-Seq data. BMC Bioinf. **16**, 350 (2015)
6. Palaniappan, S.K., et al.: Abstracting the dynamics of biological pathways using information theory: a case study of apoptosis pathway. Bioinformatics **33**(13), 1980–1986 (2017)
7. Schraiber, J.G., et al.: Inferring evolutionary histories of pathway regulation from transcriptional profiling data. PLoS Comput. Biol. **9**(10), e1003255 (2013)
8. Hanzelmann, S., Castelo, R., Guinney, J.: GSVA: gene set variation analysis for microarray and RNA-seq data. BMC Bioinf. **14**, 7 (2013)

9. Mussel, C., Hopfensitz, M., Kestler, H.A.: BoolNet–an R package for generation, reconstruction and analysis of Boolean networks. Bioinformatics **26**(10), 1378–1380 (2010)
10. Yu, X., et al.: Unravelling personalized dysfunctional gene network of complex diseases based on differential network model. J. Transl. Med. **13**, 189 (2015)
11. Liu, T.Y., et al.: An individualized predictor of health and disease using paired reference and target samples. BMC Bioinf. **17**, 47 (2016)

Discovering an Integrated Network in Heterogeneous Data for Predicting lncRNA-miRNA Interactions

Pengwei Hu[1], Yu-An Huang[1], Keith C. C. Chan[1], and Zhu-Hong You[1,2(✉)]

[1] Department of Computing, The Hong Kong Polytechnic University,
Hong Kong, Hong Kong
{csphu, csyahuang, cskcchan}@comp.polyu.edu.hk
[2] Xinjiang Technical Institute of Physics and Chemistry,
Chinese Academy of Science, Ürümqi 830011, China
zhuhongyou@ms.xjb.ac.cn

Abstract. Long noncoding RNAs (lncRNAs) belong to a class of non-protein coding RNAs, which have recently been found to potentially act as a regulatory molecule in some important biological processes. MicroRNAs (miRNAs) have been proved by many biomedical studies to be closely associated to many human diseases. Recent studies have suggested that lncRNAs could potentially interact with miRNAs to modulate their regulatory roles. Hence, predicting lncRNA–miRNA interactions is biologically significant due to their potential roles in determining the effectiveness of gene regulations. As diverse heterogeneous datasets for describing lncRNA and miRNA have been made available, it becomes more feasible for us to develop a model to describe potential interactions between lncRNAs and miRNAs. In this work, we presented a new computational pipeline, called INLMI, to predict lncRNA–miRNA interactions by integrating the expression similarity network and the sequence similarity network. Based on a measure of similarities between these networks, INLMI computes an *interaction score* for a pair of lncRNA and a miRNA. The novelty of INLMI lies in that we used network integration on two similarity networks. Using a real data set, we have shown that INLMI can be a very effective approach as the model that it has learnt can be used to very accurately predict lncRNA-miRNA interactions.

Keywords: lncRNA–miRNA interaction · Network integration
Two-way diffusion

1 Introduction

Rapidly evolving sequencing techniques are making us progressively easier to collect multiple and diverse non-coding RNAs (ncRNAs) data sets to address genomic and biological questions [1, 2]. Meanwhile, the discovery of the essential role of the microRNAs (miRNAs) and long noncoding RNAs (lncRNAs) in the regulation of gene expression has received great attention. The mutual regulation of lncRNA and miRNA

© Springer International Publishing AG, part of Springer Nature 2018
D.-S. Huang et al. (Eds.): ICIC 2018, LNCS 10954, pp. 539–545, 2018.
https://doi.org/10.1007/978-3-319-95930-6_51

constitute a complex network of molecular regulation, which has been proved to be closely related to the development of various human diseases, but the specific mechanism research is still in the preliminary stage of exploration. With a deep insight into the lncRNA-miRNA interaction, researchers can find new ways to understanding the gene regulation and disease incentive. [3, 4]. So far, the amount of lncRNA-miRNA interactions determined by biological experiments is limited by a variety of resource requirements which cannot have a very broad impact [5, 6]. Thanks to the accumulation of functional knowledge of miRNA and lncRNA in the past few years [7], if the interaction mechanism between lncRNA and miRNA can be exposed or even inferred, we can have a deeper understanding of their complex functions. Some experiments have shown that certain lncRNAs are the co-regulation of groups in the expression network, which indicates that multiple lncRNAs can regulate biological processes through synergistic interaction with specific miRNA clusters [8]. Therefore, we can assume that in different tissues and cell lines, lncRNA has a similar expression with those miRNA that interacts with each other. In reality, some new studies have shown that the sequence of lncRNAs that interact with the same miRNA tend to be more similar to those of lncRNAs that do not interact with each other [9]. Moreover, a number of machine learning methods have emerged in recent years to predict potential lncRNA-disease associations [10] small RNA regulation [11] and miRNA–target threshold effects [10, 12, 13] reveal that lncRNAs and miRNAs interact with each other because they appear in a variety of synergy together.

Previous experiments have provided a starting point for understanding mechanisms and collecting enough samples. Motivated by this discovery and known mutual regulation rules, we present here an integrated network-based prediction model for lncRNA-miRNA interactions (INLMI). Without using the limited single similarity network of lncRNAs and miRNAs, we develop INLMI to predict lncRNA–miRNA interactions by integrating the expression similarity network and the sequence similarity network. Then, we infer the interaction strength of lncRNA–miRNA interaction according to the known interaction network with matrix completion method. To evaluate the performance of proposed method, we have performed experiments using the real world data by using the dataset collected from some of the latest databases [9].

INLMI was proposed to computationally infer potential lncRNA–miRNA interactions via adopting similarity network integration with non-negative matrix factorization on the similarity networks. The proposed model can achieve AUC of 0.8517 based on 5-fold cross validation. Experimental results show that INLMI is better able to identify lncRNA–miRNA interactions more accurate when compared with the state-of-the-art algorithms due to it relies on the integrated network.

2 Methodology

2.1 lncRNA/miRNA Similarity Network

For each ncRNA, diverse profiles may have different contributions in similarity network. Hence, we collected two different types of lncRNA/miRNA information through different databases and generated two different similarity networks based on this information. To discover such kinds of similarity values, we bring in Pearson correlation coefficient (PCC) to compute the first kind of similarity network based on the expression profile of lncRNAs/miRNAs. Given p_a and p_b represent two expression profiles of two RNAs, we define:

$$PS(a,b) = \frac{\sum_{i=1}^{N}(p_{ai} - \bar{p}_a)(p_{bi} - \bar{p}_b)}{\sqrt{\sum_{i=1}^{N}(p_{ai} - \bar{p}_a)^2 \sum_{i=1}^{N}(p_{bi} - \bar{p}_b)^2}} \tag{1}$$

as the correlation coefficient score, where N denotes the number of features of the expression profile. In general, we believe that two RNAs with higher correlation scores have higher similarity. Furthermore, we compute another kind of RNA similarity by use of the sequence data. Given q_a and q_b represent two sequence data of two RNAs, the sequence similarity of lncRNAs/miRNAs $QS(a,b)$ was calculated using the Needleman-Wunsch pairwise sequence alignment [14].

2.2 Similarity Network Integration

Above similarity network from two different sources are inherently correlated, and sometimes provide complementary information to each other. As a result, network integration has been paid much attention to, which mainly aims to generate most similar representation between entities under the existing domains. Considering that expression similarity score and the sequence similarity score do not cover all the RNAs, we use either one of the available similarity scores as integrated similarity. Following the criteria defined for the desirable lncRNA and miRNA similarity networks, we formulate an integrated network by using the average value of expression similarity score and the sequence similarity score when both similarities are present in the network. Here, the integrated representation of two kinds of similarity between miRNA a and b is defined as follows:

$$IM(a,b) = \frac{PS_M(a,b) + QS_M(a,b)}{2} \tag{2}$$

the integrated similarity between lncRNA a and b is defined as follows:

$$IL(a,b) = \frac{PS_L(a,b) + QS_L(a,b)}{2} \tag{3}$$

For the two above networks, we observed that matrix is high dimensional representation. And, we assume that $IM(a,b)$ and $IL(a,b)$ should be nonnegative and a higher magnitude of it means two RNAs are more related. Since matrix factorization is efficient to look for the low-dimensional representation of network, we propose to use NMF [15] for similarity network decomposition. The target of NMF is to find an approximate factorization $V \approx WH$. The overall cost function of NMF can be defined using Kullback-Leibler divergence to measure the distance between two non-negative matrices. We use matrix M to denote one of the original networks. This matrix first approximately factorized into a $n \times r$ matrix W and a $r \times n$ matrix H, and then we consider following formulations as optimization problems:

$$minD_{KL}(V||WH)$$
$$= \sum_{i=1}^{I} \sum_{j=1}^{J} \left(v_{ij} \ln \frac{v_{ij}}{[WH]_{ij}} - v_{ij} + [WH]_{ij} \right) \tag{4}$$

subject to the constraints $W, H \geq O$.

where

$$W_{ik} \leftarrow W_{ik} \frac{\sum_{j=1}^{J} H_{kj}v_{ij}/[WH]_{ij}}{\sum_{j=1}^{J} H_{kj}} \tag{5}$$

and

$$H_{kj} \leftarrow H_{kj} \frac{\sum_{i=1}^{I} W_{ik}v_{ij}/[WH]_{ij}}{\sum_{i=1}^{I} W_{ik}} \tag{6}$$

This step is to initially insert known values for those entries, then perform NMF, producing W and H. Then, we can compute WH as our estimate of original similarity network, and now have a low-dimensional representation for the integrated similarity network. To compare the effect of the integration step, in this study we adopted two-way diffusion [9] to obtain the potential interactions. This state-of-the-art method solved the prediction task as a matrix-completion problem.

3   Results

3.1   Evaluation

In this study, the lncRNA and miRNA networks which we used to predict lncRNA–miRNA interactions come from [9]. This standard dataset collected lncRNA–miRNA interactions from the lncRNASNP database [16]. We removed duplicated entries and observed the number of known lncRNA–miRNA interactions in original dataset are 5348. In addition, there are 780 different lncRNAs and 275 different miRNAs involved in the known lncRNA–miRNA interactions. We obtained the sequence data of lncRNAs from LNCipedia database to calculate the RNA sequence-based similarity network, [17]. To obtain the expression profiles of lncRNAs, we downloaded the information from the NONCODE database [18]. The miRNAs sequence information is obtained from the miRBase database [19]. The expression profiles of microRNA were obtained from [20]. The prediction of lncRNA–miRNA interactions is a typical labeling unbalance prediction task, where the positive sample is very low, so we choose AUROC (Area under Receiver Operating Characteristic) as our evaluation criterion. AUROC has been widely used to evaluate such models, and using AUROC score can effectively avoid experimental bias. To measure the performance of the methods, we adopted 5-fold cross validation, so each fold is going to be a 20% of the data as a test data set and the other 80% as a training data set.

3.2 Compare with State-of-the-Art Approaches

To evaluate the performance of proposed prediction method, we also applied some classical methods include the Katz measure [21] and latent factor model (LFM) to predict lncRNA–miRNA interactions based on similarity network. We compared LFM because our prediction step adopted the matrix completion, so the two recommendation algorithms were introduced. To further evaluate the performance of INLMI, we also compared it with the EPLMI which is the only one initially proposed for lncRNA–miRNA interactions prediction problem. Katz measure as a special algorithm to solve the problem of network link prediction, here is also used to carry out the contrast test. Table 1 reports the AUC scores of different algorithms on the same standard dataset.

Table 1. Performance comparison among different methods

Method	AUROC
INLMI	**0.8517**
EPLMI	0.8402
KATZ	0.7435
LFM	0.8257

4 Conclusion

Since the lncRNA–miRNA interactions are very significant for dissecting various biological mechanisms, the prediction of unknown lncRNA–miRNA interaction is an important direction that researchers have been working on. So far, apart from a few tools based on a small amount of information predicting miRNA targets, there's not much work to be done to predict the lncRNA–miRNA interactions. Thanks to the generation of these multiple types of data, a network integrated approach for lncRNA–miRNA interactions prediction is proposed in this paper. We do not predict the interaction of lncRNA–miRNA interactions using one similarity network constructed of one kind of information, as does the latest method. INLMI introduces two similarity measurements, which first characterizes the valuable information of each individual network. And then, we applying similarity integration with NMF to the two networks. In addition, we adopt matrix completion for predicting lncRNA–miRNA interactions based on the integrated network. The experimental results on real-world dataset show that our method has the ability to achieve the good performance.

Acknowledgments. This work is supported by the National Natural Science Foundation of China under Grant No. 61702424, and the National Natural Science Foundation of China under Grant No. 61572506.

References

1. Kung, J.T.Y., Colognori, D., Lee, J.T.: Long noncoding RNAs: past, present, and future. Genetics **193**(3), 651–669 (2013)
2. Salmena, L., Poliseno, L., Tay, Y., Kats, L., Pandolfi, P.P.: A ceRNA hypothesis: the Rosetta Stone of a hidden RNA language? Cell **146**(3), 353–358 (2011)
3. Quinn, J.J., Chang, H.Y.: Unique features of long non-coding RNA biogenesis and function. Nat. Rev. Genet. **17**, 47–62 (2016)
4. Du, Z., Sun, T., Hacisuleyman, E., Fei, T., Wang, X., Brown, M., Rinn, J.L., et al.: Integrative analyses reveal a long noncoding RNA-mediated sponge regulatory network in prostate cancer. Nat. Commun. **7**, 10982 (2016)
5. Li, J.-H., Liu, S., Zhou, H., Qu, L.-H., Yang, J.-H.: Star base v2.0: decoding miRNA-ceRNA, miRNA-ncRNA and protein–RNA interaction networks from large-scale CLIP-Seq data. Nucleic Acids Res. **42**, D92–D97 (2013)
6. Paraskevopoulou, M.D., Hatzigeorgiou, A.G.: Analyzing miRNA–lncRNA interactions. In: Long Non-Coding RNAs, pp. 271–286. Humana Press, New York (2016)
7. Chen, X., Sun, Y.Z., Zhang, D.H., Li, J.Q., Yan, G.Y., An, J.Y., You, Z.H.: NRDTD: a database for clinically or experimentally supported non-coding RNAs and drug targets associations. Database **2017** (2017)
8. Li, J., et al.: LncTar: a tool for predicting the RNA targets of long noncoding RNAs. Brief. Bioinf. **16**, 806–812 (2015)
9. Huang, Y.-A., Chan, K.C.C., You, Z.-H.: Constructing prediction models from expression profiles for large scale lncRNA-miRNA interaction profiling. Bioinformatics (2017)
10. Chen, X., Yan, C.C., Zhang, X., You, Z.H., Huang, Y.A., Yan, G.Y.: HGIMDA: heterogeneous graph inference for miRNA-disease association prediction. Oncotarget **7**(40), 65257–65269 (2016)
11. Levine, E., Hwa, T.: Small RNAs establish gene expression thresholds. Curr. Opin. Microbiol. **11**, 574–579 (2008)
12. Mukherji, S., Ebert, M.S., Zheng, G.X., Tsang, J.S., Sharp, P.A., van Oudenaarden, A.: MicroRNAs can generate thresholds in target gene expression. Nat. Genet. **43**(9), 854 (2011)
13. Yi, H.C., You, Z.H., Huang, D.S., Li, X., Jiang, T.H., Li, L.P.: A deep learning framework for robust and accurate prediction of ncRNA-protein interactions using evolutionary information. Mol. Ther. Nucleic Acids **11**, 337–344 (2018)
14. Cock, P.J., et al.: Biopython: freely available Python tools for computational molecular biology and bioinformatics. Bioinformatics **25**, 1422–1423 (2009)
15. Lee, D.D., Seung, H.S.: Learning the parts of objects by non-negative matrix factorization. Nature **401**(6755), 788 (1999)
16. Gong, J., et al.: lncRNASNP: a database of SNPs in lncRNAs and their potential functions in human and mouse. Nucleic Acids Res. **43**, D181–D186 (2015)
17. Volders, P.-J., et al.: LNCipedia: a database for annotated human lncRNA transcript sequences and structures. Nucleic Acids Res. **41**, D246–D251 (2013)
18. Bu, D., et al.: NONCODE v3.0: integrative annotation of long noncoding RNAs. Nucleic Acids Res. **40**, D210–D215 (2011)

19. Kozomara, A., Griffiths-Jones, S.: miRBase: annotating high confidence microRNAs using deep sequencing data. Nucleic Acids Res. **42**, D68–D73 (2014)
20. Betel, D., et al.: The microRNA.org resource: targets and expression. Nucleic Acids Res. **36**, D149–D153 (2008)
21. Chen, X., Huang, Y.A., You, Z.H., Yan, G.Y., Wang, X.S.: A novel approach based on KATZ measure to predict associations of human microbiota with non-infectious diseases. Bioinformatics **33**(5), 733–739 (2016)

Transcriptomic Analysis of Flower Development in the Bamboo *Phyllostachys violascens* (Poaceae: Bambusoideae)

Yulian Jiao[1,3], Qiutao Hu[1], Yan Zhu[2], Longfei Zhu[1], Tengfei Ma[1], Haiyong Zeng[1], Qiaolu Zang[1], Xinchun Lin[1(✉)], and Xuan Li[2(✉)]

[1] State Key Laboratory of Subtropical Silviculture, Zhejiang A&F University, Lin'an, Zhejiang 311300, People's Republic of China
linxcx@163.com
[2] Key Laboratory of Synthetic Biology, Institute of Plant Physiology and Ecology, Shanghai Institutes for Biological Sciences Chinese Academy of Sciences, Shanghai 200032, China
lixuan@sippe.ac.cn
[3] Research Institute of Subtropical Forestry, Chinese Academy of Forestry, Fu'yang, Zhejiang 311400, People's Republic of China

Abstract. Bamboo flowering is a complicate phenomenon for its long period and unpredictability. In this study, three successive stages of flowering buds from flowering bamboo (Lei bamboo, *Phyllostachys violascens*) plants and the corresponding vegetative buds of non-flowering plants were collected for transcriptome analysis. By using Illumina RNA-Seq method, about 442 million clean sequence reads were generated from the above samples and assembled into 317,273 transcripts with N50 of 1,968 bp, then acquired 132,678 unigenes with N50 of 1,080 bp, about 44.18% of the unigenes annotated in at least one database. 7,266 differentially expressed genes (DEGs) were determined through GO and KEGG analysis. Some DEGs were involved in plant hormone signal transduction and circadian rhythm pathways, which were highly expressed in the former and middle flower development stages, respectively. Our paper presents a useful data and a critical sampling method for studying flowering mechanism in bamboo.

Keywords: *Phyllostachys violascens* · RNA-seq
Plant hormone signal transduction · Circadian rhythm · Flower

1 Introduction

Flowering plays a key role in the transition from vegetative stage to reproductive stage. Bamboo is one of the most important non-timber forest resources in the world. Its fast growth and strong nitrogen fixation capacity have gained much attention in economy and ecology. The flowering of bamboo is a peculiar phenomenon, usually bamboo

---

Y. Jiao and Q. Hu—Contributed equally to this work.

© Springer International Publishing AG, part of Springer Nature 2018
D.-S. Huang et al. (Eds.): ICIC 2018, LNCS 10954, pp. 546–552, 2018.
https://doi.org/10.1007/978-3-319-95930-6_52

flowers synchronously on a large scale, and caused the death of bamboo after flowering, which leads to great economic and ecological losses. Recently, the draft genome sequences of moso bamboo *(Phyllostachys heterocycla)* was reported [1], and provides useful information for genomic research in bamboo. Furthermore, transcriptomic analysis of flowering development in several bamboo species has been reported [2, 3], and many putative flowering-related genes have been identified [4, 5]. However, the flowering mechanism in bamboo is still poorly understood.

In this paper, the transcriptome of 6 samples among different flowering development stages of Lei bamboo *(Phyllostachys violascens)*, an important bamboo species for bamboo shoots, were analyzed. The goal of this study was to obtain a complete set of assembled unigenes and transcripts for Lei bamboo and to identify flowering-related genes. Our paper could provide not only useful information, but also a critical sampling method to studying flowering mechanism in bamboos.

2   Materials and Methods

Sample plants of Lei bamboo were from bamboo forests cultivated about 30 years in Lin'an, Zhejiang province, China. In this study, three successive stages of flowering buds (TF_1,2,3) were collected from flowering bamboo plants, and the corresponding vegetative buds of non-flowering (TV_1,2,3) plants from the same rhizome were sampled to ensure the same genetic background (Fig. 1).

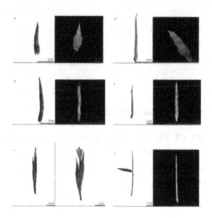

Fig. 1. The flowering and vegetable buds for sequencing. (a–c) The flowering buds sampled on March 8, March 29 and April 12 are named as TF_1, TF_2 and TF_3 respectively. (d–f) Vegetative buds sampled on March 8, March 29 and April 12 are named TV_1, TV_2 and TV_3 respectively.

In this paper, our sequencing libraries were constructed by using NEBNext® Ultra™ Directional RNA Library Prep Kit for Illumina® (NEB, USA) instruction. Gene function was annotated by using the following databases: Nr, Nt, Pfam, KOG, Swiss-Port, KO and GO. We used NCBI blast 2.2.28+ to search genes in these databases. Blast2GO v2.5 were used for GO function annotation based on the annotation of the NR and Pfam databases [6]. We calculated the gene expression by using RSEM software and transformed the read counts to Reads Per Kilobase of transcript per Million mapped reads (RPKM) [7, 8]. The DEGseq R package were used for differential expression analyses, and $\log_{10}$ (RPKM + 1) were used for heat map data.

3   Results

3.1   Illumina Sequencing and De novo Assembly

Using the Illumina HiSeq 2000 platform, we obtained two pair-end sequencing about 317,273 transcripts and 132,678 unigenes (Table 1), and analyzed the length distribution of both (Table 2).

Table 1. Transcript and unigene length intervals

Transcript length interval	200–500 bp	500–1 kbp	1–2 kbp	>2 kbp	Total reads
No. of transcripts	125980	64002	68737	58554	317273
No. of unigenes	85724	24704	13221	9029	132678

Table 2. Transcript and unigene length distribution

	Length				N50	N90	Total nucleotides
	Min	Mean	Median	Max			
No. of transcripts	201	1147	709	16212	1968	473	364054006
No. of unigenes	201	675	366	16212	1080	269	89499052

3.2   Functional Analysis of Total Unigenes

We analyzed the function of the total unigenes by annotating the unigenes with 7 databases (Table 3). Unigenes annotated in the NR, NT, KO, Swissport, PFAM, GO and KOG databases were 44,299 (33.38%), 35,526 (26.77%), 6,932 (5.22%), 28,209 (21.26%), 32,851 (24.75%), 39,043 (29.42%), and 16,232 (12.23%), respectively.

Table 3. Annotation of total unigenes in databases

Annotated in database	Databases							All databases	At least 1 database	Total unigenes
	NR	NT	KO	SwissProt	PFAM	GO	KOG			
No. of unigenes	44299	35526	6932	28209	32851	39043	16232	2931	58628	132678
Percentage (%)	33.38	26.77	5.22	21.26	24.75	29.42	12.23	2.2	44.18	100

From the KOG functional classification, the cluster for general functional prediction (17.37%) represented the largest group, followed by post-translational modification, protein turnover, chaperon (11.82%), translation (10.24%), signal transduction (6.95%), energy production and conversion (6.12%), with little for extracellular structures, nuclear structure, and cell motility (Fig. 2a). To identify biochemical pathways, we mapped the annotated sequences in the KEGG database (Fig. 2b), a classification of gene function that places emphasis on biochemical pathways. Most unigenes were classified as involved in translation, energy metabolism and carbohydrate metabolism, with the least in membrane transport, signaling molecules and degradation. These pathway assignments provide valuable information for investigating specific biochemical and development processes.

The GO classification usually includes three major aspects: molecular function, cellular component and biological process. In molecular function, functional terms highly represented among our genes were binding, catalytic activity, and then transporter activity and structural molecule activity, with the least being channel regulator activity (Fig. 2c). For cellular component, functional terms were cell, cell part, organelle, macromolecular complex, membrane, and organelle part, with the least being cell junction. For biological process, functional terms were mostly cellular process, metabolic process, single-organism process, biological regulation, and little on rhythmic process.

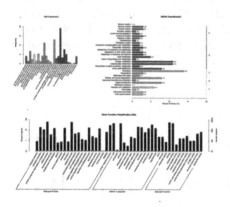

Fig. 2. The functional analysis of total unigenes. (a) KOG classification (b) KEGG classification, and (c) GO functional classification.

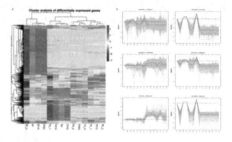

Fig. 3. Cluster analysis of differentially expressed genes (DEGs). (a) Expression values for 6 libraries are presented as RPKM normalized by transformed counts. Red and blue indicate up-regulated and down-regulated transcripts respectively; (b) Six main clusters are shown based on the heat map of DEGs. (Color figure online)

3.3 The Analysis of Differentially Expressed Genes (DEGs)

In total, 7,266 unigenes were identified as differential expressed genes (DEGs) in our six cDNA libraries from 16 comparisons, including unregulated (UP) and downregulated (DOWN) DEGs in each comparison group. We analyzed the gene expression of all 7,266 DEGs in the heat map (Fig. 3a), and the phases of TF_3 and TV_1 possessed more genes with high expression in flowering and non-flowering samples respectively. We also performed k_means clustering of all the DEGs (Fig. 3b), and 6 clusters were plotted with expression patterns. Subcluster_1 included 2,136 genes that were highly expressed in TF_3 sample and had low expression in TF_1, TF_2 and all vegetative samples. The subcluster_2 included 591 genes, subcluster_4 included 1,256 genes, and subcluster_6 included 264 genes, which all had a similar expression pattern as subcluster_1. The subcluster_3 included 2,538 genes and showed high expression in TV_1 samples but lowest expression in TF_3 samples. The subcluster_5 included 481 genes and showed high expression in vegetative samples versus flowering samples, especially in TV_1 samples. The results corresponded to the heat map of DEGs revealing the DEGs expressed mostly in TF_3 and TV_1.

3.4 Hormone and Circadian Rhythm-Related Genes in Flower Development

DEGs enriched in the path-ways of hormone signal transduction (ko04075) and circadian rhythm in plant (ko04712) were used to explore the possible key flower genes in our species (Fig. 4). For the DOWN DEGs of TF_3 versus TF_1, DOWN DEGs of TF_3 versus TF_2 and UP DEGs of TF_1 versus TV_1, the enriched KEGG pathway hormone signal transduction in plants was associated with ARF (comp139738_c1), CRE1 (comp138523_c0), BRI1 (comp130360_c0, comp139502_c2), BSK (comp137102_c0), NPR1 (comp140280_c0); AUX1 (comp138485_c1, comp 137757_c1), TGA (comp133847_c0); TIR1 (comp136544_c2); B-ARR (comp 139166_c1, comp134657_c0) and CTR1 (comp138697_c0). These genes tend to express highly in the stages of TF_1 and TF_2.

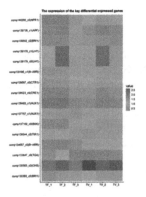

Fig. 4. The expression patterns of key genes in circadian rhythm and plant hormone pathways in Lei bamboo.

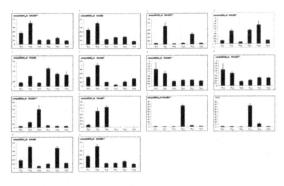

Fig. 5. Real-time quantitative RT-PCR confirmation of 12 candidate genes at the six stages, TF_1, TF_2, TF_3, TV_1, TV_2, and TV_3.

The circadian rhythm-plant was associated with *LHY* (*comp139179_c0* and *comp 139179_c1*) and *CHS* (*comp130563_c0*). Both *LHY* and *CHS* had high expression in TF_2 but low expression in the phases of TF_1 and TF_3 (Fig. 4).

3.5   Verification of the Gene Expression by RT-PCR

Transcriptional regulation revealed by RNA-seq was confirmed in a biologically independent experiment using RT-PCR, a total of 12 genes were chosen to design gene-specific primers. Each gene showed significantly different expression (P = 0.05) (Fig. 5). Moreover, 10 genes showed significant correlations (P = 0.05 & P = 0.01) between the RT-qPCR data and the RNA-seq results, for good reproducibility between transcript abundance assayed by RNA-seq and the expression profile revealed by RT-qPCR.

4   Conclusion

RNA-seq analysis were used to monitor global transcriptional changes at the three developmental stages of flower buds and vegetative buds, and it was enabled comprehensive analysis of differential transcriptional events that occurred during flower formation in Lei bamboo. In total, 132,678 unigenes were assembled, and 7266 DEGs were identified in analyzing flower buds and vegetative buds at the three stages. Most of these DEGs were expressed in TF_3 and TV_1, which corresponds to morphologic change in these stages. Most of DEGs related to plant hormones, including *NPR1*, *BRI1*, *BSK*, *CTR1*, *CRE1*, *ARR-B*, *AUX1* and *ARF*, had a high expression in flower initiation and differentiation stages (TF_1 and TF_2), so these genes may have roles in these two stages. DEGs such as *LHY* and *CHS* in the circadian rhythm pathway tended to have high expression in TF_2, which suggests that they may function in the stage of formation and identity determination of floral organs. Our results lay a foundation for future studies of the molecular mechanisms underlying bamboo flower development.

References

1. Peng, Z., Lu, Y., Li, L., Zhao, Q., Feng, Q., et al.: The draft genome of the fast-growing non-timber forest species moso bamboo (Phyllostachys heterocycla). Nat. Genet. **45**, 456–461 (2013)
2. Lin, X.C., Chow, T.Y., Chen, H.H., Liu, C.C., Chou, S.J., et al.: Understanding bamboo flowering based on large-scale analysis of expressed sequence tags. Genet. Mol. Res. **9**, 1085–1093 (2010)
3. Gao, J., Zhang, Y., Zhang, C., Qi, F., Li, X., et al.: Characterization of the floral transcriptome of Moso Bamboo (Phyllostachys edulis) at different flowering developmental stages by transcriptome sequencing and RNA-Seq analysis. PLoS ONE **9**, e98910 (2014)
4. Shih, M., Chou, M., Yue, J., Hsu, C.H., Chang, W., et al.: BeMADS1 is a key to delivery MADSs into nucleus in reproductive tissues-De novo characterization of Bambusa edulis transcriptome and study of MADS genes in bamboo floral development. BMC Plant Biol. **14**, 179 (2014)

5. Liu, S., Ma, T., Ma, L., Lin, X.: Ectopic expression of PvSOC1, a homolog of SOC1 from Phyllostachys violascens, promotes flowering in Arabidopsis and rice. Acta Physiol. Plant. **38**, 166 (2016)
6. Conesa, A., Gotz, S., Garciagomez, J.M., Terol, J., Talon, M., et al.: Blast2GO: a universal tool for annotation, visualization and analysis in functional genomics research. Bioinformatics **21**, 3674–3676 (2005)
7. Li, B., Dewey, C.N.: RSEM: accurate transcript quantification from RNA-Seq data with or without a reference genome. BMC Bioinf. **12**, 323 (2011)
8. Mortazavi, A., Williams, B.A., Mccue, K., Schaeffer, L., Wold, B.J.: Mapping and quantifying mammalian transcriptomes by RNA-Seq. Nat. Methods **5**, 621–628 (2008)

PASA: Identifying More Credible Structural Variants of Hedou12

Huiqiang Jia[1], Haichao Wei[2], Daming Zhu[1(✉)], Ruizhi Wang[1],
Haodi Feng[1], and Xiangzhong Feng[2]

[1] School of Computer Science and Technology, Shandong University,
Jinan 250101, China
dmzhu@sdu.edu.cn
[2] Key Laboratory of Soybean Molecular Design Breeding,
Northeast Institute of Geography and Agroecology,
Chinese Academy of Sciences, Changchun 130102, China

Abstract. In this paper, we devote to find structural variants including deletions, insertions, and inversions which occur in Hedou12 genome in constrast to Williams82 genome. To find as many as possible potential structural variants, we try to develop new principles to detect discordant and split read map sets supporting structural variants. Aiming to enhance the precision of structural variant detection, we propose two new sequencing characteristic based models, which use the sequencing parameters of Hedou12 paired-end reads, as well as the parameters for Hedou12 paired-end reads to be aligned onto Williams82, to evaluate the probability a potential structural variant can occur in. To remove those false members from the potential structural variants, we propose a integer linear program to describe formally on which potential structural variants it should accept to achieve as high as possible a probability summation, whose solution can help predict more credible structural variants. The feasibility and precision of our algorithm are verified by comparing with DELLY version 0.5.8 and LUMPY version 0.2.2.3. The software is available for download at https://pan.baidu.com/s/1rasmtti.

Keywords: Algorithm · Deletion · Insertion · Inversion · Structural variant

1 Introduction

Using sequencing data to detect genome structural variants has been drawing attention of many computer science and molecular biology scientists and playing an important role in computational biology and bioinformatics [1]. A typical genome structural variant (SV) can be recognized as a deletion, an insertion, a translocation, a duplication, an inversion, and so on [2]. The alignment signals such as the read coverage degree [3], the read map distribution, as well as where and in what forms the paired-end reads are mapped to the reference genome, have been being frequently used for detecting SVs [4–6]. The familiar SV predicting software tools based on mining such kinds of sequencing data properties, can be looked up in [5, 7–9].

© Springer International Publishing AG, part of Springer Nature 2018
D.-S. Huang et al. (Eds.): ICIC 2018, LNCS 10954, pp. 553–558, 2018.
https://doi.org/10.1007/978-3-319-95930-6_53

It has to indicate that the existing software tools for predicting SVs usually aim at human genomes. On predicting plant genome SVs, little has been done as far as we know. In this paper, we focus on Hedou12 genome to predict SVs in contrast to the reference genome, Williams82.

To detect sufficient SVs, those paired-end reads which have been aligned onto the reference genome abnormally, are really what we need. By finding *discordant* as well as *split* reads, which have been mapped to a specific reference genome region, a set of paired-end read maps can be reached to support a potential structural variant. This can be done repeatedly to help find a collection of potential structural variants, each of which are supported by a discordant read map set or a split read map set.

As we know so far, DELLY [8] and LUMPY [9] are the most typical software tools, which can predict SVs by means of mapping the paired-end reads onto the reference genome. Although DELLY and LUMPY can use various kinds of alignment signals to detect SVs, it is not sure for any SV they predict to occur, to really occur.

In this paper, we firstly propose criterias which can be used to collect discordant as well as split read maps into a set which may be caused by a structural variant, and is called to *support a potential* SV. Then, aiming to filter out the false members from those potential SVs, we propose two new models to evaluate the probabilities a potential discordant read map set supported SV, and respectively a potential split read map set supported SV, can really occur in. Based on the probabilities of those potential SVs, we propose to use an integer linear program to select those SVs as what we acknowledge to occur.

The validated verification shows that our algorithm can with better precision, find no less deletions, insertions and inversions than those DELLY version 0.5.8 [8] and LUMPY version 0.2.2.3 [9] can find, in the majority of those SV length ranges. The algorithm has been demonstrated with the same performance in more practices of identifying structural variants in Hedou12 genome [10] in contrast to Williams82 genome [11], where Hedou12 and Williams82 have been widely used in soybean seed design researches.

2   Extract Structural Variants

We start with mapping the paired-end reads of Hedou12 genome [10] onto the reference genome Williams82 [11]. A paired-end read is referred to as *concordant*, if for every time it is aligned onto the reference genome, its ends can both be aligned onto the reference genome in the same orientation as they are given in the read, and in the distance without going too longer or too shorter than the insert size a normal read should have. A paired-end read is referred to as *discordant*, if it is not concordant. A paired-end read is referred to as *split*, if at least one of its ends has to be split into two segments, before it can be mapped onto one reference genome chromosome.

To extract those discordant and split read maps, we have to use BWA aligner [12] and mrsFast [13] to align the paired-end reads of Hedou12 onto the reference genome. This can help develop criterias for getting a collection of discordant read map sets and split read map sets, each of which supports a potential SV. To make a credible decision on which potential SVs supported by those read map sets really occur, a preferable way is to select those which can occur in as high as possible a probability summation which can cover all those discordant and split read maps.

2.1   Discordant Read Map Cluster

Let PE denote a discordant paired-end read set, $PE_i \in PE$ for $1 \leq i \leq |PE|$. Provided PE_i can be aligned onto the reference genome with N_i maps, we denote by PE_i^m the mth map of PE_i, for m with $1 \leq m \leq N_i$. We use the so called *insert-size* of the paired-end reads, and those parameters accompanied with the discordant read maps to cluster the discordant read maps into sets, to support potential SVs (Deletion, Insertion, Inversion). Let $\Gamma_1 = \{V_1, \ldots, V_N\}$ denote a collection of discordant read map sets each of which supports a potential deletion, insertion, or inversion.

The *similarity measure* of a read map end indicates how alike the sequence map and that reference genome subsequence spanned by the sequence map are. Let $SeqL\left(PE_j^m\right)$ (resp. $SeqR\left(PE_j^m\right)$) denote the similarity measure of PE_j^m left (resp. right) end. Then the probability PE_j can be aligned onto the reference genome as PE_j^m, can be accounted by,

$$q(PE_j^m) = \frac{LR\left(PE_j^m\right) * SeqL\left(PE_j^m\right) + RR\left(PE_j^m\right) * SeqR\left(PE_j^m\right)}{LR\left(PE_j^m\right) + RR\left(PE_j^m\right)}. \tag{1}$$

Let $q(PE_j^m \in V_i | PE_j^m)$ denote the probability for PE_j^m to occur in V_i in premise that it occurs. Consequently, we set $q(PE_j^m \in V_i | PE_j^m)$ by $\frac{1}{K}$, if PE_j^m has been recognized to occur in K SVs. Thus a potential structure variant V_i supports can occur in Hedou12 genome in the evaluated probability,

$$P(V_i) = \frac{1}{|V_i|} \sum_{PE_j^m \in V_i} q\left(PE_j^m\right) * q(PE_j^m \in V_i | PE_j^m). \tag{2}$$

2.2   Split-Read Map Cluster

A split-read is referred to as *balanced*, if its split end has been split into two segments with the same number of base symbols, and *unbalanced* otherwise. In what follows, a split read map is balanced by default. We have to formalize criterias to cluster split read maps into sets to support SVs. Let $\Gamma_2 = \{R_1, \ldots, R_M\}$ denote a collection of split read map sets each of which supports a deletion, an insertion, or an inversion. Let Q $(SE_j^m) \equiv Q(SE_j)$ denote the quality value of those two segments that SE_j end has been split into [14]. We set a probability for SE_j^m to occur in R_i in premise that SE_j^m occurs as $q(SE_j^m \in R_i | SE_j^m) = \frac{1}{K}$, provided SE_j^m occurs in K (>0) potential SVs. Finally, we propose to evaluate the probability the potential structural variant supported by R_i occurs by,

$$P(R_i) = 1 - 10^{-\frac{\frac{1}{|R_i|}\sum_{SE_j^m \in R_i} Q\left(SE_j^m\right)*seq\left(SE_j^m\right)*q\left(SE_j^m \in V_i | SE_j^m\right)}{100}}. \tag{3}$$

2.3 On Throwing Away False Structural Variants

Let $\Gamma = \Gamma_1 \cup \Gamma_2 = \{V_1, \ldots, V_N\} \cup \{R_1, \ldots, R_M\}$ be the collection of all those read map sets we get in the way of Sects. 2.1 and 2.2, each of which, say V_i (resp. R_j), can support a deletion, an insertion, or an inversion with probability $P(V_i)$ (resp. $P(R_j)$), $U = V_1 \cup \ldots \cup V_N \cup R_1 \cup \ldots \cup R_M$. To exclude those false members in Γ, we select sets in Γ to cover all read maps in U, such that those structural variants supported by the selected sets can occur with as high as possible a probability summation. We can summarize it into a integer linear program. The solution of this program will provide us with those deletions, insertions, and inversions, each supported by a read map set. We name by PASA the aforementioned algorithm including the computations of discordant read map cluster (Sect. 2.1), split read map cluster (Sect. 2.2), and throwing away false structural variants (Sect. 2.3).

3 Experiments and Discussions

Let U_1 denote the set of read maps each with a similarity measure no less than 90%, and U_2 the set of read maps each with a similarity measure no less than 80%. We use those read maps in U_1 to find a collection of sets, say Γ, to support potential SVs. If there is a read map in $U_2 - U_1$ which can be appended to a set, say V in Γ to support the SV the set V supports, then the structural variant V supports will be thought of as what can be believed.

Although PASA may find less insertions than those DELLY as well as LUMPY can find, the number ratio of those $U_2 - U_1$ read map supported insertions divided by all those insertions PASA can find is always larger than that of DELLY and LUMPY in each insertion length range. PASA can always find more deletions than DELLY as well as LUMPY can find, with an exception in the length range 1000–1999 bp. It also deserves to note that the longest and shortest deletions PASA, DELLY and LUMPY can find have 50,783,338 bp/17 bp, 52,455,922/17 bp, and 52,455,922/17 bp base symbols respectively. Moreover, it can be observed that, as the deletion length increases, less deletions will occur in Hedou12 genome. The longest and shortest inversions PASA, DELLY and LUMPY can find have 47,581,392 bp/98 bp, 47,581,392 bp/98 bp, and 47,581,392 bp/98 bp base symbols respectively.

Finally, it has to specialize that PASA can always find deletions, insertions and inversions with more than 90% of them shared with those DELLY and LUMPY can find.

4   Conclusion

Better methods for aligning reads onto the reference genome wait to be developed in the future. This seems possible to help detect structural variants in better performance. Finding long length of insertions have also been looking forward to new techniques. The existing software may find structural variants which conflict to occur together. Removing the conflictions between those structural variants seems necessary and interesting.

Acknowledgment. This paper is supported by National natural science foundation of China, No. 61472222, 61732009, 61672325, 61761136017.

References

1. Jiang, Y., Wang, Y., Brudno, M.: PRISM: pair-read informed split-read mapping for base-pair level detection of insertion, deletion and structural variants. Bioinformatics **28**(20), 2576–2583 (2012)
2. Korbel, J., Abyzov, A., Mu, X., Carriero, N., Cayting, P., Zhang, Z., Snyder, M., Gerstein, M.: PEMer: a computational framework with simulation-based error models for inferring genomic structural variants from massive paired-end sequencing data. Genome Biol. **10**(2), R23 (2009)
3. Abyzov, A., Urban, A., Snyder, M., Gerstein, M.: CNVnator: an approach to discover, genotype, and characterize typical and atypical CNVs from family and population genome sequencing. Genome Res. **21**(6), 974–984 (2011)
4. Chen, K., Wallis, J., McLellan, M., Larson, D., Kalicki, J., Poh, C., McGrath, S., Wendl, M., Zhang, Q., Locke, D., Shi, X., Fulton, R., Ley, T., Wilson, R., Ding, L., Mardis, E.: BreakDancer: an algorithm for high-resolution mapping of genomic structural variation. Nat. Methods **6**(9), 677–681 (2009)
5. Fan, X., Abbott, T., Larson, D., Chen, K.: BreakDancer: identification of genomic structural variation from paired-end read mapping. Curr. Protoc. Bioinf. **45**, 15.6.1–15.6.11 (2014)
6. Karakoc, E., Alkan, C., O'Roak, B., Dennis, M., Vives, L., Mark, K., Rieder, M., Nickerson, D., Eichler, E.: Detection of structural variants and indels within exome data. Nat. Methods **9**(2), 176–178 (2012)
7. Medvedev, P., Fiume, M., Dzamba, M., Smith, T., Brudno, M.: Detecting copy number variation with mated short reads. Genome Res. **20**(11), 1613–1622 (2010)
8. Rausch, T., Zichner, T., Schlattl, A., Stütz, A., Benes, V., Korbel, J.: DELLY: structural variant discovery by integrated paired-end and split-read analysis. Bioinformatics **28**(18), i333–i339 (2012)
9. Layer, R., Chiang, C., Quinlan, A., Hall, I.: LUMPY: a probabilistic framework for structural variant discovery. Genome Biol. **15**(6), R84 (2014)
10. Song, X., Wei, H., Cheng, W., Yang, S., Zhao, Y., Zhang, H., Feng, X.: Development of INDEL markers for genetic mapping based on whole genome resequencing in soybean. G3 (Bethesda, Md) **5**(12), 2793–2799 (2015)
11. Gao, J., Yang, S., Cheng, W., Fu, Y., Leng, J., Yuan, X., Jiang, N., Ma, J., Feng, X.: GmILPA1, encoding an APC8-like protein, controls leaf petiole angle in soybean. J. Am. Soc. Plant Biologists **174**(2), 1167–11176 (2017)

12. Li, H., Durbin, R.: Fast and accurate short read alignment with burrows-wheeler transform. Bioinformatics **25**(14), 1754–1760 (2009)
13. Hach, F., Hormozdiari, F., Alkan, C., Hormozdiari, F., Birol, I., Eichler, E., Sahinalp, S.: mrsFAST: a cache-oblivious algorithm for short-read mapping. Nat. Methods **7**(8), 576–577 (2010)
14. Ewing, B., Green, P.: Base-calling of automated sequencer traces using Phred. II. Error probabilities. Genome Res. **8**(3), 186–194 (1998)

RPML: A Learning-Based Approach for Reranking Protein-Spectrum Matches

Qiong Duan[1], Hao Liang[1], Chaohua Sheng[1], Jun Wu[2], Bo Xu[1(✉)], and Zengyou He[1(✉)]

[1] School of Software, Dalian University of Technology, Dalian, China
{boxu, zyhe}@dlut.edu.cn
[2] School of Information Engineering, Zunyi Normal University, Zunyi, China

Abstract. Searching top-down spectra against a protein database has been a mainstream method for intact protein identification. Ranking true Protein-Spectrum Matches (PrSMs) over their false counterparts is a feasible method for improving protein identification results. In this paper, we propose a novel model called RPML (Rerank PrSMs based on Machine Learning) to rerank PrSMs in top-down proteomics. The experimental results on real data sets show that RPML can distinguish more correct PrSMs from incorrect ones. The source codes of algorithm are available at https://github.com/dqiong/spectra_protein_match_rerank.

Keywords: Protein identification · Protein-spectrum matches
Machine learning · Rerank method

1 Introduction

Proteomics aims at studying the complete set of proteins expressed in a given cell, tissue, or organism. One important problem of proteomics is to identify all proteins in a complex mixture effectively and sensitively within a limited time.

The current revolution in mass spectrometry instruments has made it possible to obtain high-resolution spectra of intact proteins. The top-down approach without the use of proteolytic digestion, proteins are less susceptible to instrumental biases than their short peptides. More importantly, the top-down approach is able to retain the correlation information between multiple PTMs. Therefore, the top-down approach has become a promising technique to the bottom-up method.

Over the last decade, some effective algorithms have been proposed to solve the top-down protein identification problem [9, 10, 12]. A key component in existing identification approaches is the scoring function used to evaluate the quality of protein-spectrum matches. However, current PrSM scoring algorithms are far from satisfactory since incorrect PrSMs are frequently observed in the identification results. It is quite challenging to design a PrSM scoring algorithm that is capable of distinguishing correct identifications from incorrect ones in a perfect manner. One alternative strategy is to develop new post-processing techniques by reranking PrSMs such that more true identifications are re-arranged before false identifications.

© Springer International Publishing AG, part of Springer Nature 2018
D.-S. Huang et al. (Eds.): ICIC 2018, LNCS 10954, pp. 559–564, 2018.
https://doi.org/10.1007/978-3-319-95930-6_54

To date, some reranking methods have been introduced for improving the peptide identification results in bottom-up proteomics [6, 7]. The success of the reranking methods in bottom-up proteomics motivates us to investigate if it is also feasible to rerank PrSMs so as to generate an improved top-down identification results. To the best of our knowledge, this is the first attempt that studies the PrSM reranking problem in the context of top-down protein identification.

2   Methods

2.1   Problem Definition

Let $\mathbb{C} = \{(s_1, p_1), (s_2, p_2), \ldots, (s_n, p_n)\}$ be a set of PrSMs, where s_i is a MS/MS spectra and p_i is a protein sequence. The set $\mathbb{C}$ is associated with a vector of initial ranking scores $X = (x_1, x_2, \ldots, x_n)^T$ provided by a standard top-down protein identification approach, where x_i is the score of (s_i, p_i). The type of score varies according to the baseline identification methods (e.g., E-value, the number of matching peaks). The goal of the reranking model is to assign (s_i, p_i) a new score y_i so that the new vector $Y = (y_1, y_2, \ldots, y_n)^T$ can improve the ranking results.

2.2   General Workflow of RPML

RPML consists of three main procedures: feature extraction, model construction and score integration.

2.3   Feature Extraction

In feature extraction, each PrSM is represented with eleven features (more details are available at https://github.com/dqiong/spectra_protein_match_rerank/blob/master/RPML.pdf). To make the feature values comparable, all eight numeric features extracted from data are transformed into [0,1] with the min-max normalization method. Other three features take binary values: 0 or 1. Here we emphasize that RPML uses one feature (e.g. E-value) as the target feature and the remaining ten features as dependent features.

2.4   Model Construction

The PrSMs have no labels in the initial ranking list, we have to set up the labels artificially. Generally, the baseline identification method reports an initial ranking score rs (this score is the feature value from one of the eleven extracted features) for each PrSM. There will be a positive correlation between rs and the correctness of identifi-cations. Therefore, we first sort all PrSMs based on rs. Then, we select a portion of top-ranked PrSMs as the candidate positive training set and the same portion of PrSMs at the end of the sorted list as the candidate negative training set. Finally, we randomly take a subset from both positive and negative candidate training sets to construct the training set used for model construction. Thus, we can generate multiple training data sets in parallel to build the prediction model so as to increase the stability and accuracy. It is difficult to determine the optimal size of both candidate training data and actual training data automatically. RPML select top 20% of PrSMs at the beginning of sorted list to construct the candidate positive training data. Similarly, those 20% of PrSMs with lowest initial ranks are used as the candidate negative training data. In addition, we use a parameter c to control the size of actually selected training data. In our experiments, we set $c = 80\%$.

In this paper, we choose two widely used classifiers: xgboost (eXtreme Gradient Boosting) [3] and LR (Logistic Regression) [1] to construct prediction models on the training set to predict the probabilities for all PrSMs.

2.5   Score Integration

After model training and prediction, each classifier generates k scores for each PrSM, where k is the number of training data sets. For instance, xgboost will produce a probability score vector $SC_{xgb(i)} = (sc^1_{xgb(i)}, sc^2_{xgb(i)}, \dots, sc^n_{xgb(i)})^T$ for the set of PrSMs based on the ith training data set, where n is the total number of PrSMs. As a result, there will be k new score vectors for the set of n PrSMs. In this paper, we use a simple approach to get the consensus score of each PrSM via calculating the arithmetic mean of its k scores. This means that each training data set is assigned with the same weight when considering its contribution to the final score for one classification model. For the xgboost method, the score for the ith PrSM is:

$$sc^i_{xgb} = \frac{sc^i_{xgb(1)} + sc^i_{xgb(2)} + \dots + sc^i_{xgb(k)}}{k} \tag{1}$$

Similarly, we can get the score of the ith PrSM for LR:

$$sc^i_{lr} = \frac{sc^i_{lr(1)} + sc^i_{lr(2)} + \dots + sc^i_{lr(k)}}{k} \tag{2}$$

To integrate the results from xgboost and LR, we set a parameter α to control the weights of two classifiers. The final consensus score of the ith PrSM is:

$$sc^i_{final} = \alpha sc^i_{xgb} + (1 - \alpha)sc^i_{lr} \tag{3}$$

where α is set to be 0.5 in our experiments. After the score integration step, we can order the PrSMs according to the new ranking vector SC_{final}.

3   Experiments

We implemented the RPML algorithm in Python and compared its performance with MS-Align+ [9] on four data sets.

3.1   Data Sets

In the experiments, we used four data sets to test the performance of RPML: Ecoli [2], H2A [13], ST [14] and Autopilot [4]. Each MS/MS spectrum of the four data sets was converted to a deconvoluted spectrum using MS-Deconv [8]. Only the spectra that have a precursor mass ≥ 2500 Da and at least 10 fragment peaks were retained for protein identification. The corresponding protein databases were downloaded from NCBI (https://www.ncbi.nlm.nih.gov/). The target-decoy strategy [5] was used for performance evaluation, where each decoy protein was generated by shuffling the corresponding protein sequence in the target databases. MS-Align+ was used as the baseline method with the following parameter settings: *searchType* = target + decoy, *shiftNumber* = 3, *errorTolerance* = 15, *cutoff* = 0.01, and default values are used for other parameters. Meanwhile, MS-Align+ reported only the PrSM with the best matching score for each spectrum. Table 1 shows the number of spectra and the number of target and decoy PrSMs in the initial identification results of MS-Align+.

Table 1. Data sets used in experiments and the distribution of target and decoy PrSMs in the ranking list generated from MS-Align+.

Data set	#Spectra	#Target PrSMs	#Decoy PrSMs
Ecoli	1848	925	943
H2A	3529	2992	537
ST	4460	1725	2735
Autopilot	12102	7698	4404

3.2   Performance Evaluation

To compare the performance different methods, we plot a curve for each competing approach. In such a curve, the number of true positives (target PrSMs) is used as the y-axis and the q-value [11] is used as the x-axis. If we specify a score threshold t and refer to PrSMs with scores that are better than t as accepted PrSMs, the false discovery rate (FDR) is defined as the percentage of accepted PrSMs that are incorrect. The q-value is defined as the minimal FDR threshold at which a given PrSM is accepted. Here the target-decoy strategy is utilized to calculate the q-value.

3.3   Results

We firstly compared the RPML algorithm with MS-Align+. Our goal is to correctly identify as many target PrSMs as possible for a given q-value. Therefore, in Fig. 1, we plot the number of identified target PrSMs as a function of q-value threshold. Note that RPML ensembles xgboost and LR to construct a prediction model, we also included these two classifiers in the performance comparison. That is, RPML equals xgboost and LR when the parameter $\alpha = 1$ and 0, respectively.

From the comparison results in Fig. 1 we can see that RPML has better performance than MS-Align+ on all data sets. This indicates that our method is capable of improving the initial ranking list of PrSMs generated from MS-Align+. For xgboost and LR, there is no method can always achieve better performance than the other. LR has better performance on the Ecoli and H2A data set, while xgboost exhibits good performance on the ST and Autopilot data set. This means that the results produced by different methods are very diverse. In contrast, RPML is able to obtain the best performance on the H2A, ST and Autopilot data set when the number of target PrSMs is apparently different from that of the decoy PrSMs. It is no doubt that the ensemble strategy is one key factor to the success of RPML.

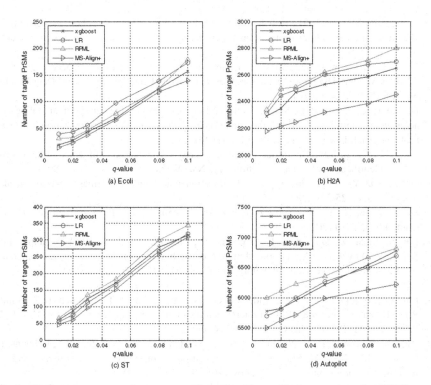

Fig. 1. Performance comparison among four algorithms. The parameters for RPML were specified as: $\alpha = 0.5$, $c = 80\%$ and $k = 100$.

4   Conclusion

Ranking true Protein-Spectrum Matches (PrSMs) over their false counterparts is a feasible strategy to improve the protein identification performance in top-down proteomics. In this paper, we propose the first PrSM reranking algorithm in the literature. Our new algorithm is called RPML, which is based on the idea of supervised ensemble learning and a novel adaptive training data construction method. The experiments on four data sets show that the proposed algorithm is capable of distinguishing more correct identifications from incorrect ones.

Acknowledgements. This work was partially supported by the Natural Science Foundation of China (Nos. 61572094, 61502071), the Fundamental Research Funds for the Central Universities (Nos. DUT2017TB02, DUT14QY07) and the Science-Technology Foundation for Youth of Guizhou Province (No. KY[2017]250).

References

1. Bishop, C.: Pattern Recognition and Machine Learning. Springer, New York (2007)
2. Cannon, J.R., Cammarata, M., Robotham, S.A., Cotham, V.C., Shaw, J.B., Fellers, R.T., Early, B.P., Thomas, P.M., Kelleher, N.L., Brodbelt, J.S.: Ultraviolet photodissociation for characterization of whole proteins on a chromatographic time scale. Anal. Chem. **86**(4), 2185–2192 (2014)
3. Chen, T., Guestrin, C.: XGBoost: a scalable tree boosting system. In: Proceedings of the 22nd ACM SIGKDD International Conference on Knowledge Discovery and Data Mining, pp. 785–794. ACM (2016)
4. Durbin, K.R., Fellers, R.T., Ntai, I., Kelleher, N.L., Compton, P.D.: Autopilot: an online data acquisition control system for the enhanced high-throughput characterization of intact proteins. Anal. Chem. **86**(3), 1485–1492 (2014)
5. Elias, J.E., Gygi, S.P.: Target-decoy search strategy for increased confidence in large-scale protein identifications by mass spectrometry. Nat. Methods **4**(3), 207–214 (2007)
6. He, Z., Yu, W.: Improving peptide identification with single-stage mass spectrum peaks. Bioinformatics **25**(22), 2969–2974 (2009)
7. He, Z., Zhao, H., Yu, W.: Score regularization for peptide identification. Asia Pac. Bioinf. Conf. **12**(1), 1–10 (2011)
8. Liu, X., Inbar, Y., Dorrestein, P.C., Wynne, C., Edwards, N., Souda, P., Whitelegge, J.P., Bafna, V., Pevzner, P.A.: Deconvolution and database search of complex tandem mass spectra of intact proteins a combinatorial approach. Mol. Cell. Proteomics **9**(12), 2772–2782 (2010)
9. Liu, X., Sirotkin, Y., Shen, Y., Anderson, G., Tsai, Y.S., Ying, S.T., Goodlett, D.R., Smith, R.D., Bafna, V., Pevzner, P.A.: Protein identification using top-down spectra. Mol. Cell. Proteomics MCP **11**(6), M111.008524 (2012)
10. Park, J., Piehowski, P.D., Wilkins, C., Zhou, M., Mendoza, J., Fujimoto, G.M., Gibbons, B. C., Shaw, J.B., Shen, Y., Shukla, A.K.: Informed-proteomics: open source software package for top-down proteomics. Nat. Methods **14**(9), 909–914 (2017)
11. Storey, J.D.: A direct approach to false discovery rates. J. R. Stat. Soc. Ser. B Stat. Methodol. **64**(3), 479–498 (2002)
12. Sun, R., Luo, L., Wu, L., Wang, R., Zeng, W., Chi, H., Liu, C., He, S.: pTop 1.0: a high-accuracy and high-efficiency search engine for intact protein identification. Anal. Chem. **88** (6), 3082–3090 (2016)
13. Tian, Z., Tolic, N., Zhao, R., Moore, R.J., Hengel, S.M., Robinson, E.W., Stenoien, D.L., Wu, S., Smith, R.D., Pasatolic, L.: Enhanced top-down characterization of histone post-translational modifications. Genome Biol. **13**(10), 1–9 (2012)
14. Tsai, Y.S., Scherl, A., Shaw, J.L., Mackay, C.L., Shaffer, S.A., Langridgesmith, P.R.R., Goodlett, D.R.: Precursor ion independent algorithm for top-down shotgun proteomics. J. Am. Soc. Mass Spectrom. **20**(11), 2154–2166 (2009)

Functional Analysis of Autism Candidate Genes Based on Comparative Genomics Analysis

Lejun Gong[1(✉)], Shixin Sun[1], Chun Zhang[1], Zhihong Gao[2], Chuandi Pan[2], Zhihui Zhang[1], Daoyu Huang[1], and Geng Yang[1]

[1] Jiangsu Key Lab of Big Data Security and Intelligent Processing, School of Computer Science, Nanjing University of Posts and Telecommunications, Nanjing 210003, China
glj98226@163.com
[2] Zhejiang Engineering Research Center of Intelligent Medicine, Wenzhou 325035, China

Abstract. In the post-genomics era, the rapid development of high-throughput technology makes data analysis become more and more important which could obtain some new biomedical knowledge, especially in understanding disease mechanism. In this work, we analyze the candidate genes related to autism by comparative genomics in two samples. We try to understand the autism disease pathology from molecular mechanism. We first select the confirmed autism susceptibility genes acting as positive sample, and the genes from biomedical literature related to autism by text mining technology acting as the unknown sample. By venn diagram analysis, the results obtain 25 autism susceptibility genes from the unknown sample. The results achieve some significant biomedical knowledge in the comparative functional analysis to the two samples, In GO analysis, we obtain that the two class of genes have some similar molecular functions including all kinds of binding functions. In the pathway analysis, VEGF signaling pathway and MAPK signaling pathway have significant enrichment about the two samples. The result also shows some genes between the two samples play a key role in the same signaling transduction pathway. It indicates that the functional analysis is helpful for candidate gene related to autism. This provides a way to study the disease from molecular mechanism.

Keywords: Bioinformatics · Biostatistics · Genomics · Functional analysis Data acquisition

1 Introduction

With the rapid development of high-throughput technology, All kinds of omics big data with the volume of biomedical annotated data and literature come out. Aiming at the biomedical big data to analyze, we would achieve some meaningful biomedical knowledge related to the disease. Effective analysis of the abovementioned data is expected to deliver novel way to understand disease's therapeutic mechanism [1].

© Springer International Publishing AG, part of Springer Nature 2018
D.-S. Huang et al. (Eds.): ICIC 2018, LNCS 10954, pp. 565–575, 2018.
https://doi.org/10.1007/978-3-319-95930-6_55

Autism is a heterogeneous neurodevelopmental syndrome with repetitive conduct, impaired ability of social behavior, and communication. Although there is a variety of studies in autism field involving experimental or computational methods [2–6], the mechanisms and etiology of autism are still unknown. Omics data analysis could discover biomedical knowledge which is verified by [1, 7]. Parker et al. [8] used a new statistical model for preserve biological heterogeneity with a permuted surrogate variable analysis for genomics batch correction. Applied in the prediction of Human Papillomavirus (HPV) status, the model result in better performance. Csala et al. [9] proposed sparse redundancy analysis for high dimensional omics data analysis, the results could show the model is able to deal with the usual high dimensionality of omics data.

Sun et al. [10] considered that genome-wide methylation data with Illumina Infinium platform can be susceptible to batch effects with profound impacts by evaluating three common normalization approaches. In this paper, we first select the two samples. A sample is from the confirmed autism susceptive genes, which is called the positive sample. Another sample is from the biomedical literature related to autism, called the unknown sample. Then by comparative genomics analysis between the two samples, we would discover some biomedical knowledge involved with autism in molecular level. This would provide a new path for research autism in molecular mechanism for understanding disease etiology. In the following sections, we would depict the work more details.

2   Materials and Methods

Comparative genomics focuses on both similarities and differences in the genes and gene products of different organisms for studying the relationship of genome structure and function across different biological species or strains. Gene finding is an important application of comparative genomics. In this paper, we used different omics data comparison to analyze the disease related genes. In this paper, we consider autism as our study object. The method is as shown in Fig. 1.

In Fig. 1, we first obtained data containing the positive samples related to a disease and unknown samples to predict the correlation of the disease. Then, we perform the functional analysis to the two class of samples separately including Pathway analysis, PPI (protein and protein interaction) analysis and Gene Ontology analysis containing three levels: Biological Process, Cell Component and Molecular Function. By comparative genomics analysis between positive samples and unknown samples, we exploited the correlation between the unknown samples and autism.

2.1   Data Acquision

In this study, we study the unknown genes related to autism from molecular level based on omics data functional analysis. Our data acquision is from the work [6]. Our positive samples are containing 292 autism genes which are collected as autism sensitive genes obtained from GeneNavi (www.cbi.seu.edu.cn/GeneNavi/autism292). The unknown genes are from biomedical literature using text mining technology including 701 genes by the de-replication.

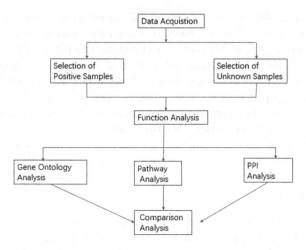

Fig. 1. Pipeline of prediction of autism genes based on comparative omics data analysis

2.2   Gene Ontology Analysis

An ontology is used to define something that we know about. The representation of things detected and directly observable, and relationships between those things are contained in "Ontologies". Gene ontology provides an ontology of defined terms representing gene product properties covering three domains: cellular component, molecular function and biological process. Gene ontology (GO) is a controlled vocabulary about gene and gene products as part of a larger classification effort, namely the Open Biomedical Ontologies (OBO) about gene and gene products. Gene Ontology analysis is a statistically enrichment analysis for the given set of genes considering to a GO term using the hypergeometric distribution statistic method of enrichment analysis as shown in (1).

$$P = 1 - \sum_{i=0}^{m-1} \frac{\binom{M}{i}\binom{N-M}{n-1}}{\binom{N}{n}} \tag{1}$$

Where N is the number of all genes of the specific organism that were annotated in GO. n is the number of query genes annotated to the GO Term; M is the number of all genes that are annotated to certain GO terms (Pop Hit); m is the number of query genes annotated to certain GO terms (count).

2.3   Pathway Analysis

Pathway depicts a model of process within a cell or tissue in molecular biology, which could trigger a chain of protein-protein or protein-small molecule interactions. Pathway analysis is related to proteins within a pathway or building pathway from the proteins of interest. It is helpful in studying differential expression of a gene in a disease and analyzing omics dataset. Pathway analysis needs bases with pathway collection and interaction networks. KEGG [11] is the most popular free public pathway collections. Pathway analysis could help to interpret omics data in the view of prior knowledges structured from the pathways graph. It also allows exploiting distinct cell processes, disease or signaling pathways based statistically associated with selection of deferentially expressed genes between two samples based on the same hypergeometric distribution as GO statistical model. Pathway analysis is often used as synonym in functional enrichment analysis and gene set analysis. The paper used the KEGG to analyze the pathway of two samples for finding signaling pathways related to autism.

2.4   PPI Analysis

Protein and protein interaction (PPI) is the physical contacts between two or more protein molecules as a result of biochemical events. Many molecular processes are carried out in that a large number of protein components organized by their PPIs, also called interact-omics. The aberrant PPIs may result in cancer because they are the basis of multiple aggregation-related disease. PPI analysis is to help the researcher discover what biological themes, and which biomolecules, are crucial to understand the phenomena under study. It also helps researcher in the identification of the biological roles of candidate genes. In this work, we performed PPI analysis to the selected samples for the molecular mechanism of autism susceptibility genes.

3   Results and Discussions

In this work, we have selected the two samples containing the positive sample and unknown sample in abovementioned method. The positive sample contained 292 autism susceptibility genes, and the unknown sample include 701 unknown genes from biomedical literature. We performed the venn diagram analysis as shown in Fig. 2.

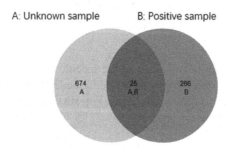

Fig. 2. Venn diagram analysis of two samples

Aiming at the two samples analysis, a gene is not find gene ID in the positive sample, and 699 unknown sample to be predict. According to the venn diagram, we found 25 intersections in the two samples. The 25 intersected genes contain these genes: HOXB1, FLT1, MET, HDAC4, PAX6, DLX2, HTR7, FABP5, PRODH, SDC2, NRP2, CNR1, CC2D1A, PLA2G6, MBD3, GTF2I, ROBO1, PTGS2, ITGB3, HSD11B1, GPX1, NTRK3, AR, MBD1 and FOXG1. Therefore, the functional analysis is performed from the three class samples containing 674 A genes, 266 B genes using comparison analysis.

3.1 GO Analysis Results

In the GO analysis, aiming at the A and B two samples, the work performs the GO analysis containing molecular function, biological process and cell component. A sample contains 7137 biological processes with 4510 ones statistically p-value, 679 cell components with 269 ones statistically p-value, and 1213 molecular functions with 542 ones of statistically p-value, and 1434 KEGG pathways with 1028 ones of statistically significant p-value. B sample contains 4799 biological processes with 2760 ones statistically significant p-value, 455 cell components with 188 ones of statistically significant p-value, 824 molecular functions with 370 ones statistically p-value, and 889 KEGG pathways with 486 ones of statistically significant p-value. The GO enrichment numbers of the two sample are shown in Fig. 3.

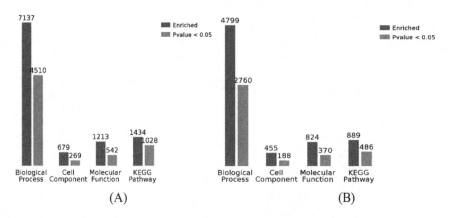

Fig. 3. (A) A sample GO enrichment numbers, (B) B sample GO enrichment numbers

By comparative analyzing GO enrichment about the genes between A sample and B sample, the two class genes primarily perform similarly all kinds of binding functions containing protein binding, enzyme binding, receptor binding, etc. Aiming at A sample, the genes are largely seated in membrane-bounded organelle, organelle, intracellular membrane-bounded organelle, nucleus, and nuclear lumen, and play a role in positive regulation of cellular process, regulation of biological process, multicellular organismal process, and development process. Aiming at B sample, these gene are largely located in cell periphery, plasma membrane, neuron part, cell projection, neuron projection, synapse part, and act a role in a series of biological processes including multicellular organismal process, developmental process, anatomical structure development, multicellular organismal development. The GO enrichments of the two samples are shown in Fig. 4.

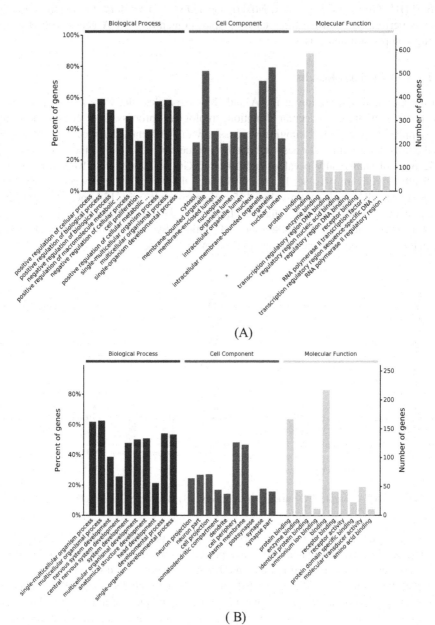

(A)

(B)

Fig. 4. (A) GO enrichment of A sample (B) GO enrichment of B sample

3.2 Pathway Analysis

In pathway analysis, aiming at the two samples, the enriched KEGG pathway has 4 classes containing metabolism, genetic information processing, environmental information processing and cellular processes as shown in Figs. 5 and 6. By the analysis according to Figs. 5 and 6, the A class sample has significant enrichment in MAPK signaling pathway, p53 signaling pathway, VEGF signaling pathway, and NF-kappa-B signaling pathway, and B class sample has significant enrichment in Calclum signaling pathway, mTOR signaling pathway, MAPK signaling pathway, and VEGF signaling pathway.

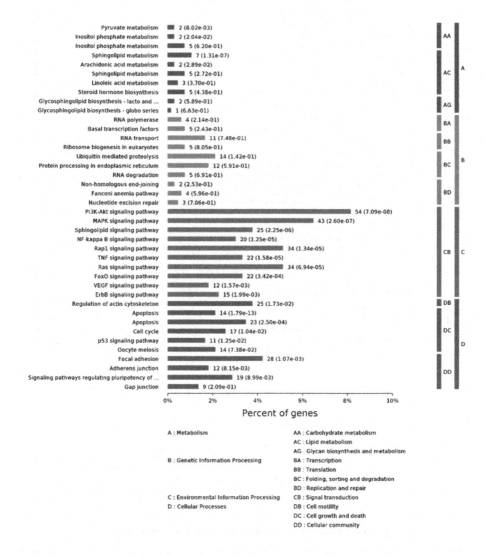

Fig. 5. Enriched KEGG pathway classifications of the A sample

3.3 PPI Analysis

PPI analysis helps researcher to identify the biological roles of candidate genes for their molecular mechanism. In PPI network of A sample in Fig. 7, MAPK8 takes signal transduction process, and interacts some proteins, for instance, STAT5A, MMP9, NOS3, MAPK11, MAPK12, etc. MAPK8 plays a key role in immune system and signaling by NGF as shown in Fig. 7.

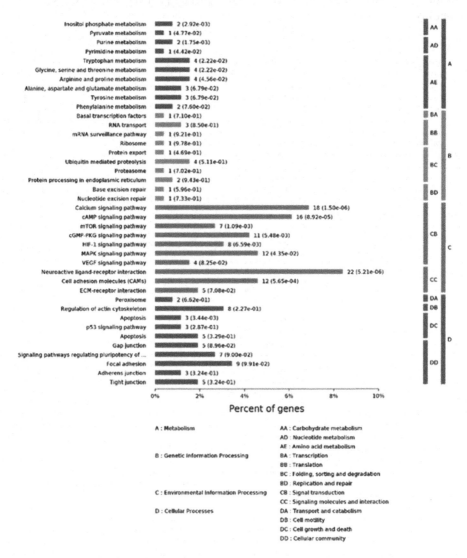

Fig. 6. Enriched KEGG pathway classifications of B sample

In Fig. 8, autism susceptibility gene STAT3 takes signal transduction pathway, interacts some proteins, for example, NF1, KRAS, CREBBP, BCL2, RPS6KA2, PTEN, etc. It also plays a key role in immune system pathway which is similar to the gene MAPK8 which is belonging to the positive sample. According to the analyzed method, we could also analyze other genes in A sample. This provides a way to Prioritization of candidate genes related to autism.

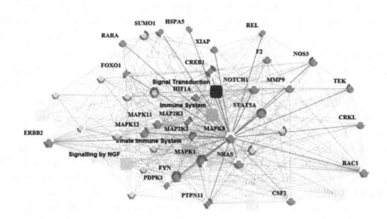

Fig. 7. PPI network of A sample

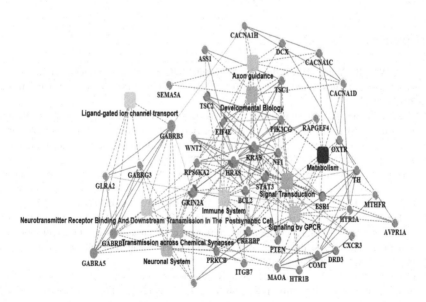

Fig. 8. PPI network of B sample

4 Conclusions

In this work, we first select the both positive and unknown samples, separately. And then we perform venn diagram analysis to the two class samples, we obtain 25 autism susceptibility genes from the unknown sample. We also perform the function analysis to both A sample and B sample. In GO analysis, we obtain that the two class genes have some similar molecular functions including all kinds of binding functions containing protein binding, enzyme binding, receptor binding. In pathway analysis, we obtain the results that the two samples have significant enrichment in VEGF signaling pathway and MAPK signaling pathway by comparative analysis. In PPI analysis, we take an example of hot spot in PPI network to analyze. The result shows the hot spot STAT3 in A sample which is belong to the unknown sample play a key role in immune system pathway. It is similar to the hot spot MAPK8 which is belonging to the positive sample. Based on analyzed results, the functional analysis is helpful for candidate gene related to autism. This provide a way to study the disease from molecular mechanism.

Acknowledgement. This research is supported by the National Natural Science Foundation of China (Grant Nos: 61502243, 61502247, 61272084, 61300240, 61572263, 61502251, 61503195), Natural Science Foundation of the Jiangsu Province (Grant Nos: BK20130417, BK20150863, BK20140895, and BK20140875), China Postdoctoral Science Foundation (Grant No. 2016M590483), Jiangsu Province postdoctoral Science Foundation (Grant No. 1501072B), Scientific and Technological Support Project (Society) of Jiangsu Province (Grant No. BE2016776), Nanjing University of Posts and Telecommunications' Science Foundation (Grant Nos: NY214068 and NY213088). This work is also supported in part by Zhejiang Engineering Research Center of Intelligent Medicine under 2016E10011.

References

1. Goh, W.W.B., Wang, W., Wong, L.: Why batch effects matter in omics data, and how to avoid them. Trends Biotechnol. **35**(6), 498–507 (2017). https://doi.org/10.1016/j.tibtech.2017.02.012
2. Cuevas-Olguin, R., Roychowdhury, S., Banerjee, A., Garcia-Oscos, F., Esquivel-Rendon, E., Bringas, M.E., Kilgard, M.P., Flores, G., Atzori, M.: Cerebrolysin prevents deficits in social behavior, repetitive conduct, and synaptic inhibition in a rat model of autism. J. Neurosci. Res. (2017). https://doi.org/10.1002/jnr.24072
3. Cheng, N., Alshammari, F., Hughes, E., Khanbabaei, M., Rho, J.M.: Dendritic overgrowth and elevated ERK signaling during neonatal development in a mouse model of autism. PLoS ONE **12**(6), e0179409 (2017). https://doi.org/10.1371/journal.pone.0179409
4. Coleman, J.R.I., Lester, K.J., Keers, R., Munafò, M.R., Breen, G., Eley, T.C.: Genome-wide association study of facial emotion recognition in children and association with polygenic risk for mental health disorders. Am. J. Med. Genet. B Neuropsychiatry Genet. (2017). https://doi.org/10.1002/ajmg.b.32558
5. Gong, L., Yan, Y., Xie, J., Liu, H., Sun, X.: Prediction of autism susceptibility genes based on association rules. J. Neurosci. Res. **90**(6), 1119–1125 (2012). https://doi.org/10.1002/jnr.23015

6. Gong, L., Yang, R., Yan, Q., Sun, X.: Prioritization of disease susceptibility genes using LSM/SVD. IEEE Trans. Biomed. Eng. **60**(12), 3410–3417 (2013). https://doi.org/10.1109/TBME.2013.2257767

7. Kim, D.H., Kim, Y.S., Son, N.I., Kang, C.K., Kim, A.R.: Recent omics technologies and their emerging applications for personalised medicine. IET Syst. Biol. **11**(3), 87–98 (2017). https://doi.org/10.1049/iet-syb.2016.0016

8. Parker, H.S., Leek, J.T., Favorov, A.V., Considine, M., Xia, X., Chavan, S., Chung, C.H., Fertig, E.J.: Preserving biological heterogeneity with a permuted surrogate variable analysis for genomics batch correction. Bioinformatics **30**(19), 2757–2763 (2014)

9. Csala, A., Voorbraak, F.P.J.M., Zwinderman, A.H., Hof, M.H.: Sparse redundancy analysis of high dimensional genetic and genomic data. Bioinformatics (2017). https://doi.org/10.1093/bioinformatics/btx374

10. Sun, Z., Chai, H.S., Wu, Y., White, W.M., Donkena, K.V., Klein, C.J., Garovic, V.D., Therneau, T.M., Kocher, J.P.: Batch effect correction for genome-wide methylation data with Illumina Infinium platform. BMC Med. Genomics **16**(4), 84 (2011). https://doi.org/10.1186/1755-8794-4-84

11. Ogata, H., Goto, S., Sato, K., Fujibuchi, W., Bono, H., Kanehisa, M.: KEGG: Kyoto encyclopedia of genes and genomes. Nucleic Acids Res. **27**(1), 29–34 (1999)

Identification of Hotspots in Protein-Protein Interactions Based on Recursive Feature Elimination

Xiaoli Lin[1,2], Xiaolong Zhang[1(✉)], and Fengli Zhou[2]

[1] Hubei Key Laboratory of Intelligent Information Processing and Real-Time
Industrial System, School of Computer Science and Technology,
Wuhan University of Science and Technology, Wuhan 430065, China
aneya@163.com, XiaolongZhang@wust.edu.cn
[2] Information and Engineering Department of City College,
Wuhan University of Science and Technology, Wuhan 430083, China
thinkview@163.com

Abstract. The study of protein-protein interactions and protein structure through computational methods is critical to understand protein function. Hot spot residues play an important role in bioinformatics to reveal life activities. However, conventional hot spots prediction methods may face great challenges. This paper proposes a hot spot prediction method based on feature selection method SVM-RFE to improve the training performance. SMOTE based over-sampling is used to adds new samples to avoid an overfitting classifier. SVM-RFE is then invoked to obtained optimal feature subset. Finally, a feature-based SVM is created to predict the hot spots. Experimental results indicate that the performance of hot spots prediction has been significantly improved compared with the previous methods.

Keywords: Protein-Protein interactions · Hot spots · SVM-RFE
Classification

1 Introduction

Proteins rarely act alone and they interact with many other molecules to form protein complexes for performing biological functions [1]. The correct 3D structure of a protein is essential to particular biological activities [2]. The folded protein allows amino acids to interact with other molecules known as targets at specific active sites. Generally, the target molecules are also proteins, DNA stretches or some other small molecules [3]. For years many models of protein-protein interaction network have been constructed to find out the biological mechanisms in living cells [4]. These physical models are based on deep understanding of amino acids that take an important action in protein-protein interactions [5]. In molecular biology, a technique is used to determine the contribution of a specific residue to the stability or function of given protein, known as alanine scanning [6] [7], which has been adopted to identify the hot spots [8]. Kortemme [9] created a simple physical model to predict the results of alanine scanning experiments. Ofran et al. [10] proposed a feature-based method called ISIS predicts the

© Springer International Publishing AG, part of Springer Nature 2018
D.-S. Huang et al. (Eds.): ICIC 2018, LNCS 10954, pp. 576–582, 2018.
https://doi.org/10.1007/978-3-319-95930-6_56

hot spots. Darnell et al. [11] created knowledge-based models to predict hot spot residues from protein complexes. Burgoyne et al. [12] used sequence conservation and some physical properties to analyze clefts corresponding to binding hot spots on the protein surface. Barata et al. [13] introduced a molecular docking approach for the identification of protein-excipient hotspots in protein excipient interfaces. Tunc bag et al. [14] applied machine learning methods to determine hot spots based on conservation, solvent accessibility ASA, and statistical pairwise residue potentials of interface residues. Keskin et al. [15] found that hot spots are clustered within locally tightly packed regions. Xia et al. [16] introduced the SVM model to predict the hot spots based on some features. In this paper, we introduce a hybrid method based on an oversampling strategy and SVM-RFE to predict the hot spot residues.

2 Method

2.1 Preprocessing of Data

We have processed the datasets and deleted the residue of the binding free energy in the interval of 0.4–2.0 kcal/mol. If the binding free energy of residues is higher than or equal to 2.0 kcal/mol, we consider these residues as hot spots. If the binding free energy of residues is less than 0.4 kcal/mol, those residues are defined as non-hot spots. The datasets come from the database ASEdb [17], SKEMPI [18], and BID [19]. In the ASEdb database, 65 hot spot residues and 90 non-hot spot residues are selected. SKEMPI database has collected 136 hot spots and 349 non-hot spots. In the BID database, 39 hot spot residues and 88 non-hot spot residues are selected.

The initial datasets obtained have unequal distribution of samples among the categories. In this paper, we use SMOTE [20] based oversampling techniques for handling imbalanced datasets. Preprocessing of datasets based on SMOTE is shown in Table 1.

Table 1. Preprocessing of datasets based on SMOTE

Input:	Training sample set
Output:	New synthetic samples

$M = (\text{int})\ (M/100)$
While $M \neq 0$
  For $i \leftarrow 1$ to S
    Randomly choose a sample X_{nn} from k nearest neighbors
    For $attr \leftarrow 1$ to $numattrs$
      $X_{new} = X_{old} + \text{rand}\ (0,1)*(X_{nn} - X_{old})$
    End For
    $\Omega \leftarrow \Omega \cup \{(X_{new}, y_{minority\ class})\}$
    $M = M - 1$
  End For
End

2.2   Feature Selection Based on SVM-REF

Feature selection is the process of selecting a subset of relevant features in model construction to avoid the curse of dimensionality, which can enhance generalization by reducing overfitting [21]. Here, a feature selection method is designed to eliminate irrelevant and redundant features, which is recursive feature elimination based on support vector machine.

Given a set of training samples (x_i, y_i), where $x_i \in R^D$, $y_i \in [-1, 1]$ is the label of x_i, D and N are the dimension and the number of samples, respectively. The objective function is defined as:

$$J = \frac{1}{2}w^T w + C \sum_{i=1}^{n} \xi_i$$
$$s.t.\, y_i\left(w^T x_i + b\right) \geq 1 - \xi_i,\ \xi_i \geq 0,\ i = 1, 2, \cdots, n \tag{1}$$

Where w and b are the weight and threshold of the hyperplane. C is a regulation parameter and ξ_i is the slack variable. C can make the balance between the minimization of misclassification and the maximization of the margin of the hyperplane. SVM-RFE is also called recursive feature elimination based on support vector machine. The weights can be used to rank representative features according to their importance with respect to classification tasks. The detail algorithm of SVM-RFE is shown in Table 2.

Table 2. Recursive feature elimination based on support vector machine

Algorithm SVM-RFE
Input: Training dataset D
Output: Ranked feature subset $Best\_D$
Begin
Initialize: The selected feature subset Current_D={1,2,...,K}, and ranked feature subset $Best\_D = \emptyset$;
Set proportion of features deleted in each step.
Repeat until Current_$D = \emptyset$:
Establish SVM model based on the current feature subset $Current\_D$, get the accuracy ACC
Compute the ranking criteria
Rank features of Current_D in descending order by weight $|w|$
Remove the features with smallest ranking
Return ranked feature subset $Best\_D$
End

3 Experiments

3.1 Evaluation Criteria

There are several measures used to estimate the performance of hot spot prediction and hot region prediction. They are respectively *Precision, Recall* and *F-Score*.

$$Precision = \frac{TP}{TP + FP} \tag{2}$$

$$Recall = \frac{TP}{TP + FN} \tag{3}$$

$$F - Score = \frac{2 * precision * Recall}{Precision + Recall} \tag{4}$$

For predicting hot spots, the true positive (*TP*) represents the quantity of hot spots which appear in both natural hot regions and predicted hot regions. The quantity of hot spots which do not exist on natural hot regions but are predicted, can be represented as the false positive (*FP*). The false negative (*FN*) of predicted hot spot represents the quantity of hot spots which exist on natural hot regions but are not predicted.

3.2 Result Analysis

In this work, we obtained features by PSAIA (Protein Structure and Interaction Analyzer) [22], which enables the calculation of protein geometric parameters and the determination of location and type for protein-protein interaction sites. The physicochemical features include ASA (accessible surface area), DI (depth index) and PI (protrusion index). In total, we selected the 32 attributes of protein residue by SVM-RFE in Table 3. Then we created the SVM model according to extracted structure-based features.

Table 3. Selected attributes by SVM-RFE

Order	Attribute	Order	Attribute	Order	Attribute	Order	Attribute
1	BtRASA	9	BtASA	17	BsRASA	25	BtmPI
2	BpRASA	10	BsASA	18	BnRASA	26	BminPI
3	BsmPI	11	BpASA	19	RctASA	27	BtmDI
4	RcsASA	12	UtRASA	20	UtmPI	28	UminPI
5	RcsmPI	13	BsmDI	21	UsmPI	29	UtASA
6	RctmPI	14	UtmDI	22	UsRASA	30	UbASA
7	UpRASA	15	BbASA	23	UsASA	31	RCtmDI
8	RcminDI	16	Prop	24	Enc	32	Na

In order to verify our method, we created a training model using mixed datasets based on ASEdb and SKEMPI, then tested our model using BID dataset for comparing against other models Robetta [23], FOLDEF [24], KFC[25], Cho's method [26] and Zhang's method [27]. The testing results in Table 4 demonstrates that our model can obtain a better performance. *Precision* of our model is same as that of Cho' method and better than others. *Recall* of our model is same as that of Zhang's method, but is better than others. *F-Score* is slightly better than Zhang's method and Cho's method, but is better than others.

Table 4. Prediction results of hot spot residues and non-hot spot residues

Model	Precision	Recall	F-Score
Robetta	0.48	0.32	0.384
FOLDEF	0.4	0.22	0.28
KFC	0.44	0.30	0.35
Cho's method	**0.53**	0.62	0.57
Zhang's method	0.5	**0.71**	0.59
Our method	**0.53**	**0.71**	**0.61**

4   Conclusion

This paper adopted a hybrid method to classify hot spot residues and non-hot spot residues in PPIs. Firstly, we adopted SMOTE to deal with the imbalance between hot spot residues and non-hot spot residues. Then, the recursive feature elimination based on support vector machine was used to select the optimal features. Finally, the SVM model based on structure features was created to predict hot spot residues. Experimental results show that the proposed method has the higher performance of predicting the hot spot residues. In our future work, we will select more useful features to train models for classifying hot spot residues and study the formation mechanism of the hot regions.

Acknowledgment. The authors thank the members of Machine Learning and Artificial Intelligence Laboratory, School of Computer Science and Technology, Wuhan University of Science and Technology, for their helpful discussion within seminars. This work was supported in part by National Natural Science Foundation of China (No. 61502356, 61273225), by Hubei Province Natural Science Foundation of China (No. 2018CFB526).

References

1. Giot, L., Bader, J.S., Brouwer, C., Chaudhuri, A., Kuang, B.: A protein interaction map of Drosophila Melanogaster. Science **302**, 1727–1736 (2003)
2. Lin, X.L., Zhang, X.L., Zhou, F.L.: Protein structure prediction with local adjust tabu search algorithm. BMC Bioinform. **5**(S15), S1 (2014)

3. Sahu, S.S., Panda, G.: Efficient Localization of hot spots in proteins using a novel S-transform based filtering approach. IEEE/ACM Trans. Comput. Biol. Bioinform. **8**(5), 1235–1246 (2011)
4. Keskin, O., Tuncbag, N., Gursoy, A.: Predicting protein-protein interactions from the molecular to the proteome level. Chem. Rev. **116**(8), 4884–4909 (2016)
5. Cho, K., Kim, D., Lee, D.: A feature-based approach to modeling protein-protein interaction hot spots. Nucl. Acids Res. **37**(8), 2672–2687 (2009)
6. Morrison, K.L., Weiss, G.A.: Combinatorial Alanine-scanning. Curr. Opin. Chem. Biol. **5** (3), 302–307 (2001)
7. Kortemme, T., Kim, D.E., Baker, D.: Computational Alanine scanning of protein-protein interfaces. Sci. STKE Signal Transduct. Knowl. Environ. (STKE) **2004**(219), pl2 (2004)
8. Bogan, A., Thorn, K.S.: Anatomy of ces. J. Mol. Biol. **280**, 1–9 (1998)
9. Kortemme, T., Baker, D.: A simple physical model for binding energy hot spots in protein-protein complexes. Proc. Nat. Acad. Sci. USA **99**(22), 14116–14121 (2002)
10. Ofran, Y., Rost, B.: ISIS: Interaction Sites Identified from Sequences. Bioinformatics **23**, e13–e16 (2006)
11. Darnell, S.J., Page, D., Mitchell, J.C.: An automated decision-tree approach to predicting protein interaction hot spots. Proteins **68**(4), 813–823 (2007)
12. Burgoyne, N.J., Jackson, R.M.: Predicting protein interaction sites: binding hot-spots in protein-protein and protein-ligand interfaces. Bioinformatics **22**(11), 1335–1342 (2006)
13. Barata, T.S., Zhang, C., Dalby, P.A., Brocchini, S., Zloh, M.: Identification of protein-excipient interaction hotspots using computational approaches. Int. J. Mol. Sci. **17**(6), 853 (2016)
14. Tuncbag, N., Gursoy, A., Keskin, O.: Identification of computational hot spots in protein interfaces: combining solvent accessibility and inter-residue potentials improves the accuracy. Bioinformatics **25**(12), 1513–1520 (2009)
15. Keskin, O., Ma, B., Nussinov, R.: Hot regions in protein- protein interaction: the organization and contribution of structurally conserved hot spot residues. J. Mol. Biol. **345** (5), 1281–1294 (2005)
16. Xia, J.F., Zhao, X.M., Song, J., Huang, D.S.: APIS: accurate prediction of hot spots in protein interfaces by combining protrusion index with solvent accessibility. BMC Bioinform. **11**, 174 (2010)
17. Thorn, K.S., Bogan, A.A.: ASEdb: a data base of Alanine mutations and their effects on the free energy of binding in protein interactions. Bioinformatics **17**(3), 284–285 (2001)
18. Moal, I.H., Fernández-Recio, J.: SKEMPI: a Structural Kinetic and Energetic database of Mutant Protein Interactions and its use in empirical models. Bioinformatics **28**(20), 2600–2607 (2012)
19. Fischer, T.B., Arunachalam, K.V., Bailey, D., Mangual: The Binding Interface Database (BID): a compilation of amino acid hot spots in protein interfaces. Bioinformatics **19**(11), 1453–1454 (2003)
20. Chawla, N.V., Bowyer, K.W., Hall, L.O., Kegelmeyer, W.P.: SMOTE: synthetic minority over-sampling technique. J. Artif. Intell. Res. **16**, 321–357 (2002)
21. Bermingham, M.L., Pongwong, R., Spiliopoulou, A., et al.: Application of high-dimensional feature selection: evaluation for genomic prediction in man. Sci. Rep. **5**, 10312 (2015)
22. Mihel, J., Sikic, M., Tomic, S., Jeren, B., Vlahovicek, K.: PSAIA-protein structure and interaction analyzer. BMC Struct. Biol. **8**(1), 21 (2008)
23. Ofran, Y., Ros, B.: ISIS: Interaction Sites Identified from Sequence. Bioinformatics **23**(2), e13–e16 (2007)

24. Guerois, R., Nielsen, J.E., Serrano, L.: Predicting changes in the stability of proteins and protein complexes: a study of more than 1000 mutations. J. Mol. Biol. **320**(2), 369–387 (2002)
25. Darnell, S.J., Legault, L., Mitchell, J.C.: KFC server: interactive forecasting of protein interaction hot spots. Nucl. Acids Res. **36**(suppl 2), W265–W269 (2008)
26. Cho, K., Kim, D., Lee, D.: A feature-based approach to modeling protein–protein interaction hot spots. Nucl. Acids Res **37**(8), 2672–2687 (2009)
27. Zhang, S.H., Zhang, X.L.: Prediction of hot spots at protein-protein interface. Acta Biophysica Sinica **29**(2), 1–12 (2013)

Compare Copy Number Alterations Detection Methods on Real Cancer Data

Fei Luo[1($\boxtimes$)] and Yongqiong Zhu[2]

[1] Wuhan University, Wuhan 430072, China
luofei@whu.edu.cn
[2] Wuhan Business University, Wuhan 430056, China
zyqzhuyongqiong@126.com

Abstract. Since the Copy Number Alterations (CNAs) are discovered to be tightly associated with cancers, accurately detecting them is an important task in the genomic structural variants research. Although a series of CNAs calling algorithms have been proposed and several evaluations made attempts to reveal their performance, their comparisons are still limited by the amount and type of experimental data and the conclusions have poor consensus. In this work, we use a large-scale real dataset from CAGEKID consortium to evaluate total 12 commonly used CNAs detection methods. This large-scale dataset comprises of SNP array data on 94 samples and the whole genome sequencing data on 10 samples. Twelve compared methods comprehensively cover the current CNAs detection scenarios, which include using SNP array data, sequencing data with tumor and normal matched samples and single tumor sample.

Keywords: Tumor · Copy Number Alterations · SNP chip · Deep sequencing

1 Introduction

A wide spectrum of genetic diseases is related to CNVs [1, 2]. Copy number change also involves in the initiation and development of cancer in a way that copy numbers are different in an individual's germline DNA and in the DNA of a clonal sub-population of cells. Such copy number change is specially called as somatic copy number alterations (CNAs) [3]. Many oncogenes and tumor suppressors behaviors are associated with the CNAs [4]. Two international research organizations TCGA (the cancer genome atlas) [5] and ICGC (international cancer genome consortium) [6] are dedicating to large-scale collecting and deeply interrogating the variants of typical cancers' genomics. In the both of them, the CNAs is a research hotspot to understand the cancer etiology.

Over decades, the SNP (single-nucleotide polymorphism) array and the aCGH (array comparative genomic hybridization) array have been widely applied to detect CNAs. Due to probes' low-resolution, these two micro-array platforms based technology are only suitable for the large CNAs detection. With the development of high-throughput DNA sequencing technology, platforms like Roche 454, Illumina solexa offer an alternative way to discover CNAs in any length. Notable, the CNVs and CNAs are two completely different biological concepts. Different from that SNP based

© Springer International Publishing AG, part of Springer Nature 2018
D.-S. Huang et al. (Eds.): ICIC 2018, LNCS 10954, pp. 583–588, 2018.
https://doi.org/10.1007/978-3-319-95930-6_57

methods that definitely distinguish their difference, few sequencing based methods integrate genotype information and it's hard for them to detect aneuploidy and LOH due to the inadequate reads coverage. If only using the read counts information, the detecting procedures for CNVs and CNAs are quite similar for deep sequencing based methods. Therefore, most deep sequencing based methods don't rigidly distinguish CNAs or CNVs.

A series of calling methods have been proposed and several attempts have been made to compare the existing methods. In order to ascertain the real features and the difference of typical methods' performance, in this work a group of representative methods will be evaluated on the renal clear cell carcinoma dataset.

2   Method

2.1   Methods Selected to Compare

According to mosén [7] assessment, GPHMM [8] and GAP [9] work best and are respectively recommended to the professional and general users. GAP and GPHMM only need tumor sample to call CNAs. Another tool called OncoSNP [10], as a special tool using both the tumor and normal samples information, is included. Hence, we take the GAP, GPHMM and OncoSNP as the representatives of SNP based methods. The ReadDepth [11], CNVnator [12], RDXplorer [13] and CNVseq [14] are mentioned in both works [15, 16]. Therefore, these four methods are also included in our work. JointSLM is discarded because it's special for calling CNAs from a group of samples. Referring to the list given by Xi [15], another five methods including BICseq [17], CNVnorm [18], FREEC [19], rSWseq [20], Varscan [21] are added into our work. Therefore, total 12 methods are put together to compare in this work.

2.2   Rank SNP Based CNAs Detection Methods

For the tumor genomes, the chromosomal aneuploidy, stromal contamination, and intratumoral heterogeneity are three major obstacles to hamper the CNAs detection. Depending on the genotype information, the SNP based methods are more reliable in estimating the aneuploidy, normal cell contamination and large-scale CNAs than the sequencing based methods. Thus GAP, GPHMM and OncoSNP are firstly compared, and then the best caller of them will be used as the golden standard to evaluate the sequencing based methods.

In order to evaluate the SNP based methods' performance, we propose two criteria to rank the three methods.

(1) correct estimation for the baseline shift,
(2) correct estimation for large length CNAs.

The aneuploidy, as a frequently appearing phenomenon in tumor genomes, always makes the LRR baseline of SNP chip result shift away from zero. Criterion one emphasizes the fact that precise LRR baseline shift estimation is the prerequisites for correctly assigning copy number state. The number of probes in a SNP chip is insufficient to discover short length CNAs, but large length CNAs take up most fraction of total CNAs length in a sample. Therefore, the second criterion can reflect SNP based methods' performance on the CNAs calling accuracy. According these two criteria, three grades are used to rank.

$$\begin{cases} \textit{if satisfying both 1 and 2, grade 1} \\ \textit{if satisfying 1 but not 2, grade 2} \\ \textit{if not satisfying 1, grade 3} \end{cases}$$

When judging whether satisfy with the criterion, all results are validated by experts' visual examination.

2.3 Evaluate Sequencing Based CNAs Detection Methods

Since results from SNP based methods are sufficient to cover most of one sample's total CNAs, the sequencing based methods' consistency with SNP based result could reflect their performance. We use the recall rate and Jaccard index to measure this consistency. The Jaccard index is a statistic used for comparing the similarity and diversity of two sets. The recall rate could measure how much fraction of the SNP based golden result is covered by the sequencing based method. It reflects the true positive rate of sequencing based method. In theory, sequencing based method could discover additional short length CNAs because of its higher resolution along the genome, but short length CNAs' length only takes up a very small fraction of total CNAs length. Therefore, if the Jaccard Index is too low, it means a high false positive rate.

3 Results

3.1 Dataset

Renal Cell Carcinoma (RCC) data from CAGEKID (the CAncer GEnomic of the KIDney) consortium, which is a part of International Cancer Genome Consortium (ICGC), are used to compare the total 12 methods. In CAGEKID project, we collect SNP array data of 94 RCC patients' samples. They are genotyped by Illumina 660 W quad BeadChip, which has more than 657,000 genetic probes. The patients' samples are also sequenced by the platform Illumina HiSeq 2000. The mean length of the read is 100 bp and paired-end sequenced. 10 patients' corresponding sequencing data are used for the deep sequencing based CNAs detection comparison. All samples' reads are aligned to NCBI 37 reference genome by the BWA.

3.2 Evaluation of SNP Based Methods

Figure 1 depicts three methods' ranking results. GPHMM gets most grade 1 on total 94 samples. We further divide the samples into two groups by their ploidy. For near diploid samples, GPHMM quite low grade 3 rate suggests that GPHMM has good performance on ploidy detection on normal ploidy samples. However, all three methods work not well on 14 over diploid samples, on which the grade 1 rate of three methods aren't more than 25%. Three SNP based methods has no enough accuracy to discriminate the small difference of the over diploid samples' complex BAF and LRR patterns, especially under the existence of noises. OncoSNP is sensitive to aneuploidy than the other two methods, but this sensitivity is at the expense of low specificity. 19% of diploid samples are mistakenly predicted as the aneuploidy by OncoSNP. On over 60% samples, at least one chromosome-scale region is given as wrong copy number by OncoSNP. OncoSNP's more baseline shift estimation leads to its over sensitivity.

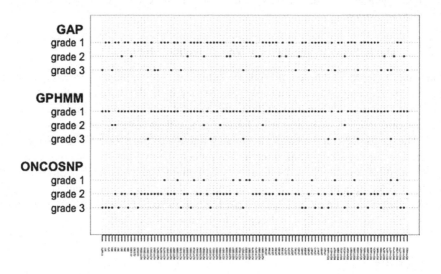

Fig. 1. SNP based methods' grades on total 94 samples.

3.3   Evaluation of Sequencing Based Methods

Use the best caller from SNP based methods as the reference. All sequencing based methods' performance are evaluated. As listed in the Tables 1 and 2, for each sample, the best two results are marked bold and italic. CNVnorm achieves best recall rate on both near diploid samples and over diploid samples, but it gets the lowest Jaccard Index. CNVnorm predicts the ploidy for 10 samples as 2, 4, 2, 3, 1, 2, 2, 5, 2 and 2. Only 4 samples' ploidy estimations are accordant with the SNP results. For total 10 samples, the average number of copy number states predicted by CNVnorm is 8. This state overestimation leads to a lot of false positive callings. High recall rate but low Jaccard rate can reflect this overestimation. Therefore, BICseq and FREEC are two superior sequencing based CNAs detection methods.

Table 1. Recall rate on 10 samples.

Sample	BICseq	CNVnorm	FREEC	CNVseq	rSWseg	Varscan	CNVnator	RDXplorer	ReadDepth
357	0.9	*0.97*	*0.97*	0.82	0.45	0.84	0.68	0	0.06
620380	0.97	*0.99*	*0.99*	0.92	0.16	0.94	0.15	0.01	0.26
K2110056	0.57	*0.98*	*0.74*	0.44	0.23	0.27	0.05	0	0.08
K2150024	*0.91*	*0.93*	0.17	0.74	0.02	0.9	0	0.01	0.05
K2310007	*0.99*	*1*	0.89	0.96	0	0.98	0.2	0.01	0.01
K2310024	*0.99*	*1*	0.88	0.96	0.52	0.97	0.36	0	0.2
K2310030	*0.99*	*0.99*	0.94	0.96	0.27	0.98	0.48	0.01	0.7
AVG	*0.91*	*0.98*	0.8	0.83	0.24	0.84	0.27	0.01	0.19
RS114527	0.17	*0.23*	*0.74*	0.18	0.01	0.17	0.01	0	0.08
K2110097	*0.39*	0.4	0.26	0.35	0	0.3	0.18	0	*0.71*
K2150051	*0.57*	*0.94*	0.55	0.52	0.01	0.46	0.04	0.01	0.04
AVG	0.38	*0.52*	*0.52*	0.35	0.01	0.31	0.08	0	0.28

Table 2. Jaccard index on 10 samples

Sample	BICSeq	CNVnorm	FREEC	CNVseq	rSWseq	Varscan	CNVnator	RDXplorer	ReadDepth
357	*0.78*	0.22	*0.79*	0.59	0.35	0.64	0.57	0	0.05
620380	0.65	0.13	*0.66*	0.6	0.1	*0.73*	0.08	0.02	0.04
K2110056	0.45	*0.6*	*0.68*	0.32	0.19	0.27	0.07	0	0.08
K2150024	*0.73*	0.23	0.05	0.55	0.01	*0.81*	0.01	0.01	0.04
K2310007	*0.6*	0.05	0.12	0.53	0	*0.7*	0.19	0.02	0.01
K2310024	*0.66*	0.06	0.14	0.52	0.41	*0.72*	0.29	0.01	0.05
K2310030	*0.73*	0.19	0.49	0.64	0.14	*0.85*	0.64	0.01	0.43
AVG	*0.66*	0.21	0.42	0.54	0.17	*0.67*	0.26	0.01	0.1
RS114527	0.13	*0.16*	*0.69*	0.15	0.02	0.14	0.02	0	0.06
K2110097	*0.38*	0.32	0.26	0.35	0	0.3	0.22	0	*0.63*
K2150051	*0.47*	0.28	*0.42*	0.41	0	0.36	0.03	0.01	0.04
AVG	*0.33*	0.25	*0.46*	0.3	0.01	0.27	0.09	0	0.24

4   Conclusion

In this work, three SNP based method are first ranked. The optimal one is revealed and used as golden standard for evaluation on six matched-pairs methods and three single sample based methods. The optimal CNAs detection methods are found.

References

1. Weischenfeldt, J., Dubash, T., Drainas, A.P., et al.: Pan-cancer analysis of somatic copy-number alterations implicates IRS4 and IGF2 in enhancer hijacking. Nat. Genet. **49**(1), 65 (2017)
2. Beroukhim, R., Zhang, X., Meyerson, M.: Copy number alterations unmasked as enhancer hijackers. Nat. Genet. **49**(1), 5 (2016)
3. Beroukhim, R., et al.: The landscape of somatic copy-number alteration across human cancers. Nature **463**(7283), 899–905 (2010)

4. Malek, J.A., et al.: Copy number variation analysis of matched ovarian primary tumors and peritoneal metastasis. PLoS ONE **6**(12), e28561 (2011)
5. Chin, L., et al.: Making sense of cancer genomic data. Genes Dev. **25**(6), 534–555 (2011)
6. Hudson, T.J., et al.: International network of cancer genome projects. Nature **464**(7291), 993–998 (2010)
7. Mosen-Ansorena, D., Aransay, A.M., Rodriguez-Ezpeleta, N.: Comparison of methods to detect copy number alterations in cancer using simulated and real genotyping data. BMC Bioinform. **13**, 192 (2012)
8. Li, A., et al.: GPHMM: an integrated hidden Markov model for identification of copy number alteration and loss of heterozygosity in complex tumor samples using whole genome SNP arrays. Nucleic Acids Res. **39**(12), 4928–4941 (2011)
9. Popova, T., et al.: Genome Alteration Print (GAP): a tool to visualize and mine complex cancer genomic profiles obtained by SNP arrays. Genome Biol. **10**(11), R128 (2009)
10. Yau, C., et al.: A statistical approach for detecting genomic aberrations in heterogeneous tumor samples from single nucleotide polymorphism genotyping data. Genome Biol. **11**(9), R92 (2010)
11. Miller, C.A., et al.: ReadDepth: a parallel R package for detecting copy number alterations from short sequencing reads. PLoS ONE **6**(1), e16327 (2011)
12. Abyzov, A., et al.: CNVnator: an approach to discover, genotype, and characterize typical and atypical CNVs from family and population genome sequencing. Genome Res. **21**(6), 974–984 (2011)
13. Yoon, S., et al.: Sensitive and accurate detection of copy number variants using read depth of coverage. Genome Res. **19**(9), 1586–1592 (2009)
14. Xie, C., Tammi, M.T.: CNV-seq, a new method to detect copy number variation using high-throughput sequencing. BMC Bioinform. **10**, 80 (2009)
15. Xi, R., Lee, S., Park, P.J.: A survey of copy-number variation detection tools based on high-throughput sequencing data. Curr. Protoc. Hum. Genet. Chapter 7: Unit7 (2012)
16. Duan, J., et al.: Comparative studies of copy number variation detection methods for next-generation sequencing technologies. PLoS ONE **8**(3), e59128 (2013)
17. Xi, R., et al.: Copy number variation detection in whole-genome sequencing data using the Bayesian information criterion. Proc. Natl. Acad. Sci. USA **108**(46), E1128–E1136 (2011)
18. Gusnanto, A., et al.: Correcting for cancer genome size and tumour cell content enables better estimation of copy number alterations from next-generation sequence data. Bioinformatics **28**(1), 40–47 (2012)
19. Boeva, V., et al.: Control-free calling of copy number alterations in deep-sequencing data using GC-content normalization. Bioinformatics **27**(2), 268–269 (2011)
20. Kim, T.M., et al.: rSW-seq: algorithm for detection of copy number alterations in deep sequencing data. BMC Bioinform. **11**, 432 (2010)
21. Koboldt, D.C., et al.: Varscan 2: somatic mutation and copy number alteration discovery in cancer by exome sequencing. Genome Res. **22**(3), 568–576 (2012)

An Integrated Method of Detecting Copy Number Variation Based on Sequence Assembly

Weiwei Liu and Jingyang Gao(✉)

College of Information Science and Technology, Beijing University of Chemical Technology, 15 East Beisanhuan Road, Beijing, China
gaojy@mail.buct.edu.cn

Abstract. The Next-generation sequencing technology is a widely used sequencing method, and many genome researches are based on its sequencing data. Currently, there are many methods of detection of genomic Structure Variation, based on NGS data. And a lot of Copy Number Variation (CNV) detection methods based on the statistical models of read depth. However, since CNV has multiple subtypes and long variant lengths, the traditional detection tools have many limitations. Therefore, this paper proposes AssCNV, a new detection method for CNV, which integrated sequence assembly strategy and read depth strategy. The subtypes of CNV considered in this paper are insertion, deletion, and duplication. Our experimental results showed that AssCNV maintains a higher level of precision and sensitivity in the simulation data of different coverage, which is much better than other available tools.

Keywords: Next-generation sequencing · Copy Number Variation
Sequence assembly · Read depth · Integrated detection

1 Introduction

As a widely used technology, the Next-generation Sequencing technology (NGS) improves the sequencing speed and maintains the high precision (less than 1%), although it obtains a shorter sequence length. The 1000 Genomes Project released 1092 human genetic sequence data in 2012. This greatly facilitates researchers. There are three types of genome variations: Single Nucleotide Variant (SNV), short Insertion/Deletion (INDEL), and Structure Variant (SV) [1]. SNV refers to a single base variation. INDEL refers to short insertion and deletion which less than 50 bp in length. And SV includes deletion, insertion, inversions, translocation, duplication variations and so on, which more than 50 bp in length. Copy Number Variation (CNV) is an important part of SV. It refers to submicroscopic variations in the size of genomic fragment, ranging from kb to Mb [2]. This paper, however, only considers insertion, deletion, and duplication, which more than 50 bp in length. Being one of the most important pathogenic factors of human diseases, CNV has been confirmed to be associated with many complex disorders [3]. Moreover, CNV is one of the most important types of variations in the

© Springer International Publishing AG, part of Springer Nature 2018
D.-S. Huang et al. (Eds.): ICIC 2018, LNCS 10954, pp. 589–594, 2018.
https://doi.org/10.1007/978-3-319-95930-6_58

cancer genome [4]. Therefore, the detailed detection for CNV within the whole genome is of great significance.

At present, the methods of detecting CNV based on NGS data are mainly based on the read depth by establishing statistical models. Typical tools include ReadDepth [5], CNVnator [6]. In addition to the read depth strategy, there are Paired-End Mapping, split read alignment, sequence assembly for detecting SV. Such as BreakDancer [7], Pindel [8]. But each strategy has its own limitations, so many researchers have begun to develop integrated detection tools, such as, DELLY [9], LUMPY [10], and so on.

2   Methods

Considering that CNV has multiple subtypes and long variant lengths, this study proposes an integrated tool, namely AssCNV, for detecting CNV. Firstly used the sequence assembly strategy to obtain high-quality soft-clipped sites, then integrated read depth strategy. At present, AssCNV is only used to detect CNV in autosomal.

2.1   Multiple Tools for Detecting CNV to Obtain the Variant Breakpoints

Experiments were performed using the currently popular tools for detecting CNV, include BreakDancer, CNVnator, Control-FREEC [11], DELLY, SpeedSeq [12], ReadDepth, Pindel. The precision and sensitivity of each tool was calculated, and two thresholds were set: e = 0.4 and m = 1000, indicating that the deviations of the two detected breakpoints with the benchmark data did not exceed 1000 bp and 40% of the variation length; such breakpoints were considered as true positive. Under these two thresholds, if there were at least three tools with sensitivity greater than 0.85 and precision greater than 0.6, merged their detected results. Otherwise, the program would divided the genome with fixed length to obtain variant breakpoint candidates.

2.2   Local Sequence Assembly

After obtaining the local breakpoint candidates, the high quality reads in the range were extracted from the BAM file to perform the subsequent sequence assembly. The filter conditions were as follows: a. The MAPQ of the normal matching read should greater than 1. b. High quality soft-clipped reads. First, the MAPQ should greater than 1. Second, the proportion of soft-clipped in the read should be greater than 25% of the read length. Third, the high-quality base ratio of the soft-clipped portion should be greater than 0.8. The base with quality greater than 20 is considered as high-quality base. c. Keep the unmatched reads in the entire BAM file. The MAPQ and CIGAR field can be find in the fifth and sixth row of BAM file.

The filtered high-quality reads were cut into short reads of fixed length, 25 bp. Then the repeat length of every two read was calculated, considering that the suffix of a read overlaps with the prefix of the other. In addition, in order to simulate SNP, an additional parameter, fault tolerance rate, was set with 1%. Then obtained the maximum repeatability of each read and the No. of subsequent reads. If the result was less than 15% of read length, this record will be canceled. Then a directed graph was constructed. For the branches and rings situations, the path compatibility strategy was used to choose the suitable paths. Finally, contigs longer than 50 bp were reserved.

2.3   Integrated Detection Method

Obtained the High-quality Soft-clipped Sites. When multiple reads have soft-clipped sites at the same aligned position, this is often the breakpoint corresponding to SV. Therefore, the contigs obtained by local assembly were aligned to the reference genome, and then the high quality soft-clipped reads were filtered. For the extracted results, the Blat tool was used for realignment to refine the CIGAR field. If the new CIGAR analyzed that the read was not a soft-clipped read, the new CIGAR wound replace the original CIGAR. Finally, obtained the high quality soft-clipped sites.

Combined Read Depth Strategy. The base coverage at a site can be obtained using the depth command of SAMtools. We considered 200 bp as a window, and 20 bp was overlapped between two adjacent windows to calculate the average coverage depth. If the result of a series of connected windows was close, connected these continuous windows into one window. The Qualimap [13] was used to obtain the mean coverage about the original BAM file, so that these windows can be tagged as normal, deletion, or duplication. Then, the high quality soft-clipped sites was combined with the tagged windows. For all high-quality soft-clipped sites in the normal windows, an insertion variation breakpoint was sought and insert length was calculated. Combining the remaining candidate breakpoints with the labeled window information, the precise start and end sites for deletion and duplication variation could be identified.

3   Results

The detected results were compared to the benchmark data, and evaluated by Precision (Pre), Sensitivity (Sen) and F1-score (F1). The precision is the correct ratio of the detected result, and the sensitivity is the correct ratio of the benchmark data. F1-score is a comprehensive indicator of Pre and Sen. And the formulas are:

$$Pre = TP/(TP + FP) \tag{1}$$

$$Sen = TP/(TP + FN) \tag{2}$$

$$F1-score = (2 * Pre * Sen)/(Pre + Sen) \tag{3}$$

Because there is a direct relationship between detected results of different tools and benchmark data, and variant lengths. Therefore, we generated simulated sequencing data for experiments. Selected the chrom11 of hg19 published by the 1000 Genomes Project as reference genome, and used SVsim to generate simulated FASTA files, which contained accurate variant information. Three variant subtypes of CNV were simulated: deletion, insertion and duplication, and their variant length contains 5 different ranges: 50–500 bp, 500–1000 bp, 1000–10000 bp, 10000–50000 bp, and

Table 1. Results of different tools for detecting CNV in simulated data.

INS

Cov	Tool	Result	Benchmark	TP	Pre	Sen	F1
20x	BreakDancer	5767	350	122	0.021	0.349	0.0396
	DELLY	1676|0		0	0	0	0
	Pindel	489106|0		0	0	0	0
	SpeedSeq	221		65	0.294	0.186	0.2279
	AssCNV	434		289	0.666	0.826	0.7374
30x	BreakDancer	6324	350	135	0.021	0.386	0.0398
	DELLY	1710|2		1	0.5	0.003	0.006
	Pindel	491019|6		4	0.667	0.011	0.0216
	SpeedSeq	243		72	0.296	0.206	0.2429
	AssCNV	456		307	0.673	0.877	0.7616
50x	BreakDancer	7460	350	164	0.022	0.47	0.042
	DELLY	2390|16		9	0.563	0.026	0.0497
	Pindel	534389|32		22	0.679	0.063	0.1153
	SpeedSeq	301		102	0.339	0.291	0.3132
	AssCNV	465		322	0.692	0.92	0.7899

DEL

Cov	Tool	Result	Benchmark	TP	Pre	Sen	F1
20x	BreakDancer	934	350	88	0.094	0.251	0.1368
	CNVnator	657		45	0.068	0.129	0.0891
	DELLY	2082		102	0.049	0.291	0.0839
	Pindel	231321		97	0.0004	0.277	0.0008
	SpeedSeq	312		106	0.34	0.303	0.3204
	AssCNV	396		317	0.801	0.906	0.8503
30x	BreakDancer	1011	350	97	0.096	0.277	0.1426
	CNVnator	827		62	0.075	0.177	0.1054
	DELLY	2373		121	0.051	0.346	0.0889
	Pindel	251158		119	0.0005	0.34	0.001
	SpeedSeq	324		112	0.346	0.32	0.3325
	AssCNV	399		323	0.81	0.923	0.8628
50x	BreakDancer	1258	350	122	0.097	0.349	0.1518
	CNVnator	1050		83	0.079	0.237	0.1185
	DELLY	2834		153	0.054	0.409	0.0954
	Pindel	297302		164	0.0006	0.469	0.0012
	SpeedSeq	331		131	0.396	0.374	0.3847
	AssCNV	401		326	0.813	0.931	0.868

DUP

Cov	Tool	Result	Benchmark	TP	Pre	Sen	F1
20x	CNVnator	178	350	73	0.41	0.209	0.2769
	DELLY	267		84	0.315	0.24	0.2724

(continued)

Table 1. (*continued*)

DUP							
Cov	Tool	Result	Benchmark	TP	Pre	Sen	F1
	FREEC	34		14		0.04	0.0729
	ReadDepth	14603		233	0.016	0.666	0.0312
	Pindel	98672		186	0.002	0.531	0.004
	SpeedSeq	368		214	0.582	0.611	0.5961
	AssCNV	382		310	0.812	0.886	0.8474
30x	CNVnator	203	350	84	0.414	0.24	0.3039
	DELLY	296		92	0.311	0.263	0.285
	FREEC	58		28	0.483	0.08	0.1373
	ReadDepth	15792		250	0.016	0.714	0.0313
	Pindel	110023		216	0.002	0.617	0.004
	SpeedSeq	353		211	0.598	0.603	0.6005
	AssCNV	388		316	0.814	0.903	0.8562
50x	CNVnator	275	350	116	0.422	0.331	0.371
	DELLY	397		113	0.285	0.323	0.3028
	FREEC	62		38	0.613	0.109	0.1851
	ReadDepth	17165		273	0.016	0.78	0.0314
	Pindel	130180		245	0.002	0.7	0.004
	SpeedSeq	372		223	0.599	0.666	0.6307
	AssCNV	396		323	0.815	0.923	0.8656

>50000 bp, with 60, 100, 100, 60, and 30 variations. Then used art_illumina of ART to generate FASTQ files. Finally generated three groups data with different coverage: 20x, 30x, 50x. Table 1 shows the detected results of those tools in simulated data.

Analyzing the above experimental results, we can find that the detection Pre and Sen of each tool had a certain increase as the coverage increased of the three types variations. For the insertion detection results, DELLY/Pindel/BreakDancer mostly detected short insertions. SpeedSeq had higher Pre and Sen, as it had a long variant length. AssCNV had the best detection performances with a Pre over 65%. For the deletion detection results, AssCNV had the best detection performances with a Pre over 80%. SpeedSeq had the second Pre. Pindel's detected results included many incorrect results. For duplication detection results, FREEC had a high Pre. AssCNV had the best performance with a F1 over 85%. SpeedSeq had the second F1 over 60%.

4   Conclusion

Based on the NGS data, this paper proposes a new detection method for CNV, namely AssCNV, combined sequence assembly strategy and read depth strategy. Sequence assembly helps to detect longer CNV, and then combined read depth information and the high-quality soft-clipped sites from assembled reads can detect deletion and duplication variations effectively, and identify insertion variant breakpoints.

At present, AssCNV only implements the experiments in autosomal sequencing data. Our experimental results showed that the Pre and Sen of AssCNV was higher than that of other detection tools, and maintained them at a high level. And with the increase of data coverage, the ability of AssCNV to detect the three subtypes of CNV increased. Therefore, AssCNV has obvious advantages in the detection of CNV. However, there are some details that need to be improved and perfected in AssCNV, and we will continue to work hard to achieve better results in future researches.

Acknowledgement. Project supported by the National Natural Science Foundation of China (Grant No. 61472026) and Beijing Natural Science Foundation (5182018).

References

1. Eichler, E.E., Nickerson, D.A., Altshuler, D., et al.: Completing the map of human genetic variation. Nature **447**(7141), 161–165 (2007)
2. Redon, R., Ishikawa, S., Fitch, K.R., et al.: Global variation in copy number in the human genome. Nature **444**(7118), 444–454 (2006)
3. Thuresson, A.C., Van Buggenhout, G., Sheth, F., et al.: Whole gene duplication of SCN2A and SCN3A is associated with neonatal seizures and a normal intellectual development. Clin. Genet. **91**(1) (2017)
4. Hackmann, K., Kuhlee, F., Betcheva-Krajcir, E., et al.: Ready to clone: CNV detection and breakpoint fine-mapping in breast and ovarian cancer susceptibility genes by high-resolution array CGH. Breast Cancer Res. Treat. **159**(3), 585–590 (2016)
5. Miller, C.A., Hampton, O., Coarfa, C., et al.: ReadDepth: a parallel R package for detecting copy number alterations from short sequence reads. Plos One **6**(1), e16327 (2011)
6. Abyzov, A., Urban, A.E., Snyder, M., et al.: CNVnator: an approach to discover, genotype, and characterize typical and atypical CNVs from family and population genome sequence. Genome Res. **21**(6), 974–984 (2011)
7. Chen, K., Wallis, J.W., McLellan, M.D., et al.: BreakDancer: an algorithm for high-resolution mapping of genomic structural variation. Nat. Methods **6**(9), 677–681 (2009)
8. Ye, K., Schulz, M.H., Long, Q., et al.: Pindel: a pattern growth approach to detect break points of large deletions and medium sized insertions from paired-end short reads. Bioinformatics **25**(21), 2865–2871 (2009)
9. Rausch, T., Zichner, T., Schlattl, A., et al.: DELLY: structural variant discovery by integrated paired-end and split-read analysis. Bioinformatics **28**(18), i333–i339 (2012)
10. Layer, R.M., Chiang, C., Quinlan, A.R., et al.: Lumpy: a probabilistic framework for structural variant discovery. Genome Biol. **15**(6), R84 (2014)
11. Boeva, V., Popova, T., Bleakley, K., et al.: Control-FREEC: a tool for assessing copy number and allelic content using next-generation sequence data. Bioinformatics **28**(3), 423–425 (2012)
12. Chiang, C., Layer, R.M., Faust, G.G., et al.: SpeedSeq: ultra-fast personal genome analysis and interpretation. Nat. Methods **12**(10), 966 (2015)
13. Garcíaalcalde, F., Okonechnikov, K., Carbonell, J., et al.: Qualimap: evaluating next-generation sequence alignment data. Bioinformatics **28**(20), 2678 (2012)

RNA Secondary Structure Prediction Based on Long Short-Term Memory Model

Hongjie Wu[1], Ye Tang[1], Weizhong Lu[1,2(✉)], Cheng Chen[1],
Hongmei Huang[1], and Qiming Fu[1,2]

[1] School of Electronic and Information Engineering,
Suzhou University of Science and Technology, Suzhou 215009, China
luwz@usts.edu.cn
[2] Jiangsu Key Laboratory of Intelligent Building Energy Efficiency,
Suzhou 215009, Jiangsu, China

Abstract. RNA secondary structure prediction is an important issue in structural bioinformatics. The difficulty of RNA secondary structure prediction with pseudoknot is increased due to complex structure of the pseudoknot. Traditional machine learning methods, such as support vector machine, markov model and neural network, have been tried and their prediction accuracy are also increasing. The RNA secondary structure prediction problem is transferred into the classification problem of base in the sequence to reduce computational complexity to a certain extent. A model based on LSTM deep recurrent neural network is proposed for RNA secondary structure prediction. Subsequently, comparative experiments were conducted on the authoritative data set RNA STRAND containing 1488 RNA sequences with pseudoknot. The experimental results show that the SEN and PPV of this method are higher than the other two typical methods by 1% and 11%.

Keywords: RNA secondary structure prediction · Recurrent neural network
Pseudoknots · Classification

1 Introduction

RNA (ribonucleotide) is not only a carrier of genetic information, but also its structure plays a crucial role in gene maturation, regulation and function [1]. Studying the relationship between RNA function and structure, and determining the form and frequency of RNA folding are important to reveal the role of RNA molecules in the life process [2, 4]. The most common way to manipulate RNA structures algorithmically is to reduce them to base pairs, the so-called secondary structures, abstracted from the actual spatial arrangement of nucleotides. For a valid secondary structure, each base

This paper is supported by the National Natural Science Foundation of China (61772357, 61502329, 61672371), Jiangsu 333 talent project and top six talent peak project (DZXX-010), Suzhou Foresight Research Project (SYG201704, SNG201610) and Postgraduate Research & Practice Innovation Program of Jiangsu Province (SJCX17_0680).

D.-S. Huang et al. (Eds.): ICIC 2018, LNCS 10954, pp. 595–599, 2018.
https://doi.org/10.1007/978-3-319-95930-6_59

i can only interact with at most one other base *j*, and form one base pair (i, j), where *i* and *j* are index of bases in sequence [5, 6].

Pseudoknot is a substructure of RNA secondary structure, it is the crossed structure between two base pairs (i, j) and (k, l) in the sequence, where $i < k < j < l$. RNA secondary structure prediction without pseudoknots has been fully studied by the dynamic programming algorithms of Zuker [7], Mathews [8, 9] and its improved algorithms such as mfold [10] and GTfold [11]. RNA secondary structure prediction with pseudoknots has been proved to be an NP-based optimization problem [12].

After translating the RNA secondary structure prediction problem into the classification problem of bases in the sequence, machine learning method can have lower computational complexity than thermodynamics method. However, there is still a difficulty: to figure out the secondary structure of a sequence with pseudoknots, bases cannot be distinguished by only using three categories of 'unpaired bases', 'paired bases near the head' and 'paired bases near the end'.

As a research hotspot in the field of machine learning, deep learning can mine deeper hidden features from data [13, 14]. Recurrent neural network is a sequence-oriented neural network model for deep learning. It has excellent performance in natural language processing, image recognition and speech recognition [15].

2   Method

2.1   RNA Secondary Structure Prediction

After translating the RNA secondary structure prediction problem into the classification problem of base pairings in the sequence, the purpose of this problem is to figure out the pairing result of every base in the sequence when the primary structure of the sequence is given in the form of $S = s_1, s_2, \ldots s_n$, where s_1 is the base near the 5' side, s_n is the base near the 3' side, s_i is the *i*-th base in sequence S and $s_i \in \{A, G, C, U\}$.

Considering the existence of pseudoknots, the secondary structure of a sequence cannot be completely calculated when its bases are distinguished by 3 categories, in the pairing results of this paper, every base of a sequence with *n* bases can be divided into $n + 1$ categories: the category of 'unpaired base' and categories of 'paired base that paired with the *k*-th base $(k = 1, 2 \ldots n)$'.

2.2   LSTM Model for RNA Secondary Structure Prediction

The structure of an RNA secondary structure prediction model based on LSTM is shown in Fig. 1 (the reshape layer for data shape transformation between the function layers is omitted in the figure). This model can predict the pseudoknots and allow base deletion. It consists of input layer, LSTM layer, fully connected layer and output layer. The bidirectional LSTM layer is used to extract the features of the RNA sequence, and the fully connection layer is used for further feature extraction and classification.

The model uses the cross-entropy commonly used in the classification problem as the loss function. For each input sequence, its total loss function is equal to the mean of all loss prediction functions of bases. The calculation formula is as followed:

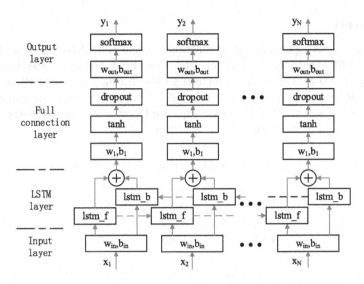

Fig. 1. LSTM model framework.

$$L(y, y') = \frac{1}{n} \sum_{i=1}^{n} \sum_{j=1}^{m} y'[i,j] * \log(y[i,j]) \tag{1}$$

Where n is the length of the sequence, m is the number of categories, y and y' are two-dimensional arrays respectively representing the predicted result of the model and the real label of the sample. After the total loss function of the sequence is got, the batch gradient descent algorithm is used to update the weight and bias parameters in the network model to make the total loss function as small as possible.

3 Experiments and Discussion

3.1 Evaluation Metrics

Sensitivity (*SEN*) and specificity (*PPV*) [16] are evaluation metrics commonly used in RNA secondary structure prediction, they are calculated as follows:

$$SEN = \frac{TP}{TP + FN}, \ PPV = \frac{TP}{TP + FP} \tag{2}$$

Where *TP* (true positives) means the number of correctly predicted bases, *FN* (false negatives) means the number of bases that are not correctly predicted, *FP* (false negatives) means the number of unpaired bases that predicted to be paired, *TN* (true negatives) means the number of correctly predicted unpaired bases [16].

3.2 Dataset

The dataset of this paper comes from Authoritative dataset RNA STRAND, including 3 subsets: TMR (The tmRNA website), RFA (The RNA family database), and ASE (RNase P Database). The 3 datasets including 1488 sequences are randomly and evenly divided into 10 parts, one of them is selected as testing data, others are training data. The number of sequences, the average sequence length, the minimum length and the maximum length of the 3 data sets are shown in Table 1.

Table 1. Datasets

Dataset	Number	Average	Max	Min
TMR	721	361.1	463	102
ASE	454	332.6	486	189
RFA	313	118.9	553	40

3.3 Experiment Results

Table 2 shows the experimental results of LSTM and other classic RNA secondary structure prediction methods: ProbKnot and cylofold. It can be seen from the table that the *SEN* and *PPV* of the LSTM algorithm are higher than other algorithms by at least 1% and 11% respectively on all 3 data sets, which can prove that LSTM model is an effective method for RNA secondary structure prediction.

Table 2. Experiment results of 3 methods

Dataset	TMR		ASE		RFA	
Method	SEN	PPV	SEN	PPV	SEN	PPV
LSTM	0.71	0.56	0.85	0.67	0.83	0.72
ProbKnot	0.63	0.44	0.72	0.56	0.82	0.58
Cylofold	0.45	0.33	0.51	0.44	0.66	0.51

4 Conclusions

By translating the RNA secondary structure prediction problem into the classification problem of base pairings in the sequence, an RNA secondary structure prediction method based on LSTM was proposed. This model consists of input layer, bidirectional LSTM layer, fully connected layer and output layer, which can extract deep features from RNA primary sequences and classify them. The experimental results show that this method can effectively improve the accuracy of the prediction of RNA secondary structure. The average *SEN* and *PPV* of this model are 81% and 65% respectively, which is higher than the classical algorithm ProbKnot on the 3 testing sets.

Acknowledgements. This paper is supported by the National Natural Science Foundation of China (61772357, 61502329, 61672371), Jiangsu 333 talent project and top six talent peak project (DZXX-010), Suzhou Foresight Research Project (SYG201704, SNG201610) and Postgraduate Research & Practice Innovation Program of Jiangsu Province (SJCX17_0680).

References

1. Anderson-Lee, J., Fisker, E., Kosaraju, V., et al.: Principles for predicting RNA secondary structure design difficulty. J. Mol. Biol. **428**(5), 748 (2016). Part A
2. Dong, H., Liu, Y.N.: A new method for RNA secondary structure prediction based on hidden markov model. J. Comput. Res. Dev. **49**(4), 812–817 (2012)
3. Wu, J.S., Zhou, Z.H.: Sequence-based prediction of microRNA-binding residues in proteins using cost-sensitive Laplacian support vector machines. IEEE/ACM Trans. Comput. Biol. Bioinf. **10**(3), 752–759 (2013)
4. Bai, Y., Dai, X., Harrison, A., et al.: Toward a next-generation atlas of RNA secondary structure. Brief. Bioinform. **17**(1), 63–77 (2016)
5. Lorenz, R., Wolfinger, M.T., Tanzer, A., et al.: Predicting RNA secondary structures from sequence and probing data. Methods **103**, 86 (2016)
6. Wu, H.J., Lv, Q., Quan, L.J., et al.: Structural topology modeling of GPCR transmembrane helix and its prediction. Chin. J. Comput. **36**(10), 2168–2178 (2013)
7. Wu, H.J., Lv, Q., Wu, J.Z., et al.: A parallel ant colony method to predict protein skeleton and its application in CASP8/9. Scientia Sinica Informationis **42**(8), 1034–1048 (2012)
8. Mathews, D.H., Turner, D.H., Watson, R.M.: RNA secondary structure prediction. BMC Bioinform. **11**(1), 129 (2007)
9. Mathews, D.H., Turner, D.H., Watson, R.M.: RNA secondary structure prediction. In: Current Protocols in Nucleic Acid Chemistry, pp. 345–363. Wiley, Hoboken (2016)
10. Zuker, M.: Mfold web server for nucleic acid folding and hybridization prediction. Nucleic Acids Res. **31**(13), 3406–3415 (2003)
11. Mathuriya, A., Bader, D.A., Heitsch, C.E., et al.: GTfold: a scalable multicore code for RNA secondary structure prediction. In: ACM Symposium on Applied Computing, pp. 981–988. ACM (2009)
12. Do, C.B., Woods, D.A., Batzoglou, S.: CONTRAfold: RNA secondary structure prediction without physics-based models. Bioinformatics **22**(14), e90 (2006)
13. Wu, H.J., Wang, K., Lu, L.Y., et al.: A deep conditional random field approach to transmembrane topology prediction and application to GPCR three-dimensional structure modeling. IEEE/ACM Trans. Comput. Biol. Bioinform. **PP**(99), 1 (2016)
14. Wu, H.J., Cao, C.Y., Xia, X.Y., et al.: Unified deep learning architecture for modeling biology sequence. IEEE/ACM Trans. Comput. Biol. Bioinform. **PP**(99), 1 (2017)
15. Reuter, J.S., Mathews, D.H.: RNA secondary structure prediction. BMC Bioinform. **9**(17), 873 (2013)
16. Mathews, D.H.: Using an RNA secondary structure partition function to determine confidence in base pairs predicted by free energy minimization. RNA **10**(8), 1178 (2004)

Protein-Protein Docking with Improved Shape Complementarity

Yumeng Yan and Sheng-You Huang[✉] ⓘ

Institute of Biophysics, School of Physics, Huazhong University of Science
and Technology, Wuhan 430074, Hubei, China
huangsy@hust.edu.cn
http://huanglab.phys.hust.edu.cn/

Abstract. Protein-protein docking is a useful computational tool for predicting
the complex structure and interaction between proteins. As the most basic
ingredient of scoring functions, shape complementarity plays a critical role in
protein-protein docking. In this study, we have presented a new pairwise scoring
function to consider long-range interactions in shape complementarity (LSC) for
protein-protein docking. Our docking program with LSC was tested on the
protein docking benchmark 4.0 of 176 diverse protein-protein complexes, and
compared with four other shape-based docking approaches, ZDOCK2.1,
MolFit/G, GRAMM, and FTDock/G. It was shown that our LSC significantly
improved the docking performance in binding mode predictions in both success
rate and number of hits per complex, compared to the other four approaches.
The software is freely available as part of our HDOCK web server at http://
hdock.phys.hust.edu.cn/.

Keywords: Protein-protein docking · Shape complementarity
Protein- protein interactions · Scoring function · Fast-Fourier transformation

1 Introduction

Protein is one of the most important biological macromolecules in life, whose structure
is determined by its sequence of amino acids [1, 2]. Many cellular processes in living
organisms like signal transduction, intracellular trafficking, and immune recognition
involve the interactions between proteins. One such example is the interactions
between histone proteins H2A, H2B, H3 and H4 in the nucleosome core [3]. Therefore, determination of the three-dimensional (3D) structures of protein-protein complexes is crucial to investigate the molecular mechanism of protein-protein interactions
and thus develop therapeutic drugs targeting the interactions [4, 5]. In this process,
protein-protein docking has played an important role in complementing experimental
methods for the determination of protein-protein complex structures [4–7]. Given two

Supported by the National Natural Science Foundation of China (No. 31670724), the National
Key R&D Program of China (Nos. 2016YFC1305800 and 2016YFC1305805), and the startup
grant of Huazhong University of Science and Technology (No. 3004012104).

individual protein structures, protein-protein docking samples possible binding modes of one protein relative to the other, and then ranks them with an energy scoring function [4].

All docking algorithms require a scoring function for ranking sampled binding poses. As the most basic ingredient of scoring functions [8–10], shape complementarity plays an important role in not only guiding the search process of putative binding modes [4, 11, 12] but also ranking the sampled binding modes as a basic component of scoring functions [4, 7]. This is especially critical for fast-Fourier transformation (FFT)-based docking algorithms, in which shape representation is not only a basic scoring element but also affecting the discretization of other energy terms on grids [4]. Here, we have developed a new pairwise scoring function to consider the long range effects in shape complementarity, which is referred to as LSC, for protein-protein docking. Our docking program with LSC can serve as an efficient initial stage of hierarchical docking algorithms, and has an implementation in the development of accurate docking/scoring algorithms.

2 Materials and Method

2.1 FFT-Based Docking with LSC

The binding modes of one protein relative to the other are globally sampled through an exhaustive search in six-dimensional (3 translational + 3 rotational) space, where the translational search is accelerated by an FFT-based method.

FFT-Based Translational Search. For an exhaustive search in the translational space, both the receptor and ligand proteins were first mapped onto a 3D grid of $N \times N \times N$ grid points [9, 14]. The grid points within the van der Waals (VDW) radius of any protein atom were considered inside the molecule; otherwise, the grid points were considered as outside the protein. Here, the VDW radii for standard protein atoms were taken from the study by Li and Nussinov [15]. Then, the inside-protein grid points were divided into three parts: surface layer, near-surface layer, and core region. It is defined that a grid point belongs to the surface layer if any of its neighboring grid points is outside the protein. Similarly, a grid point will belong to the near-surface layer if any of its neighbors is the surface-layer grid point. All the other grid points except the surface and near-surface layers inside the protein are defined as the core region. Then, each grid point for the receptor (R) and the ligand (L) was assigned a complex value as follows:

$$
R(l, m, n) = \begin{cases}
-\sum_{i,j,k} exp\left[-(r-1)^2\right] + J & \text{for surface layer} \\
-1 + 2J \times \sum_{i,j,k} exp(-r^2) & \text{for near surface layer} \\
-1 + 10J & \text{for the core} \\
0 & \text{outside the protein}
\end{cases}
\tag{1}
$$

$$
L(l,m,n) =
\begin{cases}
1 - J & \text{for surface layer} \\
1 - 2J \times \sum_{i,j,k} exp(-r^2) & \text{for near surface layer} \\
1 - 10J & \text{for the core} \\
0 & \text{outside the protein}
\end{cases}
\tag{2}
$$

where $J^2 = -1$, l, m and n are the indices of the 3D grid $(l, m, n = 1, \cdots, N)$, and r is the distance between the grid points of (i, j, k) and (l, m, n). For the surface layer, the long-range interactions are considered by setting $i \in [l - 3, l + 3]$, $j \in [m - 3, m + 3]$ and $k \in [n - 3, n + 3]$. For the near-surface layer, we set the normal ranges of $i \in [l - 1, l + 1]$, $j \in [m - 1, m + 1]$ and $k \in [n - 1, n + 1]$.

After mapping the proteins on the grid, the long-range shape complementarity (LSC) score of a relative translation between receptor and ligand can be generally expressed by the following equation [9, 14]

$$
E(o, p, q) = \text{Re}\left[\sum_{l=1}^{N} \sum_{m=1}^{N} \sum_{n=1}^{N} R(l, m, n) \times L(l+o, m+p, n+q) \right]
\tag{3}
$$

where Re[ ] denotes the real part of a complex function, and o, p and q stand for the number of grid points by which the ligand (L) is shifted with respect to the receptor (R) in each of three translational dimensions, respectively. The correlation of Eq. (3) can be calculated by an FFT-based algorithm. A higher correlation means a better shape complementarity between the receptor and ligand grids for a relative translation of (o, p, q).

Rotational Sampling. For an exhaustive search in the rotational space, the receptor is fixed and the ligand is rotated by an interval of Euler angles $(\Delta\theta, \Delta\phi, \Delta\psi)$. For each rotation of the ligand, an FFT-based algorithm, as described above, is used to calculate the correlations between the receptor and ligand grids in terms of shape complementarity. During the docking calculations, an angle interval of 15° was used for rotational sampling, and a grid spacing of 1.2 Å was adopted for FFT-based translational search. Evenly distributed Euler angles were used for exhaustive rotational search, resulting in a total of 4392 rotations in the rotational space. One binding mode, which corresponds to the best shape complementarity in an FFT-based translational search, was kept for each rotation of the ligand, yielding a total of 4392 ligand binding modes for a docking run.

2.2   Test Set

We evaluated our LSC on the protein-protein docking benchmark version 4.0 constructed by the Weng's group [13]. The benchmark includes a total of 176 diverse protein-protein complexes including 52 enzyme-inhibitor (EI), 25 antibody-antigen (AA), and 99 cases of other types (OT). The benchmark has been widely used to assess the performance of docking programs and scoring functions [7].

2.3 Evaluation Criterion

Similar to that used in previous studies [7], we adopted the ligand root-mean- square-deviation (RMSD) to evaluate the quality of a binding mode. Here, the ligand RMSD (L_{rmsd}) was calculated based on the C_α atoms of the ligand protein. If a predicted binding mode has an L_{rmsd} of less than 10 Å, it is defined as a successful prediction or a 'hit'. The docking performance was measured by the success rate, the fraction of test cases with at least one hit when a certain number of top predictions were considered.

3 Results and Discussion

We tested our LSC-implemented docking program on the bound (co-crystallized) structures of 176 protein-protein complexes in the protein docking benchmark 4.0 [13]. Docking with bound structures (bound docking) serves as a primary evaluation for the performance of a docking/scoring algorithm, as no conformational change is involved in the individual structures. A reasonable scoring function should be able to give a good performance in bound docking.

Figure 1 shows the success rate and the average number of hits per complex by our LSC method for bound docking on the 176 protein-protein complexes of the benchmark. The corresponding success rates and average number of hits for several specific numbers of top predictions were shown in Table 1. As a reference, Fig. 1 and Table 1 also give the corresponding results of four other shape-based docking algorithms, ZDOCK2.1 [9], MolFit/G [16], GRAMM [17], and FTDock/G [18]. Here, GRAMM and FTDock/G use a grid-based shape complementarity (GSC) scoring function, and ZDOCK2.1 adopts a pairwise shape complementarity (PSC) function in docking. It can be seen from Fig. 1a that our LSC method achieved a significantly better performance in the success rate than the other four docking/scoring methods. When the top 1, 10, 100, and 1000 predictions were considered, our LSC obtained a success rate of 34.09, 51.71, 69.32, and 87.50%, respectively, compared to 25.57, 42.61, 61.36, and 83.52% for ZDOCK2.1, 24.43, 33.52, 51.14, and 79.55% for MolFit/G, 9.09, 15.91, 38.64, 67.61% for GRAMM, and 4.55, 14.21, 43.18, 77.27 for FTDock/G (see Table 1).

Our LSC also performed significantly better than the other four scoring methods in the average number of hits and yielded an average of 3.21, 6.26, 8.38, and 11.64 hits per complex, compared to 2.18, 4.51, 6.38, and 8.97 hits for ZDOCK2.1, 1.96, 4.36, 6.37, and 9.47 hits for MolFit/G, 0.76, 1.73, 2.54, and 3.66 hits for GRAMM, and 1.05, 2.43, 3.87, and 5.82 hits for FTDock/G when the top 100, 500, 1000, and 2000 predictions were considered (see Fig. 1b and Table 1). The significantly better performance of our LSC scoring method compared to the other four scoring functions for bound docking suggests an advantage of LSC in characterizing shape complementarity for initial docking stage.

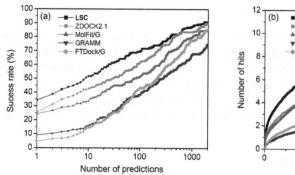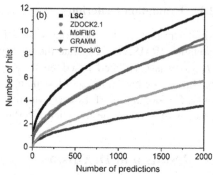

Fig. 1. The success rate (a) and the average number of hits per target (b) as a function of the number of top predictions by our FFT-based docking program with LSC and four other approaches (ZDOCK2.1, MolFit/G, GRAMM, and FTDock/G) for bound docking on the protein-protein docking benchmark 4.0. The results of MolFit/G, GRAMM, and FTDock/G were taken from our previous study [7].

Table 1. The success rates (%) and average number of hits per complex predicted by our docking program with LSC and four other docking programs with shape-based scoring functions on the protein docking benchmark 4.0 when a certain number of top predictions were considered.

Method	Success rate (%)				Number of hits			
	1	10	100	1000	100	500	1000	2000
LSC	**34.09**	**51.71**	**69.32**	**87.50**	**3.21**	**6.26**	**8.38**	**11.64**
ZDOCK 2.1	25.57	42.61	61.36	83.52	2.18	4.51	6.38	8.97
MolFit/G	24.43	33.52	51.14	79.55	1.96	4.36	6.37	9.47
GRAMM	9.09	15.91	38.64	67.61	0.76	1.73	2.54	3.66
FTDock/G	4.55	14.21	43.18	77.27	1.05	2.43	3.87	5.82

4   Conclusion and Future Remarks

We have developed a new pairwise shape complementarity scoring function to consider long-range interactions in grid-based protein-protein docking. The long-range shape-based complementarity (LSC) function is characterized by an exponentially decaying interaction potential. Our docking program with LSC was tested on the 176 protein-protein complexes in the protein docking benchmark 4.0 by the Weng's group. It was shown that our LSC significantly improved the docking performance in both success rate and number of hits in binding mode predictions, compared to four other shape-based docking approaches, ZDOCK2.1, MolFit, GRAMM, and FTDock. The results suggest the efficacy of our method in characterizing shape complementarity. Extended evaluation on realistic docking will be conducted to further validate the robustness of LSC in a future study.

References

1. Wang, J., Wang, W.: A computational approach to simplifying the protein folding alphabet. Nat. Struct. Biol. **6**, 1033–1038 (1999)
2. Zuo, Y., Li, Y., Chen, Y., Li, G., Yan, Z., Yang, L.: PseKRAAC: a flexible web server for generating pseudo K-tuple reduced amino acids composition. Bioinformatics **33**, 122–124 (2017)
3. Luger, K., Mader, A.W., Richmond, R.K., Sargent, D.F., Richmond, T.J.: Crystal structure of the nucleosome core particle at 2.8 °A resolution. Nature **389**, 251–260 (1997)
4. Huang, S.Y.: Search strategies and evaluation in protein-protein docking: principles, advances and challenges. Drug Discov. Today **19**, 1081–1096 (2014)
5. Schreiber, G., Fleishman, S.J.: Computational design of protein-protein interactions. Curr. Opin. Struct. Biol. **23**, 903–910 (2013)
6. Smith, G.R., Sternberg, M.J.: Prediction of protein-protein interactions by docking methods. Curr. Opin. Struct. Biol. **12**, 28–35 (2002)
7. Huang, S.Y.: Exploring the potential of global protein-protein docking: an overview and critical assessment of current programs for automatic ab initio docking. Drug Discov. Today **20**, 969–977 (2015)
8. Norel, R., Petrey, D., Wolfson, H.J., Nussinov, R.: Examination of shape complementarity in docking of unbound Proteins. Proteins **36**, 307–317 (1999)
9. Chen, R., Weng, Z.: A novel shape complementarity scoring function for protein-protein docking. Proteins **51**, 397–408 (2003)
10. Shentu, Z., Al Hasan, M., Bystroff, C., Zaki, M.J.: Context shapes: efficient complementary shape matching for protein-protein docking. Proteins **70**, 1056–1073 (2008)
11. Kuroda, D., Gray, J.J.: Shape complementarity and hydrogen bond preferences in protein-protein interfaces: implications for antibody modeling and protein-protein docking. Bioinformatics **32**, 2451–2456 (2016)
12. Nicola, G., Vakser, I.A.: A simple shape characteristic of protein-protein recognition. Bioinformatics **23**, 789–792 (2007)
13. Hwang, H., Vreven, T., Janin, J., Weng, Z.: Protein-protein docking benchmark version 4.0. Proteins **78**, 3111–3114 (2010)
14. Katchalski-Katzir, E., Shariv, I., Eisenstein, M., Friesem, A.A., Aflalo, C., Vakser, I.A.: Molecular surface recognition: determination of geometric fit between proteins and their ligands by correlation techniques. Proc. Natl. Acad. Sci. USA **89**, 2195–2199 (1992)
15. Li, A.J., Nussinov, R.: A set of van der Waals and coulombic radii of protein atoms for molecular and solvent-accessible surface calculation, packing evaluation, and docking. Proteins **32**, 111–127 (1998)
16. Heifetz, A., Katchalski-Katzir, E., Eisenstein, M.: Electrostatics in protein-protein docking. Protein Sci. **11**, 571–587 (2002)
17. Vakser, I.A.: Evaluation of GRAMM low-resolution docking methodology on the hemagglutinin-antibody complex. Proteins Suppl. **1**, 226–230 (1997)
18. Gabb, H.A., Jackson, R.M., Sternberg, M.J.: Modelling protein docking using shape complementarity, electrostatics and biochemical information. J. Mol. Biol. **272**, 106–120 (1997)

Carbon-Efficient Scheduling of Blocking Flow Shop by Hybrid Quantum-Inspired Evolution Algorithm

You-Jie Yao[1], Bin Qian[1(⊠)], Rong Hu[1], Ling Wang[2],
and Feng-Hong Xiang[3]

[1] School of Information Engineering and Automation, Kunming University
of Science and Technology, Kunming 650500, China
bin.qian@vip.163.com
[2] Department of Automation, Tsinghua University, Beijing 10084, China
[3] Computer Center, Kunming University of Science and Technology,
Kunming 650500, China

Abstract. In this paper, a hybrid quantum-inspired evolution algorithm
(HQEA) is proposed to solve the blocking flow shop scheduling problem
(BFSP) with the objectives of makespan and carbon-efficient. First, depending
on the characteristics of quantum, we provided a feasible coding and decoding
method for HQEA. Then, a mechanism intended to update the quantum probability matrix. Meanwhile, new individuals are generated through the quantum
probability matrix and have a specified probability of cataclysm. In addition,
some local search operators are utilized to improve the non-dominated solutions.
Finally, the effectiveness of HQEA in solving the BFSP is demonstrated by
experiments and comparisons.

Keywords: Blocking flow shop scheduling problem · Carbon-efficient
Hybrid quantum-inspired evolution algorithm · Local search

1 Introduction

With the intense global greenhouse effect, carbon emissions have become an important
indicator of enterprise production. Industrial production should consider the protection
of the environment while considering the economic effect. In addition, manufacturing
enterprises have become a major source of global carbon emissions. The blocking flow
shop scheduling problems (BFSP) as a typical industrial production model, which is
NP-hard and has attracted more attention in recent years. Due to its 'no buffer' characteristics, many modern manufacturing industries can be modeled as the blocking flow
shop scheduling problems, such as chemical industry [1], iron and steel industry [2],
and serial manufacturing [3]. Therefore, it is necessary to research in the engineering
fields.

Under the situation of the global warming, there is much research about low-carbon
and energy-saving. In [4], two mixed integer programming formulations was proposed
for flow shop scheduling problem with a restriction on peak power consumption.

© Springer International Publishing AG, part of Springer Nature 2018
D.-S. Huang et al. (Eds.): ICIC 2018, LNCS 10954, pp. 606–617, 2018.
https://doi.org/10.1007/978-3-319-95930-6_61

Luo et al. [5] presented a new ant colony optimization to minimize the objectives of makespan and electric power cost of hybrid flow shop. Liu et al. [6] studied a batch-processing machine scheduling and a hybrid flow shop scheduling with economic and environmental criteria. To solve flow shop scheduling with the objectives of minimizing the total carbon emissions and the makespan, Song et al. [7] developed a multi-objective optimization. Deng et al. [8] proposed a competitive memetic algorithm for carbon-efficient scheduling of distributed flow shop, which has better performance.

Quantum-inspired evolutionary algorithm (QEA) is the product of the integration of quantum computing and evolutionary computation [9]. Due to its simple structure, easy implementation, quick convergence, QEA has been applied to the solution of many scheduling problems. Niu et al. [10] proposed a quantum immune algorithm for hybrid flow shop scheduling problems with mean flow time. In order to improve the efficiency, Latif et al. [11] combined the QEA with the estimation of distribution algorithm for the permutation flow shop scheduling problem. To solve the travelling salesman problem, Ma et al. [12] presented a new quantum ant colony algorithm. Deng et al. [13] developed a co-evolutionary quantum genetic algorithm for the no-wait flow shop scheduling problem with the criterion to minimize makespan. Singh et al. [14] used quantum-behaved particle swarm optimization to flexible job shop scheduling problem. Wu et al. [15] proposed an elitist quantum-inspired evolutionary algorithm for the flexible job-shop scheduling problem. As far as we know, there is no any published paper in QEA for BFSP with carbon emissions and makespan.

In this paper, we proposed a quantum-inspired evolution algorithm (HQEA) for the multi-objective BFSP with carbon emissions and makespan. The BFSP with carbon and makespan is described in the Sect. 2. In Sect. 3, the framework of HQEA for solving it is proposed. Experiments and comparisons are provided to demonstrate the effectiveness of HQEA. Finally, we put an end to the paper with some conclusions.

2 Problem Description

The BFSP is described as follows. There are n jobs and m machines. Each of n jobs will be processed through m machines with the same permutation. The processing time of each job on each machine is predefined. Each machine can process at most one job. Each job can be just processed on one machine at a time. Preemption and interruption are forbidden at any time. There is no buffer between two consecutive machines, which mean the finished job cannot leave its machine until the next machine is available for processing.

2.1 BFSP with Carbon and Makespan

Let $\pi = \{\pi_1, \pi_2, \cdots \pi_n\}$ denote the permutation sequence, π_i $(i \in 1, 2, \cdots, n)$ the ith job in π. Denote p_{ij} the processing time of job i on machine j, v_{ij} the processing speed of job i on machine j $(v_{ij} \in S, S = \{v_1, v_2, \cdots, v_s\})$. The departure time of π_i on machine j is denoted as $D_{\pi_i,j}$, $D_{\pi_1,0}$ denotes the departure time of π_1 on virtual machine 0. Then $D_{\pi_i,j}$ can be calculated as follows.

$$D_{\pi_1,0} = 0 \tag{1}$$

$$D_{\pi_1,j} = D_{\pi_1,j-1} + p_{\pi_1,j}/v_{\pi_1,j} \; j = 1,2,\cdots m-1 \tag{2}$$

$$D_{\pi_i,0} = D_{\pi_{i-1},1} \; i = 2,3,\cdots n \tag{3}$$

$$D_{\pi_i,j} = \max\{D_{\pi_i,j-1} + p_{\pi_i,j-1}/v_{\pi_i,j-1}, D_{\pi_{i-1},j+1}\} \; i = 2,3,\cdots n \; j = 1,2,\cdots m-1 \tag{4}$$

$$D_{\pi_i,m} = D_{\pi_i,m-1} + p_{\pi_i,m}/v_{\pi_i,m} \; i = 1,2,\cdots n \tag{5}$$

Then, the first objective of permutation π is $C_{max}(\pi) = D_{\pi_n,m}$. The other objective of BFSP is to minimize the total carbon emissions (TEC) during the processing period. Obviously, the machine is either processing or standby. So, $x_{j,v}(t)$ and $y_j(t)$ be the following binary variables:

$$x_{j,v}(t) = \begin{cases} 1, & \text{if machine } j \text{ is in processing at speed } v \text{ at time } t \\ 0, & \text{otherwise} \end{cases} \tag{6}$$

$$y_j(t) = \begin{cases} 1, & \text{if machine } j \text{ is in standby at time } t \\ 0, & \text{otherwise} \end{cases} \tag{7}$$

Thus, the total carbon emissions (TCE) can be calculated by the following formula:

$$TCE = \int_0^{C_{max}} \left(\sum_{v=1}^{s} \sum_{j=1}^{m} PP_{j,v} \cdot x_{j,v}(t) + \sum_{j=1}^{m} SP_j \cdot y_j(t) \right) dt \tag{8}$$

$$TEC = \varepsilon \cdot TCE \tag{9}$$

$PP_{j,v}$ denotes the energy consumption per unit time when machine j is run at speed v. SP_j denotes the energy consumption per unit time when machine j is run at standby mode. Where is the total energy consumption and ε refers to the carbon emissions per unit of consumed energy. According to [16], it is assumed that when a machine is processing at higher speed, the power consumption increases while its processing time decreases. It means that:

$$\forall u > v(u, v \in S) \tag{10}$$

$$p_{\pi_i,j}/u < p_{\pi_i,j}/v \tag{11}$$

$$PP_{j,u} \cdot p_{\pi_i,j}/u > PP_{j,v} \cdot p_{\pi_i,j}/v \tag{12}$$

The TCE of π_1 and π_2 can be calculated as follows:

$$TCE(\pi) = TEC^S(\pi) + TEC^P(\pi) \tag{13}$$

Where $TEC^S(\pi)$ and $TEC^P(\pi)$ denote TCE in standby and in processing. Based on the assumption and property of FSP in [17], consider two solutions, π_1 and π_2. If $v_{i,j}(\pi_1) = v_{i,j}(\pi_2)$, then $TEC^P(\pi_1) = TEC^P(\pi_2)$, while $C_{max}(\pi_1) > C_{max}(\pi_2)$, then $TCE(\pi_1) > TCE(\pi_2)$. That is $\pi_1 \prec \pi_2$. If $C_{max}(\pi_1) = C_{max}(\pi_2)$, and $v_{i,j}(\pi_1) \geq v_{i,j}(\pi_2)$, $TCE(\pi_1) > TCE(\pi_2)$. That is $\pi_1 \prec \pi_2$.

To illustrate the multi-objective BFSP introduced above, the gantt of a sequence with 3 jobs and 3 machines is presented in Fig. 1. The blocking indicates that the job blocked on the current machine and the machine is in standby mode. The real-time power consumption curve for the BFSP is provided in Fig. 2. At time t_0, machine 2 and machine 3 are in processing mode, machine 1 is in standby mode. Then, the system's power consumption at time t_0 is calculated. Similarly, the power consumption of the system at different time point can be calculated. Finally, TCE is computed according to Eq. 13.

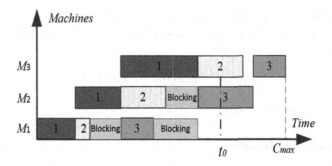

Fig. 1. Gantt of BFSP with 3 jobs and 4 machines

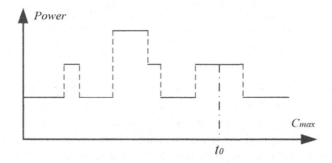

Fig. 2. Real-time power consumption of BFSP

3 HQEA for BFSP with Carbon and Makespan

In this section, HEQA is presented to solve the BFSP with the objectives of makespan and carbon-efficient. The framework of HEQA is illustrated. following the feasible coding and decoding method, qubit updating mechanism, quantum cataclysm and local search scheme.

In the theory of quantum, the quantum state $|\varphi >$ can be expressed as a linear superposition of the ground state of $|0 >$ and $|1 >$. It can be described as follows:

$$|\varphi > = \alpha|0 > + \beta|1 > \tag{14}$$

Where α and β denote a pair of complex numbers, called the probability of quantum state and satisfied:

$$|\alpha|^2 + |\beta|^2 = 1 \tag{15}$$

$|\alpha|^2$ denote the probability of making $|\varphi >$ collapsed to $|0 >$. Meanwhile, $|\beta|^2$ denote the probability of making $|\varphi >$ collapsed to $|1 >$. Thus, it can be expressed as means of trigonometric functions:

$$|\varphi > = cos\theta|0 > + sin\theta|1 > \tag{16}$$

That is: $\alpha = cos\theta, \beta = sin\theta$.

3.1 Encoding and Decoding

We design a two-dimensional quantum probability matrix for encoding qubit in this paper [18], it is crucial for the HQEA. It can be written as:

$$G = \begin{bmatrix} |\beta_{1,1}|^2 & |\beta_{1,2}|^2 & \cdots & |\beta_{1,n}|^2 \\ |\beta_{1,2}|^2 & |\beta_{2,2}|^2 & \cdots & |\beta_{2,n}|^2 \\ \vdots & \vdots & |\beta_{i,j}|^2 & \vdots \\ |\beta_{n,1}|^2 & |\beta_{n,2}|^2 & \cdots & |\beta_{n,n}|^2 \end{bmatrix}_{n \times n} \tag{17}$$

In accordance with the theory of quantum, where G is a qubit observation model and $|\beta_{i,j}|^2$ represent the probability of 1 in row i and column j. By observing G, the 0–1 observation matrix X can be obtained. Since the matrix X contains more than one 1 per row, it can't be determined the sequence of jobs. Then, we translate X into a matrix C with only one 1 per row and per column. Let E_1 denote the set of 1 column numbers, E_0 denote the set of 0 column numbers, We can get E_1 and E_0 through the statistics of X. The larger the value in the probability matrix, the greater the probability that the location will be selected. The way to generate the permutation can be described as follows:

Step 1: set $i \leftarrow 1$, $j \leftarrow 1$.
Step 2: for $i = 1, \cdots, n$
   Step 2.1: set $E_1 = [\,]$, $E_1 = [\,]$;
   Step 2.2: for $j = 1, \cdots, n$
      If $X_{i,j}=1$ then $E_1 = E_1 + [j]$.
      If $X_{i,j}=0$ then $E_0 = E_0 + [j]$.
      end for j;
   Step 2.3: If $E_1 \neq [\,]$ then $e \leftarrow arg\,\max_{j \in E_1}(G_{i,j})$
                  else $e \leftarrow arg\,\max_{j \in E_0}(G_{i,j})$.
      for $j = 1, \cdots, n$
        $X_{je} = -1$.
      end for j
   Step 2.4: $C_{ie} = 1$.
   end for i.

Where argmax denote the maximum value in the set. For example, when we get the matrix C, we can convert it to permutation π.

$$C = \begin{bmatrix} 0 & 1 & 0 \\ 1 & 0 & 0 \\ 0 & 0 & 1 \end{bmatrix} \rightarrow \pi = [2, 1, 3] \tag{18}$$

3.2 Update Probability

The qubit probabilistic update is the main part of the HQEA. The proper updating mechanism not only keeps the diversity of the population, but also doesn't make the algorithm converge rapidly. We propose an updated pattern with learning mechanisms to learn better solutions in this paper. The probability matrix G can be updated according to the following formula.

$$G_{i,j}(gen + 1) = (1 - \alpha)G_{i,j}(gen) + (1 - \alpha \times L) \tag{19}$$

$$L = sign\left(\left(f_1(b\_\pi) - f_1(\pi)\right) \times \left(f_2(b\_\pi) - f_2(\pi)\right) \times \left(b\_x_{ij} - x_{ij}\right)\right) \tag{20}$$

$$sign(x) = \begin{cases} 1, & if\ x > 0 \\ 0, & if\ x = 0 \\ -1, & if\ x < 0 \end{cases} \tag{21}$$

Where $sign$ is a function, it can be described as Eq. (21). Where b_π denote the better solutions, which is randomly selected from the set of non-dominated solution. b_x_{ij} is the value of matrix X of b_π. Then the value of L is 0 when one of the objectives is equal.

3.3   Quantum Cataclysm

In order to improve the effectiveness of the algorithm, the probability of cataclysm was proposed when the population was updated. It is beneficial to escape from the local optimum. The probability of cataclysm to escape from the local optimum is designed to enhance the global search of the algorithm. When generating new individuals, we designed a cataclysm probability. If the individual cataclysm occurs, set the quantum probability matrix $G_{ij} = 0.5$ and generate a random permutation.

3.4   Local Search

It is well-known that the local search is efficient in improving the solutions. Two local search operators are used to enhance the local exploitation around the non-dominated solution in this paper. Two local search operators can be described as follows.

   Interchange: randomly select two different locations of the sequence, replace the corresponding jobs in the two locations.

   Insert: randomly select two different locations of the sequence and insert the back one before the front one.

   These operators are performed sequentially in the given order to generate another solution, and then the new solution replaces the old one if it dominates the old. The above procedure is designed as follows:

Step 1: Convert QEA's individual to a job permutation π according to decoding rule. $\pi' = \pi$.
Step 2: Randomly select u and v, where $\pi'_u = \pi'_v$, $\pi'' = interchange(\pi', u, v)$;
Step 3: Set $loop \leftarrow 0$.
    Repeat
    k=0;
      Repeat
      Randomly select u and v, where $\pi''_u = \pi''_v$
      If $k = 0$ then $\pi''' = interchange(\pi'', u, v)$.
      If $k = 1$ then $\pi''' = insert(\pi'', u, v)$.
      If $\pi''' \succ \pi''$ then $k = 0$, $\pi'' = \pi'''$
             else $k \leftarrow k + 1$.
    Until $k = 2$;
    $loop \leftarrow loop + 1$;
    Until $loop = max\_loop$.
Step 4: If $\pi'' \succ \pi$, then $\pi = \pi''$.
Step 5: Updating the quantum probability matrix of the current individual.

With the above design, the procedure of HQEA for solving the BFSP with carbon and makespan is illustrated in Fig. 3.

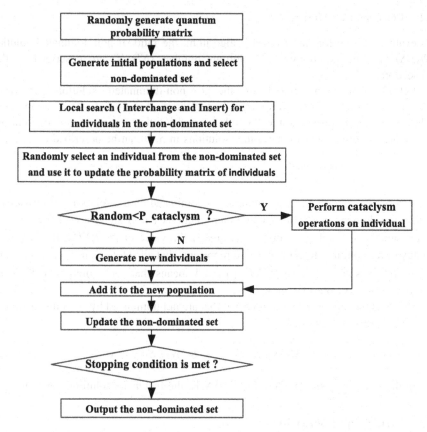

Fig. 3. Framework of HQEA

4 Simulation Result and Comparisons

4.1 Experimental Setup

To test the performances of HQEA, we compare with the multi-objective genetic local search (IMMOGLS) [19] and the simulated annealing genetic algorithm (SAGA) [20] in different scales. IMMOGLS is an efficient multi-objective algorithm based on random weight. Its performance is superior to the famous algorithm NSGAII in solving the multi-objective scheduling problems.

HEQA and IMMOGLS are coded in Delphi 2010 and run on a PC with Intel 7700HQ 2.80 GHz. The proposed HEQA's parameters are set as follows: the population size $P = 50$, the learning rate $\alpha = 0.2$, the probability of cataclysm $p_c = 0.1$, $v = \{1, 1.5, 2\}$. IMMOGLS's population size is also set to 50 and its other parameters are set to the same values in [19, 20]. For fair comparison, HEQA and IMMOGLS are run the same time and 30 times.

4.2   Performance Metrics

To evaluate the performance of each algorithm, the ratio of non-dominated solution (*RNDS*) and the non-dominated solutions number (*NDSN*) are described in this subsection.

Let S denote the union of the K non-dominated solution sets (i.e., $S = S_1 \cup \cdots \cup S_k$). A straightforward performance measure of the nondominated solution set S_j with respect to the K nondominated solution sets is the ratio of solutions S_j that are not dominated by any other solutions in S. It can be described as

$$R\_NDS(S_j) = |S_j - x \in S_j| \exists y \in S : y \prec x| / |S_j| \tag{22}$$

Where $y \prec x$ means that the solution x is dominated by the solution y. In the above numerator, dominated solutions x by other solutions y in S are removed from S_j. $|S_j|$ denote numbers of solutions in S_j. The greater the value of the *RNDS*, the better the quality of the solution. $R\_NDS(S_j) = 0$ represents that all solutions in S_j are dominated by the others solutions in S. $R\_NDS(S_j) = 1$ means that each solution in S_j is not dominated by any solutions in S.

NDSN is the number of solutions in S_j that are not dominated by any other solutions in S. This measure is written as

$$NDSN(S_j) = |S_j - x \in S_j| \exists y \in S : y \prec x| \tag{23}$$

In the same way, the larger value *NDSN* is, the better the solutions set S_j is.

4.3   Results and Comparison

Considering the 14 different size instances, we compare the HQEA with the IMMOGLS and SAGA. For each instance, we run both the same time and obtain the RNDS and the NDSN. Table 1 summarizes the results grouped by each instance. Where $S = S\_HQEA \cup S\_IMMOGLS \cup S\_SAGA$, RNDS_HQEA, RNDS_SAGA, RNDS_IM, NDSN_HQEA, NDSN_IM, NDSN_SAGA denote the average ratio of S_HQEA, the average ratio of S_IMMOGLS, the average ratio of S_SAGA, the average value of NDSN (S_HQEA), the average value of NDSN (S_IMMOGLS), the average value of NDSN (S_SAGA).

From Table 1, it can be seen that the results obtained by HQEA are much better than those of IMMOGLS and SAGA for most instances. For the large-sized instances, most of the solution of IMMOGLS and SAGA are dominated by the solution of HQEA. Thus, it is concluded that HQEA has the better performance for BFSP with carbon and makespan in the given time.

Figure 4 is on the scale of 20 jobs 5 machines after the output of nondominated solutions set. It can be seen from the diagram that each solution is not dominated by each other. HQEA runs any generation, and other nondominated solution sets are similar to Fig. 4.

Table 1. Comparisons of HQEA and IMMOGLS

Instance n_m	IMMOGLS		SAGA		HQEA	
	RNDS_IM	*NDSN_IM*	*RNDS_SAGA*	*NDSN_SAGA*	*RNDS_QEA*	*NDSN_QEA*
20_5	0.17	1.10	0.09	0.63	**1.00**	**2.03**
20_10	0.19	**1.27**	0.07	0.17	**0.97**	0.97
30_5	0.15	0.85	0.01	0.05	**1.00**	**1.13**
30_10	0.09	0.13	0.02	0.13	**1.00**	**1.07**
30_20	0.04	0.03	0.00	0.00	**1.00**	**1.27**
50_5	0.00	0.00	0.00	0.00	**1.00**	**1.33**
50_10	0.00	0.00	0.00	0.00	**1.00**	**1.03**
50_20	0.00	0.00	0.00	0.00	**1.00**	**1.00**
70_5	0.00	0.00	0.00	0.00	**1.00**	**1.57**
70_10	0.00	0.00	0.00	0.00	**1.00**	**1.13**
70_20	0.00	0.00	0.00	0.00	**1.00**	**1.00**
100_10	0.00	0.00	0.00	0.00	**1.00**	**1.33**
100_20	0.00	0.00	0.00	0.00	**1.00**	**1.57**
100_40	0.00	0.00	0.00	0.00	**1.00**	**1.13**

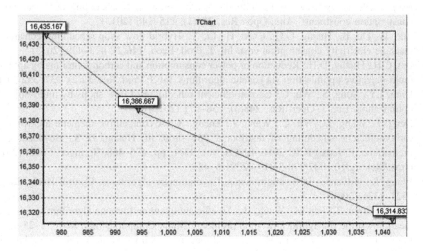

Fig. 4. The nondominated solution set of HQEA

5 Conclusion

A hybrid quantum-inspired evolution algorithm (HQEA) is proposed to solve the blocking flow shop scheduling problem (BFSP) with the objectives of makespan and carbon-efficient in this paper. First, we provided a feasible coding and decoding method for HQEA. To update the quantum probability matrix, an update mechanism is developed. Then, new individuals are generated through the quantum probability matrix and have a certain probability of cataclysm. In addition, some local search operators are used to improve the non-dominated solutions. Finally, the effectiveness of HQEA in solving the BFSP is demonstrated by experiments and comparisons. Future work includes developing another version of HQEA and using it for other scheduling problems. Especially for the construction of quantum model, it is an important part of my future work.

Acknowledgements. This research is partially supported by the National Science Foundation of China (51665025), the Applied Basic Research Foundation of Yunnan Province (2015FB136), and the National Natural Science Fund for Distinguished Young Scholars of China (61525304).

References

1. Merchan, A.F., Maravelias, C.T.: Preprocessing and tightening methods for time-indexed MIP chemical production scheduling models. Comput. Chem. Eng. **84**, 516–535 (2016)
2. Gong, H., Tang, L., Duin, C.W.: A two-stage flow shop scheduling problem on a batching machine and a discrete machine with blocking and shared setup times. Comput. Oper. Res. **37**(5), 960–969 (2010)
3. Koren, Y., Wang, W., Gu, X.: Value creation through design for scalability of reconfigurable manufacturing systems. Int. J. Prod. Res. **55**(5), 1227–1242 (2016)
4. Fang, K., Uhan, N.A., Zhao, F., Sutherland, J.W.: Flow shop scheduling with peak power consumption constraints. Ann. Oper. Res. **206**(1), 115–145 (2013)
5. Luo, H., Du, B., Huang, G.Q., Chen, H., Li, X.: Hybrid flow shop scheduling considering machine electricity consumption cost. Int. J. Prod. Econ. **146**(2), 423–439 (2013)
6. Liu, C.H., Huang, D.H.: Reduction of power consumption and carbon footprints by applying multi-objective optimisation via genetic algorithms. Int. J. Prod. Res. **52**(2), 337–352 (2014)
7. Ding, J.Y., Song, S., Wu, C.: Carbon-efficient scheduling of flow shops by multi-objective optimization. Eur. J. Oper. Res. **248**(3), 758–771 (2015)
8. Deng, J., Wang, L., Wu, C., Wang, J., Zheng, X.: A competitive memetic algorithm for carbon-efficient scheduling of distributed flow-shop. In: Huang, D.-S., Bevilacqua, V., Premaratne, P. (eds.) ICIC 2016. LNCS, vol. 9771, pp. 476–488. Springer, Cham (2016). https://doi.org/10.1007/978-3-319-42291-6_48
9. Wang, L.: Advances in quantum-inspired evolutionary algorithms. Control Decis. **23**(12), 1321–1326 (2008)
10. Niu, Q., Zhou, T., Fei, M., Wang, B.: An efficient quantum immune algorithm to minimize mean flow time for hybrid flow shop problems. Math. Comput. Simul. **84**(6), 1–25 (2012)
11. Latif, M.S., Zhou, H., Amir, M.: A hybrid Quantum Estimation of Distribution Algorithm (Q-EDA) for Flow-Shop Scheduling (2013)
12. Ma, Y., Tian, W.J., Fan, Y.Y.: Improved quantum ant colony algorithm for solving TSP problem. In: 2014 IEEE Workshop on Electronics, Computer and Applications, pp. 453–456. IEEE (2014)
13. Deng, G., Wei, M., Su, Q., Zhao, M.: An effective co-evolutionary quantum genetic algorithm for the no-wait flow shop scheduling problem. Adv. Mech. Eng. **7**(12) (2015)
14. Singh, M.R., Mahapatra, S.S.: A Quantum Behaved Particle Swarm Optimization for Flexible Job Shop Scheduling. Pergamon Press Inc., Tarrytown (2016)
15. Wu, X., Wu, S.: An elitist quantum-inspired evolutionary algorithm for the flexible job-shop scheduling problem. J. Intell. Manuf. **28**(6), 1441–1457 (2017)
16. Fang, K., Uhan, N.A., Zhao, F., Sutherland, J.W.: Flow shop scheduling with peak power consumption constraints. Ann. Oper. Res. **206**(1), 115–145 (2013)

17. Ding, J.Y., Song, S., Wu, C.: Carbon-efficient scheduling of flow shops by multi-objective optimization. Eur. J. Oper. Res. **248**, 758–771 (2016)
18. Zhao, J. X., Qian, B., Hu, R., Zhang, C. S., & Li, Z. H.: An improved quantum-inspired evolution algorithm for no-wait flow shop scheduling problem to minimize makespan. In: International Conference on Intelligent Computing, pp. 536–547 (2016)
19. Ishibuchi, H., Yoshida, T., Murata, T.: Balance between genetic search and local search in memetic algorithms for multiobjective permutation flowshop scheduling. IEEE Trans. Evol. Comput. **7**(2), 204–223 (2003)
20. Mokhtari, H., Hasani, A.: An energy-efficient multi-objective optimization for flexible job-shop scheduling problem. Comput. Chem. Eng. **104**, 339–352 (2017)

Hybrid Discrete Teaching-Learning-Based Optimization Algorithm for Solving Parallel Machine Scheduling Problem with Multiple Constraints

Yu-Jie He[1], Bin Qian[1(✉)], Bo Liu[2], Rong Hu[1], and Chao Deng[3]

[1] School of Information Engineering and Automation,
Kunming University of Science and Technology, Kunming 650500, China
bin.qian@vip.163.com
[2] Academy of Mathematics and Systems Science,
Chinese Academy of Sciences, Beijing 100190, China
[3] School of Mechanical and Electrical Engineering,
Kunming University of Science and Technology, Kunming 650500, China

Abstract. This paper proposes a hybrid discrete teaching-learning-based optimization algorithm (HDTLBO) for solving the parallel machine scheduling problem with arrival time, multiple operations and process restraints (PMSP_AMP), which widely exists in the various manufacturing process. The criterion is to minimize the maximum completion time (i.e., makespan). Firstly, a discretization method is designed to remold the standard teaching-learning-based optimization algorithm, which enhances its global exploration ability and makes it can execute the global search directly in the discrete solution space. Then, the *swap*-based and the *insert*-based neighborhood are utilized to construct the local search for performance improvement. Simulation results and comparisons based on s set of random instances demonstrate the effectiveness and searching quality of the presented HDTLBO.

Keywords: Parallel machine scheduling problem · Multiple operations
Arrival times · Discrete teaching-learning-based optimization

1 Introduction

Parallel machine scheduling problem with arrival time, multiple operations and process restraints (PMSP_AMP) is a typical discrete optimization problem with strong engineering background in many manufacturing industries, which have gained increasing attention in recent years [1]. To minimize the makespan of unrelated parallel machine scheduling problem, Avalos-Rosales et al. [2] proposed a meta-heuristic algorithm based on multi-start and variable-domain reduction. Diana [3] proposed a hybrid immune algorithm based on variable-domain reduction. Veronique et al. [4] proposed a hybrid tabu search and a truncated branch-and-bound. Abdoul et al. [5] proposed a memetic algorithm. Joo and Kim [6] proposed three kinds of hybrid genetic algorithms with dispatching rules. Damodaran et al. [7] proposed a PSO algorithm to solve the

© Springer International Publishing AG, part of Springer Nature 2018
D.-S. Huang et al. (Eds.): ICIC 2018, LNCS 10954, pp. 618–627, 2018.
https://doi.org/10.1007/978-3-319-95930-6_62

problem of unrelated parallel batch machine problem. Obviously, PMSP_AMP is a typical NP-Complete problem in computational complexity, various parallel machine scheduling problems can be reduced to such problems. Therefore, it has great engineering significance and theoretical value to study the effective algorithm of solving PMSP_AMP.

Teaching-learning-based algorithm (TLBO) is a continuous intelligence optimization algorithm, which was first proposed by Rao et al. [8] in 2011. Recently, Shao et al. [9, 10] proposed a hybrid discrete optimization algorithm based on teaching-probabilistic learning mechanism, they construct an individual updating mechanism by Insert operation and two-dimensional probability model. They also proposed an extended teaching-learning based algorithm based on Gaussian distribution for no-wait flow shop scheduling problem. Li et al. [11] proposed a kind of DTLBO algorithm for rescheduling problem, which include four kinds of discrete teaching-learning operations.

This paper proposed a hybrid discrete teaching-learning-based optimization (HDTLBO) algorithm for PMSP_AMP, which aims to minimize the maximum completion time. HDTLBO redefined the individual updating mechanism by discrete bit-based operation, which makes the algorithm can execute directly in the discrete solution space. Firstly, we use the job permutation as individuals in the population and randomly initialize the population, which can ensure the diversity of the population. Secondly, we proposed discrete teaching phase and discrete learning phase, we construct a discrete individual updating mechanism through the discretization of standard TLBO algorithm, which enhances the efficiency of the global search. Finally, we design the neighborhood search based on Swap and Insert operation to enhance the local search ability. Simulation results and comparisons on different instances demonstrate the effectiveness and robustness of the presented HDTLBO.

The remainder of the paper is structured as follows. In Sect. 2, PMAP_AMP is stated. In Sect. 3, HDTLBO is proposed and described in detail. Simulation results and comparisons are provided in Sect. 4. Conclusions and future work are presented in Sect. 5.

2 Problem Description

The following notions will be used to describe the PMSP_AMP.

n: total number of jobs; m: total number of machines; Π: the set of all schedule; m_i: The total number of processing operations on the i-th machine; $\pi_k^{T(i)}$: the job processing permutation based on sequence in the machine i, $k = 1, 2, \cdots, m_i$; stg_j: the number of processing operations of job j; $S\left(\pi_k^{T(i)}\right)$: the start processing time of $\pi_k^{T(i)}$, $S\left(\pi_0^{T(i)}\right) = 0$; $P\left(\pi_k^{T(i)}\right)$: the processing time of $\pi_k^{T(i)}$, $P\left(\pi_0^{T(i)}\right) = 0$; $R\left(\pi_k^{T(i)}\right)$: the release time of $\pi_k^{T(i)}$; $pre\_k\left(\pi_k^{T(i)}\right)$: the position of $\pi_k^{T(i)}$ in the permutation of last processing machine, if $\pi_k^{T(i)}$ is the first time for processing, then $pre\_k\left(\pi_k^{T(i)}\right) = 0$;

$pre\_m\left(\pi_k^{T(i)}\right)$: the machine used to the last processing of $\pi_k^{T(i)}$, if $\pi_k^{T(i)}$ is the first time for processing, then $pre\_m\left(\pi_k^{T(i)}\right) = 0$; $C_{\max}(\pi)$: the maximum completion time, i.e. makespan.

The problem is described as follows. There are n identical jobs are to be processed on m machines. Each job needs to be processed for stg_j times. Different processes of one job need to be processed in sequence. The job in any permutation is assigned to machines from left to right according to the certain rules and processing constraints. All processes can only be processed by the machine that meets the processing constraints. The processing time of each job is related to the processing machine, any machine can only be processed one job at the same time. Machines will not be shut down before all jobs are completed. The makespan $C_{\max}(\pi)$ can be calculated as follows:

$$C_{\max}(\pi) = \max\left\{S\left(\pi_{m_1}^{T(1)}\right) + P\left(\pi_{m_1}^{T(1)}\right), S\left(\pi_{m_2}^{T(2)}\right) + P\left(\pi_{m_2}^{T(2)}\right), \cdots, S\left(\pi_{m_j}^{T(i)}\right) + P\left(\pi_{m_j}^{T(i)}\right)\right\} \quad (1)$$

$$S\left(\pi_k^{T(i)}\right) = \begin{cases} \max\left\{S\left(\pi_{k-1}^{T(i)}\right) + P\left(\pi_{k-1}^{T(i)}\right), R\left(\pi_k^{T(i)}\right)\right\}; & if\ pre\_k\left(\pi_k^{T(i)}\right) = 0 \\ \max\{S\left(\pi_k^{T(i)}\right) + P\left(\pi_{k-1}^{T(i)}\right), S(\pi_{pre\_k(\pi_k^{T(i)})}^{T(pre\_m(\pi_k^{T(i)}))}) + P(\pi_{pre\_k(\pi_k^{T(i)})}^{T(pre\_m(\pi_k^{T(i)}))}); & else \end{cases}$$

$$(2)$$

$$C_{\max}(\pi^*) = \min_{m \subset \Pi} C_{\max}(\pi) \quad (3)$$

$$\pi^* = \arg\{C_{\max}(\pi)\} \rightarrow \min, \forall \pi \in \Pi \quad (4)$$

Equations (1) and (2) is calculation formula of the earliest completion time, the Eqs. (3) and (4) is to find the best job sequence in all solutions, which can get the minimum of $C_{\max}$.

3   HDTLBO for PMSP_AMP

In this section, we propose a hybrid discrete teaching-learning-based optimization algorithm (HDTLBO) for PMSP_AMP. The proposed HDTLBO includes four phases: solution representation and population initialization, discrete teaching phase, discrete learning phase and local search.

3.1   Solution Representation and Population Initialization

Since some jobs in PMSP_AMP need to be processed in multiple processes, HDTLBO adopts an operation-based encoding mechanism to represent solutions. For example, $\pi = [1, 3,2,5,4,1,3,1,3,4]$ is an individual when the scale of problem n is set to 5, m is set to 3 and the job processing operations set is set to {3,1,3,2,1}, which convey that the first processing job is job 1, it has three processing operations, and the last processing job in the sequence is job 4, it has two processing operations totally. Similarly, decoding for PMSP_AMP is the Early Completion Time (ECT) rule. In this paper,

HDTLBO adopt the random initialization method to produce individuals as the initial population, which not only guarantees the diversification the population, but also can realize the fair comparison for HDTLO with other algorithms.

3.2 Discrete Teaching Phase

Let X_i^{gen} denote the i-th individual in the gen-th generation, X_t^{gen} the teacher, which is the best individual in the current population at the gen-th generation, X_m^{gen} the individual which in the middle of the current population at the gen-th generation, TF the teaching factor. Let $r[a,b]$ denote randomly select integer a and b to constitute interval $[a,b]$, $\left[X_t^{gen}, X_m^{gen}\right]^{r[a,b]}$ denote the crossover operation based on interval $[a,b]$ between X_t^{gen} and X_m^{gen}, $r[c]$ denote randomly select an integer $c(1 \leq c \leq n)$. In teaching phase, each individual in a class learns from the difference between the mean individual of students and the best individual currently (i.e., teacher). According to the above definition, we propose the individual update formula in teaching phase as it shown in Eq. (5).

$$X_i^{gen+1} = \left\{X_i^{gen}, TF \odot \left[X_t^{gen}, X_m^{gen}\right]^{r[a,b]}\right\}^{r[c]} \tag{5}$$

Obviously, there are two core operations in Eq. (5), which are $TF \odot \left[X_t^{gen}, X_m^{gen}\right]^{r[a,b]}$ and $\left\{X_i^{gen}, TF \odot \left[X_t^{gen}, X_m^{gen}\right]^{r[a,b]}\right\}^{r[c]}$, the details will be introduced in Sects. 3.2.1 and 3.2.2.

3.2.1 Description of $TF \odot \left[X_t^{gen}, X_m^{gen}\right]^{r[a,b]}$

The specific operation of $TF \odot \left[X_{teacher}^{gen}, X_{mean}^{gen}\right]^{r[a,b]}$ can denote as Eq. (6).

$$X_{m\_new}^{gen} = TF \odot \left[X_t^{gen}, X_m^{gen}\right]^{r[a,b]} = \begin{cases} \left[X_t^{gen}, X_m^{gen}\right]^{r[a,b]}, & TF = 2 \\ X_m^{gen}, & else \end{cases} \tag{6}$$

This operation can be understood as an updating of the X_m^{gen}, that is, X_m^{gen} will improve after all individual improves their own performance through continuous learning. If $TF = 2$, we will adopt the Order-Based Crossover (OBX) method [12] to achieve $\left[X_t^{gen}, X_m^{gen}\right]^{r[a,b]}$. For example, suppose X_t^{gen} = [1 3 2 5 4 1 3 1 3 4], X_m^{gen} = [4 3 1 5 3 4 2 1 3 1], $a = 3$ and $b = 6$. Firstly, take the sub-array [2 5 4 1] from the third position to the sixth position in X_t^{gen}, and set the position which first appeared 2,5,4,1 to 0 from left to right in X_m^{gen}, then we can get the temporary array $temp1$ = [0 3 0 0 3 4 0 1 3 1]. Secondly, let 2,5,4,1 replace the element 0 in $temp1$ in order from left to right. Finally, we can generate the new mean individual $X_{m\_new}^{gen}$ = [2 3 5 4 3 4 1 1 3 1] after crossover. After this operation, we can transfer the Eq. (5) to the Eq. (7).

$$X_i^{gen+1} = \left\{X_i^{gen}, X_{m\_new}^{gen}\right\}^{r[c]} \tag{7}$$

3.2.2   Description of $\left\{ X_i^{gen}, X_{m\_new}^{gen} \right\}^{r[c]}$

This operation is the last operation of the discrete teaching phase in Eq. (7), which can be understood as the generating procedure of the new population. This operation can be exemplified. Suppose X_i^{gen} = [2 3 5 4 3 4 1 1 3 1], $X_{m\_new}^{gen}$ = [5 4 1 3 3 4 2 1 3 1] and $c = 3$. Firstly, set the element 3 to zero in X_i^{gen}, we can get the temporary array $temp1$ = [2 0 5 4 0 4 1 1 0 1]. Secondly, set the element which not equal 3 to zero from left to right in $X_{m\_new}^{gen}$, then we can get the second temporary array $temp2$ = [0 0 0 3 3 0 0 0 3 0]. Finally, let the element which is not equal to zero in $temp1$ replace the element 0 in $temp2$ in order from left to right, then we can generate the new individual X_i^{gen+1} = [2 5 4 3 3 4 1 1 3 1] in the next population. At this point, we finished all operations in the discrete teaching phase.

3.3   Discrete Learning Phase

In this phase, several learners evolve through learning from each other. For any student, we will choose another learning object randomly. let X_i^{gen} and X_j^{gen} denote two individuals which are selected from current population randomly. According to the above definition, we propose the updating formula as it shown in Eq. (8) for learning phase.

$$X_i^{gen+1} = r_i \odot \left\{ X_i^{gen}, X_j^{gen} \right\}^{r[c]} = \begin{cases} \left\{ X_i^{gen}, X_j^{gen} \right\}^{r[c]}, & rand[0,1] < r_i \\ X_i^{gen}, & else \end{cases} \qquad (8)$$

This operation is the last step of the learning phase, it can be understood as the updating process of new populations. If $rand < r_i$, execute the crossover operation $\left\{ X_i^{gen}, X_j^{gen} \right\}^{r[c]}$, this operation is similar to the last crossover operation in the discrete teaching phase, which described in Sect. 3.2.2. At this point, the entire DTLBO algorithm is completed.

3.4   Local Search

To enhance the search ability of the algorithm, a more detailed local search must be performed on the neighborhoods of good solutions. Two effective neighborhoods are often used in the literature. (i) swap the job at the u dimension and the job at the dimension of job permutation $\pi(swap(\pi, u, v))$, (ii) remove the job at the $u-th$ dimension and insert it in the $v-th$ dimension of the job permutation $\pi(insert(\pi, u, v))$. Thus, we employ swap and insert here as neighborhood for local search. The procedure of the local search is given as follows:

Step 1: Input the best individual π_{best}, let $\pi_{i\_0} = \pi_{best}$.

Step 2: Perturbation phase.

      Set $km = 1$;

      Do

            Randomly select u and v, where $u \neq v$; $\pi_{i\_1} = swap\left(\pi_{i\_0}, u, v\right)$;

            if $f\left(\pi_{i\_1}\right) < f\left(\pi_{i\_0}\right)$, then $\pi_{i\_0} = \pi_{i\_1}$;

            km++;

      While $km < 5$.

Step 3: Exploitation phase.

      Set $flag = 1$;

      Do

            Randomly select u and v, where $u \neq v$;

            $\pi_{i\_2} = insert\left(\pi_{i\_1}, u, v\right)$;

            if $f\left(\pi_{i\_2}\right) < f\left(\pi_{i\_1}\right)$, then $\pi_{i\_1} = \pi_{i\_2}$;

            $flag$++;

      While $flag < n$.

Step 4: if $f\left(\pi_{i\_1}\right) < f\left(\pi_{i\_0}\right)$, then $\pi_{i\_0} = \pi_{i\_1}$.

Step 5: Output the best individual π_{best}.

3.5 Procedure of HDTLBO

The procedure framework of HDTLBO is shown in Fig. 1. It can be seen from Fig. 1 that teaching phase and learning phase are directly responsible for global search. Then, we applied the local search to perform exploitation for the best individual to improve solution's quality, which can make the algorithm to achieve a good balance between global and local search.

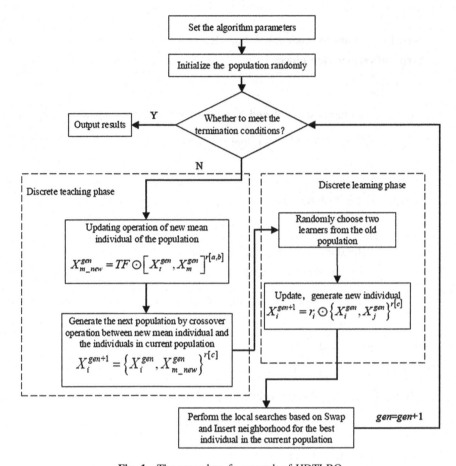

Fig. 1. The procedure framework of HDTLBO

4   Simulation Result and Comparisons

4.1   Experimental Setup

To test the performance of proposed HDTLBO, a set of instances under differential scales is generated randomly. That is, the $n \times m$ combinations include $10 \times 5, 20 \times 5, 30 \times 5, 40 \times 5, 50 \times 5, 40 \times 10, 50 \times 10, 60 \times 10, 70 \times 10, 80 \times 10, 80 \times 20, 90 \times 30, 100 \times 20$. The processing operations of each job are generated from a uniform distribution [1, 3], the processing time and release time are generated from a uniform distribution [1,100]. Let AVG denote the average value of the $C_{\max}$, BST the best value of $C_{\max}$, WST the worst value of $C_{\max}$. In our tests, HDTLBO use the parameters: the population size $popsize = 30$, the mutation probability $r_m = 0.9$, the teaching factor $TF=2$. Each instance is independently run for 20 times for every algorithm at the same runtime. The running time of every algorithm is $n \times m \times 0.2$ second. All algorithms used in the comparisons are re-implemented by ourselves and are coded in Delphi 2010 and are executed on an Intel I5-5200U 2.2 GHz with 4 GB RAM.

4.2   Comparison of DTLBO and TLBO, DTLBO-I

To verify the effectiveness of DTLBO (i.e., the HDTLBO without local search), we carry out simulations to compare our DTLBO with two scheduling algorithms in international journals for global search performance, which are a new teaching-learning-based optimization algorithm (TLBO) [8] and a discrete teaching-learning-based algorithm with TP-I and LP-I (DTLBO-I) [11]. The test results are shown in Table 1.

Table 1. Comparison of *BST, WST, AVG* of TLBO, DTLBO-I and DTLBO

Instance	TLBO			DTLBO-I			DTLBO		
$n \times m$	*BST*	*WST*	*AVG*	*BST*	*WST*	*AVG*	*BST*	*WST*	*AVG*
10 × 5	**192**	**192**	**192**	**192**	**192**	**192**	**192**	**192**	**192**
20 × 5	399	431	417.1	426	446	436.9	**382**	**410**	**392.5**
30 × 5	493	521	508.95	521	539	532.3	**476**	**495**	**482.55**
40 × 5	504	530	517.9	531	562	545.1	**487**	**503**	**495.25**
50 × 5	784	838	817.45	830	863	849.85	**757**	**793**	**774.65**
40 × 10	193	213	203.4	213	223	218.85	**180**	**203**	**187.95**
50 × 10	249	262	254.4	260	271	266.15	**216**	**241**	**231.3**
60 × 10	327	346	337.8	347	360	354.15	**306**	**330**	**316.95**
70 × 10	415	430	422.7	438	454	446.35	**387**	**407**	**396.9**
80 × 10	374	390	380.55	395	406	400.6	**350**	**379**	**362.55**
80 × 20	163	171	166.9	161	175	170.15	**148**	**160**	**153.45**
90 × 30	112	117	114.15	116	120	117.8	**105**	**114**	**110.4**
100 × 20	202	202	202	193	204	199.25	**170**	**181**	**175.3**

It can be seen from Table 1 that the BST, WST and AVG values obtained by DTLBO are obviously better than those obtained by TLBO and DTLBO-I, which not only shows the superiority of DTLBO but also validates the effectiveness of utilizing discrete bit-based operation in standard TLBO.

4.3   Comparisons of HDTLBO and GA_DR_C, PSO

To further test the overall performance of HDTLBO, this section compares HDTLBO with some effective algorithms in important international journals, which are the hybrid genetic algorithms with dispatching rules (GA_DR_C) [6] and a particle swarm optimization algorithm (PSO) [7]. The results are shown in Table 2.

Table 2. Comparison of *BST, WST and AVG* of PSO, GA_DR_C and HDTLBO

Instance	PSO			GA_DR_C			HDTLBO		
$n \times m$	BST	WST	AVG	BST	WST	AVG	BST	WST	AVG
10 × 5	**192**	**192**	**192**	**192**	**192**	**192**	**192**	**192**	**192**
20 × 5	**382**	410	398.2	393	417	405.45	**382**	**400**	**387.9**
30 × 5	476	505	487.85	487	511	496.45	**473**	**489**	**481**
40 × 5	484	506	496.95	498	525	506.75	**483**	**503**	**491.8**
50 × 5	777	799	787.75	785	813	797.85	**762**	**795**	**776.9**
40 × 10	186	203	193.8	186	203	194.05	**181**	**191**	**186.9**
50 × 10	225	243	236.1	229	250	238.1	**219**	**241**	**229.35**
60 × 10	315	341	326.85	316	335	325.2	**307**	**326**	**317.35**
70 × 10	397	425	410.15	387	419	405.45	**385**	**408**	**399.65**
80 × 10	361	385	370.8	356	384	368.35	**344**	**368**	**359.55**
80 × 20	152	162	158.05	151	161	155.3	**143**	**155**	**151.1**
90 × 30	108	118	111.75	105	112	107.8	**105**	**110**	**107.5**
100 × 20	172	192	182.35	172	190	178.35	**170**	**182**	**174.4**

As it can be seen from Table 2, the BST, WST and AVG values obtained by HDTLBO are better than those obtained by PSO and GA_DR_C for all instances, which not only verifies the necessity of adopting the local search in DTLBO but also shows that the effective and robustness of HDTLBO.

In conclusion, HDTLBO can solve PMSP_AMP very effectively.

5   Conclusion and Future Work

This paper proposes a hybrid discrete teaching-learning-based optimization algorithm (HDTLBO) for parallel machines scheduling problem with arrival time, multiple processes and processing constraints (PMSP_AMP) to minimize the makespan. This is the first report on the application of the discrete TLBO-based algorithm for PMSP_AMP. In the global search of HDTLBO, the discretization method is designed to deal with the standard TLBO algorithm, which makes it can execute the global search directly in discrete solution space. In the local search, a neighborhood search based on *Swap* and *Insert* is constructed to perform a more detailed search on the high-quality region obtained by the global search, which can achieve a better trade-off between global and local search, thus enhancing the overall search capability of the algorithm. Depending on the computational results and statistical analysis, the HDTLBO outperforms all other considered algorithms with the effectiveness and robustness for PMSP_AMP on different test problems. Our future work is to develop some effective TLBO-based algorithms to deal with some green scheduling problems in Intelligent manufacturing.

Acknowledgments. This research is supported by the National Science Foundation of China (No. 51665025), and the Applied Basic Research Foundation of Yunnan Province (No. 2015FB136).

References

1. Pinedo, M.: Scheduling: Theory, Algorithms, and Systems. Springer, Heidelberg (2012)
2. Avalos-Rosales, O., Angel-Bello, F., Alvarez, A.: Efficient metaheuristic algorithm and reformulations for the unrelated parallel machine scheduling problem with sequence and machine-dependent setup times. Int. J. Adv. Manuf. Technol. **76**(9–12), 1705–1718 (2015)
3. Diana, R.O.M.: An immune-inspired algorithm for an unrelated parallel machines' scheduling problem with sequence and machine dependent setup-times for makespan minimisation. Neurocomputing **163**(C), 94–105 (2015)
4. Sels, V., Coelho, J., Dias, A.M., et al.: Hybrid tabu search and a truncated branch-and-bound for the unrelated parallel machine scheduling problem. Comput. Oper. Res. **53**, 107–117 (2015)
5. Bitar, A., Yugma, C., Roussel, R.: A memetic algorithm to solve an unrelated parallel machine scheduling problem with auxiliary resources in semiconductor manufacturing. J. Sched. **19**(4), 367–376 (2016)
6. Joo, C.M., Kim, B.S.: Hybrid genetic algorithms with dispatching rules for unrelated parallel machine scheduling with setup time and production availability. Comput. Ind. Eng. **85**(C), 102–109 (2015)
7. Damodaran, P., Diyadawagamage, D.A., Ghrayeb, O.: A particle swarm optimization algorithm for minimizing makespan of nonidentical parallel batch processing machines. Int. J. Adv. Manuf. Technol. **58**(9–12), 1131–1140 (2012)
8. Rao, R.V., Savsani, V.J., Vakharia, D.P.: Teaching- learning-based optimization: a novel method for constrained mechanical design optimization problems. Comput. Aided Des. **43**(3), 303–315 (2011)
9. Shao, S.W., Pi, D.C., Shao, Z.: A hybrid discrete optimization algorithm based on teaching–probabilistic learning mechanism for no-wait flow shop scheduling. Knowl. Based Syst. **107**, 219–234 (2016)
10. Shao, S.W., Pi, D.C., Shao, Z.: An extended teaching-learning based optimization algorithm for solving no-wait flow shop scheduling problem. Appl. Soft Comput. **61**, 193–210 (2017)
11. Li, J.Q., Pan, Q.K., Mao, K.: A discrete teaching-learning-based optimization algorithm for realistic flowshop rescheduling problems. Eng. Appl. Artif. Intell. **37**, 279–292 (2015)
12. Abdoun, O., Abouchabaka, J.: A Comparative study of adaptive crossover operators for genetic algorithms to resolve the traveling salesman problem. Comput. Sci. **31**(11), 49–57 (2011)

Hybrid Estimation of Distribution Algorithm for Blocking Flow-Shop Scheduling Problem with Sequence-Dependent Setup Times

Zi-Qi Zhang[1,2], Bin Qian[1,2(✉)], Bo Liu[3], Rong Hu[1],
and Chang-Sheng Zhang[1]

[1] School of Information Engineering and Automation,
Kunming University of Science and Technology, Kunming 650500, China
bin.qian@vip.163.com
[2] School of Mechanical and Electrical Engineering,
Kunming University of Science and Technology, Kunming 650500, China
[3] Academy of Mathematics and Systems Science,
Chinese Academy of Sciences, Beijing 100190, China

Abstract. This paper presents an innovative hybrid estimation of distribution algorithm, named HEDA, for blocking flow-shop scheduling problem (BFSP) with sequence-dependent setup times (SDSTs) to minimize the makespan criterion, which has been proved to be typically NP-hard combinatorial optimization problem with strong engineering background. Firstly, several efficient heuristics are proposed according to the property of BFSP with SDSTs. Secondly, the genetic information of both the order of jobs and the promising blocks of jobs are concerned to generate the guided probabilistic model. Thirdly, after the HEDA-based global exploration, a reference sequence-based local search with path relinking technique is developed and incorporated into local exploitation to escape from local optima and improve the convergence property. Due to the reasonable balance between EDA-based global exploration and sequence dependent local exploitation as well as comprehensive utilization of the speedup evaluation method, the BFSP with SDSTs can be solved effectively and efficiently. Finally, computational results and comparisons with the existing state-of-the-art algorithms are carried out, which demonstrate the superiority of the proposed HEDA in terms of searching quality, robustness, and efficiency.

Keywords: Estimation of distribution algorithm
Blocking flow-shop scheduling problem · Sequence-dependent setup times
Path relinking

1 Introduction

Flow shop scheduling problems (FSSPs) have attracted much attention in the manufacturing systems of contemporary enterprises and have widely potential applications in the research field of both computer science and operation research. The blocking flow-shop scheduling problem (BFSP) is a special case of FSSPs, which considers that the buffer capacity between two machines may be absent. Many modern production and

© Springer International Publishing AG, part of Springer Nature 2018
D.-S. Huang et al. (Eds.): ICIC 2018, LNCS 10954, pp. 628–637, 2018.
https://doi.org/10.1007/978-3-319-95930-6_63

manufacturing systems can be modelled as the BFSP when no buffers exist between consecutive machines, e.g. serial manufacturing processes, chemical industry, iron and steel industry and robotic cell, etc. In BFSP, there are no infinite buffer capacity between consecutive machines. In other words, under the no buffers circumstances, the completed jobs on a machine have to be blocked on its machine until its next machine is available. In addition, each machine can handle no more than one job at a time and all jobs have to visit each machine exactly once. To the best of our knowledge, almost all of the works about BFSP usually focused on assuming the setup times on each machine as the part of processing times. However, the setup times are very important and need to be explicitly treated in some actual production process [1], which should be separated from the processing times and consider independently (e.g., cleaning, fixing or releasing parts to machines, changing tools, and adjustments to machines). Therefore, the main motivation of this paper is that we consider a kind of setup times in BFSP, i.e., sequence-dependent setup times (SDSTs), where SDSTs depend upon both current job being processed and the next job in the sequence.

For the computational complexity of the BFSP, Hall and Sriskandarajah proved that the BFSP with more than two machines was NP-hard in the strong sense [2]. That is to say, the BFSP with exponential computational complexity, which becomes more complicated and is difficult to solve completely with the increase of problem size. Because of the BFSP is proven to be NP-hard, let alone with SDSTs, many state-of-the-art approaches which include constructive heuristics and metaheuristics evolutionary algorithms (EAs) have been developed recent years to address BFSP with different objectives including makespan, total flow time and total tardiness, etc. Constructive heuristics utilize some rules to determine the relative priority of each job to construct a specific scheduling permutation, which usually consume less time but obtain unsatisfied performance. However, metaheuristics start from previous generated solutions and iteratively improve performance with domain knowledge through different evolutionary strategies, which usually obtain fairly satisfactory solutions but the processes are always time-consuming. For BFSP with makespan, Wang et al. [3] proposed a hybrid discrete differential evolution (HDDE), where several crossover and mutation operators were employed. Meanwhile, a speedup method was also proposed to evaluate the insertion neighborhood, which largely reduced the complexity of HDDE. Pan and Wang [4] introduced the concept of weight value into PF and presented two simple constructive heuristics, i.e., profile fitting (wPF) and the Pan-Wang (PW) heuristic. Then, based on wPF and PW three improved constructive heuristics, i.e., PF-NEH, wPF-NEH and PW-NEH were proposed. Wang et al. [5] developed a three-phase algorithm (TPA) in which a priority rule based on the average value and standard deviation of the processing time, a variant of the NEH-insert procedure and a modified simulated annealing algorithm were utilized to three phases, respectively. Ding et al. [6] investigated some new blocking properties of the BFSP and proposed an iterated greedy algorithm (B-IG) based on these priorities. Han et al. [7] presented a modified fruit fly optimization (MFFO) algorithm in which three key operators, i.e., a problem-specific heuristic, neighborhood strategy and a speedup insert neighborhood based local search were employed. Tasgetiren et al. [8] proposed two iterated greedy algorithms, i.e., IG_{IJ} and IG_{RLS}, which combined a constructive heuristic and two well-known speedup methods for both insertion and swap neighborhood structures.

For BFSP with total flow time, Wang et al. [9] developed three harmony search algorithms, i.e., hybrid harmony search (hHS) algorithm, hybrid global best harmony search (hgHS) algorithm and hybrid modified global best harmony search (hmgHS) algorithm. Fernandez-Viagas et al. [10] conducted a comprehensive evaluation of the available heuristics for the BFSP and proposed an efficient constructive heuristic based on the beam search. For BFSP with total tardiness, Ronconi and Henriques [11] proposed a constructive heuristic (FPDNEH) and a GRASP-based heuristic. Nagano et al. [12] presented an evolutionary clustering search (ECS) algorithm combined with a variable neighborhood search (VNS), which NEH-based procedure to generate an initial solution, the genetic algorithm to generate solutions and VNS to improve the solutions. For the parallel blocking flow shop scheduling problem, Ribas et al. [13] proposed some constructive and improvement heuristics to address parallel or distributed blocking flow-shop problem recently, which included an iterated local search (ILS) and an iterated greedy algorithm (IGA), both of which are combined with a variable neighborhood search (VNS). From the previous literature reviews, it can be seen that researches on this concerned problem with SDSTs are considerably scarce. Therefore, it is meaningful to develop and propose more effective and efficient approaches for addressing variety of blocking flow-shop scheduling problems.

As a statistical learning based stochastic optimization, Estimation of Distribution Algorithm (EDA) [14] has attracted increasing attention in the field of evolutionary computation during recent years. Unlike GA, which apply mutation and crossover operators, EDA generates offspring through building and sampling explicit probabilistic models from superior candidate solutions. Owing to its outstanding global exploration and inherent parallelism, EDA has already been extensively applied to deal with production scheduling problems. Ceberio et al. [15] provided an overview of EDA in combinatorial optimization problems, which is of great importance of EDA for researchers. Pan and Ruiz [16] proposed a probabilistic model by taking into account of both the order of the jobs in the sequence and the promising similar blocks of jobs and employed the efficient NEH initialization and local search to enhance the performance of EDA. Recently, Wang et al. [17] presented an EDA with critical path based local search scheme to solve the flexible job-shop scheduling problem. Moreover, Wang et al. [18] developed an effective estimation of distribution algorithm (EEDA) to solve the distributed permutation flow-shop scheduling problem (DPFSP). From the mechanism of EDA, it can be seen that the tradeoff between the proper probability model with updating mechanism and local search are crucial to develop effective metaheuristics, which should achieve a better balance between global exploration and local exploitation. Consequently, to the best of our knowledge, there is no published work on EDA for the BFSP with SDSTs. In this work, EDA is employed to guide the promising search direction and find superior solutions, and the local intensification methods are designed to emphasize the exploitation from those promising regions.

The remainder of this work are organized as follows. Section 2 briefly introduces the BFSP with SDSTs. Section 3 elaborates on the proposed HEDA, such as HEDA global exploration, and reference sequence based local exploitation, respectively. Computational comparisons are given and analyzed in Sect. 4. Finally, Sect. 5 gives some concluding remarks and suggestions of future research.

2 Problem Statement

The BFSP with SDSTs, which denoted as $Fm|blocking, ST_{sd}|C_{max}$, can be described as follows. There are n jobs and m machines. Each of n jobs will be continuously processed without intermediated buffers. Since there are no intermediate buffers between consecutive machines, a job has to be blocked on current machine until the next machine has prepared for processing it. The sequence-dependent setup times are separable from the processing times and dependent upon both current job being processed and the adjacent job on each machine. Note that other common flow-shop assumptions are considered. Let $\pi = [\pi_1, \pi_2, \ldots, \pi_n]$ denote a permutation sequence. Let $p_{\pi_i, l}$ denotes the processing time of job π_i on machine l, $d_{\pi_i, l}$ denotes the departure time of job π_i on machine l, and $d_{\pi_i, 0}$ denotes the time job π_i starts its processing on the first machine. $s_{\pi_{i-1}, l\pi_i, l}$ represents the setup times between two consecutive jobs π_{i-1} and π_i on machine l. Let $p_{\pi_0, l} = 0$, $l = 1, \ldots, m$ for initial job and $s_{\pi_0, \pi_i, l}$ represents the initial setup times of job π_i. Then $d_{\pi_i, l}$ can be calculated as follows:

$$d_{\pi_0, l} = 0, \ l = 1, \ldots, m \tag{1}$$

$$d_{\pi_i, 0} = d_{\pi_{i-1}, 1} + s_{\pi_{i-1}, \pi_i, 1}, \ i = 1, 2, \ldots, n \tag{2}$$

$$d_{\pi_i, l} = \max\{d_{\pi_{i-1}, \pi_i, l+1} + s_{\pi_{i-1}, \pi_i, l+1}, d_{\pi_i, l-1} + d_{\pi_i, l}\}, \ i = 1, 2, \ldots, n; \ l = 1, 2, \ldots, m-1 \tag{3}$$

$$d_{\pi_i, m} = d_{\pi_i, m-1} + d_{\pi_i, m}, \ i = 1, 2, \ldots, n \tag{4}$$

Therefore, the makespan of the $\pi = [\pi_1, \pi_2, \ldots, \pi_n]$ can be given by $C_{max}(\pi) = d_{\pi_n, m}$. The aim of this paper is to find a schedule π^* in the set of all schedules Π such that

$$C_{max}(\pi^*) = \min_{\pi \in \Pi} C_{max}(\pi) \tag{5}$$

3 HEDA for BFSP with SDSTs

In this section, the HEDA for $Fm|blocking, ST_{sd}|C_{max}$ is proposed after explaining the solution representation and population initialization, probabilistic model, updating mechanism and new population generation and local intensification, i.e. a reference sequence based local search and a path relinking based local search, respectively.

3.1 Solution Representation and Population Initialization

In this work, we use the job permutation based representation, that is, each individual presents a feasible solution. Because of the properties of BFSP, the longer total processing time of job is, the higher probability of blocking occurs [6]. Therefore, an effective heuristic [4], i.e. PF-NEH, is employed to initialize and intensify the quality of the partial initial population. In addition, other initial solutions are randomly generated to maintain the diversity of the population. Note that the speedup method [3] is also utilized to evaluate the generated sequences and reduce the computational efforts.

3.2 Probabilistic Model

Probabilistic model is utilized to investigate the distribution information of the superb solutions, which guides the promising evolutionary directions. Therefore, the model has a key influence on the performances of HEDA. Obviously, an appropriate model can greatly enhance the efficiency and effectiveness of the presented algorithms. In this study, the probabilistic model of HEDA is denoted by $\mathbf{P(gen)}$ as follows:

$$\mathbf{P(gen)} = \begin{bmatrix} \mathbf{P_1(gen)} \\ \vdots \\ \mathbf{P_n(gen)} \end{bmatrix}_{n \times 1} = \begin{bmatrix} P_{1,1}(gen) & \cdots & P_{1,n}(gen) \\ \vdots & \ddots & \vdots \\ P_{n,1}(gen) & \cdots & P_{n,n}(gen) \end{bmatrix}_{n \times n} \tag{6}$$

where $\mathbf{P_i(gen)} = [P_{i,1}(gen), P_{i,2}(gen), \cdots, P_{i,n}(gen)]$ denote the ith vector of $\mathbf{P(gen)}$ in gen generation. Especially, $P_{i,j}(gen)$ is the probability of job j appearing in the ith position of π at gen. It's clear that the probability matrix is a random matrix, where each $P_{i,j}(gen)$ is a probability value for a certain event. In addition, for a certain position, the total probability of all jobs in this position of a solution sequence inevitable to 1, i.e., $\sum_{j=1}^{n} P_{i,j}(gen) = 1$, $gen \geq 0$. Note that the values in $\mathbf{P(gen)}$ reflect the priority of promising jobs in candidate sequences. Therefore, the elements of probability matrix affect the selection of the superior individuals. The higher value of $P_{i,j}(gen)$ $(i,j = 1, 2, \cdots, n)$ is, the higher possibility for job j is selected in position i.

According to the above definitions, HEDA generates offspring by sampling from the probability matrix proposed in Eq. (6), which means each individual is generated based on $\mathbf{P(gen)}$ by roulette sampling. Therefore, the values in $\mathbf{P(gen)}$ determine the composition of the generated population. Note that we set $P_{i,j}(0) = 1/(n \times n)$ $(i,j = 1, 2, \cdots, n)$ when $gen = 0$, which can guarantee the diversity of initial population, and accumulate more quality information appropriately at the initial phase.

3.3 Updating Mechanism

Probability model guides the searching direction and an effective update mechanism has a great influence on the searching performance. To make the $\mathbf{P(gen)}$ accurately estimate and effectively update the probability distribution of the superior population, and guide HEDA to promising search directions, two matrixes are proposed in this section, which record the total information of the order of jobs and the similar blocks, respectively. Therefore, the two matrixes $\mathbf{\eta(gen)}$ and $\mathbf{\xi(gen)}$ are defined as follows:

$$\mathbf{\eta(gen)} = \begin{bmatrix} \mathbf{\eta_1(gen)} \\ \vdots \\ \mathbf{\eta_n(gen)} \end{bmatrix}_{n \times 1} = \begin{bmatrix} \eta_{1,1}(gen) & \cdots & \eta_{1,n}(gen) \\ \vdots & \ddots & \vdots \\ \eta_{n,1}(gen) & \cdots & \eta_{n,n}(gen) \end{bmatrix}_{n \times n} \tag{7}$$

$$\xi(\mathbf{gen}) = \begin{bmatrix} \xi_1(\mathbf{gen}) \\ \vdots \\ \xi_n(\mathbf{gen}) \end{bmatrix}_{n \times 1} = \begin{bmatrix} \xi_{1,1}(gen) & \cdots & \xi_{1,n}(gen) \\ \vdots & \ddots & \vdots \\ \xi_{n,1}(gen) & \cdots & \xi_{n,n}(gen) \end{bmatrix}_{n \times n} \tag{8}$$

where each $\eta_{i,j}(gen)$ in $\eta(\mathbf{gen})$ records the information of the number of times that job j appears before or in position i in the set of superior individuals. Additionally, $\eta(\mathbf{gen})$ can efficiently reduce the rate of convergence and effectively prevent premature convergence. The ordinal matrix $\eta(\mathbf{gen})$ can be calculated as follows:

$$\eta_{i,j}(gen) = \sum_{s=1}^{Sbest} X_{i,j}^s(gen), \ i,j = 1, 2, \ldots, n \tag{9}$$

where $X_{i,j}^s$ is a binary variable and $Sbest$ represents the number of elite solution. In order to make the HEDA not trapped into local optimum and efficiently increase the diversity of population, the value of $X_{i,j}^s$ is set as 1 from i to n when job j appears in position i, otherwise 0. Then the indicative function is expressed as follows:

$$X_{i,j}^s(gen) = \begin{cases} 1, & \text{if job } j \text{ appears before or in position } i \\ 0, & \text{else} \end{cases} \tag{10}$$

Using the ordinal matrix can save the historical information of selected elite solutions. Note that the normalization should be performed to $\eta(\mathbf{gen})$ after it is updated.

Inspired by the schema theorem and the hypothesis of building blocks, the dependency matrix, i.e., $\xi(\mathbf{gen})$, is proposed to record the number of times that job j' appears immediately after job j when job j is in the previous position of job j' placed. Then each element $\xi_{j',j}(gen)$ in $\xi(\mathbf{gen})$ can be calculated as follows:

$$\xi_{j',j}(gen) = \sum_{s=1}^{Sbest} Y_{j',j}^s(gen), \ j',j = 1, 2, \ldots, n \tag{11}$$

$$Y_{j',j}^s(gen) = \begin{cases} 1, & \text{if job } j \text{ appears immediately after job } i \\ 0, & \text{else} \end{cases} \tag{12}$$

It is note that the normalization is also required for the dependency matrix $\xi(\mathbf{gen})$ after updating, which can slow down the convergence and adjust the searching space. Then, $\eta(\mathbf{gen})$ and $\xi(\mathbf{gen})$ which present the importance of the job order and the similar blocks can be obtained by Eqs. (7)–(12). Therefore, the probabilistic model can be updated by utilizing the information of $\eta(\mathbf{gen})$ and $\xi(\mathbf{gen})$, which can effectively balance the historical and current information of superior population. The incremental learning probabilistic model based on Heb-rule can be established as Eq. (13).

$$P_{i,j}(gen+1) = \begin{cases} rP_{i,j}(gen) + (1-r)\eta_{i,j}(gen+1), \ i = 1 \\ rP_{i,j}(gen) + \frac{(1-r)}{\mu}\left[\delta_1\eta_{i,j}(gen+1) + \delta_2\xi_{[i-1],j'}(gen+1)\right], \ i = 2, 3, \ldots, n \end{cases} \tag{13}$$

Note that $[i-1]$ presents the selected job placed in position $i-1$ and r $(0 < r \leq 1)$ is learning rate. Especially, if $r = 0$, the $\mathbf{P(gen)}$ can update directly without inertia. Note that $\mu = \sum_{j=1}^{n} \left(\delta_1 \eta_{ij}(gen+1) + \delta_2 \xi_{i-1,j}(gen+1) \right)$ $(i = 2, 3, \ldots, n)$, where δ_1 and δ_2 are two user-defined parameters, which indicate the importance of the two matrix $\boldsymbol{\eta(gen)}$ and $\boldsymbol{\xi(gen)}$ and the diversity of the population in each iteration. It has obvious that the updating mechanism is more advantageous to enhance the promising genetic information, which efficiently improve the guidance of the global exploration.

3.4 New Population Generation and Local Intensification

Sampling to generate new population means that all jobs are selected dependent on the probabilistic model $\mathbf{P(gen)}$. The sampling procedure is described as follows:

Step 1: Set control parameter $p = 1$ and randomly generate a probability value r where $r \in \left[0, \sum_{l=1}^{n} P_{il}(gen)\right)$.

Step 2: Get a candidate job j_c by using roulette selection scheme.

    Step 2.1: If $r \in [0, P_{i1}(gen))$ $(i \in \{1, \ldots, n\})$, then set $j_c = 1$, and go to Step 3.

    Step 2.2: If $r \in \left[\sum_{l=1}^{h_c-1} P_{il}(gen), \sum_{l=1}^{h_c} P_{il}(gen)\right)$ and $(i \in \{1, \ldots, n\}, h_c = 2, \ldots, n)$, then select the candidate job $j_c = h_c$, and go to Step 3.

Step 3: If the candidate job j_c do not repeat with the selected jobs, then return j_c.

Step 4: Put the selected j_c into the corresponding position and execute the sampling method independently and repeatedly until generating a feasible solution.

Step 5: Set $p = p + 1$. If $p \leq popsize$, go to Step 2. Otherwise output $\mathbf{pop(gen)}$.

To enhance the exploitation, two local search methods based on the speedup methods, i.e., a reference sequence based local search (RLS) [9] and a path relinking based local search, are employed and embedded in the EDA for stressing exploitation.

4 Simulation Results and Comparisons

4.1 Experimental Setup

Some randomly generated instances are adopted to test the performance of HEDA. The $n \times m$ instances are generated with $\{20, 30, 50, 70, 100, 200, 500\} \times \{5, 10, 20\}$. Each group consists of 10 instances. Therefore, we have a total of 210 different instances. The processing time $p_{j_i,l}$ and the setup time $s_{j_{i-1}j_i,l}$ are generated from two different uniform distributions $[1, 100]$ and $[0, 100]$, respectively. To evaluate the performance of each algorithm, the average relative percentage deviation (ARPD) is computed to measure the quality of the solutions. The ARPD is calculated as follows:

$$ARPD = \frac{1}{R}\sum_{i=1}^{R}\frac{C_{max}^{i} - C_{max}^{best}}{C_{max}^{best}} \times 100 \tag{14}$$

where C_{max}^{i} is the solution obtained by a specific algorithm in ith experiment and C_{max}^{best} is the best solutions obtained by all of the compared algorithms and R is the number of repetitions. Obviously, the less the value of ARPD is, the better performance of the algorithm is. Note that a full factorial design of experiment is carried out to determine the best parameters for the compared algorithms. The parameters for HEDA are set as follows: $popsize = 100$, $r = 0.75$, $Sbest = 10$, $\delta_1 = 0.7$ and $\delta_2 = 0.3$. All algorithms have been re-implemented in Embarcadero Studio XE-10. Computational experiments independently run 30 times on a server with 2.60 GHz Intel® Xeon® E5-4620 v2 processors (32 cores) and 64 GB RAM running Windows 10 Operating System. Therefore, the results are impartial and completely comparable.

4.2   Computational Results and Comparisons

To evaluate the effectiveness of the proposed HEDA, we have re-implemented the existing state-of-the-art algorithms, i.e. HDDE [3], TPA [5], B-IG [6], MFFO [7], IG_{RLS} [8], hmgHS [9], ECS [12], PEDA [16], and EEDA [18], for comprehensive comparisons as all details given by the original papers. Since no algorithms were proposed for the problem under consideration until now, we have adapted them by using the makespan criterion presented in Sect. 2. The statistical results of *ARPD* with the same maximum elapsed time limit of $t = 30 \times n \times (m/2)$ milliseconds as a termination criterion are reported in Table 1, where *Average* is a statistical average and the best, the second-best, and the third-best values in each cell are represented by using the bold, the bold with underlined, and the italic with underlined fonts, respectively.

It is clear through Table 1 that HEDA performs steadily and well in terms of the overall ARPD values, and that it outperforms the others in different scale problems. However, IG_{RLS} is a potential competitive algorithm better than HDDE and hmgHS.

Table 1. Computational results of compared algorithms and HEDA.

Instance	HDDE	TPA	B-IG	MFFO	IG_{RLS}	hmgHS	ECS	PEDA	EEDA	HEDA
20 × 5	0.016	0.137	0.018	*0.009*	**0.007**	0.064	0.562	0.078	0.027	**0.005**
20 × 10	*0.011*	0.034	0.012	*0.011*	**0.006**	0.036	0.457	0.034	0.012	**0.004**
20 × 20	**0.017**	0.047	0.025	*0.021*	**0.019**	0.083	0.239	0.076	0.023	**0.017**
30 × 5	**0.398**	0.784	*0.578*	0.512	*0.438*	0.815	1.283	0.645	0.536	**0.364**
30 × 10	**0.346**	0.732	0.507	0.493	*0.412*	0.726	1.184	0.535	0.482	**0.337**
30 × 20	**0.489**	0.836	0.612	0.587	*0.535*	0.923	1.326	0.624	0.546	**0.478**
50 × 5	1.924	1.536	*0.878*	1.382	**0.763**	1.185	2.651	1.436	1.147	**0.563**
50 × 10	1.967	1.464	*0.988*	1.402	**0.787**	1.231	2.649	1.312	1.347	**0.546**
50 × 20	1.783	1.215	*0.936*	1.117	**0.683**	1.189	2.276	1.183	1.382	**0.524**
100 × 5	2.163	1.687	*1.105*	1.782	**0.871**	1.624	2.853	1.382	1.515	**0.787**
100 × 10	2.231	1.602	*1.138*	1.927	**0.938**	1.633	2.589	1.236	1.674	**0.538**
100 × 20	2.562	1.275	*1.107*	1.776	**0.783**	1.687	2.611	1.312	1.647	**0.534**
200 × 10	2.216	1.131	**0.683**	1.132	1.028	1.732	1.521	0.979	*0.762*	**0.383**
200 × 20	1.874	1.137	**0.638**	1.213	1.126	1.734	1.527	0.837	*0.762*	**0.376**
500 × 10	1.457	1.038	**0.586**	1.178	*1.032*	1.637	1.426	0.787	*0.721*	**0.368**
500 × 20	0.957	0.794	**0.499**	0.793	*0.535*	0.937	0.871	0.668	0.639	**0.276**
Average	1.276	0.966	*0.644*	0.958	**0.623**	1.077	1.627	0.820	0.826	**0.381**

To be specific, for small-scaled instances ($n = 20$, 30), MFFO and HDDE obtain the best solutions, except for 20×5. For the medium-scaled instances ($n = 50$, 100), IG_{RLS} obtains the best performance. HEDA outperforms other algorithms at the considered margin for large-scaled instances ($n = 200$, 500). Meanwhile, some meta-heuristics, i.e. HDDE, MFFO and hmgHS, are highly depend on the parameters for different scale instances. The superiority of HEDA owes to some aspects as follow: (1) The PF-NEH heuristic provides excellent initial solutions. (2) With a well-designed probability model and the suitable updating mechanism, it is helpful to explore the promising area in the solution space effectively. (3) With the path relinking based local search, it is helpful to further exploit these regions and enhance the exploitation capability. (4) With the suitable calibration of parameters, it is helpful to obtain satisfactory schedules. With the above merits, it is safely concluded that the HEDA is a new state-of-the-art algorithm for $Fm|blocking, ST_{sd}C_{max}$ with excellent quality and robustness.

5 Conclusions and Future Research

To the best of the current authors' knowledge, this is the first report on the application of the HEDA for blocking flow-shop scheduling problem (BFSP) with sequence-dependent setup times (SDSTs) to minimize the maximum completion time. The speedup method is designed to reduce the computing complexity successfully. Both the effectiveness of searching solutions and the efficiency of evaluating solutions were stressed, and the influence of parameter setting was investigated as well. Due to reasonably hybridize and balance the global exploration and local exploitation, HEDA's search behavior can be enriched and its search performance can be greatly enhanced, which outperforms other considered algorithms with the effectiveness and robustness for the BFSP with SDSTs. Our future work is to develop some effective metaheuristics to solve the flexible distributed energy efficient scheduling problems.

Acknowledgments. The authors are sincerely grateful to the anonymous referees. This research is partially supported by the National Science Foundation of China (51665025), and the Applied Basic Research Foundation of Yunnan Province (2015FB136).

References

1. Allahverdi, A.: The third comprehensive survey on scheduling problems with setup times/costs. Eur. J. Oper. Res. **246**, 345–378 (2015)
2. Hall, N.G., Sriskandarajah, C.: A survey of machine scheduling problems with blocking and no-wait in process. Oper. Res. **44**(3), 510–525 (1996)
3. Wang, L., Pan, Q.K., Suganthan, P.N., Wang, W.H., Wang, Y.M.: A novel hybrid discrete differential evolution algorithm for blocking flow shop scheduling problems. Comput. Oper. Res. **37**(3), 509–520 (2010)
4. Pan, Q.K., Wang, L.: Effective heuristics for the blocking flowshop scheduling problem with makespan minimization. Omega **40**(2), 218–229 (2012)

5. Wang, C., Song, S., Gupta, J.N.D., Wu, C.: A three-phase algorithm for flowshop scheduling with blocking to minimize makespan. Comput. Oper. Res. **39**(11), 2880–2887 (2012)
6. Ding, J.Y., Song, S., Gupta, J.N.D., Wang, C., Zhang, R., Wu, C.: New block properties for flowshop scheduling with blocking and their application in an iterated greedy algorithm. Int. J. Prod. Res. **54**(16), 1–14 (2016)
7. Han, Y., Gong, D., Li, J., Zhang, Y.: Solving the blocking flow shop scheduling problem with makespan using a modified fruit fly optimisation algorithm. Int. J. Prod. Res. **54**(22), 6782–6797 (2016)
8. Tasgetiren, M.F., Kizilay, D., Pan, Q.K., Suganthan, P.N.: Iterated greedy algorithms for the blocking flowshop scheduling problem with makespan criterion. Comput. Oper. Res. **77**(C), 111–126 (2017)
9. Wang, L., Pan, Q.K., Tasgetiren, M.F.: Minimizing the total flow time in a flow shop with blocking by using hybrid harmony search algorithms. Expert Syst. Appl. **37**(12), 7929–7936 (2010)
10. Fernandez-Viagas, V., Leisten, R., Framinan, J.M.: A computational evaluation of constructive and improvement heuristics for the blocking flow shop to minimise total flowtime. Expert Syst. Appl. **61**, 290–301 (2016)
11. Ronconi, D.P., Henriques, L.R.S.: Some heuristic algorithms for total tardiness minimization in a flowshop with blocking ☆. Omega **37**(2), 272–281 (2009)
12. Nagano, M.S., Komesu, A.S., Mi, H.H.: An evolutionary clustering search for the total tardiness blocking flow shop problem. J. Intell. Manuf. 1–15 (2017)
13. Ribas, I., Companys, R., Tort-Martorell, X.: Efficient heuristics for the parallel blocking flow shop scheduling problem. Expert Syst. Appl. **74**, 41–54 (2017)
14. Larranaga, P., Lozano, J.A.: Estimation of Distribution Algorithms: A New Tool for Evolutionary Computation. Springer, Netherlands (2002). https://doi.org/10.1007/978-1-4615-1539-5
15. Ceberio, J., Irurozki, E., Mendiburu, A., Lozano, J.A.: A review on estimation of distribution algorithms in permutation-based combinatorial optimization problems. Prog. Artif. Intell. **1** (1), 103–117 (2012)
16. Pan, Q.K., Ruiz, R.: An estimation of distribution algorithm for lot-streaming flow shop problems with setup times. Omega Int. J. Manag. Sci. **40**(2), 166–180 (2012)
17. Wang, L., Wang, S.Y., Xu, Y., Zhou, G., Liu, M.: A bi-population based estimation of distribution algorithm for the flexible job-shop scheduling problem. Comput. Ind. Eng. **62**, 917–926 (2012)
18. Wang, S.Y., Wang, L., Liu, M., Xu, Y.: An effective estimation of distribution algorithm for solving the distributed permutation flow-shop scheduling problem. Int. J. Prod. Econ. **145**, 387–396 (2013)

Salp Swarm Algorithm Based on Blocks on Critical Path for Reentrant Job Shop Scheduling Problems

Zai-Xing Sun[1], Rong Hu[1($\boxtimes$)], Bin Qian[1], Bo Liu[2], and Guo-Lin Che[1]

[1] School of Information Engineering and Automation,
Kunming University of Science and Technology, Kunming 650500, China
ronghu@vip.163.com
[2] Academy of Mathematics and Systems Science,
Chinese Academy of Sciences, Beijing 100190, China

Abstract. In this paper, salp swarm algorithm based on blocks on critical path (SSA_BCP) is presented to minimize the makespan for reentrant job shop scheduling problem (RJSSP), which is a typical NP-complete combinational optimization problem. Firstly, the mathematical model of RJSSP based on the disjunctive graph is established. Secondly, the extended reentrant-smallest-order-value (RSOV) encoding rule is designed to transform SSA's individuals from real vectors to job permutations so that SSA can be used to perform global search for finding high-quality solutions or regions in the solution space. Thirdly, four kinds of neighborhood structures are defined after defining the insert operation based on block structure, which can be used to avoid search in the invalid regions. Then, a high-efficient local search integrating multiple neighborhoods is proposed to execute a thorough search from the promising regions found by the global search. Simulation results and comparisons show the effectiveness of the proposed algorithm.

Keywords: Reentrant job shop scheduling problem · Makespan
Salp swarm algorithm · Block structure

1 Introduction

Job shop scheduling problem (JSSP) is one of the most typical and most difficult combinatorial optimization problems [1]. Garey et al. [2] has proved that JSSP is a NP-complete problem when the number of machines exceeds two. Its algorithm research has always been an important topic of common concern in academia and engineering. JSSP is limited to machining or manufacturing parts that can only be machined once on a single machine. In the production systems of high-tech industries such as semiconductor manufacturing and computer networks, some parts are required to be machined many times by the same machine. Therefore, the research of the reentrant job shop scheduling problem (RJSSP) has important engineering value and practical significance. Since JSSP has NP-complete property, and JSSP can be reduced to RJSSP, RJSSP is also NP-complete. Therefore, the research of RJSSP for minimizing the makespan has high practical and theoretical value, which can provide feasible guidance and help for relevant production enterprises.

As far as JSSP is concerned, the design of neighborhood structure is very important for the performance of algorithm, and it also has some universality. Therefore, the research is of great significance. Some scholars have studied the neighborhood structure of JSSP with the goal of minimizing makespan. Zhang et al. [3] were respectively proposed the neighborhoods N7 on the basis of predecessors. Zhang and Wu [4] proposed a neighborhood structure to solve the JSSP with the goal of minimizing total weighted tardiness. And using the property of the dual model of the problem, [4] was proved that the proposed neighborhood structure can not improve the target value when satisfying certain constraints, and effectively enhances the local search efficiency of the algorithm. Kuhpfahl and Bierwirth [5] proposed six new neighborhood structures, and the simulation experiments compared the performance and search capability of different structures and combinations. Then Bierwirth and Kuhpfahl [6] derived six further neighborhoods from [5]. By adopting a suitable neighborhood structure, the abovementioned algorithms have achieved good performance. However, there is no relevant research on the effective neighborhood structure of RJSSP.

For RJSSP, Topaloglu and Kilincli [7] proposed a modified shifting bottleneck heuristic to solve the RJSSP with makespan minimization, which shows that the proposed algorithm is very effective compared with the simulation of other algorithms. Qian et al. [8] proposed a hybrid differential evolution algorithm for solving multi-objective RJSSP with total machine idleness and maximum tardiness criteria. Simulation results and comparisons show the effectiveness of the proposed algorithm. Thus it is very important to develop effective algorithms for this kind of problem.

Mirjalili et al. [9] first proposed a new meta-heuristic swarm intelligence optimization salp swarm algorithm (SSA) based on the swarming behaviour of salps when navigating and foraging in oceans. SSA is novel algorithm that attracts the attention in many applications as engineering design problem [9] and Electrical Engineering problem [10]. In Ref. [9], the authors indicate that the cropped results in comparison to other challenging optimizers demonstrate the value of the SSA in solving optimization problems with hard and unidentified search spaces. On top of that, to the author's knowledge, this is a novel application and first time to use the SSA in solving the scheduling problem, so it is very important to carry out relevant research.

In this paper, a SSA based on blocks on critical path (SSA_BCP) is proposed to minimize makespan of the RJSSP. In SSA_BCP, SSA is used to obtain promising regions over the solution space, and a problem-dependent local search with different neighborhoods is applied to perform exploitation from these regions. Simulation results and comparisons demonstrate the effectiveness of the proposed SSA_BCP. The rest of this paper is arranged as follows.

In Sect. 2, the RJSSP description using disjunctive graph representation is briefly introduced. In Sect. 3, a brief review of SSA is provided. In Sect. 4, SSA_BCP is proposed and described in detail. In Sect. 5, simulation results and comparisons are provided. Finally, we end the paper with some conclusions and future work in Sect. 6.

2　Problem Description Using Disjunctive Graph Representation

The RJSSP can be described as follows: given the processing time p_v of operation v, each of n jobs will be processed r times on m machines. At any time, each machine can process at most one job and each job can be processed on at most one machine.

The disjunctive diagram was proposed by Roy and Sussman in 1964 to describe the process sequence constraints and machine uniqueness constraints in the scheduling problem. RJSSP can be represented by the disjunctive graph $G = (N, A, E, Re)$. A simple disjunctive graph for RJSSP is shown in Fig. 1. $N = O \cup \{0\} = \{0, 1, 2, \cdots, TO\}$ is the set of nodes. Node 0 stands for a dummy operation which starts before all the real operations, while the starting time and the processing time of this dummy operation are both zero. $TO = n \times m \times r$ is the total number of operations. A is the set of unidirectional conjunctive arcs, which describes the order constraints of the same job process route. E is the set of disjunctive arcs, where represents the disjunctive arcs related with one machine. The disjunctive arc orientation is not determined before scheduling, for bi-directional. Re is a set of reentrant arcs that is first proposed for reentrant properties which describes the processing constraints of the same job on the same machine and is unidirectional. U is the unidirectional tail arc, describing the last operation of each job pointing to node 0.

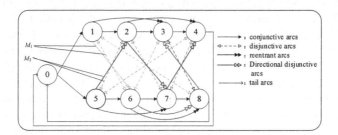

Fig. 1. The disjunctive graph of RJSSP with $2 \times 2 \times 2$

Let s_v be the starting processing time of the operation v, in which $s_0 = 0$. p_v is the processing time of the operation v, in which $p_0 = 0$. μ_j is the last operation of job j. (v, w) is the arc with processing constraints. $\langle v, w \rangle$ is a disjunction arc in an undefined direction. '$\vee$' is a logical OR, describing the operation of the arc in the disjunction. $C_{\max}$ is the makespan. The disjunctive planning model of RJSSP is described as follows:

$$Min \quad C_{\max} \tag{1}$$

$$s.t. \quad s_v + p_v \leq s_w \qquad \forall (v, w) \in A \tag{2}$$

$$(s_v + p_v \leq s_w) \vee (s_w + p_w \leq s_v) \quad \forall \langle v, w \rangle \in E \tag{3}$$

$$s_v + p_v \leq s_w \qquad \forall (v, w) \in Re \tag{4}$$

$$C_{\max} \geq s_{\mu_j} + p_{\mu_j} \qquad \forall \mu_j \in U(O) \tag{5}$$

In the above model, $s_v \geq 0 \,(\forall v \in O)$ can be obtained by $s_0 = p_0 = 0$ and formula (2), so this constraint does not need to be given separately.

The important scheduling decision is to determine for each pair of operations that require the same machine, whether one is performed before the other, or the other way around. This decision is often referred to as a selection: for each of the disjunctive constraints (3), either $s_v + p_v \leq s_w$ or $s_w + p_w \leq s_v$ is selected. A selection, denoted by σ, transforms RJSSP into a linear programming problem. Let σ be the collection of ordered pairs of operations (v, w) that require the same machine, and for which it has been decided that v must be performed before w. The linear program is then as follows.

$$Min \quad C_{\max} \tag{6}$$

$$s.t. \quad s_v + p_v \leq s_w \qquad \forall (v, w) \in A \tag{7}$$

$$s_v + p_v \leq s_w \qquad \forall (v, w) \in \sigma \tag{8}$$

$$s_v + p_v \leq s_w \qquad \forall (v, w) \in Re \tag{9}$$

$$C_{\max} \geq s_{\mu_j} + p_{\mu_j} \qquad \forall \mu_j \in U(O) \tag{10}$$

3 Brief Review of SSA

SSA is a branch of meta-heuristic proposed by Mirjalili et al. [9] for optimization problems over continuous domains. To mathematically model the salp chains, the population is first divided to two groups: leader and followers. The leader is the salp at the front of the chain, whereas the rest of salps are considered as followers. As the name of these salps implies, the leader guides swarm and the followers follow each other (and leader directly of indirectly). Similarly to other swarm-based techniques, the position of salps is defined in an n-dimensional search space where n is the number of variables of a given problem. Therefore, the position of all salps are stored in a two-dimensional matrix called X. It is also assumed that there is a food source called F in the search space as the swarm's target. The details of SSA can be found in Subsect. 4.4.

4 SSA_BCP for RJSSP

In this section, we will present SSA_BCP for RJSSP after explaining the solution representation and decoding scheme, problem-dependent neighborhood structure, problem-dependent local search.

4.1 Solution Representation and Decoding Scheme

Because of the continuous characters of the individuals in SSA, a canonical SSA cannot be directly applied to the considered problem in this paper. So, we extend the reentrant-smallest-order-value (RSOV) [8] encoding rule to convert SSA's ith individual $X_i = [x_{i,1}, x_{i,2}, \cdots, x_{i,TO}]$ to the job permutation vector $\pi_i = [\pi_{i,1}, \pi_{i,2}, \cdots, \pi_{i,TO}]$ and disjunctive graph operation permutation vector $\varphi_i = [\varphi_{i,1}, \varphi_{i,2}, \cdots, \varphi_{i,TO}]$, where $\varphi_{i,j} = (\pi_{i,j} - 1) \times m \times r + z_{\pi_{i,j}}$, $z_{\pi_{i,j}}$ is the $z_{\pi_{i,j}}$th occurrence of job $\pi_{i,j}$ in π_i. Table 1 illustrates the representation of vector X_i in SSA for a simple problem ($n = 3, m = 2, r = 2$).

Table 1. Solution representation

Dimension k	1	2	3	4	5	6	7	8	9	10	11	12
$x_{i,k}$	1.51	2.83	3.25	2.30	1.69	2.57	3.83	0.61	2.40	3.08	0.11	1.03
$y_{i,k}$	4	9	11	6	5	8	12	2	7	10	1	3
$\pi_{i,k}$	1	3	3	2	2	2	3	1	2	3	1	1
$\varphi_{i,k}$	1	9	10	5	6	7	11	2	8	12	3	4

According to the properties of RJSSP, the set of active schedules is a subset of the semi-active ones and an optimal solution must be an active schedule to minimize the makespan. Therefore, when decoding we check the possible idle interval before appending an operation at the last position, and fill the first idle interval before the last operation (called active decoding). In order to adapt to the operation of the block structure, when the scheduling is completed, the process is sorted by the completion time from the smallest to the largest, and the obtained new solution replaces the old solution as the current solution output.

4.2 Problem-Dependent Neighborhood Structure

The design of the neighborhood structure is an important research issue in the field of combinatorial optimization [11]. The effective neighborhood structure can significantly speed up the convergence of the algorithm and improve the search performance of the algorithm.

Definition 1 (Critical path): based on the Machine Predecessor First (MPF) rule [4], the longest path from the last completed operation to the virtual node 0.

Definition 2 (Block structure): a sequence of operations in a critical path is called a block structure if it contains at least two operations and not all belongs to one job.

Definition 3 (Insert operation based on block structure): in disjunctive graph operation permutation vector φ_i, when the operation v in the block structure is inserted before (after) the operation w, all the operations between v and w that belong to the same job as v are inserted before (after) the operation w.

Neighborhood Structure 1 (NS1): moving the first operation of the block structure into the internal operation within the block structure.

Neighborhood Structure 2 (NS2): moving the last operation of the block structure into the internal operation within the block structure.

Neighborhood Structure 3 (NS3): in disjunctive graph operation permutation vector, the first operation on the block structure is inserted to its previous random operation.

Neighborhood Structure 4 (NS4): in disjunctive graph operation permutation vector, the last operation on the block structure is inserted in its later random operation.

The four neighborhood structures are illustrated in Fig. 2.

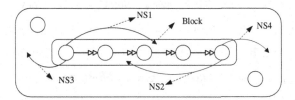

Fig. 2. The four neighborhood structures

4.3   Problem-Dependent Local Search

We design a problem dependent local search based on the neighborhood structure in the above subsection to enrich the search behavior and enhance the search ability. The perturbation phase is based on the following two neighborhoods which are often employed in the literature. (1) remove the job at the vth dimension and insert it in the wth dimension of the solution $\varphi(insert(\varphi, v, w))$, (2) invert the subsequence between the vth dimension and the wth dimension of the solution $\varphi(inverse(\varphi, v, w))$. Let *Block_num* be the numbers of block structure. *Block_n_i* is the numbers of operation in ith block structure. *Block$_{i,j}$* is ith operation in jth block structure. ***LBest_φ*** is the optimal solution currently searched. The procedure of SSA_BCP's local search is given as follows:

Step 1: Convert SSA's individual X_i to disjunctive graph operation permutation vector φ_{i_0} according to the RSOV rule;

Step 2: Set $kk = 0$, $\varphi_{i\_pp} = \varphi_{i\_0}$ and *kick-move strength* (ks) = 10;

  Do

    Randomly select v and w, where $v \neq w$;

    If *random* < 0.8 then $\varphi_i = insert(\varphi_{i\_pp}, v, w)$, $\varphi_{i\_pp} = \varphi_i$;

    Else $\varphi_i = inverse(\varphi_{i\_pp}, v, w)$, $\varphi_{i\_pp} = \varphi_i$;

    $kk = kk + 1$;

  While $kk < ks$;

Step 3: Statistics *Block_num*, *Block_n* and *Block* of φ_i, let $\varphi_{i\_1} = \varphi_i$, *LBest_$\varphi$* $= \varphi_i$, $x = 1$;

Step 4: the xth block structure in φ_{i_1} search using NS1 to get φ_{i_2}, if $\varphi_{i\_2} <$ *LBest_φ*, then *LBest_φ* $= \varphi_{i\_2}$;

let $\varphi_{i\_1} = \varphi_i$, the xth block structure in φ_{i_1} search using NS2 to get φ_{i_2}, if $\varphi_{i\_2} <$ *LBest_φ*, then *LBest_φ* $= \varphi_{i\_2}$;

let $\varphi_{i\_1} = \varphi_i$, the xth block structure in φ_{i_1} search using NS3 to get φ_{i_2}, if $\varphi_{i\_2} <$ *LBest_φ*, then *LBest_φ* $= \varphi_{i\_2}$;

let $\varphi_{i\_1} = \varphi_i$, the xth block structure in φ_{i_1} search using NS4 to get φ_{i_2}, if $\varphi_{i\_2} <$ *LBest_φ*, then *LBest_φ* $= \varphi_{i\_2}$;

Step 5: Set $x = x+1$. If $x \leq$ *Block_num*, then go to **Step 4**;

Step 6: Output *LBest_φ* and its objective value. Convert *LBest_φ* back to X_i.

As can be seen from the above procedure, step 2 is the perturbation phase, which is used to escape local optima and guide the search to a different region, and step 3–5 executes a thorough four neighborhood structures exploitation from the region obtained by step 2.

4.4   SSA_BCP

Based on the above subsections, the procedure of SSA_BCP is proposed as follows:

Step 1: Let G denotes a generation, P a population of size PS, and $x_i(t)$ the ith individual of dimension TO in population P in generation t, $x_{i,k}(t)$ the kth dimension of individual $x_i(t)$, random$(0,1)$ the random value in the interval $[0,1]$, and t_max the maximum number of iteration;

Step 2: Input, m n, r, PS. Initial bounds: $lb_k = -2$, $ub_k = 2$, $k = 1,2,\cdots TO$;

Step 3: Set $t=1$. Initialize the salp population x_i considering low and upper. Calculate the fitness of each search agent (salp). F = the best search agent;

Step 4: Update $c1$ by Eq.(11). Set $i=1$;

Step 5: Set $c3$ = random$(0,1)$. If $c3 < 0.5$ or $i=1$, then update the position of the leading salp by Eq.(12); else update the position of the follower salp by Eq.(13).

$$c1 = 2e^{-(4t/t\_max)^2} \tag{11}$$

$$x_{i,k} = \begin{cases} F_k + c1((ub_k - lb_k)c2 + lb_k) & c3 < 0.5 \\ F_k - c1((ub_k - lb_k)c2 + lb_k) & c3 \geq 0.5 \end{cases} \tag{12}$$

$$x_{i,k} = \frac{1}{2}(x_{i-1,k} + F_k) \tag{13}$$

Where $c2$ and $c3$ are random numbers;

Step 6: Amend the salps based on the upper and lower bounds of variables;

Step 7: Set $i=i+1$. If $i \leq PS$, then go to **Step 5**;

Step 8: Calculate the fitness of each search agent (salp). F = the best search agent.

Step 9: Apply the problem-dependent local search to F;

Step 10: Set $t=t+1$. If $t \leq t\_max$, then go to **Step 4**;

Step 11: Output F.

The steps of the algorithm show that SSA used and found high quality solutions for global search in the solution space of regional problems, at the same time into local search for different neighborhood structures for high quality solutions for regional implementation further detailed search, the SSA_BCP in the global and local search to achieve a reasonable balance. SSA_BCP is expected to be an effective algorithm for solving this problem. The fifth section will verify the effectiveness of the proposed algorithm by simulation experiments and algorithm comparisons.

5 Simulation Results and Comparisons

In this section, we first perform experimental analysis based on the Design of Experiment (DOE) for the key parameters of the SSA_BCP to determine the optimal combination of the algorithm parameters. Then by comparing SSA_BCP with other effective algorithms, SSA_BCP is an effective algorithm for solving RJSSP. A set of instances under different scales is randomly generated. The $n \times m \times r$ combinations include $10 \times 10 \times 3$, $20 \times 5 \times 3$, $10 \times 20 \times 3$, $20 \times 10 \times 3$, $20 \times 15 \times 3$, $50 \times 6 \times 3$, $20 \times 20 \times 3$, $40 \times 10 \times 3$, $50 \times 10 \times 3$, $100 \times 5 \times 3$. The processing time is generated from a uniform distribution [1,99]. All algorithms are coded in Delphi10.2 and are executed on Intel Core I7-7700 k 4.20 GHz PC with 8 GB memory.

For the purpose of evaluating the effectiveness of SSA_BCP, we carry out simulations to compare our SSA_BCP with NIMGA [12]. NIMGA is an effective algorithm for JSSP. The performance of NIMGA is superior to a variety of effective algorithms.

For each instance, each algorithm is run 20 times independently, running at the same time each time. SSA_BCP's parameters are set as follows: the population size $PS = 10$, and the maximum number of iteration $t\_max = 200$. Other parameters of SSA or NIMGA set reference [9, 12]. *Min, Max, Avg* and *Std* are the best makespan, worst makespan, average makespan, and standard derivation of the 20 times. The optimal results for each problem are shown in bold, and the test results are shown in Table 2.

Table 2. Comparison of SSA_BCP and NIMGA

$n \times m \times r$	SSA_BCP				NIMGA			
	Min	*Max*	*Avg*	*Std*	*Min*	*Max*	*Avg*	*Std*
$10 \times 10 \times 3$	**2275**	2420	**2347.25**	41.22	2299	**2406**	2348.2	**23.57**
$20 \times 5 \times 3$	**3250**	3314	3262.75	18.53	**3250**	**3270**	**3253.1**	**6.24**
$10 \times 20 \times 3$	**3996**	4267	**4096.1**	69.55	4090	**4184**	4135.75	**28.02**
$20 \times 10 \times 3$	**3543**	**3757**	**3646.75**	54.46	3671	3845	3768.1	**44.91**
$20 \times 15 \times 3$	**4529**	**4760**	**4643.75**	57.08	4759	4975	4836.05	**51.8**
$50 \times 6 \times 3$	**8085**	**8085**	**8085**	**0**	**8085**	8094	8085.55	1.96
$20 \times 20 \times 3$	**5086**	**5400**	**5230.55**	80.89	5353	5595	5477.35	**58.98**
$40 \times 10 \times 3$	**6767**	**6887**	**6842.25**	**32.73**	6912	7115	7023.85	53.78
$50 \times 10 \times 3$	**8175**	**8323**	**8237.59**	41.26	8340	8453	8394.35	**26.70**
$100 \times 5 \times 3$	**15592**	**15592**	**15592**	**0**	**15592**	**15592**	**15592**	**0**

From Table 2, it can be seen that SSA_BCP performs much better than NIMGA with respecting to solution quality. The values of *Min* and *Avg* obtained by SSA_BCP are obviously better than that obtained by NIMGA for all instances except $20 \times 5 \times 3$. For *Max*, SSA_BCP is better than NIMGA on large-scale problems. For *Std*, although SSA_BCP is larger than NIMGA on several issues, its *Avg* is superior, so SSA_BCP is effective. For $50 \times 6 \times 3$ and $100 \times 5 \times 3$, their *Std* is 0, because the block structure of the current solution is only one and only on one machine, so the solution is the optimal solution. So, it can be concluded that SSA is an effective algorithm for the RJSSP. Moreover, the test results also manifest neighborhood structure based on blocks on critical path is more suitable for guiding the search to the promising regions in the solution space of RJSSP.

6   Conclusions and Future Research

This paper presented a salp swarm algorithm based on blocks on critical path (SSA_BCP) for the reentrant job shop scheduling problem (RJSSP). To the best of the current authors' knowledge, this is the first report on the application of salp swarm algorithm (SSA) approach to solve the problem considered. Firstly, considering that the problem has reentrant features, a reentry arc is added to the disjunctive graph model of Job Shop Scheduling Problem (JSSP), and the RJSSP mathematics planning model based on disassembly graph is established. Secondly, the extended reentrant-smallest-order-value (RSOV) encoding rule is designed so that SSA can be used to perform global search for finding high-quality solutions or regions in the solution space. Then, a high-efficient local search integrating multiple neighborhoods is proposed to execute a thorough search from the promising regions found by the global search. Simulation results and comparisons demonstrated the effectiveness of SSA_BCP. Future work includes developing of multi-objective SSA for the uncertain multi-objective RJSSP.

Acknowledgements. This research is partially supported by the National Science Foundation of China (51665025), and the Applied Basic Research Foundation of Yunnan Province (2015FB136).

References

1. Quanke, P., Ling, W., Liang, G., et al.: Differential evolution algorithm based on blocks on critical path for job shop scheduling problems. J. Mech. Eng. **46**(22), 182–188 (2010)
2. Garey, M.R., Johnson, D.S., Sethi, R.: The complexity of flowshop and jobshop scheduling. INFORMS (1976)
3. Zhang, C.Y., Li, P.G., Guan, Z.L., et al.: A tabu search algorithm with a new neighborhood structure for the job shop scheduling problem. Comput. Oper. Res. **34**(11), 3229–3242 (2007)
4. Zhang, R., Wu, C.: A simulated annealing algorithm based on block properties for the job shop scheduling problem with total weighted tardiness objective. Comput. Oper. Res. **38**(5), 854–867 (2011)

5. Kuhpfahl, J., Bierwirth, C.: A study on local search neighborhoods for the job shop scheduling problem with total weighted tardiness objective. Elsevier Science Ltd. (2016)
6. Bierwirth, C., Kuhpfahl, J.: Extended GRASP for the job shop scheduling problem with total weighted tardiness objective. Eur. J. Oper. Res. **261**, 835–848 (2017)
7. Topaloglu, S., Kilincli, G.: A modified shifting bottleneck heuristic for the reentrant job shop scheduling problem with makespan minimization. Int. J. Adv. Manuf. Technol. **44**(7–8), 781–794 (2009)
8. Qian, B., Li, Z.H., Hu, R., et al.: A hybrid differential evolution algorithm for the multi-objective reentrant job-shop scheduling problem, pp. 485–489 (2013)
9. Mirjalili, S., Gandomi, A.H., et al.: Salp Swarm Algorithm: a bio-inspired optimizer for engineering design problems. Adv. Eng. Softw. **114**, 163–191 (2017)
10. Elfergany, A.A., Kalogirou, S.A., Christodoulides, P.: Extracting optimal parameters of PEM fuel cells using Salp Swarm Optimizer. Renew. Energy **119**, 641–648 (2018)
11. Abdel-Basset, M., Gunasekaran, M., El-Shahat, D., et al.: A hybrid whale optimization algorithm based on local search strategy for the permutation flow shop scheduling problem. Future Gener. Comput. Syst. **85**, 129–145 (2018)
12. Kurdi, M.: An effective new island model genetic algorithm for job shop scheduling problem. Comput. Oper. Res. **67**, 132–142 (2016)

The Hybrid Shuffle Frog Leaping Algorithm Based on Cuckoo Search for Flow Shop Scheduling with the Consideration of Energy Consumption

Ling-Chong Zhong, Bin Qian[✉], Rong Hu, and Chang-Sheng Zhang

School of Information Engineering and Automation,
Kunming University of Science and Technology, Kunming 650500, China
bin.qian@vip.163.com

Abstract. Green manufacturing requires full consideration of energy-related optimization objective. This paper presents a hybrid shuffle frog leaping algorithm based on the cuckoo search algorithm (HFLCS), and the algorithm for solving multi-objective based green flow shop scheduling problem (MOPFS), the optimization objectives are the maximum completion time and energy consumption. Since the traditional flow shop scheduling problem (PFS) is a typical NP-hard combinatorial problem, MOPFS is also a NP-hard combinatorial problem. Firstly, the levy flight update formula in cuckoo algorithm is based on the update formula of global search, and its nature can make the search out of the local optimum and generate the disturbance; secondly, a global search mechanism based on shuffled frog leaping algorithm and levy flight (LACS) is designed to explore the solution space; thirdly, a multi-neighborhood local search is proposed to search potential solutions in better space. With the application of global search and local search, we can prevent the algorithm iteration from getting into local optimum and find high quality solutions so that we can solve the problem of MOPFS. Finally, the simulation results and comparisons demonstrate the superiority of HFLCS in terms of search quality, robustness, and efficiency.

Keywords: Hybrid shuffle frog-leaping algorithm · Cuckoo search algorithm
Multi-objective flow-shop scheduling · Energy consumption

1 Introduction

The problem of scheduling in permutation flow shop has been extensively investigated by many researchers. Exact and heuristic algorithm have been proposed over the years for solving the static permutation flow shop scheduling problems with the multi-objective of minimizing the make-span and energy consumption (MOPFS). The make-span is the final time required to complete the all jobs' processing. The energy consumption produced during the processing is believed to be a critical reason that causes the waste of resources. In this research, which aims at obtaining solutions, with the consideration of those double objectives. These solutions are called Pareto-optimal

© Springer International Publishing AG, part of Springer Nature 2018
D.-S. Huang et al. (Eds.): ICIC 2018, LNCS 10954, pp. 649–658, 2018.
https://doi.org/10.1007/978-3-319-95930-6_65

solutions [1]. There is a survey on multi-objective scheduling problems is given by T'kindt and Billaut [2]. According to the survey, the DOPFS is NP-hard.

Recently, metaheuristic methods have attracted wide research attention, including such topics as cuckoo search algorithm (CS) (Yang et al. 2013), shuffled frog-leaping algorithm (SFLA) (Lei et al. 2015), hybrid differential evolution (HDE) (Wang et al. 2012), genetic algorithm (GA) (Chen et al. 1996), hybrid particle swarm optimization (HPSO) (Liu et al. 2007). Over the past 10 years, frog-leaping algorithm (FLA) and cuckoo search algorithm (CS) have been a hot topic in the fields of scheduling research. Some recent work demonstrates that CS and FLA can yield promising results for solving combinatorial [3] and engineering problems [4, 5]. As for multi-objective scheduling problem, Yin and Li proposed a multi-objective genetic algorithm based on a simplex lattice design to solve the mixed-integer programming model effectively [6]; Lei and zheng designed a multi-objective frog-leaping algorithm (MOFLA) with the local search method, which used a concept of Pareto dominance that is used to rank the population [7]; Yang XS devised a multi-objective cuckoo search for design optimization, which could outperform many classical algorithms such as PSO [8]. In addition, metaheuristic algorithms start to emerge as a major player for multi-objective optimization. Many new algorithms are emerging with combinational optimization.

In this paper, we will combine FLA and CS to solve multi-objective problems and formulate a hybrid frog-leaping algorithm based on cuckoo search algorithm. In the proposed HFLCS, FLA is used to find the promising solutions or regions over the solution space, and then a local search based on the landscape, after the FLA and local search find the Pareto solutions, levy flights is conceived to breaking away from the local optimum and increasing diversity.

The remaining contents are organized as follows. In Sect. 2, the MOPFS is introduced. In Sect. 3, HFLCS is proposed after FLA-based both global and local search, and levy flights overstepping the local optimum. In Sect. 4, experimental results and comparisons are presented and analyzed. Finally, we end the paper with some conclusions and future work.

2 MOPFS

The PFS can be described mathematically as follows. There are N jobs and M machines. The machines are ordered so that a job cannot start on machine k until it is completed on machine $k - 1$, for $k = [2, \ldots, M]$. In addition, each job needs to be processed on each machine in M. Each machine should be completed without interruption once it started. Suppose all the jobs are available at time zero. And at any time, each job can be processed at most one machine, and each machine can just process at most one job. For the MOPFS with the make-span criterion $C_{\max}$ and energy consumption TEC criterion, it needs to find the Pareto solutions in the set of all solutions to obtain a balance between the two criterions [9].

Let $\pi = [j_1, j_2, \ldots, j_N]$ denote a permutation of all job numbers to be processed, for example, $\pi = [2, 5, 1, 3, 4]$, it represents job 2, job 5, job 1, job 3, and job 4 processed in sequence on each machine. Π is a set of different π. $p_{j_i, k}$ the processing time of job j_i on the machine k, $i = [1, \ldots, N]$, $k = [1, \ldots, M]$. There is a finite and discrete set of d

different speeds $S = [v_1, \ldots, v_d]$ for each machine that affect the energy consumption. And the speed of machine cannot be changed during its execution of a job. $V_{j_i,k}$ the speed of machine k when it process the job j_i. $P_{j_i,k}/V_{j_i,k}$ denotes the real processing time when the job j_i processed with speed $V_{j_i,k}$ on machine k. PP_{k,v_l} is the energy consumption per unit time when machine k at speed v_l, $l = [1, \ldots, d]$. SP_k means the energy consumption per unit time when machine k is in standby state. $x_{k,v_l}(t)$ is equal to 1 if the machine k with speed v_l is working at t time, and 0 otherwise. $y_k(t)$ is equal to 1 if the machine k is in standby state, and 0 otherwise. $C(j_i, k)$ denotes the complete time of job j_i processed on machine k.

The make-span C_{max} can be computed as follows:

$$C(j_1, 1) = P_{j_1,1}/V_{j_1,1} \tag{1}$$

$$C(j_i, 1) = C(j_{i-1}, 1) + P_{j_i,1}/V_{j_1,1} \tag{2}$$

$$C(j_1, k) = C(j_1, k-1) + P_{j_1,k}/V_{j_1,k} \tag{3}$$

$$C(j_i, k) = \max\{C(j_{i-1}, k), C(j_i, k-1)\} + P_{j_i,k}/V_{j_i,k} \tag{4}$$

$$C_{max} = C(j_n, m) \tag{5}$$

The total energy consumption TEC can be computed as follows:

$$TEC = \int_0^{C_{max}} \left(\sum_{v=1}^{s} \sum_{k=1}^{m} PP_{k,v} \bullet x_{kv}(t) + \sum_{k=1}^{m} SP_k \bullet y_k(t) \right) dt \tag{6}$$

Table 1 shows an example, the entries of which are the processing times. The speed set $S = [1, 2, 3]$.

Table 1. Processing requirement $p_{j_i,k}$ on machines when N = 4, M = 2.

Job	M1	M2	Job	M1	M2
J_1	12	9	J_3	14	16
J_2	7	20	J_4	2	9

3 HFLCS for MOPFS

In this section, we will present HFLCS for MOPFS after explaining the Pareto front, levy flight, LACS-based global search, and local search.

3.1 Pareto Front

A solution vector, $u = (u_1, \ldots, u_n)^T \in F$, is said to dominate another vector $v = (v_1, \ldots, v_n)^T \in F$ if and only if $u_i \leq v_i$ for $\forall i \in \{1, \ldots, n\}$ and $\exists i \in \{1, \ldots, n\} : u_i < v_i$. We can define another dominance relationship $u \prec v$. A point $x \in F$ is called a non-dominated solution if no solution can be found that dominates. For example, suppose individual 1 and individual 2 are two individuals based on a problem scale of 5×3, and the individual 1 represents the following: $\pi_1 = [3, 1, 5, 2, 4]$, $f_{11} = 589$, $f_{12} = 34789$, the individual 2 represents the following: $\pi_2 = [3, 4, 1, 2, 5]$, $f_{21} = 582$, $f_{22} = 34770$. It can be seen that $f_{21} < f_{11}, f_{22} < f_{21}$. So we can conclude that individual 1 dominates the individual 2.

The Pareto front PF of a multi-objective can be defined as the set of non-dominated solutions so that $PF = \{s \in S | \exists s' \in S : s' \prec s\}$.

3.2 Levy Flights in Cuckoo Search

In the original cuckoo search for k different objectives, there are two idealized rules. And the first rule is: each cuckoo lays k eggs at a time, and dumps them in a randomly chosen nest. Egg k corresponds to the solution to the kth objective. In cuckoo search algorithm, levy flight is used to generate solutions randomly. Mathematically speaking, a new solution can be randomly generated either by a random walk or by levy flight, in other word, levy flights ensure the good diversity of the solutions.

When generating new solutions x_i^{t+1} for, say cuckoo i, a levy flight is performed:

$$x_i^{t+1} = x_i^t + \alpha \oplus \text{Lévy}(\lambda) \tag{7}$$

In the formula (7), the x_i^{t+1} is the $(t+1)th$ generation's solution and the x_i^t is the $(t)th$ generation's solution. Where $\alpha > 0$ is the step size which should be related to the scales of the problem of interest. In most cases, we can use $\alpha = O(1)$. The $\oplus$ means entry-wise multiplications. Levy flights provide a random walk while the random step obeys levy distribution. Then $\text{Lévy}(\lambda)$ can be written as follows:

$$\text{Lévy}(\lambda) \sim u = t^{-\lambda}, (1 < \lambda \leq 3) \tag{8}$$

Here the walks obey a power-law step-length distribution with a heavy tail. The nature of levy's flight can increase the diversity of the population, making it easier to jump out of the local optimum. In this paper, the following calculation is used to calculate the levy random number:

$$\text{Lévy}(\lambda) = \frac{\phi \times \mu}{|v|^{1/\beta}} \tag{9}$$

In the formula, $\mu \sim N(0,1)$, $v \sim N(0,1)$; β is a constant, and the value range is 1–2. And the calculation formula of ϕ is as follows:

$$\phi = \left\{ \frac{\Gamma(1+\beta) \times \sin(\frac{\pi \times \beta}{2})}{\Gamma(\frac{1+\beta}{2}) \times \beta \times 2^{\frac{\beta-1}{2}}} \right\}^{1/\beta} \tag{10}$$

3.3 LACS-Based Global Search

The procedure of LACS-based global search is given as follows:

Step1: Generate FLA's individuals with random values.
Step2: Calculate C_{max} and TEC of FLA's population, and then select Pareto solutions by non-dominant principle.
Step3: Update LACS's individuals using LACS-based scheme.
Step4: Calculate C_{max} and TEC of FLA's population and update Pareto front.
Step5: Remix the memeplexes into next population.
Step6: Apply levy flights to a random number of dominant to increasing the diversity.
Step7: Set t = t + 1. If $t \leq time\_consume$, then go to step3.
Step8: Output final Pareto solutions.

As we can find that, step 6 is a levy flight perturbation phase, which is used to escape local optima and guide the search to a more different region. In addition, a multi-neighborhoods local search is easy to incorporate into LACS to develop effective hybrid algorithm. Next, we will present a local search to perform exploitation.

3.4 Local Search

For the MOPFS, a multi-neighborhoods local search is presented to exploit the excellent sub-regions in the algorithm. This local search is composed of two parts: part A consists of *(i)* and *(ii)*, and apply part A to each Pareto solutions in each memeplexes; part B just includes *(iii)*, a finite applies of part B to x_b and x_w in each memeplexes.

(i) Interchange the job at the uth dimension and the job at the vth dimension of the job solution $\pi(\text{int } erchange(\pi, u, v))$ [10, 11]

(ii) Remove the job at the uth dimension and insert it in the vth dimension of the job solution $\pi(insert(\pi, u, v))$.

(iii) If a random number α less than a constant number δ, the first job J_1 of x_b is selected and then delete job J_1 of x_b from x_b and x_w; else the first job J_1 of x_w is chosen, at the same time delete the first job J_1 of x_w from x_b and x_w. The above step is repeated until the x_b and x_w are empty; as a result, a new permutation is obtained. Where δ is a probability, x_b is one of the Pareto solution in each memeplexes, x_w is one of the dominant solution in each memeplexes. $H - select(\pi_b, \pi_w)$ [12].

Denote $x_i^{(t)}$ is the ith individual at generation t, $\pi_i^{(t)}$ is the permutation of individual $x_i^{(t)}$. The procedure of the local search is given as follows:

Step1: Set loop=1

Do

Randomly select x_b and x_w in each memeplexes;

$\pi_{new} = H - select(\pi_b, \pi_w)$;

If $x_{new} \prec x_b$,

then $x_b = x_{new}$, Loop++, continue

Else

begin

$\pi_{new} = insert(\pi_b, u, v)$;

If $x_{new} \prec x_b$

then $x_b = x_{new}$, Loop++, continue

Else

begin

$\pi_{new} = int\,erchange(\pi_b, u, v)$;

If $x_{new} \prec x_b$, Loop++, continue

then $x_b = x_{new}$;

End;

End;

End;

While loop<11

As can be seen from the above procedure, there is an exploitation from the region obtained by global search. The multi-neighborhoods local search can improve the search ability of FLCS to some extent.

3.5   HFLCS

Based on the above solution representation, the Pareto front, the levy flight, the LACS-based global search, and local search, and you end up with the procedure framework of HFLCS in Fig. 1. It can be seen from Fig. 1, that does HFLCS apply FLCS includes levy flight to execute exploration for all individuals to find promising regions. And then uses a multi-neighborhoods local search to perform exploitation for the non-dominant solutions to improve the solutions' quality. The methods above are expected to achieve good results for MOPFS. In the next section, we will investigate the performances of the proposed HFLCS.

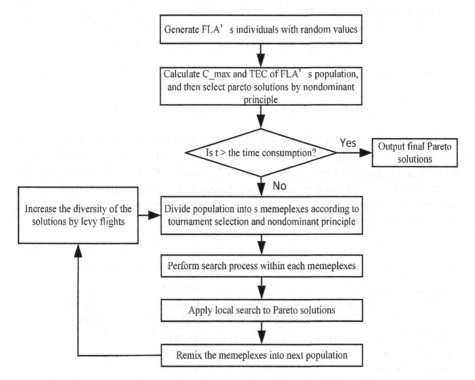

Fig. 1. The procedure framework of HFLCS

4   Test and Comparisons

In order to prove that HFLCS is more effective than the original standard algorithm and to prove that the algorithm is superior in solving the multi-objective problem, we compared HFLCS with CS, FLA, and NSGA-II at the same time below. We used 40, 50, 60, 70, 80, 90, 100 for N, and 10, 20, 30 for M. $V_{j_i,k}$ and $p_{j_i,k}$ in every instance are randomly generated Numbers. And the $V_{j_i,k}$ has three different values, the speed set $S = [1, 2, 3]$. The other parameters are set as follows: the population size of the four algorithms popsize $= 20$, each memeplex of FLA $s = 10$, the control factor of CS step $\alpha = 0.01$, the mutation probability of NSGA-II $P_m = 0.3$, the crossover probability $P_c = 0.9$, each memeplex of HFLCS $s = 10$, the step control factor of HFLCS $\alpha = 0.01$. Four algorithms were used to test the MOPFS problem of 10 problems, each of which was independently tested 20 times, and the test time was 50 * N, where N was the number of works of each problem size.

Denote S_j the non-dominated solutions set, S the union of the K non-dominated solution sets (i.e. $S = S_1 \cup \cdots \cup S_r \cup \cdots S_K$), $R\_NDS(S_j)$ the S_j with respect to the K non-dominated solution sets is the ratio of solutions in S_j that are not dominated by any other solutions in S. $NDS\_NUM(S_j)$ the number of solutions S_j are not dominated by y. $R\_NDS(S_j)$ is written as [13]:

$$R\_NDS(S_r) = |S_r - \{x \in S_r | \exists y \in S : y \prec x\}| / |S_r| \tag{11}$$

$NDS\_NUM(S_j)$ is written as:

$$NDS\_NUM(S_j) = |S_j - \{x \in S_j | \exists y \in S : y \prec x\}| \tag{12}$$

The statistical results of R_NDS and NDS_NUM produced by the compared algorithms are shown in Tables 2, 3 and 4, respectively. In this paper, 10 problems are given. Can be seen from the results of Tables 2, 3 and 4, HFLCS gain value is superior to other three kinds of algorithms, it shows that under the same computer time, HFLCS algorithm are superior, and can quickly converge to the approximate optimal solution, shows that this algorithm has good convergence.

Table 2. Comparison of R_NDS and NDS_NUM of HFLCS and NSGA-II

Instance	HFLCS		NSGA-II	
N, M	R_NDS_HFLCS	NDS_NUM_HFLCS	R_NDS_NSGA-II	NDS_NUM_NSGA-II
40,10	**0.925**	1	0.6304167	2.8
50,10	**0.9**	0.9	0.476369	2.45
60,10	**0.95**	1.05	0.275456	1.25
70,10	**1**	1.05	0.495357	2.3
80,10	**0.85**	0.85	0.2875	1
80,20	1	1	0.065	0.3
90,20	1	1	0.0333	0.1
90,30	1	1	0	0
100,20	1	1.05	0	0
100,30	1	1.05	0	0

All the algorithms and test programs are implemented with Delphi10.2, the operating system is Win10, the processor is Intel(R) Core(TM) i5-4210U 1.70 GHZ, and the memory is 4 GB.

In conclusion, the above comparisons show that HFLCS is an effective approach with excellent quality and robustness for MOPFS and HFLCS has good convergence.

Table 3. Comparison of *R_NDS* and *NDS_NUM* of HFLCS and FLA

Instance	HFLCS		FLA	
N, M	R_NDS_HFLCS	NDS_NUM_ HFLCS	R_NDS_ FLA	NDS_NUM_ FLA
40,10	**1**	1	0	0
50,10	**1**	1	0	0
60,10	**1**	1	0	0
70,10	**1**	1	0	0
80,10	**1**	1	0	0
80,20	**1**	1	0	0
90,20	**1**	1	0	0
90,30	**1**	1	0	0
100,20	**1**	1	0	0
100,30	**1**	1	0	0

Table 4. Comparison of *R_NDS* and *NDS_NUM* of HFLCS and CS

Instance	HFLCS		CS	
N, M	R_NDS_HFLCS	NDS_NUM_ HFLCS	R_NDS_ CS	NDS_NUM_ CS
40,10	**1**	1	0	0
50,10	**1**	1	0	0
60,10	**1**	1	0	0
70,10	**1**	1	0	0
80,10	**1**	1	0	0
80,20	**1**	1	0	0
90,20	**1**	1	0	0
90,30	**1**	1	0	0
100,20	**1**	1	0	0
100,30	**1**	1	0	0

5 Conclusions and Future Research

This paper presented a hybrid shuffle frog-leaping algorithm based on cuckoo search (HFLCS) for the flow shop scheduling with the consideration of energy consumption (MOPFS). It is the first report on the application of combine FLA and CS. The developed algorithm considers both of global search and local search. The simulation results and comparisons with some classical algorithms demonstrated the effectiveness and robustness of our proposed algorithms. For the future, we apply ourselves to study CS and FLA for other kinds of scheduling problems, such as no-wait flow-shop.

Acknowledgements. This research is partially supported by the National Science Foundation of China (51665025), and the Applied Basic Research Foundation of Yunnan Province (2015FB136).

References

1. Pasupathy, T., Rajendran, C., Suresh, R.K.: A multi-objective genetic algorithm for scheduling in flow shop to minimize the makespan and total flow time of jobs. Int. J. Adv. Manuf. Technol. **27**, 804–815 (2006)
2. T'kindt, V., Billiaut, J.-C.: Multicriteria scheduling problems: a survey. RAIRO Oper. Res. **35**, 143–163 (2001)
3. Wang, L.J., Yin, Y.L., Zhong, Y.W.: Cuckoo search with varied scaling factor. Front. Comput. Sci. China **9**(4), 623–635 (2015)
4. Aziz, O., Belaïd, A., Yang, X.: Discrete Cuckoo search applied to job shop scheduling problem. Recent Adv. Swarm Intell. Evol. Comput. **585**, 121–137 (2015)
5. Xu, Y., Wang, L., Wang, S.Y., Liu, M.: An effective shuffled frog-leaping algorithm for solving the hybrid flow-shop scheduling problem with identical parallel machines. Eng. Optim. **45**, 1409–1430 (2013)
6. Yin, L., Li, X., Gao, L., Lu, C., Zhang, Z.: A novel mathematical model and multi-objective method for the low-carbon flexible job shop scheduling problem, considering productivity, energy efficiency and noise reduction. Sustain. Comput. Inf. Syst. **13**(3), 15–30 (2017)
7. Lei, D., Zheng, Y., Guo, X.: A shuffled frog-leaping algorithm for flexible job shop scheduling with the consideration of energy consumption. Int. J. Prod. Res. **55**, 3126–3140 (2016)
8. Yang, X.S.: Multiobjective cuckoo search for design optimization. Comput. Oper. Res. **40**(6), 1616–1624 (2013)
9. Yin, L., Li, X., Lu, C., Gao, L.: Energy-efficient scheduling problem using an effective hybrid multi-objective evolutionary algorithm. Sustainability **8**(12), 1268–1301 (2016)
10. Wang, L., Qian, B.: Hybrid Differential Evolution and Scheduling Algorithm. Tsinghua University Press, Beijing (2012)
11. Wang, S.Y., Wang, L., Liu, M., Xu, Y.: An order-based estimation of distribution algorithm for stochastic hybrid flow-shop scheduling problem. Int. J. Comput. Integr. Manuf. **28**(3), 307–320 (2015)
12. Lei, D.M.: Minimizing makespan for scheduling stochastic job shop with random breakdown. Appl. Math. Comput. **218**(24), 11851–11858 (2012)
13. Qian, B., Li, Z.H., Hu, R., Zhang, C.S.: A hybrid differential evolution algorithm for the multi objective reentrant job-shop scheduling problem. In: IEEE International Conference on Control and Automation (ICCA) Hangzhou, China, 12–14 June 2013, pp. 485–489 (2013)

Multi-objective Flexible Job Shop Scheduling Problem with Energy Consumption Constraint Using Imperialist Competitive Algorithm

Chengzhi Guo and Deming Lei[✉]

School of Automation, Wuhan University of Technology, Wuhan 430070, China
deminglei11@163.com

Abstract. In this paper, multi-objective flexible job shop scheduling problem (MOFJSP) with energy consumption constraint is investigated and a novel imperialist competitive algorithm (ICA) is used to optimize makespan and total tardiness under a constraint that total energy consumption doesn't exceed a given threshold. The flow of ICA consists of two parts. In the first part, a MOFJSP is obtained by adding total energy consumption as objective and optimized, all generated feasible solutions are stored and updated to build a population of the second part; in the second part, the original MOFJSP is solved by starting with the population. New strategies are applied to build initial empires twice to adapt to the two-part structure and imperialist's reinforced search is added. The computational results show that the new approach to constraint is effective and ICA is a very competitive algorithm.

Keywords: Flexible job shop scheduling · Energy consumption constraint
Imperialist competitive algorithm

1 Introduction

Flexible job shop scheduling problem (FJSP) has been extensively existed in many industries such as automobile assembly, textile, chemical material process and semi-conductor manufacturing. There often exits several conflicting objectives in real life flexible job shop and it is necessary to consider multi-objective flexible job shop scheduling problem (MOFJSP). Many works have been done on MOFJSP by using various meta-heuristics, such as genetic algorithm (GA [1–8]), PSO [9, 10], harmony search algorithm [11], artificial bee colony [12], tabu search [13], variable neighborhood search (VNS [14]), shuffled frog leaping algorithm (SFLA [15]) and estimation of distribution algorithm [16].

In recent years, energy-efficient scheduling problem has been considered extensively [17–23]. Regarding energy-efficient FJSP, it is often MOFJSP because of the inclusion of energy related objective and a number of low carbon strategies have been obtained [18–23].

© Springer International Publishing AG, part of Springer Nature 2018
D.-S. Huang et al. (Eds.): ICIC 2018, LNCS 10954, pp. 659–669, 2018.
https://doi.org/10.1007/978-3-319-95930-6_66

In this study, A novel ICA is proposed to solve MOFJSP with energy consumption constraint to minimize total tardiness and makespan under the constraint that total energy consumption doesn't exceed a given threshold. The energy constraint is not met frequently in the decoding procedure of meta-heuristics. To solve the problem, ICA is divided into two parts. In the first part, a MOFJSP is obtained by adding total energy consumption as objective and optimized and all obtained feasible solutions are utilized to build a population of the second part; in the second part, the original MOFJSP is solved by starting with the population.

2   Problem Description

MOFJSP with energy consumption constraint is described as follows. There are a set of jobs $J=\{J_1, J_2, \cdots, J_n\}$ and a set of machines $M=\{M_1, M_2, \cdots, M_m\}$. Job J_i consists of h_i operatations. Operation o_{ij} is the jth operation of job J_i and can be processed on a set S_{ij} of compatible machines, $S_{ij} \subset M$. Each machine M_k exists two modes: processing mode and stand-by mode. E_k is the energy consumption per unit time in processing mode and SE_k is the energy consumption per unit idle time.

There are several constraints on jobs and machines, such as, each machine can process at most one operation at a time, no jobs may be processed on more than one machine at a time, operations cannot be interrupted and setup times and remove times are included in the processing times etc.

A new constraint is added below.

$$TEC = \int_0^{C_{max}} \left(\sum_{i=1}^n \sum_{j=1}^{h_i} \sum_{k=1}^m E_k y_{ijk}(t) + \sum_{k=1}^m SE_k z_k(t) \right) dt \leq Q_{EC} \qquad (1)$$

Where TEC denotes total energy consumption. $y_{ijk}(t)$ is a binary variable. If machine $M_k \in S_{ij}$ is in processing mode at time t, then $y_{ijk}(t)$ is equal to 1; otherwise $y_{ijk}(t)$ is 0. $z_k(t)$ is 1 if M_k is in stand-by mode at time t and 0 otherwise. C_{max} indicates the maximum completion time of all jobs. Q_{EC} is a given energy consumption threshold.

The goal of the problem is to minimize simultaneously the following two objectives under the condition that constraints are all met.

$$f_1 = C_{max} \qquad (2)$$

$$f_2 = \sum_{i=1}^n \max\{C_i - D_i, 0\} \qquad (3)$$

3 ICA for MOFJSP with Energy Consumption Constraint

ICA is composed of two parts.

3.1 The First Part

In general, ICA starts with an initial population P with N solutions. Each solution is represented with a scheduling string $[(\theta_1, r_1), (\theta_2, r_2), \cdots, (\theta_i, r_i), \cdots, (\theta_h, r_h)]$ and a machine assignment string $[q_{11}, q_{12}, \cdots, q_{1h_1}, \cdots, q_{nh_n}]$, where $h = \sum_{i=1}^{n} h_i$ is the total number of operations. In the first string, a doublet (θ_i, r_i) indicates an operation $o_{\theta_i r_i}$, as a result, the whole string corresponds to an ordered operation list $[o_{\theta_1 r_1}, o_{\theta_2 r_2}, \cdots,$ $o_{\theta_i r_i}, \cdots, o_{\theta_h r_h}]$, $\theta_i \in \{1, 2, \cdots, n\}, 1 \le r_i \le h_{\theta_i}$. In the second string, a gene $q_{ij} \in S_{ij}$ represents a compatible machine for operation o_{ij}.

The above coding method has also been used in the previous works [20, 24] and can guarantee all constraints of MOFJSP except $TEC \le Q_{EC}$ are met, so only $TEC \le Q_{EC}$ is required to be considered specially.

To effectively deal with the constraint $TEC \le Q_{EC}$, we delete the constraint from the considered MOFJSP and add TEC as f_3, as a result, the problem is transformed into FJSP with f_1, f_2, f_3.

In this study, a simple strategy is adopted to construct initial empires according to the above feature and the non-dominated sorting [25]. The detailed steps are shown in Algorithm 1, where F_k indicates the total number of countries in empire k and $\lfloor x \rfloor$ indicates the biggest integer being smaller than x.

If $\eta < N_{im}$, there are two kinds of empires. η non-dominated solutions are selected as the imperialist of initial empires $k = 1, 2, \cdots, \eta$, the second kind of empires are empires $k = \eta + 1, \cdots, N_{im}$, which are the dominated solutions with least rank value. If $\eta \ge N_{im}$, N_{im} randomly chosen non-dominated solutions are set to be imperialists.

For each country in an empire, if it is a colony, its assimilation is done, if it is an imperialist, the reinforced search is executed.

Algorithm 1. InitialEmpires

> Decide the number η of non-dominating solutions
>
> Determine the imperialist for each empire
>
> Let $A = \lfloor N/N_{im} \rfloor$
>
> **For** $k = 1$ to N_{im}
>
> if $k < N_{im}$, $F_k = A + u$; else $F_{N_{im}} = N - \sum_{k=1}^{N_{im}-1} F_k$
>
> Assign $F_k - 1$ randomly chosen countries into empire k
>
> **End For**

For each colony λ in the empire of imperialist v, a random number *rand* is generated, if $rand < \xi$, then the first crossover on the scheduling string is executed between colony λ and imperialist v, otherwise, the second one is done on the machine assignment string of countries λ and v. A new solution z is obtained and compared with the colony λ, if the solution z dominates or is non-dominated with the colony λ, then the colony λ is replaced with the solution z and the set Ω is updated with z. Where *rand* follows uniform distribution on [0, 1] and the set Ω is used to store the non-dominated solutions.

For the imperialist v, if a set $\{y \in \Omega | y \neq v\}$ is not empty, a member is randomly chosen from the set and crossover is executed between the member and v in the same way as done between λ and v in the assimilation.

The detailed descriptions on two crossovers are shown in [20]. We set ξ to be 0.6 based on a number of experiments.

Algorithm 2. Revolution

For $k = 1$ to N_{im}

   $\alpha \leftarrow 1$

   For $\tau = 1$ to F_k

      If *rand* $< U_R$, **then** $\alpha \leftarrow \alpha + 1$

      All colonies in empire k are compared according to Pareto dominance

      Assign each colony λ with w_λ

      If $\alpha > 1$, **then**

      For $l = 1$ to α

         Decide a colony λ with the least w_λ. Let $g \leftarrow 1$

         For $\tau = 1$ to R

            if $g = 1$, *insert* is executed on colony λ, **else** *change* is done on colony λ

            A new solution z is obtained

            if z is non-dominated with or dominates λ,

               then colony λ is replaced and Ω is updated;

               else $g \leftarrow g + 1$ and $g \leftarrow 1$ if $g = 3$

         End For

      End For

      End If

   End For

The set Ω is updated as follows. The new solution is added into Ω and all solutions in Ω are compared in terms of Pareto dominance; the non-dominated members are remained and the dominated ones are removed from the set Ω.

The detailed procedure of revolution is listed in Algorithm 2, where R is an integer and U_R is the revolution probability. Because only good colonies are the revolution object in this study, so we use a smaller U_R of 0.1.

In the above procedure, comparisons among colonies are first done. For a colony λ in an empire k, if there are w_λ colonies in the same empire dominating the colony λ, the colony λ is assigned a value w_λ. Only some colonies with the smallest w_λ have chance for revolution because the revolution of good colonies can generate probably some better colonies or a new imperialist.

In this study, two neighborhood structures *insert* and *change* are used. *insert* produces new solution by inserting an element (θ_i, r_i) into position $k \neq i$ and assigning new values of all r_i again. *change* is described below: a set $\Theta = \{q_{ij} \| S_{ij}| > 1\}$ is first constructed, some elements are randomly selected from the set Θ, suppose that q_{ij} is chosen, then q_{ij} is replaced with a selected machine from S_{ij} in a random way.

Exchange step is described as follows. For each colony of an empire, if it is non-dominated with or dominates its imperialist, then the colony substitutes for its imperialist and becomes the new imperialist.

For a set H_l of solutions with the same rank l, all solutions except boundary solutions are assigned a distance by using the method of Deb et al. [25]; then the maximum value ψ of these distances are obtained; finally, a boundary solution is given a distance, which is $\psi + \alpha$, where $\alpha \geq 1$ is an integer. The usage of α is to distinguish boundary solutions from other solutions. Let $dist_i$ be the crowding distance of a solution $i \in H_l$. we define cost as $1 + dist_i \big/ \sum_{f \in H_l} dist_f$ for each solution $i \in H_l$. Obviously, so the cost is in interval $[1, N+1]$.

After the cost is defined for all solutions, imperialist competition is done in the same way the basic ICA [24]. ζ is set to be 0.1 to calculate TC_k. In this study, we directly use this setting because good results are gained in the experiments.

In the first part, a set Ψ is used to store feasible solutions generated by assimilation, reinforced search and revolution. The set Ψ is updated in the following way. If $|\Psi| \leq N$, the newly obtained solution is directly added into the set Ψ; otherwise, for a new feasible solution z, if a set $\Xi = \{y \in \Psi | z \succ y\}$ is not empty, then we choose a worst member $y \in \Xi$ and replace y with the new solution z. The worse member is decided by the dominance relation among members in Ξ, for example, if there are two members y_1, y_2, y_1, y_2 are dominated by three and four members of Ξ, respectively; obviously, y_2 is the worst member.

Algorithm 3. FirstPart

Randomly produce an initial population and construct an initial set Ω

Construct initial empires,

While termination condition of the first part is not met **do**

  Assimilation, reinforced search and Ξ update

  Revolution and Ξ update

  Exchange colony and imperialist if possible

  Imperialist competition

End While

The detailed steps of ICA are shown in Algorithm 3. The main goal of first part is to generate and store enough feasible solutions. Two kinds of dominance are used. In all steps except the updating of Ξ, Pareto dominance based on f_1, f_2, f_3 is applied. When Ξ is updated, dominance based on f_1, f_2 is utilized because all members of Ξ are feasible.

Let max_it_1 be the number of objective vector evaluations. When a new solution is generated, its objective vector is evaluated once. The first part is finished after max_it_1 solutions are produced.

3.2   The Second Part

TEC is limited and not optimized. Obviously, Q_{EC} must be greater than the optimal value of *TEC* and feasible solutions can be produced after the first part; moreover, the updating of $\varXi$ is useful to obtain more feasible solutions, thus, after the first part is finished, we should test if there are enough feasible solutions in the set $\varXi$.

If sufficient feasible solutions are stored in the set $\varXi$, then the second part begins. In this part, the original MOFJSP is considered. When all solutions in population are feasible, it is easy to form initial solution and execute imperialist competition, so we first construct an initial population $\bar{P}$ with members in the set $\varXi$ and define $\bar{P} = \varXi$ and $\bar{N} = |\varXi|$. $\bar{N}$ is the size of $\bar{P}$. Obviously, $\bar{N} \leq N$.

To form the initial empires of the second part, we first decide $\bar{N}_{im}$ by

$$\bar{N}_{im} = \begin{cases} 1 & 1 < \bar{N} \leq 3 \\ 2 & 3 < \bar{N} < 10 \\ \max\{N_{im}, 3 + \lfloor \bar{N}/10 \rfloor\} & \bar{N} \geq 10 \end{cases} \tag{4}$$

Where $\bar{N}_{im}$ indicates the number of imperialists. In general, $N_{im} \geq 2$, so $\bar{N}_{im} \leq N_{im}$.

After $\bar{P}$ is built and $\bar{N}_{im}$ is decided, we form initial empires using the same way shown in the first part. In the previous works [26–29], initial empires are seldom formed twice. We also implement assimilation, exchange and imperialist as done in the first part; however, there are no set $\varXi$ and its updating, different ways are used to execute revolution, compare solutions and renew the set Ω.

In the revolution procedure, computation on g is removed from algorithm 3, line 12 is revised below: if $l \leq \beta$, *insert* is executed on colony λ, else *change* is done on colony λ, where $\beta < R$ is an integer.

A novel principle is developed to decide if x can be updated with the new one z, which is shown below. For x and z, if $TEC(z) \leq Q_{EC}$ and z dominates x a or non-dominated with x according to objectives f_1, f_2, then x is replaced with z.

Each solution of $\bar{P}$ is feasible, so a new one z must be feasible if it can substitute for x and will be directly thrown away if it is infeasible. On the contrary, if not all solutions in $\bar{P}$ are feasible, three cases exist for x and z: (1) both x and z are feasible, (2) both x and z are infeasible; (3) one of them is feasible. Obviously, when $\bar{P}$ just consists of feasible solutions, solution comparison and constraint handling become easier.

When a solution $x \in \bar{P}$ is replaced with a new one z, the set Ω is renewed as follows. If all members of Ω are feasible, then add x into the set Ω and compare all solutions in the set Ω according to Pareto dominance on f_1, f_2 and remove the dominated ones from the Ω. If not all members of Ω are feasible, then replace an infeasible member with x.

The first part is about FJSP with f_1, f_2, f_3, so after the first part is ended, it is likely that some infeasible solutions exist in Ω. If not all solutions in $\bar{P}$ are feasible, there are two cases: x is feasible or infeasible; obviously, the updating procedure is simplified because each member of $\bar{P}$ is feasible. *max_it* denotes the maximum number of objective vector evaluations. When the total number of the obtained solutions is equal to *max_it*, the second part and the search of ICA are terminated.

4   Computational Experiments

Extensive experiments are conducted on a set of problems to test the performance of ICA for the considered FJSP. All experiments are implemented by using Microsoft Visual C++ 6.0 and run on 4.0G RAM 1.70 GHz CPU PC.

We choose MK1-15 [30], which are extended by adding energy consumption information $E_k \in [2, 4]$ and $SE_k = 1$. Duedate is calculated in the following:

$$D_i = \delta \sum_{j=1}^{h_i} \max_{k=1,2,\cdots,m} \left\{ p_{ijk} \right\} \tag{5}$$

Where δ is a random real number in an interval and shown in Table 1.

In this study, *TEC* is limited to Q_{EC} to meet special requirements such as on-time delivery. We decide Q_{EC} based on the results of *TEC* in the first part of ICA. The results on Q_{EC} are presented in Table 2.

Table 1. Setting on δ

δ	Instances	δ	Instances
[0.5, 0.7]	MK01,03-05,	[0.3, 0.5]	MK02,06,08-10
[1.2, 1.6]	MK11-15	[0.7, 0.9]	MK07

Metric DI_R [31] is often used to measure the convergence performance by computing the distance of the non-dominated set Ω_l relative to a reference set Ω^*.

$$DI_R(\Omega_l) = \frac{1}{|\Omega^*|} \sum_{y \in \Omega^*} min\left\{ \sigma_{xy} | x \in \Omega_l \right\} \tag{6}$$

The reference set Ω^* is composed of the non-dominated solutions in $\bigcup_l \Omega_l$.

Metric C (Zitzler and Thiele [32]) is used to compare the approximate Pareto optimal set respectively obtained by three algorithms.

Table 2. Computational results of four algorithms on metric DI_R

Ins.	Q_{EC}	ICA	NSGA-II	VNS	Ins.	Q_{EC}	ICA	NSGA-II	VNS
MK01	682.4	0.0000	65.908	8.3619	MK09	9082	11.52	73.482	46.02
MK02	550.7	11.784	41.213	7.9463	MK10	9728	2.261	39.311	20.42
MK03	3597	0.8720	69.355	51.090	MK11	12890	11.37	45.872	17.76
MK04	1191	5.9387	31.738	19.798	MK12	14933	2.338	47.186	44.68
MK05	2748	5.6780	62.398	4.9910	MK13	14202	5.056	59.165	59.16
MK06	2349	1.568	36.129	8.1942	MK14	16345	9.471	75.800	44.05
MK07	1728	0.1856	43.663	25.777	MK15	15846	3.167	77.379	49.11
MK08	10156	2.1517	31.662	31.662					

Table 3. Comparisons among ICA, NSGA-II and VNS on metric C

Ins	$C(I,N2)$	$C(N2,I)$	$C(I,V)$	$C(V,I)$	$C(V,N2)$	$C(N2,V)$
MK01	1.0000	0.0000	1.0000	0.0000	1.0000	0.0000
MK02	1.0000	0.0000	0.0000	0.3333	1.0000	0.0000
MK03	1.0000	0.0000	1.0000	0.0000	1.0000	0.0000
MK04	1.0000	0.0000	1.0000	0.0000	0.8333	0.1429
MK05	1.0000	0.0000	0.5714	0.1111	1.0000	0.0000
MK06	0.8334	0.0000	1.0000	0.0000	0.9000	0.0000
MK07	1.0000	0.0000	1.0000	0.0000	1.0000	0.0000
MK08	1.0000	0.0000	1.0000	0.0000	1.0000	0.0000
MK09	1.0000	0.0000	1.0000	0.0000	1.0000	0.0000
MK10	1.0000	0.0000	1.0000	0.0000	1.0000	0.0000
MK11	1.0000	0.0000	0.2000	0.0000	1.0000	0.0000
MK12	1.0000	0.0000	1.0000	0.0000	0.4000	0.0000
MK13	1.0000	0.0000	1.0000	0.0000	1.0000	0.0000
MK14	1.0000	0.0000	1.0000	0.0000	1.0000	0.0000
MK15	1.0000	0.0000	1.0000	0.0000	1.0000	0.0000

In this study, NSGA-II [7] and VNS [14] are chosen. We extended NSGA-II to our FJSP in the following way: the parts on machine breakdown are first deleted, rank value and crowding distance of all feasible solutions are obtained using the original non-sorting method and crowding distance assignment, all infeasible solutions are assigned into the same front and their rank is bigger than the rank of all feasible solutions, crowding distance of an infeasible solution x is set to be $\omega_{max} - TEC(x)$, where ω_{max} is a big enough positive number. The other parts of NSGA-II [23] are directly kept.

Bagheri and Zandieh [14] introduced a VNS. This VNS can be applied to solve our MOFJSP by using a usual constraint handling method: for the current solution x and a new solution z, if both x and z are infeasible, they are compared according to TEC of them; if one of them is feasible, the feasible one is better than the infeasible one; if they are feasible, they are compared based on Pareto dominance on f_1, f_2. After the method is adopted, VNS can be directly used to solve the considered MOFJSP.

The parameter settings of ICA are listed below: $N = 80$, $N_{im} = 6$, $R = 8$. $\beta = 5$, $max\_it_1 = 30000$, max_it is set to be 10^5. The parameters $\bar{N}$, $\bar{N}_{im}$ are decided by the size of Ξ and the Eq. (4).

For NSGA-II, population scale of 100, crossover probability of 0.8, maximum generation of $max\_it/100$ and mutation probability of 0.1.

For VNS [14], $n_{max} = 350$, max_it is 10^5 for all instances to compare VNS with other algorithms fairly. $n_{max} = 350$ is decided by Bagheri and Zandieh [19] and we adopt it directly.

Table 4. Computational times of ICA, NSGA-II and VNS

Ins	Running time (s)			Ins	Running time (s)		
	ICA	NSGA-II	VNS		ICA	NSGA-II	VNS
MK01	3.358	4.068	3.136	MK09	14.82	14.84	14.87
MK02	3.325	4.037	3.358	MK10	14.74	13.16	14.82
MK03	8.149	7.595	8.498	MK11	14.94	12.27	12.39
MK04	4.794	5.141	4.895	MK12	15.54	13.44	14.93
MK05	7.175	5.335	6.600	MK13	15.68	13.62	15.08
MK06	7.583	6.037	7.419	MK14	15.08	16.77	18.95
MK07	6.317	6.072	5.325	MK15	15.71	16.60	18.65
MK08	15.70	13.98	15.31				

Tables 2, 3 and 4 show the computational results and times of ICA, NSGA-II and VNS, where "N2" indicates NSGA-II and "V" denotes VNS. We can observe that ICA gets better results than other two algorithms on most of instances. The solutions of NSGA-II and VNS are always far away from those of ICA on nearly all instances; moreover, all non-dominated solutions of NSGA-II are dominated by those of ICA on nearly all instances. The non-dominated solutions of ICA dominates all solutions of VNS on 12 instances, thus, it can be concluded that ICA is a very competitive method for MOFJSP with energy consumption constraint.

5 Conclusions

In this study, a new ICA with two-part structure is proposed to minimize total tardiness and makespan under the constraint that total energy consumption doesn't exceed a given threshold. Total energy consumption constraint is added. In the first part, the problem is converted into FJSP with makespan, total tardiness and total energy consumption, initial empires are formed in a new way, imperialist's reinforced search is added and data on feasible solutions are stored and updated. In the second part, the original problem is considered, ICA starts with a set of feasible solutions obtained in the first part and some new strategies are adopted to compare solutions and update the non-dominated set Ω. We conduct a number of experiments and the computational results show that ICA can provide better results than the methods from the literature.

Acknowledgement. This work is supported by the National Natural Science of Foundation of China (61573264)

References

1. Kacem, I., Hammadi, S., Borne, P.: Pareto-optimality approach for flexible job-shop scheduling problems: hybridization of evolutionary algorithms and fuzzy logic. Math. Comput. Simul. **60**(3–5), 245–276 (2002)
2. Gao, J., Gen, M., Sun, L., Zhao, X.: A hybrid of genetic algorithm and bottleneck shifting for multi-objective flexible job shop scheduling problems. Comput. Ind. Eng. **53**(1), 149–162 (2007)
3. Yuan, Y., Xu, H.: Multiobjective flexible job shop scheduling using memetic algorithms. IEEE Trans. Autom. Sci. Eng. **12**(1), 336–353 (2015)
4. Rohaninejad, M., Kheirkhah, A., Fattahi, P., Vahedi-Nouri, B.: A hybrid multi-objective genetic algorithm based on the ELECTRE method for a capacitated flexible job shop scheduling problem. Int. J. Adv. Manuf. Technol. **77**(1), 51–66 (2015)
5. Rohaninejad, M., Sahraeian, R., Nouri, B.V.: Multi-objective optimization of integrated lot-sizing and scheduling problem in flexible job shop. PAIRO Oper. Res. **50**(3), 587–609 (2015)
6. Li, J., Huang, Y., Niu, X.: A branch population genetic algorithm for dual-resource constrained job shop scheduling problem. Comput. Ind. Eng. **102**(1), 113–131 (2016)
7. Ahmadi, E., Zandieh, M., Farrokh, M., Emami, S.M.: A multi objective optimization approach for flexible job shop scheduling problem under random machine breakdown by evolutionary algorithm. Comput. Oper. Res. **73**(1), 56–66 (2016)
8. Shen, X.N., Han, Y., Fu, J.Z.: Robustness measures and robust scheduling for multi-objective stochastic flexible job shop scheduling problems. Soft Comput. (2018, in press)
9. Moslehi, G., Mahnam, M.: A Pareto approach to multi-objective flexible job-shop scheduling problem using particle swarm optimization and local search. Int. J. Prod. Econ. **129**(1), 14–22 (2011)
10. Singh, M.R., Singh, M., Mahapatra, S.S., Jagadev, N.: Particle swarm optimization algorithm embedded with maximum deviation theory for solving multi-objective flexible job shop scheduling problem. Int. J. Adv. Manuf. Technol. **85**(9), 2353–2366 (2016)
11. Gao, K.Z., Suganthan, P.N., Pan, Q.K., Chua, T.J., Cai, T.X., Chong, C.S.: Pareto-based grouping discrete harmony search algorithm for multi-objective flexible job shop scheduling. Inf. Sci. **289**(1), 76–90 (2014)
12. Li, J.Q., Pan, Q.K., Tasgetiren, M.F.: A discrete artificial bee colony algorithm for the multi-objective flexible job-shop scheduling problem with maintenance. Appl. Math. Model. **38**(3), 1111–1132 (2014)
13. Jia, S., Hu, Z.H.: Path-relinking tabu search for the multi-objective flexible job shop scheduling problem. Comput. Oper. Res. **47**(1), 11–26 (2014)
14. Bagheri, A., Zandieh, M.: Bi-criteria flexible job-shop scheduling with sequence-dependent setup times-variable neighborhood search approach. J. Manuf. Syst. **30**(1), 8–15 (2011)
15. Li, J.Q., Pan, Q.K., Xie, S.X.: An effective shuffled frog-leaping algorithm for multi-objective flexible job shop scheduling problems. Appl. Math. Comput. **218**(18), 9353–9371 (2012)

16. Wang, L., Wang, S.Y., Liu, M.: A Pareto-based estimation of distribution algorithm for the multi-objective flexible job-shop scheduling problem. Int. J. Prod. Res. **51**(12), 3574–3592 (2013)

17. Li, J.Q., Sang, H.Y., Han, Y.Y., Wang, C.G., Gao, K.Z.: Efficient multi-objective optimization algorithm for hybrid flow shop scheduling problems with setup energy consumptions. J. Cleaner Prod. **181**, 584–598 (2018)

18. He, Y., Li, Y.F., Wu, T., Sutherland, J.W.: An energy-responsive optimization method for machine tool selection and operation sequence in flexible machining job shops. J. Cleaner Prod. **87**(1), 245–254 (2015)

19. Yin, L.J., Li, X.Y., Gao, L., Lu, C., Zhang, Z.: A novel mathematical model and multi-objective method for the low-carbon flexible job shop scheduling problem. Sustain. Comput. Inf. Syst. **13**, 15–30 (2017)

20. Lei, D.M., Zheng, Y.L., Guo, X.P.: A shuffled frog leaping algorithm for flexible job shop scheduling with the consideration of energy consumption. Int. J. Prod. Res. **55**(11), 3126–3140 (2017)

21. Mokhtari, H., Hasani, A.: An energy-efficient multi-objective optimization for flexible job shop scheduling. Comput. Ind. Eng. **104**, 339–352 (2017)

22. Lei, D.M., Li, M., Wang, L.: A two-phase meta-heuristic for multi-objective flexible job shop scheduling problem with total energy consumption threshold. IEEE Trans. Cybern. (2018, in press)

23. Lei, D.M, Yang, D.J.: Research on flexible job shop scheduling problem with total energy consumption constraint. ACTA Autom. Sinica (2018, in press). (in Chinese)

24. Lei, D.M.: Simplified multi-objective genetic algorithm for stochastic job shop scheduling. Appl. Soft Comput. **11**(8), 4991–4996 (2011)

25. Deb, K., Pratap, A., Agarwal, S., Meyarivan, T.: A fast and elitist multiobjective genetic algorithm: NSGA-II. IEEE Trans. Evol. Comput. **6**(1), 182–197 (2002)

26. Atashpaz-Gagari, E., Lucas, C.: Imperialist competitive algorithm: an algorithm for optimization inspired by imperialist competition. In: IEEE Congress on Evolutionary Computation, Singapore, pp. 4661–4667 (2007)

27. Goldansaz, S.M., Jolai, F., Anaraki, A.H.Z.: A hybrid imperialist competitive algorithm for minimizing makespan in a multi-processor open shop. Appl. Math. Model. **37**(23), 9603–9616 (2013)

28. Naderi, B., Yazdani, M.: A model and imperialist competitive algorithm for hybrid flow shops with sublots and setup times. J. Manuf. Syst. **33**(4), 647–653 (2014)

29. Ghasemishabankareh, B., Shahsavari-Pour, N., Basiri, M.A., Li, X.D.: A hybrid imperialist competitive algorithm for the flexible job shop problem. Ray, T., et al. (eds.) ACALCI 2016, LNAI 9592, pp. 221–233 (2016)

30. Brandimarte, P.: Routing and scheduling in a flexible job shop by tabu search. Ann. Oper. Res. **41**(1), 157–183 (1993)

31. Knowles, J.D., Corne, D.W.: On metrics for comparing non-dominated sets. In: Proceedings of 2002 Congress on Evolutionary Computation, Honolulu, 12–17 May, pp. 711–716 (2002)

32. Zitzler, E., Thiele, L.: Multi-objective evolutionary algorithms: a comparative case study and the strength Pareto approach. IEEE Trans. Evol. Comput. **3**(4), 257–271 (1999)

Single-Machine Green Scheduling to Minimize Total Flow Time and Carbon Emission

Hong-Lin Zhang[1], Bin Qian[1]($\boxtimes$), Zai-Xing Sun[1], Rong Hu[1], Bo Liu[2], and Ning Guo[1]

[1] School of Information Engineering and Automation,
Kunming University of Science and Technology, Kunming 650500, China
bin.qian@vip.163.com
[2] Academy of Mathematics and Systems Science,
Chinese Academy of Sciences, Beijing 100190, China

Abstract. In this paper, single-machine scheduling with carbon emission index is studied. The objective function is to minimize the sum of total flow time and carbon emission. Firstly, the problem is shown to be NP-hard by Turing reduction. Then mathematical programming (MP) model is established. A pseudo-time algorithm based on dynamic programming (DPA) is proposed for small scale. And a Bird Swarm Algorithm (BSA) is proposed to compete with DPA. In addition, simulation experiments are used to compare the proposed algorithms. DPA is shown to be more efficient for small scale problem, and BSA is better for large scale problem.

Keywords: Single-Machine Scheduling · Flow time · Carbon emission
Dynamic programming · Mathematical Programming · Bird Swarm Algorithm

1 Introduction

As the most typical scheduling problem, Single-machine scheduling prollem (S-MSP) is instructive for parallel machine scheduling problem. As a kind of most difficult combinatorial optimization problem, it is shown to be NP-hard, and attracts much attention in decades. Since Baker et al. [1] studied single machine scheduling for the first time, many scholars carry further research on this problem. Brucker et al. [2] prove that some single machine scheduling and parallel machine scheduling is NP-hard. Mahdavi et al. [3] study on single-machine scheduling problem to minimize total flow time and delivery cost, and it's shown to be NP-hard. As for single machine scheduling with unavailability intervals, Lee et al. [4] overview researches on this problem. Chu et al. [5] study on single machine scheduling with unavailability interval to minimize to minimize weighted completion time, a dynamic programming algorithm and a branch and bound algorithm is proposed. Yang et al. [6] propose a heuristic algorithm to minimize max completion time. Yin et al. [7] propose a fully polynomial time algorithm scheduling(PFTAS) based on dynamic programming (DPA) to minimize total flow time and delivery cost in single machine scheduling with unavailability intervals. But the study on single machine scheduling with cooling-standby intervals is very limited, and it's meaningful to research on this problem both in academic field and application field.

With the grim situation of global warming, green manufacturing to decrease carbon emission attracts more and more attention worldwide. As a modern manufacturing model, green manufacturing aims to achieve the simultaneous optimization of economic and green indexes. That is, guarantee the quality and function of products, reduce the manufacturing profit, and at the same time to decrease environmental pollution and energy waste. Green scheduling, as an important part of green manufacturing, is much more difficult than traditional scheduling problems. Many scholars study on this subject. Wang et al. [8] overview advances in green shop scheduling and optimization. Yildirim et al. [9] study on single machine green scheduling to minimize total completion time and energy consumption, and propose a multi-objective genetic algorithm to solve it. Liu et al. [10] study single machine scheduling with total completion time, energy consumption and carbon emission base on the research of Yildirm's, and an improved genetic algorithm is proposed to solve this problem. Fang et al. [11] presents a new mathematical programming model of the flow shop scheduling problem that considers peak power load, energy consumption, and associated carbon footprint in addition to cycle time. As shown above, research on green scheduling problem with both cooling-standby interval and carbon emission consideration is limited.

In this paper, a single-machine green scheduling to minimize the sum of total flow time and total carbon emission index is studied. The rest of this paper is arranged as follow: In Sect. 2, the S-MGS-FC problem is described and a mathematical programming model is established. In Sect. 3, a dynamic programming algorithm (DPA) is proposed to optimize the objective function. In Sect. 4, a bird swarm algorithm (BSA) is proposed to optimize S-MGS-FC. In Sect. 5, experiment results and comparisons of DPA and BSA is provided. And we end this paper with conclusion and acknowledgement in Sect. 6.

2   Problem $1|\text{CS-I}| \sum F_j + \sum C_j$

2.1   Problem Description and the Complexity

S-MGS-FC is described as follows: a job set which contains n independent jobs is supposed to be processed on single machine. Only one job can be processed at the same time, and preemption is forbidden while processing. A cooling-standby interval (C-SI) is set to cool down the machine, on the safe side, accordingly. The C-SI lasts times, which begins at moment T_1 and ends at T_2. Job j requires a processing time of p_j, and is available at time zero. All jobs follow shortest processing time(SPT) order. The objective function is to seek an optimal job sequence to minimize the sum of total flow time and total carbon emission. The problem is donated as $1|\text{CS-I}| \sum F_j + \sum C_j$ by using three-field notation.

And when it comes to the complexity, Turing reduction is used to prove that S-MGS-FC is NP-hard. Leave out of account of carbon emission, our problem is reduced to $1|C - SI| \sum F_j$, which is shown to be NP-hard by Yin et al. [12]. And S-MGS-FC, therefore, is also NP-hard.

2.2 Problem Formulation

As shown in Sect. 2.1, the objective function consists of two parts: total flow time and total carbon emission. Flow time equals to the sum of each job's flow time. And total carbon emission is included with three parts: total processing carbon emission, cooling-standby carbon emission, and turn-on(off) carbon emission. The function model based on mathematical programming is established as follows.

Parameters and variables:

a_j: arrival time of job j;
p_j: processing time of job j;
x_j: time to start processing of job j;
n: number of jobs in set N;
e_1: energy consumption of processing per minitue;
e_2: energy consumption of C-SI per minitue;
e_3: energy consumption of turn-on(off) per minitue;
E_1: energy consumption of processing;
E_2: energy consumption of C-SI;
E_3: energy consumption of turn-on(off);
φ: energy-carbon conversion coefficient
t_1: time of C-SI;
t_2: time of turn-on(off);
k: number of jobs processed before C-SI;
d_j: delivery time of job j.

Mathematical programming model:

$$\min Z(j) = \sum F_j + \sum C_j \tag{1}$$

$$s.t. p_j \geq a_j, \forall j = 1, 2, \ldots, n \tag{2}$$

$$p_j \leq p_{j+1}, \forall j = 1, 2, \ldots, n - 1 \tag{3}$$

$$x_j + p_j \leq d_j, \forall j = 1, 2, \ldots, n \tag{4}$$

$$x_{j+1} \geq x_j + p_j, \forall j = 1, 2, \ldots, n - 1 \tag{5}$$

$$\sum F_j = np_1 + (n-1)p_2 + \ldots + p_n + (n-k)t_1, \forall j = 1, 2, \ldots, n \tag{6}$$

$$\sum C_j = \sum p_j \cdot e_1 \cdot \varphi + t_1 \cdot e_2 \cdot \varphi + t_2 \cdot e_3 \cdot \varphi, \forall j = 1, 2, \ldots, n \tag{7}$$

Function (1) is to minimize the sum of total flow time and total carbon emission of the entire progress. Constraint (2) ensures the processing of job j js no earlier than its arrival time. Constraint (3) ensures all jobs follow SPT order. Constraint (4) ensures the completion time of job j is no earlier than its delivery time. Constraint (5) ensures the beginning of job $(j+1)$'s processing is no earlier than job j's completion time. Function (6) and (7) are detailed explanation of total flow time and total carbon emission.

3   Dynamic Programming Algorithm (DPA) for S-MGS-FC

In this section, a dynamic programming algorithm (DPA) is proposed to solve the MP model established in 2.3. Firstly, SPT is to be shown optimal for S-MGS-FC. Then, three kinds of reasonable states are tobe analyzed. And finally, a dynamic programming algorithm (DPA) for S-MGS-FC is proposed.

Theorem 3.1. SPT is the optimal order for S-MGS-FC.

Proof. For any processing sequence $S = (c_1, c_2, \ldots, c_k, c_l, \ldots, c_n)$ sorted by SPT, i.e. $p_k < p_l$. Suppose there exists a better sequence $S' = (c_1, c_2, \ldots, c_l, c_k, \ldots, c_n)$ which is to reduce the function value. S', however, increases the function value, which contradicts the optimality of S. It implies that $p_k < p_l$ holds for all jobs in an optimal sequence, and SPY is shown to be the optimal scheduling order for S-MGS-FC.

In what follows, we develop DPA based on Theorem 3.1. There are n steps in total. The jth step cares about job j's state $R_j(l, c_1, c_2, t)$. Meaning of the variables in R_j is given below:

l: completion time of the last job processed before T_1.
c_1:   number of jobs processed before T_1.
c_2:   number of jobs processed after T_2.
t: total flow time and cabon emission of all jobs processed from $(1, 2, \ldots, j)$.

Let Z be the optimal value and U be a large number satisfying $Z \le U$. The DPA is to be described as follows.

DPA:

Step 1. [intialization]
$$R_1 = \{(p_1,1,0,p_1),(0,0,1,T_2+p_1)\}.$$

Step 2. [from R_{j-1} to R_j]

For $j = 2$ to n do
$$R_j = \varnothing$$

For $(l,c_1,c_2,t) \in R_{j-1}$

If $l + p_j < T_1$ and $c_1 > 1$

Then $R_j = (l+p_j, c_1+1, c_2, t+l+(c_1+1)\cdot p_j)$

If $l + p_j < T_1$

Then $R_j = (l+p_j, 1, c_2, t+l+p_j)$

If $c_2 > 1$

Then $R_j = (l, c_1, c_2+1, t+T_2+\sum p_{k-l}+c_2\cdot p_j)$

Else then
$$R_j = (l, c_1, c_2+1, t+T_2+\sum p_{k-l})$$

Step 3. [elimination]

For $(l,c_1,c_2,t),(l,c_1,c_2',t') \in R_j, t < t'$

Eliminate (l,c_1,c_2',t') from R_j.

End

Step 4. [optimization]
$$Z = \min(l,c_1,c_2,t) \in R_j$$

4   Bird Swarm Algorithm (BSA) for S-MGS-FC

4.1   Brief Review of BSA

BSA is proposed by Meng et al. based on the simulation of bird swarm's foraging behavior, migration behavior and vigilance behavior. And it's efficient for optimization problems.

Forgaing behavior is described as follows:

$$\begin{aligned} bird_{i,j}^{t+1} = bird_{i,j}^{t} + (person_{i,j} - bird_{i,j}^{t}) \cdot C \cdot rand(0,1) \\ + (G_{i,j} - bird_{i,j}^{t}) \cdot S \cdot rand(0,1) \end{aligned} \tag{8}$$

Variables in formula (8) are described as follows: $bird_{i,j}^t$ is the jth dimension of the ith bird in tth bird swarm generation. And $person_{i,j}$ is the best position of the ith bird's jth dimension. And $G_{i,j}$ is the best position of current bird swarm, **C** and **S** are positive constant.

Vigilance behavior is described as follows:

$$bird_{i,j}^{t+1} = bird_{i,j}^t + A1 \cdot (mean_j - bird_{i,j}^t) \cdot C \cdot rand(0,1)$$
$$+ A2 \cdot (p_{k,j} - bird_{i,j}^t) \cdot rand(-1,1) \tag{9}$$

$$A1 = \alpha \cdot \exp(-\frac{pfit_i}{sumfit + \xi} \cdot popsize) \tag{10}$$

$$A2 = \beta \cdot \exp((\frac{pfit_i - pfit_k}{|pfit_i - pfit_k| + \xi}) \cdot \frac{popsize \cdot pfit_k}{sumfit + \xi}) \tag{11}$$

Meanings of variables in formula (9)–(11) are as follows: k is a random integer satisfying $k \in (1, popsize)$, $pfit_i$ is the best fitness of the jth bird swarm, *sumfit* is the sum of bird swarms in generation, ξ is a small positive number, $mean_j$ is the average of all swarm's *j*th dimension in generation.

Migration behavior is described as follows:

$$bird_{i,j}^{t+1} = bird_{i,j}^t + bird_{i,j}^t \cdot randn(0,1) \tag{12}$$

$$bird_{i,j}^{t+1} = bird_{i,j}^t + (bird_{k,j}^t - bird_{i,j}^t) \cdot FL \cdot rand(0,1) \tag{13}$$

Meanings of variables in formula (12) and (13) are as follows: k is a random integer satisfying $k \in (1, popsize)$, FL is a constant satisfying $FL \in [0,1]$, $randn(0,1)$ is standard Gaussian number.

4.2 LOV Encoding

Because of the continuous character of the individuals in BSA, canonical BSA cannot be directly applied to S-MGS-FC. So, the largest-order-value (LOV) encoding rule based on random key is proposed to convert BSA's individual $X_i = [x_{i,1}, x_{i,2}, \ldots, x_{i,n}]$ to job permutation $\pi_{i,j} = [\pi_{i,1}, \pi_{i,2}, \ldots, \pi_{i,n}]$. Table 1 illustrates the representation of vector X_i in BSA for a simple problem.

Table 1. LOV encoding sample

Dimension k	1	2	3	4	5	6
$x_{i,k}$	1.36	3.85	2.55	0.63	2.68	0.82
$\varphi_{i,k}$	4	1	3	6	2	5
$\pi_{i,k}$	2	5	3	1	6	4

4.3 Pseudo Code of BSA

The pseudo code of BSA is described as follows:

Algorithm 1 Bird Swarm Algorithm (BSA)

Input : N : the number of individuals (birds) contained by the population

       M : the maximum number of iteration

       FQ : the frequency of birds' migration behaviours

       P : the probability of forging for food

       C, S, a1, a2, FL : five constant parameters

$t = 0$; Initialize the population and define the related parameters

Evaluate the N individuals' fitness value, and find the best solution

While $(t < M)$

    If $(t\%FQ \neq 0)$

      For $i = 1 : N$

      If rand(0,1) < P

      Birds forge for food

        Else

          Birds keep vigilance

        End if End for

   Else

Divide the swarm into two parts: producers and scroungers.

For $i = 1 : N$

  If i is a producer

    Producing

  Else

    Scrounging

    End if End for

End if Evaluate new solutions

If the new solutions are better than their previous ones, update them

Find the current best solution

   $t = t + 1$; End while

    Output : the individual with the best objective function value in the population

---

5 Experiment Results and Comparison

In this section, 10 sets of random number are adopted to test the performance of DPA and BSA, respectively. That is, n numbers are carried out with $\{10, 20, 30, \ldots, 100\}$. The setup of machine is as follows: The cooling-standby interval (C-SI) is set per 4 h and lasts 30 min. Processing is forbidden during C-SI (e.g., the last job before C-SI will not be processed if its completion time is later than the beginning of C-SI, and is to be processed after C-SI. Processing power is set as 63.4^{kWh} per minute, C-SI power is

37.6^{kWh} per minute, turn-on(off) power is 15^{kWh} per minute and the turn-on(off) time is 45 min. Energy-carbon conversion rate is 0.7559.

Table 2. Comparison of DPA and BSA on 10 sets of jobs

Instance	DPA		BSA	
n	Avg	Std	Aver	Std
10	**16622.07**	**0**	16629.27	4.78
20	**36198.07**	**0**	36312.37	27.35
30	**57642.06**	**0**	58131.76	45.11
40	**79800.48**	**0**	80830.68	97.78
50	–	–	**107009.62**	**138.94**
60	–	–	**135902.65**	**152.69**
70	–	–	**172763.67**	**302.71**
80	–	–	**204287.77**	**321.21**
90	–	–	**249829.34**	**426.52**
100	–	–	**296174.67**	**497.07**

Both algorithms are coded by Delphi 7.0 and run on PC with Intel processor i7-7700 k (4.2 GHz) and 16 GB memory in 10 min. Results and comparisons are shown at Table 2.

Apparently, from Table 2, DPA performs better than BSA for small scale problem, and BSA is better for large scale problem. For $n = \{10, 20, 30, 40\}$, DPA get the optimal solution. Because dynamic programming is an exact algorithm. For large scale problem, however, exact algorithm can not get any solution because of dimension explosion. BSA, as an intelligent algorithm, is an approximate algorithm, can get approximate solutions.

6 Conclusions and Further Research

Scheduling problem is one of the most difficult combinational optimization problem. Single-machine scheduling is the basis of all scheduling problems, and it's instructive to parallel machine scheduling. Green scheduling, as core of green manufacturing, adapts to the world's development situation. This paper studies single-machine green scheduling with carbon emission index. Some researchers study on green scheduling with peak load, carbon footprint, power consumption et al. to the best of the authors' knowledge, green scheduling has good prospects for development. As for the algorithms, the optimal solution is to be obtained by exact algorithm in theory. When it comes to large scale problem, however, it becomes non solvable for exact algorithm. Intelligent algorithm, as one of the approximate algorithms, has advantages in solving large scale problems, but it's easy to get trapped in local optimality. The following research is to combine both exact algorithm and approximate algorithm, solving complex single-machine green scheduling problem.

Acknowledgement. This research is partially supported by the National Science Foundation of China (51665025), and the Applied Basic Research Foundation of Yunnan Province (2015FB136).

References

1. Baker, K.R.: Heuristic procedures for scheduling job families with set-ups and due dates. Naval Res. Logist. **46**, 978–991 (1999)
2. Brucker, P., Gladky, A., Hoogeveen, H., et al.: Scheduling a batching machine. J. Sched. **1**, 31–54 (1998)
3. Mazdeh, M.M., Sarhadi, M., Hindi, K.S.: A branch-and-bound algorithm for single-machine scheduling with batch delivery and job release times. Comput. Oper. Res. **35**, 1099–1111 (2008)
4. Lee, C.Y., Lei, L., Pinedo, M.: Current trends in deterministic scheduling. Ann. Oper. Res. **70**, 1–41 (1997)
5. Ma, Y., Chu, C.B., Yang, S.L.: Minimizing total weighted completion time in semiresumable case of single-machine scheduling with an availability constraint. Syst. Eng. Theory Pract. **29**(2), 134–143 (2009)
6. Ma, Y., Yang, S.L., Chu, C.B.: Minimizing makespan in semiresumable case of single-machine scheduling with an availability constraint. Syst. Eng. Theory Pract. **29**(4), 128–134 (2009)
7. Yin, Y., Wang, Y., Cheng, T.C.E., Wang, D.J., Wu, C.C.: Two-agent single-machine scheduling to minimize the batch delivery cost. Comput. Ind. Eng. **92**, 16–30 (2016)
8. Ling, W., Wang, J., Wu, C.: Advances in green shop scheduling and optimization [J/OL]. http://kns.cnki.net/kcms/detail/21.1124.TP.20170904.0922.001.html
9. Yildirim, M.B., Mouzon, G.: Single-machine sustainable production planning to minimize total energy consumption and total completion time using a multiple objective genetic algorithm. IEEE Trans. Eng. Manage. **59**(4), 585–597 (2012)
10. Liu, C., Yang, J., Lian, J., et al.: Sustainable performance oriented operational decision-making of single machine systems with deterministic product arrival time. J. Clean. Prod. **85**, 318–330 (2014)
11. Fang, K., Uhan, N., Zhao, F., et al.: A new approach to scheduling in manufacturing for power consumption and carbon footprint reduction. J. Manuf. Syst. **30**(4), 234–240 (2011)
12. Yin, Y., Cheng, T.C.E., Hsu, C.J., Wu, C.C.: Single-machine batch delivery scheduling with an assignable common due window. Omega **41**, 216–225 (2013)

The Model of Flight Recovery Problem with Decision Factors and Its Optimization

Zhurong Wang$^{(\boxtimes)}$, Feng Wang, Xinhong Hei, and Haining Meng

School of Computer Science and Engineering, Xi'an University of Technology,
NO. 5 South Jinhua Road, Xi'an, Shaanxi, China
wangzhurong@xaut.edu.cn

Abstract. For different scenarios of flight recovery problems, a mathematical model for flight recovery is constructed while considering the factors of decision variables and scenarios. The model deals with the problem of optimized objective functions and the constraints based on actual requirements. The mathematical optimization model M1 with determining recovery time is established to minimize the objective function of aircraft swapping, flight cancellation and flight delay cost. For the case of uncertain recovery time, or uncertain opening and closing time of the airport due to weather conditions, the mathematical optimization model M2 is used to make simple adjustments to the results obtained by the model M1, including flight rearrangements, flight delays or cancellations when curfew is encountered, and to minimize the total cost. For actual flight recovery time, the stochastic model M of the flight recovery problem is established by the combination of the constructed model M1 and M2. Finally, the lingo9 is applied to solve the optimization problem. The computational results indicate that the proposed model can handle actual target requirements of the flight recovery problem, and has the characteristics of good flexibility and scalability.

Keywords: Flight recovery · Stochastic modeling · Aircraft route
Optimization

1 Introduction

Air travel has become an increasingly popular choice for passengers in our lives. However, delays and cancellations of flights can cause huge economic losses to the airlines and lead to great inconvenience for passengers. Delays in some flights may even cause social unrest. According to the 2016 Civil Aviation Development Statistics Bulletin [1] issued by the civil aviation administration of China (CAAC), the national civil aviation transport airport completed 9.235 million trips, a total of 3.679 million flights, of which 2.824 million were regular flights, and the average regular rate was 76.76% this year. For this reason, the study of flight recovery issues is of great significance.

© Springer International Publishing AG, part of Springer Nature 2018
D.-S. Huang et al. (Eds.): ICIC 2018, LNCS 10954, pp. 679–690, 2018.
https://doi.org/10.1007/978-3-319-95930-6_68

In 1967, Jedlinsky studied the optimization of flight recovery. Afterwards, some scholars studied on single resources such as aircraft, aircrew and passengers, and restoration of resource combinations and restoration of overall resources. For example, Teodorovic et al. in [2] established a mathematical model that minimizes the total time for passenger delays through using the flight swaps and delays with no changing on the numbers of the flights, and the branch-and-bound method is used to solve the problem while no considering the cost of flight cancellation. Arguello et al. in [3] applied the Greedy randomized adaptive search procedure (GRASP) to solve the aircraft rearrangement problem with considering the aircraft flow balance constraint in the case of aircraft shortage. Rosenberger, et al. in [4] proposed the aircraft selection heuristic algorithm in which the flight recovery problem is converted into a set coverage problem with time windows and time slots. Froyland et al. in [5] converted the recovery study from "post-event" to "pre-event" and simplified aircraft recovery by establishing an easy-to-recover aircraft robust assignment. Jafari et al. in [6] converted flights into aircraft rotation to determine the recovery scope to reduce the size of the problem, and add some manual decision factors to study the problem of simultaneous recovery of flights and passengers. Yan and Yang in [7] proposed a basic framework for dispatchers to carry out aircraft recovery in the event of an aircraft failure, through modifying the model by changing the objective function and boundary constraints. Zhou in [8] proposed a real-time integrated recovery framework including flight plans, aircraft, crews, and passenger trips, and designed a pseudo-genetic algorithm to solve the problem. Bai and Zhu et al. in [9] established a multi-commodity-goods network flow mathematical model while considering both the shortage of aircraft resources and the closure of airport fields, and the column generation algorithm is designed to solve the problem. Considering the flight recovery problem in the case of an uncertain recovery time, Zhu et al. in [10] established a two-stage recovery model, and design a random algorithm framework combining Greedy Simulated Annealing (GSA) method to solve the model.

In summary, most of the above studies only concern flight recovery problems at a given time interval. Although the uncertain flight model is studied in [10], the model do not deal with complex scenarios such as aircraft swap, flight rotation constraints, etc. And all of the above studies have attributed delays or cancellations of flights to aircraft failures. In this study, delays or cancellations of flights due to temporary closure of the airport due to uncertain weather conditions will be considered and researches will be conducted under this new scenario.

In this paper, a stochastic model for flight recovery is established. The proposed model could handle the actual cases of certainty and uncertainty aircraft recovery time with respect to determining and uncertain opening and closing time of the airport. In the proposed model flight delays, flight cancellations, aircraft swap, airport closures due to weather conditions and some decision strategy factors are considered. Furthermore, in order to verify the correctness and effectiveness of the proposed model, the software Lingo is used to solve the proposed model. Test data demonstrates the correctness and effectiveness of the proposed model.

2   Flight Recovery Model (M1) with Determining Recovery Time

In this section, we focus on analyzing flight recovery model (M1) with determining recovery time. Several assumptions and analysis are given firstly, and then objective function and constraints are discussed in the model.

2.1   Assumptions and Analysis

According to actual requirements of the flight, we give the following assumptions.

a. Some time-related changes are given. For example, the maximum delay time, the minimum airplane interval, and the minimum connection time, etc.
b. All the flights must return to be normal while the airport is opened with the closing time being end. For example, for a given closing time FCT = [apt1s, apt1e], all flights which are affected by FCT could recover immediately after apt1e. In addition, closing time could be temporary time gap and deterministic time gap.
c. All flights cannot be reached in advance and can arrive on time or be delayed.

According to the above assumptions, we assume that the maximum delay time is 5 h for a single flight; the minimum airplane interval is 45 min, the minimum connection time is 10 min, and the airport closing time FCT = [18:00, 21:00] due to weather reason.

Figure 1 is the routes of the aircraft F1 and F2. Figure 2 is flight rotation related to Fig. 1. Figure 3 demonstrates recovery scope of flight.

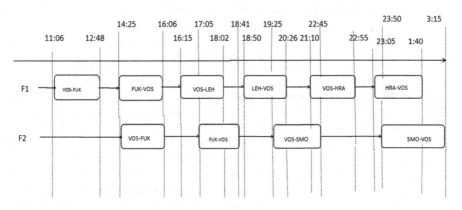

Fig. 1. Aircraft routes

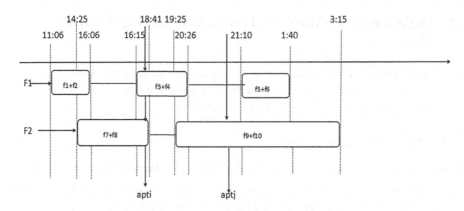

Fig. 2. Flight rotation

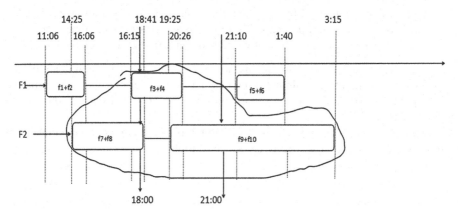

Fig. 3. Recovery scope

When the aircraft fails, there are many recovery strategies such as flight delays, flight cancellations, flight swaps, etc. which can make the flight back to be normal. However, in general, as shown in Fig. 1, while considering that the VOS airport is closed at FCT = [18:00, 21:00] due to the weather, flights on the F1 and F2 routes will be treated as delayed flights. This will increase problem complex processing since some flights in the routes are not delayed. Here we introduce the concept of aircraft rotation [6], firstly converting flights into the concept of aircraft rotation, then determining the restoration scope on the basis of aircraft rotation, which can reduce the scale of recovery problems. As shown in Fig. 2, the flights performed by the flight of F1 and F2 are divided into 3 rotations and 2 rotations respectively according to the position of the starting and ending airport. For the given airport closing time [18:00, 21:00], it can be seen that the delayed flight rotations are f3 + f4, f7 + f8, f9 + f10, so the restoration scope can be determined, as shown in Fig. 3, covered by the curve is the recovery scope.

2.2   Variables and Parameters

The variables to the model M1 are as follows.

F_s: the set of flights.

K_s: the set of aircraft.

RO_s: the set of rotations.

AP: the set of airports.

AR: the set of airlines.

$x_{k,f}$: flag variable. $x_{k,f} = 1$ if aircraft k is assigned to flight f, and $x_{k,f} = 0$ otherwise.

z_f: flag variable. $z_f = 1$ if flight f is canceled, and $z_f = 0$ otherwise.

$t_{df}, t_{df_{ro(i)}}$: actual departure time of flight $f, f_{ro(i)}$.

$t_{af_{ro(j)}}$: actual arrival time of flight $f_{ro(j)}$.

$C_{k,f}$: cost of assigning aircraft k to flight f.

CC_f: cost of canceling flight f.

CD_f: cost of delay of flight f.

T_f: the scheduled departure time of flight f.

$x_{k,f_{ro(i)}}$: flag variable. $x_{k,f_{ro(i)}} = 1$ if aircraft k is assigned to flight $f_{ro(i)}$ which is the first flight within $ro(i)$, and $x_{k,f_{ro(i)}} = 0$ otherwise.

$x_{k,f_{ro(i)}^1}, x_{k,f_{ro(i)}^2}$: flag variable. $x_{k,f_{ro(i)}^1}, x_{k,f_{ro(i)}^2}$ are set to 1 if aircraft k is assigned to flight $f_{ro(i)}^1, f_{ro(i)}^2$, $f_{ro(i)}^1, f_{ro(i)}^2$ is the first and second flight within $ro(i)$, and 0 otherwise.

$y_{k,r}$: flag variable. $y_{k,r} = 1$ if aircraft k is assigned to airline r, and $y_{k,r} = 0$ otherwise.

U: Minimum connection time between two connected flights before and after.

$\psi_{r,ap}$: 1 if airline r passes through airport ap, airport ap belongs to a subset of closed airport set AP, and 0 otherwise.

$ap_{f,r}, ap_{k,r}$: The starting airport when flight f is arranged to execute airline r and aircraft k is arranged to execute airline r.

2.3   The Mathematics Model M1

While considering the influence of airport closing time FCT, the model M1 can be described as below.

$$\min W_1 = \left(\sum_i \mu(x,z)\omega_i \right)$$

$$= \min \left(\sum_{f \in F_s} \sum_{k \in K_s} C_{k,f} x_{k,f} + \sum_{f \in F_s} CD_f \left(1 - z_f\right)\left(t_{df} - T_f\right) + \sum_{f \in F_s} CC_f z_f \right) \tag{1}$$

s.t.

$$x_{k,f_{ro(i)}} + \sum_{i \neq j} x_{k,f_{ro(j)}} \leq 1, \forall f_{ro(i)}, f_{ro(j)} \in F_s, \forall ro(i), ro(j) \in RO_s \tag{2}$$

$$1 - z_f = \sum_{k \in K_s} x_{k,f} \quad \forall f \in F_s \tag{3}$$

$$x_{k,f_{ro(i)}^2} - x_{k,f_{ro(i)}^1} = 0 \ \forall k \in K_s, \forall f_{ro(i)}^1, f_{ro(i)}^2 \in F_s, \forall ro(i) \in RO_s \tag{4}$$

$$\sum_{r \in AR} y_{k,r} \leq 1 \quad \forall k \in K_s \tag{5}$$

$$td_{f_{ro(i)}} \geq ta_{f_{ro(j)}} x_{k,f_{ro(j)}} + U \ \forall f \in F_s, \forall ro(i), ro(j) \in RO_s \tag{6}$$

$$td_{f_{i+1}} \geq ta_{f_i} + U \quad \forall f_i, f_{i+1} \in F_s \tag{7}$$

$$\sum_{r \in AR} \sum_{f \in F_s} \psi_{r,ap} y_{r,f} = 0 \quad \forall ap \in AP^{CLOSE}[apt1, apt2] \tag{8}$$

$$\sum_{r \in AR} ap_{f,r} = \sum_{r \in AR} ap_{k,r} \quad \forall f \in F_s, \forall k \in K_s \tag{9}$$

$$x_{k,f}, y_{r,f}, z_f, \psi_{r,ap} = \{0, 1\} \ td_f, ta_f \in Z^+ \tag{10}$$

The objective function (1) minimizes the cost of aircraft assignment, flight delay and cancellation. Equation (2) represents that each aircraft can only be used for one flight at the same time; Eq. (3) indicates that if the flight is cancelled, no aircraft will be scheduled to perform this flight; Eq. (4) confines that the aircraft is flying within the same aircraft rotation; Eq. (5) means that each aircraft is arranged with a maximum of one route; Eq. (6) indicates the departure time of the second flight rotation cannot be earlier than the sum of the real arrival time and the minimum connection time of the first flight rotation when the two aircraft rotations are executed by the same aircraft; Eq. (7) ensures that the flight departure time cannot be earlier than the sum of the flight arrival time and the minimum connection time; Eq. (8) indicates that when the airport is closed, there is no take-off and landing task for the flight that is related to the airport shutdown; Eq. (9) indicates that the departure of the flight is the same as the departure of the aircraft from the airport; Eq. (10) confines that decision variable x, y, z, ψ are binary variables, t_{df}, t_{af} is a positive integer in twenty-four hour form.

3   Flight Recovery Mathematical Model (M2) with Uncertain Recovery Time

In this section, we address analyzing flight recovery model (M2) with uncertain recovery time. Several assumptions and analysis are given firstly, and then the objective function and constraints are discussed in the model.

3.1   Assumptions and Analysis

The model of flight recovery discussed in the second part is under the condition with determining recovery time. Because the recovery time is given by the dispatcher based on the empirical value, it is difficult to accurately determine the airport recovery time under normal circumstances. And this value is rarely the same as the actual value, which may lead to a poor or infeasible recovery plan. The mathematical model of uncertain recovery time is based on the recovery time given by the dispatcher, first establishing a model for determining the recovery time, and then a simple adjustments (rearrange flights, delays or cancellations when encountering a curfew) is made so that the total cost caused by the uncertain recovery time is minimized.

3.2   Variables and Parameters

The variables to the model M2, which are not the same as listing in the model M1, are as follows.

cp_f: flight f one minute delay cost.

$\Delta_{f,\zeta}$:   flight f estimated delays under scenario ζ.

Δ_f: delay time of flight f calculated from optimization on model M1.

cx_f: the cost of flight f broken the curfew.

$v_{f,\zeta}$:   flag variable. $v_{f,\zeta} = 1$ if flight f broken the curfew under scenario ζ, and $v_{f,\zeta} = 0$ otherwise.

$p(\zeta)$:   the probability of scenario ζ.

$td_{f,\zeta}, rtd_{f,\zeta}$:   flight f departure time and scheduled departure time under the scenario ζ.

$q(f)$:   predecessor flight of flight f in the same airline after optimization on the model M1.

$ra_{q(f),\zeta}$:   the arrival time of flight $q(f)$ under the scenario ζ.

g_k: minimum turnaround time of aircraft k.

Ω: the set of scenarios variable.

3.3   The Mathematical Model M2

In case of uncertain recovery time, the recovery time of the maximum probability is brought into the model M1 to obtain the initial arrangement of the flight. Then the model M2 is used to make simple adjustments to the results obtained by the model M1, including flight rearrangements, flight delays or cancellations when curfew is encountered, and minimize the total cost. Model M2 is described as follows:

$$\min W_2 = \min \sum_{\xi} \vartheta(x, z, \xi) p(\xi)$$

$$= \min \sum_{\xi} \left(\sum_{f} cp_f(\Delta_{f,\xi} - \Delta_f) + cx_f v_{f,\xi} \right) p(\xi) \tag{11}$$

s.t.

$$td_{f,\xi} - rtd_f = \Delta_{f,\xi} \quad \forall f \in F_s, \forall \xi \in \Omega \tag{12}$$

$$x_{k,f_{ro(i)}^2} - x_{k,f_{ro(i)}^1} = 0 \quad \forall k \in K_s, \forall f_{ro(i)}^1, f_{ro(i)}^2 \in F_s, \forall ro(i) \in RO_s \tag{13}$$

$$td_{f,\xi} - ra_{q(f),\xi} \geq g_k \sum_{r} a_{f,r} x_{k,r} \quad \forall k \in K_s, \forall f \in F_s, \forall \xi \in \Omega, \forall r \in AR \tag{14}$$

$$td_{f_{ro(i)}} \geq ta_{f_{ro(j)}} x_{k,f_{ro}} + U \quad \forall k \in K_s, \forall f_{ro(i)}, f_{ro(j)} \in F_s, \forall ro(i), ro(j) \in RO_s \tag{15}$$

$$td_{f_{i+1}} \geq ta_{f_i} + U, \quad \forall f_i \in F_s \tag{16}$$

$$td_{f,\xi}, ra_{f,\xi}, \Delta_{f,\xi} \geq 0, \quad \forall f \in F_s, \forall \xi \in \Omega, v_{f,\xi} = 0, 1 \tag{17}$$

The objective function (11) indicates that for each scenario ζ, on the basis of the model M1, due to the uncertainty of aircraft recovery time, the sum of the flight time rearrangement costs and the risk costs incurred when encountering the curfew is minimized. Equation (12) is the relationship between aircraft delays; Eq. (13) confines that the aircraft is flying within the same aircraft rotation; Eq. (14) is minimum turn-around time for adjacent flights on an aircraft route; Eq. (15) represents that when the flight rotation is executed by the same aircraft, the departure time of the second flight rotation cannot be earlier than the sum of the real arrival time and the minimum connection time of the first flight rotation; Eq. (16) means that the flight departure time cannot be earlier than the sum of the flight arrival time and the minimum connection time; Eq. (17) is non-negative constraints and $v_{f,\zeta}$ is 1 if the curfew under scenario ζ is broken, and 0 otherwise.

4 Stochastic Model M of Flight Recovery

While considering the cases of recovery time being determining and uncertainty, a stochastic model M for flight recovery is proposed. The stochastic recovery model M combines the recovery time determination model M1 and the recovery time uncertainty model M2. When the recovery time is determined, the parameter λ is set to 0. At this time, the stochastic model M is the recovery time determination model; When the recovery time is indeterminate, the parameter λ is set to 1. In this case, the stochastic model is M1+M2, which means that based on the recovery time given by the dispatcher, the initial situation of the flight arrangement is calculated by taking the recovery time of the maximum probability as the determination of the recovery time., and then a simple adjustment is made so that the total cost caused by the uncertain recovery time is minimized. The mathematical model M is as follows:

$$\min \quad W = \min W_1(x, z) + \lambda * W_2(x, z, \xi) \tag{18}$$

$$s.t.$$
$$A_1(x, z, \xi) = C_2 \tag{19}$$

$$A_2(x, z, \xi) \le C_3 \tag{20}$$

$$x, z = 0, 1 \quad \forall \xi \in \Omega \tag{21}$$

Equation (18) is the objective function of the stochastic model and is divided into two parts as W_1 and W_2. For the recovery time determination, let $\lambda = 0$; For uncertain recovery times, let $\lambda = 1$; Eqs. (19) and (20) are the equality constraints and inequality constraints in the formula of (2)–(10) and (12)–(17); Eq. (21) gives that values of decision variables x, z and scenario variable ξ.

From the above model M with the objective functions and constraints, M is a minimum function optimization problem with equality constraints and inequality constraints.

5 Computational Test

Experimental data is obtained from a challenge cup, and 2 aircraft and 10 flights are selected for computational test. The original flight data is shown in Table 1. Decision variable and constants are presented in Table 2. Due to the weather factor, the airport closing time FCT is set to [18:00, 21:00]. However, when the bad weather has passed, the airport will be opened immediately, Therefore usually the opening time of airport is uncertain. According to historical data, the probability distribution of opening the airport is shown in Table 3.

When the airport opening time is determined, the recovery time of the flight is also determined, Let $\lambda = 0$. The data from Tables 1 and 2 is used in the model M, and the lingo9.0 is used to solve the model M. The minimum flight swapping, delay, and cancellation cost are 80 in this case. At this time, the flights in the recovery scope should be arranged in such a way that flights 3, 4, 7 and 8 are delayed, flight 9,10 are canceled, and flights 1, 2, 5, 6 are not within the range of recovery scope and they are normal flights.

When the opening time of the airport is uncertain, the recovery time of the flight is uncertain. For example, flight reservations are scheduled to open at 21:00, because of the abrupt end of the weather, the airport opening time is put in advance to 20:30, or the bad weather continues for a long time, and the airport opening time is delayed to 21:30. In the case of uncertain recovery time, first the recovery time of the maximum probability is taken as the deterministic time in the model M1 to obtain the flight schedule, Then the recovery time of other probabilities is taken into account. Afterwards, the resulting flight schedule is adjusted to minimize total costs due to uncertain recovery times. Let $\lambda = 1$. The data from Tables 1, 2 and 3 is used in the model M and solved by lingo9.0. If the delay flight is 3, 4, 7, and 8, the flight 9, 10 are canceled, flight cancellation is equivalent to 5 h of flight delay, at this time lingo9.0 calculated the minimum cost of 3656.3; if the flight 9, 10 also as a delay, that is, the flight delay 3, 4, 7, 8, 9, 10; The calculation result is 734.3. In two above cases, the flight execution table can be rescheduled. The flights within the recovery scope should be arranged in this way that flight 1, 2, 5, and 6 are in normal flight and flights 3, 4, 7, and 8, 9, 10 delays. The experimental results are shown in Table 4.

When the recovery time is determined through the analysis of the actual situation, the decision weights for swapping, delay and flight cancellation are 10, 1 and 20, respectively. Within the scope of recovery, arrangements are made for delays on flights 3, 4, 7, 8, and cancellations for the flight 9, 10 because the delay time for the flights 9,10 are more than two hours. In order to minimize the total cost, the flight 9 and 10 are canceled and the minimum cost is 80. When the recovery time is uncertain, due to the specific delay cancellation cost of the flight, further optimization is performed on the recovery time, and the flights are re-calculated after balancing the cancellation and delay. And if there is no simple adjustment, only the flight schedule according to the model M1, the minimum cost is 3656.3, but after a simple adjustment of the model M2, the minimum cost can be reduced to 734.3, from this, it can be seen that the model M is

Table 1. The flight data

Aircraft	Departure airport	Arrival airport	Scheduled time of departure	Scheduled time of arrival
9	VOS	FUK	11:06	12:48
9	FUK	VOS	14:25	16:06
9	VOS	LEH	16:15	18:02
9	LEH	VOS	18:50	20:36
9	VOS	HRA	21:10	22:55
9	HRA	VOS	23:05	1:40
32A	VOS	FUK	14:25	16:05
32A	FUK	VOS	17:05	18:41
32A	VOS	SMO	19:25	22:45
32A	SMO	VOS	23:50	3:15

Table 2. Decision variable and constants

Swap decision value	Cancellation decision value	Delay decision value	Minimum connection time	Unit cost of flight delays	The cost of breaking curfew	Minimum station time
C	CC	CD	U	cp	cx	g
10	20/100	1	10 min	5 ¥/ min	10000 ¥	45 min

Table 3. The probability of opening the airport

Open time	20:30	21:00	21:30	22:00
Probability	0.14	0.6	0.15	0.11

Table 4. The experimental results

	Cost	Delaying flight	Cancellation flight	Normal flight
Recovery time determining	80	f3, f4, f7, f8	f9, f10	f1, f2, f5, f6
Recovery time uncertain	734.3	f3, f4, f7, f8, f9, f10	Null	f1, f2, f5, f6

effective in solving uncertain recovery time. The calculation result of the schedule can satisfy the problem that flight restore operation and target requirements. The above test results verify the correctness and feasibility of the proposed model of the flight recovery problem proposed in this study.

6 Conclusion

In this paper, a stochastic model for flight recovery is established considering the time of flight recovery is determinate or uncertain. The manual decision factor and recovery scope are taken into account. The temporary closure of the airport due to weather conditions was also considered, and the delay and cancellation of flights was attributed to the temporary closure of the airport rather than the failure of the aircraft itself. This model can simultaneously resolve the two cases of flight recovery time determination and uncertainty. Through using optimization software lingo9.0, the experiment shows that the model can better deal with the two situations. This is our initial attempt to study this problem model, and there are still some deficiencies in this paper. The model only considers the recovery of small-scale flights. It does not take into account the issues of large-scale flight recovery and flight recovery with passengers. In the near future we consider the model with more complex scenarios and strategies, and seek evolutionary algorithms with heuristics strategies for solving the model. And we also try to apply the proposed model and algorithm to practical problems.

Acknowledgments. The research presented is supported in part by the National Natural Science Foundation (No: U1334211, 61773313, 61602375), Shaanxi Province Key Research and Development Plan Project (No: 2015KTZDGY0104, 2017ZDXM-GY-098), The Key Laboratory Project of Shaanxi Provincial Department of Education (No: 17JS100).

References

1. Civil Aviation Industry Development Statistics Bulletin 2016. China Civil Aviation Administration, Beijing (2017)
2. Teodorović, D., Guberinić, S.: Optimal dispatching strategy on an airline network after a schedule perturbation. Eur. J. Oper. Res. **15**(2), 178–182 (2007)
3. Argüello, M.F., Bard, J.F., Yu, G.: A GRASP for aircraft routing in response to groundings and delays. J. Comb. Optim. **1**(3), 211–228 (1997)
4. Rosenberger, J.M., Johnson, E.L., Nemhauser, G.L.: Rerouting aircraft for airline recovery. Transp. Sci. **37**(4), 408–421 (2003)
5. Froyland, G., Maher, S.J., Wu, C.L.: The recoverable robust tail assignment problem. Transp. Sci. **48**(3), 351–372 (2011)
6. Jafari, N., Zegordi, S.H.: Simultaneous recovery model for aircraft and passengers. J. Franklin Inst. **348**(7), 1638–1655 (2011)
7. Yan, S., Yang, D.H.: A decision support framework for handling schedule perturbation. Transp. Res. Part B Methodol. **30**(6), 405–419 (1996)
8. Zhizhong, Z.: Research on Real-time Optimization of Flight Operation Control. Beijing University of Aeronautics and Astronautics, Beijing (2001)
9. Feng, B.A.I., Jin-fu, Z.H.U., Qiang, G.A.O.: Disrupted airline schedules dispatching based on column generation methods. Syst. Eng. Theory Pract. **30**(11), 2036–2045 (2010)
10. Zhu, B., Zhu, J.-F., Gao, Q.: A Stochastic programming approach on aircraft recovery problem. Math. Probl. Eng. **2015**(1), 1–9 (2015)

Mining Overlapping Protein Complexes in PPI Network Based on Granular Computation in Quotient Space

Jie Zhao and Xiujuan Lei[(⊠)]

School of Computer Science, Shaanxi Normal University, Xi'an 710062, China
xjlei@snnu.edu.cn

Abstract. Proteins complexes play a critical role in many biological processes. The existing protein complex detection algorithms are mostly cannot reflect the overlapping protein complexes. In this paper, a novel algorithm is proposed to detect overlapping protein complexes based on granular computation in quotient space. Firstly, problems are expressed by quotient space and different quotient space embodies the quotient set of different granular. Then the method estimates the relationship between particles to make up for the inadequacy of data in combination with the PPI data and Gene Ontology data, deals with the network based on quotient space theory. Graining the network to construct the quotient space and merging the particles layer by layer. The final protein complexes is obtained after purification. The experimental results on Saccharomyces cerevisiae and Homo sapiens turned out that the proposed method could exploit protein complexes more accurately and efficiently.

Keywords: Protein complexes · Gene Ontology · Quotient space
Granular computation · Clustering

1 Introduction

Mining protein complexes can help to unfold the structure of PPI networks, to predict protein functions and explain biological processes [1]. It is costly and time-consuming to detect protein complexes based on experimental methods [2], so researchers pay much attention to propose more computational methods based on the topological property of PPI network and structural information of proteins. Bader and Hogue [3] proposed MCODE method based on the proteins' connectivity values. Nepusz et al. developed ClusterOne [4] algorithm to detect overlapping protein complexes. Gavin et al. [5] demonstrated that protein complexes was made up of core and additional attachment proteins or protein modules. Leung et al. [6] designed CORE algorithm which calculated the p-value for all pairs of proteins to detect cores. Wu et al. [7] proposed COACH algorithm which detected dense subgraphs as protein-complex cores. Some classical clustering algorithms such as Markov Clustering (MCL) [8] and fuzzy clustering [9] were also developed to detect protein complexes. Lei et al. [10] proposed F-MCL clustering model based on Markov clustering and firefly algorithm which automatically adjusted the parameters by introducing the firefly algorithm.

© Springer International Publishing AG, part of Springer Nature 2018
D.-S. Huang et al. (Eds.): ICIC 2018, LNCS 10954, pp. 691–696, 2018.
https://doi.org/10.1007/978-3-319-95930-6_69

Although the above methods have been shown to identify protein complexes effectively, the result data was highly false positive and false negative due to the small world and scale-free properties of PPI network.

In recent years, quotient space theory has been applied to cluster. Zhang [11] defined the fuzzy equivalence relation and stratified hierarchical structure. Xu [12] proposed fuzzy clustering method based on Gaussian function. In this paper, a novel algorithm named Granular computation in Quotient space Clustering (GQC) is introduced to mine overlapping protein complexes. Firstly, the proposed method grains the relationship between particles through PPI data and Gene Ontology (GO) data. Then, granulating the network to construct the quotient space and merging the particles layer by layer. Finally, saving the particles un-clustered each time to increase the diversity of particles, the final complexes are obtained after purification. The algorithm is tested on two species Saccharomyces cerevisiae (DIP [13] database) and Homo sapiens (HPRD [14] database). The simulation results illustrate that GQC algorithm has a higher performance and outweighs than other algorithms.

2   Methods

In this paper, it computes the similarity between nodes according to GO data, then grains each node through graining function, finally, merges particle according to the granularity synthesis coefficient and gets the clustering results. It describes the network using $G = (X, F, T)$ in the quotient space theory [15]. Domain X refers to a collection of all nodes, F is the attribute set of X, T is the structure of X, in this paper, it is on behalf of the relation of each node in the X.

For PPI network G, it defines an equivalence relation R for domain X to construct the fuzzy quotient set [X], then defines the granularity synthesis coefficient to control the fuzzy degree, in the granularity space, ([X], [F], [T]) is called quotient space and it is a granularity classification results for the network.

The related definition is as follows:

Definition 1. The similarity between the proteins

$$SIM_{v_i, v_j} = \frac{\left| GO_{v_i} \cap GO_{v_j} \right|}{min\left(\left| GO_{v_i} \right|, \left| GO_{v_j} \right| \right)} \tag{1}$$

where GO_{v_i} and GO_{v_j} are the GO annotation set of node v_i and v_j respectively, $\left| GO_{v_i} \cap GO_{v_j} \right|$ represents the number of the same annotation between GO_{v_i} and GO_{v_j}.

Definition 2. The graining function

$G' = ([X], [F], [T]) = GrFitness(G) = \{GrFitness(v_i) | v_i \in X\}$ expresses to grain for each node, and gets the granularity space.

For $\forall v_i, v_j \in X$, if there was a side (v_i, v_j), that is, $(v_i, v_j) \in T$, and the similarity of two nodes is greater than the threshold sm, the $(v_i, v_j) \in GrFitness(v_i)$, $GrFitness(v_i) \in [X]$, $F \in [F]$, $(GrFitness(v_i), GrFitness(v_j)) \in [T]$, and the quotient space G is grained into $G' = ([X], [F], [T])$.

Definition 3. The granularity synthesis coefficient

In the granulating process, the graining result of node v_i is called GrF_i. Granularity synthesis space of network G is called $G' = \{GrF_i\}$, then merging the particles layer by layer, The nth layer of granularity synthesis space can be expressed as GrF. The degree of granularity thickness increases gradually.

The particle granularity synthesis coefficient is used to control the fuzzy degree, its formula is as follows:

$$k\left(GrF_i^n \cap GrF_j^n\right) = \frac{\left|GrF_i^n \cap GrF_j^n\right|}{\left|GrF_i^n\right| + \left|GrF_j^n\right|} \tag{2}$$

where $\left|GrF_i^n \cap GrF_j^n\right|$ is the number of the same proteins of Gr_i^n and GrF_j^n. The basic steps of the GQC algorithm can be summarized as the pseudo code as Table 1.

Table 1. Pseudo code of GQC algorithm

Algorithm GQC
Input : $G = (X, F, T) = \{PPI\ network\}$, Gene Ontology data, similarity threshold sm, granularity synthesis coefficient α.
Output : the complexes
1 : Calculate SIM_{v_i, v_j} of node v_i and v_j;
2 : for each $v_i \in X$
3 :   if $\exists (v_i, v_j) \in T$ and $SIM_{v_i, v_j} > sm$
4 : $(v_i, v_j) \in GrFitness(v_i)$;
5 :   end if
6 : end for
7 : $GrF^n = GrFitness(G)$;
8 : $\forall GrF_i^n, GrF_j^n$, calculate $k\left(GrF_i^n, GrF_j^n\right)$;
9 : if $k\left(GrF_i^n, GrF_j^n\right) > \alpha$
10 : $GrF' = GrF_i^n \cup GrF_j^n$;
11 :   Delete GrF_i^n and GrF_j^n from GrF^n;
12 : $GrF^{n+1} = GrF^n + GrF'$;
13 : end if
14 : Repeat 8-13, until $\forall GrF_i^n, GrF_j^n, k\left(GrF_i^n, GrF_j^n\right) < \alpha$
15 : $\forall GrF_i^n \in GrF^n$
16 : if $\exists GrF_j^n \subseteq GrF_i^n$
17 :   Delete GrF_i^n from GrF^n;
18 : end if
19: Output the protein complexes (GrF^n)

3   Implementation and Results

3.1   Experimental Data Set and Evaluation Metrics

In this study, the GQC method was applied to two species: Saccharomyces cerevisiae (DIP [13] (version of 20160114) database) and Homo sapiens (HPRD [14] database). After pretreatment, DIP included 5028 proteins and 22302 interactions, HPRD contained 9454 proteins and 36868 interactions. The GO annotation data was extracted from GO-slims dataset [1] and HPRD database, respectively. CYC2008 [16] is used to evaluate Saccharomyces cerevisiae clustering results, which includes 408 protein complexes. The standard protein complexes of Homo sapiens were extracted from the HPRD database, which includes 1514 protein complexes.

Three commonly used metrics sensitivity (SN), specificity (SP) and F-measure [17] are used to measure the efficiency of the proposed GQC algorithm and evaluate the performance of the clustering results.

$$SN = \frac{TP}{TP+FN}, \quad SP = \frac{TP}{TP+FP}, \quad F-measure = \frac{2 \times SN \times SP}{SN+SP} \tag{3}$$

The overlapping score OS is also used to evaluate the match quality of a predicted protein complex and standard protein complex.

$$OS(pc, sc) = \frac{\left|V_{pc} \cap V_{sc}\right|^2}{\left|V_{pc}\right| \times \left|V_{sc}\right|} \tag{4}$$

where Vpc and Vsc denote the node sets of predicted protein complex pc and standard protein complex sc, respectively. Usually we set the threshold for 0.2 [18].

3.2   Clustering Results and Analysis

This paper adopts the thought of graining to identify protein complex, it takes similarity threshold value 0.6 and granularity synthesis parameter 0.4 for performance. This paper compares the GQC algorithm with four classics algorithms MCODE, MCL, CORE, ClusterONE, COACH, the results are shown in Table 2.

Where PC is the total number of predicted protein complexes, AS represents the average size of the predicted protein complexes. The MCODE and ClusterONE are run using Cytoscape [19] and the parameters are set to the default setting. In Table 2, it is clear that GQC performs better than other five methods in terms of *SN* and *F-measure*. The *F-measure* of GQC is the highest on the DIP and HPRD datasets, it is proved that the GQC algorithm is stable and effective.

In order to clearly show the GQC algorithm results, we visualize the 408th standard protein complex of HPRD 'COM_2313' in Fig. 1. As shown in Fig. 1(a), there are 5 proteins in this standard protein complex. The GQC could mine overlapping protein complexes, as shown in Fig. 1(b), there are two protein complexes detected by GQC are considered to match standard protein complex 'COM_2313', the two protein complexes in the red dashed oval and black dashed oval, respectively.

Table 2. Comparisons of clustering results with different algorithms

Dataset	Algorithms	PC	AS	SN	SP	F-measure
DIP	MCODE	59	13.59	0.07	0.47	0.12
	MCL	623	6.57	0.36	0.38	0.3717
	CORE	1707	3.02	0.47	0.11	0.1785
	ClusterONE	371	4.9	0.36	0.42	0.39
	COACH	902	9.18	0.63	0.35	0.45
	GQC	**897**	**6.85**	**0.7444**	**0.3735**	**0.4974**
HPRD	MCODE	102	11.38	0.0663	0.2647	0.1061
	MCL	1811	4.92	0.4232	0.1171	0.1834
	CORE	2689	3.16	0.4330	0.0840	0.1408
	ClusterONE	787	4.50	0.4118	0.2668	0.3238
	COACH	1738	8.25	0.764	0.3297	0.4606
	GQC	**1983**	**6.79**	**0.649**	**0.4704**	**0.5454**

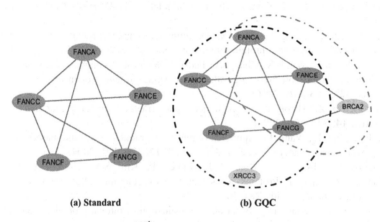

(a) Standard   (b) GQC

Fig. 1. Visualization of the 408th standard protein complex 'COM_2313' of HPRD

4   Conclusion and Discussion

This paper proposes a new method to mine overlapping protein complexes based on the natural commonalities of granular computing in quotient space. The experiments on DIP and HPRD turned out that, the proposed method is more likely to identify the clusters with large size, and has high accuracy. Though GQC performs well and achieves good results, there is certainly room for improvement. We should further study the dynamic characteristics of the PPI network.

Acknowledgement. This paper is supported by the National Natural Science Foundation of China (61672334, 61502290, 61401263) and the Fundamental Research Funds for the Central Universities, Shaanxi Normal University (GK201804006).

References

1. Zhao, J., Lei, X., Wu, F.X.: Predicting protein complexes in weighted dynamic PPI networks based on ICSC. Complexity **2017**, 1–11 (2017)
2. Lei, X., Ding, Y., Fujita, H., Zhang, A.: Identification of dynamic protein complexes based on fruit fly optimization algorithm. Knowl. Based Syst. **105**, 270–277 (2016)
3. Bader, G.D., Hogue, C.W.: An automated method for finding molecular complexes in large protein interaction networks. BMC Bioinf. **4**, 2 (2003)
4. Nepusz, T., Yu, H., Paccanaro, A.: Detecting overlapping protein complexes in protein-protein interaction networks. Nat. Meth. **9**, 471 (2012)
5. Gavin, A.C., Aloy, P., Grandi, P., Krause, R., Boesche, M., Marzioch, M., Rau, C., Jensen, L.J., Bastuck, S., Dümpelfeld, B.: Proteome survey reveals modularity of the yeast cell machinery. Nature **440**, 631–636 (2006)
6. Leung, H.C., Xiang, Q., Yiu, S.M., Chin, F.Y.: Predicting protein complexes from PPI data: a core-attachment approach. J. Comput. Biol. J. Comput. Mol. Cell Biol. **16**, 133 (2009)
7. Min, W., Li, X., Kwoh, C.K., Ng, S.K.: A core-attachment based method to detect protein complexes in PPI networks. BMC Bioinf. **10**, 1–16 (2009)
8. Van Dongen, S.: Graph clustering by flow simulation. Ph.D. thesis University of Utrecht (2000)
9. Wu, H., Gao, L., Dong, J., Yang, X.: Detecting overlapping protein complexes by rough-fuzzy clustering in protein-protein interaction networks. PLoS ONE **9**, e91856 (2014)
10. Lei, X., Wang, F., Wu, F.X., Zhang, A., Pedrycz, W.: Protein complex identification through Markov clustering with firefly algorithm on dynamic protein–protein interaction networks. Inf. Sci. **329**, 303–316 (2016)
11. Ling, Z., Bo, Z.: Theory of fuzzy quotient space (methods of fuzzy granular computing). J. Softw. **14**, 770–776 (2003)
12. Xu, F., Zhang, L., Wang, L.: Approach of the fuzzy granular computing based on the theory of quotient space. Pattern Recognit. Artif. Intell. **17**, 424–429 (2004)
13. Xenarios, I., Salwinski, L., Duan, X.J., Higney, P., Kim, S.M., Eisenberg, D.: DIP, the Database of interacting Proteins: a research tool for studying cellular networks of protein interactions. Nucleic Acids Res. **30**, 303 (2002)
14. Mathivanan, S., Periaswamy, B., et al.: An evaluation of human protein-protein interaction data in the public domain. BMC Bioinf. **7**, S19 (2006)
15. Zhao, S., Wang, K.E., Chen, J., et al.: Community detection algorithm based on clustering granulation. J. Comput. Appl. **34**, 2812–2815 (2014)
16. Pu, S., Wong, J., Turner, B., Cho, E., Wodak, S.J.: Up-to-date catalogues of yeast protein complexes. Nucleic Acids Res. **37**, 825 (2009)
17. Baldi, P., Brunak, S., Chauvin, Y., Nielsen, H.: Assessing the accuracy of prediction algorithms for classification: an overview. Bioinformatics **16**, 412 (2000)
18. Lei, X., Ding, Y., Wu, F.X.: Detecting protein complexes from DPINs by density based clustering with Pigeon-Inspired Optimization Algorithm. Sci. China Inf. Sci. **59**, 070103 (2016)
19. Tang, Y., Min, L.I.: A cytoscape plugin for visualization and clustering analysis of protein interaction networks. Chin. J. Bioinf. (2014)

Prediction of Protein-Protein Interaction Sites Combing Sequence Profile and Hydrophobic Information

Lili Peng[1,2,3], Fang Chen[6], Nian Zhou[1,2,3], Peng Chen[4], Jun Zhang[5], and Bing Wang[1,2,3(✉)]

[1] School of Electronics and Information Engineering, Tongji University, Shanghai, China
{1631719, zhounian}@tongji.edu.cn, wangbing@ustc.edu
[2] The Advanced Research Institute of Intelligent Sensing Network, Tongji University, Shanghai, China
[3] The Key Laboratory of Embedded System and Service Computing, Tongji University, Shanghai, China
[4] Institute of Health Sciences, Anhui University, Hefei, Anhui, China
pchen.ustc10@gmail.com
[5] College of Electrical Engineering and Automation, Anhui University, Hefei, Anhui, China
wwwzhangjun@gmail.com
[6] School of Information and Electromechanical Engineering, Shanghai Normal University, Shanghai, China
chenfangshnu@163.com

Abstract. Identification of the residues in protein-protein interaction sites has an important impact in a lot of biological problems. We propose an extra-trees method to identify protein interaction sites in hetero-complexes by combing profile and hydrophobic information based on extra-trees. The efficiency and the effectiveness of our proposed approach are verified by its better prediction performance compared with other methods. The experiment is performed on the 1250 non-redundant protein chains. Without using any structure data, we only use profile and a binary profile hydrophobic attribute as input vectors.

Keywords: Protein-protein interaction sites · Hydrophobic · Binary profile

1 Introduction

Protein-protein interactions play an important role in living biological cells, which control the proteins performing their specific biological functions [1, 2]. Mapping these interactions to functional sites or interactive sites will allow us to understand this biological process of how proteins recognize other molecules and provide us with clues to understanding their functions at a cellular and organism level [3, 4]. Identification of these significant interaction sites will provide useful help in drug design [5]. Current prevalent experimental methods for identifying protein-protein interaction sites are through site-directed mutagenesis like alanine scanning are costly and time-consuming

© Springer International Publishing AG, part of Springer Nature 2018
D.-S. Huang et al. (Eds.): ICIC 2018, LNCS 10954, pp. 697–702, 2018.
https://doi.org/10.1007/978-3-319-95930-6_70

[6, 7]. This prompted researchers to find effective computational methods to determine the role of protein functional residues.

In the past few decades, a growing number of machines learning computational methods used for the prediction of protein-protein interaction sites. These computational methods are used to guide experimental studies to identify interface residues and explore the functions and mechanisms of protein–protein interactions [8, 9]. Some methods are trained on a set of features derived from the sequences and/or structures of proteins with known interface sites [10–12]. Sequence-based methods mainly use these features: sequence hydrophobicity, residue composition, residue propensity and sequence alignment, sequence conservation information. Structure features have include the size of interfaces, solvent accessible surface area buried upon association, shape of interfaces, clustering of interface atoms, B-factor, electrostatic potential, spatial distribution of interface residues, and others. Among them, many machine learning methods have been developed or adopted [13–18].

Although a number of structure-based methods or hybrid methods have been proposed for the study and the effect has been verified is almost better, structure-based methods rely on structural features extracted from the structure and a lot of protein complexes is lacking the structure information. Therefore, this target of our paper is to propose a method to predict the binding sites based on sequence alone. We present a method that combines the sequence profile and the binary hydrophobic profile attributes using extra-trees algorithm. Finally, a predictor is built to predict protein-protein interaction sites. The dataset includes 194415 surface residues and positive samples accounted for 34.3%. For consequence, the predictor was used to achieve accuracy 74% and F1 55%.

2 Materials and Methods

2.1 Dataset Preparation

The complexes used in this work were extracted from the 3dComplex database, which is a database for automatically generating non-redundant sets of complexes. Only those proteins in hetero-complexes with sequence identity $\leq 30\%$ were selected in this work. Meanwhile, proteins and molecules with fewer than 30 residues were excluded from our dataset. Protein chains which are not available in HSSP database (http://swift. cmbi.ru.nl/gv/hssp/) were also removed. As a result, our dataset contains 1250 protein chains in 606 complexes. In our case, a residue is considered to be an interface residue if the difference of its solvent accessible area in unbound and bound form is $>1\text{Å}^2$. As a result, we obtained 66780 interface residues (positive samples) and 127635 non-interface residues (negative samples).

2.2 Predictor Construction

Generation of Features. It is well known that hydrophobic force is often a major driver to binding affinity. We integrate a hydrophobic scale and sequence profile in the identification of protein-protein interaction residues. In this work, Kyte-Doolittle (KD) hydrophobic scale of 20 common types of amino acids is used. In fact, we convert sequence profiles from HSSP into binary profiles with a probability threshold. Here we use a mean of the multiplication of binary profile and hydrophobic information to measure the fluctuation of residue i in its evolutionary context with respect to hydrophobicity binary profile attribute MH. Then for residue i in a protein is shown as the following form:

$$MH_i = \frac{1}{n}\sum_{k=1}^{n}\left(BP_i^k \times KD_i^k\right) \tag{1}$$

where BP_i^k and KD_i^k denote the k-th value of BP_i and KD_i for residue i, respectively, and $BP_i \times KD_i$ denotes the mean value of vector BP × KD. Note that Equation is an unbiased estimation of $BP_i^k \times KD_i^k$. In addition BP_i^k and KD_i^k represent the same amino acid type. For instance, KD_i^8 and BP_i^8 all represent residue ALA.

Predictor Construction. ET or Extra-Trees (Extremely randomized trees) is very similar to the random forest algorithm. ET is obtained using all the training samples for each decision tree, that is, each decision tree applies all the same training samples and ET is completely random to get the attribution value. We constructed five extra-trees predictors and one RF predictor, one DT predictor using residue sequence profile, hydrophobic, binary hydrophobic profile or a combination of these attributes. For the hydrophobic scale and binary hydrophobic scale based predictor, each input vector is the mean of 20 score to amino acid position. Each residue is represented by a 220-component vector in the predictor based on the residue spatial sequence profile, and by an 11-component vector in the hydrophobic scale-based predictor. For the predictor which combines residue sequence profile with hydrophobic mean score or binary hydrophobic mean, a 231-component vector is required for each amino acid residue.

2.3   Evaluation Measures of Predictor Performance

To assess our method objectively, another three indices are introduced in this paper, namely sensitivity (Sn), accuracy (ACC), precision (PPV), F-measure (F1).

(a) sensitivity or recall, measures the proportion of the known interface residues that are correctly predicted as interface residues: Sn = TP/(TP + FN);
(b) accuracy, the proportion of the known residues that are correctly predicted in all predictions: ACC = (TP + TN)/(TP + TN + FP + FN);
(c) precision, measures the proportion of the residues predicted as interface that are known interface residues: PPV = TP/(TP + FP);
(d) F-measure, combines precision and recall into their harmonic mean F1 = 2 × Sn × PPV/(Sn + PPV);

3 Results

3.1 Performance of the Predictor

We trained the predictor using a 10-fold cross validation procedure. This was carried out by splitting the dataset into ten subsets, almost equal in size. The 231 features vectors including profile and hydrophobic binary profile attribute with 11 sliding windows of the 9 train subsets input the predictor, then to predict the test subsets. The follow figure (Fig. 1) shows the results of 10 experiments. It can be seen that, compared to random forest and decision tree predictor, for imbalance data, extra-tree predictor has the better F1-measure and accuracy, which represent the stability and effect.

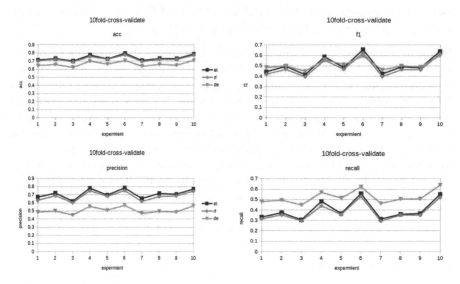

Fig. 1. The detailed performance measures across 10fold-cross-validate. Blue corresponds to extra-tree-based predictor; red denotes random-forest-based predictor; and yellow denotes the decision tree.

Table 1. The scoring efficiency of the average performing predictor in the testing phase.

Features	Sn	PPV	ACC	F1
Profile	0.34	0.68	0.72	0.34
Hydro-binary	0.28	0.61	0.70	0.38
Hydro	0.30	0.62	0.70	0.40
Profile+hydro	0.30	0.64	0.70	0.40
Profile+hydro-binary	0.47	0.71	0.74	0.55

We also construct different features-based predictor to verify the combination of profile and hydrophobic binary profile. The following table (Table 1) in this section are the average performance of different features-based predictor. It can be seen that the predictor using both residue sequence profile and hydrophobic binary profile mean as feature vectors outperforms others.

3.2 Conclusion

The experimental results show that our experimental method will yield progress for studying protein–protein interactions. Although the experimental datasets cannot contain all proteins, the experimental results are not as good as some current studies using protein structure information for prediction. The paper implies a direction for the study of prediction methods for protein interaction sites only by sequence information. The experiment verified that the hydrophobic binary profile can improve the predictor efficiency. Additionally imbalanced data of interface residues and non-interface residues is a very challenging issue, which always causes classifier over-fitting. The extra-trees algorithm may be a feasible pathway to balance training data.

References

1. Clackson, T., Wells, J.A.: A hot spot of binding energy in a hormone-receptor interface. Science **267**, 383–386 (1995)
2. Chothia, C., Janin, J.: Principles of protein-protein recognition. Nature **256**(5520), 705 (1975)
3. Kortemme, T., Baker, D.: A simple physical model for binding energy hot spots in protein-protein complexes. Proc. Natl. Acad. Sci. U.S.A. **99**, 14116–14121 (2002)
4. Keskin, O., Ma, B., Nussinov, R.: Hot regions in protein-protein interactions: The organization and contribution of structurally conserved hot spot residues. J. Mol. Biol. **345**, 1281–1294 (2005)
5. Chelliah, V., Chen, L., Blundell, T.L., Lovell, S.C.: Distinguishing structural and functional restraints in evolution in order to identify interaction sites. J. Mol. Biol. **342**, 1487–1504 (2004)
6. Williams, N.E.: Immunoprecipitation procedures. Meth. Cell Biol. **62**, 449–453 (1999)
7. Wells, J.A.: Systematic mutational analyses of protein-protein interfaces. Meth. Enzymol. **202**, 390–411 (1991)
8. Fernandezrecio, J.: Prediction of protein binding sites and hot spots. Wiley Interdisc. Rev. Comput. Mol. Sci. **1**(5), 680–698 (2011)
9. Esmaielbeiki, R., Krawczyk, K., Knapp, B., Nebel, J.C., Deane, C.M.: Progress and challenges in predicting protein interfaces. Brief Bioinf. **17**, 117–131 (2016)
10. Lise, S., Buchan, D., Pontil, M., Jones, D.T.: Predictions of hot spot residues at protein-protein interfaces using support vector machines. PLoS ONE **6**, e16774 (2011)
11. Lise, S., Archambeau, C., Pontil, M., Jones, D.T.: Prediction of hot spot residues at protein-protein interfaces by combining machine learning and energy-based methods. BMC Bioinf. **10**, 365 (2009)
12. Wang, L., Liu, Z.P., Zhang, X.S., Chen, L.: Prediction of hot spots in protein inter faces using a random forest model with hybrid features. Protein Eng. Des. Sel. **25**, 119–126 (2012)

13. Wang, B., Chen, P., Zhang, J.: Protein interface residues prediction based on amino acid properties only. In: Huang, D.-S., Gan, Y., Premaratne, P., Han, K. (eds.) ICIC 2011. LNCS, vol. 6840, pp. 448–452. Springer, Heidelberg (2012). https://doi.org/10.1007/978-3-642-24553-4_59

14. Chen, P., Wong, L., Li, J.: Detection of outlier residues for improving interface pre diction in protein heterocomplexes. IEEE/ACM Trans. Comput. Biol. Bioinf. **9**, 1155–1165 (2012)

15. Wang, B., Huang, D.S., Jiang, C.: A new strategy for protein interface identification using manifold learning method. IEEE Trans. Nanobiosci. **13**(2), 118–123 (2014)

16. Chen, H., Zhou, H.: Prediction of interface residues in protein-protein complexes by a consensus neural network method: test against NMR data. Proteins **61**, 21–26 (2005)

17. Wang, B., Chen, P., Zhang, J., et al.: Inferring protein-protein interactions using a hybrid genetic algorithm/support vector machine method. Protein Pept. Lett. **17**(9), 1079 (2010)

18. Bystroff, C., Krogh, A.: Hidden Markov models for prediction of protein features. In: Zaki, M.J., Bystroff, C. (eds.) Protein Structure Prediction. Methods in Molecular Biology, vol. 413, pp. 173–198. Humana Press, Totowa (2007)

Prediction of Indoor PM2.5 Index
Using Genetic Neural Network Model

Hongjie Wu[1,2(✉)], Cheng Chen[1], Weisheng Liu[3], Ru Yang[1],
Qiming Fu[1,2], Baochuan Fu[1,2], and Dadong Dai[1]

[1] School of Electronic and Information Engineering, Suzhou University
of Science and Technology, Suzhou 215009, China
Hongjie.wu@qq.com
[2] Jiangsu Province Key Laboratory of Intelligent Building Energy Efficiency,
Suzhou University of Science and Technology, Suzhou 215009, China
[3] Suzhou Municipal Hospital (North Area), Suzhou 215000, China

Abstract. Since people spend more than 80% of the daytime in indoor environment every day, the effect on people's health of the indoor PM2.5 is much greater than outdoor PM2.5. This paper proposes a method based on genetic neural network to predict the indoor PM2.5. We use seven features including indoor ventilation rate, air temperature, relative humidity and others to train the model. The experiment results showed that the relative error is 5.60%, which is 7.55% lower than the traditional artificial neural network, 5.98% lower than the support vector regression method, 8.36% lower than the Random Forest.

Keywords: Indoor PM2.5 · Genetic algorithm · Ventilation rate

1 Introduction

The prediction of air quality has become an important requirement for people's daily life. In fact, people stay in indoor environment for more than 80% of the whole day [1], and indoor air pollution always causes much more harm to the human body than outdoor air pollution. Indoor environment is closely related to people's health. So the monitoring and prediction of indoor PM2.5 concentration is of great significance.

Many factors may affect the concentration of pollutants in the air like temperature, humidity [2]. So the concentration of pollutants in air is highly complex and uncertain. Many researchers have begun to use machine learning methods to do research and exploration on air quality prediction. The existing research results show that the air quality predict model based on machine learning algorithm can get higher prediction accuracy [3]. The main reason is that machine learning method has better fitting for nonlinear mapping. However, most of the current researches focused on the prediction of outdoor air pollutants, while the prediction of indoor air pollutants by machine learning is rarely mentioned [4].

Compared with outdoor PM2.5, indoor PM2.5 is affected by more factors. Therefore, indoor PM2.5 prediction is facing more difficulties which causes almost no research for prediction of indoor PM2.5 at present [5]. In general, indoor PM2.5 mainly comes from the PM2.5 of outdoor and various types of combustion in the indoor air.

© Springer International Publishing AG, part of Springer Nature 2018
D.-S. Huang et al. (Eds.): ICIC 2018, LNCS 10954, pp. 703–707, 2018.
https://doi.org/10.1007/978-3-319-95930-6_71

Besides the direct contribution of indoor and outdoor pollution sources, indoor ventilation status, season, temperature and humidity level will also directly affect the pollution level of indoor PM2.5 [6]. People only know that there is a nonlinear mapping relationship between the concentration of indoor PM2.5 and the above factors, but the specific relationship formula can not be calculated. However, neural network can fit any continuous mapping with any accuracy [7]. Based on this, this paper proposes a real-time indoor PM2.5 concentration prediction method based on genetic neural network. Genetic algorithm is applied to the indoor air quality predict novelly, and ventilation rate, air temperature, relative humidity and other 7 features which are closely related to indoor PM2.5 were selected for modeling.

2   Data and Method

2.1   Dataset and Feature

The monitoring points were set in the room and outside the room. We collected data in summer, autumn and winter. The collected data amount to 28800. The prediction model can be defined as follow:

$$l = G(t, p_{out}, w_{in}, w_{out}, s_{in}, s_{out}, v) \tag{1}$$

The current indoor PM2.5 concentration is related to 7 parameters which correspond to the 7 inputs of the neural network. 9000 data in each dataset is used as the training set of the network, and 600 data is used as the test set. The characteristics of the network are as follows (Table 1):

Table 1. Features for the neural network

Feature	Unit	Range	Explanation
t	h	0–24	Current time point
p_{out}	µg/m³	0–150	Outdoor PM2.5
w_{out}	°C	−20–50	Outdoor temperature
w_{in}	°C	−20–50	Indoor temperature
s_{out}	%	−20–50	Outdoor humidity
s_{in}	%	0–100%	Indoor humidity
r	/	0–100	Indoor ventilation rate

r is the indoor ventilation rate in the room, and the calculate formula is as follow:

$$r = S * |w_{out} - w_{in}| \tag{2}$$

In (2), S is the open area of window, w_{out} is outdoor temperature, and w_{in} is indoor temperature. Before the experiment, the data is normalized as follow:

$$x^* = (x - \mu)/\sigma \tag{3}$$

μ is the mean of all sample data, and σ is the standard deviation of all sample data.

2.2   Genetic Neural Network Model

The neural network of 3 layers can reproduce any continuous function arbitrarily. The neural network has the characteristic of self-adaptive, self-organizing and real-time learning. In general, the network errors are usually used to update the network parameters by error derivatives. It is not a global optimization algorithm from the perspective of principle. If the network weights and bias are initialized randomly, the output result will fall into the local minimum point.

Genetic algorithm is a global optimization algorithm. Different from the traditional optimization algorithm, genetic algorithm can deal with different individuals in a population and guide the search direction according to the uncertainty principle. The algorithm expands the coverage of solutions and the diversity of the search direction greatly so that it can reduce the possibility of falling into local minimum point. Therefore, we can use genetic algorithm to select the initial weights and bias of the network so as to reduce the possibility of obtaining the local minimum value, and improve the predict performance of neural network (Fig. 1).

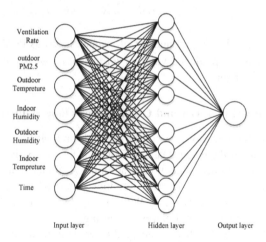

Fig. 1. Block diagram of neural network model

In the model, evaluation function of genetic algorithm is defined as follow:

$$\text{SSE} = \sum_{i=1}^{n} \left(y_i - y_i' \right)^2 \tag{4}$$

$$\text{Fitness} = 1/\text{SSE} \tag{5}$$

In (4), y_i represents the real output of ith sample, y_i' represents the predict output.

3 Experiments and Discussion

3.1 Evaluating Metrics

Many indexes can be used to evaluate the performance of the model, but the fundamental purpose is to analyze the difference between the real value and predictive value. For regression problems, this difference can be evaluated from two aspects: the degree of deviation and consistency between the real and the predicted values. Therefore, this paper uses 3 indicators to evaluate the predicted result respectively: relative error (RE), root mean square error (RMSE) and IA (Index of Agreement). RE and RMSE reflect the deviation between predicted and true value, IA reflecting the consistency between the predicted value and the real value. The calculation formulas are as follows:

$$RE = \sqrt{\frac{\sum_{i=1}^{n} \frac{|obs_i - pred_i|}{obs_i}}{n}} \times 100\% \tag{6}$$

$$RMSE = \sqrt{\frac{\sum_{i=1}^{n}(obs_i - pred_i)^2}{n}} \tag{7}$$

$$IA = 1 - \frac{\sum_{i=1}^{n}(obs_i - pred_i)^2}{\sum_{i=1}^{n}\left(|pred_i - \overline{obs}| + |obs_i - \overline{obs}|\right)^2} \tag{8}$$

In formula above, n is the total number of instances. obs_i is PM2.5 real value of ith instance, and $pred_i$ is the predicted value of ith instance. $\overline{obs}$ is the average value the actual value of the test set.

3.2 Comparison Between GA and Other 3 Prediction Methods

According to the definition above, as for the RE and the RMSE, the smaller the better. As for the IA, the larger the better. Data in Table 2 is the comparison of predict results of Genetic Artificial (GA) neural network, traditional Artificial Neural Network(ANN), Support Vector Regression (SVR) and Random Forest(RF). We had 6-fold cross validation in our experiments. We have 9600 pieces data for each season and we use 8000 pieces for training and 1600 pieces for testing.

From Table 2 we can see that for air quality prediction, compared with the traditional ANN, SVR and RF method, GA neural network shows better predicting performance on 3 datasets. The average RE are decreased by 7.55%, 5.98% and 8.36%. RMSE are decreased by 0.04, 0.07 and 0.07. IA are increased by 0.13, 0.08 and 0.17. Indoor PM2.5 is affected by many factors and has more complex mapping relationship. Neural networks have the inherent ability to adjust their own synaptic weights to adapt to the changes in the environment. It allows the neural network to find some approximate solutions to the complex problems that are currently difficult to deal with.

Table 2. Prediction results comparison between GA, ANN, SVR and RF

Dataset	Summer			Autumn			Winter			Average		
Method	RE (%)	RMSE	IA	RE (%)	RMSE	IA	RE (%)	RMSE	IA	RE	RMSE	IA
GA	6.09	0.03	0..95	7.69	0.02	0.82	3.03	0.23	0.95	5.60	0.02	0.91
ANN	16.6	0.69	0.91	13.01	0.06	0.66	9.85	0.06	0.78	13.15	0.06	0.78
SVR	11.38	0.09	0.86	12.95	0.10	0.85	10.42	0.07	0.78	11.58	0.09	0.83
RF	15.38	0.14	0.81	6.36	0.06	0.54	10.15	0.09	0.87	13.96	0.09	0.74

4 Conclusions

Based on the experiment and analysis of the GA model above, it shows that taking factors such as time, temperature and humidity, ventilation rate and outdoor PM2.5 concentration as network inputs can accurately predict indoor PM2.5 concentration with the model we proposed. Finally, we collected 28800 data in different seasons from the hospital room for experimental verification. The results showed that the relative error was 5.60%, which was 7.55% lower than that of the traditional ANN, 5.98% lower than that of the SVR, 8.36% lower than that of the RF method. This result is feasible for the occurrence of early warning pollution in the hospital room.

Acknowledgements. This paper is supported by the National Natural Science Foundation of China (61772357, 61502329, 61672371), Jiangsu 333 talent project and top six talent peak project (DZXX-010), Suzhou Foresight Research Project (SYG201704, SNG201610) and Postgraduate Research & Practice Innovation Program of Jiangsu Province (SJCX17_0680).

References

1. Ma, Z.W., Hu, X.F.: Estimating ground-level PM2.5 in China using satellite remote sensing. Environ. Sci. Technol. **48**(13), 7436–7441 (2014)
2. Cincinelli, A., Martellini, T.: Indoor air quality and health. Int. J. Environ. Res. Public Health **14**(11), 4535–4564 (2017)
3. Phala, K.S.E., Kumar, A., Hancke, G.P.: Air quality monitoring system based on ISO/IEC/IEEE 21451 standards. IEEE Sens. J. **16**(12), 5037–5045 (2016)
4. Hong, B., Qin, H.: Prediction of wind environment and indoor/outdoor relationships for PM2.5 in different building–tree grouping patterns. Atmosphere **9**(2), 39–43 (2018)
5. Huang, Y., Yuan, X.: Present situation and development of indoor PM2.5 pollution control. Shanxi Archit. **11**(2), 85–90 (2017)
6. Kuang, C.L.: Influence of relative humidity on real-time measurement of indoor PM2.5 concentration. Environ. Sci. Technol. **40**(1), 107–111 (2017)
7. Anders, U., Korn, O., Schmitt, C.: Improving the pricing of options: a neural network approach. J. Forecast. **17**(5), 369–388 (2016)

A Two-Phase Variable Neighborhood Search for Flexible Job Shop Scheduling Problem with Energy Consumption Constraint

Chengzhi Guo and Deming Lei[✉]

School of Automation, Wuhan University of Technology, Wuhan 430070, China
demingleill@163.com

Abstract. This paper investigates flexible job shop scheduling problem (FJSP) with energy consumption constraint, the goal of which is to minimize makespan and total tardiness under the constraint that total energy consumption doesn't exceed a given threshold. Energy consumption constraint is not always met and a new method for this constraint is proposed. A two-phase variable neighborhood search (TVNS) is presented. In the first phase, the problem is converted into FJSP with makespan, total tardiness and total energy consumption and a VNS is applied for the new problem. In the second phase, another VNS is for the original problem by strategies for comparing solutions and updating the non-dominated set Ω of the first phase. The current solution of TVNS is replaced with a member of Ω every a prefixed number of iterations to improve solution quality. Extensive experiments are conducted and computational results validate the effectiveness and advantages of TVNS for the considered FJSP.

Keywords: Flexible job shop scheduling problem
Energy consumption constraint · Variable neighborhood search

1 Introduction

There often exits several conflicting objectives in real life flexible job shop and it is necessary to consider multi-objective flexible job shop scheduling problem (MOFJSP).

Many works have been done on MOFJSP by using various meta-heuristics, such as genetic algorithm (GA [1–8]), PSO [9, 10], harmony search algorithm [11], artificial bee colony [12], tabu search [13], variable neighborhood search (VNS [14]), shuffled frog leaping algorithm (SFLA [15]) and estimation of distribution algorithm [16].

In recent years, energy-efficient scheduling problem, which has objective of improving energy efficiency, has been considered extensively. Regarding energy-efficient FJSP, it is often MOFJSP because of the inclusion of energy related objective and a number of low carbon strategies and optimization methods have been obtained [17–23].

In this study, we consider MOFJSP with energy consumption constraint, the goal of which is to minimize total tardiness and makespan under the constraint that total energy consumption doesn't exceed a given threshold. A new strategy is applied to deal with energy consumption and two-phase variable neighborhood search (TVNS) is proposed.

© Springer International Publishing AG, part of Springer Nature 2018
D.-S. Huang et al. (Eds.): ICIC 2018, LNCS 10954, pp. 708–716, 2018.
https://doi.org/10.1007/978-3-319-95930-6_72

In the first phase, the problem is converted into FJSP with makespan, total tardiness and total energy consumption and a VNS is applied to generate feasible solutions of the original problem by solving the new problem. In the second phase, some ways are first adopted to compare solutions and update the non-dominated set Ω of the first phase and then another VNS is used to solve the original problem. The current solution of TVNS is replaced with a member of Ω when the prefixed condition is met. We examine the impact of the new strategies of TVNS on its performance and compare TVNS with other methods from the literature and the results validate that TVNS is an effective and efficient method for multi-objective FJSP with energy consumption constraint.

2 Problem Description

MOFJSP with energy consumption constraint is described as follows. There are a set of jobs $J = \{J_1, J_2, \cdots, J_n\}$ and a set of machines $M = \{M_1, M_2, \cdots, M_m\}$. Job J_i consists of h_i operations. Operation o_{ij} is the jth operation of job J_i and can be processed on a set S_{ij} of compatible machines, $S_{ij} \subset M$. Each machine M_k exists two modes: processing mode and stand-by mode. E_k is the energy consumption per unit time in processing mode and SE_k is the energy consumption per unit idle time.

There are several constraints on jobs and machines, such as, each machine can process at most one operation at a time, no jobs may be processed on more than one machine at a time, operations cannot be interrupted and setup times and remove times are included in the processing times etc.

A new constraint is added below.

$$TEC = \int_0^{C_{\max}} \left(\sum_{i=1}^n \sum_{j=1}^{h_i} \sum_{k=1}^m E_k y_{ijk}(t) + \sum_{k=1}^m SE_k z_k(t) \right) dt \leq Q_{EC} \quad (1)$$

Where TEC denotes total energy consumption. $y_{ijk}(t)$ is a binary variable. If machine $M_k \in S_{ij}$ is in processing mode at time t, then $y_{ijk}(t)$ is equal to 1; otherwise $y_{ijk}(t)$ is 0. $z_k(t)$ is 1 if M_k is in stand-by mode at time t and 0 otherwise. $C_{\max}$ indicates the maximum completion time of all jobs. Q_{EC} is a given energy consumption threshold.

The goal of the problem is to minimize simultaneously the following two objectives under the condition that constraints are all met.

$$f_1 = C_{\max} \quad (2)$$

$$f_2 = \sum_{i=1}^n \max\{C_i - D_i, 0\} \quad (3)$$

3 TVNS for FJSP with Energy Consumption Constraint

3.1 Initialization

For our FJSP, a solution is represented by a machine assignment string $[q_{11}, q_{12}, \cdots,$ $q_{1h_1}, \cdots, q_{nh_n}]$, a scheduling string $[(\theta_1, r_1), (\theta_2, r_2), \cdots, (\theta_i, r_i), \cdots, (\theta_h, r_h)]$ $h =$ $\sum_{i=1}^{n} h_i$ is the total number of operations. In the first string, a gene $q_{ij} \in S_{ij}$ represents a compatible machine for operation o_{ij}. In the second string, a doublet (θ_i, r_i) indicates an operation $o_{\theta_i r_i}$ and the whole string corresponds to an ordered operation list $[o_{\theta_1 r_1}, o_{\theta_2 r_2}, \cdots, o_{\theta_i r_i}, \cdots, o_{\theta_h r_h}]$, $\theta_i \in \{1, 2, \cdots, n\}$, $1 \le r_i \le h_{\theta_i}$. The decoding proce- dure is provided as follows. (1) Convert $[o_{\theta_1 r_1}, o_{\theta_2 r_2}, \cdots, o_{\theta_i r_i}, \cdots, o_{\theta_h r_h}]$ into an ordered operation list $[o_{\theta_1 r_1}, o_{\theta_2 r_2}, \cdots, o_{\theta_i r_i}, \cdots, o_{\theta_h r_h}]$. (2) Start with the first operation in the ordered operation list, allocate each operation $o_{\theta_i r_i}$ on its assigned machine $q_{\theta_i r_i}$ in the best available time.

The previous works [20, 23] often apply the above decoding procedure and have shown that all constraints of our FJSP except $TEC \le Q_{EC}$ are met, so only $TEC \le Q_{EC}$ is required to be considered specially. An initial solution x is randomly generated and an initial non-dominated set Ω is constructed.

3.2 The First Phase

In this phase, TVNS tries to generate enough feasible solutions. To implement the above purpose, energy constraint is deleted from the considered FJSP and TEC is added as f_3, as a result, the problem is transformed into FJSP with f_1, f_2 and f_3.

In this study, three neighborhood structures are utilized, which are shown as follow. Neighborhood structure *change* is performed on the first string $[q_{11}, q_{12}, \cdots q_{1h_1}, \cdots, q_{n1}, \cdots, q_{nh_n}]$, which is described below: a set $\Theta = \{q_{ij} || S_{ij}| > 1\}$ is first constructed, some elements are randomly selected from the set Θ, suppose that q_{ij} is chosen, then q_{ij} is replaced with a selected machine from S_{ij} in a random way. Neighborhood structure *insert* and *swap* are designed for scheduling string.

insert produces new solution by deciding an element (θ_i, r_i) and a position $k \ne i$, inserting the element into the position and assigning new values of all r_i again. *swap* generates new solutions by exchanging some pairs of doublets from the scheduling string and reassigning new value of r_i for each θ_i [22].

$\mathcal{N}_1, \mathcal{N}_2, \mathcal{N}_3$ indicate *insert*, *change* and *swap*, respectively and $\mathcal{N}_i(x)$ represents the neighborhood of x by using $\mathcal{N}_i$.

When a new solution $z \in \mathcal{N}_i(x)$ is produced and compared with x based on Pareto dominance on f_1, f_2, f_3, if z dominates x or is non-dominated with x, then x is replaced with z.

The set Ω is renewed in the following way: x is added into Ω, then all solutions in Ω are compared based on Pareto dominance on f_1, f_2, f_3 and the dominated solutions are removed from the set Ω.

The detailed steps of the first phase are listed below.

(1) Transform the problem into FJSP with f_1, f_2, f_3.
(2) $w \leftarrow 1, g \leftarrow 1$.
(3) Randomly generate a solution $z \in \mathcal{N}_g(x)$.
(4) If z dominates or non-dominated with x, then replace x with z, update the set Ω
     and $g \leftarrow 1$.
     else $g \leftarrow g + 1$ and let $g \leftarrow 1$ if $g = 3$.
(5) If w is exactly divided by an integer β, then select a solution $y \in \Omega$ and substitute
     for x.
(6) $w \leftarrow w + 1$, if $w \leq \max\_it_1$, then go to (3); otherwise stop the search of the first
     phase.
     Where $\max\_it_1$ is the number of iterations in the first phase.

When energy consumption constraint is met, it is necessary to optimize f_1, f_2 further
to produce high quality solutions, so after enough feasible solutions are generated, the
search of the first phase should be terminated. This is also the reason that two-phase
structure is used.

3.3   The Second Phase

After the first phase, if we continue the optimization of FJSP with f_1, f_2 and f_3, the
improvement on *TEC* of feasible solutions is not necessary, so the original FJSP with
f_1, f_2 and $TEC \leq Q_{EC}$ should be considered to improve the obtained solutions of the first
phase further.

In our problem, *TEC* is limited and not optimized. Obviously, Q_{EC} must be greater
than the optimal value of *TEC* and feasible solutions can be produced after the first
phase. A solution is stochastically chosen from the set Ω and acts as the current solution
x of VNS in the second phase.

The main steps of the second phase are shown below.

(1) Randomly choose a member of Ω as the current solution x.
(2) $g \leftarrow 1, w \leftarrow 1 + \max\_it_1$.
(3) If $g = 1$, randomly generate a new solution $z \in \mathcal{N}_3(x)$; if $g = 2$, produce a
     solution $z \in \mathcal{N}_1(x)$ randomly; if $g = 3$, stochastically provide a solution
     $z \in \mathcal{N}_2(x)$.
(4) If x can be replaced with z according to a principle, then replace x with z, update
     the set Ω and $g \leftarrow 1$; else $g \leftarrow g + 1$ and let $g \leftarrow 1$ if $g = 4$.
(5) If w is exactly divided by an integer β, then select a solution $y \in \Omega$ and substitute
     for x.
(6) $w \leftarrow w + 1$, if $w \leq \max\_it$, then go to (3); otherwise stop the search of TVNS.
     Where $\max\_it$ is the maximum number of iterations.

The principle is developed to decide if x can be updated with the new one z, which
is shown as follows. if one of the following three conditions is met, then x is replaced
with z.

(1) $TEC(x) > Q_{EC}$ and $TEC(z) \leq Q_{EC}$;
(2) $TEC(x) \leq Q_{EC}$, $TEC(z) \leq Q_{EC}$ and z is non-dominated with or dominates x according to Pareto dominance on objectives f_1, f_2;
(3) $TEC(x) > Q_{EC}$, $TEC(z) > Q_{EC}$, $TEC(z) \leq TEC(x)$.

In the first phase, the set Ω is utilized to store non-dominated solutions decided by f_1, f_2, f_3. In the second phase, the set Ω may consist of some infeasible solutions and a group of non-dominated feasible solutions.

When the current solution x is replaced with the new one z, the set Ω is renewed as follows.

(1) **If** x is infeasible

If there is at least one infeasible solution in the set Ω and $TEC(x) < TEC(y)$, **then** x substitutes for $y \in \Omega$, which is an infeasible member of Ω.

(2) **If** x is feasible

If all members of Ω are feasible, then
If the solution x have the same f_1, f_2 as a member of Ω and less TEC than the member, the member is directly replaced with x;

otherwise add x into the set Ω and compare all solutions in the set Ω according to Pareto dominance on f_1, f_2 and remove the dominated ones from the Ω.

If not all members of Ω are feasible, then replace a randomly chosen infeasible member with x.

In the above procedure, at most an infeasible solution is removed from the set Ω when Ω is updated once to keep enough members in Ω for substituting for the current solution with more choices.

3.4   Algorithm Description

After the initial solution x is generated and the initial set Ω is constructed, the first phase and the second phase are executed sequentially. The notable characteristic of our FJSP is that $TEC \leq Q_{EC}$ is frequently not met. The structure of TVNS effectively adapts to this feature and a new way is applied to deal with energy consumption constraint; moreover, the search of TVNS often re-starts from a new solution and this feature can intensify its exploration ability, thus, TVNS is likely to be an appropriate method for solving multi-objective FJSP with energy consumption constraint.

4 Computational Experiments

Extensive experiments are conducted on a set of problems to test the performance of TVNS for the considered FJSP. All experiments are implemented by using Microsoft Visual C++ 6.0 and run on 4.0G RAM 1.70 GHz CPU PC.

We choose 12 instances DP1-12 [25], which are designed for testing FJSP. We extend them by adding energy consumption information $E_k \in [2, 4]$ and $SE_k = 1$. The detailed data are shown in appendix. Duedate is calculated in the following:

$$D_i = \delta_i \sum_{j=1}^{h_i} \max_{k=1,2,\cdots,m} \{p_{ijk}\} \tag{4}$$

where δ_i is a real number and also listed in the appendix, $i = 1, 2, \cdots, n$.

In this study, *TEC* is limited to Q_{EC} to meet special requirements such as on-time delivery. We decide Q_{EC} based on the results of the first phase of TVNS on FJSP with f_1, f_2, f_3. The results on Q_{EC} are presented in Table 1.

Many performance metrics have been applied to evaluate the results of multi-objective optimization algorithms. The following two metrics are chosen in this study.

Metric DI_R [26] is often used to measure the convergence performance by computing the distance of the non-dominated set Ω_l relative to a reference set Ω^*.

$$DI_R(\Omega_l) = \frac{1}{|\Omega^*|} \sum_{y\in\Omega^*} \min\{\sigma_{xy}|x \in \Omega_l\} \tag{5}$$

where σ_{xy} is the distance between a solution x and a reference solution y in the normalized objective space, The reference set Ω^* consists of the non-dominated solutions of $\bigcup_{l=1}^{A} \Omega_l$, A is the total number of algorithms.

The smaller the value of $DI_R(\Omega_l)$ is, the better the solutions of Ω_l are.

Metric ρ_l [24] indicates the ratio of number of the elements in the set $\{x \in \Omega_l | x \in \Omega^*\}$ to $|\Omega^*|$.

In this study, NSGA-II [7] and VNS [14] are chosen.

We extended this version of NSGA-II [7] to our FJSP in the following way: the parts on machine breakdown are first deleted, rank value and crowding distance of all feasible solutions are obtained using the original non-sorting method and crowding distance assignment, all infeasible solutions are assigned into the same front and their rank is bigger than the rank of all feasible solutions, crowding distance of an infeasible solution x is set to be $\omega_{\max} - TEC(x)$, where $\omega_{\max}$ is a big enough positive number. The other parts of NSGA-II are directly kept.

Bagheri and Zandieh [14] introduced a VNS for FJSP with the aggregate objective value of total tardiness and makespan. This VNS can be applied to solve our FJSP by using the principle used to decide if the current solution can be replaced with a new one, and the updating method of Ω in Sect. 3.3.

Parameter settings of TVNS is as follows: β is 5000, *max_it* is 10^5 and *max_it$_1$* is 20000.

For NSGA-II, population scale of 100, crossover probability of 0.8, maximum generation of *max_it*/100 and mutation probability of 0.1.

For VNS, $n_{max} = 350$, *max_it* is same as TVNS fairly. $n_{max} = 350$ is decided by Bagheri and Zandieh [14] and we adopt it directly.

Tables 1, 2 and 3 describe the results and the average running time of TVNS, VNS and NSGA-II. It can be found from Tables 1, 2 and 3 that TVNS performs better than VNS and NSGA-II on two metrics using less time. TVNS provides better DI_R than VNS and NSGA-II on 10 instances and bigger ρ_l than VNS and NSGA-II on 10 instances. The comparison among TVNS2, VNS and NSGA-II presents that our TVNS has advantages even it is directly applied to solve the considered problem.

Table 1. Results of three algorithms on metric DI_R

Ins.	Q_{EC}	TVNS	NSGA-II	VNS	Ins.	Q_{EC}	TVNS	NSGA-II	VNS
DP1	34534	0.000	23.69	9.694	DP7	62373	0.000	36.38	9.172
DP2	40332	3.254	40.13	9.991	DP8	52422	15.23	39.67	19.51
DP3	36428	1.064	46.93	13.00	DP9	64539	2.356	62.38	29.88
DP4	36880	5.456	35.71	13.55	DP10	60350	0.000	64.70	30.50
DP5	40085	13.18	49.85	0.000	DP11	57613	11.31	43.74	21.78
DP6	37091	0.000	47.27	10.21	DP12	60711	13.90	62.12	11.78

Table 2. Results of three algorithms on metric ρ

Instance	TVNS	NSGA-II	VNS	Instance	TVNS	NSGA-II	VNS
DP1	1.000	0.000	0.000	DP7	1.000	0.000	0.000
DP2	0.940	0.000	0.060	DP8	0.572	0.000	0.428
DP3	0.466	0.000	0.000	DP9	0.875	0.000	0.125
DP4	0.750	0.000	0.250	DP10	1.000	0.000	0.000
DP5	0.000	0.000	1.000	DP11	0.550	0.050	0.400
DP6	1.000	0.000	0.000	DP12	0.350	0.000	0.650

Table 3. Running times of TVNS, NSGA-II and VNS

Ins.	Average running time (s)			Ins.	Average running time (s)		
	TVNS	NSGA-II	VNS		TVNS	NSGA-II	VNS
DP1	11.9	10.8	13.0	DP7	17.2	18.1	19.7
DP2	12.7	11.4	14.5	DP8	16.7	17.5	19.8
DP3	10.9	10.9	14.6	DP9	16.2	17.8	18.8
DP4	10.6	11.5	11.8	DP10	17.3	17.8	20.2
DP5	9.60	11.9	11.7	DP11	16.7	18.0	19.7
DP6	9.47	11.2	11.5	DP12	17.8	17.9	19.8

The notable characteristics of TVNS include its two-phase structure, new strategies for comparing solutions and updating the set Ω and the updating of current solution every β iterations. TVNS first explores in objective space of the problem with f_1, f_2, f_3 and then concentrates on the objective space of the original problem. The updating of current solution with a member of Ω diminishes the possibility of falling local optimum, as a result, the good balance between exploration and exploitation is made, thus, TVNS is an effective and competitive method for FJSP with energy consumption threshold.

5   Conclusions

Multi-objective FJSP has been extensively investigated in the past decade; however, multi-objective FJSP with energy consumption constraint is seldom considered. In this study, a novel algorithm named TVNS is proposed to minimize makespan and total tardiness under the constraint that total energy consumption doesn't exceed a given threshold. In the first phase, the problem is converted into FJSP with makespan, total tardiness and total energy consumption and a VNS is presented for the new problem. In the second phase, some strategies are applied to compare solutions and update the set Ω of the first phase and another VNS is used for the original problem. The current solution of TVNS is replaced with a member of Ω every β iterations to improve solution quality. We conduct a number of experiments and the computational results show that TVNS can provide better results than the methods from the literature.

Acknowledgement. This work is supported by the National Natural Science of Foundation of China (61573264).

References

1. Kacem, I., Hammadi, S., Borne, P.: Pareto-optimality approach for flexible job-shop scheduling problems: hybridization of evolutionary algorithms and fuzzy logic. Math. Comput. Simul. **60**(3–5), 245–276 (2002)
2. Gao, J., Gen, M., Sun, L., Zhao, X.: A hybrid of genetic algorithm and bottleneck shifting for multi-objective flexible job shop scheduling problems. Comput. Ind. Eng. **53**(1), 149–162 (2007)
3. Yuan, Y., Xu, H.: Multiobjective flexible job shop scheduling using memetic algorithms. IEEE Trans. Autom. Sci. Eng. **12**(1), 336–353 (2015)
4. Rohaninejad, M., Kheirkhah, A., Fattahi, P., Vahedi-Nouri, B.: A hybrid multi-objective genetic algorithm based on the ELECTRE method for a capacitated flexible job shop scheduling problem. Int. J. Adv. Manufact. Technol. **77**(1), 51–66 (2015)
5. Rohaninejad, M., Sahraeian, R., Nouri, B.V.: Multi-objective optimization of integrated lot-sizing and scheduling problem in flexible job shop. PAIRO Oper. Res. **50**(3), 587–609 (2015)
6. Li, J., Huang, Y., Niu, X.: A branch population genetic algorithm for dual-resource constrained job shop scheduling problem. Comput. Ind. Eng. **102**(1), 113–131 (2016)
7. Ahmadi, E., Zandieh, M., Farrokh, M., Emami, S.M.: A multi objective optimization approach for flexible job shop scheduling problem under random machine breakdown by evolutionary algorithm. Comput. Oper. Res. **73**(1), 56–66 (2016)
8. Shen, X.N., Han, Y., Fu, J.Z.: Robustness measures and robust scheduling for multi-objective stochastic flexible job shop scheduling problems. Soft Comput. (2018, in press)
9. Moslehi, G., Mahnam, M.: A Pareto approach to multi-objective flexible job-shop scheduling problem using particle swarm optimization and local search. Int. J. Prod. Econ. **129**(1), 14–22 (2011)
10. Singh, M.R., Singh, M., Mahapatra, S.S., Jagadev, N.: Particle swarm optimization algorithm embedded with maximum deviation theory for solving multi-objective flexible job shop scheduling problem. Int. J. Adv. Manuf. Technol. **85**(9), 2353–2366 (2016)

11. Gao, K.Z., Suganthan, P.N., Pan, Q.K., Chua, T.J., Cai, T.X., Chong, C.S.: Pareto-based grouping discrete harmony search algorithm for multi-objective flexible job shop scheduling. Inf. Sci. **289**(1), 76–90 (2014)

12. Li, J.Q., Pan, Q.K., Tasgetiren, M.F.: A discrete artificial bee colony algorithm for the multi-objective flexible job-shop scheduling problem with maintenance. Appl. Math. Model. **38** (3), 1111–1132 (2014)

13. Jia, S., Hu, Z.H.: Path-relinking tabu search for the multi-objective flexible job shop scheduling problem. Comput. Oper. Res. **47**(1), 11–26 (2014)

14. Bagheri, A., Zandieh, M.: Bi-criteria flexible job-shop scheduling with sequence-dependent setup times-variable neighborhood search approach. J. Manuf. Syst. **30**(1), 8–15 (2011)

15. Li, J.Q., Pan, Q.K., Xie, S.X.: An effective shuffled frog-leaping algorithm for multi-objective flexible job shop scheduling problems. Appl. Math. Comput. **218**(18), 9353–9371 (2012)

16. Wang, L., Wang, S.Y., Liu, M.: A Pareto-based estimation of distribution algorithm for the multi-objective flexible job-shop scheduling problem. Int. J. Prod. Res. **51**(12), 3574–3592 (2013)

17. Li, J.Q., Sang, H.Y., Han, Y.Y., Wang, C.G., Gao, K.Z.: Efficient multi-objective optimization algorithm for hybrid flow shop scheduling problems with setup energy consumptions. J. Cleaner Prod. **181**, 584–598 (2018)

18. He, Y., Li, Y.F., Wu, T., Sutherland, J.W.: An energy-responsive optimization method for machine tool selection and operation sequence in flexible machining job shops. J. Cleaner Prod. **87**(1), 245–254 (2015)

19. Yin, L.J., Li, X.Y., Gao, L., Lu, C., Zhang, Z.: A novel mathematical model and multi-objective method for the low-carbon flexible job shop scheduling problem. Sustain. Comput. Inf. Syst. **13**, 15–30 (2017)

20. Lei, D.M., Zheng, Y.L., Guo, X.P.: A shuffled frog leaping algorithm for flexible job shop scheduling with the consideration of energy consumption. Int. J. Prod. Res. **55**(11), 3126–3140 (2017)

21. Mokhtari, H., Hasani, A.: An energy-efficient multi-objective optimization for flexible job shop scheduling. Comput. Ind. Eng. **104**, 339–352 (2017)

22. Lei, D.M., Li, M., Wang, L.: A two-phase meta-heuristic for multi-objective flexible job shop scheduling problem with total energy consumption threshold. IEEE Trans. Cybern. (2018, in press)

23. Lei, D.M., Yang, D.J.: Research on flexible job shop scheduling problem with total energy consumption constraint. ACTA Autom. Sinica (2018, in press). (in Chinese)

24. Lei, D.M.: Simplified multi-objective genetic algorithm for stochastic job shop scheduling. Appl. Soft Comput. **11**(8), 4991–4996 (2011)

25. Dauzère-Pérès, S., Paulli, J.: An integrated approach for modeling and solving the general multiprocessor job-shop scheduling problem using tabu search. Ann. Oper. Res. **70**(2), 281–306 (1997)

26. Knowles, J.D., Corne, D.W.: On metrics for comparing non-dominated sets. In: Proceedings of 2002 Congress on Evolutionary Computation, Honolulu, 12–17 May, pp. 711–716 (2002)

Research on Airport Refueling Vehicle Scheduling Problem Based on Greedy Algorithm

Zhurong Wang[✉], You Li, Xinhong Hei, and Haining Meng

School of Computer Science and Engineering, Xi'an University of Technology,
No. 5 South Jinhua Road, Xi'an, Shaanxi, China
{wangzhurong,Heixinhong}@xaut.edu.cn

Abstract. Aiming at airport refueling vehicle scheduling problem (ARVSP), the mathematical model with time window constraints is established. Firstly, considering the distance between refueling vehicle and flight, and in view of the influence of flight refueling service time windows on service flight selection, an evaluation function is designed to achieve the least total vehicle distance and the minimum required vehicle. And then based on the evaluation function, a greedy algorithm is proposed to solve airport refueling vehicle scheduling problem. Finally, the correctness and effectiveness of the proposed model are verified by a practical case of airport refueling vehicle scheduling problem.

Keywords: Airport refueling vehicle scheduling · Greedy algorithm
Evaluation function

1 Introduction

With the rapid development of the civil aviation industry and airport scale enlargement in China, the problem of the low efficiency of the airport is becoming more and more prominent; thus how to respond quickly to basic business in airport operation has become a key issue to achieve efficient operation of airports. As a key step in ground support services, ground support vehicles with improper scheduling are likely to cause flight delays and affect the subsequent missions, which might bring negative effects to passengers and airports. Therefore, it is economic and socially significant to study the problem of ground guarantee vehicle scheduling.

Because of the characteristics of computation complexity, multiple objectives, multiple constraints and dynamic randomness, the scheduling of actual airport ground support vehicles is a combinatorial optimization problem with NP-hard feature. In recent years, scholars have been widely concerned the problem, and present many heuristic algorithms. For example, Norin et al. in [1] present the scheduling model of scheduling de-icing vehicles, and design a greedy randomized algorithm which minimizes the delay of flights due to de-icing, and the travel distance of the de-icing vehicles. The experiment demonstrates the effectiveness of the optimal scheduling model, and the dynamic optimization method is further demonstrated that the optimized scheduling model can effectively reduce the waiting time of deicing. Cheung et al. in

[2] aim to minimize the total flow time of the vehicle, and carry out the separate dispatch of the tractor, the clean water car and the cleaning car. Due to the low efficiency of airport ground support service, Jie et al. in [3] considered the time window constraints of the ground service guarantee vehicle scheduling and the different resource requirements of the vehicle during the service process, and then proposed the multi-objective mathematical programming model and the two stage heuristic algorithm for solving in the problem. Li et al. in [4] developed an ant colony optimization (ACO)-based hyperheuristic (ABH) for intercell scheduling with single processing machines and batch processing machines.

There are mainly two kinds of optimization methods for vehicle scheduling problem. One is mathematical exact algorithm, the other is heuristic algorithm based on knowledge information and stochastic characteristics. The most commonly used exact algorithms include branch and bound algorithm, enumeration method based on Disjunctive graph model, mixed integer programming model, Lagrange relaxation algorithm and priority rule scheduling algorithm. Although these algorithms can guarantee the global optimal solution of the resource scheduling problem, they can only solve the problems with small scale instead of the problems with large scale in a given time. For this reason, many scholars study the heuristic algorithms with random characteristics to solve large scale problems, such as greedy algorithm, simulated annealing, ant colony algorithm, genetic algorithm [5], particle swarm optimization algorithm [6] and so on.

The greedy algorithm is an approximate method to solve the optimization problem [7]. Its basic idea is to refine and improve the quality of the solution of the problem based on gradient descent direction and heuristic information, which is difficult to search from the global viewpoint and generally makes local improvement. The execution process of a greedy choice is an improvement for the solution, and finally the global solution or approximate optimal solution is obtained through a series of local optimal selection.

Greedy algorithm is widely used in scheduling problems because it has the characteristics of simple procedure, high efficiency and low time complexity. For example, a hybrid iterated greedy algorithm for the distributed no-wait flow shop scheduling problem [8]. A task scheduling algorithm combining greedy algorithm with granularity control is designed for scheduling problem of structured parallel control mechanism on cluster system [9].

While considering the influence between the waiting time on the arrival of the vehicle and the distance between the vehicle and the flight, the optimization model is given, and an evaluation value function with respect to two greedy strategy is set up so that the flight with a minimal evaluation value can be taken as the next service object and the best service process can be searched.

The rest of the paper is as follows. In the second section, the mathematical model of airport vehicles scheduling optimization is established while considering the total distance and the number of the vehicles for the scheduling problem of the airport vehicle. Section three gives the scheduling algorithm based on greedy algorithm. The fourth part in this paper shows the experimental study and comparative analysis. The last part outlines the conclusion of the paper and discusses future works.

2 Airport Refueling Vehicle Scheduling Problem (ARVSP)

2.1 Problem Description

Airport refueling vehicle scheduling problem can be described as follows: There are N vehicles providing refueling services for M flights. The vehicles starts from the parking lot and returns to the parking lot when the number of service flights is reached to the maximum number of the vehicles allowed. Vehicles number set is set to $U = \{1, 2, 3\ldots N\}$, flights number set is $P = \{1, 2, 3\ldots M\}$. The capacity of oil required for the flight i is Q_i, and the maximum allowable capacity of each vehicle is Q. Thus the aim of ARVSP is to plan the service flight sequence of each vehicle. At the same time, the following conditions are required to meet, a vehicle is only allocated to a flight anytime, and a flight is only serviced by a vehicle. All flights must be serviced in the specified time window.

2.2 Mathematical Model

The mathematical model of ARVSP is to determine the scheduling strategy of vehicle while considering the constraint conditions, so that the total distance of the vehicle flight service and the number of vehicles to be dispatched are all minimum. The objective function and constraint conditions in the model are described as follows.

$$Min \quad f(x) = [f_1(x), f_2(x)] \tag{1}$$

$$f1(x) = \sum_{k \in U} \sum_{k \in U} \sum_{k \in U} d_{ij} y_{i,j,k} \tag{2}$$

$$f2(x) = \sum_{k \in U} \sum_{j \in P} y_{o,j,k} \tag{3}$$

$$s.t. \ \min \sum_{k \in U} \left(\sum_{i \in P} y_{i,j,k} - \frac{M}{N} \right) \tag{4}$$

$$\min \sum_{k \in U} \left(\sum_{i \in P} Q_i x_{ik} - \frac{Q}{D} \right), Q = \sum_{i \in P} Q_i \tag{5}$$

$$\sum_{j \in P} y_{o,j,k} + \sum_{i \in P} \sum_{j \in P} y_{i,j,k} + \sum_{i \in P} y_{i,o,k} \le E, \forall k \in U \tag{6}$$

$$\sum_{i \in P} \sum_{j \in P} Q_{i,j,k} < Q, \forall k \in U \tag{7}$$

$$A_i < s_{ik} < B_i, \forall i, j \in P, k \in U \tag{8}$$

$$s_{ik} + t_i + t_{ij} < s_{jk}, \forall i, j \in \mathrm{P}, k \in \mathrm{U} \tag{9}$$

$$y_{i,o,k} \in \{0, 1\}, \forall i \in \mathrm{P}, k \in \mathrm{U} \tag{10}$$

$$y_{o,j,k} \in \{0, 1\}, \forall j \in \mathrm{P}, k \in \mathrm{U} \tag{11}$$

$$x_{ik} \in \{0, 1\}, \forall j \in \mathrm{P}, k \in \mathrm{U} \tag{12}$$

$$y_{i,j,k} \in \{0, 1\}, \forall i, j \in \mathrm{P}, k \in \mathrm{U} \tag{13}$$

Where d_{ij} is the distance between adjacent flight served. $y_{i,j,k}$ represents the vehicle k has finished service for the previous flight i and then service for the next flight j. $y_{o,j,k}$ means that a new vehicle k is sent out from parking lot, and begins service for the flight j. $y_{i,o,k}$ means that the vehicle k completes service for the flight i, and exits service. D is the number of the vehicles sent out. x_{ik} is a flag variable. x_{ik} is set to 1 if the vehicle k refuels for the flight i, else x_{ik} is 0. E represents the maximum number of service times for each vehicle. A_i indicates the earliest start time for the service on the flight i. B_i indicates the latest time for the service on the flight i. s_{ik} indicates the time when the vehicle k arrives at the stop of the flight i. t_i is time how long it takes for to the vehicle complete refueling service for the flight i. t_{ij} indicates the time of the vehicle from the stop of the flight i to the stop of the flight j. $f(x)$ is the objective function in (1). Equation (2) indicates the total distance of the refueling vehicle flight service are minimum. Equation (3) indicates the minimum number of dispatched vehicles. Equation (4) represents the difference in the number of tasks between each of the refueling vehicle is minimize. Equation (5) represents the minimum difference for each vehicle task. Equation (6) means the restraint of service times for each vehicle. Equation (7) indicates capacity constraint of the refueling vehicle. Equation (8) indicates that all flights must receive refueling service within its specified time window. Equation (9) translates the end of the previous flight is earlier than the start of the next flight. Equation (10) indicates the range of $y_{i,o,k}$ value. $y_{i,o,k}$ is 1 if the vehicle k starts from the parking lot for the flight i service. Equation (11) means the range of $y_{o,j,k}$ value. $y_{o,j,k}$ is 1 if the vehicle k return to the parking lot after serves flight j. Equation (12) indicates the range of x_{ik} value. x_{ik} is equals 1 if the vehicle k provides service for the flight i. Equation (13) indicates the range of $y_{i,j,k}$ value. $y_{i,j,k}$ is 1 if the vehicle k serves the flight i firstly, and then serves the next flight j.

3   Airport Vehicle Scheduling Algorithm Based on Greedy Strategy (AVSAGS)

This section firstly gives the idea of the algorithm, then defines the evaluation function used in the algorithm, and finally describes the process of the algorithm.

3.1   The AVSAGS Idea

The refueling service of different flights is connected, and the flight service sequences of the vehicles are planned. In the process of looking for the optimal service sequence, an evaluation function (VF) is designed while considering the two factors, i.e. the distance between the vehicles and flight, and the arrival time difference between the arrival time of the tanker and the upper limit of the flight service time window. A flight is served by a candidate vehicle with the minimum VF value. In this way we could get the vehicle scheduling order so that the smallest total travel distance of the vehicle is obtained. By limiting the maximum number of refueling vehicles, the mission of the refueling vehicle is balanced, thus the number of refueling vehicles reaches the smallest value.

3.2   Evaluation Function Index

Two goals of airport vehicle scheduling are considered for the proposed algorithm. The first goal is to minimize the sum of service distances for the fligh, and the second one is that times of the vehicles to be sent out is required to be minimized. To make the vehicle as early as possible for the flight service, it will enable the vehicle to serve more flights for a period of time, so as to achieve the target of fewer vehicles. The two goals are relevant, the nearest flight is always served by the same vehicle while considering time window. Because the two goals are very important which affecting the vehicle scheduling in actual situations, we propose the evaluation function (VF) to choose the flight to serve as early as possible while considering time window of flight service. The evaluation function (W) is defined as follows.

$$W = d_{jk} + \alpha \times T_w + \beta \times T_d \tag{14}$$

$$T_w = \begin{cases} A_i - s_{ik}, & T_w > 0 \\ 0, & T_w < 0 \end{cases} \tag{15}$$

$$T_d = \begin{cases} s_{ik} - A_i, & T_d > 0 \\ 0, & T_d < 0 \end{cases} \tag{16}$$

Where dkj represents the distance between the refueling vehicle and the flight, α represents the evaluation factor waiting for the refueling vehicle, β represents the evaluation factor for the delay of the refueling vehicle. Tw indicates the wait time of the vehicle, Td is the delay time of the vehicle Ai means the earliest time for flight to be served. Ski represents the time for the vehicle to get to the flight; Eq. (14) defines evaluation function (W) that considers two factors, i.e., vehicle traveling distance and delay time. W is used as the standard for evaluating the quality of scheduling strategy. The smaller of the VF (W) value is, the sooner the selected flight to be served is. Equations (15) and (16) give computation formula of Tw and Td.

3.3   The AVSAGS Description

Based on the above analysis, the **AVSAGS** can be described as follows.

Step 1. The departure flights are arranged on the timetable to form the sequence of services (L_1).

Step 2. Taking time window constraints as the criterion, the first flight which has not been served is selected to check. Calculate the time when the vehicle arrives at the flight, and if the refueling vehicles do not satisfy the time window $[A_i,B_i]$, then the new one is dispatched; and flight service is joined into the vehicle transmission chain; and go to Step 4. Else continue.

Step 3. If only one vehicle meets the time window $[A_i,B_i]$, then flight service is joined into the vehicle transmission chain; Else there are multiple vehicles meet the time window constraint simultaneously, the evaluation function is calculated according to Eq. (14), and the vehicle with the smallest VF value is added.

Step 4. In accordance with step two and step three, refueling tasks of flight service is added one by one.

Step 5. When the capacity of the transmission chain of the vehicle reaches the maximum allowable service number, the new task is terminated, and the vehicle returns to the parking lot. After a period of rest, the vehicle can be dispatched again.

Step 6. Repeat the above steps to the vacant flight to be served.

The flow chart of the algorithm is shown in Fig. 1.

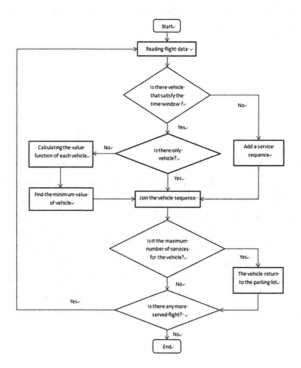

Fig. 1. Flow chart of the AVSAGS

4   Results and Discussion

This paper selects the actual flight data of an airport in January 22nd of 2017 year to perform experiment for the proposed algorithm. The airport's daily arrivals and flights amount to about 250 sorties according to the airport data. As the incoming flights do not need refueling service, 115 departure flights are selected to test.

4.1   Refueling Time

The refueling time required for a flight is related to the aircraft type. According to the number of aircraft passenger seats, China's civil aviation aircraft is divided into large, medium and small types. Flight with the number of passengers seats below 100 seats belongs to small type; flight with 100 to 200 seats belongs to medium type; and flight with more than 200 seats belongs to large type. After analyzing the flight data, there are Airbus A380 series belonging to large type; A320, A319 series belong to medium type; CRJ-900 series, ERJ-190 series, ARJ21 belong to small type. Based on the flight history refueling data, average refueling time is given as the time required for each flight type, as shown in Table 1.

Table 1. Refueling time of flight type

	Model	Refueling time (minutes)
Large	A380	35
Medium	A320, A319, Boeing 737	25
Small	CRJ-90, ERJ-190, ARJ21	15

4.2   Airport Stand Distance Matrix

The airport stand is the location of the flight at the airport. The distance between different stops affects the driving time of the vehicle. There are 37 stands in the airport, which are distributed in 3 regions of T1, T2 terminal and parking apron. According to the field calculation, the distance between adjacent parking stands is 40 m. In order to ensure the safety of the airport, the civil aviation authority stipulates that all ground support vehicles at the airport must travel along the prescribed route and cannot enter the other airport areas. The 101 to 108 stands are located in the T1 terminal, and the 109–116 stands are located in the T2 terminal. The remaining is distributed on the parking apron, and the airport stands are shown in Fig. 2.

According to Fig. 2, the distance matrix among airport stands can be obtained, in which the parking lot is expressed using O. The distance matrix of the part of the stands is shown in Table 2 as shown below.

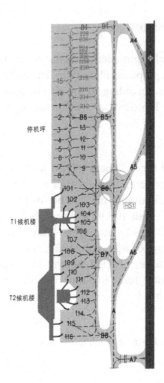

Fig. 2. Airport stands distribution

Table 2. Distance matrix of partial stands

	212	214	216	218	O	219	217
212	0	40	80	120	160	200	240
214	40	0	40	80	120	160	200
216	80	40	0	40	80	120	160
218	120	80	40	0	40	80	120
O	160	120	80	40	0	40	80
219	200	160	120	80	40	0	40
217	240	200	160	120	80	40	0

4.3   Time Window for Refueling Service

The time window of the refueling is a time range between the earliest start of the refueling service and the latest start of the refueling service for the flight. If the vehicle arrives before the earliest start of the refueling time, it needs to wait. If the vehicle arrives after the flight starts at the latest, the flight may be delayed.

According to the relevant documents [10], civil aviation flight must complete fuel refueling operations five minutes before the passengers start boarding. In special circumstances, fuel injection should be completed five minutes before the aircraft is expected to leave. Passengers generally start boarding ahead of 30 min.

The departure flights are divided into two types, i.e. originating flight and passing flight. Their time windows are also different. The former should be served ahead of time before the passengers start boarding; the latter needs be served immediately. The two types of flights time windows of refueling service are as follows.

Service time window for passing flight:

$$A_i = t_{arr} \tag{17}$$

$$B_i = t_{arr} + 10 \tag{18}$$

Service time window for originating flight:

$$A_i = t_d - t_i - 35 \tag{19}$$

$$B_i = t_d - t_i - 5 \tag{20}$$

where t_{arr} is arrival time of passing flight, t_d is the departure time of the flight plan, t_i represents the time that service for the flight i. Equation (17) means the time window lower limit of the passing flight is at the arrival time, Eq. (18) means the time window upper limit of the passing flight, Eq. (19) represents the time window upper limit of the originating flight, Eq. (20) represents the time window lower limit of the originating flight. Part of flight service time window for part of the flight, as shown in Table 3.

Table 3. Part of flight service time window

Flight number	Planned landing time	Stands	Ai	Bi
103	0830	102	0830	0840
104	0835	108	0835	0845
105	0855	107	0855	0905
106	0910	109	0910	0930
107	0910	115	0910	0920
108	0910	112	0910	0920
109	0925	111	0925	0935

4.4 Experimental Data and Analysis

The experimental data are obtained by using the proposed algorithm designed for solving the ARVSP with the above data. Service sequence for a vehicle starts and ends using 0 which means the vehicle starting from the parking lot, providing refueling service for the flights, and returning to the parking lot. The time table for flight service sequence is shown in Table 4. The distance of the driving path and the service number for the vehicles are shown in Table 5.

Table 4. Flight service sequence

Number	Service sequence	Start time	End time
1	0,101,104,106,112,116,0	0755	1055
2	0,102,103,105,107,111,0	0805	1020
3	0,108,109,110,114,118,0	0910	1115
4	0,113,,117,119,125,136,0	1010	1215
5	0,115,120,123,129,141,0	1015	1235
6	0,121,127,137,142,147,0	1100	1305
7	0,122,126,135,143,148,0	1100	1305
8	0,124,128,151,163,170,0	1310	1535
9	0,130,138,145,153,156,0	1335	1540
10	0,131,139,144,152,160,0	1335	1520
11	0,132,140,146,162,166,0	1335	1555
12	0,133,149,155,157,159,0	1335	1525
13	0,134,150,154,158,161,0	1400	1635
14	0,164,169,172,174,177,0	1410	1710
15	0,165,167,168,171,178,0	1415	1720
16	0,173,175,176,179,181,0	1540	1805
17	0,180,183,187,193,199,0	1725	1930
18	0,182,184,189,208,0	1735	2035
19	0,185,190,196,200,201,0	1800	2005
20	0,186,191,209,213,0	1800	2220
21	0,188,192,197,202,204,0	1810	2025
22	0,194,198,203,206,0	1835	2035
23	0,195,205,207,211,212,0	1835	2205
24	0,210,214,215,0	2015	2300

Since no flight need s to be refueled before 7:55, so the flight service sequence was empty from 0:00 AM to 7:55 AM. At 7:55 AM, the first flight 101 needs refueling service, and the 101 flight is added to the first flight service sequence. Because it is a passing flight, The service sequence starts at 7:55 and ends at 8:20. From 8:05 to 8:35, there are 102, 103 and 104 flights which are in service waiting flights. While considering time watch for 102 flight, a new vehicle needs to assign, and flights are added to different service sequences. The results of other service sequences are no longer stated in detail.

The total distance of the refueling vehicle is 61.920 km, which is 30% less than that traditional manual Scheduling (88.560 km) by the results of the experiment. The number of total vehicle dispatched are 24 times that decrease 79.1% compared with the traditional manual scheduling of 115 times. And only five vehicles are needed to provide service for the flights.

Table 5. The distance of the driving path and the service number for the vehicles

Number	Driving distance (Meter)	Start time
1	2800	0755
2	2720	0805
3	2080	0910
4	3200	1010
5	3520	1015
6	3760	1100
7	2640	1100
8	3920	1310
9	1280	1335
10	2160	1335
11	2240	1335
12	2240	1335
13	3040	1400
14	2240	1410
15	4160	1415
16	2480	1540
17	2640	1725
18	2560	1735
19	2960	1800
20	2000	1800
21	2000	1810
22	1200	1835
23	2080	1835
24	2000	2015

5   Conclusion

In this paper, the airport refueling vehicle scheduling problem is analyzed, and its mathematical model is established. A scheduling algorithm based on greedy strategy is designed for solving the model. The test results show that the proposed algorithm has a significant improvement in aircraft vehicle scheduling and service cost reduction.

In the future research, more scenarios and strategies are to be considered so that the model and the algorithm could handle more complex cases such as flight delays and dynamic scheduling. Meanwhile, we apply the proposed algorithm into an actual practice.

Acknowledgments. The research presented is supported in part by the National Natural Science Foundation (NO:U1334211, 61773313,61602375), Shaanxi Province Key Research and Development Plan Project (NO:2015KTZDGY0104, 2017ZDXM-GY-098). The Key Laboratory Project of Shaanxi Provincial Department of Education (NO:17JS100).

References

1. Norin, A., Värbrand, P.: Scheduling de-icing vehicles within airport logistics: a heuristic algorithm and performance evaluation. J. Oper. Res. Soc. **63**(8), 1116–1125 (2012)
2. Cheung, A., Ip, W.H., Lu, D., Lai, C.L.: An aircraft service scheduling model using genetic algorithms. J. Manufact. Technol. Manage. **16**(1), 109–119 (2005)
3. Jie, Z., Gao, J.: Research on problems in airport ground service scheduling. J. Hebei North Univ. **24**(6), 60–62 (2016)
4. Li, D., Li, M., Meng, X., et al.: A hyperheuristic approach for intercell scheduling with single processing machines and batch processing machines. IEEE Trans. Syst. Man Cybern. Syst. **45**(2), 315–325 (2015)
5. Chang, H.-C., Chen, Y.-P., Liu, T.-K., Chou, J.-H.: Solving the flexible job shop scheduling problem with makespan optimization by using a hybrid Taguchi-Genetic algorithm. IEEE Comput. Soc. **3**, 1740–1754 (2015)
6. Liu, B., Wang, L., Jin, Y.-H.: An effective PSO-based memetic algorithm for flow shop scheduling. IEEE Trans. Syst. Man Cybern. Part B (Cybern.) **37**(1), 18–27 (2007)
7. Sarkar, U.K., Chakrabarti, P.P., Ghose, S., et al.: A simple 0.5-bounded greedy algorithm for the 0/1 knapsack problem. Inf. Process. Lett. **42**(42), 173–177 (1992)
8. Shao, W., Pi, D., Shao, Z.: A hybrid iterated greedy algorithm for the distributed no-wait flow shop scheduling problem. In: Evolutionary Computation, pp. 9–16. IEEE Computer Society (2017)
9. Zhang, H., Fang, B., Hu, M.: An algorithm on task scheduling in structural parallel control mechanism. J. Softw. **12**(5), 706–710 (2001)
10. Civil Aviation Administration of China. Civil aviation, Normal Operation Standard of Airline Flight (execution). Civil Aviation Bureau General Division, Beijing (2013)

Improved Running Time Analysis of the (1+1)-ES on the Sphere Function

Wu Jiang[1(✉)], Chao Qian[1], and Ke Tang[2]

[1] Anhui Province Key Lab of Big Data Analysis and Application,
University of Science and Technology of China, Hefei 230027, China
{jwl992, chaoqian}@mail.ustc.edu.cn
[2] Shenzhen Key Lab of Computational Intelligence,
Southern University of Science and Technology, Shenzhen 518055, China
tangk3@sustc.edu.cn

Abstract. During the last two decades, much progress has been achieved on the running time analysis (one essential theoretical aspect) of evolutionary algorithms (EAs). However, most of them focused on discrete optimization, and the theoretical understanding is largely insufficient for continuous optimization. The few studies on evolutionary continuous optimization mainly analyzed the running time of the (1+1)-ES with Gaussian and uniform mutation operators solving the sphere function, the known bounds of which are, however, quite loose compared with the empirical observations. In this paper, we significantly improve their lower bound, i.e., from $\Omega(n)$ to $\Omega(e^{cn})$. Then, we study the effectiveness of 1/5-rule, a widely used self-adaptive strategy, for continuous EAs using uniform mutation operator for the first time. We prove that for the (1+1)-ES with uniform mutation operator solving the sphere function, using 1/5-rule can reduce the running time from exponential to polynomial.

Keywords: Running time analysis · Continuous optimization
Evolution strategies

1 Introduction

Evolutionary algorithms (EAs) are widely used in real-world applications and have achieved great success in solving both continuous and discrete optimization problems. As a class of general-purpose optimization algorithms, EAs are designed to search for the optimum without the information of gradients or Hessian matrix. Thus, they have been regarded as a major approach when the optimization problems are non-differential, multi-modal, or black-box. However, a great gap lies between the numerous practice and weak theoretical foundation. Moreover, most of the theoretical works focus on the discrete domain, such as pseudo-Boolean functions [4, 10–12, 14, 15], while only a few works deal with continuous optimization problems [1, 7, 8].

Note that the theoretical analysis of EAs in discrete domain cannot be directly extended to continuous domain. In the following, we briefly summarize the difference between continuous EAs and discrete EAs.

© Springer International Publishing AG, part of Springer Nature 2018
D.-S. Huang et al. (Eds.): ICIC 2018, LNCS 10954, pp. 729–739, 2018.
https://doi.org/10.1007/978-3-319-95930-6_74

- Optimization problems they are designed to solve originally. Continuous EAs are designed to solve continuous optimization problems, while discrete EAs are more often used to solve discrete optimization problems, such as combinatorial optimization problems, etc.
- Coding they use to represent the individuals. Individuals are represented by a real vector in continuous EAs in contrast to a binary vector in discrete EAs.
- Mutation operators. In discrete EAs, an offspring solution is generated by flipping the bits of the parent solution. In continuous EAs, we first sample a stochastic vector from a given distribution. A new candidate solution is generated by adding the preceding stochastic vector to the solution.
- Stopping criteria. The searching space is infinite and uncountable in continuous domain, which means the probability of finding the exact optimum is 0. Thus, we focus on the number of steps that are needed to halve the approximation error, i.e., the distance from the optimum in continuous domain. While in discrete domain, the evolution would not stop until the exact optimum or an acceptable suboptimal solution is found.

Most of the previous theoretical works in continuous EAs analyze the expected running time of evolution strategies (ES) on the sphere function. Jägersküpper [7] proved that the (1+1)-ES with Gaussian mutation operator has a polynomial lower bound $\Omega(n)$ on certain function scenarios (including the sphere function), and derived an upper bound $O(n)$ when 1/5-rule is introduced. These two results together confirmed the effectiveness of 1/5-rule. Moreover, Akimoto et al. proved that the expected running time of the (1+1)-ES with Gaussian mutation operator adapted by 1/5-rule on the sphere function is $\Theta\big((\log\frac{d}{\epsilon})n\big)$, where d denotes the distance of the initial individual to the global optimum and ϵ is the parameter of the stopping criteria [2]. Uniform mutation operator was then analyzed by Agapie et al., with a lower bound $\Omega(n)$ and an upper bound $O(e^{cn})$ [1].

In this paper, we first prove that the general lower bound on the running time of the (1+1)-ES with uniform mutation operator inside a hypersphere is $\Omega(e^{cn})$ which largely improves the known bound $\Omega(n)$ which was given in [1]. Second, we derive a similar result when the mutation operator is replaced by Gaussian mutation, i.e. $\Omega(e^{cn})$, in contrast to $\Omega(n)$ given by Jägersküpper in [7]. Third, we study the effectiveness of 1/5-rule for the (1+1)-ES using uniform mutation inside a hypersphere for the first time. We prove that the running time of the (1+1)-ES using uniform mutation inside a hypersphere has polynomial upper bound $O(n)$ after the incorporation of 1/5-rule.

The rest of the paper is organized as follows. Section 2 introduces some preliminaries, including the studied problem and algorithm, and also the analysis tools that will be used. Section 3 derives the exponential lower bound of the (1+1)-ES using uniform and Gaussian mutation operators. Section 4 analyzes the effectiveness of 1/5-rule for the (1+1)-ES using uniform mutation operator inside a hypersphere. Section 5 concludes the paper.

2 Preliminaries

In this section, we first introduce the problem and algorithm studied in this paper, respectively. Then, we present the analysis tools that we use throughout the paper.

2.1 Problem

We consider minimizing the sphere function in this paper. Its formulation is as follows,

$$f(x) = \sqrt{\sum_{i=1}^{n} x_i^2} \tag{1}$$

The sphere function has two properties:

- A minimum exists.
- $f(x) < f(y)$ if $|x| < |y|$ for any two points x and y, which means that a mutant closer to the optimum is always accepted.

2.2 (1+1)-ES and 1/5-Rule

Jägersküpper analyzed the time complexity of the $(1 + \lambda)$-ES on the sphere function and proved that the offspring population helps to reduce the running time of ES [8]. However, parent population does not help for the hypersphere function, and even makes the expected running time larger [6]. Our theoretical analysis will focus on the (1+1)-ES, which maintains only one parent solution and one offspring solution in each generation. The (1+1)-ES replaces the current individual by a better candidate offspring, and discards a worse one in every iteration. The process of the (1+1)-ES using uniform mutation inside a sphere is as follows,

Initialization: Create a random initial individual ξ_0 which satisfies $d(\xi_0, o) = R$, where R denotes a positive constant and o is the global optimum.

Mutation: Create a new search point $\xi' = \xi_t + \sigma m$, where σ is scaling factor and m is a n-dimensional vector that uniformly distributed in a sphere of fixed radius r (or sampled from n-variate standard Gaussian distribution).

Evaluation: If $f(\xi') \le f(\xi_t)$, $\xi_{t+1} = \xi'$, else $\xi_{t+1} = \xi_t$.

Stopping criteria: $f(\xi_t) \le R/2$.

Among all the self-adaptive strategies for the coefficients of mutation operators in real-world applications and research areas, 1/5-rule is the most commonly used one [3, 13]. This heuristic strategy has shown excellent performance by controlling the mutation strength of ES. The 1/5-rule works as follows: the optimization process is observed for n steps without changing σ; if more than one fifth of the steps in this observed phase have been successful, σ is doubled; otherwise, σ is halved.

2.3   Analysis Tools

It is not hard to notice the common property between the evolving of population in ES and a Markov Chain $\{\xi_t\}_{t \geq 0}$, since the future state/population is independent of the past states/populations and only effected by the current state/population. However, it is nearly impossible to construct a transition probability matrix for ES. Thus, a new effective model, renewal process, is introduced for ES. For each $t = 0, 1, 2, \ldots$, let ξ_t be the individual of t-th generation in ES and $d(\xi_t)$ be the distance from ξ_t to the optimum 0. Define the stochastic process $\{\rho_t\}_{t \geq 0}$ by $\rho_t = d(\xi_t) - d(\xi_{t+1})$, then $\{\rho_t\}_{t \geq 0}$ is a renewal process.

Since ES is a class of stochastic algorithms, the first hitting time is a random variable, either. We take the expectation of first hitting time (FHT) as the measure of time complexity of ES. The definition of FHT is as follows,

First Hitting Time. Assume that the initial state set is χ_I, and the final state set is χ^*, the first hitting time T is defined as $T = \min\{t | \xi_t \in \chi^* \wedge \xi_0 \in \chi_I\}$. In our paper, particularly $\chi_I = \{\xi \in \mathbb{R}^n | d(\xi) = R\}$ and $\chi^* = \{\xi \in \mathbb{R}^n | d(\xi) < R/2\}$.

Several analysis tools have been proposed with different ideas and principles in discrete domain. All of these methods have shown great power in the running time analysis of EAs. Drift analysis, constructs a distance function and derives the expected running time by bounding the one-step progress [4]. Fitness level method cuts the search space into different sets or levels based on the fitness values and derives the expected running time by bounding the transition probability of the solutions moving from the current level to other levels [14]. Switch analysis compares the evolutionary process with another one which has been well analyzed and derives the expected running time by estimating the one-step differences of the two [15]. However, these three effective methods cannot be applied to continuous optimization directly since they are all based on Markov Chain with finite states.

Similar to the drift analysis theorem, which was first introduced to analyze the running time of EAs by He and Yao [4], the analysis approach for ES introduced by Jägersküpper also focuses on the difference of the two adjacent generations, defined as progress rate ρ_t [7].

Progress Rate. Assume that the t-th generation is ξ_t, the distance of ξ_t from the optimizer is $d(\xi_t)$, then $\rho_t = d(\xi_t) - d(\xi_{t+1})$ is called the progress rate of the t-th generation.

Theorem 1 (Lower bound theorem [7]). Let $\rho_1, \rho_2, \ldots$ denote random variables with bounded range and T be the random variable defined by $T = \min\{t | \sum_{i=0}^{t} \rho_i \geq g\}$ for a given $g > 0$. If $\mathrm{E}(T)$ exists and $\mathrm{E}(\rho_i | T \geq i) \leq u$, then $\mathrm{E}(T) \geq g/u$.

Theorem 1 reveals that if the progress rate has an upper bound, we could derive a lower bound for the expected running time.

3   Exponential Lower Bound of the (1+1)-ES

In this section, we prove that the expected running time of the (1+1)-ES using uniform mutation operator inside a hypersphere and Gaussian mutation operator has exponential lower bound when used for minimizing the sphere function, presented in Theorems 2 and 3, respectively.

3.1   Uniform Mutation Inside a Hypersphere

Lower bound of the (1+1)-ES minimizing the sphere function derived by Agapie et al. is not tight enough, which could be testified by simulation experiments [1]. The super-polynomial bound by Huang et al., however, is concerned with uniform mutation inside a hypercube and based on a very strong assumption on the initialization and mutation operator [5]. In this subsection, we give out the lower bound which is tighter than the state-of-the-art bound. Besides, our method is more general, which has no assumption on the coefficients of the mutation operator and initialization.

Theorem 2. For the (1+1)-ES minimizing the sphere function in n-dimensional Euclidian space using uniform mutation in a hypersphere and elitist selection, the expected number of generations T to half the approximation error is $\Omega(e^{cn})$.

Proof. Denote the global optimum as o and the current individual as c, i.e. ξ_t. Supposing that the distance between o and c is $d(o, c) = R$ and the radius of the mutation operator is r. Due to symmetry, we assume that c is on axes ox_1.

As shown in Fig. 1. Intersection of fitness sphere and mutation sphere, we call the left sphere $S_o = \{\xi \in \mathbb{R}^n | d(\xi, o) \leq R\}$ fitness sphere, since the all points inside S_o are better than the current population c. Accordingly, we call the right sphere $S_c = \{\xi \in \mathbb{R}^n | d(\xi, c) \leq r\}$ mutation sphere, since all the candidate offsprings scatter in it. A is denoted as the intersection of the two spheres.

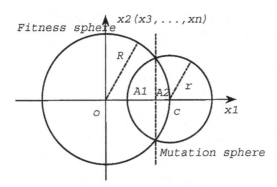

Fig. 1. Intersection of fitness sphere and mutation sphere

According to the definition of progress rate, we get

$$E(\rho_t) = \frac{1}{C_n r^n} \int (R - d(\xi_{t+1})) d\xi_{t+1} \leq \frac{1}{C_n r^n} \cdot R \cdot V(A) \tag{2}$$

Where $C_n r^n = \frac{\pi^{n/2}}{\Gamma(\frac{n}{2}+1)} r^n$ is the volume of the n-dimensional sphere with radius r and $V(A)$ is the volume of the intersection of the two hyperspheres. We begin our deduction in an algebraic way. By solving the equations below

$$\begin{cases} \sum_{i=1}^{n} x_i^2 = R^2 \\ (x_1 - R)^2 + \sum_{i=2}^{n} x_i^2 = r^2 \end{cases} \tag{3}$$

We get

$$x_1 = R - \frac{r^2}{2R} \tag{4}$$

Which is the x_1-coordinate of the hyperplane P.

1. $r > R$. In this case, we have

$$V(A) \leq V(S_o) = C_n R^n = C_n r^n \cdot O(e^{c_1 \cdot n}) \tag{5}$$

Where $c_1 < 0$ is a negative constant.

2. $r \leq R$. In this case, the intersection A can be viewed as two spherical caps.

$$V(A) = V(A_1) + V(A_2)$$

$$= C_n r^n \cdot \frac{1}{2} I_{\left(1 - \frac{r^2}{4R^2}\right)} \left(\frac{n+1}{2}, \frac{1}{2}\right) + C_n R^n \frac{1}{2} I_{\left(\frac{r^2}{R^2} - \frac{r^4}{4R^4}\right)} \left(\frac{n+1}{2}, \frac{1}{2}\right)$$

$$< C_n r^n \cdot \frac{\frac{R}{r} \left(1 - \frac{r^2}{4R^2}\right)^{\frac{n+1}{2}}}{\frac{n+1}{2} \cdot B\left(\frac{n+1}{2}, \frac{1}{2}\right)} + C_n R^n \cdot \frac{\frac{R^2}{2R^2 - r^2} \left(\frac{r^2}{R^2} - \frac{r^4}{4R^4}\right)^{\frac{n+1}{2}}}{\frac{n+1}{2} \cdot B\left(\frac{n+1}{2}, \frac{1}{2}\right)}$$

$$= C_n r^n \cdot \frac{\frac{2R^3}{r(2R^2 - r^2)} \left(1 - \frac{r^2}{4R^2}\right)^{\frac{n+1}{2}}}{\frac{n+1}{2} \cdot B\left(\frac{n+1}{2}, \frac{1}{2}\right)}$$

$$= C_n r^n \cdot O(e^{c_2 \cdot n})$$

Where $c_2 < 0$ is a negative constant. The second equation derived by applying the formula of volume of spherical cap $V(C_{a,b}) = C_n a^n \cdot \frac{1}{2} I_{\frac{2ab-b^2}{a^2}}\left(\frac{n+1}{2}, \frac{1}{2}\right)$ by Li, where a is the radius of the hypersphere that the spherical cap is cut from and b is the height of the spherical cap [9]. The first inequality is based on

$$
\begin{aligned}
I_x\left(\frac{n+1}{2}, \frac{1}{2}\right) &\equiv \frac{B\left(x; \frac{n+1}{2}, \frac{1}{2}\right)}{B\left(\frac{n+1}{2}, \frac{1}{2}\right)} = \frac{\int_0^x t^{\frac{n-1}{2}}(1-t)^{-\frac{1}{2}}dt}{B\left(\frac{n+1}{2}, \frac{1}{2}\right)} \\
&< \frac{(1-x)^{-\frac{1}{2}}\int_0^x t^{\frac{n-1}{2}}dt}{B\left(\frac{n+1}{2}, \frac{1}{2}\right)} = \frac{(1-x)^{-\frac{1}{2}}x^{\frac{n+1}{2}}}{\frac{n+1}{2} \cdot B\left(\frac{n+1}{2}, \frac{1}{2}\right)}
\end{aligned}
\tag{6}
$$

The last equation is based on

$$
B\left(\frac{n+1}{2}, \frac{1}{2}\right) = \frac{\Gamma\left(\frac{n+1}{2}\right)\Gamma\left(\frac{1}{2}\right)}{\Gamma\left(\frac{n}{2}+1\right)} = \Theta\left(\left(\frac{n}{2}+1\right)^{-\frac{1}{2}}\right) = \Theta\left(n^{-\frac{1}{2}}\right)
\tag{7}
$$

Summing up the two cases, we get

$$
E(\rho_t) = \frac{1}{C_n r^n}\int(R - d(\xi_{t+1}))d\xi_{t+1} \le \frac{1}{C_n r^n} \cdot R \cdot V(A) = O(e^{c_3 \cdot n})
\tag{8}
$$

Where $c_3 = \max\{c_1, c_2\} < 0$ is a negative constant. By lower bound theorem, we have

$$
E(T) \ge \frac{R - R/2}{O(e^{c_3 \cdot n})} = \Omega(e^{cn})
\tag{9}
$$

Where $c = -c_3 > 0$ is a positive constant.

3.2 Gaussian Mutation

The analysis by Huang et al. for Gaussian mutation is based on a very strong assumption on the initialization [5]. In this section, we give out a lower bound which is much tighter than the bound $\Omega(n)$ given by Jägersküpper [7]. Moreover, our analysis is more general than [5] since we have no assumption on the initialization.

Theorem 3. For the (1+1)-ES minimizing the sphere function in n-dimensional Euclidian space using Gaussian mutation and elitist selection, the expected number of generations T to half the approximation error is $\Omega(e^{cn})$.

Proof. Denote the global optimum as o and the current individual as c, i.e. ξ_t. Supposing that the distance between o and c is $d(o,c) = R$. Due to symmetry, we assume that c is on axes ox_1. Similarly, we call the left sphere $S_o = \{\xi \in \mathbb{R}^n | d(\xi, o) \le R\}$ fitness sphere, since the all points inside S_o are better than the current population c.

According to the definition of progress rate, we get

$$E(\rho_t) = \int \varphi(\xi_{t+1})(R - d(\xi_{t+1}))d\xi_{t+1} \le \int \varphi(\xi_{t+1})Rd\xi_{t+1} \tag{10}$$

Where $\varphi(\cdot)$ is the probability density function of the n-variate standard Gaussian distribution. Since the $\varphi(\xi_{t+1}) \le \max\{\varphi(m)|m \in \mathbb{R}^n\} = (2\pi)^{-n/2}$, we have

$$E(\rho_t) \le \int (2\pi)^{-\frac{n}{2}}Rd\xi_{t+1} = (2\pi)^{-\frac{n}{2}} \cdot R \cdot V(S_o) = (2\pi)^{-\frac{n}{2}} \cdot \frac{\pi^{n/2}R^{n+1}}{\Gamma(n/2+1)} \tag{11}$$

Where $V(S_o)$ is the volume of the fitness sphere S_o. Since $R/2 \le d(\xi_i) \le R$ for $\forall i \in [0, T]$, we have $E(\rho_t) \le \frac{R^{n+1}}{\Gamma(n/2+1) \cdot 2^{n/2}}$. By lower bound theorem, we have

$$E(T) \ge \frac{R - \frac{R}{2}}{\frac{R^{n+1}}{\Gamma(\frac{n}{2}+1) \cdot 2^{\frac{n}{2}}}} = \frac{2^{\frac{n}{2}-1}}{R^n} \cdot \Gamma\left(\frac{n}{2}+1\right) = \Omega(e^{cn}) \tag{12}$$

Where $c > 0$ is a positive constant, since $R > 0$ is a positive constant.

4   On the Effectiveness of 1/5-Rule

Jägersküpper proved that the upper bound of the (1+1)-ES minimize the sphere function using Gaussian mutations adapted by 1/5-rule is $O(n)$ [7]. In this section, we study the effectiveness of 1/5-rule for the (1+1)-ES using uniform mutation operator.

Lemma 1 [7]. For a n-dimensional vector m with each component independently standard normal distributed, the expectation $l_E = E(|m|)$ exists, and $P\{||m| - l_E| \ge \delta \cdot l_E\} \le \frac{\delta^{-2}}{2n-1}$.

Similar result holds for uniform mutation inside a hypersphere. Lemma 2 reveals that the length of the n-dimensional vector uniformly sampled from the unit hypersphere is $\Theta(1)$ w.o.p., with overwhelming probability.

Lemma 2. For a n-dimensional vector m uniformly distributed in a unit hypersphere, the expectation $l_E = E(|m|)$ exists, and $P\{||m| - l_E| \ge \delta \cdot l_E\} \le \frac{\delta^{-2}}{n^2+2n}$.

Proof. According to the definition of expectation in mathematics, we have

$$E(|m|) = \int_0^1 rA_n(r) \cdot \frac{1}{V_n(1)}dr = \frac{n}{n+1} \tag{13}$$

Where $V_n(1) = \frac{\pi^{n/2}}{\Gamma(\frac{n}{2}+1)}$ is the volume of the unit hypersphere and $A_n(r) = \frac{2\pi^{n/2}}{\Gamma(\frac{n}{2})} \cdot r^{n-1}$ is the surface area of the n-dimensional hypersphere of radius r. Similarly, we have

$$E\left(|m|^2\right) = \int_0^1 r^2 A_n(r) \cdot \frac{1}{V_n(1)} dr = \frac{n}{n+2} \tag{14}$$

Consequently, we get

$$\text{Var}(|m|) = E\left(|m|^2\right) - E(|m|)^2 = \frac{n}{(n+1)^2(n+2)} \tag{15}$$

Thus, we get

$$P\{||m| - l_E| \geq \delta \cdot l_E\} \leq \frac{\delta^{-2}}{n^2 + 2n} \tag{16}$$

by applying the Chebyshev's inequality.

Intuitively, we partition a run of the (1+1)-ES adapted by 1/5-rule into phases each of which lasts n steps. Thus, in each phase, the radius of the mutation sphere is fixed. Denote the radius and the scaling factor in the t-th phase as r_t and δ_t. Jägersküpper find the relation between the radius of the mutation sphere and the progress rate, for Gaussian mutation though, but could be extended to uniform mutation similarly.

Lemma 3 [7]. Let the (1+1)-ES minimize the sphere function using Gaussian mutations adapted by the 1/5-rule and elitist selection. Then

1. If $r_t = \Theta(d(\xi_t/\sqrt{n}))$, then w.o.p. $d(\xi_{t+1}) = d(\xi_t) - \Omega(d(\xi_t))$, i.e., the distance to the optimum is reduced by a constant fraction in the t-th phase.
2. If δ is doubled after the t-th phase, then $r_t = O(d(\xi_t/\sqrt{n}))$ w.o.p.;
3. If δ is halved after the t-th phase, then $r_{t+1} = \Omega(d(\xi_{t+1}/\sqrt{n}))$ w.o.p.;

Since the proof of Lemma 3 is merely based on the isotropy of Gaussian mutation and Lemma 1, we could derive the following result by simply applying the isotropy of uniform mutation and Lemma 2.

Lemma 4. The (1+1)-ES minimize the sphere function using uniform mutations inside a hypersphere adapted by the 1/5-rule and elitist selection. Then

1. If $r_t = \Theta(d(\xi_t/\sqrt{n}))$, then w.o.p. $d(\xi_{t+1}) = d(\xi_t) - \Omega(d(\xi_t))$, i.e., the distance to the optimum is reduced by a constant fraction in the t-th phase.
2. If δ is doubled after the t-th phase, then $r_t = O(d(\xi_t/\sqrt{n}))$ w.o.p.;
3. If δ is halved after the t-th phase, then $r_{t+1} = \Omega(d(\xi_{t+1}/\sqrt{n}))$ w.o.p.;

Theorem 4 [7]. Let the (1+1)-ES minimize the sphere function in n-dimensional Euclidian space using Gaussian mutation adapted by 1/5-rule and elitist selection. Given that the initialization ensures $d(\xi_0)/\delta_0 = \Theta(n)$, the expected number of generations T to half the approximation error is w.o.p. $O(n)$.

Since the proof of Theorem 4 is merely based on the isotropy of Gaussian mutation and Lemma 3, we could derive the following result by simply applying the isotropy of uniform mutation and Lemma 4.

Theorem 5. Let the (1+1)-ES minimize the sphere function in n-dimensional Euclidian space using uniform mutation in a hypersphere adapted by 1/5-rule and elitist selection. Given that the initialization ensures $d(\xi_0)/\delta_0 = \Theta(\sqrt{n})$, the expected number of generations T to half the approximation error is w.o.p. $O(n)$.

5 Conclusion

In this paper, we first derive a tighter lower bound, i.e., $\Omega(e^{cn})$, for (1+1)-ES using uniform mutation inside a hypersphere on the sphere function in contrast to the state-of-the-art bound $\Omega(n)$. Second, we extend the result to the case when the mutation operator is Gaussian mutation. Third, we study the effectiveness of 1/5-rule for (1+1)-ES using uniform mutation inside a hypersphere for the first time, and prove that the incorporation of 1/5-rule reduces the time complexity from exponential to polynomial.

Uniform mutation operator that distributed in a hypercube is also used by some researchers. It does not belong to isotropic distribution like uniform distribution inside a hypersphere do. The effectiveness of 1/5-rule on it has not been studied yet. We leave it as our future work.

References

1. Agapie, A., Agapie, M., Rudolph, G., Zbaganu, G.: Convergence of evolutionary algorithms on the n–dimensional continuous space. IEEE Trans. Cybern. **43**(5), 1462–1472 (2013)
2. Akimoto, Y., Auger, A., Glasmachers, T.: Drift theory in continuous search spaces: expected hitting time of the (1+1)-ES with 1/5 success rule. arXiv:1802.03209 (2018)
3. Beyer, H.-G.: The Theory of Evolution Strategies. Springer, New York (2001). https://doi.org/10.1007/978-3-662-04378-3
4. He, J., Yao, X.: A study of drift analysis for estimating computation time of evolutionary algorithms. Nat. Comput. **3**(1), 21–35 (2004)
5. Huang, H., Xu, W., Zhang, Y., Lin, Z., Hao, Z.: Runtime analysis for continuous (1+1) evolutionary algorithm based on average gain model. SCIENTIA SINICA Informationis **44**(6), 811–824 (2014)
6. Jägersküpper, J., Witt, C.: Rigorous runtime analysis of a $(\mu + 1)$ ES for the sphere function. In: 7th International Proceedings of Genetic and Evolutionary Conference, Washington, D. C., pp. 849–856. ACM (2005)
7. Jägersküpper, J.: Algorithmic analysis of a basic evolutionary algorithm for continuous optimization. Theor. Comput. Sci. **379**(3), 329–347 (2007)
8. Jägersküpper, J.: Probabilistic runtime analysis of $(1 +, \lambda)$ ES using isotropic mutations. In: 8th International Proceedings of the Genetic and Evolutionary Computation Conference, Seattle, WA, USA, pp. 461–468. ACM (2006)
9. Li, S.: Concise formulas for the area and volume of a hyperspherical cap. Asian J. Math. Stat. **4**(1), 66–70 (2011)

10. Neumann, F.: Expected runtimes of a simple evolutionary algorithm for the multi-objective minimum spanning tree problem. Eur. J. Oper. Res. **181**(3), 1620–1629 (2007)
11. Qian, C., Bian, C., Jiang, W.,Tang, K.: Running time analysis of the (1+1)–EA for onemax and leadingones under bitwise noise. In: 19th International Proceedings of the Genetic and Evolutionary Computation Conference, Berlin, German, pp. 1399–1406. ACM (2017)
12. Qian, C., Yu, Y., Jin, Y., Zhou, Z.-H.: On the effectiveness of sampling for evolutionary optimization in noisy environments. In: Bartz-Beielstein, T., Branke, J., Filipič, B., Smith, J. (eds.) PPSN 2014. LNCS, vol. 8672, pp. 302–311. Springer, Cham (2014). https://doi.org/10.1007/978-3-319-10762-2_30
13. Tang, K., Yang, P., Yao, X.: Negatively correlated search. IEEE J. Sel. Areas Commun. **34**(3), 542–550 (2016)
14. Wegener, I.: Methods for the analysis of evolutionary algorithms on pseudo–Boolean functions. In: Sarker, R., Mohammadian, M., Yao, X. (eds.) Evolutionary Optimization, pp. 349–369. Springer, Boston (2003). https://doi.org/10.1007/0-306-48041-7_14
15. Yu, Y., Qian, C., Zhou, Z.-H.: Switch analysis for running time analysis of evolutionary algorithms. IEEE Trans. Evol. Comput. **19**(6), 777–792 (2015)

DTAST: A Novel Radical Framework for de Novo Transcriptome Assembly Based on Suffix Trees

Jin Zhao[1], Haodi Feng[1(✉)], Daming Zhu[1], Chi Zhang[2], and Ying Xu[3]

[1] School of Computer Science and Technology, Shandong University,
Shun Hua Road, Jinan 250101, Shandong, China
fenghaodi@sdu.edu.cn
[2] Department of Medical and Molecular Genetics, Indiana University,
Indianapolis, USA
[3] Department of Biochemistry and Molecular Biology, University of Georgia,
Athens, USA

Abstract. In this article, we develop a novel radical framework for de novo transcriptome assembly based on suffix trees, called DTAST. DTAST extends contigs by reads that have the longest overlaps with the contigs' terminuses. These reads can be found in linear time of the length of the reads through a well-designed suffix tree structure. Besides, DTAST proposes two strategies to extract transcript-representing paths: a depth-first enumeration strategy and a hybrid strategy based on length and coverage. Experimental results showed that DTAST performs more competitive than the other compared state-of-the-art de novo assemblers. The software with choice for either strategy is available at https://github.com/Jane110111107/DTAST.

1 Introduction

In this paper, we focus on de novo transcriptome assembly in which case we have no reference genome and we assemble the transcripts mainly based on the overlaps of the reads and the information hidden in the reads' structure.

Existing strategies for transcriptome assembly usually adopt the following scheme: first constructing graphs based on the RNA-seq reads, and then extracting paths from the graphs to represent plausible transcripts. Various algorithms are employed to recover transcript-representing paths [1–4]. DTAST developed two new strategies: depth-first enumeration strategy, DTAST-E for short, aiming to distinguish as many as possible transcripts and a hybrid strategy based on length and coverage, DTAST-H for short, trying to target candidates more accurately (Sect. 2).

Historically, the de novo assembly approaches mostly rely on the pioneering works on de Bruijn graphs [5], including Trinity [4], SOAPdenovo-Trans [6], Oases [7],

This work is supported by National Natural Science Foundation of China under No. 61672325, No. 61472222, No. 61732009, and No. 61761136017.

IDBA-Tran [8], and Trans-AByss [9]. Recently, Bridger [2], BinPacker [10], and IsoTree [11] applied splicing graphs [12] to represent alternative splicing. Both de Bruijn graphs and splicing graphs are usually constructed by extending contigs with k-mers (k-character substrings of reads) while the latter usually contain much less vertices and thus more applicable.

Since a k-mer may originate from quite a few different reads which may easily lead to a wrong extension, these strategies deny making full use of the information of the whole nucleotides arrangement in each read. Hence, both the accuracy and sensitivity of k-mer-based strategies are still far from meeting the requirement. In this work, we introduce a more straightforward contig extension strategy that extends a contig by the read that has the longest overlap with the contig's terminus based on suffix trees (Sect. 2).

2 Methods

2.1 Suffix Tree Construction

To facilitate the contig extension from 5′ to 3′, DTAST builds a suffix tree, called right extension suffix tree (REST). Although the beginning of the maximum overlap between the current contig's 3′ terminus and the candidate reads is unknown, the ending of the maximum overlap is known. The maximum overlap is actually a prefix of the candidate reads. Consequently, DTAST reverses all the read sequences and constructs a suffix tree for the $l \sim L - 1$-character suffixes of all the reverse reads, where l denotes the predefined minimum overlap length and L represents the read length. If the x-character ($l \leq x \leq L - 1$) path from the root node to the node v represents a suffix of the reverse of read r, DTAST stores the read ID r in the node v (as shown in Fig. 1).

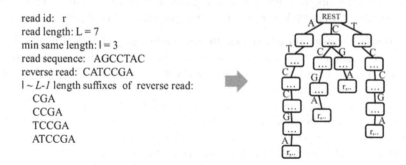

Fig. 1. An example for adding the suffixes of a reverse read to the right extension suffix tree.

Similarly, DTAST constructs a left extension suffix tree (LEST) to facilitate the contig extension from 3′ to 5′. In this case, the maximum overlap is actually a suffix of the candidate reads. Through the suffix trees constructed with the above method, DTAST gets the candidate reads that hold the longest overlaps with the current contig's terminus in $O(L)$ time by traversing the left extension tree (for left extension, right extension is processed similarly) along the edges marked with the characters as in the prefix of the contig as far as possible and then moving back to find the read IDs stored in the nearest node.

2.2 Splicing Graph Construction

The splicing graph used by DTAST is similar to that defined in Bridger [2], BinPacker [10], and IsoTree [11].

Briefly, DTAST constructs splicing graphs as follows: DTAST first sets the reads whose coverage exceeds the average as seeds, and then selects an unused seed as the main contig and extends the contig with help of the suffix trees. When the contig cannot be extended in either direction, DTAST makes the branch extensions (extensions from some possible positions of the contig other than the left and right ends) to construct splicing variants. Each subsequence between two splicing points (ending points are processed as splicing points) is defined as a node, and an edge is added between two nodes besides a common splicing point. Finally, DTAST trims fake nodes and edges in the graph. DTAST repeats the above steps until all seeds have been checked.

2.3 Transcripts Assembly

In the following discussion, let $G(V, E)$ denote the splicing graph. DTAST adds a source vertex s and a sink vertex t into the graph, and connects s (or t) with the vertices without incoming edges (or outgoing edges).

Depth-First Enumeration Strategy. DTAST-E enumerates the paths from the source vertex s to the sink vertex t with a depth-first search strategy. The algorithm starts from the source vertex s, and iteratively traverses all the edges leaving the current vertex in a depth-first manner. Once it encounters the sink vertex t, the algorithm will output the path. Since the graph is a directed acyclic graph, this procedure can be easily implemented with a stack in $O((V + E)p)$ time, where p is the number of paths starting from s.

The Hybrid Strategy Based on Length and Coverage. DTAST-H iteratively searches in the remaining graph the longest path from the paths with the maximum coverage by using a dynamic procedure until the candidate path meets a pre-given empirical condition. Note that the coverage of a path is defined as the minimum coverage of all the edges and vertices in the path. Once such a path is derived, it will be deleted from the splicing graph, i.e., the coverage of all the edges and vertices along the path will be reduced by the coverage of the path. Let $capacity(v)$ denote the maximum coverage among all the paths that start from vertex s and end at vertex v. Then, for each vertex v in topological order, $capacity = \max_{(u,v) \in E}\{min\{capacity(u), w(u, v), \ capacity(v)\}\}$. With computing $capacity(v)$, we can simultaneously compute the length of the thickest path from s to v. When processing vertex t, we can pick the longest path from s to t with the maximum coverage. The time consumed by this strategy is $O((V + E)E)$.

3   Results

We compared DTAST-E and DTAST-H with six state-of-the-art de novo assemblers including IsoTree, Trinity, BinPacker, SOAPdenovo-Trans, IDBA-Tran, and Oases on both simulated and real datasets.

3.1   Evaluation Criteria

We define the full-length reconstructed transcript as an assembled transcript that holds at least 95% sequence identity to some annotated transcripts. The full-length identified transcript represents an annotated transcript with at least 95% sequence covered by an assembled transcript. On the simulated datasets, we applied recall and precision to measure the performances of the de novo assemblers. As the annotated transcripts of the real datasets are usually not the ground truth expressed transcripts, for real datasets we use the number of full-length identified transcripts to represent the recall, and we measure the precision by comparing the number of full-length reconstructed transcripts and the number of assembled transcripts.

3.2   Simulated Data

In order to explore the sensitivity of assemblers on the length of reads, we used FluxSimulator [13] to simulate five samples with read lengths of 50 bp, 75 bp, 100 bp, 125 bp, and 150 bp, respectively. Each sample contains 0.1 million paired-end reads that are generated from 100 isoform transcripts originated from 41 different genes. The de novo assemblers' performances on these samples are shown in Fig. 2.

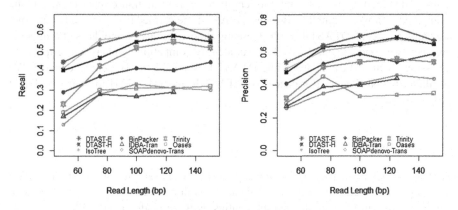

Fig. 2. Impact of the length of read on the performance of assemblers.

Figure 2 shows that DTAST performed more competitive than the other compared de novo assemblers especially with precision measure. The average precision obtained by DTAST-E on these five simulated samples was 0.66, which had 34.7%, 24.5%, 6.5%, 73.7%, 88.6%, 88.6%, and 6.5% increase over that achieved by Trinity (0.49),

BinPacker (0.53), IsoTree (0.62), SOAPdenovo-Trans (0.38), Oases (0.35), IDBA-Tran (0.35), and DTAST-H (0.62), respectively. For the recall measure, DTAST performed better than the other de novo assemblers except that IsoTree leads a little with two read lengths 75 bp and 150 bp. DTAST-E obtained the global best value of 0.63.

3.3   Real Data

We retrieved a dog dataset from NCBI SRA database with Accession Code of SRX295047. In the dog dataset, there are totally 30968059 paired-end reads with length of 50 bp. The numbers of full-length identified transcripts, full-length reconstructed transcripts, and candidate transcripts collected by the de novo assemblers are shown in Table 1. The computational demands (running time and peak memory) of the assemblers are also given in Table 1.

Table 1. The de novo assemblers' performances on the dog dataset.

Assemblers	Trinity	BinPacker	IDBA-Tran	SOAPdenovo	Oases	IsoTree	DTAST-E	DTAST-H
Identified	1017	1149	598	1005	530	1354	1504	1370
Reconstructed	1633	2601	1011	1006	957	2974	3916	2514
Candidates	96018	73419	69757	85028	113361	81597	103461	71356
Memory (G)	64	34	12	16	22	38	106	106
Time (min)	1088	162	43	33	50	410	211	207

From Table 1, we observed that DTAST-E obtained the most numbers of full-length identified transcripts and full-length reconstructed transcripts among all the compared de novo assemblers. Besides, DTAST-H performed better than other de novo assemblers except DTAST-E with guessing almost the least candidates. From these numbers we conclude that DTAST-H can identify competitively many transcripts with guessing much less candidates while DTAST-E is suitable to the general demand. We attribute the excellent performances of DTAST-E and DTAST-H to their read-based extension strategy.

4   Conclusion

In this article, we first proposed a new approach for contig extension. We applied suffix trees of reads to quickly find the candidate reads that have the longest overlaps with contigs' terminuses, and extended the contigs by these reads directly. We also developed two strategies to extract the transcript-representing paths in the splicing graphs: a depth-first enumeration strategy and a hybrid strategy based on length and coverage. The experimental results provide a whole picture of the superior performance of DTAST with the cost of comparable memory. Our future work includes developing smaller-sized data structure while keeping the searching speed and transcript extracting algorithms with higher recall and precision.

References

1. Trapnell, C., Williams, B.A., Pertea, G., Mortazavi, A., Kwan, G., van Baren, M.J., Salzberg, S.L., Wold, B.J., Pachter, L.: Transcript assembly and abundance estimation from RNA-seq reveals thousands of new transcripts and switching among isoforms. Nat. Biotechnol. **28**(5), 511 (2010)
2. Chang, Z., Li, G., Liu, J., Zhang, Y., Ashby, C., Liu, D., Cramer, C.L., Huang, X.: Bridger: a new framework for de novo transcriptome assembly using RNA-seq data. Genome Biol. **16**(1), 30 (2015)
3. Pertea, M., Pertea, G.M., Antonescu, C.M., Chang, T.C., Mendell, J.T., Salzberg, S.L.: Stringtie enables improved reconstruction of a transcriptome from RNA-seq reads. Nat. Biotechnol. **33**(3), 290–295 (2015)
4. Grabherr, M.G., Haas, B.J., Yassour, M., Levin, J.Z., Thompson, D.A., Amit, I., Adiconis, X., Fan, L., Raychowdhury, R., Zeng, Q., et al.: Full-length transcriptome assembly from RNA-seq data without a reference genome. Nat. Biotechnol. **29**(7), 644–652 (2011)
5. Pevzner, P.A., Tang, H., Waterman, M.S.: An Eulerian path approach to DNA fragment assembly. Proc. Natl. Acad. Sci. **98**(17), 9748–9753 (2001)
6. Xie, Y., Wu, G., Tang, J., Luo, R., Patterson, J., Liu, S., Huang, W., He, G., Gu, S., Li, S., et al.: Soapdenovo-trans: de novo transcriptome assembly with short RNA-seq reads. Bioinformatics **30**(12), 1660–1666 (2014)
7. Schulz, M.H., Zerbino, D.R., Vingron, M., Birney, E.: Oases: robust de novo RNA-seq assembly across the dynamic range of expression levels. Bioinformatics **28**(8), 1086–1092 (2012)
8. Peng, Y., Leung, H.C., Yiu, S.M., Lv, M.J., Zhu, X.G., Chin, F.Y.: IDBA-tran: a more robust de novo de bruijn graph assembler for transcriptomes with uneven expression levels. Bioinformatics **29**(13), i326–i334 (2013)
9. Robertson, G., Schein, J., Chiu, R., Corbett, R., Field, M., Jackman, S.D., Mungall, K., Lee, S., Okada, H.M., Qian, J.Q., et al.: De novo assembly and analysis of RNA-seq data. Nat. Methods **7**(11), 909–912 (2010)
10. Liu, J., Li, G., Chang, Z., Yu, T., Liu, B., McMullen, R., Chen, P., Huang, X.: Binpacker: packing-based de novo transcriptome assembly from RNA-seq data. PLoS Comput. Biol. **12**(2), e1004772 (2016)
11. Zhao, J., Feng, H., Zhu, D., Zhang, C., Xu, Y.: IsoTree: de novo transcriptome assembly from RNA-Seq reads. In: Cai, Z., Daescu, O., Li, M. (eds.) ISBRA 2017. LNCS, vol. 10330, pp. 71–83. Springer, Cham (2017). https://doi.org/10.1007/978-3-319-59575-7_7
12. Heber, S., Alekseyev, M., Sze, S.H., Tang, H., Pevzner, P.A.: Splicing graphs and EST assembly problem. Bioinformatics **18**(suppl 1), S181–S188 (2002)
13. Griebel, T., Zacher, B., Ribeca, P., Raineri, E., Lacroix, V., Guigó, R., Sammeth, M.: Modelling and simulating generic RNA-seq experiments with the flux simulator. Nucleic Acids Res. **40**(20), 10073–10083 (2012)

Energy-Efficient Single Machine Total Weighted Tardiness Problem with Sequence-Dependent Setup Times

M. Fatih Tasgetiren[1(✉)], Hande Öztop[2], Uğur Eliiyi[3],
Deniz Türsel Eliiyi[2], and Quan-Ke Pan[4]

[1] Department of International Logistics Management, Yasar University,
Bornova, Turkey
fatih.tasgetiren@yasar.edu.tr
[2] Department of Industrial Engineering, Yasar University, Bornova, Turkey
{hande.oztop,deniz.eliiyi}@yasar.edu.tr
[3] Department of Computer Science, Dokuz Eylül University, İzmir, Turkey
ugur.eliiyi@deu.edu.tr
[4] State Key Laboratory, Huazhong University of Science and Technology,
Wuhan, People's Republic of China
panquanke@hust.edu.cn

Abstract. Most of the problems defined in the scheduling literature do not yet take into account the energy consumption of manufacturing processes, as in most of the variants with tardiness objectives. This study handles scheduling of jobs with due dates and sequence-dependent setup times (SMWTSD), while minimizing total weighted tardiness and total energy consumed in machine operations. The trade-off between total energy consumption (TEC) and total weighted tardiness is examined in a single machine environment, where different jobs can be operated at varying speed levels. A bi-objective mixed integer linear programming model is formulated including this speed-scaling plan. Moreover, an efficient multi-objective block insertion heuristic (BIH) and a multi-objective iterated greedy (IG) algorithm are proposed for this NP-hard problem. The performances of the proposed BIH and IG algorithms are compared with each other. The preliminary computational results on a benchmark suite consisting of instances with 60 jobs reveal that, the proposed BIH algorithm is very promising in terms of providing good Pareto frontier approximations for the problem.

Keywords: Energy efficient scheduling · Multi-objective optimization
Heuristic optimization · Sequence-dependent setup times · Weighted tardiness

1 Introduction

This paper mainly focuses on employing an energy-efficient approach for a well-known scheduling problem. The single machine scheduling problem with its many variants are among the most studied in the related literature, yet most of these studies do not consider the energy consumption of the machine. Modern manufacturing policies

© Springer International Publishing AG, part of Springer Nature 2018
D.-S. Huang et al. (Eds.): ICIC 2018, LNCS 10954, pp. 746–758, 2018.
https://doi.org/10.1007/978-3-319-95930-6_76

involve the reduction of the associated carbon footprints corresponding to lower negative impacts on both the environment and financial structures [1]. Energy consumption, by directly affecting carbon emissions and the incurred costs, is also associated with sustainability issues. One alternative for the production companies is to obtain more energy-efficient machines for reducing their energy utilization [2]. However, the investments required, make this option inapplicable for most of the sector. In this study, energy efficiency is tackled with an operational standpoint for a single machine setting with sequence dependent setup times and arbitrary due dates.

Although the energy-efficiency topic attracts a lot of public interest, the studies in the context of scheduling are rare. Energy consumption problem is studied for two different cases, namely for total tardiness in single machine and flowshop settings [3, 4]. Idle and operating times of the machines are investigated. For example, Nawaz et al. [5] tackled with energy minimization for permutation flowshop problem by considering the weighted sum of idle times of the machines. Mouzon et al. [2] dealt with energy consumption and total completion time of the tasks on a single machine. By turning the idle machines off, energy savings are put into effect. A similar approach is employed for total energy consumption and total tardiness objectives on a single machine, via a metaheuristic algorithm [3]. Energy savings during machine operation are not considered in these studies, just the energy consumptions of idle machines. Diaz et al. [6] claimed that consumed energy during machine operation has much more significance than machine idling, where the machines' processing speeds affect operating time that dominates energy consumption. Moreover, higher machine speeds result in more energy costs due to increased consumption [7]. In this paper, we consider also energy savings during machine operation by employing a speed scaling strategy.

Energy utilization with respect to the speeds of machines and job completion times are considered, but they did not include the setup times between jobs in [1]. Flowshop scheduling with setup times are investigated and explained in detail as well as including the recent trends in [8]. In their extensive survey [8], they remarked the importance of considering setup times separately from the processing times of tasks, as the assumption of sequence-independent setup times does not represent the actual conditions in many real-world cases. The authors studied a bi-objective two-machine flowshop problem to show the tradeoff between makespan and energy consumption in [9]. They also considered sequence dependent setup times between tasks to cope with real-life cases.

The single machine scheduling problem with sequence-dependent setup times and weighted tardiness minimization is NP-hard, as a result of the particular case in which all setup times are zero [10]. Tanaka and Araki [11] proposed an exact algorithm for this problem, which solved most of the 120 instances of the Cicirello set [12] to optimality. Despite being able to quickly solve most instances of size up to 60 jobs and solve instances with up to 85 jobs, their algorithm still requires too much computation time for some of the largest instances in the set [12].

In this paper, a speed scaling strategy is proposed for the SMWTSD, where the machine can operate at varying speed levels. Thus, different speed levels create a clash between processing time and total energy consumption (TEC), namely higher speed level causes the energy consumption to increase, while the processing time is reduced. A novel multi-objective mixed integer linear programming model (MILP) model is proposed for SMWTSD that employs speed-scaling strategy. An effective multi-objective BIH and a multi-objective IG are also developed, as this new problem is also NP-hard. To the best of our knowledge, due dates and sequence-dependent setup times are not studied in a single machine setting with multi-criteria including the energy-efficiency objective.

2 Problem Definition

In energy-efficient single machine total weighted tardiness problem with sequence-dependent setup times (EE_SMWTSD), a set of n jobs $J = \{1, \ldots, n\}$ must be processed on a single machine to minimize two conflicting objectives: total energy consumption (TEC) and total weighted tardiness ($\sum_j w_j T_j$). Each job $j \in J$ has an uninterrupted processing time p_j, a due date $d_j \geq 0$ and a weight for tardiness $w_j \geq 0$. Setup time s_{ij} occurs if job j processed immediately after job i. Even if job j is the first job in the schedule, a setup time (s_{0j}) is required. For this reason, we define a dummy job (j_0) with $p_0 = 0; w_0 = 0; d_0 = 0; s_{0j} \geq 0$ and $s_{j0} = 0; \forall j \in J$, to handle the setup of the first job in the schedule. The tardiness of a job j is calculated as $T_j = max\{C_j - d_j, 0\}$, where C_j denotes the completion time of job j. All of the jobs and the machine are available at time 0. The machine can process one job at a time and job preemption is not permitted.

Following the standard scheduling notation, the studied problem can be referred as $1|s_{ij}| \sum_j w_j T_j, TEC$. Unlike the standard SMWTSD, the machine can process jobs with different speed levels in EE_SMWTSD. Therefore, processing time of a job can vary according to chosen speed level $l \in L$. It is assumed that setup times resulting from speed changes are included in the processing times of jobs. The necessary parameters and decision variables are given in Table 1. Following the TEC calculation of [9], MILP model of the problem is given below:

Table 1. Problem notation

Parameters		Decision Variables	
p_j	Processing time of job $j \in J$	y_{jl}	1 if job j is processed with speed level l, 0 otherwise
w_j	Weight of job $j \in J$	x_{ij}	1 if job i precedes job j, 0 otherwise
d_j	Due date of job $j \in J$	z_j	Start time of job j
s_{ij}	Sequence dependent setup time for changing from job i to job j	T_j	Tardiness of job j
v_l	Speed factor of speed level $l \in L$	θ	Idle time on machine
λ_l	Conversion factor for speed level $l \in L$	C_{max}	Maximum completion time
φ	Conversion factor for idle time	TEC	Total energy consumption
μ	Power of machine		
U	A very large number		

$$Minimize \sum_{j \in J} w_j T_j \tag{1}$$

$$Minimize\ TEC \tag{2}$$

$$\sum_{i \in J} x_{ij} = 1 \quad \forall j \in J, i \neq j \tag{3}$$

$$\sum_{j \in J} x_{ij} = 1 \quad \forall i \in J, i \neq j \tag{4}$$

$$\sum_{l \in L} y_{jl} = 1 \quad \forall j \in J \tag{5}$$

$$z_j + \sum_{l \in L} \frac{p_j}{v_l} y_{jl} - d_j \leq T_j \quad \forall j \in J \tag{6}$$

$$z_j + \sum_{l \in L} \frac{p_j}{v_l} y_{jl} \leq C_{max} \quad \forall j \in J \tag{7}$$

$$z_i + \sum_{l \in L} \frac{p_i}{v_l} y_{il} + s_{ij} + U(x_{ij} - 1) \leq z_j \quad \forall i \in J, j \in J/j_0\ i \neq j \tag{8}$$

$$\theta = C_{max} - \sum_{j \in J} \sum_{l \in L} \frac{p_j}{v_l} y_{jl} - \sum_{j \in J} \sum_{k \in J: k \neq j} s_{jk} x_{jk} \tag{9}$$

$$TEC = \sum_{j \in J} \sum_{l \in L} \frac{\mu p_j \lambda_l}{60 v_l} y_{jl} + \frac{\varphi \mu}{60} \left(\sum_{j \in J} \sum_{k \in J: k \neq j} s_{jk} x_{jk} + \theta \right) \tag{10}$$

$$z_0 = 0 \tag{11}$$

$$x_{ij} \in \{0,1\} \quad \forall i,j \in J \quad y_{jl} \in \{0,1\} \quad \forall j \in J, l \in L\ T_j \geq 0 \quad z_j \geq 0 \quad \forall j \in J \tag{12}$$

The objective functions (1) and (2) minimize the total weighted tardiness and TEC, respectively. Constraints (3) and (4) state that there are one immediate predecessor and one immediate successor for each job. Constraint (5) ensures that exactly one speed level is selected for each job. Constraint (6) calculates the tardiness of each job. Constraint (7) computes the makespan, since it is used in calculation of TEC. Constraint (8) guarantees that the next job in the sequence can be started only after preceding job has been completed. Constraint (9) computes the idle time on the machine. Constraint (10) calculates the total energy consumption in kilowatt/hour as proposed in [9]. Constraint (11) fixes the dummy job j_0 as first job. Constraint (12) defines the decision variables.

As mentioned above, in our study, we aim to minimize two conflicting objectives of total weighted tardiness and TEC by considering the sequence dependent setup times of the jobs in a single machine setting. Therefore, there is no single optimal solution for this multi-objective problem. However, a set of non-dominated solutions, which are feasible solutions that are not dominated by any other feasible solution, can be obtained. There are two types of dominance concepts in multi-objective optimization. For a minimization problem, a feasible solution $\vec{a}$ dominates a feasible solution $\vec{b}$

$(\vec{a} \succ \vec{b})$ iff $\forall i \in F;\ f_i(\vec{a}) \leq f_i(\vec{b})$ and $\exists i;\ f_i(\vec{a}) < f_i(\vec{b})$, where F is the number of objectives. Namely, $\vec{a}$ dominates $\vec{b}$ if no objective of $\vec{a}$ is larger than the corresponding objective of $\vec{b}$ and at least one objective of $\vec{a}$ is smaller. As another dominance concept, it can be said that a feasible solution $\vec{a}$ weakly dominates another feasible solution $\vec{b}$ $(\vec{a} \succeq \vec{b})$ iff $\forall i \in F;\ f_i(\vec{a}) \leq f_i(\vec{b})$.

3   Energy-Efficient SMWTSD Problem

In the traditional IG algorithm [13], the initial solution is constructed by the NEH insertion heuristic [5]. Then, destruction and construction (DC) procedure is employed to generate new solutions. The destruction phase is involved in removing a number ds of jobs randomly from the current solution. Then, these ds jobs are re-inserted into partial solution in the order they are removed to obtain a complete solution. An insertion-based local search is applied to the solution after the DC procedure. An acceptance criterion is used to accept the new solution after a local search. Note that acceptance criterion is used with a constant temperature, which is suggested by [14], as follows: $T = \sum_{j=1}^{n} p_j / 10n \times \tau P$, where τP is a parameter to be adjusted. These simple steps are repeated until a stopping criterion is satisfied. Recently, an IG_{ALL} algorithm is presented for the PFSP, where the IG_{ALL} algorithm applies an additional local search to partial solutions after destruction, which substantially improves solution quality [15]. The pseudo-code of the IG_{ALL} algorithm is given in Fig. 1 and details can be found in [15].

Procedure IG_{ALL}
    π_0 = GenerateInitial Solution
    π = LocalSearch(π_0)
    do
        π' = Destruction(π, ds)
        π' = ApplyLocalSearchToPartialSolution(π')
        π' = Construction(π')
        π'' = ApplyLocalSearchToCompleteSolution(π')
        π = AcceptionCriterion(π'', π)
    while (Termination criterion is met)
    endprocedure

Fig. 1. IG_{ALL} algorithm.

Recently, block move-based search algorithms are presented for single machine scheduling problems in literature [16–18]. We employ a modified variant of the block insertion heuristic (VBIH) algorithm used in [19]. In the proposed BIH algorithm in this paper, it begins with a constructive heuristic $(PF\_NEH(x))$, and the block size is fixed to $b = 2$. It simply removes a block $b = 2$ of jobs from the current solution, and then it makes a number $n - b + 1$ of block insertion moves sequentially in the partial solution. Then, the best one from the block moves is retained in order to undergo a local search procedure. If the new solution obtained after the local search is better than the current solution, it replaces the current solution. Otherwise, a simple simulated annealing-type of acceptance criterion is used with a constant temperature, which is suggested by [14]. The outline of the BIH algorithm is given in Fig. 2.

Procedure BIH
    π_0 = GenerateInitial Solution
    π = LocalSearch(π_0)
    do
        π' = *Remove a block b from π randomly*
        π' = ApplyLocalSearchToPartialSolution(π')
        π' = *Best insertion of block b in all positions in π'*
        π'' = ApplyLocalSearchToCompleteSolution(π')
        π = AcceptionCriterion(π'', π)
    while (Termination criterion is met)
    endprocedure

Fig. 2. Block insertion heuristic.

In this paper, we develop a constructive heuristic denoted as $PF\_NEH(x)$ for the SMWTSD problem, which provides an initial solution to IG_{ALL} and BIH algorithms. The initial order of jobs is obtained by ATCS heuristic [20]. The number x of new solutions can be generated from the initial order of jobs as follows. The first job of the initial order is taken as the first job and the $PF\_NEH(x)$ heuristic is applied to generate a new solution. Then, the second job of the initial order is taken as the first job and the $PF\_NEH(x)$ heuristic is applied to generate another new solution. This is repeated $x = n$ times and the number x of new solutions will be generated. Of course, the best one among them is chosen as the final solution. In the proposed constructive heuristic, suppose that π_{i-1} jobs are scheduled, thus, ending up with a partial sequence, $\pi = \{\pi_1, \pi_2, \ldots, \pi_{i-1}\}$. Obviously, job π_i will be the next job to be inserted into the partial solution π_{i-1}. Job π_i can be any job from the set U of unscheduled jobs. To determine job π_i, a cost function is necessary. For the SMWTSD problem, we propose a cost function as follows: We first calculate the completion time of job π_i (To ease the notation, job i refers to job π_i): $C_i = C_{i-1} + p_i + s_{i-1,i}$. Then, we also consider the weighted total tardiness wT_i of job π_i: $wT_i = w_i \max\{C_i - d_i, 0\}$. Now, we define our cost function as: $cF_i = (n - i - 1) \times C_i + 0.1 \times wT_i$. Then, the job with the smallest sum of cF_i amongst all jobs in U is determined as the ith job to be inserted to the partial solution π_{i-1}. Finally, we apply the NEH heuristic to only the last $\delta = 10$ jobs.

In this study, a job-based speed-scaling strategy is developed for the energy-efficient EBIH and EIG_{ALL} algorithms. Instead of acceptance criterion, dominance rule is employed when comparing two solutions in both algorithms. The solution representation is a multi-chromosome structure, consisting off a permutation of n jobs and a speed vector with three levels. We assume that machines can operate with three speed levels corresponding to fast, normal and slow speed levels, respectively. The solution representation for an individual s_i is given in Fig. 3, where an individual $s_i(\pi_{ij}, v_{ij})$

indicates that job $\pi_{i1} = 3$ has a normal speed level; job $\pi_{i2} = 2$ has a fast speed level ($v_1 = 1$) and so on.

π	3	2	5	.	n
$s_i(\pi, v)$ v	2	1	3	.	3

Fig. 3. Solution representation.

3.1   Initial Population

The initial the population with size NP is constructed as follows: Once we obtain a solution after $PF\_NEH(x)$ heuristic, this solution is taken as an initial solution for the IG_{ALL} algorithm with total weighted tardiness (wT) minimization. Ten percent of the total CPU time budget is devoted to the IG_{ALL} with wT minimization in order to obtain a good starting point for both energy-efficient EBIH and EIG_{ALL} algorithms. Once the best solution π_{best} is found by the single objective IG_{ALL} algorithm with wT minimization, the first individual in population is established by assigning fast speed level; the second individual by assigning normal speed level; and third individual by assigning slow speed level to each job in the best solution π_{best}. The rest of the population is obtained by assigning random speed levels between 1 and 3 to each job in the best solution π_{best}. The archive set Ω is initially empty and it is updated.

3.2   Destruction and Construction Procedure

The destruction-construction procedure is a core part of IG_{ALL} algorithms. In the destruction step, ds jobs with their speeds are randomly removed from the solution. A first improvement insertion local search is applied to the partial solution by considering speed levels. Then, random speed levels $v_{ij} = rand()\%l, \forall i \in 1, .., NP$ and $j \in 1, .., ds$ are assigned to the removed jobs, where l is the number of speed levels. These ds jobs are reinserted into the partial solution with their respective speed levels sequentially until a complete solution of n jobs is established. The dominance rule ($\succ$) in multi-objective optimization is used when two solutions are compared. Partial solutions are evaluated based on partial dominance. The complete solution is obtained by choosing the non-dominated solution among n solutions after the last removed job is inserted for n positions.

3.3   Block Insertion Procedure

The procedure randomly removes a block $b = 2$ of jobs with their speeds from the current solution π. Then, we apply a local search to the partial solution as in EIG_{ALL} algorithm before carrying out a block insertion. Then, the procedure carries out $n - b + 1$ block insertion moves. In other words, block $s^b(\pi^b, v^b)$ is inserted in the partial solution $s^p(\pi^p, v^p)$ sequentially. It should be noted that dominance rule ($\succ$) in multi-objective optimization will be used when two solutions and/or partial solutions are compared. So, partial solutions are evaluated based on the partial dominance rule.

3.4   First Improvement Insertion Local Search

In both algorithms, we employ a very effective first-improvement insertion local search for each individual i in the population, as seen in Fig. 4. Job π_{ij} and speed v_{ij} are removed from position j of solution $s_i(\pi_{ij}, v_{ij})$. A new speed level is randomly assigned to position j by $v_{ij} = rand()\%l$. Then; the local search inserts job π_{ij} and speed v_{ij} into all possible positions of the incumbent solution. Job π_{ij} and speed v_{ij} is inserted into the best insertion position that dominates the incumbent solution, and the archive set Ω is updated. This is repeated for all job and speed pairs. If any non-dominated solution is found, the local search is invoked again until any non-dominated solution cannot be obtained.

---

for $i = 1$ to NP do
  for $j = 1$ to n do
    $(\pi^*, v^*) = $ Remove (π_{ij}, v_{ij}) from $s_i(\pi_{ij}, v_{ij})$. Assign $v_{ij} = $ rand()%l
    $s^*(\pi^*, v^*) = $ Insert (π^*, v^*) in best position in $s(\pi_{ij}, v_{ij})$
    if $\left(f\big(s^*(\pi^*, v^*)\big) \succ f\big(s_i(\pi_{ij}, v_{ij})\big) \right)$ then do
      $s_i(\pi_{ij}, v_{ij}) = s^*(\pi^*, v^*)$
      $\Omega \in x^*(\pi^*, v^*)$ if dominates any solution in Ω
    end if
  end for
end for

Fig. 4. First improvement insertion neighborhood.

3.5   Energy Minimization

The energy-efficient EBIH and EIG_{ALL} algorithms given above are extremely effective for wT minimization. In order to obtain more energy-efficient schedules, we propose a local search algorithm based on uniform crossover operator for speed levels only. In other words, after applying energy-efficient EBIH or EIG_{ALL} algorithm to each individual in the population, we keep the same permutation for each individual in the population and make a uniform crossover on speed levels only as follows: For each individual s_i in the population, select another individual from population randomly, say s_k. Then; we generate offspring by making a uniform crossover as follows:

$$s(\pi_{ij}, v_{ij}) = \begin{cases} \pi_i(v_{ij}) & \text{if } r_{ij} \leq CR[i] \\ \pi_k(v_{kj}) & \text{otherwise} \end{cases} \quad \forall i \in 1, .., NP \text{ and } j \in 1, .., n \quad (13)$$

where r_{ij} is a uniform random number in $U(0, 1)$ and $CR[i]$ is the crossover probability, which is drawn from unit normal distribution $N(0.5, 0.1)$ with mean 0.5 and standard deviation 0.1. If s_{new} dominates x_i (i.e., $s_{new} \succ x_i$), s_i is replaced by s_{new} and the archive set Ω is updated. This is repeated for all individuals in the population.

After crossover local search, we mutate the speed levels (implicitly, we lower the speed levels) of jobs with a small mutation probability as follows:

$$s_i(\pi_{ij}, v_{ij}) = \begin{cases} s_i(v_{ij} = 1 + rand()\%l - 1) & if\ r_{ij} \leq MR[i] \\ s_i(v_{ij}) & otherwise \end{cases} \quad \forall i \in 1,..,NP; j \in 1,..,n$$

$$(14)$$

where r_{ij} is a uniform random number in $U(0,1)$ and $MR[i]$ is the mutation probability, which is drawn from unit normal distribution $N(0.05, 0.01)$ with mean 0.05 and standard deviation 0.01 for each individual s_i in the population. When we update the archive set Ω, we use a simple method as follows: When a new individual is generated, we check the archive set Ω in such a way that if the new individual dominates any individual in the archive set Ω, we remove the dominated solutions and add the non-dominated solution. In addition, we also check each pair (i, j) of individuals. We remove individual j if it is weakly dominated. We provide the outline of the energy-efficient EBIH and EIG_{ALL} algorithm in Fig. 5. Note that in weighted tardiness minimization of Step 3, either energy-efficient EIG_{ALL} algorithm or EBIH algorithm can be used. In other words, we generated two algorithms just changing from EIG_{ALL} to EBIH algorithm.

Step1. Initialize parameters

- For the single objective IG_{ALL} algorithm, destruction size and temperature parameter for acceptance criterion are taken as $d = 4$ and $\tau = 0.4$. For EBIH algorithm, block size is taken as $b = 2$ and for EIG_{ALL} algorithm destruction size is taken as $d = 4$.
- For both algorithms, the population size is taken as $NP = 100$. The archive size is taken as $|\Omega| = n \times 5$.

Step2. Initialize population

- Find the PF_NEH(x) solution, employ it in single objective IG_{ALL} as an initial solution and run IG_{ALL} for ten percent of the total CPU time budget in order to find the best solution π_{best}.
- Assign $v_l = 1, 2, 3$ to π_{best} and construct the first three solutions in the population
- Construct the rest of the population by assigning $s_i(\pi_{ij}, v_{ij}) = s(\pi_{best}, v_{ij})$
  $v_{ij} = rand()\%l, \forall i = 1,.., NP, \forall j = 1,.., n$.
- Evaluate population and update the archive set Ω.

Step3. Repeat following steps until a termination criterion

- **Weighted tardiness minimization**
  a. For each individual $s(\pi_{ij}, v_{ij})$ in the population, apply either energy-efficient EIG_{ALL} or EBIH algorithm and obtain a new individual.
  b. If the new individual dominates the incumbent individual $s_i(\pi_{ij}, v_{ij})$, replace it with the new individual and update the archive set Ω.

- **Energy minimization**
  a. For each individual $s_i(\pi_{ij}, v_{ij})$ in the population, keep the same permutation from previous stage.
  b. Apply uniform crossover operator by using two individuals s_i and s_k and generate a new individual.
  c. If the new individual dominates the incumbent individual $s_i(\pi_{ij}, v_{ij})$, replace it with the new individual and update the archive set Ω.

Step4. Check Termination

- If termination criterion is not satisfied, go to step 3.
- Otherwise, report non-dominated solutions from the archive set Ω

Fig. 5. Energy-Efficient EBIH and EIG_{ALL} Algorithms.

4   Computational Results

The most famous benchmark for the $1|s_{ij}|\sum_j w_jT_j$ problem is the Cicirello set proposed in [12]. It consists of 120 problem instances with 60 jobs, where there are 12 problem settings with 10 instances. These instances can be downloaded from the web site [http://loki.stockton.edu/~cicirelv/benchmarks.html]. In this paper, the benchmark suite of [12] is employed to compare the performances of the proposed EBIH and EIG_{ALL} algorithms. In computation of TEC, we use the same speed and energy parameters of [9]. There are three speed levels (slow, normal and fast) with speed factors $v = \{0.8, 1, 1.2\}$ and conversion factors $\lambda = \{0.6, 1, 1.5\}$. The machine power is assumed to be 60 kw and the conversion factor for idle time is taken as 0.05.

As the single objective SMWTSD is known to be NP-hard [11], the multi-objective EE_SMWTSD is also NP-hard. Therefore, exact optimization methods are not applicable to solve these instances with 60 jobs. Since the solution times of mathematical model increase exponentially even for the single objective problem [11], the proposed multi-objective MILP model cannot be solved within reasonable computation times for these instances. Consequently, we obtained the non-dominated solution sets for all instances employing the BIH and EIG_{ALL} algorithms. The both algorithms are coded in C++ programming language on Microsoft Visual Studio 2013 and all instances are solved on a Core i5, 3.20 GHz, 8 GB RAM computer. In both algorithms, ten replications are made for each instance. In each replication, the algorithm is run for 50 s. Note that after ten replications, we report only the non-dominated solutions amongst 1000 (Population size x Number of replications).

We compare the non-dominated solution sets of EBIH and EIG_{ALL} with each other in terms of coverage, spacing and cardinality performance metrics:

- Cardinality: the number of non-dominated solutions found.
- Spacing (S) [21]: $S_H = \left[\frac{1}{|H|}\sum_{i\in H}(d_i - \bar{d})^2\right]^{1/2}/\bar{d}$

  where $\bar{d} = \sum_{i\in H} d_i/|H|$, d_i is the minimum Euclidean distance between solution i and its nearest neighbor in set H. If spacing value is low, it can be said that the solutions of H are more evenly distributed.
- Coverage of Two Sets (C) [22]: $C(H,P) = |p \in P; \exists h \in H : h \succcurlyeq p\}|/|P|$

  where $C(H,P)$ is 1 if some solutions of H weakly dominate all solutions of P. However, $C(H,P)$ is not always equal to $1 - C(P,H)$, as some solutions in H and P sets may not be dominated by each other.

The average results over 10 instances are reported for each problem setting in Table 2. As shown in the table, EBIH generates approximately 2.5 times as many non-dominated solutions than EIG_{ALL}. In terms of coverage, EBIH clearly performs much better than the EIG_{ALL}, as the solutions generated by EBIH weakly dominate 95% of those found by EIG_{ALL}. While, solutions of EIG_{ALL} algorithm weakly dominate only 3% of those from EBIH. Especially, in 35 instances, solutions of EBIH weakly dominate all solutions found by EIG_{ALL}. This result is expected as the BIH algorithm usually performs well for the weighted tardiness objective as in [16–18]. In terms of spacing, both methods have low spacing values. However, solutions generated by the

EIG_{ALL} algorithm are slightly more even distributed than those solutions found by EBIH. Consequently, as EBIH generates many good quality solutions in reasonable computation times, we can say that EBIH performs better than EIG_{ALL}.

Table 2. Computational results

Inst.	|EBIH|	|EIG|	C(EBIH, EIG_{ALL})	C(EIG_{ALL}, EBIH)	S_{EBIH}	S_{EIGALL}
1–10	339.80	127.10	0.95	0.03	3.84	3.11
11–20	278.10	126.70	0.97	0.01	5.72	3.02
21–30	186.60	82.80	0.95	0.04	3.37	3.36
31–40	95.90	88.20	1.00	0.00	4.41	2.96
41–50	310.30	103.30	0.95	0.02	1.89	1.84
51–60	303.10	91.20	0.91	0.05	2.13	2.26
61–70	233.20	91.70	0.95	0.03	2.61	1.82
71–80	217.50	92.20	0.98	0.02	2.57	2.12
81–90	218.70	72.00	0.89	0.05	1.84	1.32
91–100	213.70	71.70	0.97	0.02	1.98	2.35
101–110	220.30	72.50	0.97	0.02	1.96	1.66
111–120	208.20	68.60	0.94	0.04	1.81	1.73
Average	235.45	90.67	0.95	0.03	2.84	2.29

5 Conclusion

In this study, we consider the trade-off between total energy consumption and total weighted tardiness for single machine scheduling problem by employing a job-based speed-scaling strategy. In order to reflect real practice, we consider due dates and sequence-dependent setup times as well. To the best of our knowledge, due dates and sequence-dependent setup times are not studied previously in the literature for the energy-efficient, multi-objective, single-machine scheduling problem. Initially, we propose a novel bi-objective MILP model for the problem. Due to the NP-hardness of the problem, multi-objective EBIH and EIG_{ALL} algorithms are also proposed. The performances of these algorithms are compared with each other using well-known benchmarks consisting of instances with 60 jobs [12]. The computational results reveal that, the proposed EBIH algorithm clearly outperforms the EIG_{ALL} algorithm in terms of quality of the solutions. The solutions generated by EBIH weakly dominate 95% of those found by EIG_{ALL}. In terms of cardinality, EBIH generates approximately 2.5 times as many non-dominated solutions than EIG_{ALL}. In further studies, different multi-objective metaheuristics such as NSGA-II can be developed for the problem, and compared with the proposed EBIH and EIG_{ALL} algorithms.

References

1. Fang, K., Uhan, N., Zhao, F., Sutherland, J.W.: A new approach to scheduling in manufacturing for power consumption and carbon footprint reduction. J. Manuf. Syst. **30**(4), 234–240 (2011)
2. Mouzon, G., Yildirim, M.B., Twomey, J.: Operational methods for the minimization of energy consumption of manufacturing equipment. Int. J. Prod. Res. **45**(18–19), 4247–4271 (2007)
3. Mouzon, G., Yildirim, M.B.: A framework to minimise total energy consumption and total tardiness on a single machine. Int. J. Sustain. Eng. **1**(2), 105–116 (2008)
4. Liu, G.S., Zhang, B.X., Yang, H.D., Chen, X., Huang, G.Q.: A branch-and-bound algorithm for minimizing the energy consumption in the PFS problem. Math. Probl. Eng. 1–6 (2013)
5. Nawaz, M., Enscore Jr., E.E., Ham, I.: A heuristic algorithm for the m-machine, n-job flowshop sequencing problem. OMEGA **11**(1), 91–95 (1983)
6. Diaz, N., Redelsheimer, E., Dornfeld, D.: Energy consumption characterization and reduction strategies for milling machine tool use. In: Hesselbach, J., Herrmann, C. (eds.) Glocalized Solutions for Sustainability in Manufacturing, pp. 263–267. Springer, Heidelberg (2011). https://doi.org/10.1007/978-3-642-19692-8_46
7. Fang, K.T., Lin, B.M.T.: Parallel-machine scheduling to minimize tardiness penalty and power cost. Comput. Ind. Eng. **64**(1), 224–234 (2013)
8. Allahverdi, A., Ng, C.T., Cheng, T.C.E., Kovalyov, M.Y.: A survey of scheduling problems with setup times or costs. Eur. J. Oper. Res. **187**(3), 985–1032 (2008)
9. Afshin Mansouri, S., Aktas, E., Besikci, U.: Green scheduling of a two-machine flowshop: trade-off between makespan and energy consumption. Eur. J. Oper. Res. **248**(3), 772–788 (2016)
10. Lenstra, J., Rinnooy Kan, A., Brucker, P.: Complexity of machine scheduling problems. Ann. Discret. Math. **1**, 343–362 (1977)
11. Tanaka, S., Araki, M.: An exact algorithm for the single-machine total weighted tardiness problem with sequence-dependent setup times. Comput. Oper. Res. **40**(1), 344–352 (2013)
12. Cicirello, V.A., Smith, S.F.: Enhancing stochastic search performance by value-biased randomization of heuristics. J. Heuristics **11**(1), 5–34 (2005)
13. Ruiz, T., Stützle, T.: A simple and effective iterated greedy algorithm for the permutation flowshop scheduling problem. Eur. J. Oper. Res. **177**(3), 2033–2049 (2007)
14. Osman, I., Potts, C.: Simulated annealing for permutation flow-shop scheduling. OMEGA **17**(6), 551–557 (1989)
15. Dubois-Lacoste, J., Pagnozzi, F., Stützle, T.: An iterated greedy algorithm with optimization of partial solutions for the makespan permutation flowshop problem. Comput. Oper. Res. **81**, 160–166 (2017)
16. Subramanian, A., Battarra, M., Potts, C.: An iterated local search heuristic for the single machine total weighted tardiness scheduling problem with sequence-dependent setup times. Int. J. Prod. Res. **52**(9), 2729–2742 (2014)
17. Xu, H., Lu, Z., Cheng, T.: Iterated local search for single-machine scheduling with sequence-dependent setup times to minimize total weighted tardiness. J. Sched. **17**(3), 271–287 (2014)
18. González, M.A., Palacios, J.J., Vela, C.R., Hernández-Arauzo, A.: Scatter search for minimizing weighted tardiness in a single machine scheduling with setups. J. Heuristics **23** (2–3), 81–110 (2017)

19. Tasgetiren, M.F., Pan, Q.-K., Ozturkoglu, Y., Chen, A.H.L.: A memetic algorithm with a variable block insertion heuristic for single machine total weighted tardiness problem with sequence dependent setup times. IEEE Congr. Evol. Comput. CEC **2016**, 2911–2918 (2016)
20. Lee, Y.H., Bhaskaran, K., Pinedo, M.: A heuristic to minimize the total weighted tardiness with sequence-dependent setups. IIE Trans. **29**(1), 45–52 (1997)
21. Tan, K.C., Goh, C.K., Yang, Y., Lee, T.H.: Evolving better population distribution and exploration in evolutionary multi-objective optimization. Eur. J. Oper. Res. **171**, 463–495 (2006)
22. Zitzler, E.: Evolutionary Algorithms for Multi Objective Optimization: Methods and Applications. Shaker, Ithaca (1999)

Optimal Chiller Loading by MOEA/D for Reducing Energy Consumption

Yong Wang[1], Jun-qing Li[2,1,3,4($\boxtimes$)], Mei-xian Song[1], Li Li[1], and Pei-yong Duan[2($\boxtimes$)]

[1] School of Computer, Liaocheng University, Liaocheng 252059, China
lijunqing.cn@gmail.com
[2] School of Information, Shandong Normal University, Jinan 250014, China
duanpeiyong@sdnu.edu.cn
[3] China Key Laboratory of Computer Network and Information Integration, Ministry of Education, Southeast University, Nanjing 211189, People's Republic of China
[4] State Key Laboratory of Synthetical Automation for Process Industries, Northeastern University, Shenyang 110819, China

Abstract. A modified multi-objective evolutionary algorithm based on decomposition (MOEA/D) is used to solve the optimal chiller loading (OCL) problem. In a multi-chiller system, the chillers are usually partially loaded for most of the running time. If the chillers are unreasonably managed, their consumption noticeably increases. To reduce power consumption, the partial load ratio (PLR) of each chiller must be adjusted. The system must meet the system cooling load (CL), so, it is a constrained optimization problem. This study uses a multi-objective method to solve the constrained optimization. The constraint condition is changed to a new objective, so, the problem can be solved as a multi-objective problem. Comparison with the experimental results in the literature proved the effectiveness and performance of the modified algorithm, which can be fully applied in air conditioning system operations.

Keywords: Optimal chiller loading · Decomposition strategy
Constrained optimization

1 Introduction

In an HVAC(Heating Ventilation Air Conditioning) system, the chiller consumes a large amount of energy to provide a cooling load. In this system, optimal loading of the cooling load is important for saving energy, and a better performance coefficient is good for the system. For a system cooling load, all the chillers provide the load. To increase energy conservation, the best combination of chillers should be determined. For the cooling load conditions, the optimal chiller loading (OCL) problem is setting the partial load ratio (PLR) of the chillers to reduce system power consumption. Heuristic optimization methods can be used to solve the problem, including the branch and bound method [1], Lagrangian method [2], Genetic Algorithm(GA) [3], and general algebraic modeling system [4]. The OCL must meet the system cooling load, and the constrained condition is

© Springer International Publishing AG, part of Springer Nature 2018
D.-S. Huang et al. (Eds.): ICIC 2018, LNCS 10954, pp. 759–768, 2018.
https://doi.org/10.1007/978-3-319-95930-6_77

transferred to a third objective function. Multi-objective optimization algorithms have been researched and applied for solving many types of industrial problems [5–24]. Further, many meta-heuristics have been increasingly researched [25–61].

The remainder of this paper is organized as follows. Section 2 describes literature on MOEA/D. Section 3 describes the OCL problem. Section 4 presents modified algorithms to solve OCL. Section 5 is the experimental results which show that the performance of the proposed algorithm is better than algorithms from the literature. The final section is the conclusion.

2   Literature Review

In recent years, multi-objective evolutionary algorithms (MOEAs), such as NSGA-II [5, 6], multi-objective PSO [7] and MOEA/D [8] have been proposed and applied in many fields. MOEA/D is widely used for the knapsack problem [9], the job shop scheduling [10], the traveling salesman problem [11], the test task scheduling problem [12], and for wireless sensor networks [13]. The experimental results show better performance using MOEA/D for MOPs.

In the MOEA/D, a MOP is decomposed into many sub-problems which are solved simultaneously by EA. Then, the algorithm uses a weight vector for the neighborhood relationship to increase the speed of the evolution. Based on this algorithm, research has been conducted in the literature to either: (a) overcome the limitations of parts of the components, or (b) improve the performance of MOEA/D, or (c) present a novel decomposition method or (d) use MOEA/D to solve different types of problems. For example, MOEA/D-DE is a new implementation of MOEA/D with a differential evolution operator [14], MOEA/D-DRA proposes a strategy for allocating computational resources [15, 16] proposes a two-way local search strategy and a new selection, adaptive replacement strategies [17], an adaptive weight adjustment [18], a non-uniform weight vector [19], and a modified Tchebycheff decomposition approach [20].

3   OCL Problem

The purpose of optimization control in an air conditioning water system is meeting the system load demand and ensuring stability. To reduce the energy consumption of the system, the performance of the water chiller must be understood. As shown in Fig. 1, the proposed system requires several chiller loads to meet the requirements. Since the power and performance of each chiller is different, the demand load allocation problem is complicated. For each part of the cold water machine load, the energy consumption is also different. These relationships are not positively correlated, so an optimization algorithm is necessary. In an HVAC system with multiple chillers in parallel, each chiller operates independently to provide the load [3]. In the system, each chiller can operate under partial-load conditions to provide the best efficiency. Figure 1 shows the structure of the system, which includes multiple chillers.

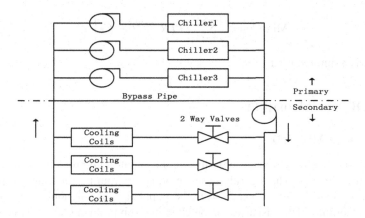

Fig. 1. Structure of multi-chiller system

PLR is the ratio of the partial cooling load to the design capacity for the chiller. In the system, the power consumption of the chiller is a convex function of its PLR value for a given temperature, as follows:

$$kW_i = a_i + b_i PLR_i + c_i PLR_i^2 + d_i PLR_i^3 \tag{1}$$

where a_i, b_i, c_i and d_i are coefficients of interpolation for consumed power versus PLR of the ith chiller in the multi-chiller system. Moreover, PLR is obtained by

$$PLR = \frac{chiller\ load}{chiller\ capacity} \tag{2}$$

In the OCL problem, the PLR of the chillers is set to reduce the sum power of the chillers and the objective function in the problem is defined as

$$Min\quad P = \sum_{i=1}^{N} kW_i \tag{3}$$

where N is the number of chillers in the cooling system, and kW_i is the power energy of the ith chiller.

In the OCL problem, the sum load of all chillers should meet the system requirements. Therefore, the following constraint:

$$\sum_{i=1}^{l} PLR_i * \overline{RT_i} = CL \tag{4}$$

where the cooling load demand of the system is CL. Usually, a penalty method replaces the constraint, but the penalty function should control a penalty parameter [21, 22], so the constraint is changed to an objective function.

$$\text{Min Con} = \left| \sum_{i=1}^{N} PLR_i * \overline{RT} - CL \right| + \varepsilon \tag{5}$$

where ε is a minimum value.

4 MOEA/D Algorithm

4.1 General MOEA/D

MOEA/D decomposes a multi-objective problem into sub-problems and optimizes them simultaneously. The algorithm uses information from neighboring sub-problems to optimize a sub-problem and evolves the population of solutions using an evolutionary algorithm. The distances of neighboring sub-problems are between weight vectors. The three decomposition approaches commonly used are the Tchebycheff approach, weighted sum approach and PBI approach. The framework of MOEA/D is shown in Fig. 2.

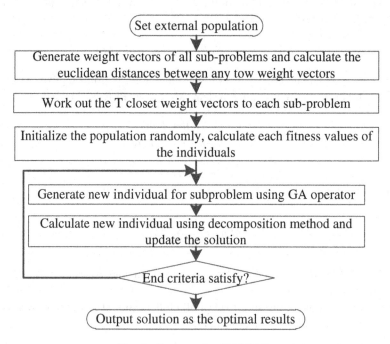

Fig. 2. Framework of MOEA/D

4.2 Modified MOEA/D Approach to OCL Problem

This paper applies MOEA/D with a modified weighted sum decomposition approach and a differential evolution (DE) operator to solve the OCL problem. The modified decomposition approach and evolutionary operator are introduced. Usually, the value ranges of objective functions for the OCL problem are different and each objective function has a different weight for the fitness value. To overcome these deficiencies for solving the OCL problem, we present a modified weight sum approach.

In the modified weight sum approach, the j-th single objective optimization subproblem is:

$$\text{maximize } g^{ws}(x|\lambda^j) = \sum_{i=1}^{m} \lambda_i^j \frac{f_i(x) - f_i^{min}}{|f_i^{max} - f_i^{min} + \varepsilon|} \tag{6}$$

$$\text{subject to } x \in \Omega$$

where $\lambda_1^j, \lambda_2^j, \ldots, \lambda_m^j$, ε is a small positive value, and f_i^{max} and f_i^{min} are the maximum and minimum value of the j-th objective function.

Let $\lambda^1, \ldots, \lambda^n$ be a set of uniformly distributed weight vectors and z^* be the reference point. The weight vectors of the j-th subproblem are $\lambda^j = (\lambda_1^j, \ldots, \lambda_m^j)^T$.

In the DE operator, each element $\bar{y}_k$ is generated as follow in $\bar{y} = (\bar{y}_1, \bar{y}_2, \ldots, \bar{y}_3)$:

$$\bar{y}_k = \begin{cases} x_k^{r_1} + F \times (x_k^{r_2} - x_k^{r_3}) & P(CR) \\ x_k^{r_1} & P(1 - CR) \end{cases} \tag{7}$$

where x^{r_1}, x^{r_2}, x^{r_3} are randomly selected from the solution. CR, F are control parameters.

5 Examples and Results

MOEA/D is used to solve the OCL problem. The cooling system consists of four chillers which have two 450 RT units, two 1000 RT units and a total capacity of 2800 RT. The data for this case is shown in Table 1.

Table 1. Coefficients and capacity of chillers

System	Chiller	a_i	b_i	c_i	d_i	Capacity (RT)
Case	CH-1	104.09	166.57	−430.13	512.53	450
	CH-2	−67.15	1177.79	−2174.53	1456.53	450
	CH-3	384.71	−779.13	1151.42	−63.20	1000
	CH-4	541.63	413.48	−3626.50	4021.41	1000

The generation number is set to 100, which determines when the algorithm stops iterating. The population size is 50, so weight vectors $C_{50+3-1}^{3-1} = 1326$. All the parameters are shown in Table 2.

Table 2. Control parameters for case of MOEA/D

Population size	50
Weight size	1326
Neighbor size	10
Generation number	100
Chromosome dimension	40
F	0.5
CR	1.0

The testing software used is written in IntelliJ IDEA and run on an Intel Core i7/3.6 GHz personal computer. In Fig. 3, the relationship between Pow and CON is shown. For different loads (60%–90%), the results in Table 3 show that some PLR is produced for each chiller to minimize power, the result of solving OCL using MOEA/D is better than GA.

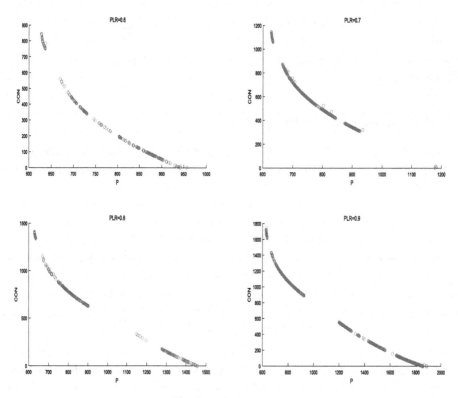

Fig. 3. Solution results

Table 3. Optimal chiller loading for case

Load (RT)	Chiller	MOEA/D			GA	
		PLR	P	CON	PLR	P
2610 (90%)	1	0.9898	1857.2986	2.2461E-07	0.9925	1862.18
	2	0.9074			0.9487	
	3	1.0000			1.0000	
	4	0.7562			0.7366	
2320 (80%)	1	0.8273	1455.6739	4.4417E-15	0.8611	1457.23
	2	0.8071			0.8132	
	3	0.8974			0.8809	
	4	0.6871			0.6859	
2030 (70%)	1	0.7272	1178.1404	6.2869-10	0.6592	1183.80
	2	0.7413			0.7605	
	3	0.7205			0.7557	
	4	0.6487			0.6360	
1740 (60%)	1	0.7471	942.06063	4.4417E-15	0.5956	907.72
	2	0.0000			0.6982	
	3	0.7487			0.5710	
	4	0.6551			0.5874	

6   Conclusion

This paper proposes an optimal chiller loading problem to reduce the electrical power consumption of systems. For this aim, we use MOEA/D to solve the OCL problem. The obtained results show that MOEA/D produces better results for the OCL problem simultaneously, many constraints can be easily solved using many objective optimization algorithms. As the algorithm handles the constraint optimization problem, the difficulty is how to solve constraints. We can convert constraints to a new problem and solve these problems simultaneously. Other performance indicators of the system can be optimized using MOEA/D in our future work.

Acknowledgments. This research is partially supported by National Science Foundation of China under Grant 61773192, 61773246 and 61503170, Shandong Province Higher Educational Science and Technology Program (J17KZ005), Key Laboratory of Computer Network and Information Integration (Southeast University), Ministry of Education (K93-9-2017-02), and State Key Laboratory of Synthetical Automation for Process Industries (PAL-N201602).

References

1. Chang, Y.C., Lin, F.A., Lin, C.H.: Optimal chiller sequencing by branch and bound method for saving energy. Energy Convers. Manage. **46**(13–14), 2158–2172 (2005)
2. Chang, Y.C.: A novel energy conservation method—optimal chiller loading. Electr. Pow. Syst. Res. **69**(2), 221–226 (2004)

3. Chang, Y.C.: Genetic algorithm based optimal chiller loading for energy conservation. Appl. Therm. Eng. **25**(17–18), 2800–2815 (2005)

4. Salari, E., Askarzadeh, A.: A new solution for loading optimization of multi-chiller systems by general algebraic modeling system. Appl. Therm. Eng. **84**(4), 429–436 (2015)

5. Jeyadevi, S., Baskar, S., Babulal, C.K.: Solving multiobjective optimal reactive power dispatch using modified NSGA-II. Int. J. Elec. Power. **33**(2), 219–228 (2011)

6. Deb, K., Pratap, A., Agarwal, S.: A fast and elitist multiobjective genetic algorithm: NSGA-II. IEEE Trans. Evolut. Comput. **6**(2), 182–197 (2002)

7. Coello, C.A.C., Pulido, G.T., Lechuga, M.S.: Handling multiple objectives with particle swarm optimization. IEEE Trans. Evolut. Comput. **8**(3), 256–279 (2004)

8. Zhang, Q., Li, H.: MOEA/D: a multiobjective evolutionary algorithm based on decomposition. IEEE Trans. Evol. Comput. **11**(6), 712–731 (2007)

9. Li, H., Landa-Silva, D.: An adaptive evolutionary multi-objective approach based on simulated annealing. Evol. Comput. **19**(4), 561–595 (2014)

10. Zhao, F., Chen, Z., Zhang, C.: A modified MOEA/D with adaptive mutation mechanism for multi-objective job shop scheduling problem. J. Comput. Inform. Syst. **11**(8), 2833–2840 (2015)

11. Souza, M.Z.D., Pozo, A.T.R.: A GPU implementation of MOEA/D-ACO for the multiobjective traveling salesman problem. IEEE Intell. Syst., 324–329 (2014)

12. Lu, H., Zhu, Z., Wang, X., Yin, L.: A variable neighborhood MOEA/D for multiobjective test task scheduling problem. Math. Probl. Eeg. **2014**(3), 1–14 (2014)

13. Konstantinidis, A., Yang, K.: Multi-objective energy-efficient dense deployment in wireless sensor networks using a hybrid problem-specific MOEA/D. Appl. Soft Comput. **11**(6), 4117–4134 (2011)

14. Li, H., Zhang, Q.: Multiobjective optimization problems with complicated Pareto sets, MOEA/D and NSGA-II. IEEE Trans. Evol. Comput. **13**(2), 284–302 (2009)

15. Zhang, Q., Liu, W., Li, H.: The performance of a new version of MOEA/D on CEC09 unconstrained MOP test instances. In: 2009 IEEE Congress on Evolutionary Computation, pp. 203–208. IEEE Press (2009)

16. Lin, S., Lin, F., Chen, H., Zeng, W.: A MOEA/D-based multi-objective optimization algorithm for remote medical. Neurocomputing **220**, 5–16 (2016)

17. Wang, Z., Zhang, Q., Zhou, A., Gong, M., Jiao, L.: Adaptive replacement strategies for MOEA/D. IEEE Trans. Cybern. **46**(2), 474–486 (2017)

18. Qi, Y., Ma, X., Liu, F., Jiao, L., Sun, J., Wu, J.: MOEA/D with adaptive weight adjustment. Evol. Comput. **22**(2), 231–264 (2014)

19. Lu, H., Zhang, M., Fei, Z., Mao, K.: Multi-objective energy consumption scheduling based on decomposition algorithm with the non-uniform weight vector. Appl. Soft Comput. **39**(C), 223–239 (2016)

20. Zhang, J., Tang, Q., Li, P., Deng, D., Chen, Y.: A modified MOEA/D approach to the solution of multi-objective optimal power flow problem. Appl. Soft Comput. **47**(C), 494–514 (2016)

21. Meng, Z., Shen, R., Jiang, M.: A penalty function algorithm with objective parameters and constraint penalty parameter for multi-objective programming. Am. J. Oper. Res. **4**(6), 331–339 (2014)

22. Chekir, N., Bellagi, A.: Performance improvement of a Butane/Octane absorption chiller. Energy **36**(10), 6278–6284 (2011)

23. Li, J.Q., Sang, H.Y., Han, Y.Y., Wang, C.G., Gao, K.Z.: Efficient multi-objective optimization algorithm for hybrid flow shop scheduling problems with setup energy consumptions. J. Clean. Prod. **181**, 584–598 (2018)

24. Zheng, Z.X., Li, J.Q.: Optimal chiller loading by improved invasive weed optimization algorithm for reducing energy consumption. Energy Buildings **161**, 80–88 (2018)
25. Duan, P.Y., Li, J.Q., Wang, Y., Sang, H.Y., Jia, B.X.: Solving chiller loading optimization problems using an improved teaching-learning-based optimization algorithm. Optim. Contr. Appl. Met. **39**(1), 65–77 (2018)
26. Li, J.Q., Pan, Q.K., Tasgetiren, M.F.: A discrete artificial bee colony algorithm for the multi-objective flexible job-shop scheduling problem with maintenance activities. Appl. Math. Model. **38**(3), 1111–1132 (2014)
27. Li, J.Q., Pan, Q.K.: Chemical-reaction optimization for solving fuzzy job-shop scheduling problem with flexible maintenance activities. Int. J. Prod. Econ. **145**(1), 4–17 (2013)
28. Li, J.Q., Pan, Q.K., Mao, K.: A discrete teaching-learning-based optimisation algorithm for realistic flowshop rescheduling problems. Eng. Appl. Artif. Intell. **37**, 279–292 (2015)
29. Li, J.Q., Pan, Q.K., Gao, K.Z.: Pareto-based discrete artificial bee colony algorithm for multi-objective flexible job shop scheduling problems. Int. J. Adv. Manuf. Tech. **55**(9–12), 1159–1169 (2011)
30. Li, J.Q., Pan, Q.K., Mao, K.: A hybrid fruit fly optimization algorithm for the realistic hybrid flowshop rescheduling problem in steelmaking systems. IEEE Trans. Autom. Sci. Eng. **13** (2), 932–949 (2016)
31. Li, J.Q., Pan, Q.K., Chen, J.: A hybrid pareto-based local search algorithm for multi-objective flexible job shop scheduling problems. Int. J. Prod. Res. **50**(4), 1063–1078 (2012)
32. Li, J.Q., Pan, Q.K., Suganthan, P.N., Chua, T.J.: A hybrid tabu search algorithm with an efficient neighborhood structure for the flexible job shop scheduling problem. Int. J. Adv. Manuf. Tech. **59**(4), 647–662 (2011)
33. Li, J.Q., Pan, Y.X.: A hybrid discrete particle swarm optimization algorithm for solving fuzzy job shop scheduling problem. Int. J. Adv. Manuf. Tech. **66**(1–4), 583–596 (2013)
34. Li, J.Q., Pan, Q.K., Duan, P.Y.: An improved artificial bee colony algorithm for solving hybrid flexible flowshop with dynamic operation skipping. IEEE Trans. Cybern. **46**(6), 1311–1324 (2016)
35. Li, J.Q., Pan, Q.K.: Solving the large-scale hybrid flow shop scheduling problem with limited buffers by a hybrid artificial bee colony algorithm. Inf. Sci. **316**(20), 487–502 (2015)
36. Li, J.Q., Pan, Q.K., Xie, S.X.: An effective shuffled frog-leaping algorithm for multi-objective flexible job shop scheduling problems. Appl. Math. Comput. **218**(18), 9353–9371 (2012)
37. Li, J.Q., Pan, Q.K., Liang, Y.C.: An effective hybrid tabu search algorithm for multi-objective flexible job shop scheduling problems. Comput. Ind. Eng. **59**(4), 647–662 (2010)
38. Li, J.Q., Pan, Q.K., Mao, K., Suganthan, P.N.: Solving the steelmaking casting problem using an effective fruit fly optimisation algorithm. Knowl.-Based Syst. **72**(12), 28–36 (2014)
39. Li, J.Q., Pan, Q.K.: Chemical-reaction optimization for flexible job-shop scheduling problems with maintenance activity. Appl. Soft Comput. **12**(9), 2896–2912 (2012)
40. Li, J.Q., Pan, Q.K., Wang, F.T.: A hybrid variable neighborhood search for solving the hybrid flow shop scheduling problem. Appl. Soft Comput. **24**, 63–77 (2014)
41. Li, J.Q., Pan, Q.K., Xie, S.X., Wang, S.: A hybrid artificial bee colony algorithm for flexible job shop scheduling problems. Int. J. Comput. Commun. Control **6**(2), 267–277 (2011)
42. Li, J.Q., Pan, Q.K., Xie, S.X.: A hybrid variable neighborhood search algorithm for solving multi-objective flexible job shop problems. ComSIS Comput. Sci. Inf. Syst. **7**(4), 907–930 (2010)
43. Li, J.Q., Wang, J.D., Pan, Q.K., Duan, P.Y., Sang, H.Y., Gao, K.Z., Xue, Y.: A hybrid artificial bee colony for optimizing a reverse logistics network system. Soft Comput. **21**(20), 6001–6018 (2017)

44. Zhang, P., Liu, H., Ding, Y.H.: Dynamic bee colony algorithm based on multi-species co-evolution. Appl. Intell. **40**, 427–440 (2014)
45. Hu, C.Y., Liu, H., Zhang, P.: Cooperative co-evolutionary artificial bee colony algorithm based on hierarchical communication model. Chin. J. Elec. **25**, 570–576 (2016)
46. Liu, Y., Jiao, Y.C., Zhang, Y.M., Tan, Y.Y.: Synthesis of phase-only reconfigurable linear arrays using multiobjective invasive weed optimization based on decomposition. Int. J. Antenn. Propag. **2014** (2014)
47. Zheng, X.W., Lu, D.J., Wang, X.G., Liu, H.: A cooperative coevolutionary biogeography-based optimizer. Appl. Intell. **43**, 95–111 (2015)
48. Liu, H., Zhang, P., Hu, B., Moore, P.: A novel approach to task assignment in a cooperative multi-agent design system. Appl. Intell. **43**, 162–175 (2015)
49. Zhang, Z.J., Liu, H.: Social recommendation model combining trust propagation and sequential behaviors. Appl. Intell. **43**, 695–706 (2015)
50. Wang, J.L., Gong, B., Liu, H., Li, S.H.: Model and algorithm for heterogeneous scheduling integrated with energy-efficiency awareness. T.I. Meas. Control. **38**, 452–462 (2016)
51. Wang, J.L., Gong, B., Liu, H., Li, S.H.: Multidisciplinary approaches to artificial swarm intelligence for heterogeneous computing and cloud scheduling. Appl. Intell. **43**, 662–675 (2015)
52. Wang, J.L., Gong, B., Liu, H., Li, S.H., Yi, J.: Heterogeneous computing and grid scheduling with parallel biologically inspired hybrid heuristics. T.I. Meas. Control **36**, 805–814 (2014)
53. Bai, J., Liu, H.: Multi-objective artificial bee algorithm based on decomposition by PBI method. Appl. Intell. **45**(4), 976–991 (2016)
54. Dong, X., Zhang, H., Sun, J., Wan, W.: A two-stage learning approach to face recognition. J. Vis. Commun. Image R. **43**, 21–29 (2017)
55. Jia, W., Zhao, D., Zheng, Y., Hou, S.: A novel optimized GA–Elman neural network algorithm. Neural Comput. Appl. **6**, 1–11 (2017)
56. Zheng, X., Yu, X., Yan, L., Liu, H.: An enhanced multi-objective group search optimizer based on multi-producer and crossover operator. J. Inf. Sci. Eng. **37**(1), 33–50 (2017)
57. Liu, H., Chen, Z.H., Zheng, X.W., Hu, B., Lu, D.J., Chen, Z.H.: Energy-efficient virtual network embedding in networks for cloud computing. Int. J. Web Grid Serv. **13**(1–1), 75 (2017)
58. Xiao, X., Zheng, X., Zhang, Y.: A multidomain survivable virtual network mapping algorithm. Secur. Commun. Netw. **2017**(10), 1–12 (2017)
59. Han, Y.Y., Gong, D.W., Jin, Y.C., Pan, Q.K.: Evolutionary multiobjective blocking lot-streaming flow shop scheduling with machine breakdowns. IEEE T. Cybern. **PP**(99), 1–14 (2017)
60. Han, Y.Y., Gong, D.W., Sun, X.Y.: An improved NSGA-II algorithm for multi-objective lot-streaming flow shop scheduling problem. Int. J. Prod. Res. **52**(8), 2211–2231 (2014)
61. Gong, D.W., Han, Y.Y., Sun, J.Y.: A novel hybrid multi-objective artificial bee colony algorithm for the blocking lot-streaming flow shop scheduling problems. Knowl.-Based Syst. **148**, 115–130 (2018)

An Enhanced Migrating Birds Optimization for the Flexible Job Shop Scheduling Problem with Lot Streaming

Tao Meng[1,2], Quan-ke Pan[1(✉)], and Qing-da Chen[3]

[1] School of Mechatronic Engineering and Automation, Shanghai University, Shanghai 200072, People's Republic of China
panquanke@qq.com
[2] School of Mathematical Science, Liaocheng University, Liaocheng 252059, People's Republic of China
[3] State Key Laboratory of Synthetic Automation for Process Industries, Northeastern University, Shenyang 110819, People's Republic of China

Abstract. This paper presents an enhanced migrating birds optimization (enMBO) for the flexible job shop scheduling problem with the consideration of lot streaming and the goal is minimizing total flowtime. In enMBO, to explore the solution space efficiently, we design a search scheme which is capable of adjusting the search radius with the increase of iteration. In addition, MBO concentrates too much on local search and hence is easily trapped in local optimum. To handle this, a special mechanism that based on precedence operation crossover is developed and incorporated into the evolutionary framework. We conduct simulations on well-known benchmarks with different scales and results verify the significance of schemes designed above. Moreover, by comparing with recent algorithms, the proposed enMBO shows its high performance for the considered problem.

Keywords: Flexible job shop scheduling problem
Migrating birds optimization · Lot streaming · Crossover operation
Meta-heuristic

1 Introduction

The job shop scheduling problem (JSP) is one of the most usual problems in modern production systems and has constituted a highly active research field in the past decades [1]. In the JSP, n jobs should be processed on m unrelated machines. Each job contains a sequence of consecutive operations must be handled on a pre-assigned machine without interruption. The JSP needs to find a processing sequence of jobs on machines so that the required objective is optimized.

The flexible job shop scheduling problem (FJSP), first addressed by Brucker and Schlie [2], is an extension to the classical JSP. The difference is that, in FJSP, operations are permitted to be performed on any machines in a candidate set, rather than a pre-assigned one in JSP. As a result, compared with JSP, FJSP should consider an additional problem that is assigning a suitable machine out of an available set to each operation. It has been indicated that FJSP belongs to the NP-hard class [3].

© Springer International Publishing AG, part of Springer Nature 2018
D.-S. Huang et al. (Eds.): ICIC 2018, LNCS 10954, pp. 769–779, 2018.
https://doi.org/10.1007/978-3-319-95930-6_78

Although FJSP is tougher than JSP, it is closer to the situations of actual manufacturing systems and thus more useful in reality [3].

After the starting work, FJSPs have captured the interests of many researchers. Due to the NP-hard nature of FJSP, exact algorithms are often not applicable for large scale instances. Therefore, meta-heuristics have been a trend in recent years. Many novel algorithms such as differential evolution algorithms (DE) [4], genetic algorithm (GA) [5], harmony search (HS) [3], artificial bee colony (ABC) [6], teaching-learning-based optimization algorithm (TLBO) [7], fruit fly (FFO) [8] and so on, have been utilized and proved effective for the FJSP. Although meta-heuristics technologies may not guarantee the global best result, the computational time is more reasonable than exact methods.

It is usually assumed on studies of FJSP that, the next operation of a job cannot start before the current operation is totally finished. But, in modern manufacturing systems, a job (or called lot) is usually divided into many small parts (i.e. sublots), which can be processed independently [9]. In these industries, a single sublot is moved to the next processing stage upon its completion and so that different sublots of a job can be parallel performed on different stages, which is often called lot streaming in related studies. Apparently, lot streaming technology contributes to increase the machine utilization and is helpful for efficiency improvement. Although being of great importance in practical production, the studies on FJSP with lot streaming (FJSPLS) is still at the initial stage. Until now, we can only find a few researches on this subject [9–12].

By mimicking the V formation flight of migrating birds, Duman et al. [13] proposed migrating birds optimization (MBO) to optimize quadratic assignment problem. MBO is a simple yet powerful meta-heuristic technique. After this study, it has been adapted to handle various of academic and application problems such as flow shop scheduling problem [14], traveling salesman problem [15], flexible manufacturing systems [16] and so on. In this paper, we proposed an enhanced MBO (enMBO) to solve the FJSPLS for total flowtime minimization. In enMBO, a new search method with adaptive radius is introduced to probe the solution space more efficiently. Besides, aiming to strengthen the global search ability, we present an effective crossover operation and merge it to the search framework.

The remainder of the paper is organized as below. The considered problem is defined in Sect. 2. Next, the basic MBO procedure is described briefly in part 3. Then, the developed enMBO is discussed in Sect. 4. In Sect. 5, extensive numerical simulations are conducted based on benchmark instances and comparisons with state of the art are also provided. At last, Sect. 6 concludes the whole study.

2   Problem Description

The FJSPLS can be defined as follows. We have n independent jobs $J = \{J_1, J_2, \ldots J_n\}$ that must be manufactured on m unrelated machines $M = \{M_1, M_2, \ldots, M_m\}$. Each job J_i consists of a sequence of precedence constrained operations $O_{i,1} \rightarrow O_{i,2} \rightarrow \ldots \rightarrow O_{i,n_i}$ that should be performed in the given order. Each operation $O_{i,j}$ (the jth operation of job J_i) could be processed on any machine within a compatible set $M_{i,j} \subset M$. The processing time of each operation is machine dependent. Let P_{ijk} to be the processing time of operation $O_{i,j}$ on machine M_k. In the studied environment, each job is split into S_i sublots with equal size and each sublot can be treated independently. Denote C_i the completion time of the last sublot of job J_i on the last machine. The objective lies in obtaining the minimum total flowtime measured as $TF_{min} = min\{\sum_{i=1}^{n} C_i\}$. A more detailed description of this model can be found in [9].

3   Introduction to the Basic MBO

MBO is a swarm-based stochastic search technique which consists of a number of solutions regarding as birds in the flock aligned with V shape. In MBO, the leading bird is firstly improved by searching around it that its best neighbor will replace it if a better fitness is obtained. Regarding other solutions in the V shape, some neighbors together with some best unused neighbors from its partner ahead will be assessed together and the best one of them will replace the current solution when an improvement occurs. After some iterations of above operations, the leader goes to the tail of the left or the right line. Then, one following bird turns into the new leader, followed by a new iteration.

The control parameters in MBO are shown below:
N the population (flock) size
K the number of neighbors generated
X the number of neighbors shared with the next partner
M number of iterations before selecting a new leader
K the maximum evaluation number of fitness

The procedure of the basic MBO is depicted following in Fig. 1 [13]:

Generate N initial solutions randomly and place them in a V formulation arbitarily
for i=1 to K do
　　for $j = 1$ to m do
　　　　Try to improve leader solution though generating and evaluating k neighbors
　　　　$i = i + k$
　　　　for each solution s_r in the flock(except the leader)
　　　　　　Try to improve s_r by evalualting $k - x$ neighbor of it and x unused best
　　　　　　neighbors from the solution in the front
　　　　　　$i = i + (k - x)$
　　　　endfor
　　endfor
　　Move the leader solution to the end and forward one of the solution following it
　　to the leader position
endfor

Fig. 1.   Pseudo code of the basic MBO

4   The Proposed enMBO

4.1   Encoding Scheme and Decoding Scheme

In enMBO, we use two-vector encoding scheme as in [9] where the solution is comprised of two parts: Machine selection (MS) vector determines the selected machine for each operation and operation sequence (OS) part lists sequence of operations on machines. Suppose there are two jobs and three machines, and each job contains three operations. The processing time of operations on machines is provided in Table 1 where the notation "/" means the machine cannot handle the corresponding operation. A solution is shown in Fig. 2. For MS part, we can see that the second integer is 2, which denotes the second machine in eligible set (i.e. M_3 in Table 1) is assigned to operation O_{12}. With respect to OS part, it can be find that the third integer (i.e. 2) represents the third scheduled operation belongs to job J_2 and further it actually denotes the second operation of job J_2 (i.e. O_{22}) because index 2 occurs here for the second time.

Table 1. Processing time of operations on machines

Job	Operation	M_1	M_2	M_3
J_1	O_{11}	10	15	/
	O_{12}	20	/	21
	O_{13}	10	12	15
J_2	O_{21}	10	/	10
	O_{22}	/	12	13
	O_{23}	/	10	/

Decoding is to find a feasible schedule according to information provided by the current solution. From the theoretical side, we can get infinite number of schedules since idle time can be added between two consecutive operations. Taking account of the total flowtime objective, the global left shift scheme is utilized in this work, where each operation should be shifted left as compact as possible without putting off operations in its front. Because of the reason of limited space, this scheme is not detailed here and a fully description is available in [9].

4.2   New Search Scheme (NSS)

As stated above, in MBO, each individual produces k-x (k for the leader) neighbors around it within a loop. For MS part, an often used neighborhood structure is selecting an operation randomly and then re-assigning it to another machine available [9]. Regarding OS part, insert and swap are two common operators in the literature [17]. In enMBO, to detect the search space from different directions, above three structures are all adopted.

Meanwhile, in early iterations, the individuals are usually far from the optimum. So, a larger search radius is useful to converge quickly. In the later iterations, however,

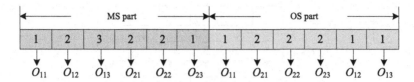

Fig. 2. Encoding scheme of enMBO.

individuals will be close to the optimal site. So, a smaller search radius is necessary for finding a better solution. In enMBO, a new search scheme (NSS) is introduced to adjust the search scope dynamically. In NSS, previous mentioned three neighbor structures will be conducted for p times to build a new candidate and p is calculated in accordance with the iteration number as in Eq. (1)

$$p = \text{int}\left(p_{\max} - (p_{\max} - p_{\min}) \cdot \frac{iter}{iter_{\max}}\right) \quad (1)$$

where $p_{\max}$ is the maximum performed times and we find that $p_{\max} = n/2$ is suitable according to experimental results. Recall that n represents the number of jobs in the problem. $p_{\min}$ is the minimum times and $p_{\min} = 1$. $iter$ is the current iteration number and $iter_{\max}$ is the maximum loop number. Finally, the result is converted to an integer by the int function which simply cuts off the decimal part.

4.3 Effective Crossover Operation (ECO) and Integration with enMBO

One of the most notable challenges in designing effective meta-heuristic algorithms is how to make a suitable compromise between local search and global search. MBO pays more attention to exploitation while its global exploration capability is relatively limited [17]. To enhance the global exploration, we propose an effective crossover operation (ECO) in this part. The detailed process is summarized as follows:

Step 1: Select two individuals, X_1 and X_2, in the current population randomly.

Step 2: Perform crossover on the OS part of two selected solutions (i.e. OS_1 and OS_2,) and generate two children OS_{child1} and OS_{child2}. We use the improved precedence operation crossover (IPOX) here to be the operator, which is described below:

Step 2.1: Randomly select part of jobs in job set J and put them into a new set $S1$.

Step 2.2: Copy the operations of OS_1 (OS_2) that belongs to jobs in $S1$ to OS_{child1} (OS_{child2}) and keep their original positions.

Step 2.3: Copy the remaining operations from OS_2 (OS_1) to fill vacant slots of OS_{child1} (OS_{child2}) and keep their order.

An example is demonstrated in Fig. 3 where job J_2 is picked out to set $S1$.

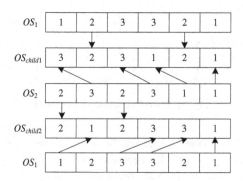

Fig. 3. An example of crossover operator

Step 3: Evaluate two children. Note that only OS part of two children are generated in step2. So, to compare their quality, we need to determine the MS part of them firstly. The procedure is devised as below: from left to right, for each operation, we assess all available machines and corresponding completion times. Then, the operation will be assigned to the machine that is able to finish it at the earliest time. That is, we make a greedy selection for MS part. After evaluation, the better one X_{child} is chosen for the next step.

Step 4: Pick out the worst solution in the current population X_{worst}. If X_{child} is better than X_{worst}, let $X_{worst} = X_{child}$.

It should be noted that, in enMBO, ECO will be carried out for R times at the end of each loop.

4.4   Search Framework of enMBO

With design above, the entire procedure of enMBO is illustrated in Fig. 4.

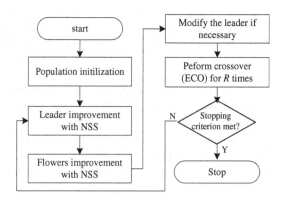

Fig. 4. The framework of enMBO

5 Numerical Tests and Comparisons

To test the performance of the presented enMBO, in this part, a well-known benchmark set (BCdata) is adopted that includes 21 problems with different scales. Interested readers may refer to the work of Barnes and Chambers [18] for further information about this dataset. In addition, as in [9], we set that all jobs are divided into 10 sublots with equal size. We implement all algorithms in C++ and run them on a PC with Intel i7-4790 CPU @ 3.6 GHZ and Windows 7 operating system.

In addition, after preliminary experiments, the parameters of enMBO are set as follows: $N = 301$, $k = 3$, $X = 1$, $m = 20$ and $R = 80$.

5.1 Effectiveness of NSS and ECO

Firstly, we show the influence of NSS and ECO in enMBO. For this aim, we compare three algorithms including enMBO$_{NSS}$, enMBO$_{ECO}$ and enMBO. In enMBO$_{NSS}$ and enMBO$_{ECO}$, we remove the NSS and ECO from enMBO respectively. All algorithms adopt the identical CPU time of $T = 100 \times m \times n$ milliseconds as the stopping criterion. Each instance is performed 20 times independent and for each run, we measure the relative percentage increase (RPI) as follows:

$$RPI(TF_i) = \frac{TF_i - TF_{\min}}{TF_{\min}} \times 100 \tag{2}$$

where TF_i is the total flowtime obtained by a given algorithm in the ith run, $TF_{\min}$ is the minimum total flowtime obtained by all compared algorithms in all runs. Clearly, a smaller value represents a solution with higher quality. The average relative percentage increase (ARPI) in 20 runs is used to evaluate the performance of compared methods as in many other studies on scheduling problems.

We report the ARPI values of all problems in Table 2 where the first column is instance names, the second column is instance sizes, the next three columns lists the ARPI of 20 replications for each instance and the last column displays the CPU time needed for each run in seconds.

It is obvious from Table 2 that enMBO is the best method, which generates the smallest ARPI value for all 21 instances. In addition, a multifactor analysis of variance (ANOVA) is conducted to check the difference of these results in a statistical way. ANOVA is a very powerful approach that provides a statistical test of whether or not the means of several groups are equal, and it has been frequently adopted in works of scheduling problems such as Ruiz and Stutzle [19], Pan and Ruiz [20], and many others. Different algorithms with the least significant difference (LSD) intervals (at the 95% confidence level) are listed in Fig. 5. We can see from Fig. 5 that enMBO yields significantly better results than other two methods, which demonstrates the effectiveness of NSS and ECO designed for enMBO.

Table 2. Comparisons of enMBO$_{NSS}$, enMBO$_{ECO}$ and enMBO (the best value are in bold)

Instance	n × m	enMBO$_{NSS}$	enMBO$_{ECO}$	enMBO	Time (s)
mt10c1	10 × 11	4.10	4.23	**2.55**	11.0
mt10cc	10 × 12	4.86	5.47	**3.25**	12.0
mt10x	10 × 11	4.75	6.56	**2.72**	11.0
mt10xx	10 × 12	5.88	6.56	**2.97**	12.0
mt10xxx	10 × 13	4.45	5.31	**2.96**	13.0
mt10xy	10 × 12	4.67	6.00	**2.75**	12.0
mt10xyz	10 × 13	6.32	10.08	**5.36**	13.0
setb4c9	10 × 11	8.13	7.03	**2.50**	11.0
setb4cc	10 × 12	4.88	4.88	**2.21**	12.0
setb4x	10 × 11	8.69	7.73	**3.38**	11.0
setb4xx	10 × 12	7.86	7.99	**3.05**	12.0
setb4xxx	10 × 13	8.25	6.82	**2.80**	13.0
setb4xy	10 × 12	5.02	5.34	**2.72**	12.0
setb4xyz	10 × 13	6.37	6.18	**2.83**	13.0
seti5c12	15 × 16	6.46	6.21	**2.79**	24.0
seti5cc	15 × 17	6.85	6.78	**2.85**	25.5
seti5x	15 × 16	8.98	8.22	**4.58**	24.0
seti5xx	15 × 17	9.10	8.10	**4.78**	25.5
seti5xxx	15 × 18	8.97	6.92	**3.22**	27.0
seti5xy	15 × 17	6.48	5.23	**2.03**	25.5
seti5xyz	15 × 18	9.00	8.75	**4.58**	27.0
Mean		6.67	6.68	**3.19**	16.5

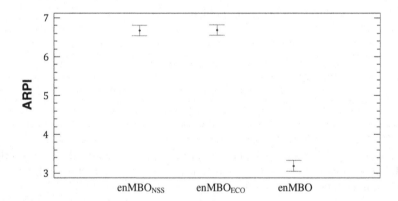

Fig. 5. Means with 95% LSD intervals for enMBO$_{NSS}$, enMBO$_{ECO}$ and enMBO

5.2   Effectiveness of the Proposed enMBO Algorithm

We now test the performance of the proposed enMBO for solving the FJSPLS with total flowtime criterion. Two recent algorithms for FJSPLT, GA [9] and MA [11], are compared in this section. Note that an adaption work has been made that GA and MA shared the same objective function as in FJSPLT. As above experiment, we use the maximum elapsed CPU time limit of $T = 100 \times m \times n$ milliseconds as a termination criterion for all algorithms and each benchmark instance is run for 20 times. We reported ARPI results in Table 3.

Table 3. Comparisons of GA, MA and enMBO (the best value are in bold)

Instance	n × m	GA	MA	enMBO	Time (s)
mt10c1	10 × 11	12.36	4.39	**3.18**	11.0
mt10cc	10 × 12	11.32	4.53	**3.25**	12.0
mt10x	10 × 11	11.51	4.74	**2.72**	11.0
mt10xx	10 × 12	12.99	4.45	**2.42**	12.0
mt10xxx	10 × 13	9.90	4.01	**2.96**	13.0
mt10xy	10 × 12	12.46	3.95	**2.75**	12.0
mt10xyz	10 × 13	14.57	5.64	**5.36**	13.0
setb4c9	10 × 11	15.24	8.23	**2.50**	11.0
setb4cc	10 × 12	11.66	5.80	**2.21**	12.0
setb4x	10 × 11	16.03	8.94	**3.38**	11.0
setb4xx	10 × 12	15.46	9.07	**3.05**	12.0
setb4xxx	10 × 13	16.41	8.40	**2.80**	13.0
setb4xy	10 × 12	11.90	6.35	**2.72**	12.0
setb4xyz	10 × 13	12.64	7.33	**2.83**	13.0
seti5c12	15 × 16	14.77	7.60	**2.79**	24.0
seti5cc	15 × 17	15.44	7.81	**2.85**	25.5
seti5x	15 × 16	16.47	9.89	**4.58**	24.0
seti5xx	15 × 17	17.38	9.15	**4.78**	25.5
seti5xxx	15 × 18	15.87	7.50	**3.22**	27.0
seti5xy	15 × 17	13.11	7.06	**2.03**	25.5
seti5xyz	15 × 18	17.32	10.26	**4.58**	27.0
Mean		14.04	6.91	**3.19**	16.5

It can be seen from Table 3 that the presented enMBO is the best algorithm in terms of solution quality since it catches the smallest APRI for all problems. Moreover, the multifactor ANOVA results in Fig. 6 verify that enMBO is significantly better than other two compared algorithms. Additionally, to provide an overall picture of the performance of competing algorithms, we also make a convergence graph in pace with CPU time. And we find that the presented enMBO converges faster than other two algorithms which further show the superiority of the proposed approach. Because of the limited space, we do not show the graph here but it can be obtained by emailing the Corresponding Author.

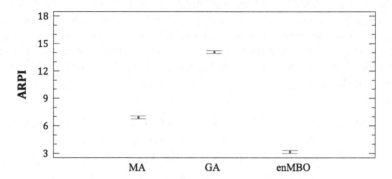

Fig. 6. Means with 95% LSD intervals for MA, GA and enMBO

6 Conclusions

The flexible job shop scheduling has enormous applications in modern industrial systems. In this study, we discussed an enhanced migrating birds optimization (enMBO) to solve the flexible job shop scheduling problem with lot-streaming and total flowtime minimization criterion. In enMBO, advanced methods are integrated to improve the solution quality including new search scheme with adaptive search radius and an effective crossover operation under the consideration of poor global search ability of MBO. Experimental results demonstrated the effectiveness of these designs, and comparisons with state-of-the-arts identified the superiority of the proposed enMBO. To further improve the performance, future work may combine MBO with other algorithms such as harmony search (HS) [21].

Acknowledgments. This research is partially supported by the National Science Foundation of China 51575212 and 61174187, and Shanghai Key Laboratory of Power station Automation Technology.

References

1. Gonzalez, M., Vela, C., Varela, R.: Scatter search with path relinking for the flexible job shop scheduling problem. Eur. J. Oper. Res. **245**(1), 35–45 (2015)
2. Brucker, P., Schlie, R.: Job-shop scheduling with multi-purpose machines. Computing **45**(4), 369–375 (1990)
3. Yuan, Y., Xu, H., Yang, J.D.: A hybrid harmony search algorithm for the flexible job shop scheduling problem. Appl. Soft Comput. **13**(7), 3259–3272 (2013)
4. Yuan, Y., Xu, H.: Flexible job shop scheduling using hybrid differential evolution algorithms. Comput. Ind. Eng. **65**, 246–260 (2013)
5. Wang, J.F., Du, B.Q., Ding, H.M.: A genetic algorithm for the flexible job-shop scheduling problem. Comput. Oper. Res. **35**(10), 3202–3212 (2008)

6. Gao, K.Z., Suganthan, P.N., Pan, Q.K., Chua, T.J., Chong, C.S., Cai, T.X.: An improved artificial bee colony algorithm for flexible job-shop scheduling problem with fuzzy processing time. Expert Syst. Appl. **65**(C), 52–67 (2016)
7. Xu, Y., Wang, L., Wang, S.Y., Liu, M.: An effective teaching-learning-based optimization algorithm for the flexible job-shop scheduling problem with fuzzy processing time. Neurocomputing **148**, 260–268 (2015)
8. Liu, Q., Zhan, M.M., Chekem, F.O., Shao, X.Y., Ying, B.S., Sutherland, J.W.: A hybrid fruit fly algorithm for solving flexible job-shop scheduling to reduce manufacturing carbon footprint. J. Clean. Prod. **168**(1), 668–678 (2017)
9. Demir, Y., Isleyen, S.K.: An effective genetic algorithm for flexible job-shop scheduling with overlapping in operations. Int. J. Prod. Res. **52**(13), 3905–3921 (2014)
10. Fattahi, P., Jolai, F., Arkat, J.: Flexible job shop scheduling with overlapping in operations. Appl. Math. Model. **33**(7), 3076–3087 (2009)
11. Farughi, H., Yegane, B.Y., Soltanpanah, H., Zaheri, F., Naseri, F.: Considering the flexibility and overlapping in operation in job shop scheduling based on meta-heuristic algorithms. Aust. J. Basic Appl. Sci. **5**(11), 526–533 (2011)
12. Bozek, A., Werner, F.: Flexible job shop scheduling with lot streaming and sublot size optimization. Int. J. Prod. Res. (in Press)
13. Duman, E., Uysal, M., Alkaya, A.F.: Migrating birds optimization: a new metaheuristic approach and its performance on quadratic assignment problem. Inf. Sci. **217**(24), 65–77 (2011)
14. Sioud, A., Gagne, C.: Enhanced migrating birds optimization algorithm for the permutation flow shop problem with sequence dependent setup times. Eur. J. Oper. Res. **264**(1), 66–73 (2018)
15. Tongur, V., Ülker, E.: The analysis of migrating birds optimization algorithm with neighborhood operator on traveling salesman problem. In: Lavangnananda, K., Phon-Amnuaisuk, S., Engchuan, W., Chan, J. (eds.) Intelligent and Evolutionary Systems. PALO, vol. 5, pp. 227–237. Springer, Cham (2016). https://doi.org/10.1007/978-3-319-27000-5_19
16. Niroomand, S., Hadi-Vencheh, A., Şahin, R., Vizvári, B.: Modified migrating birds optimization algorithm for closed loop layout with exact distances in flexible manufacturing systems. Expert Syst. Appl. **42**(19), 6586–6597 (2015)
17. Meng, T., Pan, Q.K., Li, J.Q., Sang, H.Y.: An improved migrating birds optimization for an integrated lot-streaming flow shop scheduling problem. Swarm Evol. Comput. **38**, 64–78 (2018)
18. Barnes, J.W., Chambers, J.B.: Flexible job shop scheduling by tabu search. Graduate Program in Operations Research and Industrial Engineering, The University of Texas, Austin, TX, Technical Report Series: ORP96-09 (1996)
19. Ruiz, R., Stutzle, T.: A simple and effective iterated greedy algorithm for the permutation flowshop scheduling problem. Eur. J. Oper. Res. **177**(3), 2033–2049 (2007)
20. Pan, Q.K., Ruiz, R.: An estimation of distribution algorithm for lot-streaming flow shop problems with setup times. Omega **40**, 166–180 (2012)
21. Wang, L., Yang, R., Xu, Y., Niu, Q., Pardalos, P.M., Fei, M.R.: An improved adaptive binary harmony search algorithm. Inf. Sci. **232**(5), 58–87 (2013)

An Improved Discrete Migrating Birds Optimization for Lot-Streaming Flow Shop Scheduling Problem with Blocking

Yuyan Han[1], Junqing Li[1,2(✉)], Hongyan Sang[1], Tian Tian[3],
Yun Bao[1], and Qun Sun[4]

[1] School of Computer Science, Liaocheng University, Liaocheng 252059, China
{hanyuyan, sanghongyan, baoyun}@lcu-cs.com
[2] School of Computer Science,
Shandong Normal University, Jinan 252000, China
lijunqing@lcu-cs.com
[3] School of Computer Science and Technology,
Shandong Jianzhu University, Jinan 250101, China
tian_tiantian@126.com
[4] School of Mechanical and Automotive Engineering,
Liaocheng University, Liaocheng 252059, China
sunqun@lcu.edu.cn

Abstract. Blocking lot-streaming flow shop (BLSFS) scheduling problems have considerable applications in various industrial systems, however, they have not yet been well studied. In this paper, an optimization model of BLSFS scheduling problems is formulated, and an improved migrating birds optimization (iMBO) algorithm is proposed to solve the above optimization problem with the objective of minimizing makespan. The proposed algorithm utilizes discrete job permutations to represent solutions, and applies multiple neighborhoods based on insert and swap operators to improve the leading solution. An estimation of distribution algorithm (EDA) is employed to obtain solutions for the rest migrating birds. A local search based on the insert neighborhood is embedded to improve the algorithm's local exploitation ability. iMBO is compared with the existing discrete invasive weed optimization, estimation of distribution algorithm and modified MBO algorithms based on the well-known lot-streaming flow shop benchmark. The computational results and comparison demonstrate the superiority of the proposed iMBO algorithm for the BLSFS scheduling problems with makespan criterion.

Keywords: Blocking · Lot-streaming flow shop · Migrating birds optimization
Estimation of distribution

1 Introduction

In many manufacturing environments, the job often refers to a set of tasks to be carried out by machines over semi-finished goods or raw materials in order to obtain a final product. Lot-steaming is a production layout in which every job can be split into a

© Springer International Publishing AG, part of Springer Nature 2018
D.-S. Huang et al. (Eds.): ICIC 2018, LNCS 10954, pp. 780–791, 2018.
https://doi.org/10.1007/978-3-319-95930-6_79

number of smaller sub-lots. When a sub-lot is completed, it can be immediately transferred to the downstream machine. By this splitting technique, the idle time on successive machines can be reduced, and thereby reducing productive cycle, accelerating the manufacturing process and enhancing the production efficiency. The goal for this problem is to find a solution (or a sequence) that optimizes a given objective function, i.e., maximum completion time minimization or makespan, the total flow time, the tardiness time and the earliness time.

For the lot-streaming problem in flow shop environment, Sarin et al. [1] presented a polynomial-time procedure to determine the number of sub-lots of a single-lot, multiple-machines flow shop lot-streaming problem. In this work, the authors minimized the unified cost-based objective function that comprised criteria pertaining to makespan, mean flow time, work-in-process, sublot-attached setup and transfer times. Pan et al. [2] presented a novel estimation of distribution algorithm (EDA) to minimize the maximum completion time, in which an estimation of a probabilistic model is constructed to direct the algorithm search towards good solutions by taking into account both job permutation and similar blocks of jobs. Defersha and Chen [3] first proposed a mathematical model for the lot-streaming problem in multi-stage flow shops where at each stage there are unrelated parallel machines, and then proposed genetic algorithm based on parallel computing platforms to solve the above problem. Numerical examples showed that the parallel implementation greatly improved the computational performance of the developed heuristic. To minimize the mean weighted absolute deviation from due dates of the lot-streaming flow shop scheduling, Yoo and Ventura developed a heuristic based on pairwise interchange strategy [4]. Chakaravarthy et al. [5] proposed a differential evolution algorithm (DE) and particle swarm optimization (PSO) to evolve the best sequence for makespan/total flow time criterion of the m-machine flow shop involved with lot-streaming and set up time. Following that Chakaravarthy et al. [6] adopted an improve sheep flock heredity algorithm and artificial bee colony algorithm, respectively, to solve lot-streaming flow shop with equal size sub-lot problems. For the same problems, Sang et al. [7] designed an effective iterated local search algorithm, in which an insertion neighborhood and a simulated annealing-typed acceptance criterion are utilized to generate good solutions.

In most practical manufacturing enterprises, there is no intermediate buffer between machines to store completed jobs. Therefore, these completed jobs have to remain in the current machine, until its following one is available for processing, which increases waiting time and the production period. Previous research has already been done to tackle a blocking flow shop (BFS) scheduling problem [8], so as to improve the production efficiency. Similarly, in a lot-streaming flow shop (LSFS) scheduling problem, each sublot will also be blocked when there is no intermediate buffer to store completed sublots. These practical scenarios encourage us to apply the blocking constraint to a LSFS scheduling problem, and form a blocking LSFS (BLSFS) scheduling problem [9].

Very recently, a new metaheuristic intelligence approach named the Migrating Birds Optimization (MBO) algorithm, which simulates the V flight formation of migrating birds, as the name implies, was presented by Duman [10]. For the scheduling problems, Tongur and Ülker [11] first applied the basic MBO algorithm to optimize the discrete flow shop sequencing problem. Following that Pan and Dong designed an improved MBO (IMBO) algorithm to minimize the total flowtime of the hybrid flow shop scheduling [12]. In this work, the authors presented a diversified method to initialize population with high quality, and constructed a mixed neighbourhood based on insertion and pairwise exchange operators to generate promising neighbouring solutions for the leader and the following birds. Similarly, Niroomand et al. [13] also proposed modified MBO (MMBO) algorithm to optimize closed loop layout with exact distances in flexible manufacturing systems, which are different from IMBO considered by Pan and Dong. In MMBO, the authors employed crossover and mutation operators to yield the neighbor regeneration.

For the above literature, the simulation experimental results have verified that the MBO algorithm is appropriate and competitive for solving continuous and discrete optimization problems. To the best of our knowledge, the MBO algorithm has not been applied to the LSFS scheduling problem with blocking. With the above motivations, we proposed an improved MBO (iMBO) algorithm to solve the BLSFS scheduling problem.

The rest of this paper is organized as follows. After this brief introduction, in Sect. 2, the description of BLSFS scheduling problem is stated. Next, Sect. 3 presents the proposed algorithm. Section 4 provides the experimental results. Finally, the paper ends with some conclusions in Sect. 5.

2   BLSFS Scheduling Problem

For each job, it can be processed at the ith machine after its front job completed at the ith machine, in which all the sub-lots of the same job should be processed continuously. At any time, for each machine, it can process at most a sub-lot, and a sub-lot can be processed on at most one machine at a time.

In the sequel, we assume that there are n jobs and m machines; denote the job sequence (solution) as $\pi = (1, 2, \ldots, n)$; and let lj be the number of sublots in job j. All sublots of the same job have to be processed on each of m machines in the same series. The processing time of each sublot of a job j on machine t is pj, t. We use Sj, t, e and Cj, t, e to represent the start and the completion time of the e-th sublot in job j on machine t, respectively, where $e = 1, 2, \ldots, lj$; dj refers to the due date of job j.

The completion time of each job on each machine can be calculated using the following equations.

$$\begin{cases} S_{1,1,1} = 0 \\ C_{1,1,1} = S_{1,1,1} + p_{1,1} \end{cases} \tag{1}$$

$$\begin{cases} S_{1,t,1} = C_{1,t-1,1} \\ C_{1,t,1} = S_{1,t,1} + p_{1,t} \end{cases} t = 2,3,\ldots,m \tag{2}$$

$$\begin{cases} S_{j,1,1} = \max\{C_{j-1,1,l_{j-1}}, S_{j-1,2,l_{j-1}}\} \\ C_{j,1,1} = S_{j,1,1} + p_{j,1} \end{cases} j = 2,3,\ldots,n \tag{3}$$

$$\begin{cases} S_{j,t,1} = \max\{C_{j,t-1,1}, S_{j-1,t+1,l_{j-1}}\} & t = 2,3,\ldots,m-1 \\ C_{j,t,1} = S_{j,t,1} + p_{j,t} & j = 2,3,\ldots,n \end{cases} \tag{4}$$

$$\begin{cases} S_{j,m,1} = \max\{C_{j,m-1,1}, C_{j-1,m,l_{j-1}}\} \\ C_{j,m,1} = S_{j,m,1} + p_{j,m} \end{cases} j = 2,3,\ldots,n \tag{5}$$

$$\begin{cases} S_{j,1,e} = \max\{C_{j,1,e-1}, S_{j,2,e-1}\} & e = 2,3,\ldots,l_j \\ C_{j,1,e} = S_{j,1,e} + p_{j,1} & j = 1,2,3,\ldots,n \end{cases} \tag{6}$$

$$\begin{cases} S_{j,t,e} = \max\{C_{j,t-1,e}, S_{j,t+1,e-1}\} & \begin{aligned} e &= 2,3,\ldots,l_j \\ t &= 2,3,\ldots,m-1 \\ j &= 1,2,3,\ldots,n \end{aligned} \\ C_{j,t,e} = S_{j,t,e} + p_{j,t} \end{cases} \tag{7}$$

$$\begin{cases} S_{j,m,e} = \max\{C_{j,m-1,e}, C_{j,m,e-1}\} & e = 2,3,\ldots,l_j \\ C_{j,m,e} = S_{j,m,e} + p_{j,m} & j = 1,2,3,\ldots,n \end{cases} \tag{8}$$

Equations (1) and (2) give the completion time of the first sublot of the first job at m machines. Equations (3–5) computes the completion time of the first sublot of job j ($j = 2, 3, \ldots, n$) at machine t ($t = 1, 2, \ldots, m$), in which the values of $S_{(j-1),2,l\ (j-1)}$ and $S_{(j-1),t+1,l\ (j-1)}$ are obtained by Eqs. (5 and 6), respectively. Equations (6–8) calculates the completion time of the e-th ($e = 2, 3, \ldots, l_j$) sublot of job j ($j = 1, 2, \ldots, n$) at machine t ($t = 1, 2, \ldots, m$).

The start time of the first sub-lot of the first job on the first machine is equal to zero, that is, $S_{1,1,1} = 0$. The makespan of the job permutation, π, is equal to the time when the last job in the processing sequence is finished at machine m. Its value can be represented according to Eq. (9).

$$C_{\max}(\pi) = C_{n,m,l_n} \tag{9}$$

3   The Proposed iMBO for the BLSFS Scheduling Problem

3.1   Initialization Population

To generate an initial population with a certain level of quality and diversity, many heuristics, i.e., NEH, MME and PFE, have been successfully adapted to initialize the seeds of the population [8]. But, they can only generate a single solution. If some good seeds in the initial population can be generated, the efficiency convergence of the whole algorithm will be enhanced. Therefore, the above idea is employed in this study. That is, a multiple-based MME initial strategy is proposed to yield β solutions with high quality, and a random method is adopted to generate the rest solutions so as to maintain the diversity of the population. The detailed process of generating β solutions is shown in Algorithm 1.

Algorithm 1. multiple-based MME

01: **Input:** the number of jobs, n
02: **Output:** β solutions

03: **Begin**
04:    Let $\pi_i = \phi$, $\pi' = \phi$, $\pi_i' = \phi$
05:    **for** i=1 to n
06:        $\pi'(i) = i$
07:    **for** i=1 to β
08:        $\pi_i \leftarrow random\_shuffle(\pi'.begin(), \pi'.end())$
09:        Set $\pi_i'(1) = \arg\min_{j \in \pi_i} p_{j,1}$; $\pi_i = \pi_i \setminus (\pi_i(1))$
10:        Set $\pi_i'(n) = \arg\min_{j \in \pi_i} p_{j,m}$; $\pi_i = \pi_i \setminus (\pi_i(n))$
11:        **for** k=3 to n

$$\theta_j = \varphi \times \sum_{i=1}^{m-1} |p_{j,i} - p_{k-1,i+1}| + (1-\varphi) \times \sum_{q=1}^{m} p_{j,q} \quad \varphi \in [0,1], \quad j \in \pi_i$$

(9)

13:            Set $\pi_i'(k) = \arg\min_{j \in S} \theta_j$; $\pi_i = \pi_i \setminus (\pi_i(k))$
14:        **end for**
15:    Pick the first two jobs of π_i', form two subsequences, $\{\pi_i'(1), \pi_i'(2)\}$ and $\{\pi_i'(2), \pi_i'(1)\}$, evaluate the quality of the two subsequences, and select the one with the minimal value of C_{max} as the current sequence, π_i^*
16:        **for** k=3 to n
17:            Pick the kth job from π_i', obtain k subsequences by inserting it into the current sequence, π_i^*, at k possible positions, and select the subsequence with the minimal value of C_{max} as the current sequence, π_i^*.
18:        **end for**
19:    **end for**
20: **End**

3.2   Improving the Leading Solution

Insertion, swap and inverse operators are commonly used to produce a promising neighboring solution, which can enhance the solution's exploitation ability by slightly disturbing the neighboring solution. For more details about the above operators, please refer to [8].

In this section, three strategies based on insert, swap, and inverse operators are proposed: (1) perform insert once; (2) apply swap one time; (3) conduct inverse once. Generally speaking, more strategies generate different solutions with a larger probability than a single strategy, and avoid the population trapping in local optima.

We randomly chose one of the above four strategies to generate solutions, in which the best neighbor solution is selected to update the leading solution, and the remaining solutions are put into two shared neighbor sets, respectively.

3.3   Improving the Other Solutions in the Population

The process of improving the other solutions in the population plays an important role, whose contribution is that it can lead the offspring to the global good solution, and improve the convergence of the algorithm. The estimation of distribution algorithm (EDA) can utilize the valued information of solutions in the population to construct a probabilistic model, and then estimate the probability distributions of good solution to build new ones. In this paper, the sequence-based discrete EDA is given to generate a number of sequences so as to improve solutions in the population.

The basic EDA mainly includes four steps [14]: First, select PS promising solutions from the original population by computing fitness value of each individual, and then put them into a candidate population $[\eta_{i,j}]_{PS \times n}$; Second, build a probability distribution model $[\xi_{i,j}]_{n \times n}$ based on the candidate population; Third, generate a new solution through learning and sampling from according to the constructed probabilistic model $[\xi_{i,j}]_{n \times n}$. Repeat the above third step for generating some new solutions. Last, update the population by evaluating the objective value of each solution in the population, and delete some bad solutions.

The detailed description of the proposed EDA is given as follows:

Algorithm 2. The estimation of the distribution algorithm

Input: the population size, *PS*, the number of jobs, *n*

Output: τ new solutions

/* **establish a candidate population** $[\eta]_{PS \times n}$ */

01. Set $[\eta]_{PS \times n} = \phi$

02. **for** $v = 1$ **to** *PS*

03.   Randomly select two solutions from the current population, evaluate their objectives, and select the best solution to put into the candidate population, $[\eta_{i,j}]_{PS \times n}$

04:       Let $v = v+1$

05: **end for**

 /*Build a probabilistic model $[\xi_{i,j}]_{n \times n}$ */

06: First, two matrixes $[\rho_{i,j}]_{n \times n}$ and $[\beta_{i,j}]_{n \times n}$ are built based on the order of the jobs in the permutation and the similar blocks of jobs.

07: Then, the probability $\xi_{i,j}$ of the each job of job sequence π is calculated according to following formulation,

$$
\xi_{i,j} = \begin{cases} \dfrac{\rho_{i,j}}{\sum_{t \in \mu(i)} \rho_{i,t}} & i = 1 \\[2em] \dfrac{\dfrac{\rho_{i,j}}{\sum_{t \in \mu(i)} \rho_{i,t}} + \dfrac{\beta_{j',j}}{\sum_{t \in \mu(i)} \beta_{j',t}}}{2} & i = 2,3..n \end{cases} \tag{10}
$$

08:   where $\mu(i)$ is the unscheduled sequence set. i is the position that job j appears in the sequence

09: /*Generate new solutions based on the probabilistic model, $[\xi_{i,j}]_{n \times n}$ */

10: Randomly select τ solutions from population, and let $s=1$

11: **while** $s < \tau$

12:         Let $i=1$

13:         Assign the *s*th solution of the population to the unscheduled sequence $\mu(i)$

14:         Randomly take 5 jobs from the unscheduled sequence $\mu(i)$, compute their probability $\xi_{i,j}$ respectively, and select the job with the largest probability $\xi_{i,j}$ as the *i*-th job of new solution π_{new}

15:         **while** $i <= n$

16:               Delete the selected job from sequence $\mu(i)$, generate a new subsequence $\mu(i+1)$, and calculate the probability of section each job in $\mu(i)$ according Eq. (11)

17:               Put the job with the largest probability $\xi_{i,j}$ into π_{mew}

18:               Let $i=i+1$

19:         **endwhile**

20:         Let $s=s+1$

21: **endwhile**

3.4   Improving the Other Solutions in the Population

In this work, the purpose of the local search is to generate a better solution from the neighborhood of a given solution. We adopt an insert-neighborhood-based local search, which has been regarded as superior to the swap or exchange neighborhood. Furthermore, we try to present a simple algorithm with few parameters, so some relative algorithms such as taboo search and simulated annealing algorithm are not applied. In this paper, we apply the local search to the solutions generated in Subsect. 3.2 with a small probability of pls. That is, a uniform random number r is generated from 0 and 1, if r < pls, the solution will employ several insertion operators. Otherwise, the solution does not perform the local search.

4   Experiments

In this section, the proposed iMBO is compared with EDA [2], DIWO [7], and MMBO [12] algorithms to evaluate the performance of the proposed algorithm. The test instance set is composed of 150 different instances, which are divided into 15 subsets and each subset consists of 10 instances with the same size. These subsets range from 10 jobs and 5 machines to 500 jobs and 20 machines [15]. Each instance is independently executed five replications. For each instance, we independentlly run each method 5 times, record the minimal makespan, and obtain the average relative percentage difference of 5 times. For all instances in a group, we obtain the above average relative percentage differences, and denote their average as ARPD. Denote the makespan of the jth instance provided by the ith algorithm in the tth run as $C_{j,t}^i$, C_j^R is the best known solution provided so far by existing algorithms for the specified problem or by our proposed algorithms. From the following Eq. (12), we can see that the smaller the average relative percentage difference APRD is, the better result the algorithm produces. Denote APRD obtained by the ith algorithm as $ARPD_i$, then $ARPD_i$ can be stated as follows.

$$ARPD_i = \frac{1}{50} \sum_{j=1}^{10} \sum_{t=1}^{5} \frac{C_{j,t}^i - C_j^R}{C_j^R} \times 100 \qquad (12)$$

All these algorithms were implemented with C++ in a PC environment with Pentium(R) Dual 2.8 GHZ and 2 GB memory. Following Yoon, Ventura, and Tseng and Liao [13], the related data for each LSFS scheduling problem are given according to the discrete uniform distributions as below. The number of solutions generated by strategies 1 and 2 are both 6, respectively, in the initialization of the population. The values of the rest parameters are set in Table 1.

Table 1. Parameter setting

Parameter	Notation	Value
Number of jobs	n	10, 30, 50, 70, 90, 110
Number of machines	m	5, 10, 20
Number of sub-lots	l	Uniform(1,6)
Processing time of a sublot of job j on machine t	$p_{j,t}$	Uniform(1,31)
Population size	PS	20
Number of initializing solutions	β	3
Local search rate	pls	0.6
Independently run times	T	5
Stopping time	$Time_{max}$	$T \times n \times m$ milliseconds

Tables 2 and 3 report APRD over each subset for computation time T = 5, 15, respectively.

It can be seen from Table 2 that the overall mean APRD value yielded by the iMBO algorithm is equal to 0.48, which is much smaller than 0.72,0.67,0.61 generated by the EDA, DIWO and MMBO algorithm. As the problem size increases, the superiority of the iMBO algorithm over EDA, DIWO and MMBO algorithms increases. On the other side, The results reported in Table 3 further justifies the superiority of the iMBO algorithm over the EDA, DIWO and MMBO algorithms for computation time T = 10. Thus, we can conclude that the presented NMBO algorithm outperforms the EDA, DIWO and MMBO algorithms for lot-streaming flowshop problems with makespan criterion.

Table 2. Performance comparison of EDA, DIWO, MMBO and NMBO algorithms (T = 5)

Instances	EDA	DIWO	MMBO	NMBO	CUP time
10 × 5	0.53	0.53	0.51	0.51	0.25
10 × 10	0.47	0.41	0.45	0.42	0.50
10 × 20	0.84	0.97	0.58	0.61	1.00
50 × 5	0.62	0.63	0.94	0.51	1.25
50 × 10	0.63	0.63	0.61	0.39	2.50
50 × 20	0.51	0.71	0.63	0.68	5.00
70 × 5	1.11	0.94	0.85	0.73	1.75
70 × 10	0.63	0.73	0.71	0.45	3.50
70 × 20	0.74	0.42	0.41	0.29	7.00
110 × 5	0.95	0.79	0.85	0.74	2.75
110 × 10	1.11	1.01	0.85	0.63	5.50
110 × 20	0.89	0.68	0.53	0.35	11.00
200 × 10	0.48	0.36	0.25	0.22	10.00
200 × 20	0.55	0.46	0.39	0.34	20.00
500 × 20	0.78	0.74	0.65	0.32	50.00
Average	0.72	0.67	0.61	0.48	8.14

Table 3. Performance comparison of EDA, DIWO, MMBO and NMBO algorithms (T = 10)

Instances	EDA	DIWO	MMBO	NMBO	CUP time
10 × 5	0.44	0.44	0.45	0.45	0.50
10 × 10	0.33	0.35	0.35	0.24	1.00
10 × 20	0.69	0.56	0.45	0.38	2.00
50 × 5	0.48	0.51	0.74	0.44	2.50
50 × 10	0.51	0.50	0.52	0.36	5.00
50 × 20	0.34	0.54	0.49	0.23	10.00
70 × 5	0.76	0.61	0.48	0.26	3.50
70 × 10	0.49	0.48	0.43	0.27	7.00
70 × 20	0.52	0.39	0.31	0.22	14.00
110 × 5	0.61	0.59	0.73	0.48	5.50
110 × 10	0.74	0.61	0.64	0.46	11.00
110 × 20	0.56	0.51	0.22	0.24	22.00
200 × 10	0.33	0.21	0.19	0.11	20.00
200 × 20	0.39	0.37	0.32	0.26	40.00
500 × 20	0.52	0.59	0.38	0.16	100.00
Average	0.51	0.48	0.45	0.30	16.27

Table 4 reports the two-side Wilcoxon rank sum tests of iMBO, EDA, DIWO and MMBO algorithms with significant level equal to 5%. In the Table 4, there are two values, i.e., p value and h value. P is the probability of observing the given result by chance if the null hypothesis is true. When h equals 1, it indicates that the results obtained by the two compared algorithms are obviously different. When h equals 0, it denotes that the difference between the two algorithms is not significant at 5% significant level. From the Table 4, the h values of the compared algorithms are equal to 1, and the p values are close to 0. Thus, it can be demonstrated that iMOB proposed in this paper is significantly different from the other compared algorithms.

Table 4. Wilcoxon two-sided rank sum test of the iMBO, EDA, DIWO and MMBO algorithms

(NMBO, EDA)		(NMBO, DIWO)		(NMBO,MMBO)	
p	h	p	h	p	h
7.42136e−057	1	3.95713e−061	1	6.96214e−0.32	1

5 Conclusions

In this paper, iMBO is proposed to minimize makespan for the BLSFS scheduling problem. In order to perform exploration for promising solutions within the entire solution space, iMBO with an effective population initialization approach is developed. A simple but effective local search algorithm was employed. To further enhance the proposed algorithm, we adopt EDA to obtain solutions for the rest migrating birds. Computational experiments are given and compared with the results yielded by the existing EDA, DIWO, and MMBO algorithms. The future work is to apply iMBO to other optimization problems and encourage us to extend the ideas proposed to the different objective functions or multi-objective in scheduling problems.

Acknowledgments. This work was jointly supported by National Natural Science Foundation of China with grant No. 61773192, 61503170, 61503220, 61603169, 61773246, and 71533001. Natural Science Foundation of Shandong Province with grant No. ZR2017BF039 and ZR2016FL13. Special fund plan for local science and technology development lead by central authority.

References

1. Sarin, S.C., Kalir, A.A., Chen, M.: A single-lot, unified cost-based flow shop lot-streaming problem. Int. J. Prod. Econ. **113**(1), 413–424 (2008)
2. Pan, Q.K., Tasgetiren, M.F., Suganthan, P.N., et al.: A discrete artificial bee colony algorithm for the lot-streaming flow shop scheduling problem. Inf. Sci. **181**(12), 2455–2468 (2011)
3. Defersha, F.M., Chen, M.: A hybrid genetic algorithm for flowshop lot streaming with setups and variable sublots. Int. J. Prod. Res. **48**(6), 1705–1726 (2010)
4. Yoon, S.H., Ventura, J.A.: An application of genetic algorithms to lot-streaming flow shop scheduling. IIE Trans. **34**(9), 779–787 (2002)
5. Chakaravarthy, G.V., Marimuthu, S., Sait, A.N.: Performance evaluation of proposed differential evolution and particle swarm optimization algorithms for scheduling m-machine flow shops with lot streaming. J. Intell. Manuf. **24**(1), 175–191 (2013)
6. Chakaravarthy, G.V., Marimuthu, S., Ponnambalam, S.G., et al.: Improved sheep flock heredity algorithm and artificial bee colony algorithm for scheduling m-machine flow shops lot streaming with equal size sub-lot problems. Int. J. Prod. Res. **52**(5), 1509–1527 (2014)
7. Sang, H.Y., Pan, Q.K., Duan, P.Y., et al.: An effective discrete invasive weed optimization algorithm for lot-streaming flowshop scheduling problems. J. Intell. Manuf. 1–13 (2015)
8. Han, Y.Y., Liang, J.J., Pan, Q.K., et al.: Effective hybrid discrete artificial bee colony algorithms for the total flowtime minimization in the blocking flowshop problem. Int. J. Adv. Manuf. Technol. **67**(1–4), 397–414 (2013)
9. Han, Y., Gong, D., Jin, Y., et al.: Evolutionary multiobjective blocking lot-streaming flow shop scheduling with machine breakdowns. IEEE Trans. Cybern. **11**(99), 1–14 (2017)
10. Duman, E., Uysal, M., Alkaya, A.F.: Migrating birds optimization: a new metaheuristic approach and its application to the quadratic assignment problem. Inf. Sci. **217**, 254–263 (2011)

11. Tongur, V., Ülker, E.: Migrating birds optimization for flow shop sequencing problem. J. Comput. Commun. **02**(4), 142–147 (2014)
12. Pan, Q.K., Dong, Y.: An improved migrating birds optimisation for a hybrid flowshop scheduling with total flowtime minimisation. Inf. Sci. **277**(2), 643–655 (2014)
13. Niroomand, S., Hadi-Vencheh, A.: Modified migrating birds optimization algorithm for closed loop layout with exact distances in flexible manufacturing systems. Expert Syst. Appl. Int. J. **42**(19), 6586–6597 (2015)
14. Han, Y.Y., Gong, D.W., Sun, X.Y., et al.: An improved NSGA-II algorithm for multi-objective lot-streaming flow shop scheduling problem. Int. J. Prod. Res. **52**(8), 2211–2231 (2014)
15. Gong, D.W., Han, Y.Y., Sun, J.Y.: A novel hybrid multi-objective artificial bee colony algorithm for the blocking lot-streaming flow shop scheduling problems. Knowl. Based Syst. **148**, 115–130 (2018)

Magnetotactic Bacteria Constrained Optimization Algorithm

Lili Liu[✉]

LiaoCheng University, Liaocheng, Shandong, China
liulili@lcu-cs.com

Abstract. Many problems encountered in the field of science and engineering can be ultimately attributed to constrained optimization problems (COPs). During the past decades, solving COPs with evolutionary algorithms have received considerable attentions among researchers. A novel approach to deal with numerical constrained optimization problems, which incorporates a Magnetotactic Bacteria Optimization Algorithm (MBOA) and an adaptive constraint-handling technique, named COMBOA, is presented in this paper. COMBOA mainly consists of an improved MBOA and archiving-based adaptive tradeoff model (ArATM) used as the constraint-handling technique. Additionally, the adaptive constraint-handling technique consists of three main situations. In detail, at each situation, one constraint-handling mechanism is designed based on current population state. It is compared with several state-of-the-art algorithms on 13 well-known benchmark functions. The experiment results show that the COMBOA is effective in solving constrained optimization problems. It shows better or competitive performance compared with other state-of-the-art algorithms referred to in this paper in terms of the quality of the solutions.

Keywords: Constrained optimization
Magnetic bacteria optimization algorithm · Constraint-handling mechanism
Adaptive tradeoff model · Individual archiving technique

1 Introduction

In the real world, the constrained optimization problems (COPs) is widely exists and difficult to be solved in practical engineering, so solving COPs has very important theoretical and practical significance. The general COPs can be expressed from [1].

Most of COPs are very difficult to be solved by traditional methods. Over the past few decades, evolutionary algorithms (EAs) and swarm intelligence (SI) algorithms have attracted more and more attentions and are widely used in solving COPs. Compared to the traditional methods, these new kinds of approaches do not need to consider the properties of optimized problems. So EAs and SI algorithms have earned more popularity over exact methods in solving optimization problems. They are also used to solve kinds of engineering or industry design problems [2–6].

Huang et al. proposed a co-evolution method adopting two subpopulations, using the penalty coefficient set to evolve solutions of the problems [7]. Coello Coello and

© Springer International Publishing AG, part of Springer Nature 2018
D.-S. Huang et al. (Eds.): ICIC 2018, LNCS 10954, pp. 792–805, 2018.
https://doi.org/10.1007/978-3-319-95930-6_80

Mezura-Montes proposed a method, which has the similar point with the niched Pareto genetic algorithm in [8].

Magnetotactic Bacteria Optimization Algorithm (MBOA) [9, 10] was proposed by Mo. It shows the potential ability of solving optimization problems and has a very fast convergence speed. In [11], an optimized support vector machine (SVM) based on MBOA is proposed to construct a high-performance classifier for motor imagery electroencephalograph-based brain-computer interface (BCI).

The MBOA has not been used to solve the COPs yet. In this paper, an improved magnetotactic bacteria algorithm (COMBOA) is proposed to solve COPs. COMBOA incorporates MBOA and an adaptive constraint-handling technique. It employs an archiving-based adaptive tradeoff model (ArATM) [12] to deal with constrains. We test it on 13 benchmark functions. The simulation results show that COMBOA is competitive with the state of the art algorithm in solving the COPs.

Organization of the rest of this paper is as follows. Section 2 briefly introduces the basic MBOA algorithm and the archiving-based adaptive tradeoff model (ArATM). Then the adaptation of the MBOA algorithm for constraint handling, including and the main process of the algorithm is presented. Experimental results are reported in Sect. 3. We compare our method with respect to the state-of-the-art approaches in constrained evolutionary optimization using 13 benchmark functions. Finally, Sect. 4 concludes this paper.

2 Magnetotactic Bacteria Optimization Algorithm and ArATM

2.1 The basic oF MBOA

(1) **Interaction distance calculation.** The distance of two cells X_i and X_r, $D_i^t = (d_{i1}^t, \ldots, d_{in}^t)$, is calculated as follows:

$$D_i^t = X_i^t - X_r^t \tag{1}$$

where i and r are mutually different integer indices from $[1, N]$, and r s randomly chosen one.

(2) **MTSs generation.** The interaction energy $E_i^t = (e_{i1}^t, \ldots, e_{in}^t)$ between two cells is defined as

$$e_{ij}^t = \left(\frac{d_{ij}^t}{1 + c_1 \times norm(D_i^t) + c_2 \times d_{pq}^t} \right)^3 \tag{2}$$

where c_1 and c_2 are constants. $norm(D_i^t)$ is the Euclidean length of vector D_i^t. d_{pq}^t is randomly selected from D_p^t. $p \in [1, N]$, $q \in [1, n]$.

The moments $M_i^t = (m_{i1}^t, \ldots, m_{in}^t)$ are generated according to (3). B is the magnetic field of the magnetosome.

$$M_i^t = \frac{E_i^t}{B} \tag{3}$$

Then the total moments of a cell is regulated as follows:

$$v_{ij}^t = x_{ij}^t + m_{ls}^t \times rand \tag{4}$$

where m_{ls}^t is randomly chosen from M_l^t. $l \in [1, N]$, $s \in [1, n]$

(3) **MTSs regulation.** After MTSs generation, evaluating the population according to cells' fitness, then the moments are regulated as follows:

If $rand > mp$

$$u_{ij}^t = v_{cbestq'}^t + (v_{cbestq'}^t - v_{iq'}^t) \times rand \tag{5}$$

Otherwise,

$$u_{ij}^t = v_{iq'}^t + (v_{cbestq'}^t - v_{iq'}^t) \times rand \tag{6}$$

V_{best} is the best cell in the current generation. $r' \in [1, N]$, $q' \in [1, n]$ mp is a parameter.

Based on (5), thus it has an enhanced local search ability. Based on (6), some randomly chosen cells which may be far from the best cell will approximate to the best one. Thus, it can enhance the ability of global search and can also increase the diversity of solutions at the same time.

(4) **MTSs replacement.** After the moments migration, evaluating the population according to cells' fitness. Some cells with worse moments are replaced according to (7) with the probability 0.5.

$$X_i^{t+1} = m_{l's'}^t \times ((rand(1, n) - 1) \times rand(1, n)) \tag{7}$$

Then the moments of remained cells are replaced by the following way:

$$x_{ij}^{t+1} = u_{ij}^t \tag{8}$$

where $m_{l's'}^t$ is randomly chosen from $M_{l'}^t$. $l' \in [1, N]$, $s' \in [1, n]$ In general, we will replace half of cells.

At last, evaluating the population according to cells' fitness after replacement and choosing the cell with the smallest fitness as the best individual.

2.2 The Constraint Handling Technique: Archiving-Based Tradeoff Model (ArATM)

In this paper, an archiving-based tradeoff model (ArATM) is employed after the MTS regulation step to handle constraints. It was proposed by Jia [12] and was incorporated with MBOA to solve the COPs. The details of ArATM are described as follows:

In this paper, two criteria are employed to compute the degree of constraint violation of a cell, which is based on the difference among the violations of different constraints.

- **The first criterion:** if $\max\limits_{j=1,2,\ldots,p} G_j^{\max} - \min\limits_{j=1,2,\ldots,p} G_j^{\max} \geq \xi$, first, $G_j^{\max}$ is found as follows:

$$G_j^{\max} = \max\limits_{i=1,2,\ldots,(N+\lambda)} G_j(\overrightarrow{x_i}),\ (j = 1,2,\ldots,p) \tag{9}$$

$$G(\overrightarrow{x_i}) = \frac{\sum_{j=1}^{p} G_j(\overrightarrow{x_i})/G_j^{\max}}{p}\ (i = 1,2,\ldots,N+\lambda) \tag{10}$$

- **The second criterion:** if $\max\limits_{j=1,2,\ldots,p} G_j^{\max} - \min\limits_{j=1,2,\ldots,p} G_j^{\max} < \xi$,

$$G(\overrightarrow{x_i}) = \sum_{j=1}^{p} G_j(\overrightarrow{x_i})\ (i = 1,2,\ldots,N+\lambda) \tag{11}$$

In general, during the process of solving COPs, the population may unavoidably experience three situations: infeasible situation, semi-feasible situation and feasible situation. Different constraint-handling mechanism should be designed for different situations, respectively.

(1) The Constraint-Handling Mechanism for the Infeasible Situation

In the infeasible situation, motivating the population toward the feasible region is the primary goal. At the early stage of optimization process, entering the feasible region promptly and maintaining the diversity in the population should be considered firstly. This paper is only concerned with those infeasible cells with fewer constraint violations in the population for that they may carry important information.

ArATM [12] is considered to achieve the above purpose. It transforms the original COPs into a bi-objective optimization problem and deals with the objective function $f(x)$ and the degree of constraint violation $G(x)$ simultaneously. ArATM is executed as follows [12]:

Step 1) If archive A is not empty, randsize individuals are randomly selected from A and put into the combined population F_t, where randsize is an integer randomly generated between 0 and $|A|$; $|A|$ denotes the number of individuals in A

Step 2) Set $A=\Phi$;

Step 3) Set $P_{t+1}=\Phi$;

Step 4) While $|P_{t+1}|<N$ /* the hierarchical nondominated individual selection scheme [13], $|P_{t+1}|$ denotes the number of individuals in P_{t+1}*/.

Step 5) The nondominated individuals in F_t are identified based on Pareto dominance;

Step 6) The nondominated individuals are sorted in ascending order based on their degree of constraint violations;

Step 7) The first half of the nondominated individuals are selected and stored into the population P_{t+1};

Step 8) These selected nondominated individuals are removed from F_t;

Step 9) End

Step 10) If $|P_{t+1}|>N$, delete the last ($|P_{t+1}|-N$) individuals in P_{t+1} and store them into F_t;

Step 11) All the individuals in F_t are stored into A.

(2) The Constraint-Handling Mechanism for the Semi-Feasible Situation

In this case, both infeasible and feasible individuals are included in the combined population. Some infeasible solutions might be promising during the optimization process, so it is not reasonable to obviate all infeasible individuals. Feasible solutions with small objective function values and infeasible solutions with slight constraint violations and small objective function values may carry important information for searching the optimal solution and should remain in the next population. The adaptive fitness transformation scheme is proposed for achieving above objective and is executed as follows [14].

Firstly, the subscripts of the two groups, i.e., the feasible group and the infeasible group, are recorded into two sets Z_1 and Z_2, respectively.

$$Z_1 = \left\{ i \middle| G(\vec{x_i}) = 0, i = 1, 2, \ldots, (N+\lambda) \right\}$$
$$Z_2 = \left\{ i \middle| G(\vec{x_i}) > 0, i = 1, 2, \ldots, (N+\lambda) \right\} \tag{12}$$

where the size of combined population F_t is NP (i.e., $NP = N + \lambda$).

Then, the best and worst feasible solutions (denoted as $\vec{x}_{best}$ and $\vec{x}_{worst}$) are found from the feasible group, respectively. Next, a converted objective function value $f'(\vec{x_i})(i \in \{1, 2, \ldots, (N + \lambda)\})$ is defined by the following form.

$$f(\vec{x}_{best}) = \min_{i \in Z_1} f(\vec{x_i}), f(\vec{x}_{worst}) = \max_{i \in Z_1} f(\vec{x_i}) \tag{13}$$

$$f'(\vec{x_i}) = \begin{cases} f(\vec{x_i}) & i \in Z_1 \\ \max\{\varphi \cdot f(\vec{x_{best}}) + (1 - \varphi) \cdot f(\vec{x_{worst}}), f(\vec{x_i})\} & i \in Z_2 \end{cases} \tag{14}$$

where φ denotes the feasibility proportion of the last population P_{t-1}.

Then, $f_{nor}(\vec{x_i})$ is obtained by the following Eq. (15):

$$f_{nor}(\vec{x_i}) = \frac{f'(x_i) - \min_{j \in Z_1 \cup Z_2} f'(x_j)}{\max_{j \in Z_1 \cup Z_2} f'(x_j) - \min_{j \in Z_1 \cup Z_2} f'(x_j)}, (i = 1, 2, \ldots, N + \lambda) \tag{15}$$

Afterward, with respect to different criterion, the constraint violation is normalized as follows:

$$G_{nor}(\vec{x_i}) = \begin{cases} 0 & i \in Z_1 \\ G(\vec{x_i}) & i \in Z_2 \text{ and criterion} = 1 \\ \dfrac{G(\vec{x_i}) - \min_{j \in Z_2} G(\vec{x_j})}{\max_{j \in Z_2} G(\vec{x_j}) - \min_{j \in Z_2} G(\vec{x_j})} & i \in Z_2 \text{ and criterion} = 2 \end{cases} \tag{16}$$

A final fitness function is obtained by adding the normalized objective function value and constraint violation together:

$$f_{final}(\vec{x_i}) = \begin{cases} f_{nor}(\vec{x_i}) + G(\vec{x_i}) & \text{if criterion} = 1 \\ f_{nor}(\vec{x_i}) + G_{nor}(\vec{x_i}) & \text{if criterion} = 2 \end{cases} (i = 1, 2, \ldots, N + \lambda) \tag{17}$$

In this situation, N individuals with the smallest final fitness function value are selected for the next population.

(3) The Constraint-Handling Mechanism for the Feasible Situation

In the feasible situation, the comparison of cells is only based on the objective function values, N individuals with the smallest objective function values are selected to constitute the next population.

2.3 The Constraint Optimization MBOA (COMBOA)

The algorithmic framework of the proposed COMBOA is as follows:

Step 1: Initialization: set $t=0$ where t denotes the generation number. Define the simple bounds, determine the parameters of the algorithm.

Step 2: Uniformly and randomly create the initial N cells in the search space S and form the initial population P_0.

Step 3: compute the constraint violations of the cells in P_0. According to the difference among the constraint violations, determine the criterion to compute the degree of the constraint violation of each cell during the evolution (see equations (9) -(11)).

Step 4: **MTSs generation:**

   Calculate interaction distances according to (1)

       for each cell do

           Calculate interaction energy according to (2)

           Obtain moments according to (3)

           MTSs generation according to (4)

       end for

   Evaluate the population according to Deb rules and choose N cells.

Step 5: **MTSs regulation**

   Generate λ offspring by executing COMBOA on all the N cells in P_t. These λ offspring form the offspring population Q_t. (see as follows for details).

Step 6: Compute the objective function value and the degree of constraint violation of each individual in Q_t.

Step 7: Combine P_t with Q_t to obtain a combined population F_t (i.e., $F_t=P_t \cup Q_t$).

step 8: Select N potential individuals from F_t to constitute the next population P_{t+1} by ArATM (see Section 2.2 for details).

Step 9: set $t=t+1$

Step10: If the termination criterion is not satisfied, go to Step 5); otherwise, stop and output the best individual xbest in P_t.

In the following, the details of the proposed COMBOA are discussed.

Compared with the classic MBOA, COMBOA also includes the MTSs generation and the MTSs regulation yet MTSs replacement is not applied. In the MTSs regulation step of COMBOA, a new population Q_t with λ cells is generated from P_t by performing the following procedures:

(1) set $Q_t = \Phi$;

(2) for each cell x_i ($i = 1, 2,..,N$) in P_t

(3) Generate the first offspring y_1 by implementing the "rand/1" strategy and "best/1" strategy similar to DE and using the binomial crossover of DE;

The details of this strategy are as follows:

If $current\_gen \leq threshold\_gen$

$$v_i = x_i + F \times (x_{r1} - x_{r2}) \tag{18}$$

else

$$v_i = x_{best} + F \times (x_{r1} - x_{r2}) \tag{19}$$

In the early stage of the evolutionary process, if *current_gen* <= *threshold_gen*, the "rand/1" strategy is used to generate the mutant vector v_i with the aim of enhancing the exploration ability of the algorithm. In the middle or the last stage of the evolutionary process, if *current_gen* > *threshold_gen*, the solutions around the best individuals should be searched to improve the exploitation ability of the algorithm. the "best/1" strategy is used to generate the mutant vector v_i for x_i.

(4) Generate the second offspring y_2 by implementing a mutation strategy (named "current-to-rand/best/1") proposed in [12] and the improved breeder genetic algorithm (BGA) mutation [15].

If *current_gen* > *threshold_gen*, the "current-to-rand/best/1" strategy is used to generate the mutant vector v_i for x_i. Afterward, the improved BGA mutation described by Eq. (20) is applied to the mutant vector v_i with a probability pm for producing the offspring y_2, with the aim of increasing the diversity of the population

$$y_{2,j} = \begin{cases} v_{i,j} \pm rang_i \times \sum_{s=0}^{15} \alpha_s 2^{-s} & rand < 1/n \\ v_{i,j} & \text{otherwise} \end{cases}, \quad j = 1, 2, \ldots, n \qquad (20)$$

$rang_i$ defines the mutation range and is set to $(U_i - L_i) \times (1 - current\_gen/total\_gen)^6$, the sign + or − is chosen with a probability of 0.5, and $\alpha_s \in \{0,1\}$ is randomly generated with the probability $p_r(\alpha_s = 1) = 1/16$. In addition, if *current_gen* <= *threshold_gen*, the "current to rand/1" strategy is used to generate the mutant vector v_i and the improved BGA mutation is not implemented (i,e., in this case $y_2 = v_i$).

(5) $Q_t = Q_t \cup y_1 \cup y_2$
(6) end

3   Experimental Results and Comparisons

To evaluate the performance of COMBOA, we used the 13 well-known test functions described in [16]. The details are shown in Table 1. In fact, this study aims to prove that the proposed COMBOA is a competitive and the efficient method by comparing with the most recent constrained optimization methods.

In all cases, 240,000 fitness function evaluations (FEs) are conducted before stopping the experiment. Therefore, the total generation is up to $240,000/(N + \lambda)$, because in the MTSs regulation step, the algorithm employs extra function evaluations for mutation purposes. The parameters in COMBOA are set as follows: $N = 70$, $\lambda = 140$, $B = 3$, $c_1 = 50$, $c_2 = 0.003$, $F = 0.8$, $CR = 0.9$, $pm = 0.05$, $\xi = 200$, $k = 0.6$, and $\delta = 0.0001$. For each test problem, 50 independent runs with different seeds are performed using the MATLAB environment.

Table 1. Details of 13 benchmark test functions

Prob	n	Type	ρ(%)	LI	NI	LE	NE	a	f(x*)
g01	13	Quadratic	0.0111	9	0	0	0	6	−15
g02	20	Nonlinear	99.9971	0	2	0	0	1	−0.8036191042
g03	10	Polynomial	0	0	0	0	1	1	−1.0005001000
g04	5	Quadratic	51.123	0	6	0	0	2	−30665.5386717834
g05	4	Cubic	0	2	0	0	3	3	5126.4967140071
g06	2	Cubic	0.0066	0	2	0	0	2	−6961.8138755802
g07	10	Quadratic	0.0003	3	5	0	0	6	24.3062090681
g08	2	Nonlinear	0.8560	0	2	0	0	0	−0.0958250415
g09	7	Polynomial	0.5121	0	4	0	0	2	680.6300573745
g10	8	Linear	0.001	3	3	0	0	6	7049.2480205286
g11	2	Quadratic	0	0	0	0	1	1	0.7499000000
g12	3	Quadratic	4.7713	0	1	0	0	0	−1
g13	5	Nonlinear	0	0	0	0	3	3	0.0539415140

3.1 Computational Results and Comparisons

The experimental results obtained by the COMBOA are shown in Table 2. The optimal results known to date for all the test problems are obtained from [16]. Table 2 gives the best, mean, worst objective function values in 50 independent runs. Gap is used to measure the solution quality. For minimization problems, $Gap = [(best\ result - optimal\ solution)/optimal\ solution] \times 100\%$.

Table 2. The results of COMBOA

Fun	f(x*)	Best	Mean	Worst	FR	Gap(%)
g01	−15	−15	−15	−15	50	0
g02	−0.80361910	−0.80361897	−0.8016037	−0.79260649	50	1.6177e-05
g03	−1.0005	−1.0005	−1.0005	−1.0005	50	0
g04	−30665.53867	**−30665.53867**	**−30665.53867**	**−30665.53867**	50	0
g05	5126.4967140	5126.496714	**5126.496714**	**5126.496714**	50	0
g06	−6961.8137558	−6961.8137558	−6961.8137558	−6961.8137558	50	0
g07	24.306209	**24.306209**	**24.306209**	**24.306209**	50	0
g08	−0.095825	−0.095825	−0.095825	−0.095825	50	0
g09	680.630057	680.630057	680.630057	680.630057	50	0
g10	7049.248	**7049.248**	**7049.248**	**7049.248**	50	0
g11	0.7499	0.7499	0.7499	0.7499	50	0
g12	−1	−1	−1	−1	50	0
g13	0.053941514	**0.053941514**	0.06933596	0.438802716	50	0

Firstly, we can observe that COMBOA is able to find the feasible solution consistently in 13 benchmark functions over 50 runs even though the feasibility ratios of most of the test functions are less than 1% and the COMBOA starts with infeasible individuals.

Secondly, the results show that the proposed COMBOA manages to obtain the "known" optimal solutions in all of these test functions with the exception of test function g02. For function g02, COMBOA found values near the optimal solutions. For function g13, the best experimental results were the same as the optimal values. Therefore, the proposed COMBOA algorithm is proven to be an effective and powerful approach to solve COPs.

Finally, the COMBOA found values near the optimal solutions for the test function g02. The results approach the "known" optimal values with small differences from 0.000016177%, which demonstrate the proposed algorithm can obtain near optimal solutions for the optimization function with several nonlinear constraints.

By further analyzing the results in Table 3, the COMBOA is incapable of obtaining the best-known solution in Table 3 for test functions g02 and g13 in all the 50 runs.

To visualize the progress of feasibility ratios, Fig. 1a and b plot the feasibility ratios for test functions g03 and g13 over all the generations by running once, respectively. In Fig. 1a, test function g03 starts with a low feasibility ratio because the problem has a small feasible region compared to its search space. Afterward, a relatively stable increase feasibility ratio gradually improves to 1.0 as the number of generations increases to 650. A moderate feasibility ratio helps the algorithm locate additional diverse solutions around the boundary of feasible space and avoid being trapped in local optima during the generation 650-800. Then the population enter the feasible situation. All the solutions are feasible. Conversely, the feasibility ratio for problem g13 (in Fig. 1b) is 0 in the beginning because of its relatively small feasible region. As the evolution progresses, the feasibility ratio increases rapidly because of the ArATM applied. Furthermore, the feasibility ratio was bound within a small range. Then the population enter the fesible situation. This phenomenon is reasonable because the ArATM plays an important role in balancing between the objective function and the constraint violations.

Table 3 shows the comparison between COMBOA and RPGA [1]. APM-ES [17], SRDE [18], DECV [19]. From the results of RPGA, it can be observed that COMBOA has distinct superiority over all the functions except g01, g05, g06, g08 and g11, on which the results of both approaches are similar. From the results of APM-ES, COMBOA performs better than APM-ES over all the functions except g02, g05, g06 and g13. COMBOA has similar results on function g02, g05 and g06. For g13, COMBOA obtains the global optimal solution and performs better than APM-ES in term of the best value, COMBOA performs worse than APM-ES in term of the mean value and the worst value. From the results of SRDE and COMBOA, COMBOA has similar performance with SRDE over all the functions except g02, g05, g07, g10 and g13. Our approach is significantly better than SRDE on g02, g05 and g13. And COMBOA is slightly better than SRDE on g07 and g10. COMBOA performs better than DECV on the functions g01, g02, g03, g07, g10 and g13 and has similar performance with DECV on the rest of the functions.

Table 3. The results of COMBOA compared with RPGA, APM-ES, DECV and SRDE

Prob.		RPGA [1]	APM-ES [17]	SRDE [18]	DECV [19]	COMBOA
g01	mean	**−15.000**	**−15.0000**	**−15.000**	−14.855	**−15**
	std	**0**	/	**0**	4.59e-01	**2.53765e-16**
	best	**−15.000**	**−15.0000**	**−15.000**	−15.0000	**−15**
	worst	**−15.000**	**−15.0000**	**−15.000**	−13.000	**−15.0000**
g02	mean	−0.794453	−0.791642	−0.796772	−0.569458	**−0.8016037**
	std	0.008188437	/	9.5e-03	9.51e-02	**3.8960e-03**
	best	−0.803619	−0.803619	−0.803619	−0.704009	**−0.80361897**
	worst	−0.780826	−0.77175	−0.7729	−0.238203	**−0.79260649**
g03	mean	−1.000	−1.000	−1.0005	−0.134	**−1.0005**
	std	8.83097e-05	/	1.7e-07	1.17e-01	**8.5880e-16**
	best	−1.000	−1.000	−1.0005	−0.461	**−1.0005**
	worst	−1.000	−1.000	−1.000499	−0.002	**−1.0005**
g04	mean	−30665.539	**−30665.538**	−30665.538672	−30665.539	−30665.538672
	std	2.07846e-05	/	0	1.56e-06	1.469965e-11
	best	−30665.539	**−30665.538**	−30665.538672	−30665.539	−30665.538672
	worst	−30665.539	**−30665.538**	−30665.538672	−30665.539	−30665.538672
g05	mean	5352.188	5126.498	5143.961035	5126.497	**5126.496714**
	std	246.1587486	/	4.13e+01	0	**1.366288e-07**
	best	5126.544	5126.498	5126.496714	5126.497	**5126.496714**
	worst	5888.510	5126.498	5288.879083	5126.497	**5126.496715**
g06	mean	**−6961.814**	**−6961.81**	**−6961.813876**	**−6961.814**	−6.961813876
	std	**1.0452e-11**	/	**0**	**0**	1.837457e-12
	best	**−6961.814**	**−6961.81**	**−6961.813876**	**−6961.814**	**−6961.813876**
	worst	**−6961.814**	**−6961.81**	**−6961.813876**	**−6961.814**	**−6961.813876**
g07	mean	24.387	24.3062	24.30621	24.794	**24.306209**
	std	0.02801179	/	3.1e-06	1.37e+00	**8.7171e-15**
	best	24.333	24.3062	24.306209	24.306	**24.306209**
	worst	24.427	24.3062	24.306227	29.511	**24.306209**
g08	mean	**−0.095825**	**−0.095825**	**−0.095825**	**−0.095825**	**−0.095825**
	std	**2.14065e-17**	/	**0**	4.23e-17	8.4415e-17
	best	**−0.095825**	**−0.095825**	**−0.095825**	**−0.095825**	**−0.095825**
	worst	**−0.095825**	**−0.095825**	**−0.095825**	**−0.095825**	**−0.095825**
g09	mean	680.634	**680.63**	**680.630057**	**680.630**	680.630057
	std	0.001665788	/	**0**	3.45e-07	4.49203e-13
	best	680.631	**680.63**	**680.630057**	**680.630**	680.630057
	worst	680.637	**680.63**	**680.630057**	**680.630**	680.630057
g10	mean	7131.084	**7049.248**	7049.248177	7103.548	**7049.24802**
	std	67.24232044	/	5.6E-04	1.48e + 02	**2.07478e-12**
	best	7131.861	**7049.248**	7049.24802	7049.248	**7049.24802**
	worst	7263.461	**7049.248**	7049.25101	7808.980	**7049.24802**

(continued)

Table 3. (*continued*)

Prob.		RPGA [1]	APM-ES [17]	SRDE [18]	DECV [19]	COMBOA
g11	mean	**0.749**[a]	**0.75**	**0.7499**	**0.75**	**0.749900**
	std	**1.16538e−07**	/	**0**	**1.12e-06**	**3.36448e-16**
	best	**0.749**[a]	**0.75**	**0.7499**	**0.75**	**0.749900**
	worst	**0.749**[a]	**0.75**	**0.7499**	**0.75**	**0.749900**
g12	mean	NA	**−1**	**−1**	**−1.0000**	**−1**
	std	NA	/	**0**	**0**	**0**
	best	NA	**−1**	**−1**	**−1.0000**	**−1**
	worst	NA	**−1**	**−1**	**−1.0000**	**−1**
g13	mean	NA	0.053924	0.260159	0.382401	0.06933596
	std	NA	/	2.1E-01	2.68e-01	7.6183e-02
	best	NA	0.053949	0.053942	0.059798	**0.053941514**
	worst	NA	0.053223[b]	0.702707	0.999094	0.438802716

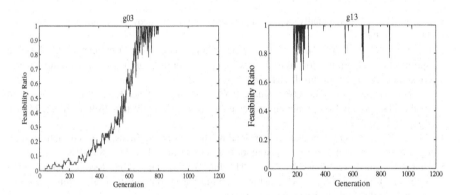

Fig. 1. Plot shows the feasibility ratios for test function (a) g03 and (b) g13

From the experimental results, we can draw a conclusion that COMBOA has a good performance on most of the test functions. In our method, we pay more attention to the exploitation of the useful information obtained from the non-dominated and feasible solutions. These two kinds of solutions describe different properties of the global optimal solutions respectively: (1) The feasibility of the global optimal solutions of COPs is embodied by the feasible solutions; and (2) the low objective function value of the global optimal solutions is reflected by the non-dominated solutions. In this way, an appropriate balance between the objective function and the constraint violations enables optimal solutions to be obtained within very few FES compared to other methods. The proposed algorithm also performs well for test problems that have very small feasible regions and more than one active constraint. For example, g07 and g10.

4   Conclusions

The MBOA deals with the unconstraint optimization problems. In this paper, COM-BOA is proposed to solve COPs and mainly consists an improved MBOA and archiving-based adaptive tradeoff model (ArATM). It adopts different constraint-handling mechanism based on the feasibility of the population. And two criteria are used to compute the degree of constraint violation of each individual in the population, according to the difference among the violations of different constraints.

The proposed algorithm was analyzed by solving COPs. The simulation results show that COMBOA performs well for test problems and has a competitive performance compared with the state-of-the-art constraint optimization algorithm.

For future work, we intent to conduct further theoretical analysis of our proposed algorithm.

References

1. Lin, C.-H.: A rough penalty genetic algorithm for constrained optimization. Inf. Sci. **241**, 119–137 (2013)
2. Li, J.Q., Sang, H.Y., Han, Y.Y., Wang, C.G., Gao, K.Z.: Efficient multi-objective optimization algorithm for hybrid flow shop scheduling problems with setup energy consumptions. J. Clean. Prod. **181**, 584–598 (2018)
3. Li, J.Q., Wang, J.D., Pan, Q.K., Duan, P.Y., Sang, H.Y., Gao, K.Z., Xue, Y.: A hybrid artificial bee colony for optimizing a reverse logistics network system. Soft. Comput. **21**(20), 6001–6018 (2017)
4. Zheng, Z., Li, J.Q.: Optimal chiller loading by improved invasive weed optimization algorithm for reducing energy consumption. Energy Build. **161**, 80–88 (2018)
5. Duan, P., Li, J.Q., Wang, Y., Sang, H., Jia, B.: Solving chiller loading optimization problems using an improved teaching-learning-based optimization algorithm. Optimal Control Appl. Methods **39**(1), 65–77 (2018)
6. Han, Y.Y., Gong, D.W., Jin, Y.C., Pan, Q.K.: Evolutionary multiobjective blocking lot-streaming flow shop scheduling with machine breakdowns. IEEE Trans. Cybern. **99**, 1–14 (2017)
7. Huang, F., Wang, L., He, Q.: An effective co-evolutionary differential evolution for constrained optimization. Appl. Math. Comput. **186**(1), 340–356 (2007)
8. Coello Coello, C.A., Mezura-Montes, E.: Constraint-Handling in genetic algorithms through the use of dominance-based tournament selection. Adv. Eng. Inform. **16**(3), 193–203 (2002)
9. Mo, H.W., Xu, L.F.: Magnetotactic bacteria algorithm for function optimization. J. Softw. Eng. Appl. **5**, 66–71 (2012)
10. Mo, H.W., Liu, L.L., Zhao, J.: A new magnetotactic bacteria optimization algorithm based on moment migration. IEEE/ACM Trans. Comput. Biol. Bioinf. **14**(1), 15–26 (2017)
11. Mo, H.W., Zhao, Y.Y.: Motor imagery electroencephalograph classification based on optimized support vector machine by magnetic bacteria optimization algorithm. Neural Process. Lett. **44**(1), 185–197 (2016)
12. Jia, G.B., Wang, Y., Cai, Z.X.: An improved $(\mu + \lambda)$-constrained differential evolution for constrained optimization. Inf. Sci. **222**, 302–322 (2013)
13. Wang, Y., Cai, Z., Zhou, Y., Zeng, W.: An adaptive tradeoff model for constrained evolutionary optimization. IEEE Trans. Evol. Comput. **12**(1), 80–92 (2008)

14. Wang, Y., Cai, Z.: Constrained evolutionary optimization by means of (μ + λ)-differential evolution and improved adaptive trade-off model. Evol. Comput. **19**(2), 249–285 (2011)
15. Wang, Y., Cai, Z., Guo, G., Zhou, Y.: Multiobjective optimization and hybrid evolutionary algorithm to solve constrained optimization problems. IEEE Trans. Syst. Man Cybern. Part B Cybern. **37**(3), 560–575 (2007)
16. Runarsson, T.P., Yao, X.: Stochastic ranking for constrained evolutionary optimization. IEEE Trans. Evol. Comput. **4**(3), 284–294 (2000)
17. Kusakci, A.O., Can, M.: An adaptive penalty based covariance matrix adaptation–evolutions strategy. Comput. Oper. Res. **40**, 2398–2417 (2013)
18. Toscano, G., Landa, R., Lárraga, G., Leguizamón, G.: On the use of stochastic ranking for parent selection in differential evolution for constrained optimization. Soft. Comput. **21**(16), 4617–4633 (2016)
19. Mezura-Montes, E., Miranda-Varela, M.E., del Carmen Gómez-Ramón, R.: Differential evolution in constrained numerical optimization: an empirical study. Inf. Sci. **180**(22), 4223–4262 (2010)

Artificial Bee Colony Algorithm Based on Ensemble of Constraint Handing Techniques

Yue-Hong Sun[1,2], Dan Wang[1], Jian-Xiang Wei[3(✉)], Ye Jin[1], Xin Xu[1], and Ke-Lian Xiao[1]

[1] School of Mathematical Sciences, Nanjing Normal University,
Nanjing 210023, China
[2] Jiangsu Key Laboratory for NSLSCS, Nanjing Normal University,
Nanjing 210023, China
[3] School of Internet of Things,
Nanjing University of Posts and Telecommunications, Nanjing 210003, China
jxwei@njupt.edu.cn

Abstract. Artificial Bee Colony (ABC) Algorithm was firstly proposed for unconstrained optimization problems. Later many constraint processing techniques have been developed for ABC algorithms. According to the no free lunch theorem, it is impossible for a single constraint technique to be better than any other constraint technique on every issue. In this paper, artificial bee colony (ABC) algorithm with ensemble of constraint handling techniques (ECHT-ABC) is proposed to solve the constraint optimization problems. The performance of ECHT-ABC has been tested on 28 benchmark test functions for CEC 2017 Competition on Constrained Real-Parameter Optimization. The experimental results demonstrate that ECHT-ABC obtains very competitive performance compared with other state-of-the-art methods for constrained evolutionary optimization.

Keywords: Artificial Bee Colony (ABC) algorithm · Ensemble
The constraint handling techniques

1 Introduction

In general, constraint optimization problems (COPs) with single-objective in minimization sense can be formulated as follows [1]:

$$
\begin{aligned}
&Min \ f(x) \\
&s.t. \ g_j(x) \leq 0, j = 1, 2, \cdots, q \\
&\qquad h_j(x) = 0, j = q+1, \ldots, m
\end{aligned}
\tag{1}
$$

where x represents a solution vector $(x_1, x_2, \cdots, x_D)$ with D dimensions. Each $x_i (i = 1, 2, \cdots, D)$ is bounded by lower and upper limits $l_i \leq x_i \leq u_i$. $f(x)$ is an objective function. $g_j(x)$ are inequality constraints, $h_j(x)$ are equality constraints. q is the number

© Springer International Publishing AG, part of Springer Nature 2018
D.-S. Huang et al. (Eds.): ICIC 2018, LNCS 10954, pp. 806–817, 2018.
https://doi.org/10.1007/978-3-319-95930-6_81

of inequality constraints and $m - q$ is the number of equality constraints. Objective functions and constraint functions can be linear or nonlinear. With a predefined tolerance parameter ε, the mean constraint violations of a solution is described as

$$CV(x) = \frac{\sum\limits_{l=1}^{q} \max(0, g_l(x)) + \sum\limits_{l=q+1}^{m} |H_l(x)|}{m} \tag{2}$$

where

$$H_l(x) = \begin{cases} |h_l(x)|, & if \ |h_l(x)| - \varepsilon \leq 0 \\ 0, & otherwise \end{cases} \tag{3}$$

Artificial Bee Colony (ABC) is a swarm intelligence algorithm with wide application. It was proposed by Karaboga and Basturk [2] with the Deb rule to deal with constraints. At the same time, two new parameters named modification rate $(M\ R)$ and scout production period (SPP) were added to accelerate local search and improve the convergence performance of the algorithm. Since then, many versions of ABC for solving COPs were proposed in succession. Brajecvic et al. proposed simple constrained ABC (SC-ABC) algorithm [3], which was applied to real engineering problems. Its main feature was each simulation run was not independent, which meant that the optimal solution in current run can be used as one initial solution in the next run. Karaboga constructed the following probability formula [4] which was related to fitness and constraint violation. The following probability of the feasible solution is larger than that of the infeasible solution, so that the feasible solution is more likely to be selected. A modified ABC [5] algorithm was proposed in 2012 by adding four mechanisms to the basic ABC: tournament selection, dynamic tolerance for equality constraints, smart flight operator, and boundary constraint-handling. Li and Yin [6] proposed a self-adaptive constrained ABC (SACABC) algorithm. It adopted different constraint handling mechanisms, including that Debs rules were used in the employed bee stage and multi-objective method was used in the onlooker bee phase. In addition, self-adaptive modification rate changed at each iteration with successful update probability. Brajevic [7] presented the crossover-based ABC (CBABC) algorithm, and the uniform crossover operator between the abandon solution and the global best solution was applied in scout phase instead of random search. The tolerance changed dynamically through iterations in order to deal with equality constraints. In 2015, Liang and Wan [8] proposed an improved ABC (IABC) algorithm. The approximate feasible solutions were introduced in order to relax Debs rules by suitably utilizing the information of the infeasible solution whose violation value was sufficiently small and objective function value was near to optimum value. For onlooker bees, the new probability selection scheme based on rank selection was also related to a new adaptive control parameter. Akay and Karaboga [9] summarized nine variants of artificial bee colony algorithm for solving COPs and compared them in terms of efficiency and stability.

In this paper, an ensemble of constraint handling techniques is embedded into ABC algorithm. The remainder of this paper is organized as follows. Five constraint handling techniques used with the EAs are given in Sect. 2. The new algorithm ECHT-ABC is described in Sect. 3. The experimental results are presented in Sect. 4 and Sect. 5 concludes the paper.

2 Constraint Handling Techniques

In order to solve the COPs, some constraint handling techniques have been developed and used in evolutionary algorithms (EAs). Recent constraint handing techniques include superiority of feasible solution (SF) [10], self-adaptive penalty (SP) [11], stochastic ranking (SR) [12], archiving-based adaptive tradeoff model (ArATM) [13], ε-constraint (EC) methods [14], ensemble of constraint-handling techniques [15], and so on.

2.1 Superiority of Feasible Solution

In SF [10], the following three conditions are considered:

(1) Any feasible solution x_i is preferable to any infeasible solutions x_j.
(2) The one having smaller objective function value (in a minimization problem) is preferred in two feasible solutions x_i and x_j.
(3) The one having smaller mean constraint violations is preferred in two infeasible solutions x_i and x_j.

2.2 Self-adaptive Penalty (SP)

An self-adaptive penalty function method [11] is proposed to solve constrained optimization problems. The fitness value is given as $F(x) = p(x) + d(x)$ where $d(x)$ is the distance value, and $p(x)$ is the penalty value. The distance value is calculated as

$$d(x) = \begin{cases} CV(x), & r_f = 0 \\ \sqrt{f''(x)^2 + CV(x)^2}, & otherwise \end{cases} \quad (4)$$

where r_f is the ratio of feasible solutions in the whole population. $f''(x) = (f(x) - f_{min})/(f_{max} - f(x))$, where f_{min} and f_{max} are the minimum and maximum value of the objection function in the current population. The penalty value is defined as $p(x) = (1 - r_f) M(x) + r_f N(x)$, where

$$M(x) = \begin{cases} 0, & r_f = 0 \\ CV(x), & otherwise \end{cases} \quad (5)$$

$$N(x) = \begin{cases} 0, & if\ x\ is\ feasible \\ f''(x), & otherwise \end{cases} \quad (6)$$

2.3 Stochastic Ranking (SR)

Stochastic Ranking [12] is a method to achieve a balance of the objective and mean constraint violation randomly. A probability factor p_f is used to judge whether the objective function value or the constraint violation value determines the rank of each individual. The basic form of SR is shown in Fig. 1.

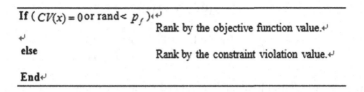

Fig. 1. Stochastic rank

2.4 Archiving-Based Adaptive Tradeoff Model (ArATM) [13]

A combined population is obtained by combining the parent population with the offspring population in ArATM. Three different constraint-handling mechanisms are designed for the three situations (infeasible, semi-feasible, and feasible situations) during the evolution of a population.

A. In the infeasible situation, the combined population contains only infeasible individuals. The objective function and the degree of constraint violation are treated as two objectives, and a bi-objective optimization problem is considered. Therefore, the hierarchical nondominated individual selection scheme [16] is adopted in ArATM to select the infeasible solution from the combined population, which makes the population enter the feasible region quickly. At the same time, the diversity of the population can be maintained by an individual archiving technique.

B. In the semi-feasible situation, the combined population contains both feasible and infeasible individuals. Because some information carried by certain infeasible individuals might be helpful to find the optimal solution, not only some feasible individuals with small objective function values but also some infeasible individuals with small degree of constraint violations and small objective function values are selected for the next generation. So an adaptive fitness transformation scheme is designed by adding the normalized objective function values and the normalized degree of constraint violation of each individual.

C. In the feasible situation, all individuals are viable. The individuals with the smallest objective function values in the combined population are selected as the parent of the next generation.

2.5 ε-Constraint (EC)

The ε-constraint handling method [14] controlled the relaxation of constraints by using the ε parameter to obtain high-quality solution for the problems with equality constraints. The ε level is updated until the generation counter k reaches the control generation T_c. The ε level is set to 0 to obtain solutions with no constraint violation after the generation counter exceeds T_c as follows

$$\varepsilon(0) = CV(x_\theta) \tag{7}$$

$$\varepsilon(k) = \begin{cases} \varepsilon(0)(1 - \frac{k}{T_c})^\infty, & 0 < k < T_c \\ 0, & k \geq T_c \end{cases} \tag{8}$$

Where x_θ is the top θth individual. The related parameter ranges are: $cp \in [2, 10]$, and $T_C \in [0.1 T_{\max}, 0.8 T_{\max}]$. The evaluation of the offspring uses superiority of feasible solution.

3 ECHT-ABC

Artificial Bee Colony algorithm (ABC) is an optimization algorithm to simulate the foraging behavior of honeybee population [17]. Karaboga and Akay [4] proposed modified ABC for solving COPs. Because a single constraint handling techniques cannot outperform all other techniques on every problem, Mallipeddi and Suganthan [15] presented an ensemble of constraint handling techniques (ECHT) with the evolutionary programming and differential evolutions in 2010. Inspired by their idea, ABC algorithm with ensemble of constraint handling techniques (ECHT-ABC) is proposed to deal with COPs in this paper.

There are three types of bees in ABC algorithms: employed bees, onlooker bees and scout bees. Different constraint handling techniques can be effective in different stages of the search process. Here, in the employed bees phase, offspring populations are evaluated by constraint handling techniques including SF, SP, SR and ArATM. In the onlooker bees phase, offspring populations are evaluated by SF, SP, SR and EC. Each constraint handling techniques has its own population. Each population corresponding to a constraint processing technology produces its offspring. The parent population with a particular constraint handling method not only competes with its own offspring, but also with the offspring produced by the other three constraint handling methods.

In the initial stage, four populations ($Foods_i, i = 1, 2, 3, 4$) are randomly generated in (9), respectively

$$x_{ij} = l_j - rand \cdot (u_j - l_j) \tag{9}$$

where $i = 1, 2, \ldots, SN, j = 1, 2, \ldots, D$, and SN is the number of food sources.

3.1 Employed Bees Phase

In employed bees phase, each parent population ($Foods_i, i = 1, 2, 3, 4$) produces offspring population ($OFFs_i, i = 1, 2, 3, 4$). An employed bee is updated [4] according to the position of the neighbour food source (solution) by (10)

$$newx_{ij} = \begin{cases} x_{ij} + \phi_{ij} * (x_{ij} - x_{kj}), & if\ R_j < MR \\ x_{ij}, & otherwise \end{cases} \tag{10}$$

Where k is an integer randomly chosen from $\{1, 2, \ldots, SN\}$ and different from i, φ_{ij} and R_j is two uniformly distributed random real numbers in the range of $(-1, 1)$ and $(0, 1)$, and MR is a control parameter. When the difference between the parameters of x_{ij} and x_{kj} decreases, the disturbance to the position decrease.

Each parent population is combined with four offspring populations to form a combined population. In a combination, such as combination 1, OFF_1 is firstly updated by competing with $OFFs_i (i = 2, 3, 4)$ generated by other populations successively. Then each individual of $Foods_1$ competes with the updated OFF_1 in order to select parent population for next cycle.

3.2 Onlooker Bees Phase

After all the employed bees completed the search process, they shared the nectar information and location information of the food source with the onlooker bees by calculating the probability value [4]

$$p(x_i) = \begin{cases} 0.5 + \dfrac{fit(x_i)}{\sum\limits_{j=1}^{SN} fit(x_j)} * 0.5, & if\ x_i\ is\ feasible \\ 0.5 - \dfrac{CV(x_i)}{\sum\limits_{j=1}^{SN} CV(x_j)} * 0.5, & if\ x_i\ is\ infeasible \end{cases} \tag{11}$$

where $fit(x_i)$ is the fitness value which is calculated as

$$fit(x_i) = \begin{cases} 1/(1 + f(x_i)), & if\ f(x_i) \geq 0 \\ 1 + |f(x_i)|, & if\ f(x_i) < 0 \end{cases} \tag{12}$$

Here we modify the process of onlooker bees phase to produce a new solution shown in Fig. 2, where x_{best} is the best individual in the employed bees phase.

Onlooker Bees Phase

Set $t = 0, i = 1$

while $t < SN$ **do**
 if $rand < p(x_i)$
  $t = t + 1$
 for $j = 1$ to D
  if $rand < MR$
   if $rand < 0.5$
    $newx_{ij} = x_{ij} + \phi_{ij}(x_{ij} - x_{kj})$
   else
    $newx_{ij} = x_{ij} + \varphi_{ij}(x_{ij} - x_{best,j})$
   end if
  else
   $newx_{ij} = x_{ij}$
  end if
 end for
 end if
 $i = i + 1$
 if $i = SN + 1$
  $i = 1$
 end if
end while

Fig. 2. Onlooker bees phase

3.3   Scout Bees Phase

Scout production period (SPP) [4] is added in order to avoid feasible solutions to be discarded quickly in the scout phase. At each SPP cycle, if a food source cannot be improved by a predetermined number (*limits*) of trial counters, it will be abandoned and the corresponding bee will become a scout bee randomly produced by Eq. (9).

4   Experimental Study and Discussion

The parameters corresponding to the constraint handling methods in ECHT-ABC are set to [15]: $MR = 0.8$, $\theta = (0.05 * SN)$, $T_c = 0.2 * T_{max}$, $cp = 5$ and p_f is linearly decreased from an initial value of $0.475 - 0.025$ in the final cycle. The tolerance parameter ε for the equality constraints is a constant 0.0001 with constraint handling techniques SF, SP, SR, ArATM, but it is adapted in EC by using (7) and (8).

To verify the performance of ECHT-ABC, this subsection compares ECHT-ABC with four state-of-the-art methods: UDE [18], LSHADE44 [19], LSHADE44-IDE [20], and CAL-SHADE [21]. CEC 2017 [22] functions are calculated by all these five methods on 25 independent runs.

Table 1. Comparison of five algorithms for 28 functions of 30D on CEC 2017

F	Criteria	UDE	LSHADE44	LSHADE44-IDE	CAL-SHADE	ECHT-ABC
C01	Median	0 (1)	0 (1)	0 (1)	0 (1)	0 (1)
	CV	0	0	0	0	0
	Mean	0 (1)	0 (1)	0 (1)	0 (1)	0 (1)
	FR	100	100	100	100	100
	vio	0	0	0	0	0
C02	Median	0 (1)	0 (1)	0 (1)	0 (1)	0 (1)
	CV	0	0	0	0	0
	Mean	0 (1)	0 (1)	0 (1)	0 (1)	0 (1)
	FR	100	100	100	100	100
	vio	0	0	0	0	0
C03	Median	7.47E+01(1)	2.06E+05(3)	6.58E+06(4)	7.36E+05(5)	1.05E+02(2)
	CV	0	0	0	1.44E-03	0
	Mean	7.33E+01(1)	3.55E+05(3)	6.70E+06(4)	1.29E+06(5)	1.63E+02(2)
	FR	100	100	100	32	100
	vio	0	0	0	2.43E-02	0
C04	Median	8.36E+01(3)	1.36E+01(2)	1.35E+01(1)	1.13E+02(4)	1.53+E02(5)
	CV	0	0	0	0	0
	Mean	8.24E+01(3)	1.36E+01(1)	1.39E+02(2)	1.15E+02(4)	1.53+E02(5)
	FR	100	100	100	100	100
	vio	0	0	0	0	0
C05	Median	0(1)	0(1)	0(1)	0(1)	1.75E−04(5)
	CV	0	0	0	0	0
	Mean	0(1)	0(1)	0(1)	7.97E−01(4)	1.11(5)
	FR	100	100	100	100	100
	vio	0	0	0	0	o
C06	Median	2.55E+02(2)	5.80E+03(4)	4.37E+03(5)	1.98+E03(3)	9.63E+01(1)
	CV	0	1.27E−02	2.55E−02	0	0
	Mean	3.03E+02(2)	4.07E+03(4)	5.53E+03(5)	3.74E+03(3)	1.00E+02(1)
	FR	100	0	0	100	100
	vio	0	1.50E−02	2.57E−02	0	0
C07	Median	−6.22E+02(2)	−1.34E-02(3)	−8.07E+01(4)	−3.26E+01(5)	−9.22E+02(1)
	CV	0	0	0	0	0
	Mean	−5.98E+02(2)	−1.09E−02(3)	−8.11E+01(4)	−2.41E+01(5)	−7.84E+02(1)
	FR	100	96	96	52	100
	vio	0	0	0	3.56E-03	0
C08	Median	−2.80E−04(3)	−2.80E−04(3)	−2.70E−04(5)	−2.84E−04(2)	−1.09E−02(1)
	CV	0	0	0	0	0
	Mean	−2.80E−04(3)	−2.80E−04(3)	−2.60E−04(5)	−2.84E−04(2)	−6.15E+02(1)
	FR	100	100	100	100	100
	vio	0	0	0	0	0

(*continued*)

Table 1. (*continued*)

F	Criteria	UDE	LSHADE44	LSHADE44-IDE	CAL-SHADE	ECHT-ABC
C09	Median	−2.67E−03(2)	−2.67E−03(2)	−2.67E−03(2)	−2.66E−03(5)	−1.84E+01(1)
	CV	0	0	0	0	0
C21	Median	9.7752(1)	28.3261(4)	28.3262(5)	9.7752(1)	9.7788(3)
	CV	0	0	0	0	0
	Mean	1.19E+01(2)	2.25E+01(5)	2.74E+01(5)	1.32E+01(3)	9.6087(1)
	FR	92	100	100	100	100
	vio	0	0	0	0	0
C22	Median	8.06E+01(1)	7.96E+02(4)	4.78E+02(3)	1.79E+05(5)	1.61E+02(2)
	CV	0	0	0	0	0
	Mean	9.41E+01(1)	1.27E+03(4)	4.78E+02(3)	3.42E+05(5)	2.26E+02(2)
	FR	100	100	100	72	100
	vio	0	0	0	2.76E−01	0
C23	Median	1.4085(2)	1.7407(4)	1.8483(5)	1.4923(3)	8.01E−01(1)
	CV	0	0	0	0	0
	Mean	1.4349(2)	1.7151(3)	1.8564(4)	1.5818(5)	9.01E−01(1)
	FR	100	100	100	84	100
	vio	0	0	0	1.60E−04	0
C24	Median	8.6393(2)	1.18E+01(3)	1.49E+01(4)	1.88E+01(5)	2.3561(1)
	CV	0	0	0	4.34E−02	0
	Mean	8.3879(2)	1.22E+01 (3)	1.39E+01(4)	2.09E+01(5)	2.4818(1)
	FR	100	100	100	20	100
	vio	0	0	0	1.47E+01	0
C25	Median	1.41E+01(1)	1.46E+02(4)	1.39E+02(3)	2.14E+02(5)	2.51E+01(2)
	CV	0	0	0	9.66E−04	0
	Mean	1.59E+01(1)	1.42E+02(4)	1.41E+02(3)	2.07E+02(5)	2.52E+01(2)
	FR	100	100	100	32	100
	vio	0	0	0	2.34E−02	0
C26	Median	1.0227(4)	9.99E−01(2)	9.84E−01(1)	1.0299(5)	1.0014(3)
	CV	1.55E+01	1.55E+01	155E+01	1.5503E+01	1.55E+01
	Mean	1.0194(4)	9.97E−01(2)	1.0123(3)	1.0353(5)	7.94E−01(1)
	FR	0	0	0	0	0
	vio	1.55E+01	1.55E+01	1.55E+01	2.15E+01	1.54E+01
C27	Median	9.04E+03(5)	1.24E+05(3)	1.15E+06(4)	2.33E+03(2)	7.0E+01(1)
	CV	2.12E+08	5.10E+06	6.83E+07	4.09E+06	3.41E+05
	Mean	11876.39(5)	9317.6(3)	52293.45(2)	3141.71(4)	83.25273(1)
	FR	0	0	0	0	0
	vio	3.41E+08	6.88E+06	4.68E+05	1.52E+07	33071.63
C28	Median	6.67E+01(3)	1.44E+02(5)	1.91E+02(4)	1.29E+02(2)	2.32E+01(1)
	CV	2.1444E+05	2.1485E+05	2.1484E+05	1.43E+05	1.42E+05
	Mean	6.43E+01(3)	1.43E+02(5)	1.53E+02(4)	1.22E+02(2)	2.61E+01(1)
	FR	0	0	0	0	0
	vio	2.1446E+05	2.1483E+05	2.1481E+05	1.43E+05	1.07E+05

In the experiments, MAXFES is set at 600000 and all other parameters are same as in the original algorithms. The results are recorded on 28 test functions for $D = 30$ in Table 1 in terms of five performance metrics (Median, CV, Mean, FR, vio). CV is the mean value of violations of all constraints at the median solution, FR is the feasibility rate of the solutions obtained in 25 runs, and vio is the mean constraint violation value of all the solutions of 25 runs. The results for UDE, LSHADE44, LSHADE44-IDE, and CAL-SHADE are directly taken from the original references respectively.

In Table 1 the numbers in parentheses represent the ranks of five algorithms based on mean values and the median solution according to the instructions in CEC 2017. The total rank value of each algorithm is recorded in Table 2. The performance of ECHT-ABC is superior to the other four algorithms because it obtains the lowest rank value.

Table 2. Ranks of five algorithm based on mean value and median solution

Algorithm	Mean rank values	Median rank values	Total rank values	Rank
UDE	76	71	147	2
LAHADE44	73	77	150	3
LSHADE44-IDE	90	94	184	4
CAL-SHADE	106	93	199	5
ECHT-ABC	**50**	**58**	**108**	**1**

5 Conclusion

Inspired by the ECHT using four constraint handling methods with the differential evolution (ECHT-DE), we propose the ECHT using five constraint handling methods (SF, SP, SR, ArATM and EC) with the Artificial bee colony (ABC) algorithm (ECHT-ABC). The final simulation results show that ECHT-ABC outperforms the other four algorithms. In the near future, we expect the ECHT can be effectively improved and more new ECHT variants with ABC algorithm can be applied to the real-world application optimization problems [23–26].

Acknowledgments. This research is partly supported by Humanity and Social Science Youth foundation of Ministry of Education of China (Grant No. 12YJCZH179), National Social Science Foundation (Grant No. 14BTQ036), and the Foundation of Jiangsu Key Laboratory for NSLSCS (Grant No. 201601). The authors thank the anonymous reviewers for providing valuable comments to improve this paper.

References

1. Qin, A.K., Huang, V.L., Suganthan, P.N.: Differential evolution algorithm with strategy adaptation for global numerical optimization. IEEE Trans. Evol. Comput. **13**(2), 398–417 (2009)
2. Karaboga, D., Basturk, B.: Artificial Bee Colony (ABC) optimization algorithm for solving constrained optimization problems. Found. Fuzzy Logic Soft Comput. **11**(3), 789–798 (2007)
3. Brajevic, I., Tuba, M., Subotic, M.: Performance of the improved Artificial Bee Colony algorithm on standard engineering constrained problems. Int. J. Math. Comput. Simul. **5**(2), 789–798 (2011)
4. Karaboga, D., Akay, B.: A modified Artificial Bee Colony (ABC) algorithm for constrained optimization problems. Soft. Comput. **11**(3), 3021–3031 (2011)
5. Mezura-Montes, E., Cetina-Domnguez, O.: Empirical analysis of a modified artificial bee colony for constrained numerical optimization. Appl. Math. Comput. **218**(22), 10943–10973 (2012)
6. Li, X., Yin, M.: Self-adaptive constrained artificial bee colony for constrained numerical optimization. Neural Comput. Appl. **24**(3–4), 723–734 (2014)
7. Brajevic, I.: Crossover-based Artificial Bee Colony algorithm for constrained optimization problems. Neural Comput. Appl. **26**(7), 1587–1601 (2015)
8. Liang, Y.S., Wan, Z.P., Fang, D.B.: An improved artificial bee colony algorithm for solving constrained optimization problems. Int. J. Mach. Learn. Cybern. **8**(3), 1–16 (2017)
9. Akay, B., Karaboga, D.: Artificial bee colony algorithm variants on constrained optimization. Int. J. Optim. Control Theor. Appl. (IJOCTA) **7**(1), 98–111 (2017)
10. Deb, K.: An efficient constraint handling method for genetic algorithms. Comput. Methods Appl. Mech. Eng. **186**(2), 311–338 (2000)
11. Tessema, B., Yen, G.G.: A self adaptive penalty function based algorithm for constrained optimization. In: IEEE Congress on Evolutionary Computation, CEC 2006, pp. 246–253. IEEE (2006)
12. Runarsson, T.P., Yao, X.: Stochastic ranking for constrained evolutionary optimization. IEEE Trans. Evol. Comput. **4**(3), 284–294 (2000)
13. Jia, G., Wang, Y., Cai, Z., et al.: An improved $(\lambda + \mu)$-constrained differential evolution for constrained optimization. Inf. Sci. **222**(4), 302–322 (2013)
14. Takahama, T., Sakai, S.: Constrained optimization by the constrained differential evolution with gradient-based mutation and feasible elites. In: Conferences, CEC 2006, pp. 1–8. IEEE (2006)
15. Mallipeddi, R., Suganthan, P.N.: Ensemble of constraint handling techniques. IEEE Trans. Evol. Comput. **14**(4), 561–579 (2010)
16. Wang, Y., Cai, Z., Zhou, Y., Zeng, W.: An adaptive tradeoff model for constrained evolutionary optimization. IEEE Trans. Evol. Comput. **12**(1), 80–92 (2008)
17. Karaboga, D.: An idea based on honey bee swarm for numerical optimization. Technical report, Engineering Faculty, Computer Engineering Department, Erciyes University (2005)
18. Trivedi, A., Sanyal, K., Verma, P., et al.: A unified differential evolution algorithm for constrained optimization problems. In: Evolutionary Computation, CEC 2017, pp. 1231–1238. IEEE (2017)
19. Polkov, R.: L-SHADE with competing strategies applied to constrained optimization. In: Evolutionary Computation, CEC 2017, pp. 1683–1689. IEEE (2017)

20. Tvrdik, J., Polakova, R.: A simple framework for constrained problems with ap- plication of L-SHADE44 and IDE. In: Evolutionary Computation, CEC 2017, pp. 1436–1443. IEEE (2017)
21. Ales, Z.: Adaptive constraint handling and Success History Differential Evolution for CEC 2017 Constrained Real-Parameter Optimization. In: Evolutionary Computation, CEC 2017, pp. 2443– 2450. IEEE (2017)
22. Wu, G., Mallipedi, R., S, P.N.: Problem definitions and evaluation criteria for the CEC 2017 competition on constrained real-parameter optimization. Technical report, CEC 2017 (2017)
23. Feng, Y., Wang, G.-G.: Binary moth search algorithm for discounted 0-1 knapsack problem. IEEE Access (2018). https://doi.org/10.1109/ACCESS.2018.2809445
24. Rizk-Allah, R.M., El-Sehiemy, R.A., Wang, G.-G.: A novel parallel hurricane optimization algorithm for secure emission/economic load dispatch solution. Appl. Soft Comput. (2018). https://doi.org/10.1016/j.asoc.2017.12.002
25. Rizk-Allah, R.M., El-Sehiemy, R.A., Deb, S., Wang, G.-G.: A novel fruit fly framework for multi-objective shape design of tubular linear synchronous motor. J. Supercomput. (2017). https://doi.org/10.1007/s11227-016-1806-8
26. Zhang, J.-W., Wang, G.-G.: Image matching using a bat algorithm with mutation. Appl. Mech. Mater. **203**(1), 88–93 (2012)

Two-Echelon Logistics Distribution Routing Optimization Problem Based on Colliding Bodies Optimization with Cue Ball

Xiaopeng Wu[1], Yongquan Zhou[1,2(✉)], Mengyi Lei[1],
Pengchuan Wang[1], and Yanbiao Niu[1]

[1] College of Information Science and Engineering,
Guangxi University for Nationalities, Nanning 530006, China
yongquanzhou@126.com
[2] Guangxi High School Key Laboratory of Complex System and Computational
Intelligence, Nanning 530006, China

Abstract. Two-echelon logistics distribution routing problem is an important optimization problem of the logistics distribution networks. It is composed of distribution center location problem and distribution routing problem. Distribution center location problem aims to find the best locations of distribution centers from all the distribution points. Meanwhile, the distribution center needs to be assigned to serve the distribution points. The goal of distribution routing problem is to decrease the total cost of delivery. In this paper, an improved version, colliding bodies optimization with cue ball (CBCBO), is proposed to tackle two-echelon logistics distribution routing problem. The new algorithm improves the lack of the colliding bodies optimization (CBO) algorithm which the number of populations must be even. The new approach based cue ball enhanced exploration ability. A strategy, elite opposition strategy, is used to promote exploitation ability. In the last, the effectiveness of the new algorithm is tested by simulation experiment. The proposed approach demonstrates its capability to optimize two-echelon logistics distribution routing problem.

Keywords: Two-echelon logistics distribution routing problem
Distribution center location problem · Distribution routing problem
Colliding bodies optimization with cue ball · Elite opposition strategy

1 Introduction

Logistics cost is an important part of the total cost of society. In recent years, logistics costs continued to grow. The costs of logistics in 2016 is 11 trillion RMB in China. Among them, the transport costs accounted for 51% of the total cost of logistics. It is the half of the total costs. Therefore, in the transport process, decreasing the cost of transportation can save a lot of money. In reality, getting a right logistics center and choosing the best route for vehicle transportation are the key to improve logistics system.

Getting a suitable logistics center is distribution center location problem which aims to find the best positions from each logistics distribution points. Usually, the

© Springer International Publishing AG, part of Springer Nature 2018
D.-S. Huang et al. (Eds.): ICIC 2018, LNCS 10954, pp. 818–830, 2018.
https://doi.org/10.1007/978-3-319-95930-6_82

distribution points are located at the somewhere of the whole city. Logistics company need to choose several distribution points from all the points and change them to distribution centers as depot. The distribution centers have two attributions. First, distribution centers itself need goods. Second, distribution centers provide goods for distribution points. But, the goods, distribution centers need, doesn't cost any money of transportation. Distribution points get goods from the distribution center which is the nearest depot to it. Choosing the best route for vehicle transportation is vehicle routing problem. There are some distribution points and distribution centers. Distribution centers provide goods for distribution demand points. Transport vehicle transports goods from distribution centers to each distribution point. Choosing the suitable route for transportation is important. The shorter route costs less transportation costs. It can save money and improve transportation efficiency.

In this paper, a variant of the Colliding Bodies Optimization is used to solve two-echelon logistics distribution routing optimization problem. Meta-heuristic optimization algorithm has attracted wide attention from scholars, because of its strong performance. Colliding Bodies Optimization is a meta-heuristic optimization algorithm. Kaveh and Mahdavi solve optimum design of truss structures with continuous variables in 2014 [1] and discrete variables in 2015 [2] respectively. Shayanfar detects bridge structures in time domain via by using enhanced colliding bodies optimization in 2016 [3]. Panda and Pani proposed multi-objective colliding bodies optimization in 2016 [4]. Kaveh also uses colliding bodies optimization and enhanced colliding bodies optimization to solve construction site layout planning in 2017 [5]. Kaveh solve modification of ground motions using enhanced colliding bodies optimization algorithm [6].

2 Two-Echelon Logistics Distribution Routing Optimization Problem

Two-echelon logistics distribution routing optimization problem can be decomposed into two problems. The first is distribution center location problems [7, 8] and the second is distribution routing problems [9–11]. Distribution center location problem is proposed to find the best locations of distribution centers from all the distribution points. The goal of distribution routing problem is to decrease the total cost of delivery.

2.1 Distribution Center Location Problem

Logistics distribution center location problem is a hot topic in the field of logistics. The problem can be described as follows: First, giving a set of distribution points that a certain number of points are selected to be established as a distribution center. Second, making the distribution center serve distribution points. These two regulations proposed so that the delivery costs of the whole logistics distribution network are minimal. The model aims to select the appropriate number of distribution points as a distribution center. This paper finally established the following mathematical model. We select the minimum conduct of the distance between the distribution center and the distribution point and the demand of each distribution. The formulation [12] shows as follows:

$$\min F_1 = \sum_{i \in NN} \sum_{j \in M_i} w_i d_{ij} Z_{ij} \qquad (1)$$

Subject to:

$$\sum_{j \in M_i} Z_{ij} = 1 \quad i \in NN \qquad (2)$$

$$Z_{ij} \leq h_j \qquad i \in NN, j \in M_i \qquad (3)$$

$$\sum_{j \in M_i} h_j = p \qquad (4)$$

$$Z_{ij}, h_j \in \{0, 1\} \quad i \in NN, j \in M_i \qquad (5)$$

$$d_{ij} \leq S \qquad (6)$$

The notations used in the formulation are listed:

Where,

$NN = \{1, 2, 3, \ldots, nn\}$ denotes the set of distribution points, and nn is the maximum number of the distribution points.

$M_i(i \in m, M_i \in NN)$ denotes potential distribution centers. m is the number of the distribution centers.

w_i denotes the demand of i th distribution point.

d_{ij} denotes the distance from distribution point i to distribution point j.

Z denotes a logical variable. Its value is 0 or 1. It can vividly express the relationship between the two forms. When the value is 1, indicating that there is a link between them, otherwise there is no contact between the two forms.

h_j is also a logical variable. Its value is 0 or 1, too, when $h_j = 1$, it denotes that j th distribution point is selected as a logistics distribution center.

S is the upper limit of the distance from the logistics distribution center to the distribution points which this logistics distribution center should serve

In these constraints, constraint Eq. (2) to ensure that each demand point can only be served by a logistics distribution center. Constraint Eq. (3) make sure that the demand for the demand point can only be distributed by distribution center. Constraint Eq. (4) sets the number of selected distribution centers that is specified as p. Constraint Eq. (5) indicates that Z_{ij} and h_j are 0–1 variables. Constraint Eq. (6) ensure that the distribution points in the range that the logistics distribution center can distribute.

2.2 Logistics Distribution Route Optimization Problem

Logistics distribution route optimization problem can be described as several vehicles transport goods from distribution center to distribution points. Each distribution points' coordinate and demand quantity are known. The capacity of vehicles is known. Vehicles transport route should be arranged reasonably to make the total distance of the transportation as short as possible. If the shortest distance in one of the centers can be computed, the whole distribution network's shortest distance can be computed easily. The shortest distance represents the least transportation cost. So, the mathematical model [13] can be described as:

$$\min F_2 = \sum_{y=1}^{M} \sum_{k=1}^{C_y} \left[\sum_{i=1}^{n_k} d_{r_{k(i-1)}r_{ki}} + d_{r_{kn_k}r_{k0}} \right] \tag{7}$$

Subject to:

$$\sum_{i=1}^{n_k} w_{r_{ki}} \leq Q \tag{8}$$

$$\sum_{i=1}^{n_k} d_{r_{k(i-1)}r_{ki}} + d_{r_{kn_k}r_{k0}} \leq D \tag{9}$$

$$0 \leq n_k \leq L_y \tag{10}$$

$$\sum_{k=1}^{C_y} n_k = L_y \tag{11}$$

$$R_k = \left\{ r_{ki} | r_{ki} \in \{1, 2, \ldots, L_y\}, i = 1, 2, \ldots, n_k \right\} \tag{12}$$

$$R_{k_1} \cap R_{k_2} = \phi, \forall k_1 \neq k_2 \tag{13}$$

The notations used in the formulation are listed:

C_y denotes the number of the vehicles in $y (y \in M)$th distribution point. M denotes the distribution centers.

Q denotes the capacity of each vehicle. It is invariable.

D denotes the longest distance that one vehicle can drive in the transportation.

L_y denotes the number of distribution points which i th distribution center need to serve and transport goods to ($y \in M$).

w_i denotes the demand quantity of i th distribution point.

d_{ij} denotes the distance from distribution point i to distribution point j. d_{0j} denotes the distance from distribution point j to distribution center.

n_k denotes the number of distribution points which k th ($k = 1, 2, \ldots, C$) vehicle transports to.

R_k denotes the set of route that k th vehicle drives.

r_{ki} denotes one distribution point. This distribution point is the i th point that will be served. r_{k0} denotes distribution center if its value is 0.

In the formulations above, Eq. (7) is the objective function of logistics distribution route optimization problem. Constraints Eq. (8) ensures that the total demand quantity of each distribution path does not exceed the capacity of vehicle. Constraints Eq. (9) ensures that the length of each distribution path does not exceed the maximum travel distance of a transport vehicle. Constraints Eq. (10) indicates that the number of distribution points on each path does not exceed the total distribution points. Constraints Eq. (11) indicates that each distribution point can be served. Constraints Eq. (12) represents each path is composed of distribution points and one distribution center. Constraints Eq. (13) limits each distribution point can only be delivered by a vehicle.

Two-echelon logistics distribution routing optimization problem is composed of these two problems. The objective function for two-echelon logistics distribution routing optimization problem can be expressed below:

$$Min\ F = F_1 + F_2 \tag{14}$$

This equation subject to the constraint Eqs. (2)–(6) and Eqs. (8)–(13).

3 Colliding Bodies Optimization with Cue Ball

This paper presents an improved version of the water cycle algorithm for constrained engineering optimization problems, i.e., the so-called dual-system (outer cycle system and inner cycle system) water cycle algorithm (DS-WCA).

3.1 Colliding Bodies Optimization

In this section, Colliding Bodies Optimization algorithm [14] is introduced briefly. The physical principle of collision is the laws of momentum and energy. The total momentum of the colliding bodies (CB) is conserved. The momentum of all objects before the collision equals the momentum of all objects after the collision. The collision happens in a one-dimensional space. The collision aims to change the velocity and position of the bodies and get a better solution. The procedure of Colliding Bodies Optimization:

1. Initialization

The bodies get a random initial location in the search space. The initial formulation is as this:

$$x_i^0 = x_{min} + random \circ (x_{max} - x_{min}),\ i = 1, 2, \ldots, n, \tag{15}$$

where x_i^0 denotes the initial position in the search space of the i th CB. x_{max} denotes upper boundary.x_{min} denote the lower boundary. *random* is a random number which its range is [0, 1]. n is the number of all CBs.

2. Defining mass

The mass is defined by this formulation:

$$m_k = \frac{\frac{1}{fit(k)}}{\sum_{i=1}^{n} \frac{1}{fit(i)}}, \quad k = 1, 2, \ldots, n, \tag{16}$$

where $fit(i)$ is the fitness function value. n is the number of agents. The CB with better values has a larger mass than the bad ones. Also, In order to facilitate the calculation, the fitness function value $fit(i)$ will be replaced by $\frac{1}{fit(i)}$.

3. Divide Groups

After Sorting the fitness function value in ascending form, all the CBs are divided to two equal groups. The lower half is named stationary CBs that their velocity that are expressed by Eq. (17) is zero before collision. The upper half of CBs are named moving CBs: These CBs move to hit the lower half. Their velocity before collision is expressed by Eq. (18):

$$v_i = 0 \quad i = 1, 2, \ldots, \frac{n}{2} \tag{17}$$

$$v_i = x_{i-\frac{n}{2}} - x_i \quad i = \frac{n}{2} + 1, \frac{n}{2} + 2, \ldots, n \tag{18}$$

where, v_i is the velocity vector of the i th CB in the moving group, and x_i is position vector. $x_{i-\frac{n}{2}}$ is the $i - \frac{n}{2}$ th CB position.

4. Criteria after the collision

When the collision end, the velocities of two groups will be calculated. The velocity of moving CBs after the collision is evaluated by Eq. (19). The new velocity of stationary group is Eq. (20):

$$v_i' = \frac{(m_i - \varepsilon m_{i-\frac{n}{2}})v_i}{m_i + m_{i-\frac{n}{2}}} \quad i = \frac{n}{2} + 1, \frac{n}{2} + 2, \ldots, n \tag{19}$$

$$v_i' = \frac{(m_{i+\frac{n}{2}} - \varepsilon m_{i+\frac{n}{2}})v_{i+\frac{n}{2}}}{m_i + m_{i+\frac{n}{2}}} \quad i = 1, 2, \ldots, \frac{n}{2} \tag{20}$$

where v_i and v'_i are the velocity of moving group bodies before and after the collision. m_i is mass of the i th CB. where ε is the coefficient of restitution (COR). It decreases linearly from unit to zero. So, it is expressed as:

$$\varepsilon = 1 - \frac{iter}{iter_{max}} \tag{21}$$

Current iteration number is expressed as iter, and $iter_{max}$ is the maximum iteration number of the optimization process.

5. Updating position

New position is determined by two factor which are the velocities after the collision and the location before the collision. The new positions of the stationary CBs are computed by Eq. (22). The new positions of every moving CBs are computed by Eq. (23):

$$x_i^{new} = x_i + rand \circ v'_i \quad i = 1, 2, \ldots, \frac{n}{2} \tag{22}$$

$$x_i^{new} = x_{i-\frac{n}{2}} + rand \circ v'_i \quad i = \frac{n}{2}+1, \frac{n}{2}+2, \ldots, n \tag{23}$$

where x_i^{new} is the new locations. x_i is previous locations. v'_i is the velocity after the collision. i denotes the i th CB. $rand$ is a random vector uniformly distributed in the range of $[-1, 1]$, and the sign 'o' denotes an element-by-element multiplication.

6. Terminal criterion check

If the number of current iteration is equal to the maximum iteration number, stop the algorithm, or the algorithm goes to step 2.

3.2   Colliding Body Encoding Scheme

According to the characteristic of the objective function, designing a CB encoding approach [15] is a key to solve emergency materials transshipment model. So, a three-dimensional CB encoding is proposed. Each CB is considered as a solution $X_i(X_i = \{X_{ijk}\}, i \in n, j = M, 1 < k < N)$. Where, n is the number of the population. M is the number of the distribution centers. The value of k denotes the number of the distribution point that j th distribution center need to serve. The value of X_{ijk} denotes the number which represents distribution point. For example, $X_{2,3,4} = 6$ indicates the third distribution center need to serve the fourth distribution point which the serial number is 6 in the second colliding body. Each CB contains two sequences of distribution points. The first sequence denotes which distribution point is changed to a distribution center. The second sequence denotes delivery order that distribution center transport goods to distribution points.

3.3   CBCBO Updating Mechanism

According to the characteristic of two-echelon logistics distribution routing optimization problem. This paper proposed a suitable updating mechanism. First, the best colliding body with best fitness function value of each iterations is preserved as a cue ball. Other CB move to cue ball and change the position. In the colliding body encoding scheme, there are two sequences. So, the suitable updating mechanism updates position of colliding body by updating the two sequences. Updating the second dimension of colliding body is trying to select different distribution points and change them to distribution centers so that the function of F_1 can get minimum value. Updating the third dimension of colliding body aims to find shortest distribution path that correspond to the sequence of the second dimension.

3.4   Colliding Bodies Optimization with Cue Balls Solving Two-Echelon Logistics Distribution Routing Optimization Problem

According to the introduction above, Colliding bodies optimization with cue ball can be detailed as follows.

Step 1: The CBCBO initialization. Utilize the colliding body encoding scheme to initialize the populations. Each CB include two sequences that are distribution center sequence and distribution route sequence. The city coordinates are generated within [1, 30000] randomly. Calculate the distance between two distribution points. Several notations are predefined:

NN	The number of the distribution points
m	The number of the distribution centers
w_i	The demand of i th distribution point ($w_i \in \{10, 20, 30, 40, 50, 60, 70, 80, 90, 100\}$).
Q	The capacity of each vehicle. It is invariable.
D	The longest distance that one vehicle can drive in the transportation.
n	The population of CB
$Iter_{max}$	The maximum iteration of CBCBO algorithm
CB_{cue}	The best objective function value of current iteration.

Step 2: According to the Eq. (14), calculate the objective function value of each CB. Select the best CB and change it to cue ball as CB_{cue}.

Step 3: All the CBs except cue ball are a group which named moving group. Moving group move to the cue ball and collide with cue ball.

Step 4: All the CBs including cue ball will change their position after the collision. If the objective function value of CB_{cue} is better after the collision than the value before the collision, update the position of CB_{cue}. If not, the position of CB_{cue} will not change. Changing positions of CBs in this paper according to the CBs updating scheme is to change the two sequences. Determine the exchange sequence. The exchange sequence of moving group is shown as:

$$v_i = x_i - x_{CB_{cue}} \quad i = 1, 2, \ldots n - 1 \tag{24}$$

The exchange sequence of CB_{cue} is shown as:

$$v_{CB_{cue}} = x_{CB_{cue}} - x_i \quad i = 1, 2, \ldots n - 1 \tag{25}$$

where v_i and $v_{CB_{cue}}$ are the exchange sequence of i th CB and CB_{cue}. x_i is the i th CB before the collision. $x_{CB_{cue}}$ is the position of cue ball.

Step 5: Update distribution centers and distribution path. Distribution plan is changed by exchange sequence before the collision. Two CBs choose some of the exchange sequences to change their position randomly. The new distribution plan x_i^{new}, $x_{CB_{cue}}$ is generated.

Step 6: Terminal criterion check. If $Iter_{max}$ is bigger than the current iteration, go to step 2. Otherwise, Stop the algorithm.

4   Simulation Experiments

4.1   Parameter Settings

$NN = 50$	The number of the distribution points is 50.
$m = 5$	The number of the distribution centers is 10% of NN. In this paper, there are 5 distribution centers.
w_i	The demand of i th distribution point ($w_i \in \{10, 20, 30, 40, 50, 60, 70, 80, 90, 100\}$). Different distribution points have different demand quantity. The minimum value is 10 cases. The maximum is 100 cases.
$Q = 200$	The capacity of each vehicle is 200 cases. It is invariable.
$D = 30000$	The longest distance that one vehicle can drive in the transportation is 30000 m.
$n = 50$	The population of CB is 50.
$Iter_{max} = 500$	The maximum iteration of CBCBO algorithm is 500 (Table 1).

Table 1. Location and demand of distribution points

Number	X-axis	Y-axis	Demand	Number	X-axis	Y-axis	Demand
1	8013	2496	80	26	14201	18155	70
2	23892	2624	60	27	12767	22724	80
3	16649	19064	50	28	25442	26799	60
4	25770	3297	100	29	603	1903	10
5	29999	16479	60	30	23808	23351	60
6	20136	25724	80	31	12316	4665	50
7	19629	2640	10	32	27255	10736	90
8	24474	5616	50	33	2325	19256	100
9	26457	5183	30	34	13183	23636	10
10	20810	21008	50	35	26200	15664	10
11	22340	26839	80	36	9925	3073	10
12	24784	20753	10	37	13359	7409	40
13	15892	11560	10	38	25458	1528	40
14	23420	27164	80	39	27578	25158	90
15	6109	8634	30	40	25174	28487	70
16	27953	28406	20	41	20400	11320	60
17	16612	7572	50	42	19491	10627	50
18	12033	2653	80	43	16232	21249	20
19	20778	29357	100	44	8115	6769	60
20	22941	16880	80	45	9684	21668	90
21	3488	12328	20	46	12898	6218	100
22	24190	16430	90	47	7584	4083	100
23	278	16387	70	48	15491	1994	60
24	24491	8736	70	49	18527	5145	60
25	15440	3180	60	50	2012	15592	60

4.2 Simulation Experiments

Colliding body optimization with cue ball is used to solve two-echelon logistics distribution routing optimization problem. The location of each distribution point is shown in Fig. 1. There are fifty distribution points in Fig. 1. Every green point represents a distribution point.

The results of selecting distribution centers are visualized in Fig. 2. The green point with red box is distribution centers. The bare green dot is the distribution points. The lines between two dot express the delivery relationship. In the picture it is easy to get each distribution center need to serve several distribution points.

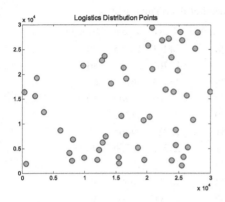

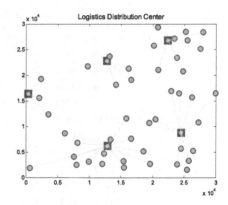

Fig. 1. The locations of distribution points (Color figure online)

Fig. 2. Logistics distribution center location (Color figure online)

The results of distribution paths are in Fig. 3. Green point with red box is distribution centers. Bare green dot is the distribution points. Red lines represent the distribution path. The transport vehicles start from distribution center, along the red line to transport the goods. Total demand quantity of ring red line is less than the capacity of vehicle. Some distribution centers only have one ring transport path. Other distribution centers have several ring transport paths. The number of ring transport path is determined by the number of distribution points and their demand quantity.

The results of two-echelon logistics distribution routing optimization problem.

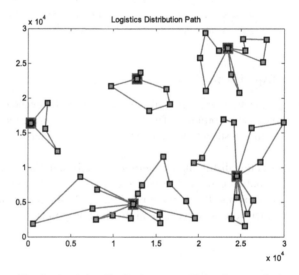

Fig. 3. Logistics distribution route (Color figure online)

5   Conclusions

In this paper, colliding body optimization algorithm with cue ball is proposed to solve two-echelon distribution routing optimization problem. The proposed algorithm does not only solve the distribution center location problem, but also deal with the distribution routing problem. Two-echelon distribution routing optimization problem improves logistics system, save the total cost of the delivery process. The CBCBO can choose suitable number and location of logistics center from the logistics system, no matter how much distribution points there are. The proposed method for enhanced the performance of Colliding Bodies Optimization. Cue ball is the goal that other colliding bodies need to collide with. So, new optimization algorithm speeds up the convergence speed and accuracy.

Acknowledgment. This work is supported by National Science Foundation of China under Grants No. 61463007; 61563008. Project of Guangxi University for Nationalities Science Foundation under Grant No. 2016GXNSFAA380264.

References

1. Kaveh, A., Mahdavai, V.R.: Colliding Bodies Optimization method for optimum design of truss structures with continuous variables. Adv. Eng. Softw. **70**, 1–12 (2014)
2. Kaveh, A., Mahdavi, V.R.: Optimal design of truss structures with discrete variables using colliding bodies optimization. Colliding Bodies Optimization: Extensions and Applications, pp. 87–104. Springer, Cham (2015). https://doi.org/10.1007/978-3-319-19659-6_4
3. Shayanfar, M.A., Kaveh, A., Eghlidos, O., et al.: Damage detection of bridge structures in time domain via enhanced colliding bodies optimization. J. Thorac. Cardio Vasc. Surg. **76** (4), 483–488 (2016)
4. Panda, A., Pani, S.: Multi-objective colliding bodies optimization. In: Pant, M., Deep, K., Bansal, J., Nagar, A., Das, K. (eds.) Proceedings of Fifth International Conference on Soft Computing for Problem Solving. Advances in Intelligent Systems and Computing, vol. 436, pp. 651–664. Springer, Singapore (2016). https://doi.org/10.1007/978-981-10-0448-3_54
5. Kaveh, A.: Construction site layout planning problem using two new meta-heuristic algorithms. Iran. J. Sci. Technol. Trans. Civ. Eng. **40**, 263–275 (2017)
6. Kaveh, A.: Modification of ground motions using enhanced colliding bodies optimization algorithm. Applications of Metaheuristic Optimization Algorithms in Civil Engineering, pp. 213–234. Springer, Cham (2017). https://doi.org/10.1007/978-3-319-48012-1_12
7. Zeng, Q., et al.: Location selection of multiple logistics distribution center based on particle swarm optimization. In: Huang, D.-S., Bevilacqua, V., Premaratne, P. (eds.) ICIC 2016. LNCS, vol. 9771, pp. 651–658. Springer, Cham (2016). https://doi.org/10.1007/978-3-319-42291-6_65
8. Wang, Y., Ma, X.L., Wang, Y.H., Mao, H.J., Zhang, Y.: Location optimization of multiple distribution centers under fuzzy environment. J. Zhejiang Univ. Sci. A **13**(10), 782–798 (2012)
9. Chen, J.: Optimization route of food logistics distribution based on genetic and graph cluster scheme algorithm. Adv. J. Food Sci. Technol. **8**(5), 359–362 (2015)
10. Xin, K.J., Qin, Z.Y.: Study on logistics distribution route optimization based on clustering algorithm and ant colony algorithm. Logistics Eng. Manag. **9**(1), 1245–1250 (2014)

11. Wang, T.: Study on optimization of logistics distribution route based on chaotic PSO. Comput. Eng. Appl. **47**(29), 218–221 (2011)

12. Hua, X., Hu, X., Yuan, W.: Research optimization on logistics distribution center location based on adaptive particle swarm algorithm. Optik Int. J. Light Electron Opt. **127**(20), 8443–8450 (2016)

13. Yin, P.Y., Chuang, Y.L.: Adaptive memory artificial bee colony algorithm for green vehicle routing with cross-docking. Appl. Math. Model. **40**(21–22), 9302–9315 (2016)

14. Kaveh, A., Mahdavai, V.R.: Colliding bodies optimization: a novel meta-heuristic method. Comput. Struct. **139**, 18–27 (2014)

15. Ma, X., Ma, X., Xu, M., et al.: Two-echelon logistics distribution region partitioning problem based on a hybrid particle swarm optimization-genetic algorithm. Expert Syst. Appl. **42**(12), 5019–5031 (2015)

Application of Ant Colony Algorithms to Solve the Vehicle Routing Problem

Mei-xian Song[1], Jun-qing Li[1,2,3,4(✉)], Li Li[1], Wang Yong[1],
and Pei-yong Duan[2(✉)]

[1] School of Computer, Liaocheng University, Liaocheng 252059, China
lijunqing.cn@gmail.com
[2] School of Information, Shandong Normal University, Jinan 250014, China
duanpeiyong@sdnu.edu.cn
[3] China Key Laboratory of Computer Network and Information Integration,
Southeast University, Ministry of Education,
Nanjing 211189, People's Republic of China
[4] State Key Laboratory of Synthetical Automation for Process Industries,
Northeastern University, Shenyang 110819, China

Abstract. Many optimization problems exist in the world. The Vehicle Routing Problem (VRP) is a relatively complex and high-level issue. The ant colony algorithm has certain advantages for solving the capacity-based vehicle routing problem (CVRP), but is prone to local optimization and high search speed problems. To solve these problems, this paper proposes an adaptive hybrid ant colony optimization algorithm to solve the vehicle routing problem with larger capacity. The adaptive hybrid ant colony optimization algorithm uses a genetic algorithm to adjust the pheromone matrix algorithm, designs an adaptive pheromone evaporation rate adjustment strategy, and uses a local search strategy to reduce computation. Experiments on some classic problems show that the proposed algorithm is effective for solving vehicle routing problems and has good performance. In the experiment, the results of different scale issues were compared with previously published papers. Experimental results show that the algorithm is feasible and effective.

Keywords: Ant colony system · Genetic algorithm · Shortest path
Vehicle routing problem

1 Introduction

In the past few decades, due to the development of global transportation and logistics, our lives have undergone major changes. For local products that are sold to other countries or cities, transportation and logistics costs are indispensable. In fact, recent research data shows that transport and logistics costs typically account for more than 20% of product value [1]; logistics systems play a growing and indispensable role in daily economic life. Despite this, many negative effects such as air pollution, noise and traffic accidents occur [2]. Despite the inevitable impact of transportation and logistics on daily life, efficient vehicle routing based on optimized algorithms can minimize negative impacts and corporate logistics costs. This is because decreasing the distance

© Springer International Publishing AG, part of Springer Nature 2018
D.-S. Huang et al. (Eds.): ICIC 2018, LNCS 10954, pp. 831–840, 2018.
https://doi.org/10.1007/978-3-319-95930-6_83

traveled by a vehicle increases the efficiency of the vehicle and the driver. In addition, the algorithm can improve customer service quality, reduce exhaust emissions, and improve vehicle scheduling efficiency [3]. Therefore, study of the vehicle routing problem is an important topic that has attracted the attention of scholars in the past decades [4]. It is of great practical significance and theoretical value to study different types of vehicle routing problems and their solving algorithms to meet the actual needs of production and operation management.

The ant colony algorithm is a kind of intelligent optimization algorithm. It is mainly an algorithm that uses the positive feedback parallel mechanism and deter-mines the shortest path of food and nest in the shortest time based on mutual cooperation among ants. Ant colony algorithm has the advantages of fast solution speed and strong parallelism. It has been widely used in water transport, railway, highway and other fields in recent years. In this paper, the ant colony algorithm is also applied to the logistics distribution vehicle path optimization problem. The mathematical model of VRP is firstly described, the solution objectives and constraints are described, and the method of solving the vehicle routing problem by using improved ant colony algorithm is proposed.

2 About Ant Colony Algorithm Related Theory

In most studies on VRP, researchers have defined some basic information about the customer location and needs, available vehicles, etc. This information is known before the service begins. However, in the actual service process, the VRP is dynamic; that is, the customer needs and arrangements gradually change over time, although some of the customer needs may be known in advance before beginning the service. In addition, VRP is an NP problem, so traditional precision algorithms (linear programming, dynamic programming, greedy algorithms, etc.) are difficult to solve under time constraints. However, modern optimization techniques (ACO [13, 14], Genetic Algorithms (GA) [15–17], Particle Swarm Optimization [18], etc.) are high quality solutions (although not exact) and are the most suitable methods for solving DVRP. Among these methods, ACO is a classic and effective heuristic algorithm. ACO is a classic bionic algorithm, which is inspired by the process of observing the foraging behavior of ant colonies. Ants exchange information by secreting pheromones (a special chemical substance) in the environment. By sensing the pheromone concentration, ants can choose the correct way to reach the food source. This behavior attracted researcher's attention and researchers created an artificial ant system to solve combinatorial optimization problems [19]. The original ACO was proposed by Marco Dorigo in 1991 and was called the ant system. The ant colony algorithm is a random search algorithm based on the study of the collective nature of the ant community, behavior of leaving to find food, simulating the entire process of the real ant colony to find food and determining the shortest foraging process of the ant colony search path. The information plays an important role through the ant pheromone. The role and approach increases, and positive feedback over time will allow more ants to cluster on the shortest possible path.

A food source and path diagram is shown in Fig. 1.

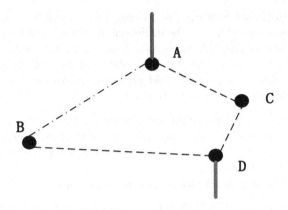

Fig. 1. Schematic diagram of ant colony foraging

In Fig. 1, the hypothesis in each Δt time interval is n ant moves from point A to point D, the ants crawl at a unit speed, the ant can be in a unit length of path to leave the element information of a unit. Assume that at moment t = 0, path A-B-D and path A-C-D information is 0, so the ant moving in the point A to point D path has the same probability as ant path A-B-D. It is also the only ants walking path A-C-D.

According to the above principle, when the ant colony has multiple feeding paths, the ant colony converges on the shortest path radially according to the change in pheromone concentration [21] until all ants pass the shortest path.

The Ant Colony Algorithm is a simulated evolutionary algorithm inspired by nature and behavior of ant colonies. In recent years, many scholars have conducted in-depth research on CVRP using the Ant Colony Algorithm and presented classic algorithms, such as the Saving Ant Algorithm [22], sweep based Multiple Ant Colonies Algorithm [23], hybrid meta-heuristics for the vehicle routing problem [24–26], and MAX–MIN Ant System [27]. These are very popular and perform well in many applications. Although the current Ant Colony Algorithms have certain advantages for solving CVRP, they are prone to local optimization and slow search speed. To overcome these deficiencies, this paper proposes an adaptive hybrid ant colony algorithm with high search performance (named AHASAC) and validates its effectiveness for solving CVRP in comparative experiments.

3   Improved Ant Colony Algorithm

3.1   Key Improvements of AHASAC

In the proposed AHASAC, the following strategies are used: First, the genetic algorithm is introduced to global convergence, the pheromone matrix of the ant colony algorithm is adjusted to avoid the local optimal solution, and the pheromone adjustment strategy is used to prevent the phenomenon of population loss and improve results. Second, a local search strategy is introduced to reduce the number of calculations and improve search speed.

Pheromone Adjustment Strategy. *Methodology of intermingling Ant Algorithm with Genetic Algorithm.* Different from the traditional algorithm in which the intermingled Ant Algorithm and Genetic Algorithm switch with each other with a fixed iteration number, AHASAC switches the Ant Algorithm and Genetic Algorithm with Ant Algorithm evolution stagnation to avoid premature convergence and local optimal cutting. The main integration process is as follows:

Step 1: Let the Ant Algorithm run for a certain number of iterations;
Step 2: Determine if the Ant Algorithm has reached the state of evolutionary stagnation. If it is in an evolutionary stagnation state, the genetic algorithm is called to optimize the selected ant;
Step 3: Update the path selected by the optimized ant into the pheromone table.

Improvement based on MMAS (1). To accelerate the convergence rate of the algorithm, only the best offspring or offspring from each generation can be allowed to establish the optimal path release pheromone. Therefore, the modified pheromone trail update rule is given by $\tau_{ij}^{new} = \rho_{ga} \tau_{ij}^{old} + (1 - \rho_{ga}) \Delta \tau_{ij}$, where ρ_{ga} is the residual parameter of the genetic algorithm for the pheromone update, and

$$\Delta \tau_{ij} = \begin{cases} \frac{1}{d_{ij}^{\frac{2}{3}}}, & \text{if } d_{ij} \text{ is the best in current generation} \\ 0, & \text{otherwise} \end{cases} \quad (1)$$

(2) The pheromone update method is based on the maximum and minimum pheromone method, and the pheromone concentration range on the path is limited. For any τ_{ij}, it holds $\lim_{t \to \infty} \tau_{ij}(t) = \tau_{ij} \leq \frac{1}{1-\rho} \frac{1}{f(s^{opt})}$, there are explicit limits τ_{min} and τ_{max} on the minimum and maximum pheromone trails such that for all pheromone trails $\tau_{ij(t)}$, $\tau_{min} \leq \tau_{ij(t)} \leq \tau_{max}$,

$$\tau_{max} = \frac{1}{(1-\rho)L_{global}} \quad (2)$$

$$\tau_{min} = \frac{\tau_{max}}{5} \quad (3)$$

After iteration, the pheromone trail must be guaranteed to respect the limits. If we have $\tau_{ij}(t) = \tau_{max}$; similarly, if $\tau_{ij}(t) \leq \pi_{min}$, we set $\tau_{ij} = (\tau_{min} + \tau_{max})/2$.

(3) The pheromone initial value is set as the reciprocal of the distance between two cities, so that the algorithm converges faster in the initial stage.

Adaptive Pheromone Evaporation Rate. Normally, updating the pheromone using the Genetic Algorithm can help avoid the local optima in the Ant Colony Algorithm, but this causes the Ant Colony Algorithm to lose its outstanding factor and destroys the original pheromone table created by the Ant Colony Algorithm. Therefore, we propose a new adaptive evaporation rate to update the pheromone. In this method, the amount of updated pheromone in the early stage of the Genetic Algorithm is decreased, so that the volatility rate of the pheromone gradually increases in the genetic algorithm. The pheromone evaporation rate ρ_{ga} formula $\rho_{ga}^{new} = \rho_{ga}^{new} + \Delta t$. In this equation, ρ_{ga}^{new} is the evaporation rate of the new pheromone, and ρ_{ga}^{new} is the evaporation rate of the old pheromone, and Δt is the increment of the pheromone evaporation rate.

3.2   Genetic Algorithm Design

Chromosome Encoding. In the proposed AHASAC, the natural number coding method is used for chromosome coding in the genetic algorithm. For example, the chromosome "0123045060" represents the route for the transportation tasks of six customers of three cars. The three paths are as follows:

Path 1: Warehouse → Customer 1 → Customer 2 → Client 3 → Warehouse
Path 2: Warehouse → Client 4 → Client 5 → Warehouse
Path 3: Warehouse → customer 6 → Warehouse

As shown in the above example, "0" is a warehouse. There are four "0"s that divide the chromosome into three sections representing the three paths.

Cross Operation. Due to CVRP constraint conditions, using ordinary cross operation will have a high probability of producing many offspring that do not comply with constraints, or result in a loss of good genes. This can cause the algorithm search results to deteriorate. To overcome this drawback, SHASA uses a modified maximum retention crossover scheme to ensure the probability of a good gene fragment, as described in the following procedure:

(1)   Randomly select two adjacent zero-crossover positions;
(2)   Place the segment between two "0"s into the header of the new offspring, then, compare the two selected cross-segments, and add the non-repeated elements to the end of the new offspring;
(3)   Exchange the position of the non-selected parts of the two cross-genes and add the non-zero elements to the end of the new offspring arranged in the original order;
(4)   Insert zero operation for two new offspring after step (3). Zero is inserted at the end of each chromosome to ensure that zero is the end element of the chromosomes.

3.3   AHASAC Algorithm

(1)   Initialize parameters. Set the iterations $N_c = 0$, and set the value for the maximum number of iterations N_{max} and the genetic algorithm iterations Nga; Place m ants in n cities, and set $\tau_{ij}^{begin} = 1 / d_{ij}$ and the initial $\Delta\tau_{ij} = 0$ and ant taboo table index number $k = 1$.
(2) $N_c = N_c + 1, k = k + 1$.
(3)   The ant selects city j with a transition probability and estimates whether it meets the load constraints. If it does, then run city j, otherwise it will go back to the warehouse.
(4)   The ant moves to the selected city which is added to the ant's taboo Table
(5)   Orderly traverse all cities.

(6) Update the pheromone based on the maximum and minimum pheromone strategy.

(7) Determine whether the Ant Algorithm has reached the status of evolution stagnation. If it is in the status of evolution stagnation go to step (8), otherwise it will go to step (12).

(8) Use roulette to select a certain ant population.

(9) The Genetic Algorithm runs to optimize the ant path. If the offspring is better than the parents after genetic operation, then replace the parents.

(10) If it meets the final conditions of the Genetic Algorithm, the Genetic Algorithm stops.

(11) Update the path selected by the optimized ant in the pheromone table.

(12) $N_c \geq N_{max}$, the algorithm ends and the result is output, otherwise the taboo table is cleared and reset $k = 1$, and goes to step (2) (See Fig. 2).

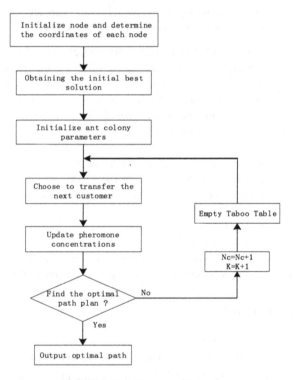

Fig. 2. Algorithm step diagram

4 Computational Experiments

In this section, experiments using AHASAC for selected CVRP problems are designed to demonstrate the effectiveness of AHASAC. Simulation environment: Windows7 operating system, 6 GB memory; Matlab R2013B simulation software. The main parameters of the experiment were set as follows: ant number = 34, pheromone = 3,

$\pi_{max} = 4$, training number = 1000, Genetic algorithm population = 40, Genetic algorithm iterations = 50, number of nearby cities = 25, and ant stagnation time = 50.

4.1 Test for CVRP Problem Presented in Literature

Table 1 shows the result of the proposed algorithm for 30 averaged runs and other results for solving the Capacitated Vehicle Routing Problem cited in references [29, 31].

Table 1. Comparison of different algorithms

Problem scale	Sweep algorithm	Gaskell's saving approach	Direct tree-search approach	Set partitioning algorithm	Location based algorithm	Genetic algorithm
50	585	532	524.9	524	534	521
100	886	851	832.9	833	832.9	840.5

As shown in Table 1, other than the set partitioning algorithm and genetic algorithm and the scanning algorithm, the proposed algorithm performs better than five of the seven algorithms mentioned in [29, 31] when the problem scale is 50. Its performance is superior to all other algorithms. The problem size of these algorithms is 100. This shows that when the problem scale is increased, the proposed algorithm performs better. Therefore, it performs better in more complex problems.

4.2 Test for International Benchmark

In this section, the proposed AHASAC method is compared with the classic ant colony algorithm experiment for solving the solomon100 problem of the VRP benchmark to show its superiority. The paper randomly selects four test cases. They are a101, ac101, b101 and b201. The results are shown in Table 2.

Table 2. Results of AHASAC for solving VRP benchmarks

Test case	Classic Ant Colony Algorithm	Proposed AHASAC
a101	885.118	880.247
ac101	1026.498	1024.36.
b101	827.946	827.787
b201	587.857	586.631

As shown in Table 2, the proposed algorithm outperforms the classic Ant Colony Algorithm for all four random test cases from the solomon100. This shows the proposed algorithm is effective for solving the benchmarks.

5 Conclusion

The artificial ant colony algorithm is a type of swarm intelligence optimization algorithm. Its greatest advantage is a positive feedback mechanism, namely, the information of the renewal process and the algorithm of the applicable conditions: an example is needed to solve the problem. Define a positive feedback process; stimulate functionality; and construct a constraint mechanism. The VRP problem coincides with this formulation. The empirical results show that the artificial ant colony algorithm is feasible for solving the VRP problem, the proposed algorithm has better convergence and a better solution. The proposed adaptive ant colony algorithm combines pheromone adjustment strategies, a selection strategy based on genetic algorithms and a node-based selection strategy, thus reducing the complexity of the search process. Finally, the computer simulation experiments confirm the correctness and validity of this method. However, due to the lack of strict mathematical proofs, this method still has disadvantages, so some parameters are determined empirically. However, this method has great potential for solving other NP-hard problems.

Acknowledgement. This research is partially supported by National Science Foundation of China under Grant 61773192, 61773246, 61603169 and 61503170, Shandong Province Higher Educational Science and Technology Program (J17KZ005), Key Laboratory of Computer Network and Information Integration (Southeast University), Ministry of Education (K93-9-2017-02), and State Key Laboratory of Synthetical Automation for Process Industries (PAL-N201602).

References

1. Braekers, K., Ramaekers, K., Nieuwenhuyse, I.V.: The vehicle routing problem: state of the art classification and review. Comput. Ind. Eng. **99**, 300–313 (2016)
2. Xu, Y., Wang, L., Yang, Y.: Dynamic vehicle routing using an improved variable neighborhood search algorithm. J. Appl. Math. **2013**(1), 1–21 (2013)
3. Lin, S., Kernighan, B.W.: An effective heuristic algorithm for the Traveling-Salesman Problem. Oper. Res. **21**(2), 498–516 (1973)
4. Hinton, G.T.: A thesis regarding the vehicle routing problem including a range of novel techniques for its solution. University of Bristol (2010)
5. Campbell, A.M., Wilson, J.H.: Forty years of periodic vehicle routing. Networks **63**(3), 2–15 (2014)
6. Ohlmann, J.W., Thomas, B.W.: A compressed-annealing heuristic for the traveling salesman problem with time windows. Inf. J. Comput. **19**(1), 80–90 (2017)
7. Wang, Z., Zhou, C.: A three-stage saving-based heuristic for vehicle routing problem with time windows and stochastic travel times. Discrete Dynamics in Nature and Society (2016)
8. Chávez, M.A.C., Martinez-Oropeza, A.: Feasible initial population with genetic diversity for a population-based algorithm applied to the vehicle routing problem with time windows. Math. Prob. Eng. **2016**(5), 1–11 (2016)
9. Prodhon, C., Prins, C.: Metaheuristics for Vehicle Routing Problems. Wiley-IEEE Press, Chichester (2016)
10. Caceres-Cruz, J., Arias, P., Guimarans, D., et al.: Rich vehicle routing problem: survey. ACM Comput. Sur. **47**(2), 32 (2015)

11. Yua, B., Yao, B.: An improved ant colony optimization for vehicle routing problem. Eur. J. Oper. Res. **196**(1), 171–176 (2009)
12. Chen, C.-H., Ting, C.-J.: An improved ant colony system algorithm for the vehicle routing problem. J. Chin. Inst. Ind. Eng. **23**(2), 115–126 (2006)
13. Hanshar, F.T., Ombuki-Berman, B.M.: Dynamic vehicle routing using genetic algorithms. Appl. Intell. **27**(1), 89–99 (2007)
14. Cheng, A., Yu, D.: Genetic algorithm for vehicle routing problem. In: International Conference on Transportation Engineering, pp. 2876–2881 (2013)
15. Abdallah, A.M.F.M., Essam, D.L., Sarker, R.A.: On solving periodic re-optimization dynamic vehicle routing problems. Appl. Soft Comput. **55**, 1–12 (2017)
16. Okulewicz, M., Mańdziuk, J.: Application of particle swarm optimization algorithm to dynamic vehicle routing problem. In: Rutkowski, L., Korytkowski, M., Scherer, R., Tadeusiewicz, R., Zadeh, Lotfi A., Zurada, Jacek M. (eds.) ICAISC 2013. LNCS (LNAI), vol. 7895, pp. 547–558. Springer, Heidelberg (2013). https://doi.org/10.1007/978-3-642-38610-7_50
17. Chandra, M.B., Baskaran, R.: Review: a survey: ant colony optimization based recent research and implementation on several engineering domain. Exp. Syst. Appl. **39**(4), 4618–4627 (2012)
18. Bonabeau, E., Dorigo, M., Theraulaz, G.: Inspiration for optimization from social insect behavior. Nature **406**(6791), 39–42 (2000)
19. Shi-Yong, L.I.: Progresses in ant colony optimization algorithm with applications. Comput. Aut. Mea. Control **11**(12), 910–911 (2003)
20. Wu, Q.H., Zhang, J.H., Xu, X.H.: An ant colony algorithm with mutation features. J. Comput. Res. Dev. **36**(10), 1240–1245 (1999)
21. Xu, H., Pu, P., Duan, F.: Dynamic vehicle routing problems with enhanced ant colony optimization. Discrete Dyn. Nat. Soc., 1–13 (2018)
22. Toth, P., Vigo, D.: The Vehicle Routing Problem. Tsinghua University Press, Siam (2011)
23. Figliozzi, M.A.: An iterative route construction and improvement algorithm for the vehicle routing problem with soft time windows. Trans. Res. Part C **18**(5), 668–679 (2010)
24. Bullnheimer, B., Hartl, R.F.: An improved Ant system algorithm for the vehicle routing problem. Ann. Oper. Res. **89**, 319–328 (1999)
25. Liu, Z., Cai, Y.: Sweep based multiple ant colonies algorithm for capacitated vehicle routing problem. In: IEEE International Conference E-business Engineering, pp. 387–394. IEEE Computing Society (2005)
26. Breedam, A.V.: An analysis of the effect of local improvement operators in genetic algorithms and simulated annealing for the vehicle routing problem (1996)
27. Jiang, D.: A study on the genetic algorithm for vehicle routing problem. Syst. Eng. Pract. **19**(6), 40–45 (1999)
28. Lin, S.W., Lee, Z.J., Ying, K.C., et al.: Applying hybrid meta-heuristics for capacitated vehicle routing problem. Expert Syst. Appl. **36**(2), 1505–1512 (2009)
29. Thomas, S., Holger, H.H.: MAX–MIN ant system. Future Gener. Comput. Syst. **16**, 889–914 (2000)
30. Li, J.Q., Pan, Q.K., Tasgetiren, M.F.: A discrete artificial bee colony algorithm for the multi-objective flexible job-shop scheduling problem with maintenance activities. Appl. Math. Model. **38**(3), 1111–1132 (2014)
31. Li, J.Q., Pan, Q.K.: Chemical-reaction optimization for solving fuzzy job-shop scheduling problem with flexible maintenance activities. Int. J. Prod. Econ. **145**(1), 4–17 (2013)
32. Li, J.Q., Pan, Q.K., Mao, K.: A discrete teaching-learning-based optimisation algorithm for realistic flowshop rescheduling problems. Eng. Appl. Artif. Intell. **37**(1), 279–292 (2015)

33. Li, J.Q., Pan, Q.K., Gao, K.Z.: Pareto-based discrete artificial bee colony algorithm for multi-objective flexible job shop scheduling problems. Int. J. Adv. Manuf. Techol. **55**(9–12), 1159–1169 (2011)

34. Han, Y.Y., Gong, D.W., Sun, X.Y., et al.: An improved NSGA-II algorithm for multi-objective lot-streaming flow shop scheduling problem. Int. J. Prod. Res. **52**(8), 2211–2231 (2014)

35. Li, J.Q., Pan, Q.K., Kun, M.: A hybrid fruit fly optimization algorithm for the realistic hybrid flowshop rescheduling problem in steelmaking systems. IEEE Trans. Autom. Sci. Eng. **13** (2), 932–949 (2016)

36. Li, J.Q., Pan, Q.K., Chen, J.: A hybrid Pareto-based local search algorithm for multi-objective flexible job shop scheduling problems. Int. J. Prod. Res. **50**(4), 1063–1078 (2012)

37. Li, J.Q., Pan, Q.K., Suganthan, P.N., Chua, T.J.: A hybrid tabu search algorithm with an efficient neighborhood structure for the flexible job shop scheduling problem. Int. J. Adv. Manuf. Techol. **59**(4), 647–662 (2011)

38. Li, J.Q., Pan, Y.: A hybrid discrete particle swarm optimization algorithm for solving fuzzy job shop scheduling problem. Int. J. Adv. Manuf. Techol. **66**(1–4), 583–596 (2013)

39. Li, J.Q., Pan, Q.K., Duan, P.Y.: An improved artificial bee colony algorithm for solving hybrid flexible flow shop with dynamic operation skipping. IEEE Trans. Cybern. **46**(6), 1311–1324 (2016)

40. Li, J.Q., Pan, Q.K.: Solving the large-scale hybrid flow shop scheduling problem with limited buffers by a hybrid artificial bee colony algorithm. Inf. Sci. **316**(20), 487–502 (2015)

41. Gong, D., Han, Y., Sun, J.: A novel hybrid multi-objective artificial bee colony algorithm for the blocking lot-streaming flow shop scheduling problems. Knowl. Sys. **148**, 115–130 (2018)

42. Li, J.Q., Pan, Q.K., Xie, S.: An effective shuffled frog-leaping algorithm for multi-objective flexible job shop scheduling problems. Appl. Mat. Comput. **218**(18), 9353–9371 (2012)

43. Li, J.Q., Pan, Q.K., Liang, Y.C.: An effective hybrid tabu search algorithm for multi-objective flexible job shop scheduling problems. Comput. Ind. Eng. **59**(4), 647–662 (2010)

44. Li, J.Q., Pan, Q.K., Mao, K., Suganthan, P.N.: Solving the steelmaking casting problem using an effective fruit fly optimisation algorithm. Knowl. Based Syst. **72**(5), 28–36 (2014)

45. Li, J.Q., Pan, Q.K.: Chemical-reaction optimization for flexible job-shop scheduling problems with maintenance activity. Appl. Soft Comput. **12**(9), 2896–2912 (2012)

46. Li, J.Q., Pan, Q.K., Wang, F.T.: A hybrid variable neighborhood search for solving the hybrid flow shop scheduling problem. Appl. Soft Comput. **24**, 63–77 (2014)

47. Han, Y., Gong, D., Jin, Y.C., Pan, Q.K.: Evolutionary multi-objective blocking lot-streaming flow shop scheduling with machine breakdowns. IEEE Trans. Cybern. **99**, 1–14 (2017)

48. Li, J.Q., Pan, Q.K., Xie, S.X., Wang, S.: A hybrid artificial bee colony algorithm for flexible job shop scheduling problems. Int. J. Comput. Commun. Control. **6**(2), 267–277 (2011)

An Mutational Multi-Verse Optimizer with Lévy Flight

Jingxin Liu and Dengxu He[(✉)]

College of Science, Guangxi University for Nationalities,
Nanning 530006, China
dengxuhe@126.com

Abstract. This paper proposes a mutational Multi-Verse Optimizer (MVO) algorithm based on Lévy flight and called LMVO algorithm. The random steps of Lévy flight enhances the ability of the search individual to escape the local optimum, and promotes the balance of exploration and exploitation for MVO algorithm. For investigate the availability of LMVO, add basic MVO algorithm and other four mainstream algorithms to compare with it on six high dimensional test functions and two fixed-dimensional test functions. Furthermore, apply it to cantilever beam design problem. These final results proved that LMVO has good convergence accuracy and stability.

Keywords: Lévy flight · Multi-Verse Optimizer · Test functions
Cantilever beam design problem

1 Introduction

In 2016, Mirjalili et al. proposed a new meta-heuristic algorithm called Multi-Verse Optimizer (MVO) [1]. MVO is inspired by three concepts in the multi verse, white holes, black holes, and wormholes. Because MVO algorithm has some characteristics, such as simple structure, less parameter adjustment, easy implementation and higher search efficiency, it has been diffusely researched by scholars in recent two years. On the application side, MVO has used in training feedforward neural networks [2], parameter extraction [3] and identifying [4], the voltage stability analysis [5], feature selection [6, 7], On the improvement side, Jangir et al. have proposed a novel hybrid PSO with MVO [8], Mirjalili et al. have used the multi verse optimization algorithm to solve optimization of problems with multiple objectives [9].

MVO as an advanced algorithm, it still has some problems in convergence precision and stability. The randomness of Levy flight [10] can improves the global search ability of algorithms and avoids the local optimization. At present, Lévy flight has been successfully applied in many intelligent optimization algorithms based on simulating animals and insects behavior, such as CS [11], PSO [12], FA [13], ABC [14]. These research studies confirm that Lévy flight can greatly improve the performance of the optimization algorithms. A novel MVO algorithm with Lévy flight called the LMVO is presented in this paper.

© Springer International Publishing AG, part of Springer Nature 2018
D.-S. Huang et al. (Eds.): ICIC 2018, LNCS 10954, pp. 841–853, 2018.
https://doi.org/10.1007/978-3-319-95930-6_84

The following part of the paper is carried out as follows. Section 2 describes the original MVO algorithm. Section 3 gives the LMVO algorithm. The experimental simulation and analysis are presented in Sect. 4. Section 5 gives a conclusion for the paper.

2 Basic Multi-Verse Optimizer (MVO)

MVO is inspired by the multi-verse theory in physics, which simulates the motion of objects in the multi-verse with the effection of white holes, black holes and wormholes. First, each universe has an inflation rate, which has major significance for the changes the position of the planets and the creation of white holes, black holes, wormholes in the universe. And then, white holes, black holes and wormholes perform their own functions, respectively: white holes send objects, black holes attract objects, and wormholes are tunnels that connect different parts of these universes to transport objects. Finally, the whole multi-verse population reached a relatively stable situation due to the interaction about three concepts. The formulaic form of this process is as follows.

First, initialize multi-verse population size and position:

$$
U = \begin{bmatrix} U_1 \\ U_2 \\ \vdots \\ U_n \end{bmatrix} = \begin{bmatrix} x_1^1 & x_1^2 & \cdots & x_1^d \\ x_2^1 & x_2^2 & \cdots & x_2^d \\ \vdots & \vdots & \vdots & \vdots \\ x_n^1 & x_n^2 & \cdots & x_n^d \end{bmatrix} \tag{1}
$$

where n denotes the number of universes (candidate solutions) and d denotes the number of objects (variables), $U_i = (x_i^1, x_i^2, \cdots, x_i^d)$ shows the i-th universe position.

$$
x_i^j = \begin{cases} x_k^j & r1 \leq NI(U_i) \\ x_i^j & r1 \geq NI(U_i) \end{cases} \tag{2}
$$

where x_i^j indicates the position of j-th object in the i-th universe, $NI(U_i)$ is normalized inflation rate of the i-th universe position, $r1$ is a random value in [0, 1], and x_k^j indicates the j-th object in k-th universe selected by a roulette wheel selection mechanism.

And then, each universe is going to be disturbed by the action of the wormholes to make objects perform movement in all universes. The implementation formula is as follows:

$$
x_i^j = \begin{cases} \begin{cases} X_j + TDR \times ((ub_j - lb_j) \times r4 + lb_j) & r3 < 0.5 \\ X_j - TDR \times ((ub_j - lb_j) \times r4 + lb_j) & r3 \geq 0.5 \end{cases} & r2 < WEP \\ x_i^j & r2 \geq WEP \end{cases} \tag{3}
$$

where X_j indicates the position of j-th object in the current optimal universe, lb_i and ub_i are the lower and upper bound of x_i^j, and r_2, r_3, r_4 are random numbers in [0, 1]. WEP (Wormhole existence probability) and TDR (Travelling distance rate) are two important parameters, their formula can be expressed as follows:

$$WEP = WEP_{min} + l \times \left(\frac{WEP_{max} - WEP_{min}}{L} \right) \tag{4}$$

$$TDR = 1 - \frac{l^{1/p}}{L^{1/p}} \tag{5}$$

where $WEP_{min} = 0.2$ $WEP_{max} = 1$, l and L indicate the current iteration and the maximum iteration. p is exploitation accuracy with typical value of 6.

3   MVO Algorithm with Lévy Flight (LMVO)

3.1   Lévy Flight

Lévy flight is a way of search path with random walk steps. The way of search path search path combines short-distance search and occasional long-distance search, which can be used to explain many random phenomena. In nature, the foraging behavior of many animals and insects is consistent with Lévy distribution. Lévy distribution can be represented by the following formula:

$$Levy(\lambda) \sim u = t^{-\lambda}, 1 < \lambda \leq 3 \tag{6}$$

Mantegna algorithm is usually used to simulate Lévy distribution in mathematical calculation. The mathematical formula of Mantegna algorithm as follows:

$$s = \frac{\mu}{|v|^{1/\beta}} \tag{7}$$

where s is the step length of the Lévy flight and $\beta = 1.5$. It's worth noting that λ in Eq. (6) follows $\lambda = 1 + \beta$. Both μ and v in Eq. (7) are obey a normal stochastic distribution:

$$\mu = N(0, \sigma_\mu^2) \tag{8}$$

$$v = N(0, \sigma_v^2) \tag{9}$$

The value of σ_μ and σ_v can be calculated by the following formulas:

$$\sigma_\mu = \left[\frac{\Gamma(1+\beta) \times \sin(\pi \times \beta/2)}{\Gamma(1+\beta/2) \times \beta \times 2^{(\beta-2)/2}}\right] \tag{10}$$

$$\sigma_v = 1 \tag{11}$$

3.2 Improvement Strategy and Flowchart

After the position of every universe is changed by the wormhole, their states are extremely unstable and come into mutation pattern. At this time, their internal objects will be perform movement with randomness and uncertainty in the entire multi verse space, not necessarily in the directions face to the current optimal universe and other

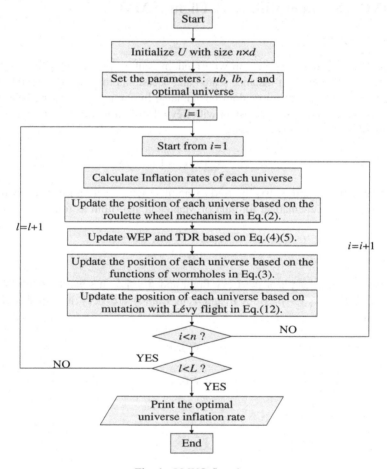

Fig. 1. LMVO flowchart

universes. In order to better balance the exploration and exploitation of MVO algorithm, we introduce Lévy flight into the location updating of each universe by these mutation.

$$U_i = U_i + \mu sign[rand - 1/2] \oplus Levy(\lambda) \tag{12}$$

where U_i shows the i-th universe position, μ is a random number that follow a uniform distribution, the product $\oplus$ denotes entrywise multiplication, and rand is a random number in $[0, 1]$. It is worth noticing that $sign[rand - 1/2]$ has only three values 1, 0 and -1.

The flowchart for the LMVO algorithm is presented in Fig. 1.

4   Experimental Results and Discussion

To evaluate the effective performance of LMVO, we selected eight test functions in Table 1. They can be classified into three categories: unimodal test functions (f_1, f_2, f_3), multimodal test functions (f_4, f_5, f_6), and fixed-dimensional multimodal test functions (f_7, f_8).

Table 1. Unimodal benchmark functions

Function	Dim	Range	f_{min}
$f_1(x) = \sum\limits_{i=1}^{n}|x_i| + \prod\limits_{i=1}^{n}|x_i|$	50	$[-10,10]$	0
$f_2(x) = \sum\limits_{i=1}^{D-1}[100(x_{i+1} - x_i^2)^2 + (x_i - 1)^2]$	50	$[-30,30]$	0
$f_3(x) = \sum\limits_{i=1}^{n}x_i^4 + random(0,1)$	50	$[-1.28,1.28]$	0
$f_4(x) = \sum\limits_{i=1}^{n}[x_i^2 - 10\cos(2\pi x_i) + 10]$	50	$[-5.12,5.12]$	0
$f_5(x) = -20\exp\left(-0.2\sqrt{\frac{1}{n}\sum\limits_{i=1}^{n}x_i^2} - \exp\left(\frac{1}{n}\sum\limits_{i=1}^{n}\cos 2\pi x_i\right)\right) + 20 + e$	50	$[-32,32]$	0
$f_6(x) = \frac{1}{4000}\sum\limits_{i=1}^{n}(x_i^2) - \prod\limits_{i=1}^{n}\cos\left(\frac{x_i}{\sqrt{i}}\right) + 1$	50	$[-600,600]$	0
$f_7(x) = \sum\limits_{i=1}^{11}[a_i - \frac{x_1(b_i^2 + b_i x_2)}{b_i^2 + b_i x_3 + x_4}]^2$	4	$[-5,5]$	0.0003075
$f_8(x) = 0.5 + \frac{\sin^2(\sqrt{x_1^2 + x_2^2}) - 0.5}{(1 + 0.001(x_1^2 + x_2^2))}$	2	$[-100,100]$	-1

4.1   Experimental Environment

Each step is performed in Windows 10 control system using Intel Core i7, 3.40 GHz, 8G RAM and Matlab R2016a.

Table 2. Results on benchmark functions

Functions	Output	Algorithms						Rank
		ABC	GSA	PSO	GWO	MVO	LMVO	
f_1	Best	15.6825	0.0272	1.5371	9.3E-13	2.558593	3.94E-85	1
	Worst	29.9950	4.7182	27.0521	6.77E-12	356153.2	1.71E-48	
	Mean	21.9370	1.2919	6.7887	2.81E-12	12039.73	**5.69E-50**	
	Std	3.6140	1.1223	6.3988	1.68E-12	64993.36	3.12E-49	
f_2	Best	258922.31	292.5174	221.4868	46.0641	148.5377	48.8683	2
	Worst	5637317.04	3742.4124	23805.13	48.6513	3368.422	49	
	Mean	2151897.65	991.3361	2147.784	**47.3814**	1010.847	48.9392	
	Std	1268517.69	660.5893	4402.57	0.8189	1008.789	0.0300	
f_3	Best	2.0973	0.1073	0.6784	0.0013	0.0277	5.16E-06	1
	Worst	14.3333	0.6401	2.6192	0.0075	0.1784	0.0015	
	Mean	4.8851	0.3315	1.4913	0.0034	0.1151	**0.0004**	
	Std	2.4966	0.1305	0.5408	0.0017	0.0301	0.0003	
f_4	Best	128.8471	34.8236	122.1591	1.14E-13	161.7884	0	1
	Worst	307.9311	86.5613	237.7097	12.4716	364.4188	0	
	Mean	217.3957	58.8197	175.8901	3.5962	268.8035	**0**	
	Std	38.5415	10.3713	30.1214	3.6812	51.6415	0	
f_5	Best	14.3207	0.0078	2.4999	1.07E-11	2.1043	8.88E-16	1
	Worst	18.1516	2.1631	19.0796	9.47E-11	4.4349	8.88E-16	
	Mean	16.1164	1.3286	5.9749	4.01E-11	3.1078	**8.88E-16**	
	Std	0.9385	0.5112	3.0135	1.94E-11	0.5081	0	
f_6	Best	9.5937	80.0843	0.3360	0	1.0490	0	1
	Worst	61.5778	181.5957	1.7970	0.0326	1.1353	0	
	Mean	29.8548	127.3535	0.8659	0.0072	1.0847	**0**	
	Std	10.2579	19.3941	0.3251	0.0111	0.0201	0	
f_7	Best	0.0006	0.0018	0.0003	0.0003	0.0004	0.0003	1
	Worst	0.0023	0.0150	0.0019	0.0204	0.0565	0.0016	
	Mean	0.0015	0.0049	0.0010	0.0038	0.0091	**0.0004**	
	Std	0.0004	0.0030	0.0003	0.0075	0.0128	0.0003	
f_8	Best	0.0004	0.0097	0	0	1.87E-07	0	1
	Worst	0.0101	0.1079	0.0097	0.0097	0.0097	0	
	Mean	0.0084	0.0255	0.0039	0.0045	0.0009	**0**	
	Std	0.0029	0.0266	0.0048	0.0049	0.0030	0	

4.2   Performance Comparison

In this paper, we used five algorithms to compare with LMVO in mainstream intelligent optimization algorithms. There are ABC, GSA, PSO, GWO and MVO, respectively. These six algorithms respective control parameters are given below.

ABC: limit = 10.
GSA: $\alpha = 20$, $G_0 = 100$.
PSO: $c1 = 1.4962$, $c2 = 1.4962$, $\omega = 0.7298$.
GWO: $\overrightarrow{a}$ Linearly decreased from 2 to 0.
MVO: $WEP_{max} = 1$, $WEP_{min} = 0.2$, $p = 6$.
LMVO: $WEP_{max} = 1$, $WEP_{min} = 0.2$, $p = 6$.

Above algorithms follow the rules: the population size is 30 and the maximum iteration number is 500. Each algorithm is run 30 times independently in the test functions, respectively. The results are shown in Table 2. In the table, the best fitness value, the worst fitness value, the average fitness value and the standard deviation are respectively replaced by Best, Worst, Mean and Std. The ranking of these algorithms is based on Mean.

4.3   Results and Analysis on Benchmark Functions

Table 2 shows all the test results, the average value of LMVO is the smallest. Among them, LMVO reached the theoretical global optimal value on f_4, f_6 and f_8. This point indicated that LMVO algorithm has good convergence accuracy. For f_2, although the average value found by GWO is superior to LMVO, LMVO gains the best stability. As can be seen from the above data, the stability of LMVO is the best compared with other algorithms. We have plotted corresponding convergent curves and anova for visually see their changing trends and stability (Figs. 2, 3, 4, 5, 6, 7, 8, 9, 10, 11, 12, 13, 14, 15, 16 and 17).

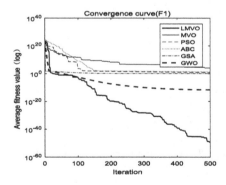

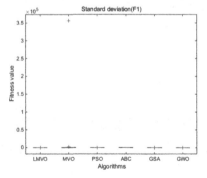

Fig. 2. Evolution curves of fitness value for f_1 **Fig. 3.** Anova test of global minimum for f_1

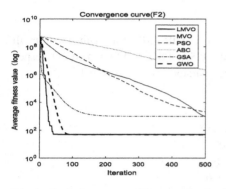

Fig. 4. Evolution curves of fitness value for f_2 **Fig. 5.** Anova test of global minimum for f_2

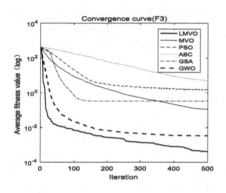

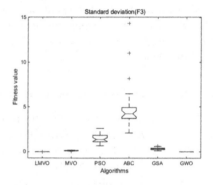

Fig. 6. Evolution curves of fitness value for f_3 **Fig. 7.** Anova test of global minimum for f_3

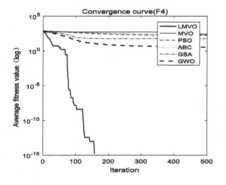

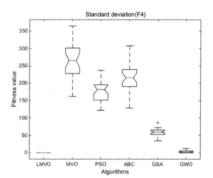

Fig. 8. Evolution curves of fitness value for f_4 **Fig. 9.** Anova test of global minimum for f_4

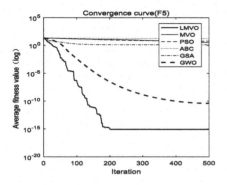

Fig. 10. Evolution curves of fitness value for f_5

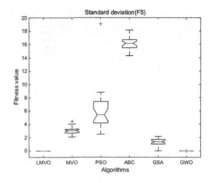

Fig. 11. Anova test of global minimum for f_5

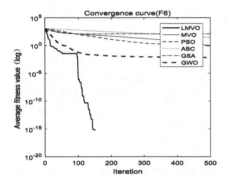

Fig. 12. Evolution curves of fitness value for f_6

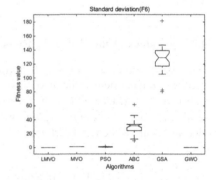

Fig. 13. Anova test of global minimum for f_6

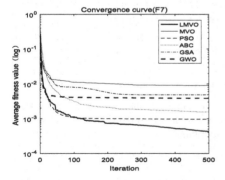

Fig. 14. Evolution curves of fitness value for f_7

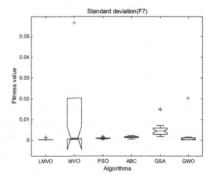

Fig. 15. Anova test of global minimum for f_7

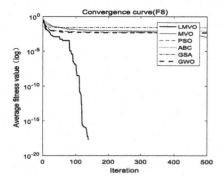

Fig. 16. Evolution curves of fitness value for f_8

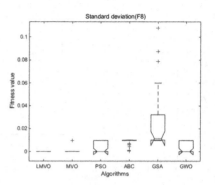

Fig. 17. Anova test of global minimum for f_8

4.4   p-Values of the Wilcoxon Rank-Sum Test

Wilcoxon's non-parametric statistical test returns a parameter value p-value that can be used to verify whether the two sets of solutions differ significantly or not in the statistical significance. Only when $p < 0.05$ can confirm the statistical significance of results. In Table 3, these $p - values$ are calculated to compare the values of other five algorithms in all the test functions.

Table 3. $p - values$ of the Wilcoxon rank-sum test results

Functions	LMVO vs ABC	LMVO vs GSA	LMVO vs PSO	LMVO vs GWO	LMVO vs MVO
f_1	3.02E−11	3.02E−11	3.02E−11	3.02E−11	3.02E−11
f_2	3.02E−11	3.02E−11	3.02E−11	3.02E−11	3.02E−11
f_3	3.02E−11	3.02E−11	3.02E−11	3.34E−11	3.02E−11
f_4	1.21E−12	1.21E−12	1.21E−12	1.21E−12	1.21E−12
f_5	1.21E−12	1.21E−12	1.21E−12	1.21E−12	1.21E−12
f_6	1.21E−12	1.21E−12	1.21Ev12	1.27E−05	1.21E−12
f_7	4.20E−10	3.02E−11	6.05E−07	0.08771	9.76E−10
f_8	1.21E−12	1.21E−12	1.23E−05	2.93E−05	1.21E−12

Noted that underline indicates $p \geq 0.05$. As can be seen from the figure: only one $p - value$ lager than 0.05 in compared with GWO, all other p-values were less than 0.05. Thus, there are significant differences between LMVO and other algorithms.

4.5   LMVO for Cantilever Beam Design Problem

In Fig. 18, cantilever structure consists of five hollow square blocks, the first block is fixed and the fifth one burdens a vertical load. Point 6 has a downward force, and point 1 has a fixed support. The objective is to minimize the weight of whole cantilever beam. The formulation of problem is as follows:

Consider $\vec{x} = [x_1 x_2 x_3 x_4 x_5]$;
Minimize $f(x) = 0.0624(x_1 + x_2 + x_3 + x_4 + x_5)$;
Subject to $g(x) = \frac{61}{x_1^3} + \frac{37}{x_2^3} + \frac{19}{x_3^3} + \frac{7}{x_4^3} + \frac{1}{x_5^3} \leq 1$;
Variable range $0.01 \leq x_1, x_2, x_3, x_4, x_5 \leq 100$

In order to verify the superiority of LMVO, Method of Moving Asymptotes (MMA) [16], Generalized Convex Approximation (GCA_I) [16], GCA_II [16], CS [17], and Symbiotic Organisms Search (SOS) [18] are used for comparison. Table 4 shows the results on 30 independent experiments.

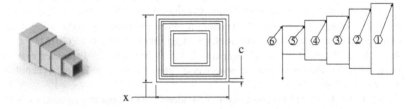

Fig. 18.   [15] Cantilever beam design problem

These data in Table 4 illustrate that LMVO can find a better optimal value than other algorithms. Therefore, LMVO algorithm is an effective tool in solving cantilever beam design problem.

Table 4. Comparison results for cantilever design problem

Algorithm	Optimal values for variables					Optimal cost
	x_1	x_2	x_3	x_4	x_5	
LMVO	5.898114	4.692344	4.554989	3.789407	2.076762	1.311125
MVO [1]	6.0239	5.3060	4.4950	3.4960	2.1527	1.33996
MMA [16]	6.0100	5.3000	4.4900	3.4900	2.1500	1.3400
GCA_I [16]	6.0100	5.3000	4.4900	3.4900	2.1500	1.3400
GCA_II [16]	6.0100	5.3000	4.4900	3.4900	2.1500	1.3400
CS [17]	6.0089	5.3049	4.5023	3.5077	2.1504	1.33999
SOS [18]	6.01878	5.30344	4.49587	3.49896	2.15564	1.33996

5 Conclusions and Future Works

This paper introduces an mutational Multi-Verse Optimizer (MVO) algorithm based on Lévy flight. Lévy flight is help to the improvement of the convergence accuracy and the ability to escape local optimal. The advantages of the proposed LMVO are proved by comparing with other algorithms on test functions. At the same time, LMVO is applied to cantilever beam design problem, which proves that LMVO is a very competitive improvement algorithm. In order to continue the works, we will try to use LMVO algorithm for solving real-world NP-hard problems.

Acknowledgment. This work is supported by Innovation Project of Guangxi Graduate Education under Grant No. gxun-chxzs2017135.

References

1. Mirjalili, S., Mirjalili, S.M., Hatamlou, A.: Multi-verse optimizer: a nature-inspired algorithm for global optimization. Neural Comput. Appl. **27**(2), 495–513 (2016)
2. Faris, H., Aljarah, I., Mirjalili, S.: Training feedforward neural networks using multi-verse optimizer for binary classification problems. Appl. Intell. **45**(2), 1–11 (2016)
3. Ali, E.E., El-Hameed, M.A., El-Fergany, A.A., El-Arini, M.M.: Parameter extraction of photovoltaic generating units using multi-verse optimizer. Sustain. Energy Technol. Assess. **17**, 68–76 (2016)
4. Fathy, A., Rezk, H.: Multi-verse optimizer for identifying the optimal parameters of PEMFC model. Energy **143**, 634–644 (2018)
5. Karthikeyan, K., Dhal, P.K., Karthikeyan, K., Dhal, P.K.: Multi verse optimization (MVO) technique based voltage stability analysis through continuation power flow in IEEE 57 bus. Energy Procedia **117**, 583–591 (2017)
6. Faris, H., Hassonah, M.A., Al-Zoubi, A.M., Mirjalili, S., Aljarah, I.: A multi-verse optimizer approach for feature selection and optimizing SVM parameters based on a robust system architecture. Neural Comput. Appl., 1–15 (2017)
7. Ewees, A.A., Aziz, M.A.E., Hassanien, A.E.: Chaotic multi-verse optimizer-based feature selection. Neural Comput. Appl. **1**, 1–16 (2017)
8. Jangir, P., Parmar, S.A., Trivedi, I.N., Bhesdadiya, R.H.: A novel hybrid particle swarm optimizer with multi verse optimizer for global numerical optimization and optimal reactive power dispatch problem. Eng. Sci. Technol. Int. J. **20**(2), 570–586 (2016)
9. Mirjalili, S., Jangir, P., Mirjalili, S.Z., Saremi, S., Trivedi, I.N.: Optimization of problems with multiple objectives using the multi-verse optimization algorithm. Knowl. Based Syst. **134**, 50–71 (2017)
10. Viswanathan, G.M., Afanasyev, V., Buldyrev, S.V., Murphy, E.J., Prince, P.A., Stanley, H. E.: Lévy flight search patterns of wandering albatrosses. Nature **381**(6581), 413–415 (1996)
11. Yang, X.S., Deb, S.: Cuckoo search via Lévy flights. In: World Congress on Nature & Biologically Inspired Computing, NaBIC 2009, vol. 71, pp. 210–214. IEEE (2010)
12. Haklı, H., Uğuz, H.: A novel particle swarm optimization algorithm with levy flight. Appl. Soft Comput. J. **23**(5), 333–345 (2014)
13. Kalantzis, G., Shang, C., Lei, Y., Leventouri, T.: Investigations of a gpu-based levy-firefly algorithm for constrained optimization of radiation therapy treatment planning. Swarm Evol. Comput. **26**, 191–201 (2016)

14. Hussein, W.A., Sahran, S., Abdullah, S.N.H.S.: Patch-levy-based initialization algorithm for bees algorithm. Appl. Soft Comput. J. **23**(5), 104–121 (2014)
15. Chickermane, H., Gea, H.C.: Structural optimization using a new local approximation method. Int. J. Numer. Meth. Eng. **39**(5), 829–846 (2015)
16. Coello, C.A.C.: Use of a self-adaptive penalty approach for engineering optimization problems. Comput. Ind. **41**, 113–127 (2000)
17. Gandomi, A.H., Yang, X.S., Alavi, A.H.: Cuckoo search algorithm: a metaheuristic approach to solve structural optimization problems. Eng. Comput. **29**(2), 245 (2013)
18. Cheng, M.Y., Prayogo, D.: Symbiotic organisms search: a new metaheuristic optimization algorithm. Comput. Struct. **139**, 98–112 (2014)

An Improved Most Valuable Player Algorithm with Twice Training Mechanism

Xin Liu[1], Qifang Luo[1,2(✉)], Dengyun Wang[1],
Mohamed Abdel-Baset[3], and Shengqi Jiang[1]

[1] College of Information Science and Engineering,
Guangxi University for Nationalities, Nanning 530006, China
l.qf@163.com
[2] Guangxi High School Key Laboratory of Complex System and Computational
Intelligence, Nanning 530006, China
[3] Faculty of Computers and Informatics, Zagazig University, El-Zera Square,
Zagazig 44519, Sharqiyah, Egypt

Abstract. The most valuable player algorithm is inspired from these players who want to win the Most Valuable Player (MVP) trophy, it have higher overall success percentage. Teaching-learning-based optimization (TLBO) simulates the process of teaching and learning. TLBO has fewer parameters that must be determined during the renewal process. This paper proposes twice training mechanism to enhance the search ability of the most valuable player algorithm (MVPA) through hybrid TLBO algorithm, and named it teaching the most valuable player algorithm (TMVPA). In TMVPA, designs two behaviors of training and abstract two training modes: pre-competition training and post-competition training. Before individual competition, join the pre-competition training to coordinated exploitation ability and the exploration ability of the original algorithm and join the post-competition training to prevent from falling into the local optimal field after the corporate competition. We test three benchmark functions and an engineering design problem. Results show that TMVPA has effectively raised algorithm accuracy.

Keywords: Most valuable player algorithm · Two training modes
Benchmark functions · Teaching-learning-based optimization
Engineering design problems

1 Introduction

Meta-heuristic algorithm is inspired by the human being's careful observation and practice of the phenomena of physics, biology and society, as well as its profound understanding of these natural laws. It is a mechanism by which human beings gradually learn from nature and imitate their rules. The resulting wisdom crystallization, so scientific and development potential is self-evident [1].

Global optimization in operations research and computer science refers to the best approximate solution to the related objective function [2]. In general, a function may have many local optimal solutions so that the global optimum can't be easily found [3]. As a result, in order to solve practical problems, many intelligent optimization

© Springer International Publishing AG, part of Springer Nature 2018
D.-S. Huang et al. (Eds.): ICIC 2018, LNCS 10954, pp. 854–865, 2018.
https://doi.org/10.1007/978-3-319-95930-6_85

algorithms have been proposed. For example, genetic algorithm (GA) [4], differential evolution (DE) [5], ant colony optimization (ACO) [6], particle swarm optimization (PSO) [7], firefly algorithm (FA) [8], bat algorithm (BA) [9], cuckoo search (CS) [10, 11], Moth-Flame Optimization (MFO) [12]. Natural intelligence algorithm can effectively solve some of the problems that traditional methods can't solve, in many ways showed excellent performance, its application is more and more widely.

In 2017, Bouchekara proposed a novel heuristic algorithm by simulating the process of player team competition. Players make up the team, and then the players keep improving in the game in order to get the best value. Unlike most of the other metaheuristics, that find the optimal solution by updating the position [13], MVPA [14] utilizes player skills. MVPA is conceptually simple and is relatively easy to implement, it converges quickly, efficient and reliable. However, it is still in its infancy.

Hence, in this article we present twice Training Mechanism MVPA cross with TLBO [15], and the goal is to increase the search breadth and accuracy of MVPA. In the teaching phase of hybrid TLBO algorithm, an improved MVPA algorithm is proposed: TMVPA. The algorithm designs two training modes. One is the precompetition training mode, giving full play to the leading role of the current best player to enable all players to actively train and enhance their ability. Second, the postcompetition training mode is promoted worst player skill level after the match, enhance the team's integrity. At this article, we test three benchmark functions. Simulation results show that TMVPA is feasible, effective and has strong robustness.

2   Preliminary

2.1   Most Valuable Player Algorithm

Like any other meta-heuristics, MVPA uses one population to evolve to the best. This population is considered as the dimension of the design variable problem corresponding to the number of skill players in a group of players with similar skills. Therefore, each player is represented as follows:

$$Player_k = \begin{bmatrix} S_{k,1} S_{k,2} \cdots S_{k,\text{Problemsize}} \end{bmatrix} \tag{1}$$

where the meaning of the player is a person who takes part in a game or sport and S stands for skill. A team composed of players, expressed as follows:

$$TEAM_i = \begin{bmatrix} Player_1 \\ Player_2 \\ \vdots \\ Player_{Playersize} \end{bmatrix} \tag{2}$$

or

$$TEAM_s^i = \begin{bmatrix} S_{1,1}, & S_{1,2}, & \cdots & S_{1.\,\mathrm{Pr}\,oblemSize} \\ S_{2,1}, & S_{2,2}, & \cdots & S_{2.\,\mathrm{Pr}\,oblemSize} \\ \vdots & \vdots & \cdots & \vdots \\ S_{\mathrm{PlayerSize},1}, & S_{\mathrm{PlayerSize},2}, & \cdots & S_{\mathrm{Playersize}.\,\mathrm{Pr}\,oblemSize} \end{bmatrix} \tag{3}$$

where the $\mathrm{Pr}\,oblemSize$ is the number of players in the league and $\mathrm{Pr}\,oblemSize$ is the dimension of the problem.

2.2 Teams Formation

The number of players in the initialization step is random generated in search space. Once the population of players is generated they are randomly distributed to form *TeamSize* teams. In order to add more agility to MVPA, the number of teams need not necessarily be a factor in the number of players. Hence, users have more freedom to choose the number of players and teams they want. Form teams, the first nT_1 teams have nP_1 players, while the remaining nT_2 teams nP_2 have players, nP_1, nP_2, nT_1 and nT_2 are calculated using the following expressions.

$$nP_1 = ceil(\frac{PlayerSize}{TeamsSize}) \tag{4}$$

$$nP_2 = nP_1 - 1 \tag{5}$$

$$nT_1 = PlayerSize - nP_2 \times TeamSize \tag{6}$$

$$nT_2 = TeamSize - nT_1 \tag{7}$$

Where, nP_1 stands for number of players of one team, nT_1 is number of teams with the same number of players. *TeamSize* is number of teams in the league.

2.3 Individual Competition

Each franchise has one franchise player (it is the best player of each team) and the league's MVP is the best player in the league. Every player wants to be the franchise player and league MVP for whom the player is a legitimate team, so he works hard to improve his skills (in training) compared to his team of chartered and league MVP. Therefore, the players' skills of the selected $TEAM_s^i$ are updated as follows:

$$TEAM_s^i(t-1) = TEAM_s^i(t-1) + rand \times \left(FranchPlayer_s^i - TEAM_s^i(t-1)\right) + 2 \\ \times rand \times \left(MVP - TEAM_s^i(t-1)\right) \tag{8}$$

where *rand* is a uniformly distributed random number between 0 and 1, the letter t represents the number of iterations and the letter s stands a skill.

2.4   Team Competition

The result of the game must be one team won another team (there are no tie games). All team fitness values are standardized in MVPA. In a team, fitness stands for strength or efficiency or rating, *fitnessN* represents normalized fitness. The fitness of that particular team can be normalized as follows:

$$finessN(TEAM_s^i) = fitness(TEAM_s^i) - \min(fitness(All\ Teams)) \qquad (9)$$

Then, the probability that $TEAM_r$ beats $TEAM_y$ is calculated using the following formula:

$$\Pr\{TEAM_r\ beats\ TEAM_y\} = 1 - \frac{(fitnessN(TEAM_s^r)^k)}{(fitnessN(TEAM_s^r))^k + (fitnessN(TEAM_s^y)^k} \qquad (10)$$

Eventually, $TEAM_j$ will compete $TEAM_i$ with if $TEAM_i$ is selected in the team competition phase, if $TEAM_i$ wins the players' skills of $TEAM_i$ are updated using the following expression:

$$TEAM_s^i = TEAM_s^i + (TEAM_s^i - FranchisePlayer_j^i) \qquad (11)$$

Otherwise, the players' skills of $TEAM_i$ are updated using the following expression:

$$TEAM_s^i = TEAM_s^i + (FranchisePlayer_j^i - TEAM_j^i) \qquad (12)$$

It is worth noting that during the competitive phase, the population will be examined to see if there are players outside the crowd's boundaries. If a skill-generated player spans the bounds of the search space it needs the value of this bound.

2.5   Procedure of MVPA

The algorithm iterates over some of the fixed devices specified by MaxNFix (the maximum number of fixed devices). In MVPA, this standard was chosen as a stop criterion, but it is clear that other stop criteria can be easily implemented by the user.

The pseudo code of the MVPA algorithm is presented as follows:

Pseudo code of MVPA
ObjFunction (objective function), ProblemSize (dimension of the problem)
1 **Inputs**   PlayersSize (number of player), TeamsSize (number of team) and MaxNFix
2 **Output** MVP
3 Initialization
4 **for** fixture=1: MaxNFix
5   **for** i=1: TeamsSize
6   $TEAM_i$ =Select the team number i from the league's teams
7   $TEAM_j$ =Randomly select another team j from the league's teams where j≠i
8   $TEAM_s^i = TEAM_s^i + rand \times (FranchisePlayer_j^1 - TEAM_s^i) + 2 \times rand \times (MVP - TEAM_s^i)$
9   **if** $TEAM_i$ wins against $TEAM_j$
10   $TEAM_s^i = TEAM_s^i + rand \times (TEAM_s^i - FranchisePlayer_j^i)$
11   **else**
12   $TEAM_s^i = TEAM_s^i + rand \times (FranchisePlayer_j^i - TEAM_s^i)$
13   **end if**
14   Check if there are players outside the search space
15   **end for**
16   Application of greediness
17   Application of elitism
18   Remove duplicates
19 **end for**

2.6   Teaching-Learning-Based Optimization (TLBO)

TLBO [15] simulates teacher teaching process to the learning process of trainees and students. TLBO from the proposed to the present seven years, has attracted the attention of many scholars, and has been well applied. The process of improving the MVPA algorithm uses the teaching phase of the TLBO algorithm to train the team players twice. Individuals with the best current fitness values are selected as teachers. During the teaching phase, each individual learns on the basis of the difference between average teacher-population values. The individual update formula is as follows:

$$X_{new}^i = X_{old}^i + rand * (X_{teacher} - T_f * Mean) \tag{13}$$

Where, $X_{teacher}$ is the optimal fitness for the current population of individuals, Mean represents the average grade of classmates, T_f represent teaching factor, $T_f = round[1 + rand(0,1)\{2 - 1\}]$ is randomly assigned a value of either 1 or 2.

3 Our Approach (TMVPA)

Based on the TLBO algorithm teaching stage, this paper presents a TMVPA algorithm. The algorithm simulates two behaviors of training and abstract two training modes: pre-competition training and post-competition training. In the TMVPA algorithm, we first initialize a random player group and evaluate the skill of each player, select the best player and the worst player in the entire player group, and calculate the overall average of the player group. Then, each time players competes in competition for the first time before the competition, that is, the players to each team's franchise player and the overall average level, to increase the depth of training. Prior to the stage of individual competition and the stage of group competition, the players conduct pre-match training, that is, the player's proximity to the franchise player and the overall average of each team increases the training depth. Before the group competition players through the pre-match training to achieve the purpose of the game, that is, the players to each team's franchise player and the overall average level of learning, increasing the depth of training. Subsequent to group competition, the second post-match training begins, where the worst player in each team learns from each franchise player and the overall average, absorbing the best of the worst player that may outperform the overall average quality, and continue to broaden the scope of training.

3.1 Pre-competition Training

In Pre-Competition mode is an important model of TMVPA. Its purpose is to rely on the current best players to guide all players to expand to a deeper skill neighborhood. The overall skill of the players is the fitness value. Each one The level of skill represents a variable of solution, and the calculation of total score is not simply the superposition of each skill level, but the solution of specific objective function. The pre-Competition mode is a process in which the best player in each team is selected and the average performance of the entire league is calculated. For skill, each student learns from the best player in each team. As follows expression:

$$TEAM_s^i(t+1) = TEAM_s^i(t) + rand \times (FranchPlayer_n^i - T_f \times Mean) \qquad (14)$$

where: T_f represent teaching factor, the letter t represents iterations.

3.2 Post-competition Training

The pre-competition training mode makes TMVPA deeply digging. The post- Competition mode absorbs the best possible skills of the worst player in the grade to further enhance TMVPA global search capabilities and prevent the algorithm from falling into the local optimal neighborhood. The specific form of post-competition training mode is as follows:

$$TEAM_s^i(t+2) = TEAM_s^i(t+1) + rand \times WorstPlayer_s^i - MF \cdot Mean \qquad (15)$$

Where, Worst Player means the worst student in every team at the moment, *MF* is Mean factor, $MF = round(2 - rand)$.

3.3 TMVPA Algorithm Description

Pseudo code of TMVPA
ObjFunction (objective function), ProblemSize (dimension of the problem)
1 **Inputs** PlayersSize (number of player), TeamsSize (number of team) and MaxNFix (maximum number of fixtures)
2 **Output** MVP
3 Initialization
4 **for** fixture=1: MaxNFix
5   **for** i=1: TeamsSize
6   $TEAM_i$ =Select the team number i from the league's teams
7   $TEAM_j$ =Randomly select another team j from the league's teams where j≠i
8   $TEAM_s^i = TEAM_s^i + rand \times (FranchisePlayer_j^1 - TEAM_s^i) + 2 \times rand \times (MVP - TEAM_s^i)$
9   $TEAM_s^i(t) = TEAM_s^i(t-1) + rand \times (FranchisePlayer_j^1 - TEAM_s^i(t-1) + 2 \times rand \times (MVP - TEAM_s^i(t-1))$
10   **if** $TEAM_i$ wins against $TEAM_j$
11   $TEAM_s^i = TEAM_s^i + rand \times (TEAM_s^i - FranchisePlayer_j^i)$
12   **else**
13   $TEAM_s^i = TEAM_s^i + rand \times (FranchisePlayer_j^i - TEAM_s^i)$
14   **end if**
15   $TEAM_s^i(t+2) = TEAM_s^i(t+1) + rand \times (WorstPlayer_j^i - MF * Mean)$
16   Check if there are players outside the search space
17   **end for**
18 Application of greediness
19   Application of elitism
20   Remove duplicates
21   **end for**

4 Simulation Experiments and Result Analysis

To verify the characteristics of the TMVPA, using a examples are considered from the literature and compared with other commonly used optimization methods. Table 1 shows the spatial dimension, the range, optimal value, and an iteration of three benchmark functions. The rest of this section is organized in the following steps: in Subsect. 4.1, the experimental setup is given; a comparison of the performance of each algorithm is shown in Subsect. 4.2.

The selected reference function can be divided into two categories. They are f_{01} for category I and $f_{02} - f_{03}$ for category II. In high-dimensional functions, f_{01} is a classical

Table 1. Benchmark test functions

Benchmark test functions	Dimension	Range	f_{min}
$f_{01}(x) = \sum\limits_{i=1}^{D-1} [100(x_{i+1} - x_i^2)^2 + (x_i - 1)^2]$	30	[−30, 30]	0
$f_{02}(x) = -20 \exp\left(-0.2\sqrt{\dfrac{1}{n}\sum\limits_{i=1}^{n} x_i^2} - \exp\left(\dfrac{1}{n}\sum\limits_{i=1}^{n}\cos 2\pi x_i\right)\right)$ $+ 20 + e$	30	[−32, 32]	0
$f_{03}(x) = 0.1\left\{\sin^2\left(3\pi x_1 + \sum\limits_{i=1}^{n}\dfrac{(x_i - 1)^2[1 + \sin^2(3\pi x_i + 1)]}{+(x_n - 1)^2[1 + \sin^2(2\pi x_n)]}\right)\right\}$ $+ \sum\limits_{i=1}^{n} u(x_i, 5, 100, 4)$	30	[−50, 50]	0

test functions. Its global minimum is located in a parabolic valley and the value of the function does not change so much in the valley that it is hard to find the global minimum. The f_{02} function has a large number of local minimum in the solution. To validate the search capability of the algorithm efficiently, the reason we chose f_{03} function, most of the functions have the characteristic of chord impulses.

4.1 Simulation Platform

All the experiments in this section were operated on computer with 3.30 GHz Intel® Core™ i5-4590 processor and 8 GB of RAM using MATLAB R2017a.

4.2 Comparison of Each Algorithm Performance

The proposed TMVPA Algorithm compared with mainstream swarm intelligence algorithms BA [9], MFO [12], PSO [7] and CS [10, 11], using standard deviation to compare their optimal performance. Set the initial value for the above algorithm control parameters. The population size all algorithm is 100. Run 30 times independently.

BA setting: A = 0.9, r = 0.5, alpa = 0.95 and gamma = 0.05, CS setting: $\beta = 1.5$ and $Pa = 0.25$, PSO setting: $C_1 = 1.4962$, $C_2 = 1.4962$, w = 0.7298 MFO setting: Constant $b = 1$, random number $r \in [-2, -1]$, $t \in [-1, 1]$.

From Table 2, in category I, TMVPA can find the optimal solution for f_{01}, and has a very strong robustness. The TMVPA algorithm finds the global optimal value, the variance value is slightly lower than the original algorithm, and the result is acceptable compared with other algorithms.

Table 2. Results of the unimodal benchmark functions.

Function		BA	MFO	PSO	CS	MVPA	TMVPA
f_{01}	Best	1.5880E+01	1.2492E+02	1.3123E+01	3.8936E+01	5.7993E−02	1.5827E−15
	Worst	5.2652E+02	9.0261E+04	7.9368E+02	1.8717E+02	3.0890E+03	2.8658E+01
	Mean	7.4914E+01	1.2550E+04	7.3939E+01	9.4782E+01	1.3922E+02	1.2383E+01
	Std	1.4096E+02	3.0975E+04	1.4542E+02	4.1581E+01	5.5797E+02	1.4403E+01
	Rank	3	6	4	2	5	1

Similarly, seen from Table 3, TMVPA can find the optimal solution for $f_{02} - f_{03}$. And variance values are optimal compared to other algorithms. The results show that this algorithm has strong search ability and strong stability and can solve the problem of high-dimensional multimodal function optimization.

Table 3. Results of multimodal benchmark function

Function		BA	MFO	PSO	CS	MVPA	TMVPA
f_{02}	Best	1.8566E+01	2.1816E−01	7.3627E+01	2.1482E+00	4.3829E+00	8.8818E−16
	Worst	1.9927E+01	1.9963E+01	4.9624E+01	3.8523E+00	1.2093E+01	8.8818E−16
	Mean	1.9066E+01	1.1329E+01	1.1298E+01	3.0173E+00	7.2144E+00	8.8818E−16
	Std	2.6009E−01	9.0084E+00	4.4914E−01	3.8002E−01	1.7996E+00	0
	Rank	2	6	4	3	5	1
f_{03}	Best	8.3871E+01	3.1668E−01	7.6046E−10	1.6999E+00	1.8745E−12	1.6971E−19
	Worst	1.1892E+02	1.4677E+01	1.0987E−02	5.3494E+00	8.5911E+00	1.7506E−09
	Mean	1.0267E+02	3.5232E+00	1.4651E−03	3.6595E+00	9.2576E−01	7.1147E−11
	Std	8.1288E+00	3.2359E+00	3.7988E−03	7.1532E−01	1.7237E+00	3.1877E−10
	Rank	6	5	2	3	4	1

Figures 1, 2 and 3 are the fitness evolution curves, Figs. 4, 5 and 6 are shows the partial variance results. Figures 1, 2 and 3 are the evolution curves of fitness value for $f_{01} - f_{03}$. From Figs. 1, 2 and 3, we can find that TMVPA converges faster than other algorithms. Figures 4, 5 and 6 is the global variance analysis. We can see that the standard deviation of TMVPA is much smaller.

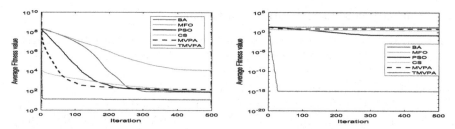

Fig. 1. Evolution curves of fitness value for f_{01} **Fig. 2.** Evolution curves of fitness value for f_{02}

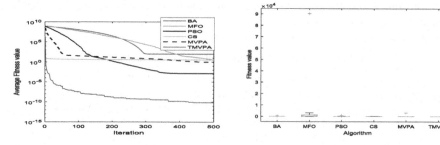

Fig. 3. Evolution curves of fitness value for f_{03} **Fig. 4.** Curves of fitness value for f_{01}

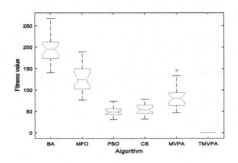

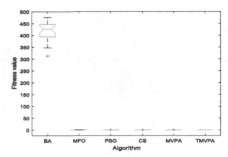

Fig. 5. Curves of fitness value for f_{02} **Fig. 6.** Curves of fitness value for f_{03}

5   TMVPA for Engineering Design Problems

5.1   Pressure Vessel Design Problem

Pressure vessel design [16] is a classic example of engineering, the main purpose of this problem is to obtain the lowest cost pressure vessel shown in Fig. 7. It shows the structural parameters of the pressure vessel design problem as the thickness of the shell (T_s), the thickness of the head (T_h), the inner radius (r) and the length of the cylindrical section without considering the head (L). The mathematical model that constrains the optimal design problem is as follows:

Consider $\vec{x} = [x_1 x_2 x_3 x_4] = [T_s T_h r L]$
Minimize $f(\vec{x}) = 0.6224 x_1 x_3 x_4 + 1.7781 x_2 x_3^2 + 3.1661 x_1^2 x_4 + 19.8 x_1^2 x_3$
Subject to $g_1(\vec{x}) = -x_1 + 0.0193 x_3 \leq 0$ $g_2(\vec{x}) = -x_3 + 0.00954 x_3 \leq 0$

$$g_3(\vec{x}) = -\pi x_3^2 x_4 - \frac{4}{3}\pi x_3^3 + 1296000 \leq 0 \quad g_4(\vec{x}) = x_4 - 240 \leq 0$$

Variable range $0 \leq x_1 \leq 99, 0 \leq x_2 \leq 99, 10 \leq x_3 \leq 200, 10 \leq x_4 \leq 200$

The result of the TMVPA on this problem are compared to PSO [7], GA [4, 17, 18], DE [5, 19], and ACO [6] in the literature. Table 4 shows the experimental results of the TMVP algorithm and other algorithms. TMPA performs exceptionally well in pressure vessel design issues and offers the best design at the lowest cost.

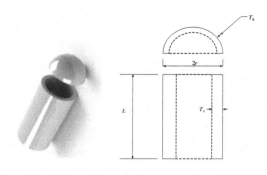

Fig. 7. Pressure vessel design problem

Table 4. Comparison results for pressure vessel design problem

Algorithms	Optimal values for variables				Optimal cost
	T_s	T_h	r	L	
PSO (He and Wang) [7]	0.812500	0.437500	42.091266	176.746500	6061.0777
GA (Coello and Montes) [17]	0.812500	0.437500	42.097398	176.654050	6059.9463
GA (Deb and Gene) [18]	0.937500	0.500000	48.329000	112.679000	6410.3811
DE (Huang et al.) [19]	0.812500	0.437500	42.098411	176.637690	6059.7340
ACO (Kaveh) [6]	0.812500	0.437500	42.103624	176.572656	6059.0888
MVPA	1.089621	0.640926	53.218026	91.1082138	6837.1603
TMVPA	0.778811	0.386771	40.36161	199.8087	5896.3688

6   Conclusions

To increase the robustness of the standard MVPA algorithm, hybridization was performed with the TLBO algorithm to generate a two-training-based MVPA for functional optimization and structural engineering design issues. The teaching phase of the TLBO algorithm increases population diversity, which helps improve its ability to explore. The pre-competition training and post-competition training designed in this paper can effectively balance the depth search and breadth search of the algorithm TMVPA algorithm balances exploration and development, addressing benchmarks function and structural engineering design issues. The results of three benchmark functions and an engineering design problem. TMVPA has faster convergence speed and higher stability.

Acknowledgment. This work is supported by National Science Foundation of China under Grant No. 61563008.

References

1. Yang, X.S.: Nature-Inspired Optimization Algorithms. Elsevier B.V., Amsterdam (2014)
2. Leung, Y.W., Wang, Y.: An orthogonal genetic algorithm with quantization for global numerical optimization. IEEE Trans. Evol. Comput. **5**(1), 41–53 (2002)
3. Ciornei, I., Kyriakides, E.: Hybrid ant colony-genetic algorithm (GAAPI) for global continuous optimization. IEEE Trans. Syst. Man Cybern. Part B Cybern. Publ. IEEE Syst. Man Cybern. Soc. **42**(1), 234 (2012)
4. Srinivas, M., Patnaik, L.M.: Adaptive probabilities of crossover and mutation in genetic algorithms. IEEE Trans. Syst. Man Cybern. **24**(4), 656–667 (2002)
5. Das, S., Suganthan, P.N.: Differential evolution: a survey of the state-of-the-art. IEEE Trans. Evol. Comput. **15**(1), 4–31 (2011)
6. Kaveh, A., Talatahari, S.: An improved ant colony optimization for constrained engineering design problems. Eng. Comput. **27**(1), 155–182 (2010)
7. He, Q., Wang, L.: An effective co-evolutionary particle swarm optimization for constrained engineering design problems. Eng. Appl. Artif. Intell. **20**(1), 89–99 (2007)
8. Yang, X.-S.: Firefly algorithms for multimodal optimization. In: Watanabe, O., Zeugmann, T. (eds.) SAGA 2009. LNCS, vol. 5792, pp. 169–178. Springer, Heidelberg (2009). https://doi.org/10.1007/978-3-642-04944-6_14
9. Yang, X.S.: A new metaheuristic bat-inspired algorithm. Comput. Knowl. Technol. **284**, 65–74 (2010)
10. Kaveh, A., Talatahari, S.: A novel heuristic optimization method: charged system search. Acta Mech. **213**(3–4), 267–289 (2010)
11. Yang, X.S., Deb, S.: Cuckoo search via lévy flights. Nat. Biol. Inspired Comput. **71**(1), 210–214 (2010)
12. Mirjalili, S.: Moth-flame optimization algorithm: a novel nature-inspired heuristic paradigm. Knowl.-Based Syst. **89**, 228–249 (2015)
13. Bouchekara, H.R.E.H.: Electromagnetic device optimization based on electromagnetism-like mechanism. Appl. Comput. Electromagnet. Soc. J. **28**(3), 241–248 (2013)
14. Bouchekara, H.R.E.H.: Most Valuable Player Algorithm: a novel optimization algorithm inspired from sport. Oper. Res. Int. J. **80**, 1–57 (2017)
15. Rao, R.V., Savsani, V.J., Vakharia, D.P.: Teaching-Learning-Based Optimization: an optimization method for continuous non-linear large scale problems. Inf. Sci. **183**(1), 1–15 (2012)
16. Gandomi, A.H., Yang, X.S., Alavi, A.H.: Cuckoo search algorithm: a metaheuristic approach to solve structural optimization problems. Eng. Comput. **29**(2), 245 (2013)
17. Coello, C.A.C., Montes, E.M.: Constraint-handling in genetic algorithms through the use of dominance-based tournament selection. Adv. Eng. Inform. **16**(3), 193–203 (2002)
18. Deb, K.: GeneAS: a robust optimal design technique for mechanical component design. In: Dasgupta, D., Michalewicz, Z. (eds.) Evolutionary Algorithms in Engineering Applications, pp. 497–514. Springer, Heidelberg (1997). https://doi.org/10.1007/978-3-662-03423-1_27
19. Li, L.J., Huang, Z.B., Liu, F., et al.: A heuristic particle swarm optimizer for optimization of pin connected structures. Comput. Struct. **85**(7–8), 340–349 (2007)

Simplex Bat Algorithm for Solving System of Non-linear Equations

Gengyu Ge, Xuexian Ruan, Pingping Chen, and Aijia Ouyang[(✉)]

School of Information Engineering, Zunyi Normal University,
Zunyi 563006, Guizhou, China
ouyangaijia@163.com

Abstract. In consideration of the fact that bat algorithm (BA) is sensitive to the initial values and simplex algorithm (SA) could often easily fall into local optimal, simplex - bat algorithm is put forward in this paper to solve system of non-linear equations based on the respective advantages of both algorithms. Such a combined algorithm does not only give full play to BAs global searching ability but also make full use of SA local searching ability. The results of simulation experiments show that this combined algorithm can be used to find the roots of all sorts of systems of non-linear equations with high accuracy, and moreover, with strong robustness and fast convergence rate, and therefore, it is indeed an effective method to solve system of non-linear equations.

Keywords: System of non-linear equations · Simplex Algorithm
Bat Algorithm · Hybrid Algorithm · Optimization

1 Introduction

Nowadays, solving system of non-linear equations is an important and basic issue, and currently, it is widely applied in many fields of scientific research such as economics, information security, network communication, engineering technology and dynamics, etc. In the final analysis, a mountain of engineering optimization problems in these fields come down to the problem of solving system of non-linear equations. However, solving system of non-linear equations is always a tough problem, although a lot of studies have been done about solving system of non-linear equations both theoretically and numerically through the ages in academic circles. In the traditional method that can be used for numerically problem solving–Newton-Raphson iteration method, the corresponding initial values for iterative procedures should be given. Moreover, this method is sensitive to the initial values for iterative procedures, and meanwhile Newton-Raphson iteration method could easily fall into the situations of calculating the derivatives of functions and matrix inversion, etc. For some functions without derivatives or whose derivatives are hard to be worked out, this method has major limitations. For this reason, we need to seek out a fast and efficient way to solve this problem.

The advantages of SA [1] lie in that it is not necessary to work out the first-order derivative matrix and Hessen matrix of the function to be solved, and there is also no

© Springer International Publishing AG, part of Springer Nature 2018
D.-S. Huang et al. (Eds.): ICIC 2018, LNCS 10954, pp. 866–873, 2018.
https://doi.org/10.1007/978-3-319-95930-6_86

need to do complex matrix operations as well; it can be used for data processing in place of the least squares method.

Cunningham [2] proposed an efficient algorithm for converting any feasible basis into a strongly feasible basis. De Wolf and Smeers [3] proposed a Simplex method for solving distributing gas through a network of pipelines is formulated as a cost minimization subject to nonlinear flow- pressure relations, material balances, and pressure bounds. Karim et al. [4] proposed a modified version of the simplex algorithm for solving a numerical procedure is presented for the inversion of leaky Lamb wave (LLW) data to determine certain material properties of a waveguide. BA [5] is a kind of newly emerging optimization algorithm of swarm intelligence, characterized by parallelism, distributivity, fast convergence rate, etc., and however, it also has several disadvantages such as that its convergence rate tends to be slow in the later stage, its convergence accuracy is not high enough, and it could often easily fall into local minimum, etc.

Rodrigues et al. [6] proposed a wrapper feature selection approach based on Bat Algorithm (BA) and Optimum-Path Forest (OPF)for solving optimizing classifier predictive performance and addressing the curse of the dimensionality problem. Sambariya and Prasad [7] proposed a bat algorithm (BA) to optimize its gain and pole-zero parameters for solving optimize conventional power system stabilizer (CPSS) gain and pole-zero parameters. Alihodzic and Tuba [8] proposed bat algorithm for solving multilevel image thresholding problem. Sathya and Ansari [9] proposed a load frequency control using dual mode Bat algorithm based scheduling of PI controllers for solving interconnected power systems. Fister et al. [10] proposed different DE strategies and applied these as a local search heuristics for improving the current best solution directing the swarm for solving towards the better regions within a search space.

In this paper, simplex algorithm - bat algorithm is introduced to solve system of non-linear equations, and for this, SA is used to optimize these points generated randomly with BA, forming a new type of combined global optimization algorithm by virtue of the advantages of both SA and BA, and this method can be widely used to solve system of non- linear equations.

2   Construction of Fitness Function

Suppose a system of nonlinear equations consists of n equations and involves m unknown quantities, in the following form:

$$
\begin{cases}
f_1(x_1, x_2, \cdots, x_m) = A_1 \\
f_2(x_1, x_2, \cdots, x_m) = A_2 \\
\cdots \\
f_n(x_1, x_2, \cdots, x_m) = A_n
\end{cases}
\tag{1}
$$

Where $f_i(X) = A_i (i = 1, 2, \ldots, n)$ is a system of nonlinear equations, $X = [x_1, x_2, \ldots, x_m]$ is the unknown vectors of this system of nonlinear equations, and $A_i (i = 1, 2, \ldots, n)$ is the constant term, a fitness function is then constructed:

$$F(X) = \sum_{i=1}^{n} |f_i(x) - A_i| \tag{2}$$

Therefore, the problem of finding the roots of this system of nonlinear equation (1) is converted to be an unconstrained optimization problem by working out the minimum value of this fitness function $F(X)$.

3 Basic Algorithms

3.1 Bat Algorithm

In 2010, a scholar at University of Cambridge named as YANG put forward a new intelligent optimization algorithm based on echolocation behavior of micro bats, which is called Bat-inspired Algorithm [11]. This algorithm is a stochastic searching optimization algorithm based on population, in which micro bats are the elementary units, and the movements of the whole population go through an evolutionary process from disorder to order in the space of problem- solving, in order to achieve the optimal solution. In BA, the solution of every optimization problem is actually a bat in the searching space, and each bat has a fitness value determined by optimization problem, all the bats search in the solution space and follow the current optimal bat by adjusting their frequency, loudness and impulse transmission rate. BA combines the major advantages of existing successful algorithms with the new features of echolocation.

A new intelligent computing mode is formed on the basis of simulating echolocation behavior of micro bats, this algorithm is based on the fundamental assumptions as follows.

(1) Bats measure distance by virtue of echolocation and can perceive the difference between prey and obstacle.
(2) Bats fly at a speed of v_i in the place of x_i, searching for prey with fixed frequency f_{min} (or λ) variable wavelength λ (or f) and loudness A_0, they adjust the transmitted pulse wavelength (or frequency) according to the distance between themselves and prey, and then adjust the frequency of transmitted pulse $r \in [1, 0]$ when they are getting closer to prey.
(3) The value of loudness is supposed to change from the maxi- mum value (positive value) A_0 to the minimum value A_{min}.

Generally speaking, the range of frequency is $[f_{min}, f_{max}]$, and the corresponding range of wavelength is $[\lambda_{min}, \lambda_{max}]$. For example, the range of frequency $[20\,kHz, 500\,kHz]$ corresponds to the range of wavelength from 0.7 mm to 17 mm. In practice, any value of wavelength can be set according to the specific problem, and then the range can be adjusted by adjusting wavelength (or frequency), the corresponding range is firstly selected and then narrowed gradually. We can assume that $f \in [0, f_{max}]$, the higher the frequency is, the shorter the wavelength and the flying range are, and the flying range is generally within a few meters.

The pulse transmission frequency is within the range of $[0, 1]$, where 0 indicates that there is no pulse and 1 implies the maximum transmission frequency.

3.2 Simplex Algorithm

Simplex algorithm (SA) [12] was first put forward by American mathematician. Dantzig in 1947, and the steps of this algorithm can be summarized as below: (1) Initialization procedure starting from a corner-point feasible solution; (2) Iteration step move to a better adjacent corner-point feasible solution (repeat this step as needed); (3) Stopping rule stop when a corner-point feasible solution is better than any other adjacent corner-point feasible solution. This corner-point feasible solution is exactly the optimal solution. The advantage of SA and its success lies in that the optimal solution can be finally achieved with less and limited number of iterations.

4 Simplex-Bat Algorithm

BA has disadvantages such as that its convergence accuracy is low, and it could easily fall into local minimum, and however, SA has the advantage that it can be used to find the optimal solution with less and limited number of iterations. Hence, on the basis of BA, SA is used to optimize the random points generated with the method of BA, hoping thereby to improve the accuracy of equation solving, enhance the local searching ability of such algorithm and further to strengthen its ability to find out the global optimal solution. For this reason, such a combined simplex-bat algorithm is designed. The advantages of this algorithm include high calculation accuracy and fast convergence rate.

4.1 Improvement Ideas

(1) BA is a kind of algorithm based on random searching, while SA is an algorithm based on deterministic searching, a mutually complementary new algorithm is designed with the combination of randomness and determinacy of these two algorithms.

(2) The gradient estimation step in the iterative process with SA can be used to determine vertexes roughly without working out the derivatives of objective equation, and the searching direction with BA can be guided by gradient estimation with SA.

(3) Both BA and SA make use of the values of fitness function to determine the evolutionary direction of algorithm without the need to work out the derivatives of fitness function, and therefore, it is very convenient to realize the transition between each other.

4.2 Algorithm Steps

Based on the analysis above, the main steps of simplex-bat algorithm can be described as follows:

Step1: Initialize parameters, objective function $f(x)$, $x = (x_1, x_2, x_3, \cdots, x_n)^T$, initial population of bats x_i ($i = 1, 2, \cdots, n$) and v_i, impulse frequency f_i at x_i, the initial pulse rate r_i and loudness of sound A_i^t;

Step2: Generate new solutions by adjusting frequency, and change the rate and location;

Step3: If ($rand_1 > r_i$), select a solution from the set of optimal solutions and then go deeper into searching with SA, and after that, a set of new points with higher accuracy can be obtained, a local solution forms in the adjacent area of this set of new points;

Step4: Generate a new solution by flying randomly;

Step5: If ($rand_2 < A^t$ & $f(x_i) < f(x_*)$), accept this new solution, increase r_i and decrease A_i^t;

Step6: Arrange bats and find out the optimal x_* at that time;

Step7: Go back to Step 2 until stopping rule is met;

Step8: Output the global optimal location.

5 Numerical Experiment and Simulation

In order to verify the performance of SABA algorithm in solving system of non-linear equations, four systems of equations in Literature [13] and four systems of equations in Literature [14, 15, 16] are used in this paper as the systems of equations to be tested (among which Examples 1 and 2 are shared by both papers), and moreover, the experimental data are compared. The comparison results show that SABA algorithm has the advantage of high calculation accuracy and it is more likely to be successful.

$$\text{Example 1} \begin{cases} f_1(x) = (x_1 - 5x_2)^2 + 40\sin^2(10x_3) = 0 \\ f_2(x) = (x_2 - 2x_3)^2 + 40\sin^2(10x_1) = 0 \\ f_3(x) = (3x_1 + x_3)^2 + 40\sin^2(10x_2) = 0 \end{cases}$$

Where: $-1 \le x_1, x_2, x_3 \le 1$, the exact solution: $x_* = (0, 0, 0)^T$.

$$\text{Example 2} \begin{cases} f_1(x) = x_1^2 - x_2 + 1 = 0 \\ f_2(x) = x_1 - \cos(0.5\pi x_2) = 0 \end{cases}$$

Where $x \in [-2, 2]$, the exact solution: $x_* = (-1/\sqrt{2}, 1.5)^T$, $x_* = (0, 1)^T$, $x_* = (-1, 2)^T$.

$$\text{Example 3} \begin{cases} f_1(x) = (x_1 + 99.7091)^2 + x_2^2 - 10000 = 0 \\ f_2(x) = \sin(5x_1) + \cos(5x_2) - 1.9932 = 0 \end{cases}$$

Where: $x \in [-2, 2]$, the exact solution: $x_* = (0.2909, 0)^T$.

$$\text{Example 4} \begin{cases} f_1(x) = x_1^2 + x_2^2 + x_3^2 - 3 = 0 \\ f_2(x) = x_1^2 + x_2^2 + x_1 x_2 + x_1 + x_2 - 5 = 0 \\ f_3(x) = x_1 + x_2 + x_3 - 3 = 0 \end{cases}$$

Where: $x \in [-1.732, 1.732]$, the exact solution: $x_* = (1, 1, 1)^T$.

$$\text{Example 5} \begin{cases} f_1(x) = x_1^3 + e^{x_1} + 2x_2 + x_3 + 1 = 0 \\ f_2(x) = -x_1 + x_2 + x_2^3 + 2e^{x_2} - 3 = 0 \\ f_3(x) = -2x_2 + x_3 + e^{x_3} + 1 = 0 \end{cases}$$

Where: $-2 \le x_1, x_2, x_3 \le 2$.

$$\text{Example 6} \begin{cases} f_1(x) = x_1 + \frac{1}{4} x_2^2 x_4 x_6 + 0.75 = 0 \\ f_2(x) = x_2 + 0.405 e^{(1 + x_1 x_2)} - 1.405 = 0 \\ f_3(x) = x_3 - \frac{1}{4} x_4 x_6 + 1.25 = 0 \\ f_4(x) = x_4 - 0.605 e^{(1 - x_3^2)} - 0.395 = 0 \\ f_5(x) = x_5 - \frac{1}{2} x_2 x_6 + 1.5 = 0 \\ f_6(x) = x_6 - x_1 x_5 = 0 \end{cases}$$

Where: $-2 \le x_1, x_2, x_3, x_4, x_5, x_6 \le 2$, the exact solution: $x_* = (-1, 1, -1, 1, -1, 1)^T$.

In this paper, SABA algorithm is used to solve the above-mentioned systems of equations, the parameters are set as follows: learning factor $c1 = 1.2, c2 = 1.2$, inertia weight factor ω changes from 0.9 to 0.4 linearly, the maximum number of iterations $MaxDT = 1000$, number of particles $N = 200$, maximum rate $vmax = 4$, the algorithm runs for 50 times, and finally the average value is worked out. The experimental results are shown in Table 1. In Example 1, SABA algorithm performs better than PSO and SCO in various aspects, and it is somewhat on a par with QPSO. In Example 2, the solution of SABA is better than that of PSO, SCO and QPSO, only two sets of solutions are found with the algorithms of PSO and SCO, while the solution $x^* = (-1, 2)T$ is not found. Although this solution is found with the algorithm of QPSO, its accuracy is not as high as that with the algorithm of SABA. In Example 3, SABA performs better than PSO but slightly worse than SCO in the second component. In Examples 4 and 6, SABA performs better than PSO, SCO and QPSO, holding all the trumps. In Example 5, SABA is somewhat on a par with QPSO, it performs slightly better in finding out one of the solutions. The success rate of searching for all these six systems of equations mentioned above is 100% with the algorithm of SABA, indicating that SABA has strong robustness and satisfying efficiency.

Table 1. Comparison results between SABA algorithm and PSO, SCO, QPSO algorithms

NO	Alg.	Times	Suc%	X1 mean X4 mean (F6)	X2 mean X5 mean (F6)	X3 mean X6 mean (F6)
F1	PSO	27	54%	0.000 001	−0.000 012	−0.000 001
	SCO	50	100%	2.595 555E−28	−1.225 309E−28	1.007 189E−28
	QPSO	50	100%	0.000 000	0.000 000	0.000 000
	SABA	50	100%	0.000 000	0.000 000	0.000 000
F2	PSO	43	86%	−0.707 724	1.500 668	26
				0.000 114	0.999 817	17
	SCO	50	100%	−0.707 106	1.500 000	12
				6.123 233E−17	1.000 000	38
	QPSO	50	100%	−0.707 106	1.500 000	31
				0.000 000	1.000 000	9
				−1.000 000	2.000 000	10
	SABA	50	100%	−0.707 106	1.500 000	30
				0.000 000	1.000 000	14
				−1	2	6
F3	PSO			0.290 9	0.001 9	
	SCO		100%	0.290 899	0.000 152	
	SABA		100%	0.290 899	0.001 910	
F4	PSO			0.996 2	1.003 8	1.000 0
	SCO		100%	0.998 771	1.001 192	0.999 885
	SABA		100%	0.999 998	1.000 000	1.000 000
F5	QPSO	50	100%	−0.767 760	0.075 330	−1.162 151
	SABA	50	100%	−0.767 760	0.075 330	−1.162 151
F6	QPSO	50	100%	−1.000 000	1.000 000	−1.000 000
	SABA	50	100%	1.000 000	−1.000 000	1.000 000
				−1	1	−1
				1	−1	1

6   Conclusion

In this paper, a combined simplex - bat algorithm is put forward based on simplex algorithm and bat algorithm. It integrates SA with BA tactfully and makes the best of the both worlds so as to enhance the performance of SABA algorithm. The simplex-bat algorithm not only improves the convergence rate with the algorithm of SABA, and but also effectively enhances the local searching ability of SABA algorithm and improves the accuracy of searching significantly improved with strong robustness. The effect is satisfying when this algorithm is used to optimize the geometric dimensions of thin-wall rectangular beam cross section.

Acknowledgements. The research was partially funded by the science and technology project of Guizhou ([2017]1207), the training program of high level innovative talents of Guizhou ([2017]3), the Guizhou province natural science foundation in China (KY[2016]018), the Science and Technology Research Foundation of Hunan Province (13C333).

References

1. Qiuruchen, K.R.: An accelerated simplex method. J. Nanjing Univ. Sci. Technol. **27**(2), 209–213 (2003)
2. Ning, X.: Graph simplex method for solution of maximum flow problem in a network. Trans. Nanjing Univ. Aeronaut. Astronaut. **28**(5), 626–630 (1996)
3. De Wolf, D., Smeers, Y.: The gas transmission problem solved by an extension of the simplex algorithm. Manage. Sci. **46**(46), 1454–1465 (2000)
4. Karim, M.R., Mal, A.K., Bar-Cohen, Y.: Inversion of leaky lamb wave data by simplex algorithm. J. Acoust. Soc. Am. **88**(1), 482–491 (1990)
5. Xiao, H.H., Duan, Y.M.: Research and application of improved bat algorithm based on de algorithm. Comput. Simul. **31**(1), 272–277 (2014)
6. Rodrigues, D., Nakamura, R.Y.M., Costa, K.A.P., Yang, X.S.: A wrapper approach for feature selection based on bat algorithm and optimum-path forest. Expert Syst. Appl. Int. J. **41**(5), 2250–2258 (2014)
7. Sambariya, D.K., Prasad, R.: Robust tuning of power system stabilizer for small signal stability enhancement using meta heuristic bat algorithm. Int. J. Electr. Power Energy Syst. **61**(61), 229–238 (2014)
8. Alihodzic, A., Tuba, M.: Improved bat algorithm applied to multilevel image thresholding. Sci. World J. **2014**, 16 (2014)
9. Sathya, M.R., Ansari, M.M.T.: Load frequency control using bat inspired algorithm based dual mode gain scheduling of pi controllers for interconnected power system. Int. J. Electr. Power Energy Syst. **64**(64), 365–374 (2015)
10. Iztok Fister, J., Fong, S., Brest, J., Fister, I.: A novel hybrid self-adaptive bat algorithm. Sci. World J. **2014**(1–2), 709738 (2014)
11. Yu, L.I., Liang, M.A., Management, S.O.: Bat-inspired algorithm: a novel approach for global optimization. Comput. Sci. **40**(9), 225–229 (2013)
12. Ouyang, A.J., Zhang, W.W.: Hybrid global optimization algorithm based on simplex and population migration. Comput. Eng. Appl. **46**(4), 29–30 (2010)
13. Sun, J.Z.: Solving nonlinear systems of equations based on social cognitive optimization. Comput. Eng. Appl. **44**(28), 42–43 (2008)
14. Zhang, A.L.: Hybrid quasi-newton/particle swarm optimization algorithm for nonlinear equations. Comput. Eng. Appl. **44**(33), 41–42 (2008)
15. Sui, Y., Zhao, W.: A quadratic programming method for solving the NSE and its application. Chin. J. Comput. Mech. **19**(2), 245–246 (2002)
16. Mo, Y., Chen, D.Z., Hu, S.: A complex particle swarm optimization for solving system of nonlinear equations. Inf. Control **35**(4), 423–427 (2006)

Earprint Based Mobile User Authentication Using Convolutional Neural Network and SIFT

Mudit Maheshwari, Sanchita Arora, Akhilesh M. Srivastava,
Aditi Agrawal, Mahak Garg, and Surya Prakash$^{(\boxtimes)}$

Discipline of Computer Science and Engineering,
Indian Institute of Technology Indore, Indore 453552, India
surya@iiti.ac.in

Abstract. Biometric verification techniques are increasingly being used in mobile devices these days with the aim of keeping private data secure and impregnable. In our approach, we propose to use the inbuilt capacitive touch-screen of mobile devices as an image sensor to collect the image of ear (earprint) and use it as biometrics. The technique produces a precision of 0.8761 and recall of 0.596 on the acquired data. Since most of the touch screens are capacitive sensing, our proposed technique presents a reliable biometric solution for a vast number of mobile devices.

Keywords: Mobile biometrics · Earprint · Convolutional neural network
SIFT

1 Introduction

There has been a tremendous amount of work to improve mobile biometrics solutions and researchers have proposed many innovative ways to improve the user experience. A lot of work that has been done in the last few years is towards providing secure and comfortable mechanism for access control and authentication in mobile devices. Many mobile and tablet manufacturers use biometrics such as fingerprint and iris pattern for authentication and access control. However, these methods make the mobile devices expensive due to use of costly hardware. Recently, there have been few attempts in which researchers have proposed the use of touch screens of mobile devices for collection of biometric data. Though the touch screens cannot be used for capturing fine features of biometric traits such as fingerprints due to low resolution of capacitive screens, it can be effectively used for capturing the geometry of biometrics such as fingers, palms, fist and ears, by placing them on the touch screen with pressure. Alan Goode in [1] has given an elaborate report of the various kinds of mobile biometrics available in the market today. A detailed analysis of ear biometrics is presented in [2] by Burge and Burger. A method is proposed in [3] by Okumura et al. which authenticates the owner during his arm sweep action using acceleration signals that are recorded with the help of the acceleration sensor in a mobile phone. In Mobile Biometrics (MoBio) [4] project, Tresadern et al. utilize face recognition in real-time coupled with voice verification for enhancing the security of mobile devices. In [5], Jillela and Ross have explored the prospects of incorporating iris recognition in mobile

© Springer International Publishing AG, part of Springer Nature 2018
D.-S. Huang et al. (Eds.): ICIC 2018, LNCS 10954, pp. 874–880, 2018.
https://doi.org/10.1007/978-3-319-95930-6_87

devices. In [6], a technique for commodity mobile devices to identify users from the biometric features collected from the body parts such as ear, fist, phalanges, palm grip and finger grip is presented. A Software Developer Kit [7] for ear recognition has been recently made available by Descartes Biometrics for the development of ear based biometric solutions on Android mobile devices.

In this paper, we propose a technique for collecting earprint of a user through touch screen of a mobile phone for access control and authentication. Since collected data from the touch screen is of low resolution and contains noise, we propose three steps to improve the quality of the image, which includes simple threshold based noise removal, super resolution using Convolutional Neural Networks (CNN) and finally overlapping of multiple scans of the earprints collected through touch screen of mobile phone. We use the final overlapped images for authentication.

Rest of the paper is organized as follows. In Sect. 2, few preliminaries used in the proposed technique are presented. In Sect. 3, the proposed technique is presented. Experimental analysis is discussed in Sect. 4. Finally, the paper is concluded in Sect. 5.

2 Preliminaries

In this section, we state all the required preliminaries which are required in developing our proposed technique. A *Capacitive Touchscreen* consists of horizontal and vertical transparent conducting wires, with tiny electrical currents running through them. The intersection points of these arrays of wires act as capacitors for the touchscreen. Overall working of capacitive touchscreen is shown in Fig. 1. The intersection points of the grid experience a change in their capacitance values whenever they come in contact with the human body. Based on these values, the touchscreen controller calculates the activated contact points. This information is passed down to the applications for gesture analysis and touch detection. *CNNs* are made up of neurons that have certain weights and biases assigned to them, which are particularly well-adapted to classify images. Each neuron receives several inputs, performs a dot product over them and optionally passes it through an activation function and responds with an output. CNN has a loss function on last layer to learn end-to-end mappings and it allows network to have fewer weights. The matching of different features across various images is done using *Scale-Invariant Feature Transform (SIFT)*, which is used to detect and describe local features in images [8] (Fig. 3).

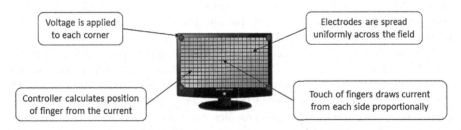

Fig. 1. Working of a capacitive touchscreen

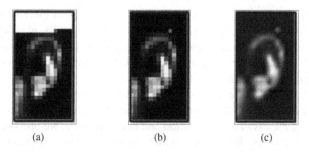

(a)	(b)	(c)

Fig. 2. (a) Raw Capacitive Image, (b) Image obtained after removal of noise from (a), (c) Resultant interpolated image using CNN

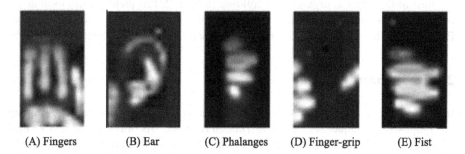

(A) Fingers	(B) Ear	(C) Phalanges	(D) Finger-grip	(E) Fist

Fig. 3. Capacitive images of various body parts

3 Proposed Technique

3.1 Data Acquisition

Applications running on the Android Operating System in a mobile phone do not have direct access to the capacitance value at each intersection node in the capacitive touchscreen. Instead, the controller computes various other attributes like x-coordinate, y-coordinate, pressure and cross-sectional area from the capacitive values of the touchscreen at a particular time instant and only these are made available to the Android Operating System. Hence, the applications can only utilize these pre-computed values and not the capacitance data in raw form. For our application, the pre-computed attributes are not enough. We need the capacitance values of each inter-section point as it is so that it can be converted to a gray-scale image that draws out the shape of the touch points on the screen. The only possible method to accomplish this is to make changes in the Android kernel itself.

We use Google Nexus 5 (code-named *Hammerhead*) which has a Synaptics ClearPad 3350 touch sensor and is based on the Android Operating System. The touchscreen of the Nexus 5 phone is composed of 27×15 intersection points. On making changes in a required file in the touchscreen driver of the Android kernel, capacitance values can be written to *proc*files which are mainly used for storing run-time system information. We use commands to obtain capacitance and coordinate values at every intersection point. These commands are called after every 150 ms in the application which we define as the *Refresh Rate*. The reason behind choosing the refresh rate to be 150 ms is to optimize the performance of our application. If we choose a higher value, this makes the application slow, while a smaller value leads to the application crashing due to excessive memory requirements. At 150 ms, a tradeoff is achieved. The capacitance values at each of these intersection points are saved as a string containing 405 (27×15) comma separated values.

3.2   Conversion of Capacitive Data to Capacitive Images

After obtaining a string of 27×15 comma separated capacitance values, we create a 2-D matrix with 27 rows and 15 columns from this data. This matrix can be directly converted to a gray-scale image by applying min-max normalization. This grey-scale image is called capacitive image. The capacitive image that we get looks similar to Fig. 2a. The obtained capacitive image is a low-resolution noisy image which needs to be enhanced to get a good quality image for feature extraction and matching.

3.3   Image Enhancement

This aims at removing noise, accentuate the contours and enhance the information in low-resolution images to enhance the prospects of obtaining the correct SIFT feature descriptor vectors for all the detected feature keypoints. For this, first *Noise Removal* is performed. The raw capacitive image, as seen in Fig. 2a, contains some unwanted and random spikes in pixel intensity values, which is eliminated by nullifying all the values which are higher than some suitable threshold (determined after some observations). Figure 2b illustrates this method carried out on the image shown in Fig. 2a. The capacitive image of low resolution obtained after applying simple threshold based noise removal is converted to a higher resolution image using a deep convolutional network. The image is upscaled using *CNN Super-Resolution Encoding* [9]. We upscale each image 8 times. Further, image is reconstructed and represented through patch extraction and non-linear mapping to obtain an upsampled higher-resolution image. A sample image obtained after super resolution is shown in Fig. 2c. Finally, we perform *Overlapping*. Even after removing the noise and scaling, we still see some unwanted regions in some images, which can be reduced by taking a composite of consecutive images obtained by the super resolution step so that the region containing noise is reduced in intensity by the ones without it. This operation also ensures that none of the features are exempted, but instead, are enhanced in the final resultant image. In our proposed technique, we take a set of five consecutive images and take a combination of the capacitance values at each pixel to get an overlapped image. Figure 4 illustrates the process of overlapping with an example.

3.4   Feature Extraction

We have used SIFT algorithm for extracting features from the enhanced capacitive images, since it provides highly distinctive features which can be correctly matched with high probability against a large set of features from images of the test data set.

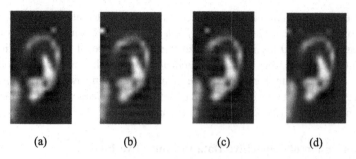

| (a) | (b) | (c) | (d) |

Fig. 4. (a), (b) and (c) are consecutive images obtained from a person, (d) Resultant superimposed image from (a), (b) and (c), which contains the features from all images

3.5   Matching

Matching of images is carried out using k-Nearest Neighbour (k-NN) matching supplied by the Fast Library for Approximate Nearest Neighbors (FLANN) in OpenCV. FLANN is a library for performing fast approximate nearest neighbor searches in high dimensional spaces. k-NN match returns the k best matches for each feature descriptor. We have chosen value of k to be 2 in our implementation.

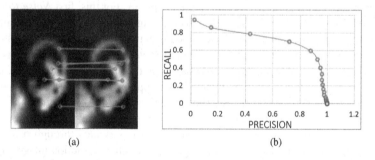

| (a) | (b) |

Fig. 5. (a) Matching of two earprints using SIFT keypoints, (b) Precision vs Recall plot for the proposed technique.

After obtaining two best matches, matching of two feature keypoint sets (obtained for two different images) is done with the help of nearest neighbor distance ratio (*nndr*) matching. The motivation behind this matching strategy is that we expect a good match to be much closer to the query feature than the second best match. There is a trade-off between high value and low value of *nndr* because a higher value of *nndr* may ignore genuine matches whereas a lower value may include some invalid matches. We chose the value of *nndr* to be 0.60 for our experiments. Figure 5a illustrates this Matching. Final match score for two images is obtained based on the number of keypoints found to be matching between them.

4 Experimental Results

Experiments are conducted on a database containing capacitive images of right ear acquired from staff members and students at IIT Indore. The current database has 40 images each from 38 different subjects. All the subjects are in the age group of 17–50 years. All the images are obtained by placing the right ear of the subject on the touchscreen of the device without changing the orientation and position of the ear. The database of 1520 images has been classified for every user with a unique identification number. The database for each subject is divided into two groups, one with 5 images (*i.e.* total 38 × 5 images used for training) and another with remaining 35 images (*i.e.* total 38 × 35 images used for testing). The resolution of each acquired image is 27 × 15 pixels and each image is available in *jpeg* format.

Recall, Precision and Rank Recognition Rate are used to measure the performance of the proposed system by testing the system for different users and keeping track of how many times it gives access to unauthorized person(s) as well as how many times an authorized person is mistakenly refrained from gaining access. This proposed system has achieved a 100% Rank-1 recognition rate. To find out the most optimized value of the threshold, we take threshold at which precision is equal to the average precision. Precision and Recall are calculated for varying threshold values (from 1 to 27). It is found that the average precision is 0.8871. Corresponding to the obtained average precision value, threshold value is found to be little more than 4 which we round up to 4. For this threshold value, precision and recall are found to be 0.8761 and 0.5960 respectively. Figure 5b shows Precision Vs. Recall plot.

5 Conclusion

In this paper, we have proposed a biometric system for the mobile devices that are not equipped with any external biometric sensors (for example, fingerprint scanner, iris scanner, camera, etc.). Our proposed technique can identify users with the help of biometric features of their ear images (earprints), and thus, increases the convenience for authentication in mobile devices. Our proposed technique utilizes the capacitive touchscreen of mobile devices as an effective input medium for capturing images and accounts for their low input resolutions, as compared to other high-end sensors. This technique has obtained a precision of 0.8761 and recall of 0.596 on the acquired data.

References

1. Goode, A.: Bring your own finger how mobile is bringing biometrics to consumers. Biometric Technol. Today **5**, 5–9 (2014)
2. Burge, M., Burger, W.: Ear biometrics. In: Jain, A.K., Bolle, R., Pankanti, S. (eds.) biometrics. Springer, Boston (1996). https://doi.org/10.1007/0-306-47044-6_13

3. Okumura, F., Kubota, A., Hatori, Y., Matsuo, K., Hashimoto, M., Koike, A.: A study on biometric authentication based on arm sweep action with acceleration sensor. In: Proceedings of International Symposium on Intelligent Signal Processing and Communication, pp. 219–222 (2006)

4. Tresadern, P., Cootes, T.F., Poh, N., Matejka, P., Hadid, A., Lvy, C., McCool, C., Marcel, S.: Mobile biometrics: combined face and voice verification for a mobile platform. IEEE Pervasive Comput. **12**(1), 79–87 (2013)

5. Jillela, R.R., Ross, A.: Segmenting iris images in the visible spectrum with applications in mobile biometrics. Pattern Recogn. Lett. **57**, 4–16 (2015)

6. Holz, C., Buthpitiya, S., Knaust, M.: Bodyprint: biometric user identification on mobile devices using the capacitive touchscreen to scan body parts. In: Proceedings of Annual Conference on Human Factors in Computing systems, pp. 3011–3014 (2015)

7. Descartes Biometrics. http://www.descartesbiometrics.com/helix-sdk/

8. Lowe, D.G.: Distinctive image features from scale-invariant keypoints. Int. J. Comput. Vis. **60**(2), 91–110 (2004)

9. Li, Y., Huang, J.-B., Ahuja, N., and Yang, M.-H.: Joint image filtering with deep convolutional Networks ArXiv e-prints, arXiv:1710.04200 (2017)

Unconstrained and NIR Face Detection with a Robust and Unified Architecture

Priyabrata Dash[1], Dakshina Ranjan Kisku[1(✉)], Jamuna Kanta Sing[2], and Phalguni Gupta[3]

[1] Department of Computer Science and Engineering,
National Institute of Technology Durgapur, Durgapur 713209, India
drkisku@cse.nitdgp.ac.in
[2] Department of Computer Science and Engineering, Jadavpur University,
Kolkata 700032, India
[3] National Institute of Technical Teachers Training and Research,
Salt Lake, Kolkata 700106, India

Abstract. This paper proposes a face detection method making use of Fast Successive Mean Quantization Transform (FSMQT) features for image representation to deal with illumination and sensor insensitive issues of the individual as well as the crowd face images. A split up Sparse Network of Winnows (SNoW) with Winnow updating rule is then exploited to speed up the original SNoW classifier. Features and classifiers are combined together with skin detection algorithm for fake face detection in crowd image and head orientation correction for near infrared faces. The experiment is performed with four databases, viz. BIOID, LFW, FDDB and IIT Delhi near infrared showing superior performance.

Keywords: Face detection · Fast SMQT · Split up SNOW classifier
Pose · Occlusion · Blur · Labeled faces · Crowd faces

1 Introduction

The aim of face detection is to find and locate a face in an image. It detects facial features and ignores anything else, such as buildings, trees and bodies in the background of cluttered scene. The face detection techniques which have been developed until now, are mainly used for frontal view and near frontal view face localization. The Viola and Jones' face detector [1] is a popular and widely accepted algorithm for visible face images. The algorithm uses Haar-like features [1] and cascaded AdaBoost classifier [1]. However, faces to be detected in unconstrained environments may not be perform well due to facial dynamics and unconstrained factors. These factors become challenging due to variations in pose, illumination and facial expression. It also includes out-of-focus blur, low resolution, infrared image and image orientation. Further, the same state-of-the-art methods are also unable to detect faces in near infrared images. A boosted exemplar-based face detection has been discussed in [2]. The work is inspired from image retrieval system where a face is detected till adequate analogous hypotheses is included into the hypotheses set, which is very large in nature,

© Springer International Publishing AG, part of Springer Nature 2018
D.-S. Huang et al. (Eds.): ICIC 2018, LNCS 10954, pp. 881–887, 2018.
https://doi.org/10.1007/978-3-319-95930-6_88

hence inefficient process. It can apply to unconstrained face images with variable appearance, but need large appearance train images of all samples which is cumbersome. The real-time high performance deformable model for unconstrained face detection [3] uses both weakly supervised settings as well as strongly supervised settings for face detection and for this reason, global root template is incorporated for low resolution face images and the parts are placed in higher resolution. It is mostly suitable for unconstrained scenario, however the computational complexity is high. A fast and accurate unconstrained face detector [4] with NPD features, Quadratic tree and Soft cascade has addressed variations in pose, occlusion and blur of a face image. NPD feature is scale invariant, rapid, memory efficient, bounded and helps to recover the input image. It uses deep quadratic tree to train the optimal subset of NPD features and a single soft-cascade classifier to meet the unconstrained facial complexities. A SMQT [5] based split up SNoW classifier [5] is proposed for illumination and structure invariant face detector. It is found to be efficient, however the face detector is less representative and computational complexity is high. In another work, aggregate channel features are for Multi-view Face Detection [6] where channel features extend the image channel to diverse types like gradient magnitude and oriented gradient histograms and therefore encodes affluent information in a simple structure. The face detector is used to detect faces in the wild. However, it needs more adjustment of training as well as test faces. A joint cascade face detection and alignment [7] method uses face alignment with unified framework using SIFT features and SVM classifier. The face detector trains the two tasks i.e. alignment and detection jointly in the same cascade framework. It is found efficient in real-time scenario. However, small changes in pose degrades the performance in large. Another cascade face detection [8] is based on histograms of oriented gradients (HOG) which uses different kinds of features and classifier to exclude non-face step by step. The face detectors which are discussed in [1–8] are able to detect faces when faces are given in either frontal view position or near frontal view positions in unconstrained environments and the cost of face detection is high. Moreover, they are also not useful for detection of near infrared face images.

This paper proposes an efficient face detection method which uses the extension of SNoW classifier [9], known as Split up SNoW [5] classifier that utilizes results of the SNoW classifier and creates a cascade of classifiers to perform a more rapid face detection. Fast SMQT features are used to represent facial features of unconstrained and NIR face images and the combination with split up SNoW classifier is found advantageous with complex facial features. Fast SMQT features are edge and gradient based transformation technique which is invariant to illumination, structure and spatial distortion. The performance of the classifier can be improved by applying advanced YCbCr color correction approach [10] to detect fake faces and head orientation correction method for NIR face images. This architecture is found to be more accurate in detecting a face both in unconstrained as well as in NIR environments.

2   Face Image Characterization with Fast SMQT

The SMQT [5] is a gradient based transformation which is invariant to structure, illumination and sensor variability. It is an advanced technique which can be compared to modified census transform [11] and it is similar to first level of SMQT in one bit structure kernel. However, the SMQT is used for any bit structural kernels on arbitrary dimensional data. As the complexity of basic SMQT solely depends on the level of quantization and the number of pixels. Selecting a lower level can reduce the processing as well as the image quality. Also the number of pixels affects the computational speed proportionally. In order to reduce the computation time without reducing image quality the Fast SMQT [12] is proposed. The Fast SMQT is invariant towards the quantity of pixels as it works on scope of pixel values rather than number of pixels and incorporating the threshold by Otsu's method [13] into the essential basic SMQT. It increases the contrast of images and also diminishes the noise. To take the advantage of Fast SMQT, both unconstrained and infrared face images are represented using this feature enhancement technique.

3   Training and Classification

The SNoW classifier [9] is a supervised and robust classifier which is based on Winnow, a multiplicative update rule algorithm, can handle substantial feature vectors toward making look-up tables for classification. The SNoW classifier is feature efficient and its architecture is a sparse network of linear units. That potentiality of the snow ensures information driven way. Thus, sparsely associated networks bring a scale free circulation. The facial features are represented using nodes over the input instance in the input layer of the sparse network. To characterize the hypothetical concept of interest over the input, target nodes are used and these target nodes are nothing but the linear units. The target nodes are features extracted from face or non-face images and an instance can map into features which are considered active in it. The outcome in the input layer propagates to the target nodes and target nodes are connected to the input features through weighted values. Face detection is very difficult task and degree of heterogeneity is more. The SNOW classifier can handle all the unconstrained conditions like variations in pose, facial expressions, illuminations and occlusions. The SNoW classifier is act as binary classifier to classify face features i.e. the Fast SMQT which is represented in sparse binary way. The speed of SNoW classifier can be increased by extending its functionalities which turn to be an efficient classifier known as Split up SNoW classifier [5]. Split up SNoW classifier requires training of only one classifier network and it is divided into several weaker classifiers in cascade. This functionality is incorporated in Split up SNoW classifier where a cascade of weak classifiers is created to overcome one class domination to input. Detail implementation of Split up SNoW classifier can be found in [5].

In order to perform face detection training with positive and negative examples, a set of 5,681 face images is collected from the Yale [14], the ORL [15], the CMU+MIT [16], the LFW-crop [17] and the BIOID [18] datasets. They have wide variations in pose, facial expression and lighting condition. For negative examples, 8765 samples from 400 images of landscapes, tree, and building are used. For Positive examples, each face image is decided to be the size of 32 × 32 pixels and is manually cropped. To reduce the effect of scale and rotation, the detection algorithm produces 20 face examples from original image. The faces are generated by rotation of images by up to 15°. To deal with illumination and sensor noise, the proposed Fast SMQT is applied.

During face detection overlapped regions are pruned by utilizing geometrical area and classification scores. Every region discovery is tried against every other detections. Given that two discoveries overlap with each other, the recognition with the most noteworthy classification score is kept and the others are expelled. This methodology is repeated until no more overlapping identifications are found. To start with the face detection process, color images are converted to grayscale images and then downscale each image by scaling factor 1.2 repeatedly. Afterward, the Fast SMQT method is applied for feature enhancement and representation on downscaled image by taking a patch of size 32 × 32, and by shifting. Now all patches are processed through Split up SNoW classifier to classify face and non-face in the image, and store pixel locations of face for localizations of facial area. The head position correction method is employed for near infrared face images. The original NIR face image is converted into binary image and clear the boarder of resultant image. Then find region filled with the most pixels which shape of binary image and finally obtain corresponding orientation of max filled region and rotate the original image with reference to orientation. It is continued till stopping criteria: either a limit of rotation (100%) is reached or face is detected.

4 Evaluation

The proposed face detection experiment is conducted on four publicly available databases such as the BIOID [18], the LFW [19], the FDDB [20], and the IIT Delhi NIR [21]. The algorithm involves detecting critical faces that are captured in both degraded indoor environments and unconstrained environments where changes in illumination, pose, facial expression, background and resolution are seen. In addition, with the same algorithm near infrared faces are also being detected from the NIR images where head pose variations are occurred.

The experimental results are determined on different databases having variety of perspectives, viz. the LFW is single face image database, the BIOID is degraded image database BioID, the IIT Delhi is NIR image database and the FDDB is crowd face image database in unconstrained scenarios. All four publicly available databases are having their own standard protocols for evaluation. These protocols differ due to variations of face image structure, representation as well as their referenced structure. The Table 1 shows the different preprocessing mechanisms and the number of samples to be processed from different databases.

Performance of the system is shown in Table 2. The datasets are shown variations on their accuracy due to their complex nature, heterogeneous characterization, unconstrained and infrared aspects. The BioID database is confined to indoor environment and small areas as well as single subject images. This dataset is limited to unconstrained characteristics and the face detection shows 98% accuracy on this database, which outperforms others. The LFW database is similar to the BioID database, however, it is found more unconstrained due to different background of same subjects or pose variations, occlusions, blur and colored. After applying the color correction method the proposed face detector gives 84% accuracy which is better than well-known face detector, The IIT Delhi NIR is another challenging database as it is not containing visible face images. On the FDDB database, we have achieved 80% accuracy for unconstrained scenario as well as for the faces present in crowd images. Results are shown in Fig. 1.

Table 1. Summarized information

Dataset	Samples	Subject	Preprocessing
BioID	1525	23	—
LFW	12777	5749	Color correction
FDDB	2845	NA	Color correction
NIR IITD	574	NA	Head-position correction

Table 2. Performance of the proposed system on four databases

Dataset	TP	FP	TPR	Accuracy	Error
BioID	1488	34	0.997	0.98	0.02
LFW	10745	1051	0.916	0.84	0.16
FDDB	2264	319	0.896	0.80	0.20
NIR IITD	418	26	0.762	0.73	0.27

Fig. 1. Face detection results on unconstrained as well as on NIR image

5 Conclusions

This paper has presented a unified and robust face detector for unconstrained as well as heterogeneous face images and the algorithm is evaluated its performance on some standard face detection benchmarks databases. An investigation is performed by increasing the size of the training set and incorporating an improved split up SNoW classifier on face detection performance. It has attained an improvement from using Split up SNoW sparse architecture as well as Fast SMQT features. Further, face detection performance is improved by employing preprocessing operations such as color correction for unconstrained faces and head position correction for IIT Delhi NIR database.

References

1. Viola, P., Jones, M.: Rapid object detection using a boosted cascade of simple features. In: Proceedings of Computer Vision and Pattern Recognition (CVPR) (2001)
2. Li, H., Lin, Z., Brandt, J., Shen, X., Hua, G.: Efficient boosted exemplar-based face detection. In: Proceedings of Computer Vision and Pattern Recognition (CVPR) (2014)
3. Yan, J., Zhang, X., Lei, Z., Li, S.Z.: Real-time high performance deformable model for face detection in the wild. In: Proceedings of International Conference on Biometrics (ICB) (2013)
4. Liao, S., Jain, A.K., Li, S.Z.: A fast and accurate unconstrained face detector. IEEE Trans. Pattern Anal. Mach. Intell. 38(2), 211–223 (2016)
5. Nilsson, M., Nordberg, J., Claesson, I.: Face detection using local SMQT features and split up SNOW classifier. In: Proceedings of IEEE International Conference on Acoustics, Speech, and Signal Processing (ICASSP), no. 2, pp. 589– 592 (2007)
6. Yang, B., Yan, J., Lei, Z., Li, S.Z.: Aggregate channel features for multi-view face detection. In: Proceedings of IEEE International Joint Conference on Biometrics (IJCB), pp. 1–8 (2014)
7. Chen, D., Ren, S., Wei, Y., Cao, X., Sun, J.: Joint cascade face detection and alignment. In: Fleet, D., Pajdla, T., Schiele, B., Tuytelaars, T. (eds.) ECCV 2014. LNCS, vol. 8694, pp. 109–122. Springer, Cham (2014). https://doi.org/10.1007/978-3-319-10599-4_8
8. Yang, H., Wang, X.A.: Cascade classifier for face detection. J. Algorithms Comput. Technol. 10(3), 187–197 (2016)
9. Roth, D., Yang, M., Ahuja, N.: A snow-based face detector. In: Proceedings of Advances in Neural Information Processing Systems (NIPS), pp. 855–861 (2000)
10. Prathibha, E., Manjunath, A., Likitha, R.: RGB to YCbCr color conversion using VHDL approach. Int. J. Eng. Res. Dev. 1(3), 15–22 (2012)
11. Fröba, B., Ernst, A.: Face detection with the modified census transform. In: Proceedings of 6th IEEE International Conference on Automatic Face and Gesture Recognition (FG), pp. 91–96 (2004)
12. https://www.toptal.com/algorithms/successive-mean-quantization-transform
13. Otsu, N.: A threshold selection method from gray-level histograms. IEEE Trans. Syst. Man Cybern. 9(1), 62–66 (1979)
14. Bellhumer, P.N., Hespanha, J., Kriegman, D.: Eigen faces vs. fisher faces: recognition using class specific linear projection. IEEE Trans. Pattern Anal. Mach. Intell. Spec. Issue Face Recogn. 17(7), 711–720 (1997)

15. Samaria, F., Harter, A.: Parameterization of a stochastic model for human face identification. In: Proceedings of 2nd IEEE Workshop on Applications of Computer Vision (WACV) (1994)
16. Heisele, B., Poggio, T., Pontil, M.: Face detection in still gray images. Technical report, Center for Biological and Computational Learning, MIT, A.I. Memo 1687 (2000)
17. Sanderson, C., Lovell, B.C.: Multi-region probabilistic histograms for robust and scalable identity inference. In: Tistarelli, M., Nixon, M.S. (eds.) ICB 2009. LNCS, vol. 5558, pp. 199–208. Springer, Heidelberg (2009). https://doi.org/10.1007/978-3-642-01793-3_21
18. https://www.bioid.com/About/BioID-Face-Database
19. Huang, B.G., Ramesh, M., Berg, T., Learned-Miller, E.: Labeled faces in the wild: a database for studying face recognition in unconstrained environments. University of Massachusetts, Amherst, Technical report 07-49 (2007)
20. Jain, V., Learned-Miller, E.: FDDB: a benchmark for face detection in unconstrained settings. Technical report, University of Massachusetts, Amherst (2010)
21. http://www.comp.polyu.edu.hk/~csajaykr/IITD/FaceIR.htm

Finding Potential RNA Aptamers for a Protein Target Using Sequence and Structure Features

Wook Lee, Jisu Lee, and Kyungsook Han[(⊠)]

Department of Computer Engineering, Inha University, Incheon, South Korea
{wooklee,22181294}@inha.edu, khan@inha.ac.kr

Abstract. Aptamers are single-stranded DNA or RNA sequences that tightly bind to a specific target molecule. This paper presents a computational method for generating potential protein-binding RNA aptamers using a random forest based on several features of protein and RNA sequences and RNA secondary structures. The results of cross validation and independent testing showed that our method can significantly reduce the initial pool of RNA sequences and that the top 10 candidates of RNA aptamers have similar secondary structures and protein-binding structures as actual RNA aptamers for a protein target. Although preliminary, our approach will be useful for constructing an initial pool of RNA sequences for experimental selection of aptamers.

Keywords: RNA aptamer · Random forest · RNA-protein interaction

1 Introduction

Aptamers are short single-stranded nucleic acid (DNA or RNA) sequences that bind to a specific target molecule with high affinity and specificity. Aptamers are usually identified *in vitro* from a random library of DNA or RNA molecules using an iterative process known as systematic evolution of ligands by exponential enrichment (SELEX) [1]. A typical SELEX process involves about 15 rounds of amplification and selection steps. Starting with a random pool of $\sim 10^{15}$ different nucleic acid sequences, only a small number of nucleic acid sequences are selected [2]. Several computational methods have been developed to find aptamers [3–9]. However, most of these methods cannot be used to find aptamers for a new target either because they are classifiers for determining the interaction between a pair of protein and nucleic acid sequences [4, 5] or because they are intended for a single target only [6–9].

In this paper, we propose a new method that predicts RNA aptamer candidates for any protein target. We identified several key features at the sequence and structure levels and built a random forest model using the features. This paper presents the method and the results of cross validation and independent testing of the method in finding RNA aptamers for several protein targets.

© Springer International Publishing AG, part of Springer Nature 2018
D.-S. Huang et al. (Eds.): ICIC 2018, LNCS 10954, pp. 888–892, 2018.
https://doi.org/10.1007/978-3-319-95930-6_89

2 Materials and Methods

We collected 35 RNA aptamer-protein complexes, which were solved by X-ray crystallography with a resolution of 5.0 Å or better from the Protein Data Bank (PDB). In addition to RNA aptamer-protein complexes, we obtained a total of 696 protein-RNA complexes from PDB, which were not included in the 35 RNA aptamer-protein complexes and contain RNA chains of 10 to 120 nucleotides. We used the 696 protein-RNA complexes to compute the interaction propensity (IP) of nucleotides with amino acids. For independent testing of our method, we obtained a benchmark dataset constructed by Li *et al.* [4] from the Apatmer Base [10].

The RNA sequence features used in our study include the IP of nucleotide triplets with amino acids [11–13], mono-nucleotide composition, di-nucleotide composition and pseudo tri-nucleotide composition (PseTNC) [14]. The protein sequence features include composition-transition-distribution of amino acid groups [15] and pseudo amino acid composition (PseAAC) [16]. Both PseTNC and PseAAC were computed with $\lambda = 1$ and $w = 0.1$.

Aptamers selected by SELEX are typically shorter than 30-mer [17, 18]. Thus, we built a random forest (RF) model for generating 27-mer RNA sequences and trained it on the feature vectors generated from the RNA aptamer-protein complexes. To handle RNA sequences of variable lengths, we used a sliding window of 27 nucleotides. As the RNA sequence is scanned with the window, a feature vector representing the window is considered positive if the middle nucleotide of the window is a protein-binding nucleotide. A feature vector is considered negative if the window contains non-binding nucleotides only. Feature vectors that are neither positive nor negative are removed from a training dataset because including them result in severely unbalanced positive and negative instances for training. Exceptionally, the first and last windows are considered as positive if they contain a protein-binding nucleotide in any position of the window. The ratio of positive to negative instances in the training dataset constructed from 35 known RNA aptamer complexes was about 1:3. When building the RF model, the number of trees was set to 35 and the number of features was set to the square root of the number of feature elements.

Potential RNA aptamers for a protein target were found in the following steps:

1. Generate 6×10^6 27-mer random RNA sequences.
2. Predict the secondary structures of the RNA sequences and their free energy.
3. Select RNAs that satisfy the four constrains on RNA secondary structures.
4. For a protein target, generate a feature vector encoding each protein-RNA pair and compute the binding probability of the RNA with the protein target by the RF model.
5. Rank the RNAs with respect to their binding probability and free energy.
6. Select the top 10 RNAs as final aptamer candidates and analyze the results of RNA-protein docking.

The four constrains on RNA secondary structures are (1) the secondary structure should be closed by at least 3 consecutive base pairs with or without a dangling end base, (2) the free energy should be lower than −5.7 kcal/mol, (3) the secondary structure should contain at least 11 unpaired bases, and (4) the number of same secondary structures in the pool should not exceed 150. RNA secondary structures were predicted by RNAfold [19], and RNA-protein docking was performed by HDOCK [20] with RNA tertiary structures predicted by RNAComposer [21].

3   Results and Discussion

From the initial pool of 6×10^6 random RNA sequences, only 38,327 RNA sequences (0.6% of the initial pool) were left after applying the four constraints on RNA secondary structures. The top 10 aptamer candidates contain at least one candidate with similar protein-binding sites as actual RNA aptamers. Figure 1 shows an example of protein-binding structures of both predicted and actual RNA aptamers for protein chain A in an RNA aptamer-protein complex (PDB ID: 3UZS).

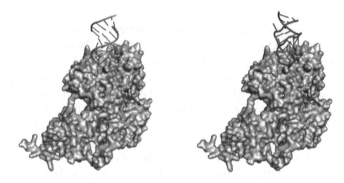

Fig. 1. The structures of an actual RNA aptamer (left) and predicted RNA aptamer (right) binding to protein chain A in a known RNA aptamer-protein complex (PDB ID: 3UZS).

The performance of the RF model was evaluated with respect to sensitivity, specificity, accuracy, positive predictive value (PPV), negative predictive value (NPV), and Matthews correlation coefficient (MCC) in 10-fold cross validation, leave-one-out cross validation and independent testing.

As shown in Table 1, the RF model showed a very good performance in both 10-fold cross validation and leave-one-out cross validation. It achieved an accuracy of 97.79% and an MCC of 0.941 in 10-fold cross validation, and an accuracy of 97.06% and an MCC of 0.922 in leave-one-out cross validation.

Independent testing of the random forest model was done using the benchmark dataset constructed by Li *et al.* [4]. 56 RNA aptamer-protein pairs from the benchmark dataset were used as positive instances and 56 random RNA sequences paired with one of the proteins in the 35 RNA aptamer-protein complexes were used as negative instances. The RF model achieved an accuracy of 71.40% and an MCC of 0.431 in independent testing.

Table 1. Performance of our random forest model. CV: cross validation.

	Sensitivity	Specificity	Accuracy	PPV	NPV	MCC
10-fold CV	95.10%	98.69%	97.79%	96.04%	98.37%	0.941
Leave-one-out CV	94.12%	98.04%	97.06%	94.12%	98.04%	0.922
Independent test	76.80%	66.10%	71.40%	69.40%	74.00%	0.431

4 Conclusion

This paper presented a new method that generates RNA aptamer candidates for a protein target. We developed a random forest model using sequence and structure features. We first generated 6×10^6 27-mer random RNA sequences and selected 38,327 27-mer sequences using constrains on RNA secondary structures. We ranked them by the random forest model and selected the top 10 RNA sequences as aptamer candidates. Structural analysis of the binding structures of the aptamer candidates with their target proteins showed that the predicted aptamers have similar protein-binding structures as actual RNA aptamers. Our method can accelerate *in vitro* selection of RNA aptamers by significantly reducing the size of an initial pool of RNA sequences for a protein target.

Acknowledgments. This work was supported by the National Research Foundation of Korea (NRF) grants (2015R1A1A3A04001243, 2017R1E1A1A03069921) funded by the Ministry of Science and ICT.

References

1. Tuerk, C., Gold, L.: Systematic evolution of ligands by exponential enrichment: RNA ligands to bacteriophage T4 DNA polymerase. Science **249**(4968), 505–510 (1990)
2. Osborne, S.E., Ellington, A.D.: Nucleic acid selection and the challenge of combinatorial chemistry. Chem. Rev. **97**(2), 349–370 (1997)
3. Chushak, Y., Stone, M.O.: In silico selection of RNA aptamers. Nucleic Acids Res. **37**(12), e87 (2009)
4. Li, B.Q., Zhang, Y.C., Huang, G.H., Cui, W.R., Zhang, N., Cai, Y.D.: Prediction of aptamer-target interacting pairs with pseudo-amino acid composition. PLoS ONE **9**(1), e86729 (2014)
5. Zhang, L., Zhang, C., Gao, R., Yang, R., Song, Q.: Prediction of aptamer-protein interacting pairs using an ensemble classifier in combination with various protein sequence attributes. BMC Bioinformatics **17**, 225 (2016)
6. Hu, W.P., Kumar, J.V., Huang, C.J., Chen, W.Y.: Computational selection of RNA aptamer against angiopoietin-2 and experimental evaluation. Biomed. Res. Int. **2015**, 658712 (2015)
7. Shcherbinin, D.S., Gnedenko, O.V., Khmeleva, S.A., Usanov, S.A., Gilep, A.A., Yantsevich, A.V., Shkel, T.V., Yushkevich, I.V., Radko, S.P., Ivanov, A.S., Veselovsky, A.V., Archakov, A.I.: Computer-aided design of aptamers for cytochrome p450. J. Struct. Biol. **191**(2), 112–119 (2015)

8. Ahirwar, R., Nahar, S., Aggarwal, S., Ramachandran, S., Maiti, S., Nahar, P.: In silico selection of an aptamer to estrogen receptor alpha using computational docking employing estrogen response elements as aptamer-alike molecules. Sci. Rep. **6**, 21285 (2016)

9. Rabal, O., Pastor, F., Villanueva, H., Soldevilla, M.M., Hervas-Stubbs, S., Oyarzabal, J.: In silico aptamer docking studies: from a retrospective validation to a prospective case study-TIM3 aptamers binding. Mol. Therapy-Nucleic Acids **5**, e376 (2016)

10. Cruz-Toledo, J., McKeague, M., Zhang, X., Giamberardino, A., McConnell, E., Francis, T., DeRosa, M.C., Dumontier, M.: Aptamer base: a collaborative knowledge base to describe aptamers and SELEX experiments. Database (Oxford) (2012)

11. Choi, S., Han, K.: Prediction of RNA-binding amino acids from protein and RNA sequences. BMC Bioinformatics **12**(Suppl. 13), S7 (2011)

12. Choi, S., Han, K.: Predicting protein-binding RNA nucleotides using the feature-based removal of data redundancy and the interaction propensity of nucleotide triplets. Comput. Biol. Med. **43**(11), 1687–1697 (2013)

13. Tuvshinjargal, N., Lee, W., Park, B., Han, K.: Predicting protein-binding RNA nucleotides with consideration of binding partners. Comput. Methods Programs Biomed. **120**(1), 3–15 (2015)

14. Chen, W., Feng, P.M., Deng, E.Z., Lin, H., Chou, K.C.: iTIS-PseTNC: a sequence-based predictor for identifying translation initiation site in human genes using pseudo trinucleotide composition. Anal. Biochem. **462**, 76–83 (2014)

15. Dubchak, I., Muchnik, I., Holbrook, S.R., Kim, S.H.: Prediction of protein folding class using global description of amino acid sequence. Proc. Natl. Acad. Sci. U.S.A. **92**(19), 8700–8704 (1995)

16. Shen, H.B., Chou, K.C.: PseAAC: a flexible web server for generating various kinds of protein pseudo amino acid composition. Anal. Biochem. **373**(2), 386–388 (2008)

17. Ruckman, J., Green, L.S., Beeson, J., Waugh, S., Gillette, W.L., Henninger, D.D., Claesson-Welsh, L., Janjic, N.: 2′-fluoropyrimidine rna-based aptamers to the 165-amino acid form of vascular endothelial growth factor (VEGF165). Inhibition of receptor binding and vegf-induced vascular permeability through interactions requiring the exon 7-encoded domain. J. Biol. Chem. **273**, 20556–20567 (1998)

18. Tasset, D.M., Kubik, M.F., Steiner, W.: Oligonucleotide inhibitors of human thrombin that bind distinct epitopes. J. Mol. Biol. **272**, 688–698 (1997)

19. Lorenz, R., Bernhart, S.H., Zu Siederdissen, C.H., Tafer, H., Flamm, C., Stadler, P.F., Hofacker, I.L.: ViennaRNA Package 2.0. Algorithms Mol. Biol. **6**(1), 26 (2011)

20. Yan, Y., Zhang, D., Zhou, P., Li, B., Huang, S.Y.: HDOCK: a web server for protein-protein and protein-DNA/RNA docking based on a hybrid strategy. Nucleic Acids Res. **45**(W1), W365–W373 (2017)

21. Popenda, M., Szachniuk, M., Antczak, M., Purzycka, K.J., Lukasiak, P., Bartol, N., Blazewicz, J., Adamiak, R.W.: Automated 3D structure composition for large RNAs. Nucleic Acids Res. **40**(14), e112 (2012)

NmSEER: A Prediction Tool
for 2'-O-Methylation (Nm) Sites
Based on Random Forest

Yiran Zhou[1] (ID), Qinghua Cui[1,2] (ID), and Yuan Zhou[1(✉)] (ID)

[1] Department of Biomedical Informatics, Department of Physiology
and Pathophysiology, MOE Key Lab of Molecular Cardiovascular Sciences,
Center for Noncoding RNA Medicine, School of Basic Medical Sciences, Peking
University, 38 Xueyuan Rd, Haidian District, Beijing 100191, China
zhouyuanbioinfo@bjmu.edu.cn
[2] Center of Bioinformatics, Key Laboratory for Neuro-Information of Ministry
of Education, School of Life Science and Technology, University of Electronic
Science and Technology of China, Chengdu 610054, China

Abstract. 2'-O-methylation (2'-O-me or Nm) is a common RNA modification, which was initially discovered in various non-coding RNAs. Recent researches also revealed its prevalence and regulatory importance in mRNA. In this work, we first demonstrate that the Nm sites can be accurately predicted by the RNA sequence features. By utilizing simple one-hot encoding scheme of positional nucleotide sequence and the random forest machine learning algorithm, we developed a computational prediction tool named NmSEER to predict Nm sites in HeLa cells, HEK293 cells or both of them. Based on our observation of the subgrouping of the Nm sites, we proposed a specialized subgroup-wise prediction strategy to further enhance the prediction performance for the Nm sites with the consensus AGAT motif. Our predictor has achieved a promising performance in both the cross-validation test and the independent test (AUROC = 0.909 and 0.928 for predicting AGAT-sites and non-AGAT sites in independent test, respectively). NmSEER is implemented as a user-friendly web server, which is freely available at http://www.rnanut.net/nmseer/.

Keywords: 2'-O-methylation · Nm site · Random forest · RNA modification
Functional site prediction

1 Introduction

With the rapid advance in the fields of genomics and molecular biology, the crucial regulatory effects of RNA modifications are being uncovered [1]. Up to now, 163 types of post-transcriptional RNA modifications are known, which endow four basic nucleotide residues with multiple additional regulations and functions [2].

2'-O-methylation (2'-O-me or Nm), the methylation modified at the 2'-OH of the ribose moiety of a nucleotide residue, occurs frequently in non-coding RNAs (ncRNAs) such as rRNA [3], tRNA [4, 5] and microRNA [6]. The Nm in ncRNAs could be catalyzed by either protein-only methyltransferases, or by enzyme fibrillarin

© Springer International Publishing AG, part of Springer Nature 2018
D.-S. Huang et al. (Eds.): ICIC 2018, LNCS 10954, pp. 893–900, 2018.
https://doi.org/10.1007/978-3-319-95930-6_90

under the guidance of C/D-box small nucleolar RNA [7, 8]. Nm plays a significant role in the biogenesis and functions of ncRNAs. For example, correct folding and assembly of rRNA require Nm modification at the specific sites [3]. Likewise, perturbation of Nm in the tRNA anticodon loop has deleterious effect on translation, and therefore is associated with non-syndromic X-linked intellectual disability [5]. Recent researches also achieved breakthroughs in the field by discovering prevalent Nm sites in mRNA. Choi et al. revealed that Nm in coding regions of mRNA has disturbing impact on the cognate tRNA selection during translation [9]. In addition, Dai and co-workers developed a novel sensitive high-throughput experimental technology termed Nm-seq to detect Nm sites at low stoichiometry with single-nucleotide resolution [10]. Nm-seq provided useful data for mining the biological features of the Nm sites, but it is expensive and labor- exhausting. Therefore, there is an urgent need of a computational prediction model to mine the sequence feature of Nm sites and identify Nm sites *in silico*. So far no available computational tool for predicting Nm sites in mRNAs has been established. Therefore, in this study, supported by the data provided by Dai et al. (which depicts Nm sites across many mRNAs and a few ncRNA molecules in HeLa's and HEK293's transcriptome), we have developed a computational tool named NmSEER for Nm site prediction. NmSEER is based on the random forest algorithm, a prominent machine learning framework for classification problems. The following sections will expatiate upon the design and performance evaluation of computational models and the implementation of NmSEER web server.

2   Materials and Methods

2.1   Datasets

We first mapped all Nm sites from Dai et al's data to human transcripts (version GRCh37 as recorded by the ENSEMBL database, queried at Nov, 2017 [11]). For each gene, the transcript which was the longest and harbored the highest number of Nm sites was selected as the representative transcript to avoid sequence redundancy. One fourth of the representative transcripts dataset was left out as the testing dataset, and the rest three fourth of the representative transcripts were assigned into the training dataset. Therefore, the sites derived from the training transcripts and testing transcripts were mutually independent. Because there are no golden standard negative samples from the experimental data, we randomly selected non-modified RNA residues as the negative samples. We conformed to an approximate 1:10 and 1:50 positive-to-negative ratio when assembling the training and test set, respectively. Note that the imbalance positive-to-negative ratio is important to evaluate the real-world performance of the predictors (since the amount of non-modified sites is ∼3148-fold more than Nm sites in natural RNA transcripts) but harmful for training the machine-learning models. In the overall consideration of computational efficiency and coverage of the dataset, we choose a compromised 1:10 positive-to-negative ratio for the training set but more rigorous 1:50 ratio for the test set. Besides, to emphasize the positional information of positive samples and avoid bias to the negative samples adjacent to Nm sites, half negative sites were selected randomly from 50 nt flanking windows in which the Nm sites were settled at the central position, and the other half were derived from the remaining positions in the transcripts randomly. As the result, we assembled 2287 positive samples and 22684 negative samples from 1658 transcripts as the training set, and 739 positive samples and 36268 negative samples from 552 transcripts as the independent test set.

After noticing that some Nm sites complied with the consensus AGAT motif, we reconstructed two pairs of mutually exclusive datasets to differentiate AGAT and non-AGAT sites (i.e., AGAT training and test sets, non-AGAT training and test sets). These datasets were also assembled by using the aforementioned dataset assembly pipeline. The only difference was that we selected negative sites as much as possible from the distal regions of AGAT sites because there were not enough non-modified AGAT sites within the 50 nt flanking window of the positive AGAT sites.

2.2 Feature Encoding Schemes

In this study, we employed two basic encoding schemes including one-hot encoding of positional nucleotide sequence [12] (one-hot encoding) and Chen's encoding where chemical property and accumulated frequency of nucleotides were considered [13] (Chen's encoding). Each encoding translated the nucleotide sequences in W nt flanking windows on each side (where the sample sites were placed at the center) into feature vectors. Exact window size W of either encoding was optimized via 5-fold cross-validation tests on the training data. The detailed introduction of encoding schemes is as hereunder mentioned.

One-Hot Encoding. This encoding aims to translate each nucleotide into a 4-dimensional binary vector. A, G, C, T and the gap character at the termini of the RNA sequence correspond to (1, 0, 0, 0), (0, 1, 0, 0), (0, 0, 1, 0), (0, 0, 0, 1) and (0, 0, 0, 0), respectively. As a result, a nucleotide sequence in the W nt flanking window was transform into a $4*(2*W + 1)$-dimensional binary feature vector under the one-hot scheme.

Chen's Encoding. This encoding extracts features by nucleotides' chemical property and frequency. According to the binary classification of four basic nucleotide based on ring structure, functional group and hydrogen bond (0 for one ring, keto group and strong bond, 1 for two rings, amino group and weak bond), each nucleotide could be presented as a three-dimensional binary vector. The accumulated frequency of the nucleotide up to its own position in sequence was further appended as the fourth dimension. Namely, A, G, C, T and the gap character can be encoded as (1, 1, 1, FreqA), (1, 0, 0, FreqG), (0, 1, 0, FreqC), (0, 0, 1, FreqT) and (0, 0, 0, 0), respectively.

One-Hot + Chen's Encoding. This encoding concatenates the feature vector of one-hot encoding with that of Chen's encoding.

One-Hot + Structure Encoding. This encoding adds several features of transcript structures in addition to the one-hot encoding. These features included the absolute and relative distances (i.e. absolute distances divided by the lengths of transcripts or exons) from sample site to the transcript start site, transcript end site, the 5'-UTR end site, the 3'-UTR start site and the nearest splicing sites (10 features in total), respectively.

2.3   Random Forest Algorithm and Performance Evaluation

Random forest machine learning framework is an outstanding supervised learning algorithm towards classification problems, and it has also been widely used in many bioinformatics studies [12, 14]. Our preliminary 5-fold cross-validation on the training data also indicated the better performance of random forest than several machine learning frameworks like decision tree, logistic regression, AdaBoost and Naïve Bayes.

The R software package *randomForest* was used to implement random forest in this study. The number of decision trees (*NT*) is the key hyperparameter of random forest, which was optimized via 5-fold cross-validation tests on the training data. Note that, for each encoding, we built three random forest models by using the different training sets as mentioned above. Namely, one full model, one AGAT specific model and one non-AGAT model were built to predict different subgroups of Nm sites. To be concise, the AGAT specific model and non-AGAT model is together referred as the detached model in the following paragraphs.

A random forest model can learn from the training data to mature numerous decision trees, and then grade for samples in the test data through voting across the trees. Namely, the random forest will give an integrated prediction score for each sample. And the sample whose score is higher than the arbitrary threshold will be classified as a positive prediction. Therefore, *TP*, *FP*, *TN* and *FN* can be calculated, which represent the counts of true positive, false positive, true negative and false negative predictions, respectively. We preliminary evaluated the encodings based on 5-fold cross-validation on the training data. To be rigorous, we further benchmarked the performance of the random forest models on the independent test sets. Sensitivity, specificity, recall and precision were chosen as the typical performance indicators as

$$Sensitivity = Recall = \frac{TP}{TP+FN}, Specificity = \frac{TN}{TN+FP}, Precision = \frac{TP}{TP+FP} \quad (1)$$

Furthermore, we plotted receiver operating characteristic curves (ROC curves) and precision-recall curves (PRC curves) to depict the overall performances of different models. The ROC curve depicts the relationship between true positive rate (sensitivity) and false positive rate (1-specificity), while the PRC curve which plots precision against recall is a more sensitive and critical approach for performance evaluation on test sets with unbalanced positive-to-negative ratio. Area under ROC curves (AUROC) and area under PRC curves (AUPRC) were calculated to quantify models' performances. For both ROC and PRC curve, a higher area under curve closer to 1 indicates more accurate prediction performance.

Once a random forest model generated, importance scores for each feature, which is a useful indicator for analyzing important biological features for prediction, can be directly observed by the *importance* function of the *randomForest* package based on Gini impurity decreases (GD). The GD of a node *d* in the decision tree and Gini of a sample set *S* are given as

$$GD(d) = \text{Gini}(S) - \frac{N_{S_1}}{N_S}\text{Gini}(S_1) - \frac{N_{S_2}}{N_S}\text{Gini}(S_2), \text{Gini}(S) = 1 - \left(\frac{N_N}{N_S}\right)^2 - \left(\frac{N_P}{N_S}\right)^2 \quad (2)$$

where sample set S is divided into subset S_1 and subset S_2 by feature d. N_S, N_{S1} and N_{S2} are the amounts of samples in set S, S_1 and S_2 respectively. In addition, N_N and N_P represent the number of negatives and positives in S. The importance score of a feature in one decision tree is defined as the total GD of nodes split from this feature. Further the importance score of this feature in random forest is averaged across all decision trees.

3   Results

3.1   Comparison Among Competitive Sequence Encoding Schemes

We first trained random forest models based on all Nm sites (i.e. building full models) with different encodings. The four full models based on different encodings turned out to have the same optimized parameters (i.e. $W = 10$, $NT = 200$). After the parameter optimization, we further tested the performances rigorously on independent test set. No sample in the independent test set was included in the feature encoding, model training and parameter optimization processes. The comparisons of ROC and PRC curves are illustrated in Figs. 1A and 1B, respectively. With acceptable false positive rate, either classifier could achieve impressive performance according the ROC curves. Even though the positive-to-negative ratio is extremely unbalanced in this test set (i.e. 1:50), the precisions of these four models, as shown in the precision-recall curves still look quite promising. The AUROC and AUPRC decreased with the order of one-hot + structure encoding, one-hot encoding, one-hot + Chen's encoding and Chen's encoding. Because the sequence pattern of Nm sites turns out to be clear and prominent, it seems that the one-hot encoding, which explicitly encodes the sequence pattern of Nm sites, is a better choice than Chen's encoding that cryptically encodes the Nm sites' sequence features. The combination of one-hot encoding with Chen's encoding did not result in better prediction performance, indicating these two encodings captured similar biological features of Nm sites. Moreover, one-hot + structure encoding scheme showed the best performance (AUROC = 0.969 and AUPRC = 0.751).

We also noticed that the transcript structure features significantly contributed to the prediction score, as indicated by their relatively high feature importance scores in the random forest model (Table 1). This result indicates that the positions and distribution of Nm sites in transcripts are non-random and helpful for Nm site prediction. However, the transcript structure feature, by definition, should take the transcript structure annotations as the additional input, but the transcript structure annotations are hard to use for non-specialist users. In consideration of the complexity of transcript structure features, we instead used the one-hot encoding model which showed the second-best performance to build our predictor.

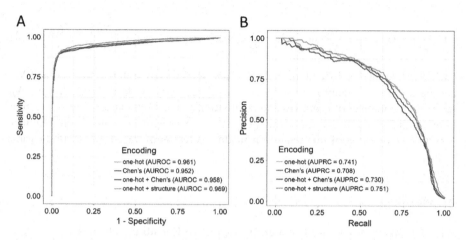

Fig. 1. Performance comparisons of four encoding schemes in independent test as illustrated by ROC and PRC curves. (A) Comparison of ROC curves of four encoding schemes. (B) Comparison of PRC curves of four encoding schemes.

Table 1. Top 10 most important features based on the one-hot + structure encoding scheme

Rank	Description	Importance
1	A at the +1 position	395.298
2	G at the +2 position	465.842
3	A at the +3 position	251.240
4	T at the +4 position	216.652
5	G at the +6 position	137.338
6	Absolute distance to transcript end site	89.401
7	C at the +5 position	87.066
8	Relative distance to 3'-UTR start site	86.433
9	Absolute distance to 3'-UTR start site	86.052
10	Absolute distance to transcript start site	84.842

3.2   The Establishment and Assessment of the AGAT-Specific Model

The most important features (Table 1), in combination, also implied a consensus AGAT motif next to the Nm sites. On the one hand, it revealed the consistency with the sequence logo observed from experimentally identified Nm sites [10]. Subsequent calculation also showed that Nm sites with the AGAT motif account for ∼43% of the total Nm sites, which highlighted a unique subgroup of Nm sites. On the other hand, by examining the prediction score of the testing samples, we noted that sample sites with the AGAT motif, no matter modified or not, were highly prone to be predicted as positive sites, indicating the presence of the AGAT sites could result in strong bias of prediction model. In order to counteract such bias, detached models were built on the detached training sets of AGAT and non-AGAT samples (i.e. we trained AGAT specific and non-AGAT models using the binary encoding, see the above 'Datasets' sub-section for detail). By 5-fold cross-validation tests, the parameters for the AGAT specific model ($W = 14$, $NT = 400$) and non-AGAT model ($W = 10$, $NT = 600$) were optimized.

Subsequently, the performance of detached models and full model (trained with all training sites) was evaluated via independent test on the AGAT test data and the non-AGAT test data, respectively. To fairly compare the detached models with the full model, same independent samples were used for performance evaluation. As a result, the detached models presented a much stronger discriminability for AGAT sites (AUROC = 0.909 *vs.* 0.890; AUPRC = 0.327 *vs.* 0.255) but slightly poorer performance than the full-model when classifying non-AGAT sites (AUROC = 0.927 *vs.* 0.928; AUPRC = 0.592 *vs.* 0.601). Thus, the AGAT specific model and full model are applied for predicting AGAT and non-AGAT sites in the NmSEER server, respectively.

3.3 NmSEER Server

According to the results of above performance assessment and comparisons, the NmSEER finally utilized the random forest models based on the one-hot encoding. For AGAT and non-AGAT sites, the AGAT-specific model and full-model were used respectively. Note that the AGAT-specific model and the full model did not share the thresholds. Therefore, we provided three pairs of pre-defined threshold values corresponding to the true positive rate of 0.2, 0.5 and 0.8 of each sub-predictor based on independent tests. These threshold pairs could be easily switched on NmSEER server by button clicking. Besides, two additional predictors for HeLa and HEK293 cells were also prepared under the same framework, using subsets of Nm sites in HeLa and HEK293 from Dai et al's data.

The NmSEER server is freely available at http://www.rnanut.net/nmseer/ with user-friendly web interface. It accepts single query RNA (or cDNA) sequence in either raw or FASTA sequence format. A successful prediction result will be shown in the result panel of webpage, which should include a task ID, the total number and basic information of predicted sites. Usually, only 5 s are required to complete a prediction task on a typical mRNA sequence (~ 4000 nt). Therefore, the NmSEER can serve as a time-efficient computational tool for screening the potential Nm sites *in silico*.

Acknowledgement. This study was supported by the National Natural Science Foundation of China (Grant Nos. 81670462 to Qinghua Cui) and Fundamental Research Funds for Central Universities (Grant Nos. BMU2017YJ004 to Yuan Zhou).

References

1. Li, S., Mason, C.E.: The pivotal regulatory landscape of RNA modifications. Ann. Rev. Genomics Hum. Genet. **15**, 127–150 (2014)
2. Boccaletto, P., Machnicka, M.A., Purta, E., Piatkowski, P., Baginski, B., Wirecki, T.K., de Crecy-Lagard, V., Ross, R., Limbach, P.A., Kotter, A., Helm, M., Bujnicki, J.M.: MODOMICS: a database of RNA modification pathways. 2017 update. Nucleic Acids Res. **46**, D303–D307 (2018)
3. Hengesbach, M., Schwalbe, H.: Structural basis for regulation of ribosomal RNA 2'-o-methylation. Angew. Chem. Int. Ed. Engl. **53**, 1742–1744 (2014)
4. Jockel, S., Nees, G., Sommer, R., Zhao, Y., Cherkasov, D., Hori, H., Ehm, G., Schnare, M., Nain, M., Kaufmann, A., Bauer, S.: The 2'-O-methylation status of a single guanosine controls transfer RNA-mediated toll-like receptor 7 activation or inhibition. J. Exp. Med. **209**, 235–241 (2012)
5. Guy, M.P., Shaw, M., Weiner, C.L., Hobson, L., Stark, Z., Rose, K., Kalscheuer, V.M., Gecz, J., Phizicky, E.M.: Defects in tRNA anticodon loop 2'-O-Methylation are implicated in nonsyndromic X-linked intellectual disability due to mutations in FTSJ1. Hum. Mutat. **36**, 1176–1187 (2015)
6. Abe, M., Naqvi, A., Hendriks, G.J., Feltzin, V., Zhu, Y., Grigoriev, A., Bonini, N.M.: Impact of age-associated increase in 2'-O-methylation of miRNAs on aging and neurodegeneration in Drosophila. Genes Dev. **28**, 44–57 (2014)
7. Somme, J., Van Laer, B., Roovers, M., Steyaert, J., Versees, W., Droogmans, L.: Characterization of two homologous 2'-O-methyltransferases showing different specificities for their tRNA substrates. RNA **20**, 1257–1271 (2014)
8. Shubina, M.Y., Musinova, Y.R., Sheval, E.V.: Nucleolar methyltransferase fibrillarin: evolution of structure and functions. Biochemistry (Mosc) **81**, 941–950 (2016)
9. Choi, J., Indrisiunaite, G., DeMirci, H., Ieong, K.W., Wang, J., Petrov, A., Prabhakar, A., Rechavi, G., Dominissini, D., He, C., Ehrenberg, M., Puglisi, J.D.: 2'-O-methylation in mRNA disrupts tRNA decoding during translation elongation. Nat. Struct. Mol. Biol. **25**, 208–216 (2018)
10. Dai, Q., Moshitch-Moshkovitz, S., Han, D., Kol, N., Amariglio, N., Rechavi, G., Dominissini, D., He, C.: Nm-seq maps 2'-O-methylation sites in human mRNA with base precision. Nat. Methods **14**, 695–698 (2017)
11. Kersey, P.J., Allen, J.E., Allot, A., Barba, M., Boddu, S., Bolt, B.J., Carvalho-Silva, D., Christensen, M., Davis, P., Grabmueller, C., Kumar, N., Liu, Z., Maurel, T., Moore, B., McDowall, M.D., Maheswari, U., Naamati, G., Newman, V., Ong, C.K., Paulini, M., Pedro, H., Perry, E., Russell, M., Sparrow, H., Tapanari, E., Taylor, K., Vullo, A., Williams, G., Zadissia, A., Olson, A., Stein, J., Wei, S., Tello-Ruiz, M., Ware, D., Luciani, A., Potter, S., Finn, R.D., Urban, M., Hammond-Kosack, K.E., Bolser, D.M., De Silva, N., Howe, K.L., Langridge, N., Maslen, G., Staines, D.M., Yates, A.: Ensembl genomes 2018: an integrated omics infrastructure for non-vertebrate species. Nucleic Acids Res. **46**, D802–D808 (2018)
12. Zhou, Y., Zeng, P., Li, Y.H., Zhang, Z., Cui, Q.: SRAMP: prediction of mammalian N6-methyladenosine (m6A) sites based on sequence-derived features. Nucleic Acids Res. **44**, e91 (2016)
13. Chen, W., Tran, H., Liang, Z., Lin, H., Zhang, L.: Identification and analysis of the N(6)-methyladenosine in the Saccharomyces cerevisiae transcriptome. Sci. Rep. **5**, 13859 (2015)
14. Wang, X., Yan, R.: RFAthM6A: a new tool for predicting m(6)A sites in Arabidopsis thaliana. Plant Mol. Biol. **96**, 327–337 (2018)

Author Index

Printed in the United States
By Bookmasters